T0138544

Geometry, Rigidity, and Group Actions

CHICAGO LECTURES IN MATHEMATICS SERIES
Editors: Spencer J. Bloch, Peter Constantin, Benson Farb,
Norman R. Lebovitz, Carlos Kenig, and J. P. May

Other *Chicago Lectures in Mathematics* titles available from the University of Chicago Press

Simplical Objects in Algebraic Topology, by J. Peter May (1967, 1993)

Fields and Rings, Second Edition, by Irving Kaplansky (1969, 1972)

Lie Algebras and Locally Compact Groups, by Irving Kaplansky (1971)

Several Complex Variables, by Raghavan Narasimhan (1971)

Torsion-Free Modules, by Eben Matlis (1973)

Stable Homotopy and Generalised Homology, by J. F. Adams (1974)

Rings with Involution, by I. N. Herstein (1976)

Theory of Unitary Group Representation, by George V. Mackey (1976)

Infinite-Dimensional Optimization and Convexity, by Ivar Ekeland and Thomas Turnbull (1983)

Commutative Semigroup Rings, by Robert Gilmer (1984)

Navier-Stokes Equations, by Peter Constantin and Ciprian Foias (1988)

Essential Results of Functional Analysis, by Robert J. Zimmer (1990)

Fuchsian Groups, by Svetlana Katok (1992)

Topological Classification of Stratified Spaces, by Shmuel Weinberger (1994)

*Unstable Modules over the Steenrod Algebra and Sullivan's Fixed
    Point Set Conjecture,* by Lionel Schwartz (1994)

Geometry of Nonpositively Curved Manifolds, by Patrick B. Eberlein (1996)

Lectures on Exceptional Lie Groups, by J. F. Adams (1996)

*Dimension Theory in Dynamical Systems: Contemporary Views
    and Applications,* by Yakov B. Pesin (1997)

A Concise Course in Algebraic Topology, by J. P. May (1999)

*Harmonic Analysis and Partial Differential Equations: Essays in Honor of Alberto
    Calderón,* edited by Michael Christ, Carlos Kenig, and Cora Sadosky (1999)

Topics in Geometric Group Theory, by Pierre de la Harpe (2000)

Exterior Differential Systems and Euler-Lagrange Partial Differential Equations, by
    Robert Bryant, Phillip Griffiths, and Daniel Grossman (2003)

Ratner's Theorems on Unipotent Flows, by Dave Witte Morris (2005)

Geometry, Rigidity, and Group Actions

Edited by Benson Farb and David Fisher

THE UNIVERSITY OF CHICAGO PRESS • CHICAGO AND LONDON

Benson Farb is professor of mathematics at the University of Chicago. He is the author of *Problems on Mapping Class Groups and Related Topics* and coauthor of *Noncommutative Algebra*. David Fisher is professor of mathematics at Indiana University.

The University of Chicago Press, Chicago 60637
The University of Chicago Press, Ltd., London
© 2011 by The University of Chicago
All rights reserved. Published 2011
Printed in the United States of America
20 19 18 17 16 15 14 13 12 11   1 2 3 4 5

ISBN-13: 978-0-226-23788-6 (cloth)
ISBN-10: 0-226-23788-5 (cloth)

Library of Congress Cataloging-in-Publication Data

Geometry, rigidity, and group actions/edited by Benson Farb and David Fisher.
        p.   cm.—(Chicago lectures in mathematics series)
      Festschrift for Robert Zimmer on the occasion of his 60th birthday.
      ISBN-13: 978-0-226-23788-6 (cloth : alk. paper)
      ISBN-10: 0-226-23788-5 (cloth : alk. paper)
      1. Rigidity (Geometry)   2. Group actions (Mathematics)   3. Mani-
    folds (Mathematics)   I. Zimmer, Robert J., 1947–   II. Farb, Benson.
    III. Fisher, David.   IV. Series: Chicago lectures in mathematics series.
    QA613.G465   2011
    516'.11—dc22   2010026631

To Bob, mentor and friend

Contents

Preface

In September 2007, a conference was held at the University of Chicago in honor of the sixtieth birthday of Robert J. Zimmer. The conference was a testament to and a celebration of Zimmer's continuing and lasting influence on the fields of geometry, rigidity, and group actions. The wide variety of papers submitted to this volume are just one indication of that influence.

This volume contains both survey papers and research papers. The common theme throughout concerns group actions. The general setup is that we have a group G, which is either a Lie group or a discrete group, acting on some space X. The examples are quite varied, and include, in (informal) order of decreasing "rigidity":

1. algebraic actions on varieties
2. isometric actions on metric spaces
3. smooth actions on manifolds
4. measurable actions on measure spaces

In each instance the goal is to relate algebraic/combinatorial properties of G to geometric/topological/measure-theoretic properties of X and the G action on X. In many cases, properties of G impose surprising restrictions on the possible choices of X and the possible actions on X.

The papers by Goldman-Xia and Lubotzky concentrate on (1). The varieties here are representation and character varieties of certain finitely generated groups Γ, with the automorphism group $G = \text{Aut}(\Gamma)$ acting by algebraic automorphisms. This setup can be viewed as a dynamical system, where one can study its dynamical properties (as in Goldman-Xia's paper), or as an algebraic system (as in Lubotzky's paper), where information about G can be gleaned from properties of the action.

A number of the papers in this volume are concerned with (2). The paper by Farb-Hruska-Thomas considers the case where X is a nonpositively curved simplicial complex and the group G is a lattice in the simplicial automorphism group of X; here the goal is to extend the classical theory of lattices in algebraic groups over fields with a discrete valuation to this more general context. Breuillard's paper provides a strengthening of the famous *Tits alternative*, and gives further applications to spectral gaps, group growth, and more. Here the G is an arbitrary, finitely generated subgroup of GL(2, **C**), and X is a projective space over various fields. The paper by Delzant-Guichard-Labourie-Mozes considers the general setup of groups acting isometrically on metric spaces. They relate the notion of a "displacing action" to the geometric property of an orbit being quasi-isometrically embedded. This setup is widely applicable, although the case of S-arithmetic groups is emphasized here. Iozzi considers isometric actions of semidirect products G of groups H with (normal) abelian groups A, acting unitarily on a Hilbert space X. The main goal is to give necessary and sufficient conditions for the pair (G, A) to have relative property (T). Taback-Wong use the large-scale geometry of wreath products to study twisted conjugacy classes in these groups. Cowling's paper develops some calculus on nilpotent groups with a view toward generalizing Mostow's rigidity theorem concerning groups acting on symmetric spaces.

The general study of smooth (usually volume-preserving) actions of (noncompact) Lie groups and their lattices on smooth manifolds goes by the name of the *Zimmer program*. The paper by Fisher gives a broad survey of this area, while the paper by Morris surveys the already rich case of actions on the circle. Burger's paper also concerns actions of various large groups on the circle and provides a unified approach to many of the major results in this area. The Feres-Ronshausen paper studies general properties of groups acting on manifolds without an invariant volume by studying a certain type of harmonic function related to these actions. The primary applications in this paper also concern groups of diffeomorphisms of the circle. Moving up one dimension, Py's paper provides a brief introduction to an interesting approach to studying volume-preserving diffeomorphisms of compact surfaces using *quasi-morphisms*. A major theme in the Zimmer program is the study of group actions that preserve geometric structures, a topic represented here in the papers by Dumitrescu and Quiroga-Barranco. While Zimmer's focus was primarily on rigidity for smooth actions, Weinberger's paper points to several interesting rigidity phenomena for actions by homeomorphisms.

Another major topic of Zimmer's research concerned cocycles over ergodic actions and orbit equivalence rigidity. Furman's survey provides a comprehensive overview of this area. The Popa-Vaes paper illustrates some recent developments deepening connections between orbit equivalence rigidity and operator algebras, and proves some striking new rigidity results. In a very different direction, Karlsson-Ledrappier write a survey on noncommutative ergodic theorems, an area in which cocycles over ergodic actions have played an interesting and important role.

Acknowledgment. We would like thank the contributors to this volume, as well as the referees, who generously gave their time to this project.

PART 1

Group Actions on Manifolds

MARC BURGER

AN EXTENSION CRITERION FOR LATTICE ACTIONS ON THE CIRCLE

TO BOB ZIMMER

1. Introduction

Let $\Gamma < G$ be a lattice in a locally compact second countable group G. The aim of this paper is to establish a necessary and sufficient condition for a Γ-action by homeomorphisms of the circle to extend continuously to G. This condition will be expressed in terms of the real bounded Euler class of this action. Combined with classical vanishing theorems in bounded cohomology, one recovers rigidity results of Ghys, Witte-Zimmer, Navas, and Bader-Furman-Shaker in a unified manner. For a survey of various approaches to the problem of classifying lattice actions on the circle we refer to the paper by David Witte Morris [25] in this volume.

Let $\mathrm{Homeo}^+(S^1)$ be the group of orientation-preserving homeomorphisms of the circle and $e \in H^2(\mathrm{Homeo}^+(S^1), \mathbb{Z})$ the Euler class; recall that e corresponds to the central extension defined by the universal covering of $\mathrm{Homeo}^+(S^1)$. The Euler class admits a representing cocycle that is bounded and this defines a bounded class $e^b \in H_b^2(\mathrm{Homeo}^+(S^1), \mathbb{Z})$ called the bounded Euler class. The relevance of bounded cohomology to the study of group actions on the circle comes from a result of Ghys [14], namely that the bounded Euler class $\rho^*(e^b) \in H_b^2(\Gamma, \mathbb{Z})$ of an action $\rho : \Gamma \to \mathrm{Homeo}^+(S^1)$ determines ρ up to quasiconjugation; a quasiconjugation is a self-map of the circle that is weakly cyclic order preserving and in particular not necessarily continuous; see Section 3 for details. If $e_{\mathbb{R}}^b$ denotes then the bounded class obtained by considering the bounded cocycle defining e^b as real valued, we call the invariant $\rho^*(e_{\mathbb{R}}^b) \in H_b^2(\Gamma, \mathbb{R})$ the real bounded Euler class of ρ. From this point of view we have the following dichotomy (see Proposition 3.2):

(E) $\rho^*(e_{\mathbb{R}}^b) = 0$: in this case, ρ is quasiconjugated to an action of Γ by rotations; as far as the extension problem is concerned, it reduces to the properties of the restriction map

$$\mathrm{Hom}_c(G, \mathbb{R}/\mathbb{Z}) \to \mathrm{Hom}(\Gamma, \mathbb{R}/\mathbb{Z}).$$

(NE) $\rho^*(e_\mathbb{R}^b) \neq 0$: in this case, ρ is quasiconjugated to a minimal unbounded action; that is, every orbit is dense and the group of homeomorphisms $\rho(\Gamma)$ is not equicontinuous.

In the first case (E) we call ρ elementary and in the second (NE) nonelementary; nonelementary actions are our main object of study in this paper.

Concerning the extension problem, an issue that has to be taken care of is the existence of a nontrivial centralizer of the action under consideration. This is illustrated by the following:

EXAMPLE 1.1. Let $\Gamma < \mathrm{PSL}(2, \mathbb{R})$ be a lattice that is nonuniform and torsion free. Since Γ is a free group we can lift the identity to a homomorphism ρ_k: $\Gamma \to \mathrm{PSL}(2, \mathbb{R})_k \subset \mathrm{Homeo}^+(S^1)$ into the k-fold cyclic covering of $\mathrm{PSL}(2, \mathbb{R})$, and this for every $k \geq 1$. In this way we get an action that is minimal, unbounded, but for $k \geq 2$ does not extend continuously to $\mathrm{PSL}(2, \mathbb{R})$. If Γ is torsion-free cocompact, this construction applies provided k divides the Euler characteristic of Γ, which is always the case for $k = 2$.

Thus given a minimal unbounded action one is led to consider its topological S^1-factors; those are easily classified and in particular there is, up to conjugation, a unique factor

$$\rho_{sp} : \Gamma \to \mathrm{Homeo}^+(S^1)$$

that is strongly proximal (Proposition 3.7).

This relies on arguments of Ghys that establish that the centralizer of $\rho(\Gamma)$ is a finite cyclic group; the strongly proximal quotient is then obtained by passing to the quotient by this cyclic group.

Our main result is:

THEOREM 1.2. *Let $\Gamma < G$ be a lattice in a locally compact, second countable group G and*

$$\rho : \Gamma \to \mathrm{Homeo}^+(S^1)$$

be a minimal unbounded action. Then the following are equivalent:

1) *The real bounded Euler class* $\rho^*(e_{\mathbb{R}}^b)$ *of* ρ *is in the image of the restriction map*

$$H_{bc}^2(G, \mathbb{R}) \to H_b^2(\Gamma, \mathbb{R}).$$

2) *The strongly proximal factor* ρ_{sp} *of* ρ *extends continuously to* G.

The main ingredient in the proof of Theorem 1.2 is a result that, for any countable group Γ, characterizes the bounded classes in $H_b^2(\Gamma, \mathbb{R})$ obtained from minimal strongly proximal actions in terms of certain cocycles defined on an appropriate Poisson boundary of Γ; see Theorem 4.5 in Section 4, where the result is proven in the more general context of locally compact, second countable groups. This leads to results of independent interest concerning the extent to which an action is determined by its real bounded Euler class and to the question of the possible values of its norm. These results are summarized in the following theorem and corollaries.

THEOREM 1.3. *Let* Γ *be a countable group.*

1) *For a homomorphism* $\rho : \Gamma \to \mathrm{Homeo}^+(S^1)$ *we have*

$$\|\rho^*(e_{\mathbb{R}}^b)\| \leq \frac{1}{2}$$

with equality if and only if ρ *is quasiconjugated to a minimal strongly proximal action.*

2) *If two minimal strongly proximal actions* ρ_1, ρ_2 *are not conjugated, then*

$$\|\rho_1^*(e_{\mathbb{R}}^b) - \rho_2^*(e_{\mathbb{R}}^b)\| = 1.$$

REMARK 1.4.

1) The first assertion in Theorem 1.3 echoes results obtained in [7] concerning tight homomorphisms with values in a Lie group of Hermitian type.
2) Let $\mathrm{ESP}(\Gamma) \subset H_b^2(\Gamma, \mathbb{R})$ denote the subset consisting of the real bounded Euler classes of minimal strongly proximal Γ-actions and $\mathbb{Z}[\mathrm{ESP}(\Gamma)]$ its $\mathbb{Z}$-span. We will show (see Section 5) that the norm takes half-integral values on $\mathbb{Z}[\mathrm{EPS}(\Gamma)]$. In this context the following question arises, namely if $\mathbb{Z}[\mathrm{SP}(\Gamma)]$ denotes the free abelian group on the set of conjugacy classes of minimal strongly proximal actions, can one determine the kernel of the homomorphism

$$\mathbb{Z}[SP(\Gamma)] \rightarrow \mathbb{Z}[ESP(\Gamma)]$$

$$\Sigma n_i[\rho_i] \rightarrow \Sigma n_i \, \rho_i(e_{\mathbb{R}}^b)$$

and what is its significance for the dynamics of Γ-actions on S^1?

The following immediate corollary is another instance of the general principle that groups whose second bounded cohomology is finite-dimensional exhibit rigidity phenomena; compare, for example, with the case of actions by isometries on Hermitian symmetric spaces (see [4], [6]).

COROLLARY 1.5. *Let Γ be a countable group and assume that $H_b^2(\Gamma, \mathbb{R})$ is finite-dimensional. Then there are, up to conjugation, only finitely many minimal strongly proximal Γ-actions on S^1.*

Together with the information concerning centralizers of minimal actions we deduce from Theorem 1.3,

COROLLARY 1.6.

1) *For a homomorphism $\rho : \Gamma \rightarrow \mathrm{Homeo}^+(S^1)$ we have*

$$\|\rho^*(e_{\mathbb{R}}^b)\| \in \{0\} \cup \left\{ \frac{1}{2k} : k \in \mathbb{N} \right\},$$

and this value equals $(2k)^{-1}$ if and only if ρ is quasiconjugated to a minimal unbounded action whose centralizer is of order k.
2) *For minimal unbounded actions ρ_1, ρ_2 we have $\rho_1^*(e_{\mathbb{R}}^b) = \rho_2^*(e_{\mathbb{R}}^b)$ if and only if, up to conjugation,*

$$\rho_2(\gamma) = h(\gamma) \, \rho_1(\gamma), \quad \gamma \in \Gamma$$

where h is a homomorphism with values in the centralizer of $\rho_1(\Gamma)$.

The extension criterion in Theorem 1.2 leads to rigidity theorems when combined with the following two ingredients, namely vanishing theorems in bounded cohomology and the description of continuous homomorphisms from a locally compact group into $\mathrm{Homeo}^+(S^1)$.

Concerning the first ingredient, we know that the restriction map $H_{bc}^2(G, \mathbb{R}) \rightarrow H_b^2(\Gamma, \mathbb{R})$ is a isomorphism in the following cases:

1) Products (see [8], [21]): $G = G_1 \times \cdots \times G_n$ is a Cartesian product of locally compact second countable groups and Γ has dense projection in every factor G_i.

2) Higher-rank Lie groups (see [8]): $G = \mathbb{G}(k)$ where $\mathbb{G}$ is a connected almost simple k-group, k is a local field and $\text{rank}_k \mathbb{G} \geq 2$.

Since it is elementary to classify continuous homomorphisms from a semi-simple Lie group over a local field into $\text{Homeo}^+(S^1)$ one obtains by combining (1) and (2),

COROLLARY 1.7. ([15], [27])

Let $\Gamma < G := \prod_{\alpha \in A} \mathbb{G}_\alpha(k_\alpha)$ be an irreducible lattice where k_α are local fields, $\mathbb{G}_\alpha$ is a connected, simply connected, almost simple k_α-group of positive rank. Assume that the sum of the k_α-ranks of $\mathbb{G}_\alpha$ is at least 2.

For a homomorphism $\rho : \Gamma \to \text{Homeo}^+(S^1)$ one of the following holds up to quasiconjugation:

1) *ρ has a finite orbit.*
2) *ρ is minimal unbounded and its strongly proximal factor ρ_{sp} extends continuously to G factoring via the projection on a factor of the form $\mathbb{G}_\alpha(k_\alpha) = SL(2, \mathbb{R})$.*

Concerning locally compact subgroups of $\text{Homeo}^+(S^1)$, one has, owing to the solution of Hilbert's fifth problem, a wealth of information and in particular, those that are connected and minimal have a simple classification; they are up to conjugation, either Rot the subgroup of rotations or $\text{PSL}(2, \mathbb{R})_k$, the k-fold cyclic covering of $\text{PSL}(2, \mathbb{R})$ (see [10], [16]). When studying continuous homomorphisms from a locally compact group G into $\text{Homeo}^+(S^1)$ one has to deal with the fact that the image is not necessarily closed. In any case we have,

THEOREM 1.8. *Let G be a locally compact group and $\pi : G \to \text{Homeo}^+(S^1)$ a continuous and minimal action. Then one of the following holds:*

1) *π is conjugated into the group Rot of rotations and has a dense image in it.*
2) *π surjects onto $\text{PSL}(2, \mathbb{R})_k$ for some $k \geq 1$, up to conjugation.*
3) *$\text{Ker}\, \pi$ is an open subgroup of G.*

From Theorem 1.8 and the vanishing result for products mentioned above we obtain,

COROLLARY 1.9. ([26], [2])

Let $G = G_1 \times \cdots \times G_n$ be a product of locally compact second countable groups and $\Gamma < G$ a lattice with dense projections on each factor G_i. Assume that $\rho : \Gamma \to \text{Homeo}^+(S^1)$ is minimal unbounded. Then the strongly proximal quotient ρ_{sp} extends continuously to G,

$$(\rho_{sp})^{\text{ext}} : G \to \text{Homeo}^+(S^1)$$

and we have one of the following:

1) $\text{Ker}(\rho_{sp})^{\text{ext}}$ is open in G.
2) Up to conjugation $(\rho_{sp})^{\text{ext}}$ factors via a projection onto some factor G_i followed by a continuous surjection onto $PSL(2, \mathbb{R}) \subset \text{Homeo}^+(S^1)$.

Finally, as shown by Bader, Furman, and Shaker, there is also in the context of actions on the circle a commensurator superrigidity theorem, which we state in a way that is somewhat different, but equivalent to their thm. C in [2].

THEOREM 1.10. ([2])
Let G be a locally compact second countable group, $\Gamma < G$ a lattice, and $\Lambda < G$ a subgroup such that $\Gamma \subset \Lambda \subset \text{Comm}_G \Gamma$ and Λ is dense in G. Let $\rho : \Lambda \to \text{Homeo}^+(S^1)$ be a homomorphism such that $\rho(\Gamma)$ is minimal and unbounded. Then the strongly proximal quotient ρ_{sp} extends continuously to G,

$$(\rho_{sp})^{\text{ext}} : G \to \text{Homeo}^+(S^1)$$

and we have one of the following:

1) $\text{Ker}(\rho_{sp})^{\text{ext}}$ is open of infinite index in G.
2) $(\rho_{sp})^{\text{ext}}$ surjects onto $PSL(2, \mathbb{R}) \subset \text{Homeo}^+(S^1)$.

Let us make the following comment about the hypothesis of Theorem 1.10. If $\rho : \Lambda \to \text{Homeo}^+(S^1)$ is a homomorphism such that $\rho(\Gamma)$ is unbounded, then either $\rho(\Gamma)$ is minimal or there is an exceptional minimal set $K \subset S^1$ (see Section 3) that is easily seen to be $\rho(\Lambda)$-invariant. Thus ρ is quasiconjugated to an action of Λ for which Γ is minimal and unbounded.

The above result follows easily from the extension criterion (Theorem 1.2) and the following general fact concerning bounded cohomology:

THEOREM 1.11. ([20])
Let $\Gamma < G$ and $\Gamma < \Lambda < \text{Comm}_G \Gamma$ be as in Theorem 1.10, in particular Λ is dense in G. Then the image of the restriction map

$$H_b^2(\Lambda, \mathbb{R}) \to H_b^2(\Gamma, \mathbb{R})$$

coincides with the image of $H_{bc}^2(G, \mathbb{R})$ in $H_b^2(\Gamma, \mathbb{R})$.

We will leave to the interested reader the exercise of deducing Theorem 1.10 from Theorem 1.11 and refer to [20] for elementary proofs of the isomor-

phism results for products and higher-rank groups mentioned above, as well
as the proof of Theorem 1.11.

ACKNOWLEDGMENTS: Thanks to Luis Hernandez for his kind invitation
to CIMAT where this work was completed. Thanks to Uri Bader and Alex
Furman for sharing with me their result on boundary maps.

2. Boundary Maps

In this section we present a general existence and uniqueness result concerning measurable equivariant maps that is due to Bader-Furman and is of general
interest in rigidity theory.

Here and in the sequel G is a second countable locally compact group,
$G \times M \to M$ is a continuous action on a compact metrisable space M, $\mu \in \mathcal{M}^1(G)$ is a spread-out probability measure on G, and (B, ν_B) is a standard
Lebesgue G-space such that ν_B is μ-stationary.

THEOREM 2.1. ([1])
Assume that

> 1) *the G-action is minimal and strongly proximal.*
> 2) *For every sequence $(g_n)_{n\geq 1}$ in G there exists a subsequence $(n_k)_{k\geq 1}$ such that*

$$M \to M$$
$$m \longmapsto g_{n_k}(m)$$

converges pointwise.

Then every measurable G-equivariant map

$$\varphi : B \to \mathcal{M}^1(M)$$

takes values in the subset of Dirac measures.

REMARK 2.2. When G is discrete countable and $M = S^1$, a result essentially
equivalent to Theorem 2.1 has been obtained by Deroin, Kleptsyn, and Navas
(see [9]).

We recall here that a G-action is strongly proximal if for every $\mu \in \mathcal{M}^1(M)$
the closure $\overline{G\mu} \subset \mathcal{M}^1(M)$ contains a Dirac mass.

Proof. The main step consists in showing that the G-space M is μ-proximal
in the sense of (2.6), def. 3 in [22] VI: this means that for every μ-stationary

measure $\nu \in \mathcal{M}^1(M)$ the map

2.1
$$\phi_\nu(\omega) := \lim_{n\to\infty} (x_1, \ldots x_n)\,\nu, \quad \omega = (x_n)_{n\geq 1}$$

with values in $\mathcal{M}^1(M)$, defined almost everywhere on the infinite product $(S, \mu_S) := \prod_{n=1}^\infty (G, \mu)$, takes values in the set of Dirac measures; this we proceed to show now. The existence of the limit (2.1) follows from the martingale convergence theorem (prop. 2.4 in [22] VI); in addition one has the following remarkable fact ([13], [17] and lemma 1.33 in [11]):

2.2
$$\lim_{n\to\infty} (x_1 \ldots x_n\, g)\,\nu = \phi_\nu(\omega)$$

for μ_S-almost every $\omega \in S$ and λ-almost every $g \in G$, where λ is the left Haar measure on G. Fix now a sequence $\omega = (x_n)_{n\geq 1}$ and a subset $E \subset G$ of full λ-measure such that (2.2) holds for ω and all $g \in E$. Passing to a subsequence we may assume that the limit

$$s(m) := \lim_{n\to\infty} (x_1 \ldots x_n)(m)$$

exists for every $m \in M$. Then the map $s : M \to M$, although in general not continuous, admits points of continuity; in fact, they form a G_δ-subset of M. Let then $m \in M$ be a point of continuity of s. Since the G-action is minimal and strongly proximal, there exists a sequence $(g_n)_{n\geq 1}$ in G such that $\lim_{n\to\infty} g_n\nu = \delta_m$. By the continuity of the G-action on M there exists for each $n \geq 1$ an open neighborhood $U_n \ni g_n$ such that whenever $(h_n)_{n\geq 1} \in \prod_{n\geq 1} U_n$, $\lim_{n\to\infty} h_n\nu = \delta_m$. Since $\lambda(U_n) > 0, \forall n \geq 1$, we may choose $h_n \in U_n \cap E$ and obtain a sequence $(h_n)_{n\geq 1}$ such that

2.3
  a) $\lim_{n\to\infty} h_n\,\nu = \delta_m$

  b) $\lim_{n\to\infty} (x_1 \ldots x_n\, h_k)\,\nu = \phi_\nu(\omega)$, for every $k \geq 1$.

Let V be a neighborhood of $s(m)$ and $W \ni m$ an open neighborhood such that $W \subset s^{-1}(V)$. Taking into account that $s : M \to M$ is a Borel map we have

$$(sh_k)\nu(V) = (h_k\,\nu)(s^{-1}(V)) \geq h_k\,\nu(W),$$

which together with $\lim_{k\to\infty} h_k\,\nu(W) = 1$ (see (2.3 a)) implies that

2.4
$$\lim_{k\to\infty} (sh_k)(\nu) = \delta_{s(m)}.$$

For every $k \geq 1$ and every continuous function $f \in C(M)$ we have

$$(sh_k)\nu(f) = \int_M f(sh_k\, x)\, d\nu(x)$$

$$= \int_M \lim_{n\to\infty} f(x_1 \ldots x_n\, h_k\, x)\, d\nu(x)$$

$$= \phi_\nu(\omega)(f)$$

where in the second equality we have used the definition of s and the continuity of f, while in the third we have used dominated convergence and (2.3 b).

Thus $sh_k\, \nu = \phi_\nu(\omega)$ for every $k \geq 1$, which together with (2.4) implies that

$$\phi_\nu(\omega) = \delta_{s(m)}\,,$$

and hence since ω could be chosen from a subset of full measure, ϕ_ν takes values in the set of Dirac masses almost everywhere; thus M is μ-proximal Let now (B, ν_B) and $\varphi : B \to \mathcal{M}^1(M)$ be as in Theorem 2.1. Then $\varphi_*(\nu_B) \in \mathcal{M}^1(\mathcal{M}^1(M))$ is a μ-stationary probability measure on $\mathcal{M}^1(M)$ and since M is μ-proximal, prop. 2.9 in [22] VI implies that the support of $\varphi_*(\nu_B)$ is contained in the subset of $\mathcal{M}^1(M)$ consisting of Dirac masses. This concludes the proof of Theorem 2.1. $\qquad\qquad\square$

In the following proposition we give examples of groups of homeomorphisms with the property in Theorem 2.1 (2).

PROPOSITION 2.3. *Let (G, M) be one of the following pairs consisting of a compact metric space M and a subgroup G of the group of homeomorphisms of M:*

1) *$M = \mathbb{P}V_k$, $G = \mathrm{PGL}(V_k)$ where V_k is a finite-dimensional vector space over a local field k.*
2) *$M = S^1$, $G = \mathrm{Homeo}^+(S^1)$.*

Then for any sequence $(g_n)_{n\geq 1}$ there is a subsequence such that $(g_{n_k})_{k\geq 1}$ converges pointwise.

Question 2.4

Let X be a proper complete CAT(0)-space, $M = X(\infty)$ its visual boundary, and $G = \mathrm{Isom}(X)$. Does (G, M) satisfy the conclusion of Proposition 2.3?

The first statement in Proposition 2.3 follows easily by recurrence on dim V_k using Furstenberg's lemma [12]. Concerning the second, we may compose g_n by an appropriate rotation r_n such that $r_n \circ g_n$ fixes a point in S^1 and, passing to a subsequence we may assume that $(r_n)_{n\geq 1}$ converges uniformly; Proposition 2.3 (2) will then follow from

LEMMA 2.5. *Let* $f_n : [0, 1] \rightarrow [0, 1]$, $n \geq 1$, *be a sequence of monotone increasing maps. Then there exists a subsequence converging pointwise.*

Proof. We may assume that $f(r) := \lim_{n \rightarrow \infty} f_n(r)$ exists $\forall r \in \mathbb{Q} \cap [0, 1]$. Then f is monotone increasing on $\mathbb{Q} \cap [0, 1]$ and thus

$$F(x) := \sup\{f(r) : r \leq x, \ r \in \mathbb{Q} \cap [0, 1]\}$$

is a well-defined function on $[0, 1]$ that is monotone increasing and extends f. The set $\mathcal{D} \subset [0, 1]$ of discontinuities of F is at most countable; passing to a subsequence we may assume then that $\lim_{n \rightarrow \infty} f_n(d)$ exists for every $d \in \mathcal{D}$.

Let now $t \notin \mathcal{D}$ be a point of continuity of F. For every $t', t'' \in \mathbb{Q}$ with $t' \leq t \leq t''$ we have $F(t') \leq \xi \leq F(t'')$ where ξ is any accumulation point of the sequence $(f_n(t))_{n \geq 1}$. Letting t', t'' approach t and using the continuity of F we get $\xi = F(t)$ and hence $\lim_{n \rightarrow \infty} f_n(t) = F(t)$. $\qquad\square$

3. Elementary Properties of Actions on the Circle

In this section we establish the basic dichotomy concerning actions on the circle in terms of their real bounded Euler class, announced in the introduction, and show the uniqueness of the strongly proximal quotient. Along the way we will take the opportunity to recall some well-known facts that will be used later on.

Let $S^1 = \mathbb{Z} \backslash \mathbb{R}$ and consider the central extension

$$0 \rightarrow \mathbb{Z} \xrightarrow{i} \text{Homeo}_{\mathbb{Z}}^+(\mathbb{R}) \xrightarrow{p} \text{Homeo}^+(S^1) \rightarrow (e)$$

where $\text{Homeo}_{\mathbb{Z}}^+(\mathbb{R})$ is the group of increasing homeomorphisms of $\mathbb{R}$ commuting with integer translations and $i(n) = T^n$, where $T(x) = x + 1, x \in \mathbb{R}$. A section of p is obtained by associating to every $f \in \text{Homeo}^+(S^1)$ the unique lift $\overline{f} : \mathbb{R} \rightarrow \mathbb{R}$ satisfying $\overline{f}(0) \in [0, 1)$. Then we have

3.1 $$\overline{fg} \ T^{c(f, g)} = \overline{f} \ \overline{g}, \ f, g \in \text{Homeo}^+(S^1)$$

and $c(f, g) \in \{0, 1\}$ is an inhomogeneous 2-cocycle called the Euler cocycle whose class $e \in H^2(\text{Homeo}^+(S^1), \mathbb{Z})$ is called the Euler class. Considering c as a $\mathbb{Z}$-valued, respectively $\mathbb{R}$-valued bounded cocycle leads respectively to the bounded Euler class $e^b \in H_b^2(\text{Homeo}^+(S^1), \mathbb{Z})$ and the real bounded Euler class $e_{\mathbb{R}}^b \in H_b^2(\text{Homeo}^+(S^1), \mathbb{R})$. Given a group Γ and a homomorphism $\rho : \Gamma \rightarrow \text{Homeo}^+(S^1)$ we obtain accordingly three invariants associated to ρ, namely, the Euler class $\rho^*(e) \in H^2(\Gamma, \mathbb{Z})$, the bounded Euler class $\rho^*(e^b) \in H_b^2(\Gamma, \mathbb{Z})$, and the real bounded Euler class $\rho^*(e_{\mathbb{R}}^b) \in H_b^2(\Gamma, \mathbb{R})$. The information contained in $\rho^*(e)$ is of purely algebraic nature; $\rho^*(e) = 0$ if and only if ρ lifts to a homomorphism with values in $\mathrm{Homeo}_{\mathbb{Z}}^{+}(\mathbb{R})$, which for example is the case when Γ is a free group. Concerning the bounded Euler class we have on the other hand,

THEOREM 3.1. ([14]) *Two actions* ρ_0, ρ_1 *are quasiconjugated if and only if* $\rho_0^*(e^b) = \rho_1^*(e^b)$.

There are several equivalent ways to look at quasiconjugations. Ghys's original definition is as follows: a quasiconjugation is a map $h : S^1 \to S^1$ induced by a monotone increasing map $\overline{h} : \mathbb{R} \to \mathbb{R}$ commuting with integer translations; observe that a quasiconjugation is not necessarily continuous. Two actions ρ_0, ρ_1 are then quasiconjugated if there is such a map h with

$$3.2 \qquad h\rho_0(\gamma) = \rho_1(\gamma)\, h, \quad \forall \gamma \in \Gamma.$$

Quasiconjugation is an equivalence relation.

Observe that if we set $(\overline{h})_+(x) = \lim_{y>x} \overline{h}(y)$, $(\overline{h})_-(x) = \lim_{y<x} \overline{h}(y)$, then $(\overline{h})_+$ and $(\overline{h})_-$ induce respectively a right continuous h_+ and a left continuous h_- quasiconjugacy and if (3.2) holds, then we have $h_{\pm}\,\rho_0(\gamma) = \rho_1(\gamma)\, h_{\pm}, \forall \gamma \in \Gamma$ as well.

An intrinsic way to define quasiconjugation rests on the consideration of the orientation cocycle $o : (S^1)^3 \to \{-1, 0, 1\}$ defined as follows:

$$3.3 \qquad o(x, y, z) = \begin{cases} 1 & \text{if } x, y, z \text{ are positively oriented,} \\ -1 & \text{if } x, y, z \text{ are negatively oriented,} \\ 0 & \text{if } x, y, z \text{ are not pairwise distinct.} \end{cases}$$

Then a map $h : S^1 \to S^1$ is a quasiconjugacy if and only if

$$o(h(x), h(y), h(z)) \geq 0 \quad \text{whenever} \quad o(x, y, z) \geq 0.$$

A third characterization is obtained as follows. Define for $x, y \in S^1$ the open interval $(x, y) := \{z \in S^1 : o(x, z, y) = 1\}$ and $[x, y]$, $(x, y]$, $[x, y)$ in the obvious way. Given a point $b \in S^1$ and a probability measure $\mu \in \mathcal{M}^1(S^1)$, the map

$$3.4 \qquad h_{b,\mu}(x) := \mu([b, x)) \bmod \mathbb{Z}$$

is a left continuous quasiconjugacy and every such quasiconjugacy is obtained this way for a unique pair (b, μ). The behavior under composition with homeomorphisms is particularly simple,

3.5 $$h_{b,\mu}(\varphi^{-1}(x)) = h_{\varphi(b),\varphi_*(\mu)}(x), \quad \varphi \in \text{Homeo}^+(S^1).$$

A very useful tool in the study of group actions on S^1 is the following trichotomy for $\rho : \Gamma \to \text{Homeo}^+(S^1)$, see ([18], chap. IV-3)

1) There is a finite orbit and all finite orbits have the same cardinality.

3.6   2) The action is minimal.

3) There is a unique proper minimal invariant closed subset $F \subset S^1$, which is a Cantor set.

In the first case ρ is quasiconjugated to an action ρ_0 by rotations; a quasiconjugacy is given by $h_{b,\mu}$ where μ is the uniform measure along a finite orbit. Then $\rho_0(\Gamma)$ is finite cyclic.

In the second case $\rho(\Gamma)$ is either bounded (equicontinuous), preserves hence a probability measure μ necessarily with full support and no atoms, and can be conjugated into the subgroup of rotations by the homeomorphism $h_{b,\mu}$; or $\rho(\Gamma)$ is unbounded.

In the third case one can collapse each connected component of $S^1 \setminus F$ to a point and obtain this way a continuous quasiconjugacy between ρ and a minimal action. In fact such a quasiconjugacy is given by $h_{b,\mu}$ where μ is a probability measure without atoms, and with support, precisely the Cantor set F.

With this at hand and denoting as usual

$$\text{rot}: \text{Homeo}^+(S^1) \to \mathbb{Z}\backslash\mathbb{R},$$

the rotation number function, we have

PROPOSITION 3.2. *Let $\rho : \Gamma \to \text{Homeo}^+(S^1)$ be a homomorphism.*

1) $\rho^*(e_{\mathbb{R}}^b) = 0$ *if and only if ρ is quasiconjugated to a subgroup of the group of rotations. When this is the case,*

$$\Gamma \to \mathbb{Z}\backslash\mathbb{R}$$

$$\gamma \longmapsto \text{rot}\,\rho(\gamma)$$

is a homomorphism and ρ is quasiconjugated to the action by rotations defined by it.

2) $\rho^*(e_{\mathbb{R}}^b) \neq 0$ *if and only if ρ is quasiconjugated to a minimal unbounded action.*

Proof. 1) Assume $\rho^*(e_{\mathbb{R}}^b) = 0$. Then there exists $\beta : \Gamma \to \mathbb{R}$ bounded with

$$d\beta(\gamma, \eta) := \beta(\gamma) - \beta(\gamma\eta) + \beta(\eta) = c(\rho(\gamma), s(\eta)), \quad \forall \gamma, \eta \in \Gamma.$$

Since $c(\,,\,)$ takes values in $\mathbb{Z}$, the map $f : \Gamma \to \mathbb{Z}\backslash\mathbb{R}$, $\gamma \longmapsto f(\gamma) = \beta(\gamma)$ mod $\mathbb{Z}$, is a homomorphism. Let $\rho_0(\gamma)(x) := x + f(\gamma)$, $x \in \mathbb{Z}\backslash\mathbb{R}$ be the corresponding action by rotations. Denoting by $\{\ \}$ the fractional part of a real number, we then have $\overline{\rho_0(\gamma)}\,(x) = x + \{\beta(x)\}$, $\forall \gamma \in \Gamma$ and hence $c(\rho_0(\gamma), \rho_0(\eta)) = d(\{\beta\})(\gamma, \eta)$, $\forall \gamma, \eta \in \Gamma$. This implies for all $\gamma, \eta \in \Gamma$:

$$c(\rho_0(\gamma), \rho_0(\eta)) + d([\beta])(\gamma, \eta) = c(\rho(\gamma), \rho(\eta))$$

where $[\beta] : \Gamma \to \mathbb{Z}$ is the bounded function associating to γ the integral part $[\beta(\gamma)]$ of $\beta(\gamma)$. As a result $\rho_0^*(e^b) = \rho^*(e^b)$ and by Ghys's theorem (Theorem 3.1) ρ and ρ_0 are quasiconjugated. Since the rotation number is an invariant of quasiconjugation we have $f(\gamma) = \operatorname{rot}\rho(\gamma)$, $\forall \gamma \in \Gamma$ and hence $\gamma \longmapsto \operatorname{rot}\rho(\gamma)$ is a homomorphism.

Conversely, if ρ is quasiconjugated to a homomorphism $\rho_0 : \Gamma \to \operatorname{Rot}$, then $\rho^*(e^b) = \rho_0^*(e^b)$ and hence $\rho^*(e_{\mathbb{R}}^b) = \rho_0^*(e_{\mathbb{R}}^b)$. But since Rot is compact, $e_{\mathbb{R}}^b|_{\operatorname{Rot}} = 0$, which implies $0 = \rho_0^*(e_{\mathbb{R}}^b) = \rho^*(e_{\mathbb{R}}^b)$.

2) If $\rho^*(e_{\mathbb{R}}^b) \neq 0$ then it follows from 1) that ρ is not quasiconjugated into Rot and hence by the trichotomy (3.6) and the discussion following it, ρ is quasiconjugated to a minimal unbounded action. $\square$

Since bounded cohomology with $\mathbb{R}$-coefficients vanishes for amenable groups we conclude that

COROLLARY 3.3. *If Γ is amenable then for any homomorphism $\rho : \Gamma \to \operatorname{Homeo}^+(S^1)$, the map*

$$\Gamma \to \mathbb{Z}\backslash\mathbb{R}$$
$$\gamma \longmapsto \operatorname{rot}\rho(\gamma)$$

is a homomorphism.

A simpler proof of the last corollary consists in observing that the rotation number with respect to a Γ-invariant measure coincides with Poincaré's rotation number and is clearly a group homomorphism.

Now we turn to a closer study of minimal unbounded actions. In this context the following definition will be useful:

DEFINITION 3.4. A nontrivial S^1-factor of an action $\rho : \Gamma \to \text{Homeo}^+(S^1)$ is a pair (φ, ρ_0) consisting of a nonconstant continuous map $\varphi : S^1 \to S^1$ and an action $\rho_0 : \Gamma \to \text{Homeo}^+(S^1)$ with $\varphi\rho(\gamma) = \rho_0(\gamma)\varphi, \forall \gamma \in \Gamma$.

Denoting by $\mathcal{Z}_{\mathcal{H}^+}(L)$ the centralizer in $\text{Homeo}^+(S^1)$ of a subgroup L we have

THEOREM 3.5. ([23], [16]) *Let $\rho : \Gamma \to \text{Homeo}^+(S^1)$ be minimal unbounded. Then the centralizer $C_\rho := \mathcal{Z}_{\mathcal{H}^+}(\rho(\Gamma))$ of $\rho(\Gamma)$ in $\text{Homeo}^+(S^1)$ is finite cyclic and the S^1-factor (φ, ρ_0), where $\varphi : S^1 \to C_\rho\backslash S^1$ is the quotient map and ρ_0 the quotient action, is strongly proximal.*

For minimal circle actions one observes that strong proximality is equivalent to the property that every proper interval can be contracted; we recall that a subset $F \subset S^1$ can be contracted if there exists a sequence $(\gamma_n)_{n \geq 1}$ such that $\lim_{n \to \infty} \text{diam}(\rho(\gamma_n)F) = 0$. The idea of the proof of Theorem 3.5 is then as follows: Since ρ is minimal unbounded, in particular not equicontinuous, every point $x \in S^1$ has a neighborhood that can be contracted. For every $x \in S^1$ the set $\{I = [x, y) | I \text{ can be contracted}\}$ is then totally ordered by inclusion, nonempty, and hence contains a unique maximal element that we denote $[x, \theta(x))$. Then one verifies that $\theta : S^1 \to S^1$ is a homeomorphism that is periodic and $C_\rho = \mathcal{Z}_{\mathcal{H}^+}(\rho(\Gamma)) = \langle \theta \rangle$. By construction, every proper interval in the quotient $C_\rho\backslash S^1$ can be contracted. We record the following immediate

COROLLARY 3.6. *Let $\rho : \Gamma \to \text{Homeo}^+(S^1)$ be minimal unbounded. Then ρ is strongly proximal if and only if $\mathcal{Z}_{\mathcal{H}^+}(\rho(\Gamma)) = (e)$.*

We now establish the uniqueness of a proximal S^1-factor:

PROPOSITION 3.7. *Let $\rho : \Gamma \to \text{Homeo}^+(S^1)$ be minimal unbounded and let (φ, ρ_0) be a nontrivial S^1-factor. If ρ_0 is strongly proximal, then φ is up to conjugation the quotient map by $\mathcal{Z}_{\mathcal{H}^+}(\rho(\Gamma))$.*

First we will need

LEMMA 3.8. *If ρ is minimal and (ρ_0, φ) is a nontrivial S^1-factor, then $\deg\varphi \neq 0$ and φ is surjective.*

Proof. Assume that $\deg\varphi = 0$ and let $\widetilde{\varphi} : S^1 \to \mathbb{R}$ be a lift of φ. Then there is for every $\gamma \in \Gamma$ a $c(\gamma) \in \mathbb{Z}$ such that

$$\widetilde{\varphi}\rho(\gamma) = T^{c(\gamma)} \overline{\rho_0(\gamma)} \, \widetilde{\varphi}.$$

Let $M = \sup_{x \in S^1} \widetilde{\varphi}(x)$ and $x_0 \in S^1$ with $M = \widetilde{\varphi}(x_0)$. Then

$$\widetilde{\varphi}(x_0) = \sup_x \widetilde{\varphi}(x) = \sup_x \widetilde{\varphi}(\rho(\gamma)x)$$

$$= c(\gamma) + \overline{\rho_0(\gamma)} \, (\sup_x \widetilde{\varphi}(x))$$

$$= c(\gamma) + \overline{\rho_0(\gamma)} \, (\widetilde{\varphi}(x_0)),$$

which implies that $\varphi(x_0) \in S^1$ is fixed by $\rho_0(\gamma)$, $\forall \gamma \in \Gamma$. The nonempty closed subset $\varphi^{-1}(\varphi(x_0))$ being $\rho(\Gamma)$-invariant equals S^1 and thus φ is constant. $\quad\square$

Proof of Proposition 3.7. Let θ be a generator of $\mathcal{Z}_{\mathcal{H}^+}(\rho(\Gamma))$ constructed as indicated in the discussion following Theorem 3.5. We claim that $\varphi\theta = \varphi$. For $x \in S^1$ we consider the closed connected subset $I_x := \varphi([x, \theta(x)])$ and, for a fixed x, distinguish two cases:

a) $I_x = S^1$: then either $\varphi(\theta(x)) = \varphi(x)$, which implies by equivariance that $\varphi\theta$ and φ coincide on $\rho(\Gamma) \cdot x$ and hence everywhere, or there is $y \in (x, \theta(x))$ with $\varphi(x) = \varphi(y)$ and $\varphi([x, y]) = S^1$. But by the construction of θ, $[x, y]$ can be contracted, which by continuity of φ implies the same for $\varphi([x, y]) = S^1$; this is absurd.

b) I_x is a proper subset of S^1: Observe that for all $\gamma \in \Gamma$, $\rho_0(\gamma)I_x = I_{\rho(\gamma)x}$. Since I_x is a proper interval and ρ_0 is strongly proximal we can choose a sequence $(\gamma_n)_{n \geq 1}$ such that diam $(\rho_0(\gamma_n)I_x)$ tends to zero for $n \to \infty$ and $(\rho(\gamma_n)x)_{n \geq 1}$ converges to some point y, which by continuity of φ implies that $\varphi(y) = \varphi\theta(y)$. Again by using equivariance and minimality we get $\varphi = \varphi\theta$.

Thus the claim is established and therefore by passing to the quotient by $\mathcal{Z}_{\mathcal{H}^+}(\rho(\Gamma))$ we may assume that ρ is strongly proximal as well. We already know (Lemma 3.8) that deg $\varphi \neq 0$ and φ is surjective. Assume then that there exists $s \neq t$ in S^1 with $\varphi(s) = \varphi(t)$. Since deg $\varphi \neq 0$ we have either $\varphi([s, t]) = S^1$ or $\varphi([t, s]) = S^1$. Assume without loss of generality that the former occurs. Since $[s, t]$ is a proper interval in S^1 it can be contracted by ρ, which clearly implies a contradiction. Thus φ is injective as well and hence a homeomorphism. $\quad\square$

4. Strongly Proximal Actions, Boundary Maps, and Cocycles

In this section we study continuous group actions of a locally compact group G on the circle using boundary maps. The main goal is Theorem 4.5, which gives a description of the space of conjugacy classes of minimal strongly proximal actions in terms of an explicit space of cocycles on appropriate Poisson boundaries of G.

4.1. Cohomological Preliminaries

For a nondiscrete group some care must be taken when defining the analogues of the various Euler classes in the continuous context. Let G be locally compact, second countable, and let A be either $\mathbb{Z}$ or $\mathbb{R}$. Then $H_{bc}^{\bullet}(G, A)$ denotes the cohomology defined via A-valued bounded Borel cochains on G; in the case $A = \mathbb{R}$ these cohomology groups coincide with those obtained by taking the subcomplex of bounded continuous cochains on G; for more details see [5], §2.3. Given a continuous action $\rho : G \to \mathrm{Homeo}^+(S^1)$ the function $(g, h) \longmapsto c(\rho(g), \rho(h))$ is a bounded Borel (inhomogeneous) 2-cocycle on G and leads to classes $\rho^*(e^b) \in H_{bc}^2(G, \mathbb{Z})$ and $\rho^*(e_{\mathbb{R}}^b) \in H_{bc}^2(G, \mathbb{R})$, which are respectively the bounded Euler class and the real bounded Euler class of the continuous action.

Recall that if (B, ν_B) is a standard Lebesgue G-space that is doubly ergodic and amenable, there is a canonical isometric isomorphism; see [24] thm. 7.5.3,

4.1
$$H_{bc}^2(G, \mathbb{R}) \simeq Z\,L_{\mathrm{alt}}^{\infty}(B^3, \mathbb{R})^G$$

where the right-hand side is the space of measurable, essentially bounded, alternating G-invariant cocycles on B^3.

PROPOSITION 4.1. *Let $\rho : G \to \mathrm{Homeo}^+(S^1)$ be a continuous action and assume that there exists a measurable G-equivariant map $\varphi : B \to S^1$. Then under the isomorphism in (4.1) the class $\rho^*(e_{\mathbb{R}}^b)$ corresponds to the cocycle*

$$B^3 \to \mathbb{R}$$

$$(x, y, z) \longmapsto -\frac{1}{2}\, o(\varphi(x), \varphi(y), \varphi(z)).$$

If in addition the essential image of φ contains at least three points, we have

$$\|\rho^*(e_{\mathbb{R}}^b)\| = \frac{1}{2}.$$

We will need the following explicit relationship between the Euler and orientation cocycle; see for instance [19] Lemma 2.1.

LEMMA 4.2. *For every $f, g \in \mathrm{Homeo}^+(S^1)$ we have*

$$2c(f,g) = -o(\dot{o}, f(\dot{o}), fg(\dot{o})) + 1$$

$$+ (\delta_{\dot{o}}(fg(\dot{o})) - \delta_{\dot{o}}(f(\dot{o})) - \delta_{\dot{o}}(g(\dot{o}))),$$

where $\dot{o} \in \mathbb{Z} \backslash \mathbb{R}$.

Proof of Proposition 4.1. This is a standard argument using [3]. Since G acts continuously on S^1, the G-complex $(B^\infty_{\text{alt}}((S^1)^{\bullet+1}, \mathbb{R}), d)$ of bounded alternating Borel cochains on S^1 is a strong resolution of the trivial G-module $\mathbb{R}$ (see [3] Prop. 2.1). Then, owing to the properness of the G-action on G and the amenability of the G-action on B, the two complexes of G-modules $(L^\infty_{\text{alt}}(G^{\bullet+1}, \mathbb{R}), d)$ and $(L^\infty_{\text{alt}}(B^{\bullet+1}, \mathbb{R}), d)$ are strong resolutions of $\mathbb{R}$ by relatively injective G-Banach modules [24] section 7.5. Then there is a morphism of G-complexes

4.2 $$c^\bullet : L^\infty_{\text{alt}}(G^{\bullet+1}, \mathbb{R}) \to L^\infty_{\text{alt}}(B^{\bullet+1}, \mathbb{R})$$

extending the identity and any two such are G-equivariantly homotopic; the canonical map induced in cohomology is then an isometric isomorphism that in degree two gives the one mentioned in (4.1).

Consider the morphisms of G-resolutions

$$\rho^{(n)} : B^\infty_{\text{alt}}(S^1)^{n+1}, \mathbb{R}) \to L^\infty_{\text{alt}}(G^{n+1}, \mathbb{R})$$

and

$$\varphi^{(n)} : B^\infty_{\text{alt}}((S^1)^{n+1}, \mathbb{R}) \to L^\infty_{\text{alt}}(B^{n+1}, \mathbb{R})$$

defined respectively for $\alpha : (S^1)^{n+1} \to \mathbb{R}$ by

$$\rho^{(n)}(\alpha)(g_0, \ldots, g_n) := \alpha(\rho(g_0)\dot{o}, \ldots, \rho(g_n)\dot{o})$$

and

$$\varphi^{(n)}(\alpha)(x_0, \ldots, x_n) := \alpha(\varphi(x_0), \ldots, \varphi(x_n)).$$

Then $c^{(\bullet)} \rho^{(\bullet)}$ and $\varphi^{(\bullet)}$ are both morphisms of G-complexes extending the identity and therefore, since $(L^\infty_{\text{alt}}(B^{\bullet+1}, \mathbb{R}), d)$ is a resolution by relatively injective G-modules, induce the same map in cohomology. Now we specialize this to degree $n = 2$. Let $[o] \in H^2(B^\infty_{\text{alt}}(S^{\bullet+1}))$ be the class defined by the orientation cocycle. Then we have by Lemma 4.2: $\rho^{(2)}([o]) = 2\rho^*(e^b_{\mathbb{R}})$ and in addition that under the isometric isomorphism $c^{(2)}$, $\rho^{(2)}([o])$ is represented by $(x, y, z) \longmapsto o(\varphi(x), \varphi(y), \varphi(z))$. In particular $2\|\rho^*(e^b_{\mathbb{R}})\|$ is the essential supremum of the latter cocycle. $\qquad \square$

We will see below that the existence of a map as in Proposition 4.1 imposes strong conditions on the type of action considered. Before we turn to this we will need

LEMMA 4.3. *Let* $\rho_0, \rho_1 : G \to \mathrm{Homeo}^+(S^1)$ *be continuous actions and* $\varphi : S^1 \to S^1$ *a degree k covering such that*

$$\rho_0(g)\,\varphi = \varphi\,\rho_1(g), \quad \forall g \in G.$$

Then we have the following equality

$$\rho_0^*(e^b) = k\,\rho_1^*(e^b)$$

in $H_{bc}^2(G, \mathbb{Z})$.

Proof. Let $p : \mathbb{R} \to \mathbb{Z}\backslash\mathbb{R}$ denote projection and $M_k : \mathbb{R} \to \mathbb{R}, x \longmapsto k \cdot x$, so that up to conjugation we have $p\,M_k = \varphi p$. Then we have $\forall g \in G$:

$$p\,M_k\,\overline{\rho_1(g)} = \varphi\,p\,\overline{\rho_1(g)} = \varphi\,\rho_1(g)\,p = \rho_0(g)\,\varphi p = \rho_0(g)\,p\,M_k = p\,\overline{\rho_0(g)}\,M_k\,.$$

Therefore there is $\alpha(g) \in \mathbb{Z}$ such that

4.3 $$M_k\,\overline{\rho_1(g)} = T^{\alpha(g)}\,\overline{\rho_0(g)}\,M_k\,.$$

Evaluation at $0 \in \mathbb{R}$ gives

$$k\,\overline{\rho_1(g)}\,(0) = \alpha(g) + \overline{\rho_0(g)}\,(0),$$

which first shows that $\alpha : G \to \mathbb{Z}$ is a Borel function and furthermore implies that $\sup_{g \in G} |\alpha(g)| \le k + 1$.

Applying (4.3) to products and using the relation defining the Euler cocycle we get

$$k\,c(\rho_1(g), \rho_1(h)) + \alpha(gh) - \alpha(g) - \alpha(h) = c(\rho_0(g), \rho_0(h)),$$

which proves the lemma. $\qquad\square$

Now we come to our first application.

COROLLARY 4.4. *Let* $\mu \in \mathcal{M}^1(G)$ *be a spread-out probability measure on G and (B, ν_B) a standard Lebesgue G-space, where ν_B is μ-stationary. Assume that B is doubly ergodic and amenable. Given a minimal unbounded action* $\rho : G \to \mathrm{Homeo}^+(S^1)$ *the following are equivalent:*

1) ρ is strongly proximal.

2) There exists a G-equivariant measurable map $\varphi : B \to S^1$.

3) $\|\rho^*(e_{\mathbb{R}}^b)\| = \frac{1}{2}$.

Proof. (1) $\implies$ (2): follows immediately from Theorem 2.1 and Proposition 2.3 (3).

(2) $\implies$ (3): follows from Proposition 4.1.

(3) $\longrightarrow$ (1): Let (ψ, ρ_0) be the strongly proximal factor of ρ. Applying the implication (1) $\implies$ (3) to ρ_0 we get $\|\rho_0^*(e_{\mathbb{R}}^b)\| = \frac{1}{2}$, and for $k = \deg \psi$, we get from Lemma 4.3 that $\rho_0^*(e_{\mathbb{R}}^b) = k \cdot \rho^*(e_{\mathbb{R}}^b)$, which implies $k = 1$ and hence that ρ is strongly proximal. $\qquad\square$

4.2. Minimal Strongly Proximal Actions and Cocycles

The main result of this section is

THEOREM 4.5. Let G be locally compact second countable and let $\mu \in \mathcal{M}^1(G)$, (B, ν_B) be as in Corollary 4.4. Then there is a bijection between

a) The set of conjugacy classes of continuous G-actions on S^1 that are minimal and strongly proximal.

b) The set of measurable functions, up to equality almost everywhere,

$$\omega : B^3 \to \mathbb{R}$$

such that

1) ω is an alternating strict cocycle.

2) $\omega(gx, gy, gz) = \omega(x, y, z)$ for every $g \in G$ and almost every $(x, y, z) \in B^3$.

3) ω takes values in $\{\pm 1\}$ almost everywhere.

This bijection is implemented by the map to which a continuous minimal strongly proximal action $\rho : G \to \text{Homeo}^+(S^1)$ associates the cocycle

$$\omega(x, y, z) = o(\varphi(x), \varphi(y), \varphi(z))$$

where $\varphi : B \to S^1$ is the map given by Corollary 4.4.

The proof of the theorem is divided into two steps. Fix $\omega : B^3 \to \mathbb{R}$ satisfying the properties (1), (2), and (3) above.

Step 1: We show that there exists a measurable map $\varphi : B \to S^1$ satisfying the following properties:

a) $\varphi_*(\nu_B) \in \mathcal{M}^1(S^1)$ has no atoms.

b) the essential image Ess Im φ of φ equals S^1.

c) $o(\varphi(x), \varphi(y), \varphi(z)) = \omega(x, y, z)$ for almost every $(x, y, z) \in B^3$.

Step 2: Given a measurable function $\varphi : B \to S^1$ satisfying (a), (b), and (c) we construct a continuous homomorphism

$$\pi_\varphi : G \to \mathrm{Homeo}^+(S^1)$$

such that $\varphi(gx) = \pi_\varphi(g)(\varphi(x))$, for every $g \in G$ and almost every $x \in B$, and conclude that π_φ is minimal strongly proximal. Finally we show that given φ, ψ satisfying (a), (b), and (c), then π_φ and π_ψ are conjugate.

Combining Corollary 4.4, Step 1, and Step 2 clearly completes the proof of Theorem 4.5.

4.3. Step 1: The Construction of the Measurable Map

We fix once and for all a measurable map $\omega : B^3 \to \mathbb{R}$ satisfying properties (1), (2), and (3) in Theorem 4.5. For every $(x, y) \in B^2$ define

$$I(x, y) = \{z \in B : \omega(x, z, y) = 1\},$$

which is measurable. The next few lemmas are intended to show that the sets $I(x, y)$ behave like intervals on S^1. We will use the notation $E \equiv F$ to indicate that two sets E, F differ by a set of measure zero; similarly $E \subset F$ means that E is contained in F up to a set of measure zero.

LEMMA 4.6.

1) $I(x, y) \cup I(y, x) \equiv B$ for a.e. $(x, y) \in B^2$.

2) $I(x, y) \cap I(y, x) = \phi, \quad \forall (x, y) \in B^2$.

3) $\forall g \in G$ and a.e. $(x, y) \in B^2$, $g(I(x, y)) \equiv I(gx, gy)$.

4) For a.e. $(x, y) \in B^2$, $0 < \nu_B(I(x, y)) < 1$.

Proof.

1) follows from the fact that ω is alternating and $\omega(x, y, z) \in \{\pm 1\}$ for a.e. $(x, y, z) \in B^3$.

2) is obvious.

3) follows from the G-invariance of ω and the fact that the G-action on B preserves the measure class of ν_B.

4) Consider

$$A_1 = \{(x, y) \in B^2 : \nu_B(I(x, y)) = 1\}$$
$$A_0 = \{(x, y) \in B^2 : \nu_B(I(x, y)) = 0\}.$$

Denoting $\wedge$ the map $(x, y) \to (y, x)$, we have from (1) that $A_1 \equiv \widehat{A_0}$; in addition we have that $A_1 \cap A_0 = \phi$ and both sets are measurable and G-invariant by (3). Thus if $v_B^2(A_1) > 0$, then $v_B^2(A_0) = v_B^2(\widehat{A_0}) = v_B^2(A_1) > 0$, which contradicts the ergodicity of the G-action on B^2. Thus A_1 and A_0 are both of measure zero. $\qquad\qquad\qquad\square$

LEMMA 4.7.

1) *For* $(a, c) \in B^2$ *and* $b \in I(a, c)$ *we have*

4.4
$$I(a, c) \supset I(a, b) \cup I(b, c)$$

4.5
$$I(a, b) \cap I(b, c) = \psi .$$

2) *For a.e.* $(a, c) \in B^2$ *and a.e.* $b \in I(a, c)$

4.6
$$I(a, c) \subseteq I(a, b) \cup I(b, c) .$$

In particular, for a.e. $(a, c) \in B^2$ *and a.e.* $b \in I(a, c)$

4.7
$$I(a, c) \equiv I(a, b) \cup I(b, c) .$$

Proof. Applying the cocycle identity to $a, x, b, c \in B$ we have

4.8
$$\omega(x, b, c) - \omega(a, b, c) + \omega(a, x, c) - \omega(a, x, b) = 0 .$$

If $b \in I(a, c)$ and $x \in I(a, b)$, we have $\omega(a, b, c) = 1$, $\omega(a, x, b) = 1$, which with (4.8) implies $\omega(x, b, c) + \omega(a, x, c) = 2$ and hence $x \in I(a, c)$. If $x \in I(b, c)$ then (4.8) gives $\omega(a, x, c) - \omega(a, x, b) = 2$ and hence $x \in I(a, c)$. This shows the inclusion (4.4). Concerning (4.5), we apply (4.8) to $x \in I(a, b) \cap I(b, c)$ to get a contradiction.

Concerning (4.6), assume that $b \in I(a, c)$ and $x \in I(a, c)$ and apply (4.8) to get

4.9
$$\omega(x, b, c) = \omega(a, x, b).$$

For a.e. $(a, b) \in B^2$ and a.e. $x \in B$ we have either $\omega(a, x, b) = 1$ and hence $x \in I(a, b)$ or $\omega(a, x, b) = -1$, which with (4.9) gives $x \in I(b, c)$. $\qquad\square$

Lemma 4.6 (4) and Lemma 4.7 imply then immediately

LEMMA 4.8. *For a.e.* $(a, x, y) \in B^3$ *we have the following dichotomy:*

1) $x \in I(a, y)$, $I(a, y) \equiv I(a, x) \cup I(x, y)$ *and* $v_B(I(a, x)) < v_B((a, y))$.
2) $y \in I(a, x)$, $I(a, x) \equiv I(a, y) \cup I(y, x)$ *and* $v_B(I(a, y)) < v_B((a, x))$.

Given $a \in B$, we define $f_a : B \to \mathbb{Z}\backslash\mathbb{R}$ by

$$f_a(x) := \nu_B(I(a, x)) \bmod \mathbb{Z}.$$

Then we have

LEMMA 4.9. *For a.e.* $a \in B$ *we have*

1) $E = \{(x, y) \in B^2 : f_a(x) = f_a(y)\}$ *is of measure zero.*
2) $(f_a)_*(\nu_B)$ *has no atoms.*

Proof. 1) If $(x, y) \in E$ then either $\nu_B(I(a, x)) = 0$ and $\nu_B(I(a, y)) = 1$, or $\nu_B(I(a, x)) = \nu_B((a, y))$; in both cases these equalities hold only for a set of (a, x, y)'s of measure zero; in the first case this follows from Lemma 4.6 (4) and in the second from Lemma 4.8.
2) follows from (1). □

LEMMA 4.10. *For a.e.* $a \in B$ *and a.e.* $(x, y, z) \in B^3$

$$o(f_a(x), f_a(y), f_a(z)) = \omega(x, y, z).$$

Proof. Observe that the left-hand side gives a measurable alternating cocycle taking values in $\{-1, 0, 1\}$; it follows from Lemma 4.9 that it takes values in $\{-1, 1\}$ almost everywhere. It therefore suffices to show that if the left-hand side equals 1, then so does the right-hand side.

If $o(f_a(x), f_a(y), f_a(z)) = 1$ then up to cyclically permuting the variables we may assume, using the definition of f_a, that $\nu_B(I(a, x)) < \nu_B(I(a, y)) < \nu_B(I(a, z))$. Together with Lemma 4.8 this implies $x \in I(a, y)$ and $y \in I(a, z)$ except possibly for a set of (a, x, y, z)'s of measure zero. Using the cocycle identity for ω applied to (a, x, y, z) we get $\omega(x, y, z) + \omega(a, x, z) = 2$ and hence $\omega(x, y, z) = 1$. □

The following lemma will be left to the reader:

LEMMA 4.11. *Let* $\xi \in \mathcal{M}^1(S^1)$ *be a probability measure without atoms and*

$$h_\xi : S^1 \to S^1$$
$$x \longmapsto \xi([\dot{o}, x)) \bmod \mathbb{Z}$$

the associated quasiconjugacy. Then,

1) h_ξ *is continuous.*
2) *the measure* $(h_\xi)_*(\xi)$ *has no atoms and its support equals* S^1.

3) *for ξ^3-almost every* $(x, y, z) \in (S^1)^3$ *we have* $o(h_\xi(x), h_\xi(y), h_\xi(z)) = o(x, y, z)$.

Combining Lemma 4.9, 4.10, and 4.11 we obtain the following lemma, which completes Step 1:

LEMMA 4.12. *For a.e. $a \in B$, let $\xi = (f_a)_*(\nu_B)$ and define $\varphi_a := h_\xi \circ f_a$. Then*

1) $(\varphi_a)_*(\nu_B)$ *has no atoms.*
2) Ess Im $\varphi_a = S^1$.
3) $o(\varphi_a(x), \varphi_a(y), \varphi_a(z)) = \omega(x, y, z)$ *for a.e.* $(x, y, z) \in B^3$.

4.4. Step 2: The Construction of the Action

The following result is the basis for our construction:

LEMMA 4.13. *Let $\varphi, \psi : B \to S^1$ be measurable maps such that*

1) $\varphi_*(\nu_B)$ *and* $\psi_*(\nu_B)$ *have no atoms.*
2) Ess Im $\varphi = $ Ess Im $\psi = S^1$.
3) $o(\varphi(x), \varphi(y), \varphi(z)) = o(\psi(x), \psi(y), \psi(z))$ *for a.e.* $(x, y, z) \in B^3$.

Then the essential image $F \subset S^1 \times S^1$ of the map

$$B \to S^1 \times S^1$$

$$x \longmapsto (\varphi(x), \psi(x))$$

is the graph of an orientation-preserving homeomorphism $h : S^1 \to S^1$ and,

$$h(\varphi(x)) = \psi(x) \text{ for a.e. } x \in B.$$

Proof. The proof is in two steps.

CLAIM 1: Given $(\xi_1, \eta_1), (\xi_2, \eta_2), (\xi_3, \eta_3)$ in F such that ξ_1, ξ_2, ξ_3 are pairwise distinct and the same holds for η_1, η_2, η_3, we have

$$o(\xi_1, \xi_2, \xi_3) = o(\eta_1, \eta_2, \eta_3).$$

Pick $V_i \ni \xi_i$ and $W_i \ni \eta_i$ open intervals such that V_1, V_2, V_3 are pairwise disjoint and the same holds for W_1, W_2, W_3. Then

$$B_i = \{x \in B : \varphi(x) \in V_i \text{ and } \psi(x) \in W_i\}$$

is of positive measure, $i = 1, 2, 3$.

We then have for a.e. $(x, y, z) \in B_1 \times B_2 \times B_3$,

$$o(\xi_1, \xi_2, \xi_3) = o(\varphi(x_1), \varphi(x_2), \varphi(x_3)) = o(\psi(x_1), \psi(x_2), \psi(x_3))$$

$$= o(\eta_1, \eta_2, \eta_3).$$

CLAIM 2: Let (ξ_1, η_1), (ξ_2, η_2) be in F. Then $\xi_1 \neq \xi_2$ iff $\eta_1 \neq \eta_2$.

Assume that $\xi_1 \neq \xi_2$. We will repeatedly use the fact that

4.10 $$(\varphi(x), \psi(x)) \in F \text{ for a.e. } x \in B.$$

Since Ess Im $\varphi = S^1$ and $\xi_1 \neq \xi_2$ the set $\varphi^{-1}((\xi_1, \xi_2))$ is of positive measure; then in virtue of (4.10) and the hypothesis that $\psi_*(\nu_B)$ has no atoms, we can pick $t \in \varphi^{-1}((\xi_1, \xi_2))$ with $(\varphi(t), \psi(t)) \in F$ and $\psi(t) \notin \{\eta_1, \eta_2\}$. Similarly we can find $s \in \varphi^{-1}((\xi_2, \xi_1))$ such that $(\varphi(s), \psi(s)) \in F$ and $\psi(s) \notin \{\psi(t), \eta_1, \eta_2\}$. Then the cocycle identity applied to $(\eta_1, \psi(t), \eta_2, \psi(s))$ gives

$$o(\psi(t), \eta_2, \psi(s)) - o(\eta_1, \eta_2, \psi(s)) + o(\eta_1, \psi(t), \psi(s)) - o(\eta_1, \psi(t), \eta_2) = 0.$$

Applying Claim 1 to the first and third terms we get

$$o(\varphi(t), \xi_2, \varphi(s)) - o(\eta_1, \eta_2, \psi(s)) + o(\xi_1, \varphi(t), \varphi(s)) - o(\eta_1, \psi(t), \eta_2) = 0$$

and by our choices we have

$$o(\varphi(t), \xi_2, \varphi(s)) = 1$$

$$o(\xi_1, \varphi(t), \varphi(s)) = 1,$$

which implies that

$$o(\eta_1, \eta_2, \psi(s)) + o(\eta_1, \psi(t), \eta_2) = 2$$

and hence $\eta_1 \neq \eta_2$.

This shows Claim 2 modulo interchanging the roles of φ and ψ.

Since $pr_i(F) = S^1$ we get from Claim 2 that F is the graph of a homeomorphism h, which by Claim 1 is orientation preserving. Finally (4.10) says precisely that $h(\varphi(x)) = \psi(x)$ for a.e. $x \in B$ and this completes the proof of the lemma. □

Now we return to our cocycle $\omega : B^3 \to \mathbb{R}$ satisfying the hypothesis of Theorem 4.5 and let $\varphi : B \to S^1$ be a measurable map as given by Lemma 4.12. Since ω is G-invariant we can apply Lemma 4.13 to φ and $\varphi g, g \in G$ to obtain an orientation-preserving homeomorphism denoted $\pi_\varphi(g) \in \text{Homeo}^+(S^1)$ satisfying

$$\pi_\varphi(g)(\varphi(x)) = \varphi(gx) \text{ for a.e. } x \in B.$$

This equivariance property implies that

$$\pi_\varphi : G \to \text{Homeo}^+(S^1)$$

is a homomorphism.

LEMMA 4.14.

1) $\pi_\varphi : G \to \text{Homeo}^+(S^1)$ *is continuous. It is minimal, unbounded, and strongly proximal.*

2) *Let* φ, ψ *be as in Lemma 4.13, and* π_φ, π_ψ *the corresponding homomorphisms. Let* $h : S^1 \to S^1$ *be the homeomorphism given by Lemma 4.13. Then,*

$$h\,\pi_\varphi(g) = \pi_\psi(g)\,h \quad \forall g \in G.$$

Proof. Assertion 2) is clear and follows from the various equivariance properties and thus we concentrate on 1).

For a.e. $x \in B$, the map $g \longmapsto \pi_\varphi(g)(\varphi(x)) = \varphi(gx)$ is measurable and hence the homomorphism $\pi_\varphi : G \to \text{Homeo}^+(S^1)$ is measurable. Since G is locally compact second countable and since $\text{Homeo}^+(S^1)$ is second countable we deduce that π_φ is continuous. Using Proposition 4.1 we see that $\pi_\varphi^*(e_{\mathbb{R}}^b) \neq 0$ and hence π_φ is nonelementary. Assume that there is an exceptional minimal set $K \underset{+}{\subset} S^1$. Let J be a connected component if $S^1 \backslash K$; since $\text{Ess Im } \varphi = S^1$, $\varphi^{-1}(J)$ is of positive measure but not of full measure in B. Since $\forall g \in B$ we have either $\pi_\varphi(g) J = J$ or $\pi_\varphi(g) J \cap J = \phi$, we conclude that either $g\varphi^{-1}(J) \equiv \varphi^{-1}(J)$ or $g\varphi^{-1}(J) \cap \varphi^{-1}(J) = \phi$. Since the G-action on B is ergodic we have $\bigcup_{g \in G} g\varphi^{-1}(J) = \varphi^{-1}(S^1 \backslash K) \equiv B$ and hence there is $g_0 \in B$ with $g_0^{-1} \varphi^{-1}(J) \cap \varphi^{-1}(J) = \phi$.

Thus $\bigcup_{h \in G} h(\varphi^{-1}(J) \times \varphi^{-1}(J))$ is of positive measure in B^2, G-invariant, and does not meet $g_0 \varphi^{-1}(J) \times \varphi^{-1}(J)$, itself also of positive measure, this contradicts the ergodicity of the G-action on $B \times B$. Thus π_φ is minimal unbounded. Finally it follows from Corollary 4.4 that π_φ is strongly proximal. $\qquad\square$

5. Proofs of Theorems 1.2 and 1.3, Remark 1.4(2), and Corollaries 1.5 and 1.6

In order to apply the results obtained so far we recall

THEOREM 5.1. ([21])

Let G be a locally compact second countable group and $\mu \in \mathcal{M}^1(G)$ a symmetric spread-out probability measure on G. Then the G-action on the associated Poisson *boundary* (B, ν_B) *is amenable, doubly ergodic; the same properties hold for the action of a lattice* $\Gamma < G$.

Using Theorem 5.1 for $\Gamma = G$ and applying Corollary 4.4, Theorem 4.5, and Proposition 4.1 implies readily Theorem 1.3. Concerning Remark 1.4 (2), if $\rho_1, \ldots, \rho_n$ are minimal strongly proximal actions and $\omega_i : B^3 \to \{-1, 1\}$ are the associated cocycle, then for any $n_i \in \mathbb{Z}$, $\sum_{i=1}^{n} n_i \omega_i$ takes integral values and hence by the isomorphism in (4.1) and Proposition 4.1 the norm of $\sum_{i=1}^{n} n_i \rho_i^*(e_{\mathbb{R}}^b)$ is half integral.

Corollary 1.6 (1) follows from Proposition 3.2, Proposition 4.1, and Lemma 4.3.

Corollary 1.6 (2) follows from Theorem 1.3 (2), which implies that ρ_1, ρ_2 have modulo conjugation the same strongly proximal quotient.

We turn now to the proof of Theorem 1.2. First assume that for the given minimal unbounded action ρ, ρ_{sp} extends continuously. Let $k = |\mathcal{Z}_{\mathcal{H}^+}(\rho(\Gamma))|$. Then we have that $(\rho_{sp})^*(e_{\mathbb{R}}^b) = k \cdot \rho^*(e_{\mathbb{R}}^b)$ (Lemma 4.3) and since $(\rho_{sp})^*(e_{\mathbb{R}}^b)$ is in the image of the restriction map, so is $\rho^*(e_{\mathbb{R}}^b)$.

Conversely, apply Theorem 5.1 and let $\mu_\Gamma \in \mathcal{M}^1(\Gamma)$ be a probability measure of full support such that ν_B is μ_Γ-stationary (see [22]). On the space (B, ν_B) the restriction map

$$H_b^2(G, \mathbb{R}) \to H_b^2(\Gamma, \mathbb{R})$$

is realized by the inclusion

5.1 $$\mathcal{Z} L_{\text{alt}}^\infty(B^3, \mathbb{R})^G \hookrightarrow \mathcal{Z} L_{\text{alt}}^\infty(B^3, \mathbb{R})^\Gamma.$$

Now if $\rho^*(e_{\mathbb{R}}^b)$ is in the image of the restriction map then so is $(\rho_{sp})^*(e_{\mathbb{R}}^b)$. Let $\omega_{\rho_{sp}} : B^3 \to \mathbb{R}$ be the cocycle associated to ρ_{sp}; using (5.1) we get that $\omega_{\rho_{sp}}$ is G-invariant, satisfies all hypotheses of Theorem 4.5 and hence corresponds to a continuous minimal strongly proximal homomorphism $\pi : G \to \text{Homeo}^+(S^1)$. Since $\omega_{\pi|_\Gamma}$ clearly coincides with $\omega_{\rho_{sp}}$ we conclude from Theorem 4.5 that $\pi|_\Gamma$ and ρ_{sp} are conjugate.

6. Locally Compact Groups Acting on S^1

In this section we prove Theorem 1.8 using some results from [16] and [10]. In the sequel, Rot denotes as usual the group of rotations of S^1 and, for every $k \geq 1$, $\text{PSL}(2, \mathbb{R})_k \subset \text{Homeo}^+(S^1)$ is the k-fold cyclic covering group of $\text{PSL}(2, \mathbb{R})$. We record the following

LEMMA 6.1. *Let H be connected, semisimple, with finite center and no compact factors, and*

6.1 $$\pi : H \to \mathrm{Homeo}^+(S^1)$$

a continuous nontrivial homomorphism. Then there exists $k \in \mathbb{N}$ such that up to conjugation, $\pi(H) = \mathrm{PSL}(2, \mathbb{R})_k$.

For the following lemma we recall that given a subgroup $L < \mathrm{Homeo}^+(S^1)$, $\mathcal{N}_{\mathcal{H}^+}(L)$ and $\mathcal{Z}_{\mathcal{H}^+}(L)$ denote respectively its normalizer and centralizer in $\mathrm{Homeo}^+(S^1)$.

LEMMA 6.2.

1) $\mathcal{N}_{\mathcal{H}^+}(\mathrm{Rot}) = \mathcal{Z}_{\mathcal{H}^+}(\mathrm{Rot}) = \mathrm{Rot}$.
2) $\mathcal{N}_{\mathcal{H}^+}(\mathrm{PSL}(2, \mathbb{R})_k) = \mathrm{PSL}(2, \mathbb{R})_k$.
3) $\mathcal{Z}_{\mathcal{H}^+}(\mathrm{PSL}(2, \mathbb{R})_k) = \mathcal{Z}(\mathrm{PSL}(2, \mathbb{R})_k)$, *which is cyclic of order k.*

Proof. 1) Let $g \in \mathcal{N}_{\mathcal{H}^+}(\mathrm{Rot})$; if $g\,r\,g^{-1} = r^{-1}$ $\forall r \in \mathrm{Rot}$, then g cannot preserve orientation; hence $g \in \mathcal{Z}_{\mathcal{H}^+}(\mathrm{Rot})$. Composing g with a rotation we may assume $g(0) = 0$, which on applying $r \in \mathrm{Rot}$ yields $g(r(0)) = r(0)$ and thus $g = e$.

2) For $g \in \mathcal{N}_{\mathcal{H}^+}(\mathrm{PSL}(2, \mathbb{R})_k)$ and $K_k < \mathrm{PSL}(2, \mathbb{R})_k$ maximal compact subgroup, $g^{-1} K_k\,g$ is maximal compact as well and hence there is $h \in \mathrm{PSL}(2, \mathbb{R})_k$ with $g\,K_k\,g^{-1} = h\,K_k\,h^{-1}$ that is $h^{-1}g \in \mathcal{N}_{\mathcal{H}^+}(K_k)$; since K_k is conjugate to Rot we deduce from 1) that $h^{-1}g \in K_k$.

3) follows from (2). □

Finally we record the following consequence of Hilbert's fifth problem:

THEOREM 6.3. ([8] thm. 3.3.3)
Let G be a locally compact group and $A(G) \lhd G$ its amenable radical. Let $G_a := G/A(G)$. Then

1) $(G_a)^0$ *is connected, semisimple, with trivial center and no compact factors.*
2) *The centralizer L of $(G_a)^0$ in G_a is totally disconnected; we have $L \cap (G_a)^0 = (e)$ and the product $L \cdot (G_a)^0$ is open of finite index in G_a.*

Proof of Theorem 1.8. Let $\pi : G \to \mathrm{Homeo}^+(S^1)$ be a continuous minimal action. We consider $\pi|_{A(G)}$ and apply our trichotomy:

Case 1: $\pi(A(G))$ has a finite orbit; then the set of finite $\pi(A(G))$-orbits with fixed cardinality is nonempty closed and $\pi(G)$-invariant, hence coincides with S^1. As a result, $\pi(A(G))$ is finite cyclic.

Case 2: An exceptional minimal set for $\pi(A(G))$ being unique would also be $\pi(G)$-invariant contradicting minimality of $\pi(G)$.

Case 3: $\pi(A(G))$ is minimal; since $A(G)$ is amenable $(\pi|_{A(G)})^*(e_{\mathbb{R}}^b) = 0$ and hence, modulo conjugating π, we may assume that $\pi(A(G))$ is a dense subgroup of Rot. In particular, $\pi(G) \subset \mathcal{N}_{\mathcal{H}^+}(\text{Rot}) = \text{Rot}$ by Lemma 6.2. This gives the first alternative in Theorem 1.8.

Now we return to Case 1 and consider $G_\pi := G/(A(G) \cap \text{Ker}\,\pi)$, which is a finite extension $q : G_\pi \to G_a$ of G_a. Observe that π factors through G_a. Let $L_\pi := q^{-1}(L)$; then $(G_\pi)^0$ is connected, semisimple with finite center and no compact factors, L_π is totally disconnected, and $L_\pi \cdot (G_\pi)^0$ is open of finite index in G_π. We then have two cases:

1) $\pi|_{(G_\pi)^0}$ is trivial. Since $\text{Homeo}^+(S^1)$ does not contain small subgroups, π sends some compact open subgroup of L_π to the identity, which implies that $\text{Ker}\,\pi$ is open.

2) $\pi|_{(G_\pi)^0}$ is nontrivial and hence by Lemma 6.1, $\pi((G_\pi)^0) = \text{PSL}(2, \mathbb{R})_k$; since $\pi(G)$ normalizes $\pi((G_\pi)^0)$ we get $\pi(G) = \text{PSL}(2, \mathbb{R})_k$ by Lemma 6.2 (2). □

References

[1]   U. Bader, A. Furman: e-mail exchange, private communication, September 9–October 6, 2006.

[2]   U. Bader, A. Furman, A. Shaker: Superrigidity, Weyl groups and actions on the circle. ArXiv: math/0605276v4.

[3]   M. Burger, A. Iozzi: Boundary maps in bounded cohomology. Appendix to: "Continuous bounded cohomology and applications to rigidity theory." *Geom. Funct. Anal.*, 12(2):281–292 (2002).

[4]   M. Burger, A. Iozzi: Bounded Kähler class rigidity of actions on Hermitian symmetric spaces. *Ann. Sci. Ecole Norm. Sup. (4)*, 37(1):77–103 (2004).

[5]   M. Burger, A. Iozzi, A. Wienhard: Surface group representations with maximal Toledo invariant. *Annals of Math.*, 172(1):517–566 (2010).

[6]   M. Burger, A. Iozzi, A. Wienhard: Hermitian symmetric spaces and Kähler rigidity. *Transform. Groups*, 12(1):5–32 (2007).

[7]   M. Burger, A. Iozzi, A. Wienhard: Tight homomorphisms and Hermitian symmetric spaces. *Geom. Funct. Anal.*, 19(3):678–721 (2009).

[8] M. Burger, N. Monod: Continuous bounded cohomology and applications to rigidity theory. *Geom. Funct. Anal.*, 12(2):219–280 (2002).

[9] B. Deroin, V. Kleptsyn, A. Navas: Sur la dynamique unidimensionnelle en régularité intermédiaire. *Acta Math.*, 199(2):199–262 (2007).

[10] A. Furman: Mostow-Margulis rigidity with locally compact targets. *Geom. Funct. Anal.*, 11(1):30–59 (2001).

[11] A. Furman: Random walks on groups and random transformations. *Handbook of dynamical systems*, Vol. 1A, 931–1014, North-Holland, Amsterdam, 2002.

[12] H. Furstenberg: A Poisson formula for semi-simple Lie groups. *Ann. of Math. (2)*, 77:335–386 (1963).

[13] H. Furstenberg: Noncommuting random products. *Trans. Amer. Math. Soc.*, 108:377–428 (1963).

[14] E. Ghys: Groupes d'homéomorphismes du cercle et cohomologie bornée. The Lefschetz centennial conference, Part III (Mexico City, 1984), 81–106, *Contemp. Math.*, 58, III, American Mathematical Society, Providence, RI, 1987.

[15] E. Ghys: Actions de réseaux sur le cercle. *Invent. Math.*, 137(1):199–231 (1999).

[16] E. Ghys: Groups acting on the circle. *Enseign. Math. (2)*, 47(3–4):329–407 (2001).

[17] Y. Guivarc'h, A. Raugi: Products of random matrices: convergence theorems. Random matrices and their applications (Brunswick, ME, 1984), 31–54, *Contemp. Math.*, 50, American Mathematical Society, Providence, RI, 1986.

[18] G. Hector, U. Hirsch: Introduction to the geometry of foliations. Part B. Foliations of codimension one. *Aspects of Mathematics*, E3. Friedr. Vieweg & Sohn, Braunschweig, 1983.

[19] A. Iozzi: Bounded cohomology, boundary maps, and rigidity of representations into $\text{Homeo}_+(S^1)$ and $\text{SU}(1, n)$. Rigidity in dynamics and geometry (Cambridge, 2000), 237–260, Springer, Berlin, 2002.

[20] A. Iozzi: Vanishing theorems in bounded cohomology, in preparation.

[21] V. Kaimanovich: Double ergodicity of the Poisson boundary and applications to bounded cohomology. *Geom. Funct. Anal.*, 13(4):852–861 (2003).

[22] G. Margulis: Discrete subgroups of semisimple Lie groups. Ergebnisse der Mathematik und ihrer Grenzgebiete (3), Springer-Verlag, Berlin, 1991.

[23] G. Margulis: Free subgroups of the homeomorphism group of the circle. *C. R. Acad. Sci. Paris Sr. I Math.*, 331(9): 669–674 (2000).

[24] N. Monod: Continuous bounded cohomology of locally compact groups. Lecture Notes in Mathematics, 1758. Springer-Verlag, Berlin, 2001.

[25] D. Witte Morris: Can lattices in $\text{SL}(n, \mathbf{R})$ act on the circle? This volume.

[26] A. Navas: Quelques nouveaux phénomènes de rang 1 pour les groupes de difféomorphismes du cercle. *Comment. Math. Helv.*, 80(2): 355–375 (2005).

[27] D. Witte, R. Zimmer: Actions of semisimple Lie groups on circle bundles. *Geom. Dedicata*, 87(1–3): 91–121 (2001).

2

SORIN DUMITRESCU

MEROMORPHIC ALMOST RIGID GEOMETRIC STRUCTURES

DEDICATED TO BOB ZIMMER

Abstract

We study the local Killing Lie algebra of meromorphic almost rigid geometric structures on complex manifolds. This leads to classification results for compact complex manifolds admitting holomorphic rigid geometric structures.

1. Introduction

Zimmer and Gromov have conjectured that "big" actions of Lie groups that preserve unimodular rigid geometric structures (for example, a pseudo-Riemannian metric or an affine connection together with a volume form) are "essentially classifiable" [1, 28, 64, 65, 66, 67].

We try to address here this general question in the framework of the complex geometry. We study holomorphic (and meromorphic) unimodular geometric structures on compact complex manifolds, and we show that in many cases the holomorphic rigidity implies that the situation is classifiable (even without assuming any isometric group action). This is a survey paper, but some results are new (see Theorem 2.1, Corollary 2.2, and Theorem 5.10).

In the sequel all complex manifolds are supposed to be smooth and connected.

Consider a complex n-manifold M and, for all integers $r \geq 1$, consider the associated bundle $R^r(M)$ of r-frames, which is a $D^r(\mathbb{C}^n)$-principal bundle over M, with $D^r(\mathbb{C}^n)$-the (algebraic) group of r-jets at the origin of local biholomorphisms of $\mathbb{C}^n$ fixing 0 [1, 28, 7, 12, 22, 63].

Let us consider, as in [1, 28], the following:

DEFINITION 1.1. A *meromorphic geometric structure* (of order r) ϕ on a complex manifold M is a $D^r(\mathbb{C}^n)$-equivariant meromorphic map from $R^r(M)$ to a

1991 Mathematics Subject Classification 53B21, 53C56, 53A55.

Keywords and phrases: complex manifolds, rigid geometric structures, transitive Killing Lie algebras

This work was partially supported by the ANR Grant Symplexe BLAN 06-3-137237.

quasiprojective variety Z endowed with an algebraic action of $D^r(\mathbb{C}^n)$. If Z is an affine variety, we say that ϕ is of *affine type*.

In holomorphic local coordinates U on M, the geometric structure ϕ is given by a meromorphic map $\phi_U : U \to Z$.

Denote by $P \subset M$ the analytic subset of poles of ϕ and by $M^* = M \setminus P$ the open dense subset of M where ϕ is holomorphic. If P is the empty set, then ϕ is a *holomorphic geometric structure* on M.

If ϕ uniquely determines a holomorphic volume form on M^*, then ϕ is called *unimodular*.

DEFINITION 1.2. The meromorphic geometric structure ϕ is called *almost rigid* if, away from an analytic subset of M of positive codimension ϕ is (holomorphic and) rigid in Gromov's sense [1, 28].

A holomorphic geometric structure on M is called *rigid* if it is rigid in Gromov's sense on all of M.

Notice that my definiton of almost rigidity is not exactly the same as in [9].

EXAMPLES On a complex connected n-manifold M the following meromorphic geometric structures are almost rigid.

- A meromorphic map ϕ from M to a projective space $P^N(\mathbb{C})$, which is an embedding on some nontrivial open subset of M. In this case M is of algebraic dimension n (see Section 2). If ϕ is a holomorphic embedding, then ϕ is rigid.
- A family $X_1, X_2, \ldots, X_n$ of meromorphic vector fields such that there exists an open (dense) subset of M where the X_i span the holomorphic tangent bundle TM. Any homogeneous manifold admits a holomorphic family of such vector fields. They pull-back after the blowup of a point (or a submanifold) in the homogeneous manifold in a meromorphic family of such vector fields.

  In the special case where $X_1, X_2, \ldots, X_n$ are holomorphic and span TM on all of M, the corresponding geometric structure is called a *holomorphic parallelization of the tangent bundle* and it is a (unimodular) holomorphic rigid geometric structure of affine type.
- A meromorphic section g of the bundle $S^2(T^*M)$ of complex quadratic forms such that g is nondegenerate on an open (dense) subset. If g is holomorphic and nondegenerate on M, then g is a holomorphic rigid geometric structure of affine type called *holomorphic Riemannian*

metric. Up to a double cover of M, a holomorphic Riemannian metric is unimodular.

A holomorphic Riemannian metric has nothing to do with the more usual Hermitian metric. It is in fact nothing but the complex version of a (pseudo-)Riemannian metric. Observe that since complex quadratic forms have no signature, there is no distinction here between the Riemannian and pseudo-Riemannian cases. This observation was at the origin of the niceuse by F. Gauß of the complexification technic of (analytic) Riemannian metrics on surfaces, in order to find conformal coordinates for them. Actually, the complexification of analytic Riemannian metrics leading to holomorphic ones is becoming a standard trick (see for instance [23]).

In the blow-up process, the pull-back of a holomorphic Riemannian metric will stay nondegenerate away from the exceptional set and will vanish on the exceptional set.

More general, if g is a meromorpic section of $S^2(T^*M)$ nondegenerate on an open dense set, then any complex manifold $\tilde{M}$ bimeromorphic to M inherits a similar section $\tilde{g}$. Moreover, if g is holomorphic, then $\tilde{g}$ is also holomorphic. This comes from the fact that indeterminacy points of a meromorphic map are of codimension at least 2 (see theorem 2.5 in [57]) and from Levi's extension principle (see [57] for the details).

- Meromorphic affine or projective connections or meromorphic conformal structures in dimension ≥ 3. For example, in local holomorphic coordinates $(z_1, \ldots, z_n)$ on the manifold M, a meromorphic affine connection ∇ is determined by the meromorphic functions Γ_{ij}^k such that $\nabla_{\frac{\partial}{\partial z_i}} \frac{\partial}{\partial z_j} = \Gamma_{ij}^k \frac{\partial}{\partial z_k}$, for all $i, j, k \in \{1, \ldots, n\}$.

  Affine connections are geometric structures of affine type, but projective connections or conformal structures are not of affine type.

- If ϕ is a meromorphic almost rigid geometric structure of order r_1 and g is a meromorphic geometric structure of order r_2, then we can put together ϕ and g in some meromorphic almost rigid geometric structure of order $max(r_1, r_2)$, denoted by (ϕ, g) [28, 1, 7, 12, 22, 63].

- The s-jet (prolongation) $\phi^{(s)}$ of a meromorphic almost rigid geometric structure ϕ is still a meromorphic almost rigid geometric structure [28, 1, 7, 12, 22, 63].

Recall that local biholomorphisms of M that preserve a meromorphic geometric structure ϕ are called *local isometries*. Note that local isometries of (ϕ, g) are the local isometries of ϕ, which also preserve g.

The set of local isometries of a holomorphic rigid geometric structure ϕ is a Lie pseudogroup $Is^{loc}(\phi)$ generated by a Lie algebra of local vector fields called *(local) Killing Lie algebra*. If the local Killing Lie algebra is transitive on M, then ϕ is called *locally homogeneous*.

2. Local Isometries and Meromorphic Functions

The maximal number of algebraically independent meromorphic functions on a complex manifold M is called *the algebraic dimension $a(M)$* of M.

Recall that a theorem of Siegel proves that a complex n manifold M admits at most n algebraically independent meromorphic functions [57]. Then $a(M) \in \{0, 1, \ldots, n\}$ and for algebraic manifolds $a(M) = n$.

We will say that two points in M are in the same *fiber of the algebraic reduction* of M if any meromorphic function on M takes the same value at the two points. There exists some open dense set in M where the fibers of the algebraic reduction are the fibers of a holomorphic fibration on an algebraic manifold of dimension $a(M)$ and any meromorphic function on M is the pull-back of a meromorphic function on the basis [57].

The following theorem shows that the fibers of the algebraic reduction are in the same orbit of the pseudogroup of local isometries for any meromorphic almost rigid geometric structure on M. This is a meromorphic version of the celebrated Gromov's open dense orbit theorem [1, 28] (see also [7, 12, 22, 63]).

THEOREM 2.1. *Let M be a connected complex manifold of dimension n, which admits a meromorphic almost rigid geometric structure ϕ. Then, there exists a nowhere-dense analytic subset S in M, such that $M \setminus S$ is $Is^{loc}(\phi)$-invariant and the orbits of $Is^{loc}(\phi)$ in $M \setminus S$ are the fibers of a holomorphic fibration of constant rank. The dimension of the fibers is $\geq n - a(M)$, where $a(M)$ is the algebraic dimension of M.*

Recall that $g' = (\phi, g)$ is still almost rigid for any meromorphic (not necessarily almost rigid) geometric structure g on M. This yields to the following:

COROLLARY 2.2. *If g is a meromorphic geometric structure on M, then, away from a nowhere dense analytic subset, the $Is^{loc}(g)$-orbits are of dimension $\geq n - a(M)$.*

In particular, if $a(M) = 0$, then g is locally homogeneous outside a nowhere dense analytic subset of M (the Killing Lie algebra $\mathcal{G}$ of $g' = (\phi, g)$ is transitive on *an open dense set). Moreover, if ϕ is unimodular and G is unimodular and simply transitive, then g is locally homogeneous on all of the open dense set where ϕ and g are holomorphic and ϕ is rigid. In this case, if ϕ is unimodular holomorphic and rigid and g is holomorphic, then g is locally homogeneous on M.*

Proof. Let ϕ be of order r, given by a map $\phi : R^r(M) \to Z$. For each positive integer s we consider the s-jet $\phi^{(s)}$ of ϕ. This is a $D^{(r+s)}(\mathbb{C}^n)$-equivariant meromorphic map $R^{(r+s)}(M) \to Z^{(s)}$, where $Z^{(s)}$ is the algebraic variety of the s-jets at the origin of holomorphic maps from $\mathbb{C}^n$ to Z. One can find the expression of the (algebraic) $D^{(r+s)}(\mathbb{C}^n)$-action on $Z^{(s)}$ in [12, 22].

Since ϕ is almost rigid, there exists a nowhere dense analytic subset S' in M, containing the poles of ϕ, and a positive integer s such that two points m, m' in $M \setminus S'$ are in the same orbit of $Is^{loc}(\phi)$ if and only if $\phi^{(s)}$ sends the fibers of $R^{(r+s)}(M)$ above m and m' on the same $D^{(r+s)}(\mathbb{C}^n)$-orbit in $Z^{(s)}$ [28, 1].

Rosenlicht's theorem (see [49]) shows that there exists a $D^{(r+s)}(\mathbb{C}^n)$-invariant stratification

$$Z^{(s)} = Z_0 \supset \cdots \supset Z_l,$$

such that Z_{i+1} is Zariski closed in Z_i, the quotient of $Z_i \setminus Z_{i+1}$ by $D^{(r+s)}(\mathbb{C}^n)$ is a complex manifold, and rational $D^{(r+s)}(\mathbb{C}^n)$-invariant functions on Z_i separate orbits in $Z_i \setminus Z_{i+1}$.

Consider the open dense $Is^{loc}(\phi)$-invariant subset U in $M \setminus S'$, such that $\phi^{(s)}$ is of constant rank above U and the image of $R^{(r+s)}(M)|_U$ through $\phi^{(s)}$ lies in $Z_i \setminus Z_{i+1}$ but not in Z_{i+1}. Then the orbits of $Is^{loc}(\phi)$ in U are the fibers of a fibration of constant rank (on the quotient of $Z_i \setminus Z_{i+1}$ by $D^{(r+s)}(\mathbb{C}^n)$). Obviously, $U = M \setminus S$, where S is a nowhere dense analytic subset in M.

If m and n are two points in U that are not in the same $Is^{loc}(\phi)$-orbit, then the corresponding fibers of $R^{(r+s)}(M)|_U$ are sent by $\phi^{(s)}$ on two distinct $D^{(r+s)}(\mathbb{C}^n)$-orbits in $Z_i \setminus Z_{i+1}$. By Rosenlicht's theorem there exists a $D^{(r+s)}(\mathbb{C}^n)$-invariant rational function $F : Z_i \setminus Z_{i+1} \to \mathbb{C}$, which takes distinct values at these two orbits.

The meromorphic function $F \circ \phi^{(s)} : R^{(r+s)}(M) \to \mathbb{C}$ is $D^{(r+s)}(\mathbb{C}^n)$-invariant and descends in a $Is^{loc}(\phi)$-invariant meromorphic function on M that takes distinct values at m and at n.

Consequently, the complex codimension in U of the $Is^{loc}(\phi)$-orbits is $\leq a(M)$, which finishes the proof. $\qquad \square$

We now deduce the corollary:

Proof. It is convenient to put together ϕ and g in some extra geometric structure $g' = (\phi, g)$. Now g' is a meromorphic almost rigid geometric structure and Theorem 2.1 shows that the complex dimension of a generic $Is^{loc}(g')$-orbit is $\geq n - a(M)$. As each $Is^{loc}(g')$-orbit is contained in a $Is^{loc}(g)$-orbit, the result follows also for g.

If $a(M) = 0$, the Killing Lie algebra $\mathcal{G}$ of g' is transitive on a maximal open dense subset U in M. Suppose now that ϕ is unimodular and also that $\mathcal{G}$ is unimodular. The other inclusion being trivial, it is enough to show that U contains the maximal subset of M where ϕ and g are holomorphic and ϕ is rigid.

Pick up a point m in the previous subset. We want to show that m is in U. Since $g' = (\phi, g)$ is holomorphic and rigid in the neighborhood of m, it follows that m admits an open neighborhood U_m in M such that any local holomorphic Killing field of g' defined on a connected open subset in U_m extends on all of U_m [48, 2].

Since $\mathcal{G}$ acts transitively on U, choose local linearly independent Killing fields $X_1, \ldots, X_n$ on a connected open set included in $U \cap U_m$. As ϕ is unimodular, it determines a holomorphic volume form vol on U_m (if necessary we restrict to a smaller U_m), which is preserved by $Is^{loc}(g')$. But $Is^{loc}(g')$ acts transitively on $U \cap U_m$ and $\mathcal{G}$ is supposed to be unimodular. This implies that the function $vol(X_1, \ldots, X_n)$ is $\mathcal{G}$-invariant and, consequently, a (nonzero) constant on $U \cap U_m$.

On the other hand, $X_1, \ldots, X_n$ extend in some holomorphic Killing fields $\tilde{X}_1, \ldots, \tilde{X}_n$ defined on all of U_m. The holomorphic function $vol(\tilde{X}_1, \ldots, \tilde{X}_n)$ is a nonzero constant on U_m: in particular, $\tilde{X}_1(m), \ldots, \tilde{X}_n(m)$ are linearly independent. We proved that $\mathcal{G}$ acts transitively in the neighborhood of m and thus $m \in U$. $\square$

3. Classification Results

Parallelizations of the Tangent Bundle

We begin with the classification of compact complex manifolds that admit a holomorphic parallelization of the tangent bundle [60].

THEOREM 3.1. (WANG) *Let M be a compact connected complex manifold with a holomorphic parallelization of the tangent bundle. Then M is a quotient $\Gamma \backslash G$, where G is a connected simply connected complex Lie group and Γ is a uniform lattice in G.*

Moreover, M is Kaehler if and only if G is abelian (and M is a complex torus).

Proof. Let n be the complex dimension of M and consider $X_1, X_2, \ldots, X_n$ global holomorphic vector fields on M that span TM. Then, for all $1 \leq i, j \leq n$, we have

$$[X_i, X_j] = f_1^{ij} X_1 + f_2^{ij} X_2 + \cdots + f_n^{ij} X_n,$$

with f_k^{ij} holomorphic functions on M. Since M is compact, these functions have to be constant and, consequently, $X_1, X_2, \ldots, X_n$ generate a n-dimensional Lie algebra $\mathcal{G}$ that acts simply transitively on M. By Lie's theorem there exists a unique connected simply connected complex Lie group G corresponding to $\mathcal{G}$.

In particular, the holomorphic parallelization is locally homogeneous, locally modeled on the parallelization given by translation-invariant vector fields on the Lie group G.

Since M is compact, the X_i are complete and they define a holomorphic simply transitive action of G on M. Hence M is a quotient of G by a cocompact discrete subgroup Γ in G.

Assume now M is Kaehler. Then, any holomorphic form on M has to be closed [27]. Consider $\omega_1, \omega_2, \ldots, \omega_n$ the dual basis with respect to X_1, $X_2, \ldots, X_n$. The one forms ω_i are holomorphic and translation-invariant on G (they descend on $\Gamma \backslash G$). The Lie-Cartan formula

$$d\omega_i(X_j, X_k) = -\omega_i([X_j, X_k]),$$

shows that the one forms ω_i are all closed if and only if G is abelian and thus M is a complex torus. $\qquad \square$

Holomorphic Riemannian Metrics

As in the real case, a holomorphic Riemannian metric on M gives rise to a covariant differential calculus, that is, a Levi-Civita (holomorphic) affine connection, and to geometric features: curvature tensors and geodesic (complex) curves [40, 41].

Locally, a holomorphic Riemannian metric has the expression $\Sigma g_{ij}(z)$ $dz_i dz_j$, where $(g_{ij}(z))$ is a complex invertible symmetric matrix depending holomorphically on z. The standard example is that of the global flat holomorphic Riemannian metric $dz_1^2 + dz_2^2 + \cdots + dz_n^2$ on $\mathbb{C}^n$. This metric is translation-invariant and thus descends to any quotient of $\mathbb{C}^n$ by a lattice. Hence, complex tori possess (flat) holomorphic Riemannian metrics. This is, however, a very special situation since, contrary to real case, *only few compact complex manifolds admit holomorphic Riemannian metrics*. In fact, Yau's proof of the Calabi conjecture shows that, up to finite unramified covers, complex tori are the only compact *Kaehler* manifolds admitting holomorphic Riemannian metrics [34].

However, very interesting examples, constructed by Ghys in [25], do exist on 3-dimensional complex non-Kaehler manifolds and deserve classification. Notice that parallelizable manifolds admit holomorphic Riemannian metrics coming from left-invariant holomorphic Riemannian metrics on G (which can be constructed by left translating any complex nondegenerate quadratic form defined on the Lie algebra $\mathcal{G}$).

Ghys's examples of 3-dimensional compact complex manifolds endowed with holomorphic Riemannian metrics are obtained by deformation of the complex structure on parallelizable manifolds $\Gamma \backslash SL(2, \mathbb{C})$ [25]. They are *non-standard, meaning they do not admit parallelizable manifolds as finite unramified covers*. Those nonstandard examples will be described later on.

A first obstruction to the existence of a holomorphic Riemannian metric on a compact complex manifold is the vanishing of its first Chern class. Indeed, a holomorphic Riemannian metric on M provides an isomorphism between TM and T^*M. In particular, the canonical bundle K is isomorphic to the anticanonical bundle K^{-1} and thus K^2 is trivial. This means that the first Chern class of M vanishes and, up to a double unramified cover, M possesses a holomorphic volume form.

The following proposition describes holomorphic Riemannian metrics on parallelizable manifolds:

PROPOSITION 3.2. *Let $M = \Gamma \backslash G$ a compact parallelizable manifold, with G a simply connected complex Lie group and Γ a uniform lattice in G. Then, any holomorphic Riemannian metric g on M comes from a nondegenerate complex quadratic form on the Lie algebra $\mathcal{G}$ of G. In particular, the pull-back of g is left invariant on the universal cover G (and g is locally homogeneous on M).*

Moreover, any compact parallelizable 3-manifold admits a holomorphic Riemannian metric of constant sectional curvature. The metric is flat exactly when G is solvable.

Proof. Consider $X_1, X_2, \ldots, X_n$ the fundamental vector fields corresponding to the locally free G-action on M. Let g be a holomorphic Riemannian metric on M and denote also by g the associated complex symmetric bilinear form. Then $g(X_i, X_j)$ is a holomorphic function on M and thus constant, for all $1 \leq i, j \leq n$. This implies that g comes from a left-invariant holomorphic Riemannian metric on G.

Assume now G is a connected simply connected complex unimodular Lie group of dimension 3. We have only four such Lie groups: $\mathbb{C}^3$, the complex Heisenberg group, the complex SOL group, and $SL(2, \mathbb{C})$ [35]. Note that the group SOL is the complexification of the affine isometry group of the Minkowski plane $\mathbb{R}^{1,1}$ or equivalently the isometry group of $\mathbb{C}^2$ endowed with its flat holomorphic Riemannian metric.

We begin with the case $G = \mathbb{C}^3$. We have seen that $\mathbb{C}^3$ admits a flat translation-invariant holomorphic Riemannian metric. The isometry group of this metric is $O(3, \mathbb{C}) \ltimes \mathbb{C}^3$. We recall here that $O(3, \mathbb{C})$ and $SL(2, \mathbb{C})$ are locally isomorphic.

Consider now the case $G = SL(2, \mathbb{C})$. The Killing form of the Lie algebra $sl(2, \mathbb{C})$ is a nondegenerate complex quadratic form that endows $SL(2, \mathbb{C})$ with a left-invariant holomorphic Riemannian metric of constant sectional curvature. Since the Killing quadratic form is invariant by the adjoint representation, the isometry group also contains all right translations. In fact, the connected component of the isometry group is $SL(2, \mathbb{C}) \times SL(2, \mathbb{C})$ acting by left and right translation. On the other hand, $\Gamma \backslash SL(2, \mathbb{C})$ doesn't admit flat torsion-free affine connections (or flat holomorphic Riemannian metrics) [18].

It is an easy exercice to exhibit in the isometry group $O(3, \mathbb{C}) \ltimes \mathbb{C}^3$ of the flat holomorphic Riemannian space copies of the complex Heisenberg group and of the complex SOL group, which act simply transitively [18]. Thus the flat holomorphic Riemannian space also admits models that are given by left-invariant metrics on the Heisenberg group and on the SOL group. One can get explicit expression of these holomorphic Riemannian metrics by complexification of flat left-invariant Lorentz metrics on the real Heisenberg and SOL groups [51, 52].

On the other hand, there exists no 3-dimensional solvable subgroups in $SL(2, \mathbb{C}) \times SL(2, \mathbb{C})$ acting with an open orbit on $SL(2, \mathbb{C})$. It follows that the compact quotients of 3-dimensional solvable groups don't admit holomorphic Riemannian metrics of nonzero constant sectional curvature. □

Ghys Nonstandard Examples

As above, for any cocompact lattice Γ in $SL(2, \mathbb{C})$, the quotient $M = \Gamma \backslash SL(2, \mathbb{C})$ admits a holomorphic Riemannian metric of nonzero constant sectional curvature. It is convenient to consider M as a quotient of $S_3 = O(4, \mathbb{C})/O(3, \mathbb{C}) = SL(2, \mathbb{C}) \times SL(2, \mathbb{C})/SL(2, \mathbb{C})$ by Γ, seen as a subgroup of $SL(2, \mathbb{C}) \times SL(2, \mathbb{C})$ by the trivial embedding $\gamma \in \Gamma \mapsto (\gamma, 1) \in SL(2, \mathbb{C}) \times SL(2, \mathbb{C})$.

New interesting examples of manifolds admitting holomorphic Riemannian metrics of nonzero constant sectional curvature have been constructed in [25] by deformation of this embedding of Γ.

Those deformations are constructed choosing a morphism $u : \Gamma \to SL(2, \mathbb{C})$ and considering the embedding $\gamma \mapsto (\gamma, u(\gamma))$. Algebraically, the action is given by:

$$(\gamma, m) \in \Gamma \times SL(2, \mathbb{C}) \to \gamma m u(\gamma^{-1}) \in SL(2, \mathbb{C}).$$

It is proved in [25] that, for u close enough to the trivial morphism, Γ acts properly (and freely) on $S_3 (\cong SL(2, \mathbb{C}))$ such that the quotient M_u is a complex compact manifold (covered by $SL(2, \mathbb{C})$) admitting a holomorphic Riemannian metric of non-zero constant sectional curvature. *In general, these examples do not admit parallelizable manifolds as finite covers.*

Note that left-invariant holomorphic Riemannian metrics on $SL(2, \mathbb{C})$ that are not right-invariant, in general, will not descend on M_u.

Let us notice that despite this systematic study in [25], there are still many open questions regarding these examples (including the question of completeness). A real version of this study is in [39, 26, 53].

Dimension 3

The classification of complex compact manifolds admitting holomorphic Riemannian metrics is simple in complex dimension 2 (see Section 5). An important step toward the classification in dimension 3 was made in [17] with the following result:

THEOREM 3.3. *Any holomorphic Riemannian metric on a compact connected complex 3-manifold is locally homogeneous. More generally, if a compact connected complex 3-manifold M admits a holomorphic Riemannian metric, then any holomorphic geometric structure of affine type on M is locally homogeneous.*

Observe that the previous result is trivial in dimension 2, since the sectional curvature is a holomorphic function and thus constant on compact complex surfaces: this implies the local homogeneity [61]. In dimension 3, the sectional curvature will be, in general, a nonconstant meromorphic function on the 2-Grassmanian of the holomorphic tangent space with poles at the degenerate planes.

Thanks to Theorem 3.3, our manifold M is locally modeled on a $(G, G/I)$-geometry in Thurston's sense [55], where I is a closed subgroup of the Lie group G such that the G-action on G/I preserves some holomorphic

Riemannian metric (see [21] for details and notice that the local Killing Lie algebra of the holomorphic Riemannian metric is the Lie algebra of G). In this context we have a *developing map* from the universal cover of M into G/I, which is a local diffeomorphism and which is equivariant with respect to the action of the fundamental group on $\tilde{M}$ by deck transformations and on G/I via the *holonomy morphism* $\rho : \pi_1(M) \rightarrow G$ [55].

Recall that the $(G, G/I)$-geometry is called *complete* if the developing map is a diffeomorphism and, consequently, $\Gamma = \rho(\pi_1(M))$ acts properly on G/I such that M is a compact quotient $\Gamma \backslash G/I$.

A second step was achieved in [21] where, in a common work with Zeghib, we proved the following result that can be seen, in particular, as a completeness result in the case where G is solvable.

THEOREM 3.4. *Let M be a compact connected complex 3-manifold that admits a (locally homogeneous) holomorphic Riemannian metric g. Then*

i) *If the Killing Lie algebra of g has a nontrivial semisimple part, then it preserves some holomorphic Riemannian metric on M with constant sectional curvature.*

ii) *If the Killing Lie algebra of g is solvable, then, up to a finite unramified cover, M is a quotient either of the complex Heisenberg group, or of the complex SOL group by a lattice.*

REMARK 3.5. If g is flat, its Killing Lie algebra corresponds to $O(3, \mathbb{C}) \ltimes \mathbb{C}^3$, which has a nontrivial semisimple part. Thus, flat holomorphic Riemannian metrics on complex tori are part of point (i) in Theorem 3.4.

The point (ii) of the previous theorem is not only about completeness, but also gives a rigidity result in Bieberbach's sense [61]: G contains a 3-dimensional closed subgroup H (either isomorphic to the complex Heisenberg group, or to the complex SOL group), which acts simply transitively (and so identifies) with G/I and (up to a finite index) the image Γ of the holonomy morphism lies in H. It follows that, up to a finite cover, M identifies with $\Gamma \backslash H$.

Since the Heisenberg and SOL groups admit left-invariant flat holomorphic Riemannian metrics this leads to:

COROLLARY 3.6. *If a compact connected complex 3-manifold M admits a holomorphic Riemannian metric, then, up to a finite unramified cover, M admits a holomorphic Riemannian metric of constant sectional curvature.*

Theorem 3.4 does not end the story, even in dimension 3, essentially because of remaining *completeness* questions, and those on the algebraic structure of the fundamental group.

It remains to classify the compact complex 3-manifolds endowed with a holomorphic Riemannian metric of constant sectional curvature.

FLAT CASE. In this case M admits a $(O(3, \mathbb{C}) \ltimes \mathbb{C}^3, \mathbb{C}^3)$-geometry. The challenge remains:

1) *Markus conjecture:* Is M complete?
2) *Auslander conjecture:* Assuming M is complete, is Γ solvable?

Note that these questions are settled in the setting of (real) flat Lorentz manifolds [13, 24], but unsolved for general (real) pseudo-Riemannian metrics. The real part of the holomorphic Riemannian metric is a (real) pseudo-Riemannian metric of signature $(3, 3)$ for which both previous conjectures are still open.

NONFLAT CASE. In this case $G = SL(2, \mathbb{C}) \times SL(2, \mathbb{C})$ and $I = SL(2, \mathbb{C})$ is diagonally embedded in the product. The completeness of this geometry on compact complex manifolds is still an open problem, despite a local result of Ghys [25]. Recall that the real analogous of this problem, that is, the completeness of compact manifolds endowed with Lorentz metrics of negative constant sectional curvature was solved in [37], but the proof cannot generalize to other signatures.

Higher Dimension

One interesting problem in differential geometry is to decide whether a given homogeneous space G/I possesses a compact quotient. A more general related question is to decide whether there exists compact manifolds locally modeled on $(G, G/I)$ (see, for instance, [6, 8, 38]).

The case $I = 1$, or more generally I compact, reduces to the classical question of existence of cocompact lattices in Lie groups. For homogeneous spaces of non-Riemannian type (i.e., I noncompact) the problem is much harder.

The case $S_n = O(n + 1, \mathbb{C})/O(n, \mathbb{C})$ is a geometric situation where these questions can be tested. It turns out that compact quotients of S_n are known to exist only for $n = 1, 3$, or 7. We discussed the case $n = 3$ above, and the existence of a compact quotient of S_7 was proved in [38]. Here, we dare ask with [38]:

CONJECTURE 3.7. [38] *S_n has no compact quotients, for $n \neq 1, 3, 7$.*

A stronger version of this question was proved in [6] for S_n, if n has the form $4m + 1$, with $m \in \mathbb{N}$.

Keeping in mind our geometric approach, we generalize the question to manifolds locally modeled on S_n. More exactly:

CONJECTURE 3.8. [21] *A compact complex manifold endowed with a holomorphic Riemannian metric of constant nonvanishing curvature is complete. In particular, such a manifold has dimension 3 or 7.*

4. Applications to Simply Connected Manifolds

Remark first that Theorem 3.3 has the following direct consequence:

COROLLARY 4.1. *A compact connected simply connected 3 manifold admits no holomorphic Riemannian metrics.*

Proof. Assume, by contradiction, (M, g) being as in the statement of the corollary. Then Theorem 3.3 implies that g is locally homogeneous. Since M is simply connected, the local Killing fields of the Killing algebra $\mathcal{G}$ extend on all of M: the unique connected simply connected complex Lie group G associated to $\mathcal{G}$ acts isometrically and transitively on M. Then M is a homogeneous space G/H. Moreover, up to a double cover, G/H admits a holomorphic volume form *vol* coming from the holomorphic Riemannian metric.

Take X_1, X_2, X_3 three global Killing fields on M that are linearly independent at some point. Since $vol(X_1, X_2, X_3)$ is a holomorphic function on M, it is a nonzero constant and, consequently, X_1, X_2, X_3 are linearly independent on M. Hence Wang's theorem implies that M is a quotient of a 3 dimensional connected simply connected complex Lie group G_1 by a discrete subgroup. Since M is simply connected, this discrete subgroup has to be trivial and M identifies with G_1. But there is no compact simply connected complex Lie group: a contradiction. $\square$

THEOREM 4.2. *A compact connected simply connected complex n-manifold without nonconstant meromorphic functions admits no holomorphic unimodular rigid geometric structures.*

Proof. Assume, by contradiction, that (M, ϕ) verifies the hypothesis. Since $a(M) = 0$, Theorem 2.1 implies ϕ is locally homogeneous on an open dense set U. As M is simply connected, elements in the Killing algebra $\mathcal{G}$ extend on all of M: the connected simply connected complex Lie group G associated to $\mathcal{G}$ acts isometrically on M with an open dense orbit. The open dense orbit U identifies with a homogeneous space G/H, where H is a closed subgroup of G.

Consider $X_1, X_2, \ldots, X_n$ global Killing fields on M that are linearly independent at some point of the open orbit U. As before, $vol(X_1, X_2, \ldots, X_n)$ is a nonzero constant, where vol is the holomorphic volume form associated to ϕ. Thus the X_i give a holomorphic parallelization of TM and Wang's theorem enables us to conclude as in the previous proof. ⊔

For nonunimodular rigid geometric structures we have the following less precise:

THEOREM 4.3. [15] *Let M be a compact connected simply connected complex n-manifold without nonconstant meromorphic functions and admitting a holomorphic rigid geometric structure ϕ. Then M is an equivariant compactification of $\Gamma \backslash G$, where Γ is a discrete noncocompact subgroup in a complex Lie group G.*

Proof. Since $a(M) = 0$, Theorem 2.1 implies ϕ is locally homogeneous on an open dense set U. As before, the extension property of local Killing fields implies U is a complex homogeneous space G/H, where G is a connected simply connected complex Lie group and H is a closed subgroup in G.

We show now that H is a discrete subgroup of G. Assume by contradiction the Lie algebra of H is nontrivial. Take at any point $u \in U$, the isotropy subalgebra $\mathcal{H}_u$ (i.e., the Lie subalgebra of Killing fields vanishing at u). Remark that $\mathcal{H}_{gu} = Ad(g)\mathcal{H}_u$, for any $g \in G$ and $u \in U$, where Ad is the adjoint representation.

The map $u \to \mathcal{H}_u$ is a meromorphic map from M to the Grassmanian of d-dimensional vector spaces in $\mathcal{G}$. But M doesn't admit any nonconstant meromorphic function and this map has to be constant. It follows that $\mathcal{H}_u$ is $Ad(G)$-invariant and H is a normal subgroup of G: a contradiction, since the G-action on M is faithful. Thus G is of dimension n and H identifies to a lattice Γ in G.

As M is simply connected, U has to be strictly contained in M and M is an equivariant compactification of $\Gamma \backslash G$. □

We don't know whether such equivariant compactifications of $\Gamma \backslash G$ admit equivariant holomorphic rigid geometric structures, but the previous result has the following application.

Recall that a well-known open question asks whether the 6-dimensional real sphere S^6 admits complex structures or not. In this context, we have the following:

COROLLARY 4.4. *If S^6 admits a complex structure M, then M admits no holomorphic rigid geometric structures.*

Proof. The starting point of the proof is a result of [11] where it is proved that M doesn't admit nonconstant meromorphic functions. If M supports holomorphic rigid geometric structures, then Theorem 4.3 implies that M is an equivariant compactification of a homogeneous space. This is in contradiction with the main theorem of [30]. □

As for *Kaehler munifolds* we have proved in [15] the following more precise:

THEOREM 4.5. *Let M be a compact connected Kaehler manifold endowed with a holomorphic unimodular rigid geometric structure ϕ of affine type. Then, up to a finite unramified cover, M is a complex torus (quotient of $\mathbb{C}^n$ by a lattice) and ϕ is translation-invariant.*

The proof is done in two steps. First we prove that ϕ is locally homogeneous. Then we use a splitting theorem [5], which asserts that such compact Kaehler manifolds with a holomorphic volume form (Calabi-Yau manifolds) are biholomorphic, up to a finite cover, to a direct product of a complex torus and a compact *simply connected* Kaehler manifold with a holomorphic volume form. Starting with ϕ and using the product structure, we construct a holomorphic unimodular rigid geometric structure on the simply connected factor that is locally homogeneous. We conclude as in the proof of Theorem 4.2 that the simply connected factor is trivial.

Recently, we adapted this proof in [19] to all holomorphic Cartan geometries of algebraic type on Calabi-Yau manifolds. A similar result was independently proved in [10].

One can also find a classification of certain holomorphic G-structures of order one on *uniruled* projective manifolds in [31].

5. Applications to Complex Surfaces

THEOREM 5.1. *Let S be a compact complex surface endowed with a holomorphic unimodular rigid geometric structure ϕ of affine type. Then the Killing Lie algebra of ϕ is nontrivial.*

The assumption on ϕ to be affine is essential:

REMARK 5.2. If S is a compact complex algebraic surface with trivial canonical bundle (for example, a complex torus or an algebraic K3 surface), the geometric structure given by a holomorphic volume form on S together with a holomorphic embedding of S in a complex projective space doesn't admit any nontrivial local isometry. However, this geometric structure is not of *affine type*.

Proof. The proof is a simple corollary of Theorems 4.5 and 2.1. Indeed, Theorem 2.1 implies the Killing Lie algebra of ϕ is trivial only if the algebraic dimension $a(S)$ equals 2. But in this case S is algebraic [4] and thus Kaehler. Then Theorem 4.5 applies and the Killing Lie algebra is transitive on S and hence of dimension at least 2. □

Recall that a complex surface is called *minimal* if it does not contain any copy of $P^1(\mathbb{C})$ with self-intersection -1 (see [4], page 91). Then we have:

THEOREM 5.3. *Let S be a compact minimal complex surface that is not biholomorphic to a nonalgebraic K3 surface or to a nonaffine Hopf surface. Then S admits holomorphic rigid geometric structures.*

REMARK 5.4. By definition, a complex algebraic manifold admits an embedding in a complex projective space, which was seen to be a holomorphic rigid geometric structure (of order zero).

Proof. By the previous remark, it remains to consider the case of nonalgebraic complex surfaces. Then the Enriques-Kodaira classification shows that, up to a finite unramified cover, any minimal nonalgebraic complex compact surface is biholomorphic to one of the following complex surfaces: a complex tori, a Hopf surface, an Inoue surface, a K3 surface, a principal elliptic principal bundle over an elliptic curve, or a principal elliptic bundle over a Riemann surface of genus $g \geq 2$, with odd-first Betti number (see [4], p. 244).

However, it is known that complex tori, Inoue surfaces, affine Hopf surfaces, principal elliptic bundles over elliptic curves, and principal elliptic bundles over a Riemann surface of genus $g \geq 2$, with odd-first Betti number, admit flat holomorphic affine connections [34, 36, 43, 54, 58, 59]. □

The classification of compact complex surfaces admitting holomorphic affine connections (see [34, 36, 43, 58, 59]) implies the following result:

THEOREM 5.5. *Let S be a compact complex surface admitting a holomorphic unimodular affine connection. Then, up to a finite unramified cover, either S is a complex torus and the connection is translation-invariant, or S is an elliptic principal bundle over an elliptic curve and the connection is locally modeled on a translation-invariant connection on a complex torus.*

Since any complex manifold endowed with a holomorphic Riemannian metric inherits a holomorphic (unimodular) affine (Levi-Civita) connection [40, 41], this easily implies the following [17]:

COROLLARY 5.6. *Let S be a complex compact surface admitting a holomorphic Riemannian metric g. Then, up to a finite unramified cover, S is a complex torus and g is (flat) translation-invariant.*

In [20] we classified the local geometry of all torsion-free holomorphic affine connections ∇ on compact complex surfaces. In particular, we proved that either ∇ is locally homogeneous, locally isomorphic to a translation-invariant connection on $\mathbb{C}^2$, or ∇ is a nonflat connection on a principal elliptic bundle over a Riemann surface of genus $g \geq 2$ with odd first Betti number and the local Killing Lie algebra of ∇ is one-dimensional generated by the fundamental vector field of the principal fibration. In all cases, ∇ is projectively flat.

As a consequence we proved in [20] the following:

THEOREM 5.7. *Normal holomorphic projective connections on compact complex surfaces are flat.*

Inoue Surfaces

Recall that Inoue surfaces are compact complex surfaces in the class VII_0, which are not Hopf, and have a vanishing second Betti number [4, 33].

In [34, 36] it is proved that any Inoue surface admits a (unique) flat torsion-free holomorphic affine connection. Here we prove the following:

THEOREM 5.8. *Holomorphic geometric structures on Inoue surfaces are locally homogeneous.*

REMARK 5.9. Inoue surfaces do not admit nonconstant meromorphic functions [4].

Proof. Let τ be a holomorphic geometric structure on an Inoue surface S. Let ∇_0 be a flat torsion-free holomorphic affine connection on S. We prove that the holomorphic rigid geometric structure $\tau' = (\tau, \nabla_0)$ is locally homogeneous. Since $a(S) = 0$, Theorem 2.1 implies τ' is locally homogeneous on some maximal open dense set $S \setminus E$, where E is a compact analytic subset of S of positive codimension.

We want to show that E is empty.

We prove first that, up to a double cover of S, the subset E is a smooth submanifold in S (this is always true if S is a finite set; but here S might have components of complex dimension 1).

Assume, by contradiction, that E is not a smooth submanifold in S.

Choose $p \in E$ a singular point in E. In particular, p is not isolated in E, but p is isolated among the singular points of E. Since $Is^{loc}(\tau')$ preserves E, it has to preserve the set of its singular points and thus it fixes p. Consequently any local Killing field defined in the neighborhood of p has to vanish at p.

Denote by $\mathcal{G}$ the Lie algebra of local Killing fields in the neighborhood of p. Since $\mathcal{G}$ acts transitively on an open set, its dimension is at least 2.

Each element of $\mathcal{G}$ preserves ∇ and fixes p. In exponential coordinates the $\mathcal{G}$-action in the neighborhood of p is linear. This gives an embedding of $\mathcal{G}$ in the Lie algebra of $GL(2, \mathbb{C})$ (the image of the isotropy representation at p). In particular, $\mathcal{G}$ is of dimension ≤ 4.

Suppose first that $\mathcal{G}$ is of dimension 2. The corresponding subgroups of $GL(2, \mathbb{C})$ are conjugated either to the group of diagonal matrices, or to one of the following subgroups $\begin{pmatrix} a & b \\ 0 & a^{-1} \end{pmatrix}$, with $a \in \mathbb{C}^*$ and $b \in \mathbb{C}$, $\begin{pmatrix} 1 & m \\ 0 & n \end{pmatrix}$, with $m \in \mathbb{C}$ and $n \in \mathbb{C}^*$, or $\begin{pmatrix} m' & n' \\ 0 & 1 \end{pmatrix}$, with $m' \in \mathbb{C}^*$ and $n' \in \mathbb{C}$.

In the first case, the invariant closed subset E lies, in exponential coordinates, in the union of the two eigen directions. In the last two cases, E locally lies in the invariant line $y = 0$. In all situations, up to a double cover of S, the analytic set E is smooth: a contradiction.

We now settle the case where $\mathcal{G}$ is of dimension 3 or 4. Then the image of $\mathcal{G}$ by the isotropy representation in p is conjugated in $GL(2, \mathbb{C})$ to one of the following subgroups: $SL(2, \mathbb{C})$, $GL(2, \mathbb{C})$, or the group of inversible upper-triangular matrices. But $GL(2, \mathbb{C})$ and $SL(2, \mathbb{C})$ don't admit invariant subsets other than p, which will be an isolated point in E: impossible.

In the last situation, E locally coincides, as before, with the unique invariant line and it is smooth.

Up to a double cover, E is a holomorphic submanifold in S. If E admits a component of dimension 1, then this component will be a union of closed curves. But Inoue surfaces contain no curves [33].

This proves that E is a finite set. Assume, by contradiction, that E is not empty and consider $p \in E$.

The previous arguments show that the Lie algebra $\mathcal{G}$ is isomorphic either to $sl(2, \mathbb{C})$ or to $gl(2, \mathbb{C})$.

Assume first $\mathcal{G} = sl(2, \mathbb{C})$. As before, the local action of $\mathcal{G}$ in the neighborhood of p is conjugated to the action of $\mathcal{G}$ on $T_p S$, which coincides with the standard linear action of $sl(2, \mathbb{C})$ on $\mathbb{C}^2$. This action has two orbits: the point p and $\mathbb{C}^2 \setminus \{p\}$.

The stabilizer H in $SL(2, \mathbb{C})$ of a nonzero vector $x \in T_p S$ is conjugated to the following one-parameter subgroup of $SL(2, \mathbb{C})$: $\begin{pmatrix} 1 & b \\ 0 & 1 \end{pmatrix}$, with $b \in \mathbb{C}$.

Observe that the action of G on G/H preserves a nontrivial holomorphic vector field. The expression of this vector field in linear coordinates (z_1, z_2) in the neighborhood of p is $z_1 \dfrac{\partial}{\partial z_1} + z_2 \dfrac{\partial}{\partial z_2}$.

Since $S \setminus E$ is locally modeled on $(G, G/H)$, this vector field is well defined on $S \setminus E$. But E is of complex codimension 2 in S and the extension theorem of Hartogs [27] shows that the vector field extends to a global nontrivial holomorphic vector field X on S. The vector field X is $\mathcal{G}$-invariant on $S \setminus E$ and hence on all of S. Since the isotropy action of $SL(2, \mathbb{C})$ at p doesn't preserve nontrivial vectors in $T_p S \simeq \mathbb{C}^2$, it follows that $X(p) = 0$. This is impossible, since Inoue surfaces don't admit *singular* nontrivial holomorphic vector fields [33].

The proof is the same in the case $\mathcal{G} = gl(2, \mathbb{C})$. $\square$

Meromorphic Affine Connections
In the following theorem we describe some meromorphic affine connections on non-Kaehler principal elliptic bundles over a compact Riemann surface of genus $g \geq 2$.

THEOREM 5.10. *Let S be a principal elliptic bundle over a compact Riemann surface Σ of genus $g \geq 2$, with odd-first Betti number.*

    i) *A meromorphic torsion-free affine connection ∇ on S is invariant by the principal fibration (i.e., the fundamental vector field of the principal fibration is a Killing field) if and only if the set of its poles intersects only a finite set of fibers.*

    ii) *The space of those previous connections admitting simple poles on a single fiber, above a point $\xi_0 \in \Sigma$, is a complex affine space of dimension $5g + 1$ and* *its underlying vector space identifies with $H^0_{\xi_0}(K^2_\Sigma) \times M^2_{2,\xi_0}$, where $H^0_{\xi_0}(K^2_\Sigma)$ is the vector space of meromorphic quadratic differentials on Σ with a single pole of order at most two at ξ_0, and M_{2,ξ_0} is the vector space of quasimodular forms of weight 2 on Σ with a single simple pole at ξ_0.*

   iii) *If ∇ is generic among the meromorphic affine connections that satisfy (ii), then the local Killing Lie algebra of ∇ is generated by the fundamental vector field of the principal fibration.*

   iv) *The meromorphic affine connections that satisfy (ii) are projectively flat away from the poles.*

Recall that *a quasimodular form of weight 2 on Σ is a holomorphic function f* defined on the upper-half-plane H such that, for some $K \in \mathbb{C}$, we have $f(\xi) = f(\gamma\xi)(c_\gamma\xi + d_\gamma)^{-2} - K(c_\gamma\xi + d_\gamma)^{-1}$, for all $\gamma = \begin{pmatrix} a_\gamma & b_\gamma \\ c_\gamma & d_\gamma \end{pmatrix} \in SL(2,\mathbb{R})$ in the fundamental group of Σ [3, 62]. If $K = 0$, then f is a classical modular form. Theorem 9 in [3] shows that the space of quasimodular forms of weight 2 on Σ that admit a simple pole in a single orbit is a complex vector space of dimension $g + 1$ (i.e., the quotient of the space of such quasimodular forms over those which are modular is 1-dimensional).

REMARK 5.11. The proof below shows that S admits a (flat) holomorphic affine connection. Theorem 2.1 applies and, since any meromorphic function on S is a pull-back of a meromorphic function on Σ [4], the orbits of the pseudogroup of local isometries of any meromorphic geometric structure on S contain the fibers of S.

Proof. i) The projection of the set of poles of ∇ on Σ is a closed analytic set that is either a finite set of points or all of Σ. If this projection is Σ and ∇ is invariant by the principal fibration, then each point of S is a pole, which is impossible. This proves the easy sense of the implication.

   We will now describe the space of connections ∇, such that only a finite set of fibers contain poles of ∇. Up to a finite unramified cover and a finite quotient, S admits holomorphic affine structures (i.e., flat torsion-free holomorphic affine connections) that can be built in the following way [36]:

   Consider Γ a discrete torsion-free subgroup in $PSL(2,\mathbb{R})$ such that $\Sigma = \Gamma\backslash H$, with H the upper-half plane. Take any holomorphic projective structure on Σ, its developing map $\tau : H \to P^1(\mathbb{C})$, and its holonomy morphism $\rho : \Gamma \to PSL(2,\mathbb{C})$. This embedding of Γ into $PSL(2,\mathbb{C})$ lifts to $SL(2,\mathbb{C})$ (this follows from the fact that orientable (real) closed 3-manifolds have a trivial second Stiefel-Whitney class [47]). Choose such a lift and consider Γ as a subgroup of $SL(2,\mathbb{C})$. Denote by $W = \mathbb{C}^2 \setminus \{0\}$ the $\mathbb{C}^*$-tautological bundle over $P^1(\mathbb{C})$. The canonical affine structure of $\mathbb{C}^2$ induces a Γ-invariant affine structure on W and hence a Γ-invariant holomorphic affine structure on the pull-back $\tau^*(W) \simeq \mathbb{C}^* \times H$. The Γ-action on $\tau^*(W)$ comes from the action by deck transformation on H and from the ρ-action on W.

The previous holomorphic affine structure on $\tau^*(W)$ is also invariant by the homotheties in the fibers (which commute with the Γ-action). Consider now $\Delta \simeq \mathbb{Z}$ a lattice in $\mathbb{C}^*$, which acts by multiplication on the fibers of $\tau^*(W)$ and takes the quotient of $\tau^*(W)$ by $\Delta \times \Gamma$. The quotient is a principal elliptic bundle over Σ, with fiber $\Delta \backslash \mathbb{C}^*$, biholomorphic to S.

The affine structure inherited by the universal cover $\mathbb{C} \times H$ of S is the pull-back of the previous affine structure on $\mathbb{C}^* \times H$ by the map

$$\mathbb{C} \times H \to \mathbb{C}^* \times H$$

$$(z, \xi) \to (e^z, \xi).$$

In the following we will consider the flat torsion-free holomorphic affine connection ∇_0 on S given by the case where τ is the standard embedding of H into $P^1(\mathbb{C})$. In this case, the action of $\gamma = \begin{pmatrix} a & b \\ c & d \end{pmatrix} \in \Gamma \subset SL(2, \mathbb{R})$ on the universal cover $\mathbb{C} \times H$ of S is easily seen to be given by [36]:

$$\gamma(z, \xi) = (z + \log(c\xi + d), \gamma\xi), \forall (z, \xi) \in \mathbb{C} \times H,$$

where $\log$ is a determination of the logarithm and the γ-action on H comes from the standard action of $SL(2, \mathbb{R})$ on H.

The difference $\nabla - \nabla_0$ is a meromorphic $(2, 1)$-tensor ω on S, or, equivalently, a $\Delta \times \Gamma$-invariant mermorphic $(2,1)$-tensor $\tilde{\omega}$ on the universal cover $\mathbb{C} \times H$. Moreover, $\tilde{\omega}$ is invariant by $t(z, \xi) = (z + 2i\pi, \xi)$.

Then we have

$$\tilde{\omega} = f_{11}(z, \xi)dz \otimes dz \otimes \frac{\partial}{\partial z} + f_{12}(z, \xi)dz \otimes d\xi \otimes \frac{\partial}{\partial z} + f_{21}(z, \xi)dz \otimes dz \otimes \frac{\partial}{\partial \xi}$$

$$+ f_{22}(z, \xi)dz \otimes d\xi \otimes \frac{\partial}{\partial \xi} + g_{11}(z, \xi)d\xi \otimes dz \otimes \frac{\partial}{\partial z} + g_{12}(z, \xi)d\xi \otimes d\xi \otimes \frac{\partial}{\partial z}$$

$$+ g_{21}(z, \xi)d\xi \otimes dz \otimes \frac{\partial}{\partial \xi} + g_{22}(z, \xi)d\xi \otimes d\xi \otimes \frac{\partial}{\partial \xi},$$

with f_{ij}, g_{ij} meromorphic functions on $\mathbb{C} \times H$, and $f_{ij}(\cdot, \xi), g_{ij}(\cdot, \xi)$ holomorphic except for ξ lying in the union of a finite number of Γ-orbits in H.

Notice that the difference between ∇_0 and the standard affine structure of $\mathbb{C} \times H$ is given by $f_{11} = f_{22} = g_{21} = 1$, the others f_{ij}, g_{ij} being trivial (see the straightforward computation in [36]).

Since $\tilde{\omega}$ is Δ-invariant and t-invariant, the functions $f_{ij}(\cdot, \xi)$ and $g_{ij}(\cdot, \xi)$ descend on an elliptic curve for all $\xi \in H$. They are constant for all $\xi \in H$ for which they are holomorphic. It follows that $f_{ij}(\cdot, \xi)$ and $g_{ij}(\cdot, \xi)$ are constant for ξ lying in a open dense subset of H and, consequently, for all $\xi \in H$. It follows that the functions f_{ij} and g_{ij} depend only on ξ. Consequently, the flow of $\frac{\partial}{\partial z}$ preserves $\tilde{\omega}$. Since the fundamental generator of the principal fibration $\frac{\partial}{\partial z}$ also preserves ∇_0, it is a Killing field for ∇.

ii) In the following we consider only torsion-free connections: $f_{12} = g_{11}$ and $f_{22} = g_{21}$.

The Γ-invariance of $\tilde{\omega}$ yields the following equations:

1) $f_{11}(\xi) = f_{11}(\gamma\xi) - cf_{21}(\gamma\xi)(c\xi + d)$

2) $f_{12}(\xi) = f_{12}(\gamma\xi)(c\xi + d)^{-2} - 2c^2 f_{21}(\gamma\xi) - cf_{22}(\gamma\xi)(c\xi + d)^{-1}$
   $+ 2cf_{11}(\gamma\xi)(c\xi + d)^{-1}$

3) $f_{21}(\xi) = f_{21}(\gamma\xi)(c\xi + d)^2$

4) $f_{22}(\xi) = 2f_{21}(\gamma\xi)c(c\xi + d) + f_{22}(\gamma\xi)$

5) $g_{12}(\xi) = g_{12}(\gamma\xi)(c\xi + d)^{-4} + c^2 f_{11}(\gamma\xi)(c\xi + d)^{-2} + cf_{12}(\gamma\xi)(c\xi + d)^{-3} -$
   $c^3 f_{21}(\gamma\xi)(c\xi + d)^{-1} - c^2 f_{22}(\gamma\xi)(c\xi + d)^{-2} - cg_{22}(\gamma\xi)(c\xi + d)^{-3}$

6) $g_{22}(\xi) = g_{22}(\gamma\xi)(c\xi + d)^{-2} + cf_{22}(\gamma\xi)(c\xi + d)^{-1} + c^2 f_{21}(\gamma\xi)$,

for all $\gamma = \begin{pmatrix} a & b \\ c & d \end{pmatrix} \in \Gamma$.

The equation (3) implies that f_{21} is a meromorphic vector field on the compact Riemann surface Σ. Since a single pole of order at most 1 is allowed, we get $f_{21} = 0$ as a direct consequence of Riemann-Roch theorem and Serre duality ([27], page 245).

The equations (1) and (4) imply then that f_{11} and f_{22} are meromorphic functions on Σ with a single pole of order at most 1. If f_{11}, f_{22} are not constant, they give a biholomorphism between Σ and the projective line $P^1(\mathbb{C})$ [27]: a contradiction, since the genus of Σ is ≥ 2.

We conclude that f_{11} and f_{22} are constants.

Then we have

2') $f_{12}(\xi) = f_{12}(\gamma\xi)(c\xi + d)^{-2} - cf_{22}(c\xi + d)^{-1} + 2cf_{11}(c\xi + d)^{-1}$

5') $g_{12}(\xi) = g_{12}(\gamma\xi)(c\xi + d)^{-4} + c^2(f_{11} - f_{22})(c\xi + d)^{-2}$
   $+ c(f_{12} - g_{22})(\gamma\xi)(c\xi + d)^{-3}$

6') $g_{22}(\xi) = g_{22}(\gamma\xi)(c\xi + d)^{-2} + cf_{22}(c\xi + d)^{-1}$.

It follows from (2′) and (6′) that f_{12} and g_{22} are quasimodular forms of weight 2 on Σ with a single simple pole. The space of those quasimodular forms is a complex vector space of dimension $g + 1$ (see [3], theorem 9).

Equation (5′) is equivalent to the Γ-invariance of the quadratic differential $w(\xi)d\xi^2$, where $w = 2g_{12} + f'_{12} - g'_{22}$. It follows that $2g_{12} + f'_{12} - g'_{22}$ is a meromorphic quadratic differential on Σ with a single pole of order at most 2.

It is classically known (as an application of Riemann-Roch theorem and Serre duality) that the space of quadratic differentials with a single pole of order at most 2 is of complex dimension $3g - 1$ (see, for example, [27]).

iii) Let $X = a(z, \xi)\frac{\partial}{\partial z} + b(z, \xi)\frac{\partial}{\partial \xi}$ be a local holomorphic Killing field of a generic connection ∇, in the neighborhood of a point where ∇ is holomorphic (a and b are holomorphic local functions on $\mathbb{C} \times H$).

The equation of the Killing field is

$$[X, \nabla_Y Z] = \nabla_{[X,Y]}Z + \nabla_Y[X, Z]$$

for all Y, Z tangents to S. It is enough to verify the equation for (Y, Z) corresponding to $\left(\frac{\partial}{\partial z}, \frac{\partial}{\partial \xi}\right)$, $\left(\frac{\partial}{\partial z}, \frac{\partial}{\partial z}\right)$, and $\left(\frac{\partial}{\partial \xi}, \frac{\partial}{\partial \xi}\right)$.

This leads to the following PDE system:

1) $a_{zz} + (1 + f_{11})a_z + 2f_{12}b_z = 0$

2) $b_{zz} + (1 + 2f_{22} - f_{11})b_z = 0$

3) $a_{z\xi} + (f_{11} - f_{22})a_\xi + g_{12}b_z + g_{11}b_\xi + \dfrac{\partial f_{12}}{\partial \xi}b = 0$

4) $b_{z\xi} + (1 + f_{22})a_z + (g_{22} - f_{12})b_z = 0$

5) $a_{\xi\xi} - g_{12}a_z + (2f_{12} - g_{12})a_\xi + 2g_{12}b_\xi + \dfrac{\partial g_{12}}{\partial \xi}b = 0$

6) $b_{\xi\xi} + 2(1 + f_{22})a_\xi - g_{12}b_z + g_{22}b_\xi + \dfrac{\partial g_{22}}{\partial \xi}b = 0$.

The general solution of the first equation is $b = \nu(\xi)e^{-\mu z} + C(\xi)$, with ν, C holomorphic functions of ξ and $\mu = 1 + 2f_{22} - f_{11}$.

We replace b_z in the first equation and we get

$$a_z = \frac{\mu}{f_{11} - f_{22}}f_{12}(\xi)\nu(\xi)e^{-\mu z} + A(\xi)e^{-(1+f_{11})z},$$

with A a holomorphic function of ξ.

Then equation (4) leads to

$$\mu\left[-\nu'(\xi) + \left(\frac{1 + f_{11}}{f_{11} - f_{22}}f_{12} - g_{22}\right)\nu(\xi)\right]e^{-\mu z} + (1 + f_{22})A(\xi)e^{-(1+f_{11})z} = 0.$$

For a generic ∇, we have $f_{11} \neq f_{22}$, thus $\mu \neq 1 + f_{11}$ and the functions $e^{-\mu z}, e^{-(1+f_{11})z}$ are $\mathbb{C}$-linearly independent. This implies $(1 + f_{22})A(\xi) = 0$ and, since for a generic ∇, $f_{22} \neq -1$, we have $A(\xi) = 0$.

We also get

$$I) \quad v'(\xi) = v(\xi) \left(\frac{1+f_{11}}{f_{11} - f_{22}} f_{12} - g_{22} \right).$$

Now we check equation (3). We replace the partial derivatives of a and b in (3) and we get the following:

$$\mu \left[\frac{f_{12}}{f_{11} - f_{22}} v'(\xi) - \left(g_{12} - \frac{f'_{12}}{f_{11} - f_{22}} \right) v(\xi) \right] e^{-\mu z} + (f_{12} C)' = 0.$$

Since generically $\mu \neq 0$ the functions $e^{-\mu z}$ and 1 are linearly independent, which yields to $(f_{12} C)' = 0$ and to

$$II) \quad v'(\xi) = v(\xi) \frac{1}{f_{12}} [(f_{11} - f_{22}) g_{12} - f'_{12}].$$

Relations (I) and (II) are compatible, for a generic connection, only if $v = 0$. This implies $b = C(\xi)$, $a = B(\xi)$.

Our PDE system becomes:

$$3') \, (f_{11} - g_{21}) a' + f'_{12} b + f_{12} b' = 0$$
$$5'') \, a'' + (2f_{12} - g_{12}) a' + 2g_{12} b' - g'_{12} b = 0$$
$$6') \, b'' + 2(1 + 2g_{21}) a' + g_{22} b' + g'_{22} b = 0.$$

Since $(f_{12} b)' = 0$, we have $a' = 0$ and thus a is a constant function. Equation $(5')$ implies then $2g_{12} b' - g'_{12} b = 0$, which, for a generic ∇, is compatible with $f_{12} b' + f'_{12} b = 0$ only if $b = 0$.

It follows that X is a constant multiple of $\frac{\partial}{\partial z}$.

iv) The projective connection associated to the affine connection ∇ is given by the following second-order ODE [14]:

$$\xi'' = K^0(z, \xi) + K^1(z, \xi) \xi' + K^2(z, \xi)(\xi')^2 + K^3(z, \xi)(\xi')^3,$$

where $K^0 = -f_{21}, = 0$, $K^1, = (1 + f_{11}) - 2(1 + f_{22})$, $K^2, = -(g_{22} - 2f_{12})$ and $K^3 = g_{12}$ (see [14]).

Liouville [42], followed by Tresse [56] and Cartan [14], proved that this projective connection is projectively flat if and only if both of the following invariants vanish:

$$L_1 = 2K^1_{z\xi} - K^2_{zz} - 3K^0_{\xi\xi} - 6K^0 K^3_z - 3K^3 K^0_z + 3K^0 K^2_\xi + 3K^2 K^0_\xi + K^1 K^2_z$$
$$- 2K^1 K^1_\xi,$$

$$L_2 = 2K^2_{z\xi} - K^1_{\xi\xi} - 3K^3_{zz} + 6K^3 K^0_\xi + 3K^0 K^3_\xi - 3K^3 K^1_z - 3K^1 K^3_\xi - K^2 K^1_\xi$$
$$+ 2K^2 K^2_z.$$

Here $K^0 = 0$, K^1 is a constant function and K^2, K^3 depend only on ξ. This implies the vanishing of both invariants L_1 and L_2. □

I would like to thank G. Chenevier and G. Dloussky for helpful conversations.

References

[1] G. D'Ambra, M. Gromov, Lectures on transformations groups: geometry and dynamics, Surveys in Differential Geometry (Cambridge), (1990), 19–111.

[2] A. M. Amores, Vector fields of a finite type G-structure, J. Differential Geom., **14(1)**, (1979), 1–6.

[3] N. Azaiez, The ring of quasimodular forms for a cocompact group, J. Number Theory, (2007).

[4] W. Barth, K. Hulek, C. Peters, A. Van De Ven, Compact complex surfaces, Ergebnisse der Mathematik, Vol. 4, 2nd Ed., Berlin, Springer, (1995).

[5] A. Beauville, Variétés Kähleriennes dont la première classe de Chern est nulle, J. Differential Geom., **18**, (1983), 755–782.

[6] Y. Benoist, Actions propres sur les espaces homogènes réductifs, Annals of Mathematics, **144**, (1996), 315–347.

[7] Y. Benoist, Orbites de structures rigides, integrable systems and foliations (Montpellier), Boston, Birkaüser, (1997).

[8] Y. Benoist, F. Labourie, Sur les espaces homogènes modèles des variétés compactes, Publ. Math. I.H.E.S., **76**, (1992), 99–109.

[9] J. Benveniste, D. Fisher, Nonexistence of invariant rigid structures and invariant almost rigid structures, Comm. Anal. Geom., **13(1)**, (2005), 89–111.

[10] I. Biswas, B. McKay, Holomorphic Cartan geometries and Calabi-Yau manifolds, arXiv math. AG/0812.3978, (2010).

[11] F. Campana, J-P. Demailly, T. Peternell, The algebraic dimension of compact complex threefolds with vanishing second Betti number, Compositio Math., **112(1)**, (1998) 77–91.

[12] A. Candel, R. Quiroga-Barranco, Gromov's centralizer theorem, Geom. Dedicata, **100**, (2003), 123–155.

[13] Y. Carrière, Autour de la conjecture de L. Markus sur les variétés affines, Invent. Math., **95**, (1989), 615–628.

[14] E. Cartan, Sur les variétés à connexion projective, Bull. Soc. Math. France, **52**, (1924), 205–241.

[15] S. Dumitrescu, Structures géométriques holomorphes sur les variétés complexes compactes, Ann. Scient. Ec. Norm. Sup., **34(4)**, (2001), 557–571.

[16] S. Dumitrescu, Métriques riemanniennes holomorphes en petite dimension, Ann. Instit. Fourier, Grenoble, **51(6)**, (2001), 1663–1690.

[17] S. Dumitrescu, Homogénéité locale pour les métriques riemanniennes holomorphes en dimension 3, Ann. Instit. Fourier, Grenoble, **57(3)**, (2007), 739–773.

[18] S. Dumitrescu, Une caractérisation des variétés complexes compactes parallélisables admettant des structures affines, C.R. Acad. Sci. Paris, Ser. I, **347**, (2009), 1183–1187.

[19] S. Dumitrescu, Killing fields of holomorphic Cartan geometries, to appear in Monatshefte fur Mathematik, (2010).

[20]   S. *Dumitrescu*, Connexions affines et projectives sur les surfaces complexes compactes, Mathematische Zeitschrift, **264(2)**, (2010), 301–316.

[21]   S. *Dumitrescu*, A. *Zeghib*, Global rigidity for holomorphic Riemannian metrics on compact threefolds, Math. Ann., **345(1)**, (2009), 58–81.

[22]   R. *Feres*, Rigid geometric structures and actions of semisimple Lie groups, Rigidité, groupe fondamental et dynamique, Panorama et synthèses, Soc. Math. France, Paris, **13**, (2002).

[23]   C. *Frances*, Une preuve du théorème de Liouville en géométrie conforme dans le cas analytique, Enseign. Math., **49(2)**, (2003), 95–100.

[24]   D. *Fried*, W. *Goldman*, Three-dimensional affine christallographic groups, Adv. Math., **47(1)**, (1983), 1–49.

[25]   E. *Ghys*, Déformations des structures complexes sur les espaces homogènes de $SL(2, \mathbb{C})$, J. Reine Angew. Math., **468**, (1995), 113–138.

[26]   W. *Goldman*, Nonstandard Lorentz space forms, J. Diff. Geom., **21(2)**, (1985), 301–308.

[27]   P. *Griffiths*, J. *Harris*, Principles of algebraic geometry, New York, Wiley-Interscience, (1994).

[28]   M. *Gromov*, Rigid transformation groups, Géométrie Différentielle, (D. Bernard et Choquet-Bruhat Ed.), Travaux en cours, Hermann, Paris, **33**, (1988), 65–141.

[29]   R. *Gunning*, Lectures on Riemann surfaces, Princeton, NJ, Princeton University Press, (1966).

[30]   A. *Huckleberry*, S. *Kebekus*, T. *Peternell*, Group actions on S^6 and complex structures on $P^3(\mathbb{C})$, Duke Math. J., **102(1)**, (2000), 101–124.

[31]   J-M. *Hwang*, N. *Mok*, Uniruled projective manifolds with irreducible reductive G-structure, J. Reine Angew. Math., **490** (1997), 55–64.

[32]   E. *Ince*, Ordinary differential equations, New York, Dover Publications, (1944).

[33]   M. *Inoue*, On surfaces of class VII_0, Invent. Math., **24**, (1974), 269–310.

[34]   M. *Inoue*, S. *Kobayashi*, T. *Ochiai*, Holomorphic affine connections on compact complex surfaces, J. Fac. Sci. Univ. Tokyo, **27(2)**, (1980), 247–264.

[35]   A. *Kirilov*, Eléments de la théorie des représentations, Moscow, M.I.R., (1974).

[36]   B. *Klingler*, Structures affines et projectives sur les surfaces complexes, Ann. Inst. Fourier, Grenoble, **48(2)**, (1998), 441–477.

[37]   B. *Klingler*, Complétude des variétés Lorentziennes à courbure sectionnelle constante, Math. Ann., **306**, (1996), 353–370.

[38]   T. *Kobayashi*, T. *Yoshino*, Compact Clifford-Klein form of symmetric spaces—revisited, Pure Appl. Math. Q., **1(3)**, (2005), 591–663.

[39]   R. *Kulkarni*, F. *Raymond*, 3-dimensional Lorentz space-forms and Seifert fiber spaces, J. Diff. Geom., **21(2)**, (1985), 231–268.

[40]   C. *Lebrun*, Spaces of complex null geodesics in complex-Riemannian geometry, Trans. Amer. Math. Soc., **278**, (1983), 209–231.

[41]   C. *Lebrun*, $\mathcal{H}$-spaces with a cosmological constant, Proc. Roy. Soc. London, Ser. A, **380(1778)**, (1982), 171–185.

[42]   R. *Liouville*, Sur les invariants de certaines équations différentielles et sur leurs applications, Journal de l'Ecole Polytechnique, **59**, (1889), 7–76.

[43]   K. *Maehara*, On elliptic surfaces whose first Betti numbers are odd, Intl. Symp. on Alg. Geom., Kyoto, (1977), 565–574.

[44]   L. *Markus*, Cosmological models in differential geometry, mimeographed notes, Univ. of Minnesota, (1962).

[45] B. McKay, Characteristic forms of complex Cartan geometries, arXiv math. DG/ 0704.2555, (2009).

[46] J. Milnor, Curvatures of left invariant metrics on Lie groups, Adv. in Math., **21**, (1976), 293–329.

[47] J. Milnor, J. Stasheff, Characteristic classes, Princeton, Princeton University Press, (1974).

[48] K. Nomizu, On local and global existence of Killing vector fields, Ann. of Math. (2), **72**, (1960), 105–120.

[49] V. Popov, E. Vinberg Invariant theory, Algebraic Geometry 4, E.M.S., **55**, (1991), 123– 280.

[50] M. Raghunathan, Discrete subgroups of Lie groups, Berlin, Springer-Verlag, (1972).

[51] S. Rahmani, Métriques de Lorentz sur les groups de Lie unimodulaires de dimension 3, J. Geom. Phys., **9**, (1992), 295–302.

[52] N. Rahmani, S. Rahmani, Lorentzian geometry of the Heisenberg group, Geom. Dedicata, **118**, (2006), 133–140.

[53] F, Salein, Variétés anti-de Sitter de dimension 3 exotiques, Ann. Inst. Fourier, Grenoble, **50(1)**, (2000), 257–284.

[54] T. Suwa, Compact quotient spaces of $\mathbb{C}^2$ by affine transformation groups, J. Diff. Geom. **10**, (1975), 239–252.

[55] W. Thurston, The geometry and the topology of 3-manifolds, Princeton, Princeton University Press, (1983).

[56] A. Tresse, Détermination des invariants ponctuels de l'équation différentielle ordinaire du second ordre $y'' = \omega(x, y, y')$, Leipzig, **87** S, gr. 8, (1896).

[57] K. Ueno, Classification theory of algebraic varities and compact complex spaces, Lect. Notes in Math., **439**, Berlin, Springer-Verlag, (1975).

[58] A. Vitter, Affine structures on compact complex manifolds, Invent. Math., **17**, (1972), 231–244.

[59] C. Wall, Geometric structures on compact complex analytic surfaces, Topology, **25(2)**, (1986), 119–153.

[60] H. Wang, Complex parallelisable manifolds, Proc. Amer. Math. Soc. **5**, (1954), 771–776.

[61] J. Wolf, Spaces of constant curvature, New York, McGraw-Hill, (1967).

[62] D. Zagier, Lecture, Cours at Collège de France, 2000–2001.

[63] A. Zeghib, On Gromov's theory of rigid transformation groups: a dual approach, Ergodic Th. Dyn. Syst., **20(3)**, (2000), 935–946.

[64] R. Zimmer, On the automorphism group of a compact Lorentz manifold and other geometric manifolds, Invent. Math., **83**, (1986), 411–426.

[65] R. Zimmer, Actions of semisimple groups and discrete subgroups, Proc. Internat. Cong. Math., Berkeley, CA, (1986), 1247–1258.

[66] R. Zimmer, Automorphisms groups of geometric manifolds, Proc. Symp. Pure Math., **54(3)**, (1993), 693–710.

[67] R. Zimmer, Ergodic theory and semisimple Lie groups, Boston, Birkhaüser, (1984).

3

RENATO FERES AND EMILY RONSHAUSEN

HARMONIC FUNCTIONS OVER GROUP ACTIONS

TO BOB ZIMMER ON HIS 60TH BIRTHDAY

Abstract

Let X be a compact space, Γ a countable group of homeomorphisms of X, and μ a probability measure on Γ. From this data one naturally defines a random walk on X and a related notion of continuous harmonic function. The question we wish to investigate is whether all continuous harmonic functions on X are Γ-invariant. If this is the case, we say that the system satisfies the topological *Liouville property*. We show that if μ is a symmetric probability measure on an arbitrary Γ and $X = S^1$ or the interval $[0, 1]$, then the Liouville property always holds. In addition, when the Poisson boundary of (Γ, μ) can be identified with the circle, we give a general construction of a non-Liouville action of Γ on S^2, generalizing an example from [FZ2].

1. Introduction

A useful idea in the study of group actions without invariant measures is to introduce and study a random $\mathbb{Z}$-action associated to a choice of probability measure on the acting group. To fix notations, let Γ denote a group of transformations of a space X and μ a probability measure on Γ. Then μ defines a random walk on Γ, which induces a random $\mathbb{Z}$-action on X. See, for example, [Fu] where this point of view is well represented. It is often assumed in this context that the Γ-space admits a μ-stationary measure (also called μ-harmonic measure—see definition below) of full support, endowing the random $\mathbb{Z}$-action with an invariant measure. This brings along the ergodic theory apparatus that relies on the existence of invariant measures.

Often, however, the Γ-space has large regions with possibly interesting dynamics that are not detected by any harmonic measure for a given μ. If this is the case, one can hope to gain some insight into the system represented by (X, Γ, μ) by investigating the space of μ-harmonic functions on X. As pointed out in Proposition 2.4 below, continuous μ-harmonic functions on compact Γ-spaces are Γ-invariant on the support of any μ-harmonic probability measure, so nontrivial (i.e., non-Γ-invariant) such functions may yield information about the regions that harmonic measures do not detect. This is the perspective motivating this paper. We have chosen here to consider compact Γ-spaces and continuous functions, although one may also try to pursue the subject on measurable Γ-spaces.

The most basic question to ask about (X, Γ, μ) in the present context is whether it satisfies a topological *Liouville property*, i.e., whether all continuous μ-harmonic functions are Γ-invariant. The main result of the paper is that the Liouville property holds for actions of countable groups on 1-dimensional spaces when μ is a symmetric probability measure.

THEOREM 1.1. *Let Γ be a countable group acting on X by homeomorphisms, where X is either the circle S^1 or the interval $[0, 1]$. Let μ be a symmetric probability measure on Γ. Then the Liouville property holds for (X, Γ, μ).*

We give at the end of the paper an example of Γ-space for which the Liouville property does not hold. More specifically, we show that if the Poisson boundary of (Γ, μ) can be identified with the circle S^1, then it is possible to construct a Γ-action on the two-sphere so that (S^2, Γ, μ) admits continuous μ-harmonic functions that are nonconstant on all orbits in the complement of a pair of fixed points. The action can be chosen to be ergodic with respect to the smooth measure class on S^2, whereas the random walk is transient in the complement of the two fixed points. A general recipe for constructing non-Liouville actions is also proposed.

The proof of Theorem 1.1 parallels the proof of the main theorem of [FFP]. That paper is about the Liouville property for codimension-one foliated Brownian motion and harmonic functions for the Laplace-Beltrami operator associated to a Riemannian metric on leaves, rather than discrete group actions. Although the two situations cannot be directly compared, our assumptions on Γ and μ are, in a sense, much more general than related assumptions on that paper. The example of a non-Liouville action on S^2 generalizes a similar construction given in [FZ2] in the context of foliations with a Riemannian metric on leaves. See, also, [FZ1] for these issues in the setting of holomorphic functions on foliations. The paper [DK] has some deep results about harmonic functions on foliated spaces in the spirit of the present discussion.

This paper was partly written during a visit by the first author to the Paris-Sud 11 University, at Orsay. He wishes to thank the group of Topology and Dynamics for their hospitality and the university for its generous support. He also wishes to thank Vadim Kaimanovich and Bertrand Deroin for helpful and illuminating discussions. In addition, we wish to express our thanks to the anonymous referee for several helpful comments.

2. Harmonic Functions on Γ-spaces

In this section we collect some basic properties and general remarks concerning harmonic functions.

Let Γ be a countable (infinite) group, μ a probability measure on Γ, and X a compact topological space on which Γ acts by homeomorphisms. It will always be assumed that μ is *non-degenerate*, that is, the support of μ generates Γ as a semigroup. In particular, $\mu(\gamma) < 1$ for all $\gamma \in \Gamma$. A continuous function $f : X \to \mathbb{R}$ is said to be μ-*harmonic* if, for all $x \in X$,

$$f(x) = \sum_{\gamma \in \Gamma} f(\gamma(x))\mu(\gamma).$$

Let $H(X, \Gamma, \mu)$ denote the set of μ-harmonic functions on X and $C(X)^\Gamma$ the subset consisting of continuous Γ-invariant functions. If $H(X, \Gamma, \mu) = C(X)^\Gamma$, we say that (X, Γ, μ) satisfies the Liouville property.

We note that if $f \in H(X, \Gamma, \mu)$ and $x \in X$, then $\tilde{f}_x(\gamma) := f(\gamma^{-1}(x))$ is a harmonic function on Γ in the following sense:

$$\tilde{f}_x(\gamma) = \sum_{\eta \in \Gamma} \tilde{f}_x(\gamma\eta)\mu(\eta^{-1}).$$

Under the conventions of [Fu], $\tilde{f}_x$ is harmonic relative to the right-$\hat{\mu}$-random walk on Γ, where $\hat{\mu}(\gamma) = \mu(\gamma^{-1})$. Thus, if $(\Gamma, \hat{\mu})$ itself has the Liouville property, that is, if bounded harmonic functions on Γ are constant, then necessarily (X, Γ, μ) has the Liouville property for all X. Therefore, the present subject is uninteresting if the Poisson boundary of $(\Gamma, \hat{\mu})$ reduces to a single point. It is known, for example, that if Γ is amenable, then there exists a symmetric $(\mu = \hat{\mu})$ nondegenerate μ for which the Poisson boundary is trivial.

A good example of a group to keep in mind, on the other hand, is the fundamental group of a closed surface of genus at least 2, represented as isometries of the Poincaré disc. It has a nontrivial boundary for many interesting probability measures. For example, a result due to Furstenberg implies that on this group there exists a probability measure μ relative to which the boundary of Γ coincides with the geometric boundary, S^1, of the disc [Fur].

Let $\Delta = P - I$ be the *Laplace operator* on the set $C(X)$ of real-valued continuous functions on X, where $(Pf)(x) := \sum \mu(\gamma)f(\gamma(x))$ is the averaging operator and I is the identity. Clearly, f is harmonic iff $\Delta f = 0$. More generally, f satisfying $\Delta f \geq 0$ (respectively, $\Delta f \leq 0$) is said to be *superharmonic* (respectively, *subharmonic*).

PROPOSITION 2.1. (MAXIMUM PRINCIPLE) *If a superharmonic $f \in C(X)$ is bounded above by C and $f(x) = C$ for an $x \in X$, then f is constant on the Γ-orbit of x.*

Proof. This trivially follows from μ being nondegenerate and the remark

$$0 \leq \sum_{\gamma \in \Gamma} (f(\gamma(x)) - f(x))\mu(\gamma) \leq \sum_{\gamma \in \Gamma} (f(\gamma(x)) - C)\mu(\gamma) \leq 0,$$

so $f(\gamma(x)) = C$ for all γ in the support of μ. $\qquad\square$

COROLLARY 2.2. *If the closure of every Γ-orbit contains a unique minimal set, then the Liouville property holds.*

Proof. Let X_0 be the closure of an orbit. The maximum and minimum values in X_0 of a harmonic function are attained on orbits where the function is constant. Since X_0 has a single minimal set, these two values must agree. $\qquad\square$

It is useful to introduce a (directed) metric d_{x_0} on the orbit Γx_0 as follows. Given x, y in Γx_0, we specify a *path* joining x to y by a sequence $\alpha = (\gamma_0, \gamma_1, \ldots, \gamma_{n-1})$, $\mu(\gamma_i) > 0$, such that $y_0 = x$, $y_n = y$, and $y_{i+1} = \gamma_i y_i$ for $i = 0, \ldots, n-1$. The *length* of α is defined as

$$L(\alpha) = \sum_{i=0}^{n-1} -\ln \mu(\gamma_i) \in (0, \infty].$$

We allow x to be joined to itself by the trivial path, defined by $n = 0$, $\alpha = \emptyset$, and $L(\alpha) = 0$. Now let $d_{x_0}(x, y) = \inf L(\alpha)$, where the infimum is taken over all n and all $(\gamma_0, \ldots, \gamma_{n-1})$ describing a path from x to y.

PROPOSITION 2.3. (HARNACK INEQUALITY) *Let f be a continuous, positive, subharmonic function on X and $x_0 \in X$ arbitrary. Then for all $x, y \in \Gamma x_0$,*

$$\frac{f(y)}{f(x)} \leq e^{d_{x_0}(x,y)}.$$

Proof. As f is subharmonic, $f(x) \geq \sum_{\gamma \in \Gamma} \mu(\gamma)f(\gamma x) \geq \mu(\eta)f(\eta x)$ for all $\eta \in \Gamma$, so $f(\eta x)/f(x) \leq e^{-\ln \mu(\eta)}$. Multiplying such inequalities along any path α from x to y yields the claim. $\qquad\square$

Given (X, Γ, μ) as above, a measure ν on X is called *harmonic* if ν is Γ-quasi invariant and $\mu * \nu = \nu$, where the convolution $\mu * \nu$ is the image of the product measure $\mu \otimes \nu$ under the action map $\Gamma \times X \to X$. It is easily shown that $\mu * \nu = \nu P$, where the averaging operator P acts on ν by duality as

$$\nu P = \sum_{\gamma \in \Gamma} \mu(\gamma)\gamma_* \nu.$$

The following useful remark is due, in the context of foliated Brownian motion, to Garnett [Gar]. The proof, which is sketched below for completeness, is adapted from Paulin [Pa].

PROPOSITION 2.4. *Let ν be a harmonic probability measure on (X, Γ, μ). Then every h in $H(X, \Gamma, \mu)$ is Γ-invariant on the support of ν.*

Proof. It may be assumed that h is non-negative. Let c be a non-negative constant and define $h_c(x)$ as the minimum of $\{h(x), c\}$. Note that $Ph_c \leq h_c$. Since ν is harmonic, $\int h_c d\nu = \int h_c d(\mu * \nu) = \int Ph_c d\nu$, therefore $\int (Ph_c - h_c)d\nu = 0$. As the functions involved are continuous, it follows that $Ph_c = h_c$ on the support of ν. Applying the maximum principle to $h_c|_{\text{supp}(\nu)}$, we conclude that $h^{-1}([c, \infty)) \cap \text{supp}(\nu)$ is Γ-invariant for all c. This implies that h is Γ-invariant on $\text{supp}(\nu)$. $\qquad\square$

Let $\mathcal{C}$ denote the largest Γ-invariant subset of X such that $h|_\mathcal{C}$ is Γ-invariant for all $h \in H(X, \Gamma, \mu)$. Clearly, $\mathcal{C}$ is closed. It is also nonempty since, by the maximum principle, it contains every minimal set in X. Thus if $\mathcal{M}$ denotes the closure of the union of all the minimal sets of X relative to the Γ-action, then $\mathcal{M} \subseteq \mathcal{C}$. Due to Proposition 2.4, the set $\mathcal{S}$, defined as the closure of the union of supports of all harmonic measures, is also contained in $\mathcal{C}$. In fact, it is easily seen that $\mathcal{M} \subseteq \mathcal{S} \subseteq \mathcal{C}$. It will be explained shortly that the random walk associated to (X, Γ, μ) (defined next) converges to $\mathcal{C}$ with probability 1.

3. Random Walk on (X, Γ, μ)

Let (X, Γ, μ) be, as above, a compact Γ-space where Γ is a countable group and μ is a probability measure on Γ. The measure induces a Markov transition kernel on X by setting $p(x, y) = 0$ if $y \notin \Gamma x$ and

$$p(x, y) = \sum_{y = \gamma x} \mu(\gamma),$$

where the sum is over all $\gamma \in \Gamma$ such that $\gamma x = y$. Let X_n denote the random walk on X with transition probabilities p and initial state x_0. For the set C, defined at the end of the previous section, we have the following:

PROPOSITION 3.1. *The random walk X_n on $X \setminus C$ is transient.*

The proposition means that for all $x_0 \in X \setminus C$ and every open U containing C, there exists a random integer N, almost surely finite, such that $X_n \in U$ for all $n \geq N$. Prior to proving this fact we need to review some information about the Poisson boundary of Γ. Our main source is Furman [Fu].

The boundary of Γ can be described as a compact Hausdorff Γ-space B endowed with a probability measure of full support, ν, satisfying the following properties:

1) ν is harmonic relative to $(\Gamma, \hat{\mu})$;
2) For almost every sample path $\bar{\gamma} = (\gamma_1, \gamma_2, \ldots)$ of a (right) $\hat{\mu}$-random walk on Γ, a limit probability $\nu_{\bar{\gamma}} = \lim_{n \to \infty} (\gamma_1 \cdots \gamma_n)_* \nu$ exists and is equal to a Dirac measure δ_{γ_∞} supported on a random point, η_∞, of B;
3) Let $H^\infty(\Gamma, \hat{\mu})$ denote the space of bounded harmonic functions on Γ relative to $\hat{\mu}$. Then the map $F : L^\infty(B, \nu) \to H^\infty(\Gamma, \hat{\mu})$ given by $F(\phi)(\gamma) = \int_B \phi \, d\gamma_* \nu$ is an isometric bijection such that, for $h = F(\phi)$,

$$\phi(\eta_\infty) = \lim_{n \to \infty} h(\gamma_1 \cdots \gamma_n).$$

It is observed in [Fu], remark 2.16(i), that if Γ is a discrete group, then, for the topological description of B given there (denoted $\overline{B}$ in Furman's article), $L^\infty(B, \nu)$ coincides with the continuous functions on B. In this case, B is nonmetrizable if Γ has a nontrivial Poisson boundary. However, measured theoretically, the Poisson boundary is always a Lebesgue space.

LEMMA 3.2. *Let $\eta_n = \gamma_1 \ldots \gamma_n$ be a right-$\hat{\mu}$-random walk starting at the unit element $e \in \Gamma$. Let $h \in H^\infty(\Gamma, \hat{\mu})$ and C a positive constant. Then, for a.e. sample* path η_n and every sequence $\xi_n \in \Gamma$ such that $d_e(\eta_n, \xi_n) < C$, the sequences $h(\eta_n)$ and $h(\xi_n)$ converge to the same value. (Here d_e is the directed metric used in Proposition 2.3.)

Proof. Define $v_\gamma = \gamma_* v$. From the above properties of the Poisson boundary and Proposition 2.3, it follows that

$$\max\left\{\frac{dv_{\eta_n}}{dv_{\xi_n}}, \frac{dv_{\xi_n}}{dv_{\eta_n}}\right\} \leq e^C$$

for all n. But since v_{η_n} converges to the Dirac measure δ_{η_∞}, the sequence v_{ξ_n} must converge to the same measure. $\square$

We now return to Proposition 3.1. Let $h \in H(X, \Gamma, \mu)$. As pointed out above, for each $x \in X \setminus C$ we obtain a harmonic function $\tilde{h}_x \in H^\infty(\Gamma, \hat{\mu})$. By Lemma 3.2, for each $C > 0$, for a.e. sample path $(x_0, x_1, \ldots)$ of the random walk on X, and for every sequence $(y_0, y_1, \ldots)$ such that $d_x(x_i, y_i) < C$ for all i, both sequences $\tilde{h}_x(x_i)$ and $\tilde{h}_x(y_i)$ converge and have the same limit. Therefore, as C is arbitrary and h is continuous, h must be constant on the orbit of every limit point of x_i. But this is to say that all limit points must lie in C. This concludes the proof of Proposition 3.1.

One further simple remark, concerning induced random walks for subgroups, is needed. Let Γ_0 be a finite-index subgroup of Γ. Let N^γ denote the first time, $n \geq 1$, at which random walk on $(\Gamma, \hat{\mu})$ starting at γ reaches Γ_0. Then N^γ is a Markov time, almost surely finite. Now define a probability measure on Γ_0 by $\hat{\mu}_0(\eta) = \text{Prob}(\gamma_{N^e} = \eta)$, that is, $\hat{\mu}_0(\eta)$ is the probability that random walk, $(\gamma_0, \gamma_1, \ldots)$, on Γ, starting at the unit element, will first return to Γ_0 at η. General properties of Markov times and martingales imply that the restriction to Γ_0 of a function in $H^\infty(\Gamma, \hat{\mu})$ is harmonic relative to $\hat{\mu}_0$.

LEMMA 3.3. *Let Γ_0 have finite index in Γ and μ_0 the image of the induced measure $\hat{\mu}_0$ under group inverse. Then $H(X, \Gamma, \mu) \subseteq H(X, \Gamma_0, \mu_0)$. In particular, if (X, Γ_0, μ_0) satisfies the Liouville property, then so does (X, Γ, μ).*

Proof. This is due to the above remarks about μ_0 and the relationship between harmonic functions h on (X, Γ, μ) and $\tilde{h}_x$ on $(\Gamma, \hat{\mu})$ or $(\Gamma_0, \hat{\mu}_0)$. (See the beginning of Section 2.) $\square$

4. Actions on the Circle

We now consider actions of discrete (countable) groups on the circle. The following lemma is a well-known group-actions version of Poincaré's classification of circle homeomorphisms.

LEMMA 4.1. *Let Γ be a countable group of homeomorphisms of S^1. Then one of the following holds:*

  1) *The action is minimal;*
  2) *The action is not minimal and there is a unique minimal set;*
  3) *There is a finite orbit.*

Proof. It suffices to prove the lemma for a subgroup of finite index. Thus we may assume that Γ is orientation preserving. Further assume that the action is not minimal and contains no finite orbit. Let Z be a minimal set and U an arbitrary connected component of $S^1 \setminus Z$. This minimal set is unique if the orbit of every point in U can be shown to accumulate on Z. Now, U and γU either are disjoint or agree for each $\gamma \in \Gamma$. If the latter, then the endpoints of U are fixed by γ. As the subgroup of Γ fixing the endpoints of U has infinite index, there are infinitely many disjoint intervals of the form γU, so we can choose a sequence $\gamma_n U$ decreasing to 0 in length. Therefore, the orbit of every point in U must limit on Z. □

By Corollary 2.2, having a unique minimal set implies the Liouville property. Thus, in order to prove Theorem 1.1, it suffices to assume that the action on S^1 contains a finite orbit. By Lemma 3.3, it can be assumed that Γ has a fixed point in S^1 and preserves orientation. This reduces the proof of Theorem 1.1 to showing the Liouville property for orientation-preserving actions on the interval [0, 1].

5. Actions on the Interval

We now restrict attention to systems $([0,1], \Gamma, \mu)$ for which the Γ-action is orientation preserving. Clearly, then, the only finite orbits are fixed points, and since the union of the fixed points is a closed subset of [0, 1], we may further restrict attention to intervals without interior finite orbits.

LEMMA 5.1. *If Γ acts by orientation-preserving homeomorphisms of [0,1] without interior finite orbits, the orbit of every $x \in (0, 1)$ must limit on both 0 and 1.*

Proof. The supremum and infimum of Γx are easily seen to be fixed points. □

LEMMA 5.2. *Suppose that the Liouville property does not hold for* $([0,1], \Gamma, \mu)$ *and the Γ-action is orientation preserving without interior finite orbits. Then,*

1) *There is a unique continuous harmonic f not Γ-invariant such that $f(0) = 0$, $f(1) = 1$;*
2) *$f(x)$ is the probability that random walk on $([0,1], \Gamma, \mu)$ starting at x converges to 1;*
3) *$\hat{f}$ is not constant on any interior orbit;*
4) *f is increasing.*

Proof. Let g be a continuous, harmonic, non-Γ-invariant function on $[0,1]$. By Lemma 5.1 and the maximum principle, $g(0) \neq g(1)$, and $g(x)$ lies in the open interval with endpoints $g(0), g(1)$ for each $x \in (0,1)$. Lemma 5.1 also implies that g cannot be constant on any interior orbit, so the set $\mathcal{C}$ of Proposition 3.1 coincides with $\{0, 1\}$. By composing g with an appropriate affine function of $\mathbb{R}$, one obtains f such that $f(i) = i$ for $i = 0, 1$ and $0 < f(x) < 1$ on interior points. Uniqueness is due to the maximum principle. By Proposition 3.1, the random walk X_n^x on $[0,1]$ starting at x must converge to a random point $X_\infty \in \{0, 1\}$ with probability 1. As f is harmonic, the expected value $E[f(X_n^x)]$ is equal to $f(x)$ and $\lim_{n \to \infty} f(X_n)$ exists almost surely. By continuity, $f(X_n^x)$ converges to either 0 or 1. Therefore,

$$f(x) = \lim_{n \to \infty} E[f(X_n^x)] = f(0)\mathrm{Prob}(X_n^x \to 0) + f(1)\mathrm{Prob}(X_n^x \to 1)$$

$$= \mathrm{Prob}(X_n^x \to 1).$$

Finally, given any $x_1 < x_2$ and a sample path $(\gamma_1, \gamma_2, \ldots)$ on Γ for the right-$\hat{\mu}$-random walk, the corresponding sample paths for the random walks on $[0,1]$ satisfy $X_n^{x_1} < X_n^{x_2}$. Therefore, the probability that $X_n^{x_2}$ converges to 1 is at least as great as the probability that $X_n^{x_1}$ converges to 1. This shows that $f(x)$ is increasing. □

We can now show how the assumption of a non-Γ-invariant harmonic function on $[0,1]$ leads to a contradiction. Let f be as in Lemma 5.2 and define a probability measure ν on $[0,1]$ by extending the definition

$$\nu((a, b]) = f(b) - f(a)$$

to the Lebesgue measurable subsets of the interval. At this point we make the further assumption that the measure μ on Γ is symmetric. Then,

$$\mu * \nu((a, b]) = \sum_{\gamma \in \Gamma} \mu(\gamma) \nu\left((\gamma^{-1}a, \gamma^{-1}b]\right)$$

$$= \sum_{\gamma \in \Gamma} \mu(\gamma) \left(f(\gamma^{-1}b) - f(\gamma^{-1}a)\right)$$

$$= \sum_{\gamma \in \Gamma} \mu(\gamma)f(\gamma b) - \sum_{\gamma \in \Gamma} \mu(\gamma)f(\gamma a)$$

$$= f(b) - f(a)$$

$$= \nu((a, b]).$$

This remark, which was shown to us by B. Dcroin, simplifies a similar but somewhat more involved argument of an earlier version of this paper. The same argument is used in the proof of proposition 5.7 of [DKN]. From this we obtain the following lemma.

LEMMA 5.3. *If μ is symmetric, ν is a harmonic probability measure.*

We can now conclude the proof of Theorem 1.1. Since by Proposition 2.4 any function in $H(X, \Gamma, \mu)$ must be Γ-invariant on the support of a harmonic probability measure, and since the above f is not constant on any interior orbit of the Γ-action on $[0, 1]$, we arrive at a contradiction. Therefore, a non-Γ-invariant, continuous harmonic function cannot exist.

6. Examples

We now give a class of examples of Γ-spaces to illustrate the way in which the Liouville property can fail to hold. We begin with a general remark that suggests a recipe for constructing examples. Let μ be a probability measure on the countable (discrete) group Γ and let $\hat{\mu}$, as before, be the image of μ under group inverse. The unit ball in $H^\infty(\Gamma, \hat{\mu})$ (the latter equipped with the supremum norm) is a compact space, which we denote by X_0. Note that Γ acts on X_0 by homeomorphisms under the definition $(\gamma, \phi) \mapsto \gamma \cdot \phi$, where $(\gamma \cdot \phi)(\eta) = \phi(\gamma^{-1}\eta)$.

The Γ-space X_0 has a tautological continuous μ-harmonic function: $f(\phi) = \phi(e)$. This is, in fact, μ-harmonic since

$$\sum_{\gamma \in \Gamma} f(\gamma \cdot \phi)\mu(\gamma) = \sum_{\gamma \in \Gamma} \phi(\gamma^{-1})\mu(\gamma) = \sum_{\gamma \in \Gamma} \phi(\gamma)\hat{\mu}(\gamma) = \phi(e) = f(\phi).$$

Furthermore, since $f(\gamma\phi) = \phi(\gamma^{-1})$, f is not constant on $\Gamma \cdot \phi$ if ϕ itself is not a constant function. This simple remark suggests the following approach to finding examples of non-Liouville actions on a Γ-space S. Suppose we can somehow construct a continuous, Γ-equivariant map $\Phi : S \to X_0$. We express equivariance by $\Phi(\gamma s) = \gamma\Phi(s)$ and write $\Phi(s) = \phi_s$. If Φ does not map S entirely into the space of constant functions, then $f \circ \Phi$ defines a continuous, μ-harmonic function on S that is not Γ-invariant. Thus, what we have shown in Theorem 1.1 amounts to the following:

PROPOSITION 6.1. *If μ is a symmetric probability measure on Γ, then every Γ-equivariant continuous $\Phi : S^1 \to X_0$ maps into the constant functions.*

We now construct a continuous Γ-equivariant map $\Phi : S^2 \to X_0$ whose image is not contained in the space of constant functions. By the above remark, this yields a non-Liouville action on S^2. The example generalizes one given in [FZ2].

Let Γ and $\hat{\mu}$ be such that the Poisson boundary of $(\Gamma, \hat{\mu})$ can be identified with the circle S^1. For example, as already noted, if Γ is a uniform group of isometries of the Poincaré disc, it is possible to find a symmetric μ for which the Poisson boundary of the group coincides with the geometric boundary of the disc [Fur]. Let ν be a harmonic measure on S^1 as in the discussion immediately after Proposition 3.1. The following general fact was pointed out to us by Kaimanovich.

LEMMA 6.2. *Let $\hat{\mu}$ be a nondegenerate probability measure on Γ such that a Poisson boundary (B, ν) of $(\Gamma, \hat{\mu})$ is nontrivial. Then the $\hat{\mu}$-harmonic measure ν has no atoms.*

Proof. Suppose for a contradiction that ν does have atoms, and let $b \in B$ be such that $\nu(b) \geq \nu(b')$ for all $b' \in B$. Since ν is $\hat{\mu}$-harmonic and $\hat{\mu}$ is nondegenerate, the maximum principle applied to $\gamma \mapsto (\gamma_*\nu)(b)$ implies $\nu(b) = \nu(\gamma b)$ for all $\gamma \in \Gamma$. On the other hand, if $(\gamma_1, \gamma_2, \ldots)$ is a random walk on $(\Gamma, \hat{\mu})$, then $(\gamma_1 \ldots \gamma_n)_*\nu$ converges to the Dirac measure δ_{g_∞}, where g_∞ is the random point on the boundary to which the random walk converges. Writing $g_n = \gamma_1 \ldots \gamma_n$, then $0 < \nu(b) = (g_n)_*\nu(b) \to \delta_{g_\infty}(b)$. This is possible only if sample paths converge to b almost surely, contradicting the assumption that the Poisson boundary is nontrivial. $\square$

We now construct a nontrivial Γ-equivariant continuous map $\Phi : S^2 \to X_0$. Let $x = (z, \theta)$ in the cylinder $S^1 \times [0, 2\pi]$ represent the interval $I_x \subset S^1$ with endpoints z and $ze^{i\theta}$. Naturally, $\gamma I_x = I_{\gamma x}$. The circle boundaries $S^1 \times \{0\}$ and $S^1 \times \{2\pi\}$ are invariant under Γ and correspond to intervals of length 0 or 2π. The quotient $(S^1 \times [0, 2\pi])/ \sim$ obtained by collapsing the boundary circles to points is homeomorphic to S^2 and the Γ-action on the cylinder defines an action on the quotient by homeomorphisms fixing two points, denoted $N = (S^1 \times \{2\pi\})/ \sim$ and $S = (S^1 \times \{0\})/ \sim$. This defines S^2 as a topological Γ-space.

Let ν be a harmonic probability measure on S^1, which is granted since S^1 is a Poisson boundary of $(\Gamma, \hat{\mu})$. Define $\Phi(x) = \phi_x$, where $\phi_x(\gamma) = \gamma_* \nu(I_x)$. Note that Φ is well defined since $\nu(I_x)$ is 0 on all intervals of length 0, and 1 on all intervals of length 2π. A simple calculation shows that ϕ_x is $\hat{\mu}$-harmonic for each $x \in S^2$ and that Φ is Γ-equivariant. Furthermore, Φ is continuous due to the fact that ν has no atoms. Since the Poisson boundary is nontrivial, Φ does not map into the set of constant functions. This shows that (S^2, Γ, μ) is non-Liouville.

In this example, the only Γ-orbits in S^2 on which $\Phi(x)$ is a constant function are the fixed points N, S, so $\mathcal{C} = \{N, S\}$, where $\mathcal{C}$ is the set defined in Proposition 3.1. The same argument used in Lemma 5.2 shows that $g = f \circ \Phi$ can be composed with an affine transformation of $\mathbb{R}$ to ensure $g(S) = 0$ and $g(N) = 1$, after which $g(x)$ is the probability that random walk on (S^2, Γ, μ) starting at $x \in S^2$ converges to N.

When (S^1, ν) coincides with the boundary of the hyperbolic disc, where Γ is a uniform lattice in $PSL(2, \mathbb{R})$, lemma 4.2 of [FZ2] can be used to show that the just-constructed action on S^2 is ergodic relative to the smooth measure class on S^2. This dynamical property of the action contrasts with the simple behavior of the random walk, which converges to N or S with probability 1.

It is interesting to regard the induced random walk on the space X_0 itself, for a general $(\Gamma, \hat{\mu})$. Let $I = [-1, 1] \subset X_0$ represent the subspace of constant functions in X_0. Then the random walk on (X_0, Γ, μ) converges toward I, that is, the random walk on $X_0 \setminus I$ is transient. On the other hand, the Γ-action on X_0 can be dynamically very complicated. For example, it is shown in [FZ2] for subgroups of $PSL(2, \mathbb{R})$ that the $\mathbb{Z}$-action on X_0 induced by a parabolic or hyperbolic element is a chaotic dynamical system.

References

[DK]   B. Deroin and V. Kleptsyn. *Random conformal dynamical systems*, GAFA 17 (2007), 1043–1105.

[DKN]   B. Deroin, V. Klepstyn, A. Navas. *Sur la dynamique unidimensionnelle en régularité intermédiaire*, Acta Mathematica Stockholm 199 (2007), no. 2, 199–262.

[FFP]   S. Fenley, R. Feres, K. Parwani. *Harmonic functions on* $\mathbb{R}$*-covered foliations*, Ergodic Theory Dynam. Systems 29 (2009), no. 4, 1141–1161.

[FZ1]   R. Feres and A. Zhegib. *Leafwise holomorphic functions*, Proc. Amer. Math. Soc. 131 (2003), no. 6, 1717–1725.

[FZ2]   R. Feres and A. Zhegib. *Dynamics on the space of harmonic functions and the foliated Liouville problem*, Ergodic Theory Dynam. Systems 23 (2003), no. 2, 303–316.

[Fu]   A. Furman. *Random walks on groups and random transformations*, Handbook of dynamical systems, Vol. 1A, chap. 12. Hasselblatt and Katok, North-Holland, 2002, 931–1014.

[Fur]   H. Furstenberg. *Random walks and discrete subgroups of Lie groups*, 1971 Advances in Probability and Related Topics, Vol. 1. New York, Dekker, 1971, 1–63.

[Gar]   L. Garnett. *Foliations, the ergodic theorem and Brownian motion*, J. Funct. Anal. 51 (1983), 285–311.

[Pa]   F. Paulin. *Analyse harmonique des relations d'équivalence mesurées discrètes*, Markov Proc. Rel. Fields. 5 (1999), 163–200.

4

DAVID FISHER

GROUPS ACTING ON MANIFOLDS: AROUND THE ZIMMER PROGRAM

TO ROBERT ZIMMER ON THE OCCASION OF HIS 60TH BIRTHDAY

Abstract

This paper is a survey on the *Zimmer program*. In its broadest form, this program seeks an understanding of actions of large groups on compact manifolds. The goals of this survey are (1) to put in context the original questions and conjectures of Zimmer and Gromov that motivated the program, (2) to indicate the current state of the art on as many of these conjectures and questions as possible, and (3) to indicate a wide variety of open problems and directions of research.

Contents

Author partially supported by NSF grants DMS-0541917 and 0643546, a fellowship from the Radcliffe Institute for Advanced Studies and visiting positions at École Polytechnique, Palaiseau and Université Paris-Nord, Villateneuse.

1. Prologue

Traditionally, the study of dynamical systems is concerned with actions of $\mathbb{R}$ or $\mathbb{Z}$ on manifolds, that is, with flows and diffeomorphisms. It is natural to consider instead dynamical systems defined by actions of larger discrete or continuous groups. For non-discrete groups, the Hilbert-Smith conjecture is relevant, since the conjecture states that non-discrete, locally compact totally disconnected groups do not act continuously by homeomorphisms on manifolds. For smooth actions known results suffice to rule out actions of totally disconnected groups; the Hilbert-Smith conjecture has been proven for Hölder diffeomorphisms [155, 207]. For infinite discrete groups, the whole universe is open. One might consider the "generalized Zimmer program" to be the study of homomorphisms $\rho : \Gamma \to \text{Diff}\,(M)$ where Γ is a finitely generated group and M is a compact manifold. In this survey much emphasis will be on a program proposed by Zimmer. Here one considers a very special class of both Lie groups and discrete groups, namely semisimple Lie groups of higher real rank and their lattices.

Zimmer's program is motivated by several of Zimmer's own theorems and observations, many of which we will discuss below. But in broadest strokes, the motivation is simpler. Given a group whose linear and unitary representations are very rigid or constrained, might it also be true that the group's representations into $\text{Diff}^\infty (M)$ are also very rigid or constrained at least when M is a compact manifold? For the higher rank lattices considered by Zimmer, Margulis's superrigidity theorems classified finite dimensional representations and the groups enjoy property (T) of Kazhdan, which makes unitary representations quite rigid.

This motivation also stems from an analogy between semisimple Lie groups and diffeomorphism groups. When M is a compact manifold, not only is $\text{Diff}^\infty (M)$ an infinite dimensional Lie group, but its connected component is simple. Simplicity of the connected component of $\text{Diff}^\infty (M)$ was proven by Thurston using results of Epstein and Herman [227, 50, 116]. Herman had used Epstein's work to see that the connected component of $\text{Diff}^\infty (\mathbb{T}^n)$ is simple and Thurston's proof of the general case uses this. See also further work on the topic by Banyaga and Mather [8, 167, 166, 168, 169], as well as Banyaga's book [9].

A major motivation for Zimmer's program came from his cocycle superrigidity theorem, discussed in detail below in Section 5. One can think of a homomorphism $\rho : \Gamma \to \text{Diff} (M)$ as defining a *virtual homomorphism* from Γ to $GL(\dim (M), \mathbb{R})$. The notion of virtual homomorphisms, now more commonly referred to in our context as cocycles over group actions, was introduced by Mackey [154].

In all of Zimmer's early papers, it was always assumed that we had a homomorphism $\rho : \Gamma \to \text{Diff}^\infty (M, \omega)$ where ω is a volume form on M. A major motivation for this is that the cocycle superrigidity theorem referred to in the last paragraph only applies to cocycles over measure preserving actions. That this hypothesis is necessary was confirmed by examples of Stuck, who showed that no rigidity could be hoped for unless one had an invariant volume form or assumed that M was very low dimensional [226]. Recent work of Nevo and Zimmer does explore the non-volume preserving case and is discussed below, along with Stuck's examples and some others due to Weinberger in Section 10.

Let G be a semisimple real Lie group, all of whose simple factors are of real rank at least 2. Let Γ in G be a lattice. The simplest question asked by Zimmer was: Can one classify smooth volume preserving actions of Γ on compact manifolds? At the time of this writing, the answer to this question is still unclear. There certainly are a wider collection of actions than Zimmer may have initially suspected and it is also clear that the moduli space of such actions is not discrete; see [12, 78, 128] or Section 9 below. But there are still few enough examples known that a classification remains plausible. And if one assumes some additional conditions on the actions, then plausible conjectures and striking results abound.

We now discuss four paradigmatic conjectures. To make this introduction accessible, these conjectures are all special cases of more general conjectures stated later in the text. In particular, we state all the conjectures for lattices in $SL(n, \mathbb{R})$ when $n > 2$. The reader not familiar with higher rank lattices can consider the examples of finite index subgroups of $SL(n, \mathbb{Z})$, with $n > 2$, rather than for general lattices in $SL(n, \mathbb{R})$. We warn the reader that the algebraic structure of $SL(n, \mathbb{Z})$ and its finite index subgroups makes many results easier for these groups than for other lattices in $SL(n, \mathbb{R})$. The conjectures we state concern, respectively, (1) classification of low dimensional actions, (2) classification of geometric actions, (3) classification of uniformly hyperbolic actions, and (4) the topology of manifolds admitting volume preserving actions.

A motivating conjecture for much recent research is the following:

CONJECTURE 1.1. (ZIMMER'S CONJECTURE) *For any $n > 2$, any homomorphism $\rho : SL(n, \mathbb{Z}) \to$ Diff (M) has finite image if* dim $(M) < n - 1$. *The same for any lattice Γ in $SL(n\mathbb{R})$.*

The dimension bound in the conjecture is clearly sharp, as $SL(n, \mathbb{R})$ and all of its subgroups act on $\mathbb{P}^{n-1}$. The conjecture is a special case of a conjecture of Zimmer's that concerns actions of higher-rank lattices on low dimensional manifolds that we state as Conjecture 4.12 below. Recently, much attention has focused on these conjectures concerning low dimensional actions.

In our second, geometric, setting (a special case of) a major motivating conjecture is the following:

CONJECTURE 1.2. (AFFINE ACTIONS) *Let $\Gamma < SL(n, \mathbb{R})$ be a lattice. Then there is a classification of actions of Γ on compact manifolds that preserve both a volume form and an affine connection. All such actions are algebraically defined in a sense to be made precise below. (See Definition 2.4 below.)*

The easiest example of an algebraically defined action is the action of $SL(n, \mathbb{Z})$ (or any of its subgroups) on $\mathbb{T}^n$. We formulate our definition of algebraically defined action to include all trivial actions, all isometric actions, and all (skew) products of other actions with these. The conjecture is a special case of a conjecture stated below as Conjecture 6.15. In addition to considering more general acting groups, we will consider more general invariant geometric structures, not just affine connections. A remark worth making is that, for the geometric structures we consider, the automorphism group is always a finite dimensional Lie group. The question of when the full automorphism group of a geometric structure is large is well studied from other points of view; see particularly [135]. However, this question is generally most approachable when the large subgroup is connected and much less is known about discrete subgroups. In particular, geometric approaches to this problem tend to use information about the connected component of the automorphism group and give much less information about the group of components particularly if the connected component is trivial.

One is also interested in the possibility of classifying actions under strong dynamical hypotheses. The following conjecture is motivated by the work of Feres-Labourie [61] and Goetze-Spatzier [105, 106] and is similar to conjectures stated in [108, 120]:

CONJECTURE 1.3. *Let* $\Gamma < SL(n, \mathbb{R})$ *be a lattice. Then there is a classification of actions of* Γ *on a compact manifold* M *that preserve both a volume form and where one element* $\gamma \in \Gamma$ *acts as an Anosov diffeomorphism. All such actions are algebraically defined in a sense to be made precise below. (Again see Definition 2.4.)*

In the setting of this particular conjecture, a proof of an older conjecture of Franks concerning Anosov diffeomorphisms would imply that any manifold M as in the conjecture was homeomorphic to an infranilmanifold on which γ is conjugate by a homeomorphism to a standard affine Anosov map. Even assuming Franks's conjecture, Conjecture 1.3 is open. One can make a more general conjecture by only assuming that γ has some uniformly partially hyperbolic behavior. Various versions of this are discussed in §7; see particularly Conjecture 7.9.

We end this introduction by stating a topological conjecture about *all* manifold admitting smooth volume preserving actions of a simple higher rank algebraic group. Very special cases of this conjecture are known and the conjecture is plausible in light of existing examples. More precise variants and a version for lattice actions will be stated below in §8.

CONJECTURE 1.4. *Let* G *be a simple Lie group of real rank at least 2. Assume* G *has a faithful action preserving volume on a compact manifold* M. *Then* $\pi_1(M)$

has a finite index subgroup Λ *such that* Λ *surjects onto an arithmetic lattice in a Lie group H where H locally contains G.*

The conjecture says, more or less, that admitting a G action forces the fundamental group of M to be large. Passage to a finite index subgroup and a quotient is necessary; see Section 9 and [84] for more discussion. The conjecture might be considered the analogue for group actions of Margulis's arithmeticity theorem.

This survey is organized on the following lines. We begin in §2 by describing in some detail the kinds of groups we consider and examples of their actions on compact manifolds. In §3 we digress with a prehistory of motivating results from rigidity theory. Then in §4, we discuss conjectures and theorems concerning actions on "low dimensional" manifolds. In this section, there are a number of related results and conjectures concerning groups not covered by Zimmer's original conjectures. Here low-dimensional is in two senses (1) compared to the group as in Conjecture 1.1, and (2) absolutely small, as in dimension being between 1 and 4. This discussion is simplified by the fact that many theorems conclude with the nonexistence of actions or at least with all actions factoring through finite quotients. In Section 5 we further describe Zimmer's motivations for his conjectures by discussing the cocycle superrigidity theorem and some of its consequences for smooth group actions. In Sections 6 and 7, we discuss, respectively, geometric and dynamical conditions under which a simple classification might be possible. In Section 8, we discuss another approach to classifying G and Γ actions using topology and representations of fundamental groups in order to produce algebraically defined quotients of actions. Then in Section 9, we describe the known "exotic examples" of the acting groups to consider. This construction reveals some necessary complexity of a high-dimensional classification. We then describe known results and examples of actions not preserving volume in Section 10 and some rather surprising group actions on manifolds in Section 11. Finally we end the survey with a collection of questions about the algebraic and geometric structure of finitely generated subgroups of Diff (M).

Some remarks on biases and omissions. Like any survey of this kind, this work is informed by its author's biases and experiences. There is an additional bias in that this paper emphasizes developments that are close to Zimmer's own work and conjectures. In particular, the study of rigidity of group actions often focuses on the low dimensional setting where all group actions are conjectured to be finite or trivial. While Zimmer did substantial work in this setting, he also proved many results and made many conjectures in more general settings where any potential classification of group actions is necessarily, due to the existence of examples, more complicated.

Another omission is that almost nothing will be said here about local rigidity of groups actions, since the author has recently written another survey on that topic [77]. While that survey could already use updating, that update will appear elsewhere.

For other surveys of Zimmer's program and rigidity of large group actions the reader is referred to [68, 144, 258]. The forthcoming book by Zimmer and Witte Morris [250] is a particularly useful introduction to ideas and techniques in Zimmer's own work. Also of interest are (1) a brief survey of rigidity theory by Spatzier with a more geometric focus [223], (2) an older survey also by Spatzier, with a somewhat broader scope [222], (3) a recent problem list by Margulis on rigidity theory with a focus on measure rigidity [164], and (4) a more recent survey by Lindenstrauss focused on recent developments in measure rigidity [147]. Both of the last two mentioned surveys are particularly oriented toward connections between rigidity and number theory, which are not mentioned at all in this survey.

Finally, while all mistakes and errors in this survey are the sole responsibility of its author, I would like to thank many people whose comments led to improvements. These include Danny Calegari, Renato Feres, Etienne Ghys, Daniel Groves, Steve Hurder, Jean-Francois Lafont, Karin Melnick, Nicolas Monod, Andrés Navas, Leonid Polterovich, Pierre Py, Yehuda Shalom, Ralf Spatzier, Dave Witte Morris, and an anonymous referee.

2. A Brief Digression: Some Examples of Groups and Actions

In this section we briefly describe some of the groups that will play important roles in the results discussed here. The reader already familiar with semisimple Lie groups and their lattices may want to skip to the Subsection 2.2 where we give descriptions of group actions. The following convention is in force throughout this paper. For definitions of relevant terms the reader is referred to the following subsections.

Convention In this article we will have occasion to refer to three overlapping classes of lattices in Lie groups that are slightly different. Let G be a semisimple Lie group and $\Gamma < G$ a lattice. We call Γ a *higher rank lattice* if all simple factors of G have real rank at least 2. We call Γ a *lattice with (T)* if all simple factors of G have property (T). Lastly, we call Γ an *irreducible higher rank lattice* if G has real rank at least 2 and Γ is irreducible.

2.1. Semisimple Groups and Their Lattices

By a simple Lie group, we mean a connected Lie group all of whose normal subgroups are discrete, though we make the additional convention that $\mathbb{R}$ and S^1 are not simple. By a semisimple Lie group we mean the quotient of a product of simple Lie groups by some subgroup of the product of their centers. Note that with our conventions, the center of a simple Lie group is discrete and is in fact the maximal normal subgroup. There is an elaborate structure theory of semisimple Lie groups and the groups are completely classified; see [115] or [134] for details. Here we merely describe some examples, all of which are matrix groups. All connected semisimple Lie groups are discrete central extensions of matrix groups, so the reader will lose very little by always thinking of matrix groups.

1) The groups $SL(n, \mathbb{R})$, $SL(n, \mathbb{C})$, and $SL(n, \mathbb{H})$ of n by n matrices of determinant 1 over the real numbers, the complex numbers, or the quaternions.

2) The group $SP(2n, \mathbb{R})$ of $2n$ by $2n$ matrices of determinant 1 that preserve a real symplectic form on $\mathbb{R}^{2n}$.

3) The groups $SO(p, q)$, $SU(p, q)$, and $SP(p, q)$ of matrices that preserve inner products of signature (p, q) where the inner product is real linear on $\mathbb{R}^{p+q}$, Hermitian on $\mathbb{C}^{p+q}$, or quaternionic Hermitian on $\mathbb{H}^{p+q}$, respectively.

Let G be a semisimple Lie group that is a subgroup of $GL(n, \mathbb{R})$. We say that G has *real rank* k if G has a k-dimensional abelian subgroup that is conjugate to a subgroup of the real diagonal matrices and no $k + 1$-dimensional abelian subgroups with the same property. The groups in (1) have rank $n - 1$, the groups in (2) have rank n, and the groups in (3) have rank min (p, q).

Since this article focuses primarily on finitely generated groups, we are more interested in discrete subgroups of Lie groups than in the Lie groups themselves. A discrete subgroup Γ in a Lie group G is called a lattice if G/Γ has finite Haar measure. The lattice is called *cocompact* or *uniform* if G/Γ is compact and *non-uniform* or *not cocompact* otherwise. If $G = G_1 \times \cdots \times G_n$ is a product, then we say a lattice $\Gamma < G$ is *irreducible* if its projection to each G_i is dense. It is more typical in the literature to insist that projections to all factors are dense, but this definition is more practical for our purposes. More generally, we make the same definition for an *almost direct product*, by which we mean a direct product G modulo some subgroup of the center $Z(G)$. Lattices in semisimple Lie groups can always be constructed by arithmetic methods; see [19] and also [175] for more discussion. In fact, one of the most important results in the theory of semisimple Lie groups is that if G is a semisimple Lie group without compact factors, then all irreducible lattices in G are arithmetic unless G is locally isomorphic to $SO(1, n)$ or $SU(1, n)$. For G of real rank at least 2, this is Margulis's arithmeticity theorem, which he deduced from his super-rigidity theorems [159, 202, 162]. For non-uniform lattices, Margulis had an earlier proof that does not use the superrigidity theorems; see [158, 160]. This earlier proof depends on the study of dynamics of unipotent elements on the space G/Γ, and particularly on what is now known as the "non-divergence of unipotent flows." Special cases of the super-rigidity theorems were then proven for $Sp(1, n)$ and F_4^{-20} by Corlette [40] and Gromov-Schoen [113], which sufficed to imply the statement on arithmeticity given above. As we will be almost exclusively concerned with arithmetic lattices, we do not give examples of non-arithmetic lattices here, but refer the reader to [162] and [175] for more discussion. A formal definition of arithmeticity, at least when G is algebraic, is

DEFINITION 2.1. Let G be a semisimple algebraic Lie group and $\Gamma < G$ a lattice. Then Γ is arithmetic if there exists a semisimple algebraic Lie group H such that

1) there is a homomorphism $\pi : H^0 \to G$ with compact kernel,
2) there is a rational structure on H such that the projection of the integer points of H to G are commensurable to Γ, that is, $\pi(H(\mathbb{Z})) \cap \Gamma$ is of finite index in both $H(\mathbb{Z})$ and Γ.

We now give some examples of arithmetic lattices. The simplest is to take the integer points in a simple (or semisimple) group G that is a matrix group, for example, $SL(n, \mathbb{Z})$ or $Sp(n, \mathbb{Z})$. This exact construction always yields lattices, but also always yields non-uniform lattices. In fact, the lattices one can construct in this way have very special properties because they will contain many unipotent matrices. If a lattice is cocompact, it will necessarily contain no unipotent matrices. The standard trick for understanding the structure of lattices in G that become integral points after passing to a compact extension is called *change of base*. For much more discussion see [162, 175, 245]. We give one example to illustrate the process. Let $G = SO(m, n)$, which we view as the set of matrices in $SL(n + m, \mathbb{R})$ which preserve the inner product

$$\langle v, w \rangle = \left(-\sqrt{2} \sum_{i=1}^{m} v_i w_i \right) + \left(\sum_{i=m+1}^{n+m} v_i w_i \right)$$

where v_i and w_i are the ith components of v and w. This form, and therefore G, are defined over the field $\mathbb{Q}(\sqrt{2})$ that has a Galois conjugation σ defined by $\sigma(\sqrt{2}) = -\sqrt{2}$. If we look at the points $\Gamma = G(\mathbb{Z}[\sqrt{2}])$, we can define an embedding of Γ in $SO(m, n) \times SO(m + n)$ by taking γ to $(\gamma, \sigma(\gamma))$. It is straightforward to check that this embedding is discrete. In fact, this embeds Γ in $H = SO(m, n) \times SO(m + n)$ as integral points for the rational structure on H where the rational points are exactly the points $(\mathbf{M}, \sigma(\mathbf{M}))$ where $\mathbf{M} \in G(\mathbb{Q}(\sqrt{2}))$. This makes Γ a lattice in H and it is easy to see that Γ projects to a lattice in G, since G is cocompact in H. What is somewhat harder to verify is that Γ is cocompact in H, for which we refer the reader to the list of references above.

Similar constructions are possible with $SU(m, n)$ or $SP(m, n)$ in place of $SO(m, n)$ and also with more simple factors and fields with more Galois automorphisms. There are also a number of other constructions of arithmetic lattices using division algebras. See [194, 175] for a comprehensive treatment.

We end this section by defining a key property of many semisimple groups and their lattices. This is property (T) of Kazhdan, and was introduced by Kazhdan in [131] in order to prove that non-uniform lattices in higher rank semisimple Lie groups are finitely generated and have finite abelianization. It has played a fundamental role in many subsequent developments. We do not give Kazhdan's original definition, but one which was shown to be equivalent by the work of Delorme [44] and Guichardet [114].

DEFINITION 2.2. A locally compact group Γ has property (T) of Kazhdan if $H^1(\Gamma, \pi) = 0$ for every continuous unitary representation π of Γ on a Hilbert space. This is equivalent to saying that any continuous isometric action of Γ on a Hilbert space has a fixed point.

REMARKS 2.3.

1) Kazhdan's definition is that the trivial representation is isolated in the Fell topology on the unitary dual of Γ.
2) If a continuous group G has property (T), so does any lattice in G. This result was proved in [131].
3) Any semisimple Lie group has property (T) if and only if it has no simple factors locally isomorphic to $SO(1, n)$ or $SU(1, n)$. For a discussion of this fact and attributions, see [43]. For groups with all simple factors of real rank at least 3, this is proven in [131].

4) No noncompact amenable group, and in particular no noncompact abelian group, has property (T). An easy averaging argument shows that all compact groups have property (T).

Groups with property (T) play an important role in many areas of mathematics and computer science.

2.2. Some Actions of Groups and Lattices

Here we define and give examples of a general class of actions. A major impetus in Zimmer's work is determining optimal conditions for actions to lie in this class. The class we describe is slightly more general than the class Zimmer termed "standard actions" in, for example, [249]. Let H be a Lie group and $L < H$ a closed subgroup. Then a diffeomorphism f of H/L is called *affine* if there is a diffeomorphism $\tilde{f}$ of H such that $f([h]) = \tilde{f}(h)$ where $\tilde{f} = A \circ \tau_h$ with A an automorphism of H with $A(L) = L$ and τ_h is left translation by some h in H. Two obvious classes of affine diffeomorphisms are left translations on any homogeneous space and linear automorphisms of tori, or more generally, automorphisms of nilmanifolds. A group action is called *affine* if every element of the group acts by an affine diffeomorphism. It is easy to check that the full group of affine diffeomorphisms Aff (H/L) is a finite dimensional Lie group and an affine action of a group D is a homomorphism $\pi : D \to$ Aff (H/L). The structure of Aff (H/L) is surprisingly complicated in general; it is a quotient of a subgroup of the group Aut $(H) \ltimes H$ where Aut (H) is a group of automorphisms of H. For a more detailed discussion of this relationship, see [81, section 6]. While it is not always the case that any affine action of a group D on H/L can be described by a homomorphism $\pi : D \to$ Aut $(H) \ltimes H$, this is true for two important special cases:

1) D is a connected semisimple Lie group and L is a cocompact lattice in H,
2) D is a lattice in a semisimple Lie group G where G has no compact factors and no simple factors locally isomorphic to $SO(1, n)$ or $SU(1, n)$, and L is a cocompact lattice in H.

These facts are [81, theorems 6.4 and 6.5] where affine actions as in (1) and (2) above are classified.

The most obvious examples of affine actions of large groups are of the following forms, which are frequently referred to as *standard actions*:

1) Actions of groups by automorphisms of nilmanifolds. That is, let N be a simply connected nilpotent group, $\Lambda < N$ a lattice (which is necessarily

cocompact), and assume a finitely generated group Γ acts by automorphisms of N preserving Λ. The most obvious examples of this are when $N = \mathbb{R}^n$, $\Lambda = \mathbb{Z}^n$, and $\Gamma < SL(n, \mathbb{Z})$, in which case we have a linear action of Γ on $\mathbb{T}^n$.

2) Actions by left translations. That is, let H be a Lie group and $\Lambda < H$ a cocompact lattice and $\Gamma < H$ some subgroup. Then Γ acts on H/Λ by left translations. Note that in this case Γ need not be discrete.

3) Actions by isometries. Here K is a compact group that acts by isometries on some compact manifold M and $\Gamma < K$ is a subgroup. Note that here Γ is either discrete or a discrete extension of a compact group.

We now briefly define a few more general classes of actions, which we need to formulate most of the conjectures in this paper. We first fix some notations. Let A and D be topological groups, and $B < A$ a closed subgroup. Let $\rho : D \times A/B \to A/B$ be a continuous affine action.

DEFINITION 2.4.

1) Let A, B, D, and ρ be as above. Let C be a compact group of affine diffeomorphisms of A/B that commute with the D action. We call the action of D on $C\backslash A/B$ a *generalized affine action*.

2) Let A, B, D, and ρ be as in (1) above. Let M be a compact Riemannian manifold and $\iota : D \times A/B \to \text{Isom}(M)$ a C^1 cocycle. We call the resulting skew product D action on $A/B \times M$ a *quasi-affine action*.

If C and D are as in (2), and we have a smooth cocycle $\alpha : D \times C\backslash A/B \to \text{Isom}(M)$, then we call the resulting skew product action of D on $C\backslash A/B \times M$ a *generalized quasi-affine action*.

Many of the conjectures stated in this paper will end with the conclusion that all actions satisfying certain hypotheses are generalized quasi-affine actions. It is not entirely clear that generalized quasi-affine actions of higher rank groups and lattices are much more general than generalized affine actions. The following discussion is somewhat technical and might be skipped on first reading.

One can always take a product of a generalized affine action with a trivial action on any manifold to obtain a generalized quasi-affine action. One can also do variants on the following. Let H be a semisimple Lie group and $\Lambda < H$ a cocompact lattice. Let $\pi : \Lambda \to K$ be any homomorphism of Λ into a compact Lie group. Let ρ be a generalized affine action of G on $C\backslash H/\Lambda$ and let M be a compact manifold on which K acts. Then there is a generalized quasi-affine action of G on $(C\backslash H \times M)/\Lambda$.

QUESTION 2.5. *Is every generalized quasi-affine action of a higher rank simple Lie group of the type just described?*

The question amounts to asking for an understanding of compact group valued cocycles over quasi-affine actions of higher rank simple Lie groups. I leave it to the interested reader to formulate the analogous question for lattice actions. For some work in this direction, see the paper of Witte Morris and Zimmer [176].

2.3. Induced Actions

I end this section by describing briefly the standard construction of an *induced* or *suspended action*. This notion can be seen as a generalization of the construction of a flow under a function or as an analogue of the more algebraic notion of inducing a representation. Given a group H, a (usually closed) subgroup L, and an action ρ of L on a space X, we can form the space $(H \times X)/L$ where L acts on $H \times X$ by $h \cdot (l, x) = (lh^{-1}, \rho(h)x)$. This space now has a natural H action by left multiplication on the first coordinate. Many properties of the L action on X can be studied more easily in terms of properties of the H action on $(H \times X)/L$. This construction is particularly useful when L is a lattice in H.

This notion suggests the following principle:

PRINCIPLE 2.6. *Let Γ be a cocompact lattice in a Lie group G. To classify Γ actions on compact manifolds it suffices to classify G actions on compact manifolds.*

The principle is a bit subtle to implement in practice, since we clearly need a sufficiently detailed classification of G actions to be able to tell which ones arise as induction of Γ actions. While it is a bit more technical to state and probably more difficult to use, there is an analogous prinicple for non-cocompact lattices. Here one needs to classify G actions on manifolds that are not compact but where the G action preserves a finite volume. In fact, one needs only to study such actions on manifolds that are fiber bundles over G/Γ with compact fibers.

The lemma begs the question as to whether or not one should simply always study G actions. While in many settings this is useful, it is not always. In particular, many known results about Γ actions require hypotheses on the

Γ action where there is no useful way of rephrasing the property as a property of the induced action. Or, perhaps more awkwardly, require assumptions on the induced action that cannot be rephrased in terms of hypotheses on the original Γ action. We will illustrate the difficulties in employing Principle 2.6 at several points in this paper.

A case where the implications of the principle are particularly clear is a negative result concerning actions of $SO(1, n)$. We will make clear by the proof what we mean by

THEOREM 2.7. *Let $G = SO(1, n)$. Then one cannot classify actions of G on compact manifolds.*

Proof. For every n there is at least one lattice $\Gamma < G$ that admits homomorphisms onto non-abelian free groups; see, for example, [150]. And therefore also onto $\mathbb{Z}$. Thus we can take any action of F_n or $\mathbb{Z}$ and induce to a G action. It is relatively easy to show that if the induced actions are isomorphic, then so are the actions they are induced from; see, for example, [78]. A classification of $\mathbb{Z}$ actions would amount to a classification of diffeomorphisms and a classification of F_n actions would involve classifying all n-tuples of diffeomorphisms. As there is no reasonable classification of diffeomorphisms, there is also no reasonable classification of n-tuples of diffeomorphisms. □

We remark that essentially the same theorem holds for actions of $SU(1, n)$ where homomorphisms to $\mathbb{Z}$ exist for certain lattices [132]. Much less is known about free quotients of lattices in $SU(1, n)$; see [148] for one example. For a surprising local rigidity result for some lattices in $SU(1, n)$ see [72, theorem 1.3].

3. Pre-history

3.1. Local and Global Rigidity of Homomorphisms into Finite Dimensional Groups

The earliest work on rigidity theory is a series of works by Calabi-Vesentini, Selberg, Calabi, and Weil that resulted in the following:

THEOREM 3.1. *Let G be a semisimple Lie group and assume that G is not locally isomorphic to $SL(2, \mathbb{R})$. Let $\Gamma < G$ be an irreducible cocompact lattice, then the defining embedding of Γ in G is locally rigid; that is, any embedding ρ close to the defining embedding is conjugate to the defining embedding by a small element of G.*

REMARK 3.2.

1) If $G = SL(2, \mathbb{R})$, the theorem is false and there is a large, well-studied space of deformation of Γ in G, known as the Teichmueller space.

2) There is an analogue of this theorem for lattices that are not cocompact. This result was proven later and has a more complicated history that we omit here. In this case it is also necessary to exclude G locally isomorphic to $SL(2, \mathbb{C})$.

This theorem was originally proven in special cases by Calabi, Calabi and Vesentini, and Selberg. In particular, Selberg gives a proof for cocompact lattices in $SL(n, \mathbb{R})$ for $n \geq 3$ in [212], Calabi-Vesentini give a proof when the associated symmetric space $X = G/K$ is Kähler in [29], and Calabi gives a proof for $G = SO(1, n)$ where $n \geq 3$ in [28]. Shortly afterward, Weil gave a complete proof of Theorem 3.1 in [231, 233].

In all of the original proofs, the first step was to show that any perturbation of Γ was discrete and therefore a cocompact lattice. This is shown in special cases in [28, 29, 212] and proven in a somewhat broader context than Theorem 3.1 in [232].

The different proofs of cases of Theorem 3.1 are also interesting in that there are two fundamentally different sets of techniques employed and this dichotomy continues to play a role in the history of rigidity. Selberg's proof essentially combines algebraic facts with a study of the dynamics of iterates of matrices. He makes systematic use of the existence of singular directions, or Weyl chamber walls, in maximal diagonalizable subgroups of $SL(n, \mathbb{R})$. Exploiting these singular directions is essential to much later work on rigidity, both of lattices in higher-rank groups and of actions of abelian groups. It seems possible to generalize Selberg's proof to the case of G an $\mathbb{R}$-split semisimple Lie group with rank at least 2. Selberg's proof, which depended on asymptotics at infinity of iterates of matrices, inspired Mostow's explicit use of boundaries in his proof of strong rigidity [179]. Mostow's work in turn provided inspiration for the use of boundaries in later work of Margulis, Zimmer, and others on rigidity properties of higher-rank groups.

The proofs of Calabi, Calabi-Vesentini, and Weil involve studying variations of geometric structures on the associated locally symmetric space. The techniques are analytic and use a variational argument to show that all variations of the geometric structure are trivial. This work is a precursor to much work in geometric analysis studying variations of geometric structures and also informs later work on proving rigidity/vanishing of harmonic forms and maps. The dichotomy between approaches based on algebra/dynamics and approaches that are in the spirit of geometric analysis continues through much of the history of rigidity and the history of rigidity of group actions in particular.

Shortly after completing this work, Weil discovered a new criterion for local rigidity [233]. In the context of Theorem 3.1, this allows one to avoid the step of showing that a perturbation of Γ remains discrete. In addition, this result opened the way for understanding local rigidity of more general representations of discrete groups than the defining representation.

THEOREM 3.3. *Let Γ be a finitely generated group, G a Lie group, and $\pi : \Gamma \to G$ a homomorphism. Then π is locally rigid if $H^1(\Gamma, \mathfrak{g}) = 0$. Here $\mathfrak{g}$ is the Lie algebra of G and Γ acts on $\mathfrak{g}$ by $\mathrm{Ad}_G \circ \pi$.*

Weil's proof of this result uses only the implicit function theorem and elementary properties of the Lie group exponential map. The same theorem is true if G is an algebraic group over any local field of characteristic zero. In [233], Weil remarks that if $\Gamma < G$ is a cocompact lattice and G satisfies the hypothesis of Theorem 3.1, then the vanishing of $H^1(\Gamma, \mathfrak{g})$ can be deduced from the computations in [231]. The vanishing of $H^1(\Gamma, \mathfrak{g})$ is proven explicitly by Matsushima and Murakami in [170].

Motivated by Weil's work and other work of Matsushima, conditions for vanishing of $H^1(\Gamma, \mathfrak{g})$ were then studied by many authors. See particularly [170] and [203]. The results in these papers imply local rigidity of many linear representations of lattices.

3.2. Strong and Super-rigidity

In a major and surprising development, it turns out that in many instances, local rigidity is just the tip of the iceberg and that much stronger rigidity phenomena exist. We now discuss major developments from the 1960s and 1970s.

The first remarkable result in this direction is Mostow's rigidity theorem; see [177, 178] and references there. Given G as in Theorem 3.1, and two irreducible cocompact lattices Γ_1 and Γ_2 in G, Mostow proves that any isomorphism from Γ_1 to Γ_2 extends to an isomorphism of G with itself. Combined with the principal theorem of [232], which shows that a perturbation of a lattice is again a lattice, this gives a remarkable and different proof of Theorem 3.1, and Mostow was motivated by the desire for a "more geometric understanding" of Theorem 3.1 [178]. Mostow's theorem is in fact a good deal stronger, and controls not only homomorphisms $\Gamma \to G$ near the defining homomorphism, but any homomorphism into any other simple Lie group G' where the image is a lattice. As mentioned above, Mostow's approach was partially inspired by Selberg's proof of certain cases of Theorem 3.1 [179]. A key step in Mostow's proof is the construction of a continuous map between the geometric boundaries of the symmetric spaces associated to G and G'. Boundary maps continue to play a key role in many developments in rigidity theory. A new proof of Mostow rigidity, at least for G_i of real rank 1, was provided by Besson, Courtois, and Gallot. Their approach is quite different and has had many other applications concerning rigidity in geometry and dynamics; see, for example, [14, 15, 39].

The next remarkable result in this direction is Margulis's superrigidity theorem. Margulis proved this theorem as a tool to prove arithmeticity of irreducible uniform lattices in groups of real rank at least 2. For irreducible lattices in semisimple Lie groups of real rank at least 2, the superrigidity theorems classifies all finite dimensional linear representations. Margulis's theorem holds for irreducible lattices in semisimple Lie groups of real rank at least 2. Given a lattice $\Gamma < G$ where G is simply connected, one precise statement of some of Margulis's results is to say that any linear representation σ of Γ *almost extends* to a linear representation of G. By this we mean that there is a linear representation $\tilde{\sigma}$ of G and a bounded image representation $\bar{\sigma}$ of Γ such that $\sigma(\gamma) = \tilde{\sigma}(\gamma)\bar{\sigma}(\gamma)$ for all γ in G. Margulis's theorems also give an essentially complete description of the representations $\bar{\sigma}$, up to some issues concerning finite image representations. The proof here is partially inspired by Mostow's work: a key step is the construction of a measurable "boundary map." However, the methods for producing the boundary map in this case are very dynamical. Margulis's original proof used Oseledec's multiplicative ergodic theorem. Later proofs were given by both Furstenberg and Margulis using the theory of group boundaries as developed by Furstenberg from his study of random walks on groups [93, 94]. Furstenberg's probabilistic version of boundary theory has had a profound influence on many subsequent developments in rigidity theory. For more discussion of Margulis's superrigidity theorem, see [159, 161, 162, 245].

Margulis's theorem, by classifying all linear representations, and not just ones with constrained images, leads one to believe that one might be able to classify all homomorphisms to other interesting classes of topological groups. Zimmer's program is just one aspect of this theory, in other directions, many authors have studied homomorphisms to isometry groups of non-positively curved (and more general) metric spaces; see, for example, [24, 98, 172].

3.3. Harmonic Map Approaches to Rigidity

In the 1990s, first Corlette and then Gromov-Schoen showed that one could prove major cases of Margulis's superrigidity theorem also for lattices in $Sp(1, n)$ and F_4^{-20} [40, 113]. Corlette considered the case of representations over Archimedean fields and Gromov and Schoen proved results over other local fields. These proofs used harmonic maps and proceed in three steps: first showing a harmonic map exists, second showing it is smooth, and third using certain special Bochner-type formulas to show that the harmonic mapping must be a local isometry. Combined with the earlier work of Matsushima and Murakami, [169], and Raghunathan [199] leads to a complete classification of linear representations for lattices in these groups [170, 203]. It is worth noting that the use of harmonic map techniques in rigidity theory had been pioneered by Siu, who used them to prove generalizations of Mostow rigidity for certain classes of Kähler manifolds [220]. There is also much later work on applying harmonic map techniques to reprove cases of Margulis's superrigidity theorem by Jost-Yau [122] and Mok-Siu-Yeung [171] among others. We remark in passing that the general problem of existence of harmonic maps for non-compact, finite volume locally symmetric spaces has not been solved in general. Results of Saper [210] and Jost-Zuo [123] allow one to prove superrigidity for fundamental groups of many such manifolds, but only while assuming arithmeticity. The use of harmonic maps in rigidity was inspired by the use of variational techniques and harmonic forms and functions in work on local rigidity and vanishing of cohomology groups. The original suggestion to use harmonic maps in this setting appears to go back to Calabi [219].

3.4. A Remark on Cocycle Super-rigidity

An important impetus for the study of rigidity of groups acting on manifolds was Zimmer's proof of his cocycle superrigidity theorem. We discuss this important result below in Section 5 where we also indicate some of its applications to group actions. This theorem is a generalization of Margulis's superrigidity theorem to the class of *virtual homomorphisms* corresponding to cocycles over measure preserving group actions.

3.5. Margulis's Normal Subgroup Theorem

I end this section by mentioning another result of Margulis's that has had tremendous importance in results concerning group actions on manifolds. This is the normal subgroups theorem, which says that any normal subgroup in a higher-rank lattice is either finite or of finite index; see, for example, [162, 245] for more on the proof. The proof precedes by a remarkable strategy.

Let N be a normal subgroup of Γ; we show Γ/N is finite by showing that it is amenable and has property (T). This strategy has been applied in other contexts and is a major tool in the construction of simple groups with good geometric properties; see [7, 25, 26, 37]. The proof that Γ/N is amenable already involves one step that might rightly be called a theorem about rigidity of group actions. Margulis shows that if G is a semisimple group of higher rank, P is a minimal parabolic, and Γ is a lattice then any measurable Γ space X that is a measurable quotient of the Γ action on G/P is necessarily of the form G/Q, where Q is a parabolic subgroup containing P. The proof of this result, sometimes called the projective factors theorem, plays a fundamental role in the work of Nevo and Zimmer on non-volume preserving actions. See Section 10.3 for more discussion. It is also worth noting that Dani has proven a topological analogue of Margulis's result on quotients [42]. That is, he has proven that continuous quotients of the Γ action on G/P are all Γ actions on G/Q.

The usual use of Margulis's normal subgroup theorem in studying rigidity of group actions is usually quite straightforward. If one wants to prove that a group satisfying the normal subgroups theorem acts finitely, it suffices to find one infinite order element that acts trivially.

4. Low-Dimensional Actions: Conjectures and Results

We begin this section by discussing results, in particular, very low dimensions; namely dimensions 1, 2, and 3.

Before we begin this discussion, we recall a result of Thurston that is often used to show that low-dimensional actions are trivial [228].

THEOREM 4.1. (THURSTON STABILITY) *Assume Γ is a finitely generated group with finite abelianization. Let Γ act on a manifold M by C^1 diffeomorphisms, fixing a point p and with trivial derivative at p. Then the Γ action is trivial in a neighborhood of p. In particular, if M is connected, the Γ action is trivial.*

The main point of Theorem 4.1 is that to show an action is trivial, it often suffices to find a fixed point. This is because, for the groups we consider, there are essentially no non-trivial low dimensional linear representations and therefore the derivative at a fixed point is trivial. More precisely, the groups usually only have finite image low dimensional linear representations, which allows one to see that the action is trivial on a subgroup of finite index.

4.1. Dimension 1

The most dramatic results obtained in the Zimmer program concern a question first brought into focus by Dave Witte Morris in [236]: can higher rank lattices act on the circle? In fact, the paper [236] is more directly concerned with actions on the line $\mathbb{R}$. A detailed survey of results in this direction is contained in the paper by Witte Morris in this volume, so we do not repeat that discussion here. We merely state a conjecture and a question.

CONJECTURE 4.2. *Let Γ be a higher rank lattice. Then any continuous Γ action on S^1 is finite.*

This conjecture is well known and first appeared in print in [101]. By the results in [236] and [146], the case of non-cocompact lattices is almost known. The paper [236] does the case of higher $\mathbb{Q}$ rank. The later work of Lifschitz and Witte Morris reduces the general case to the case of quasisplit lattices in $SL(3, \mathbb{R})$ and $SL(3, \mathbb{C})$. It follows from work of Ghys [99, 101, 102] or Burger-Monod [22, 23] that one can assume the action fixes a point, so the question is equivalent to asking if the groups act on the line. We remark that actions of these groups on the circle, which fix a point and are C^1, are easily seen to be finite by using Theorem 4.1 and the fact that all the groups in question have finite first homology. An interesting approach might be to study the induced action of Γ on the space of left orders on Γ; for ideas about this approach, we refer the reader to work of Navas [182] and Witte Morris [174].

Perhaps the following should also be a conjecture, but here we only ask it as a question.

QUESTION 4.3. *Let Γ be a discrete group with property (T). Does Γ admit an infinite action by C^1 diffeomorphisms on S^1? Does Γ admit an infinite action by homeomorphisms on S^1?*

By a result of Navas, the answer to the above question is no if C^1 diffeomorphisms are replaced by C^k diffeomorphisms for any $k > \frac{3}{2}$ [183]. A result noticed by the author and Margulis [82] and contained in a paper of Bader, Furman, Gelander, and Monod [6] allows one to adapt this proof for values of k slightly less than $\frac{3}{2}$. By Thurston's Theorem 4.1, in the C^1 case it suffices to find a fixed point for the action.

We add a remark here pointed out to the author by Navas, that perhaps justifies only calling Question 4.3 a question. In [184], Navas extends his results from [183] to groups with relative property (T). This means that no such group acts on the circle by C^k diffeomorphisms where $k > \frac{3}{2}$. However, if we let Γ be the semi-direct product of a finite index free subgroup of $SL(2, \mathbb{Z})$ and $\mathbb{Z}^2$, this group is left orderable, and so acts on S^1; see, for example, [176]. This is the prototypical example of a group with relative property (T). This leaves open the possibility that groups with property (T) would behave in a similar manner and admit continuous actions on the circle and the line.

Other possible candidate for a group with property (T) acting on the circle by homeomorphisms are the more general variants of Thompson's group F constructed by Stein in [225]. The proofs that F does not have (T) do not apply to these groups [38].

A closely related question is the following, suggested to the author by Andrés Navas along with examples in the last paragraph.

QUESTION 4.4. *Is there a group Γ that is bi-orderable and does not admit a proper isometric action on a Hilbert space? That is, it does not have the Haagerup property?*

For much more information on the fascinating topic of groups of diffeomorphisms of the circle see both the survey by Ghys [102], the more recent book by Navas [181], and the article by Witte Morris in this volume [173].

4.2. Dimension 2

Already in dimension 2 much less is known. There are some results in the volume preserving setting. In particular, we have

THEOREM 4.5. *Let Γ be a non-uniform irreducible higher rank lattice. Then any volume-preserving Γ action on a closed orientable surface other than S^2 is finite. If Γ has $\mathbb{Q}$-rank at least 1, then the same holds for actions on S^2.*

This theorem was proven by Polterovich [195] for all surfaces but the sphere and shortly afterward proven by Franks and Handel for all surfaces [87, 86]. It is worth noting that the proofs use entirely different ideas. Polterovich's proof belongs very clearly to symplectic geometry and also implies some results for actions in higher dimensions. Franks and Handel use a theory of normal forms for C^1 surface diffeomorphisms that they develop in analogy with the Thurston theory of normal forms for surface homeomorphisms with finite fixed sets. The proof of Franks and Handel can be adapted to a setting where one assumes much less regularity of the invariant measure [88].

There are some major reductions in the proofs that are similar, which we now describe. The first of these should be useful for studying actions of cocompact lattices as well. Namely, one can assume that the homomorphism $\rho : \Gamma \to \text{Diff}(S)$ defining the action takes values in the connected component. This follows from the fact that any $\rho : \Gamma \to MCG(S)$ is finite, which can now be deduced from a variety of results [17, 22, 53, 124]. This result holds not only for the groups considered in Theorem 4.5, but also for cocompact lattices and even for lattices in $SP(1, n)$ by results of Sai-Kee Yeung [239]. It may be possible to show something similar for all groups with property (T) using recent results of Andersen showing that the mapping class group does not have property (T) [2]. It is unrealistic to expect a simple analogue of this result for homomorphisms to $\text{Diff}(M)/\text{Diff}(M)^0$ for general manifolds; see Section 4.3 below for a discussion of dimension 3.

Also, in the setting of Theorem 4.5, Margulis's normal subgroup theorem implies that it suffices to show that a single infinite order element of Γ acts trivially. The proofs of Franks-Handel and Polterovich then use the existence of distortion elements in Γ, a fact established by Lubotzky, Mozes, and Raghunathan in [151]. The main result in both cases shows that $\text{Diff}(S, \omega)$ does not contain exponentially distorted elements and the proofs of this fact are completely different. An interesting result of Calegari and Freedman shows that this is not true of $\text{Diff}(S^2)$ and that $\text{Diff}(S^2)$ contains subgroups with elements of arbitrarily large distortion [30]. These examples are discussed in more detail in Section 9.

Polterovich's methods also allow him to see that there are no exponentially distorted elements in $\text{Diff}(M, \omega)^0$ for certain symplectic manifolds (M, ω). Again, this yields some partial results toward Zimmer's conjecture. On the other hand, Franks and Handel are able to work with a Borel measure μ with some properties and show that any map $\rho : \Gamma \to \text{Diff}(S, \mu)$ is finite.

We remark here that a variant on this is due to Zimmer, when there is an invariant measure supported on a finite set.

THEOREM 4.6. *Let Γ be a group with property (T) acting on a compact surface S by C^1 diffeomorphisms. Assume Γ has a periodic orbit on S_j, then the Γ action is finite.*

The proof is quite simple. First, pass to a finite index subgroup that fixes a point x. Look at the derivative representation $d\rho_x$ at the fixed point. Since Γ has property (T) and $SL(2, \mathbb{R})$ has the Haagerup property, it is easy to prove that the image of $d\rho_x$ is bounded. Bounded subgroups of $SL(2, \mathbb{R})$ are all virtually abelian and this implies that the image of $d\rho_x$ is finite. Passing to a subgroup of finite index, one has a fixed point where the derivative action is trivial of a group that has no cohomology in the trivial representation. One now applies Thurston Theorem 4.1.

The difficulty in combining Theorem 4.6 with ideas from the work of Franks and Handel is that while Franks and Handel can show that individual surface diffeomorphisms have large sets of periodic orbits, their techniques do not easily yield periodic orbits for the entire large group action.

We remark here that there is another approach to showing that lattices have no volume preserving actions on surfaces. This approach is similar to the proof that lattices have no C^1 actions on S^1 via bounded cohomology [22, 23, 99]. That Diff $(S, \omega)^0$ admits many interesting quasimorphisms, follows from the work of Entov-Polterovich [49], Gambaudo-Ghys [96], and Py [198, 200, 199]. By results of Burger and Monod [41], for any higher-rank lattice and any homomorphism $\rho : \Gamma \to$ Diff $(S, \omega)^0$, the image is in the kernel of all of these quasi-morphisms. (To make this statement meaningful and the kernel well defined, one needs to take the homogeneous versions of the quasimorphisms.) What remains to be done is to extract useful dynamical information from this fact; see the article by Py in this volume for more discussion [198].

In our context, the work of Entov-Polterovich mentioned above really only constructs a single quasi-morphism on Diff $(S^2, \omega)^0$. However, this particular quasimorphism is very nice in that it is Lipschitz in metric known as Hofer's metric on Diff $(S^2, \omega)^0$. This construction does apply more generally in higher dimensions and indicates connections between quasimorphisms and the geometry of Diff $(M, \omega)^0$, for ω a symplectic form, that are beyond the scope of this survey. For an introduction to this fascinating topic, we refer the reader to [195].

Motivated by the above discussion, we recall the following conjecture of Zimmer.

CONJECTURE 4.7. *Let Γ be a group with property (T), then any volume preserving smooth Γ action on surface is finite.*

Here the words "volume preserving" or at least "measure preserving" are quite necessary. As $SL(3, \mathbb{R})$ acts on S^2, so does any lattice in $SL(3, \mathbb{R})$ or any irreducible lattice in G where G has $SL(3, \mathbb{R})$ as a factor. The following question seems reasonable; I believe I first learned it from Leonid Polterovich.

QUESTION 4.8. *Let Γ be a higher rank lattice (or even just a group with property (T)) and assume Γ acts by diffeomorphisms on a surface S. Is it true that either (1) the action is finite or (2) the surface is S^2 and the action is smoothly conjugate to an action defined by some embedding $i : \Gamma \to SL(3, \mathbb{R})$ and the projective action of $SL(3, \mathbb{R})$ on S^2?*

This question seems quite far beyond existing technology.

4.3. Dimension 3

We now discuss briefly some work of Farb and Shalen that constrains actions by homeomorphisms in dimension 3 [55]. This work makes strong use of the geometry of 3 manifolds, but uses very little about higher rank lattices and is quite soft. A special case of their results is the following:

THEOREM 4.9. *Let M be an irreducible 3-manifold and Γ be a higher rank lattice. Assume Γ acts on M by homeomorphisms so that the action on homology is nontrivial. Then M is homeomorphic to $\mathbb{T}^3$, $\Gamma < SL(3, \mathbb{Z})$ with finite index, and the Γ action on $H^1(M)$ is the standard Γ action on Z^3.*

Farb and Shalen actually prove a variant of Theorem 4.9 for an arbitrary 3 manifold admitting a homologically infinite action of a higher rank lattice. This can be considered as a significant step toward understanding when, for Γ a higher-rank lattice and M^3 a closed 3 manifold, $\rho : \Gamma \to \text{Diff}(M^3)$ must have image in $\text{Diff}(M^3)^0$. Unlike the results discussed above for dimension 2, the answer is not simply "always." This three dimensional result uses a great deal of the known structure of 3 manifolds, though it does not use the full geometrization conjecture proven by Perelman, but only the Haken case due to Thurston. The same sort of result in higher dimensions seems quite out of reach. A sample question is the following. Here we let Γ be a higher rank lattice and M a compact manifold.

QUESTION 4.10. *Under what conditions on the topology of M do we know that a homomorphism $\rho : \Gamma \to \text{Diff}(M)$ has image in $\text{Diff}(M)^0$?*

4.4. Analytic Actions in Low Dimensions

We first mention a direction pursued by Ghys that is in a similar spirit to the Zimmer program, and that has interesting consequences for that program. Recall that the Zassenhaus lemma shows that any discrete linear group generated by small enough elements is nilpotent. The main point of Ghys's article [100] is to attempt to generalize this result for subgroups of Diff$^\omega$ (M). While the result is not actually true in that context, Ghys does prove some intriguing variants that yield some corollaries for analytic actions of large groups. For instance, he proves that $SL(n, \mathbb{Z})$ for $n > 3$ admits no analytic action on the 2 sphere. We remark that the attempt to prove the Zassenhaus lemma for diffeomorphism groups suggests that one can attempt to generalize other facts about linear groups to the category of diffeomorphism groups. We discuss several questions in this direction, mainly due to Ghys, in Section 13.

For the rest of this subsection we discuss a different approach of Farb and Shalen for showing that real analytic actions of large groups are finite. This method is pursued in [54, 56, 57].

We begin by giving a cartoon of the main idea. Given any action of a group Γ, an element $\gamma \in \Gamma$, and the centralizer $Z(\gamma)$, it is immediate that $Z(\gamma)$ acts on the set of γ fixed points. If the action is analytic, then the fixed sets are analytic and so have good structure and are "reasonably close" to being submanifolds. If one further assumes that all normal subgroups of Γ have finite index, then this essentially allows one to bootstrap results about $Z(\gamma)$ not acting on manifolds of dimension at most $n - 1$ to facts about Γ not having actions on manifolds of dimension n provided one can show that the fixed set for γ is not empty. This is not true, as analytic sets are not actually manifolds, but the idea can be implemented using the actual structure of analytic sets in a way that yields many results.

For example, we have

THEOREM 4.11. *Let M be a real analytic 4 manifold with zero Euler characteristic, then any real analytic, volume preserving action of any finite index subgroup in $SL(n, \mathbb{Z})$ for $n \geq 7$ is trivial.*

This particular theorem also requires a result of Rebelo [205], which concerns fixed sets for actions of nilpotent groups on $\mathbb{T}^2$ and uses the ideas of [100].

The techniques of Farb and Shalen can also be used to prove that certain cocompact higher rank lattices have no real analytic actions on surfaces of genus at least 1. For this result one needs only that the lattice contains an element γ whose centralizer already contains a higher rank lattice. In the paper [54] a more technical condition is required, but this can be removed using the results of Ghys and Burger-Monod on actions of lattices on the circle. (This simplification was first pointed out to the author by Farb.)

The point is that γ has fixed points for topological reasons and the set of these fixed points contains either (1) a $Z(\gamma)$ invariant circle or (2) a $Z(\gamma)$ invariant point. Case (1) reduces to case (2) after passing to a finite index subgroup via the results on circle actions. Case (2) is dealt with by the proof of Theorem 4.6.

4.5. Zimmer's Full Conjecture, Partial Results

We now state the full form of Zimmer's conjecture. In fact we generalize it slightly to include all lattices with property (T). Throughout this subsection G will be a semisimple Lie group with property (T) and Γ will be a lattice in G. We define two numerical invariants of these groups. First, for any group F, let $d(F)$ be the lowest dimension in which F admits an infinite image linear representation. We note that the superrigidity theorems imply that $d(\Gamma) = d(G)$ when Γ is a lattice in G and either G is a semisimple group property (T) or Γ is irreducible higher rank. The second number, $n(G)$ is the lowest dimension of a homogeneous space K/C for a compact group K on which a lattice Γ in G can act via a homomorphism $\rho : \Gamma \to K$. In Zimmer's work, $n(G)$ is defined differently, in a way that makes clear that there is a bound on $n(G)$ that does not depend on the choice of Γ. Namely, using the superrigidity theorems, it can be shown that n satisfies

$$\frac{n(n+1)}{2} \geq \min\{\dim_{\mathbb{C}} G' \text{ where } G' \text{ is a simple factor of } G\}.$$

The more fashionable variant of Zimmer's conjecture is the following, first made explicitly by Farb and Shalen for higher rank lattices in [54].

CONJECTURE 4.12. *Let Γ be a lattice as above. Let $b = \min\{n, d\}$ and let M be a manifold of dimension less than $b - 1$, then any Γ action on M is trivial.*

It is particularly bold to state this conjecture including lattices in $Sp(1, n)$ and F_4^{-20}. This is usually avoided because such lattices have abundant non-volume preserving actions on certain types of highly regular fractals. I digress briefly to explain why I believe the conjecture is plausible in that case by discussing how the actions on fractals arise. Namely such a lattice Γ, being a hyperbolic group, has many infinite proper quotients that are also hyperbolic groups [111]. If Γ' is a quotient of Γ by an infinite index, infinite normal subgroup N, and Γ' is hyperbolic, then the boundary $\partial \Gamma'$ is a Γ space with interesting dynamics and a good (Ahlfors regular) quasi-invariant measure class. However, $\partial \Gamma'$ is only a manifold when it is a sphere. It seems highly unlikely that this is ever the case for these groups and even more unlikely that this is ever the case with a smooth boundary action. The only known way to build a hyperbolic group that acts smoothly on its boundary is to have the group be the fundamental group of a compact negatively curved manifold M with smoothly varying horospheres. If smooth is taken to mean C^∞, this then implies the manifold is locally symmetric [14] and then superrigidity results make it impossible for this to occur with $\pi_1(M)$ a quotient of a lattice by an infinite index infinite normal subgroup. If smooth only means C^1, then even in this context, no result known rules out M having fundamental group a quotient of a lattice in $Sp(1,n)$ or F_4^{-20}. All results one can prove using harmonic map techniques in this context only rule out M with non-positive complexified sectional curvature. We make the following conjecture, which is stronger than what is needed for Conjecture 4.12.

CONJECTURE 4.13. *Let Γ be a cocompact lattice in $Sp(1,n)$ or F_4^{-20} and let Γ' be a quotient of Γ that is Gromov hyperbolic. If $\partial\Gamma'$ is a sphere, then the kernel of the quotient map is finite and $\partial\Gamma' = \partial\Gamma$.*

For background on hyperbolic groups and their boundaries from a point of view relevant to this conjecture, see [133]. As a cautionary note, we point the reader to Subsection 11.3 where we recall a construction of Farrell and Lafont that shows that any Gromov hyperbolic group has an action by homeomorphisms on a sphere.

The version of Zimmer's conjecture that Zimmer made in [249] and [250] was only for volume preserving actions. Here we break it down somewhat explicitly to clarify the role of d and n.

CONJECTURE 4.14. (ZIMMER) *Let Γ be a lattice as above and assume Γ acts smoothly on a compact manifold M preserving a volume form. Then if $\dim(M) < d$, the Γ action is isometric. If, in addition, $\dim(M) < n$ or if Γ is non-uniform, then the Γ action is finite.*

Some first remarks are in order. The cocycle super-rigidity theorems (discussed below) imply that, when the conditions of the conjecture hold, there is always a measurable invariant Riemannian metric. Also, the finiteness under the conditions in the second half of the conjecture follow from cases of Margulis's superrigidity theorem as soon as one knows that the action preserves a smooth Riemannian metric. So from one point of view, the conjecture is really about the regularity of the invariant metric.

We should also mention that the conjecture is proven under several additional hypotheses by Zimmer around the time he made the conjecture. The first example is the following:

THEOREM 4.15. *Conjecture 4.14 holds provided the action also preserves a rigid geometric structure, for example, a torsion-free affine connection or a pseudo-Riemannian metric.*

This is proven in [246] for structures of finite type in the sense of Elie Cartan; see also [250]. The fact that roughly the same proof applies for rigid structures in the sense of Gromov was remarked in [85]. The point is simply that the isometry group of a rigid structure acts properly on some higher order frame bundle and that the existence of the measurable metric implies that Γ has bounded orbits on all frame bundles as soon as Γ has property (T). This immediately implies that Γ must be contained in a compact subgroup of the isometry group of the structure.

Another easy version of the conjecture follows from the proof of Theorem 4.6. That is

THEOREM 4.16. *Let G be a semisimple Lie group with property (T) and $\Gamma < G$ a lattice. Let Γ act on a compact manifold M by C^1 diffeomorphisms where $\dim(M) < d(G)$. If Γ has a periodic orbit on S, then the Γ action is finite.*

A more difficult theorem of Zimmer shows that Conjecture 4.14 holds when the action is distal in a sense defined in [247].

4.6. Some Approaches to the Conjectures

4.6.1. DISCRETE SPECTRUM OF ACTIONS. A measure preserving action of a group D on a finite measure space (X, μ) is said to have *discrete spectrum* if $L^2(X, \mu)$ splits as the sum of finite dimensional D invariant subspaces. This is a strong condition that is (quite formally) the opposite of weak mixing; for a detailed discussion see [92]. It is a theorem of Mackey (generalizing earlier results of Halmos and Von Neumann for D abelian) that an ergodic discrete spectrum D action is measurably isomorphic to one described by a dense embedding of D into a compact group K and considering a K action on a homogeneous K-space.

The following remarkable result of Zimmer from [253] is perhaps the strongest evidence for Conjecture 4.14. This result is little known and has only recently been applied by other authors; see [83, 91].

THEOREM 4.17. *Let Γ be a group with property (T) acting by smooth, volume preserving diffeomorphisms on a compact manifold M. Assume in addition that Γ preserves a measurable invariant metric. Then the Γ action has discrete spectrum.*

This immediately implies that no counterexample to Conjecture 4.14 can be weak mixing or even admit a weak mixing measurable factor.

The proof of the theorem involves constructing finite dimensional subspaces of $L^2(M, \omega)$ that are Γ invariant. If enough of these subspaces could be shown to be spanned by smooth functions, one would have a proof of Conjecture 4.14. Here by "enough" we simply mean that it suffices to have a collection of finite dimensional D invariant subspaces that separate points in M. These functions would then specify a D equivariant smooth embedding of M into $\mathbb{R}^N$ for some large value of N.

To construct finite dimensional invariant subspaces of $L^2(M)$, Zimmer uses an approach similar to the proof of the Peter-Weyl theorem. Namely, he constructs Γ invariant kernels on $L^2(M \times M)$ that are used to define self-adjoint, compact operators on $L^2(M)$. The eigenspaces of these operators are then finite dimensional, Γ invariant subspaces of $L^2(M)$. The kernels should be thought of as functions of the distance to the diagonal in $M \times M$. The main difficulty here is that for these to be invariant by "distance" we need to mean something defined in terms of the measurable metric instead of a smooth one. The construction of the kernels in this setting is quite technical and we refer readers to the original paper.

It would be interesting to try to combine the information garnered from this theorem with other approaches to Zimmer's conjectures.

4.6.2. EFFECTIVE INVARIANT METRICS. We discuss here briefly another approach to Zimmer's conjecture, due to the author, which seems promising.

We begin by briefly recalling the construction of the space of "L^2 metrics" on a manifold M. Given a volume form ω on M, we can consider the space of all (smooth) Riemannian metrics on M whose associated volume form is ω. This is the space of smooth sections of a bundle $P \to M$. The fiber of P is $X = SL(n, \mathbb{R})/SO(n)$. The bundle P is an associated bundle to the $SL(n, \mathbb{R})$ sub-bundle of the frame bundle of M defined by ω. The space X carries a natural $SL(n, \mathbb{R})$-invariant Riemannian metric of non-positive curvature; we denote its associated distance function by d_X. This induces a natural notion of distance on the space of metrics, given by $d(g_1, g_2)^2 = \int_M d_X(g_1(m), g_2(m))^2 d\omega$. The completion of the sections with respect to the metric d will be denoted $L^2(M, \omega, X)$; it is commonly referred to as the *space of L^2 metrics on M* and its elements will be called L^2 *metrics* on M. That this space is CAT (0) follows easily from the fact that X is CAT (0). For more discussion of X and its structure as a Hilbert manifold, see, for example, [79]. It is easy to check that a volume preserving Γ action on M defines an isometric Γ action on $L^2(M, \omega, X)$. Given a generating set S for Γ and a metric g in $L^2(M, \omega, X)$, we write disp $(g) = max_{\gamma \in S} d(\gamma g, g)$.

Given a group Γ acting smoothly on M preserving ω, this gives an isometric Γ action on $L^2(M, \omega, X)$ that preserves the subset of smooth metrics. Let S be a generating set for Γ. We define an operator $P : L^2(M, \omega, X) \rightarrow L^2(M, \omega, X)$ by taking a metric g to the barycenter of the measure $\sum_S \delta_{(\gamma g)}$. The first observation is a consequence of the (standard, finite dimensional) implicit function theorem.

LEMMA 4.18. *If g is a smooth metric, then Pg is also smooth.*

Moreover, we have the following two results.

THEOREM 4.19. *Let M be a surface and Γ a group with property (T) and finite generating set S. Then there exists $0 < C < 1$ such that the operator P satisfies:*

1) disp $(Pg) < C$ disp (g), *and*
2) *for any g, the $\lim_n (P^n g)$ exists and is Γ invariant.*

A proof of this theorem can be given by using the standard construction of a negative definite kernel on $\mathbb{H}^2$ to produce a negative definite kernel on $L^2(S, \mu, \mathbb{H}^2)$. The theorem is then proved by transferring the first property from the resulting Γ action on a Hilbert space. The second property is an obvious consequence of the first and completeness of the space $L^2(M, \omega, X)$.

THEOREM 4.20. *Let G be a semisimple Lie group all of whose simple factors have property (T) and $\Gamma < G$ a lattice. Let M be a compact manifold such that $\dim(M) < d(G)$. Then the operator P on $L^2(M, \omega, X)$ satisfies the conclusions of Theorem 4.19.*

This theorem is proven from results in [80] using convexity of the distance function on $L^2(M, \omega, X)$. For cocompact lattices a proof can be given using results in [138] instead.

The problem now reduces to estimating the behavior of the derivatives of $P^n g$ for some initial smooth g. This expression clearly involves derivatives of random products of elements of Γ, that is, derivatives of elements of Γ weighted by measures that are convolution powers of equidistributed measure on S. The main cause for optimism is that the fact that disp $(P^n g)$ is small immediately implies that the first derivative of any $\gamma \in S$ must be small when measured at that point in $L^2(M, \omega, X)$. One can then try to use estimates on compositions of diffeomorphisms and convexity of derivatives to control derivatives of $P^n g$. The key difficulty is that the initial estimate on the first derivative of γ applied to $P^n g$ is only small in an L^2 sense.

4.7. Some Related Questions and Conjectures on Countable Subgroups of Diff (M)

In this subsection, we discuss related conjectures and results on countable subgroups of Diff (M). All of these are motivated by the belief that countable subgroups of Diff (M) are quite special, though considerably less special than, say, linear groups. We defer positive constructions of non-linear subgroups of Diff (M) to Section 11 and a discussion of possible algebraic and geometric properties shared by all finitely generated subgroups of Diff (M) to Section 13. Here we concentrate on groups that do not act on manifolds, either by theorems or conjecturally.

4.7.1. GROUPS WITH PROPERTY (T) AND GENERIC GROUPS. We begin by focusing on actions of groups with property (T). For a finitely generated group Γ, we recall that $d(\Gamma)$ is the smallest dimension in which Γ admits an infinite image linear representation. We then make the following conjecture:

CONJECTURE 4.21. *Let Γ be a group with property (T) acting smoothly on a compact manifold M, preserving volume. Then if $\dim(M) < d(\Gamma)$, the action preserves a smooth Riemannian metric.*

For many groups Γ with property (T), one can produce a measurable invariant metric; see [83]. In fact, in [83], Silberman and the author prove that there are many groups with property (T) with no volume preserving actions on compact manifolds. Key steps include finding the invariant measurable metric, applying Zimmer's theorem on discrete spectrum from Section 4.6, and producing groups with no finite quotients and so no linear representations at all.

A result of Furman announced in [90] provides some further evidence for the conjecture. This result is analogous to Proposition 5.1 and shows that any action of a group with property (T) either leaves invariant a measurable metric or has positive *random entropy*. We refer the reader to [90] for more discussion. While the proof of this result is not contained in [90], it is possible to reconstruct it from results there and others in [91].

Conjecture 4.21 and the work in [83] are motivated in part by the following conjecture of Gromov:

CONJECTURE 4.22. *There exists a model for random groups in which a "generic" random group admits no smooth actions on compact manifolds.*

It seems quite likely that the conjecture could be true for random groups in the density model with density more than $\frac{1}{3}$. These groups have property (T); see Ollivier's book [193] for a discussion on random groups. For the conjecture to be literally true would require that a random hyperbolic group have no finite quotients and it is a well-known question to determine whether there are any hyperbolic groups that are not residually finite, let alone generic ones. If one is satisfied by saying the generic random group has only finite smooth actions on compact manifolds, one can avoid this well-known open question.

4.7.2. UNIVERSAL LATTICES. Recently, Shalom has proven that the groups $SL(n, \mathbb{Z}[X])$ have property (T) when $n > 3$. (As well as some more general results.) Shalom refers to these groups as *universal lattices*. An interesting and approachable question is

QUESTION 4.23. *Let Γ be a finite index subgroup in $SL(n, \mathbb{Z}[X])$ and let Γ acting by diffeomorphisms on a compact manifold M preserving volume. If $\dim(M) < n$, is there a measurable Γ invariant metric?*

If one gives a positive answer to this question, one is then clearly interested in whether or not the metric can be chosen to be smooth. It is possible that this is easier for these "larger" groups than for the lattices originally considered by Zimmer. One can ask a number of variants of this question including trying to prove a full cocycle super-rigidity theorem for these groups; see below. The question just asked is particularly appealing as it can be viewed as a fixed point problem for the Γ action on the space of metrics (see Section 4.6 for a definition). In fact, one knows one has the invariant metric for the action of any conjugate of $SL(n, \mathbb{Z})$ and only needs to show that there is a consistent choice of invariant metrics over all conjugates. One might try to mimic the approach from [216], though a difficulty clearly arises at the point where Shalom applies a scaling limit construction. Scaling limits of $L^2(M, \omega, X)$ can be described using nonstandard analysis, but are quite complicated objects and not usually isomorphic to the original space.

4.7.3. IRREDUCIBLE ACTIONS OF PRODUCTS. Another interesting variant on Zimmer's conjecture is introduced in [91]. In that paper Furman and Monod study obstructions to irreducible actions of product groups. A measure preserving action of a product $\Gamma_1 \times \Gamma_2$ on a measure space (X, μ) is said to be *irreducible* if both Γ_1 and Γ_2 act ergodically on X. Furman and Monod produce many obstructions to irreducible actions; for example, one can prove from their results that

THEOREM 4.24. *Let* $\Gamma = \Gamma_1 \times \Gamma_2 \times \Gamma_3$. *Assume that* Γ_1 *has property* (T) *and no unbounded linear representations. Then there are no irreducible, volume preserving* Γ *actions on compact manifolds. The same is true for* $\Gamma = \Gamma_1 \times \Gamma_2$ *if* Γ_1 *is as above and* Γ_2 *is solvable.*

A key step in the proof of this result is to use Zimmer's result on actions with discrete spectrum to show that the Γ action is measurably conjugate to an irreducible isometric action. In particular, the second conclusion follows since no compact group can contain simultaneously a dense (T) group and a dense solvable group. This motivates the following conjecture.

CONJECTURE 4.25. *Let* $\Gamma = \Gamma_1 \times \Gamma_2$. *Assume that* Γ_1 *has property* (T) *and no unbounded linear representations and that* Γ_2 *is amenable. Then there are no irreducible, volume preserving* Γ *actions on compact manifolds.*

One might be tempted to prove this by showing that no compact group contains both a dense finitely generated amenable group and a dense Kazhdan group. This is, however, not true. The question was raised by Lubotzky in [149] but has recently been resolved in the negative by Kassabov [127].

In the context of irreducible actions, asking that groups have property (T) is perhaps too strong. In fact, there are already relatively few known irreducible actions of $\mathbb{Z}^2 = \mathbb{Z} \times \mathbb{Z}$! If one element of $\mathbb{Z}^2$ acts as an Anosov diffeomorphism, then it is conjectured by Katok and Spatzier that all actions are algebraic [130]. Even in more general settings where one assumes non-uniform hyperbolicity, there are now hints that a classification might be possible, though clearly more complicated than in the Anosov case; see [125] and references there. In a sense, that paper indicates that the "exotic examples" that might arise in this context may be no worse than those that arise for actions of higher rank lattices.

4.7.4. TORSION GROUPS AND HOMEO (M). A completely different and well-studied aspect of the theory of transformation groups of compact manifolds is the study of finite subgroups of Homeo (M). It has recently been noted by many authors that this study has applications to the study of "large subgroups" of Homeo (M). Namely, one can produce many finitely generated and even finitely presented groups that have no non-trivial homomorphisms to Homeo (M) for any compact M. This is discussed in, for example, [20, 83, 234]. As far as I know, this observation was first made by Ghys as a remark in the introduction to [100]. In all cases, the main trick is to construct infinite, finitely generated groups that contain infinitely many conjugacy classes of finite subgroups, usually just copies of $(\mathbb{Z}/p\mathbb{Z})^k$ for all $k > 1$. A new method of constructing such groups was recently introduced by Chatterji and Kassabov [38].

As far as the author knows, even if one fixes M in advance, this is the only existing method for producing groups Γ with no non-trivial (or even no infinite image) homomorphisms to Homeo (M) unless $M = S^1$. For $M = S^1$ we refer back to Subsection 4.1 and to [173].

5. Cocycle Super-rigidity and Immediate Consequences

5.1. Zimmer's Cocycle Super-rigidity Theorem and Generalizations

A main impetus for studying rigidity of group actions on manifolds came from Zimmer's theorem on superrigidity for cocycles. This theorem and its proof were strongly motivated by Margulis's work. In fact, Margulis's theorem reduces to the special case of Zimmer's theorem for a certain cocycle $\alpha : G \times G/\Gamma \to \Gamma$. In order to avoid technicalities, we describe only a special case of this result, essentially avoiding boundedness and integrability assumptions on cocycles that are automatically fulfilled in any context arising from a continuous action on a compact manifold. Let M be a compact manifold, H a matrix group, and P an H bundle over M. For readers not familiar with bundle theory, the results are interesting even in the case where $P = M \times H$. Now let a group Γ act on P continuously by bundle automorphisms, that is, such that there is a Γ action on M for which the projection from P to M is equivariant. Further assume that the action on M is measure preserving and ergodic. The cocycle superrigidity theorem says that if Γ is a lattice in a simply connected, semisimple Lie group G all of whose simple factors are noncompact and have property (T), then there is a measurable map $s : M \to H$, a representation $\pi : G \to H$, a compact subgroup $K < H$ that commutes with $\pi(G)$, and a measurable map $\Gamma \times M \to K$ such that

5.1 $$\gamma \cdot s(m) = k(m, \gamma)\pi(\gamma)s(\gamma \cdot m).$$

It is easy to check from this equation that the map K satisfies the equation that makes it into a cocycle over the action of Γ. One should view s as providing coordinates on P in which the Γ action is *almost a product*. For more discussion of this theorem, particularly in the case where all simple factors of G have higher rank, the reader should see any of [65, 66, 81, 95, 245]. (The version stated here is only proven in [81]; previous proofs all yielded somewhat more complicated statements that require passing to finite ergodic extensions of the action.) For the case of G with simple factors of the form $Sp(1, n)$ and F_4^{-20}, the results follow from work of the author and Hitchman [80], building on earlier results of Korevaar-Schoen [138, 139, 140] and Corlette-Zimmer [41].

As a sample application, let $M = \mathbb{T}^n$ and let P be the frame bundle of M, that is, the space of frames in the tangent bundle of M. Since $\mathbb{T}^n$ is parallelizable, we have $P = \mathbb{T}^n \times GL(n, \mathbb{R})$. The cocycle super-rigidity theorem then says that "up to compact noise" the derivative of any measure preserving Γ action on $\mathbb{T}^n$ looks measurably like a constant linear map. In fact, the cocycle superrigidity theorems apply more generally to continuous actions on any principal bundle P over M with fiber H, an algebraic group, and in this context produces a measurable section $s : M \to P$ satisfying equation (5.1). So in fact, cocycle superrigidity implies that for any action preserving a finite measure on any manifold the derivative cocycle looks measurably like a constant cocycle, up to compact noise. That cocycle superrigidity provides information about actions of groups on manifolds through the derivative cocycle was first observed by Furstenberg in [95]. Zimmer originally proved cocycle superrigidity in order to study orbit equivalence of group actions. For recent surveys of subsequent developments concerning orbit equivalence rigidity and other forms of superrigidity for cocycles, see [89, 197, 215].

5.2. First Applications to Group Actions and the Problem of Regularity

The following result, first observed by Furstenberg, is an immediate consequence of cocycle superrigidity.

PROPOSITION 5.1. *Let G be a semisimple Lie group with no compact factors and with property (T) of Kazhdan and let $\Gamma < G$ be a lattice. Assume G or Γ acts by volume preserving diffeomorphisms on a compact manifold M. Then there is a linear representation $\pi : G \to GL(\dim(M), \mathbb{R})$ such that the Lyapunov exponents of $g \in G$ or Γ are exactly the absolute values of the eigenvalues of $\pi(g)$.*

This proposition is most striking to those familiar with classical dynamics where the problem of estimating, let alone computing, Lyapunov exponents is quite difficult.

Another immediate consequence of cocycle superrigidity is the following:

PROPOSITION 5.2. *Let G or Γ be as above, acting by smooth diffeomorphisms on a compact manifold M preserving volume. If every element of G or Γ acts with zero entropy, then there is a measurable invariant metric on M. In particular, if G admits no nontrivial representations of* dim (M)*, then there is a measurable invariant metric on M.*

As mentioned above in Subsection 4.5, this reduces Conjecture 4.14 to the following technical conjecture.

CONJECTURE 5.3. *Let G or Γ acting on M be as above. If G or Γ preserves a measurable Riemannian metric, then they preserve a smooth-invariant Riemannian metric.*

We recall that first evidence toward this conjecture is Theorem 4.15. We remark that in the proof of that theorem, it is not the case that the measurable metric from Proposition 5.2 is shown to be smooth, but instead it is shown that the image of Γ in Diff (M) lies in a compact subgroup. In other work, Zimmer explicitly improves regularity of the invariant metric; see particularly [247].

In general the problem that arises immediately from cocycle superrigidity in any context is understanding the regularity of the straightening section σ. This question has been studied from many points of view, but still relatively little is known in general. For certain examples of actions of higher rank lattices on compact manifolds, discussed below in Section 9, the σ that straightens the derivative cocycle cannot be made smooth on all of M. It is possible that for volume preserving actions on manifolds and the derivative cocycle, σ can always be chosen to be smooth on a dense open set of full measure.

We will return to the theme of regularity of the straightening section in Section 7. First we turn to more geometric contexts in which the output of cocycle superrigidity is also often used but usually more indirectly.

6. Geometric Actions and Rigid Geometric Structures

In this section, we discuss the role of rigid geometric structures in the study of actions of large groups. The notion of rigid geometric structure was introduced by Gromov, partially in reaction to Zimmer's work on large group actions.

The first subsection of this section recalls the definition of rigid geometric structure, gives some examples, and explains the relation of Gromov's rigid geometric structures to other notions introduced by Cartan. Subsection 6.2 recalls Gromov's initial results relating actions of simple groups preserving rigid geometric structures on M to representations of the fundamental group of M, and extensions of these results to lattice actions due to Zimmer and the author. Subsection 6.3 concerns a different topic, namely the rigidity of connection preserving actions of a lattice Γ, particularly on manifolds of dimension not much larger than $d(G)$ as defined in Section 4. Subsection 6.4 recalls some obstructions to actions preserving geometric structures, particularly a result known as Zimmer's geometric Borel density theorem and some recent related results of Bader, Frances, and Melnick. We discuss actions preserving a complex structure in Subsection 6.5 and end with a subsection 6.6 on questions concerning geometric actions.

6.1. Rigid Structures and Structures of Finite Type

In this subsection we recall the formal definition of a rigid geometric structure. Since the definition is somewhat technical, some readers may prefer to skip it and read on keeping in mind examples rather than the general notion. The basic examples of rigid geometric structures are Riemannian metrics, pseudo-Riemannian metrics, conformal structures defined by either type of metric, and affine or projective connections. Basic examples of geometric structures that are not rigid are a volume form or symplectic structure. An intermediate type of structure that exhibits some rigidity but that is not literally rigid in the sense discussed here is a complex structure.

If N is a manifold, we denote the kth order frame bundle of N by $F^k(N)$, and by $J^{s,k}(N)$ the bundle of k-jets at 0 of maps from $\mathbb{R}^s$ to N. If N and N' are two manifolds, and $f : N \to N'$ is a map between them, then the k-jet $j^k(f)$ induces a map $J^{s,k}N \to J^{s,k}N'$ for all s. We let $D^k(N)$ be the bundle whose fiber D_p^k at a point p consists of the set of k-jets at p of germs of diffeomorphisms of N fixing p. We abbreviate $D_0^k(\mathbb{R}^n)$ by D_n^k or simply D^k; this is a real algebraic group. For concreteness, one can represent each element uniquely, in terms of standard coordinates $(\xi_1, \ldots, \xi_n)$ on $\mathbb{R}^n$, in the form

$$(P_1(\xi_1, \ldots, \xi_n), \ldots, P_n(\xi_1, \ldots, \xi_n))$$

where $P_1, P_2, \ldots, P_n$ are polynomials of degree $\leq k$. We denote the vector space of such polynomial maps of degree $\leq k$ by $\mathcal{P}_{n,k}$.

The group D_n^k has a natural action on $F^k(N)$, where n is the dimension of N. Suppose we are given an algebraic action of D_n^k on a smooth algebraic variety Z. Then following Gromov [112], we make the following definition:

DEFINITION 6.1.

1) An *A-structure* on N (of order k, of type Z) is a smooth map $\phi : F^k(N) \to Z$ equivariant for the D_n^k actions.

2) With notation as above, the *rth prolongation* of ϕ, denoted ϕ^r, is the map $\phi^r : F^{k+r}(N) \to J^{n,r}(Z)$ defined by $\phi^r = j^r(\phi) \circ \iota_k^{r+k}$ where $\iota_k^{r+k} : F^{k+r}(N) \to J^{n,r}(F^k(N))$ is the natural inclusion and $j^k(h) : J^{n,r}(F^k(N)) \to J^{n,r}(Z)$ is as before; this is an A-structure of type $J^{n,r}(Z)$ and order $k+r$.

Equivalently, an A-structure of type Z and order k is a smooth section of the associated bundle $F^k(N) \times_{D^k} Z$ over N. Note that an A-structure on N defines by restriction an A-structure $\phi|_U$ on any open set $U \subset N$.

REMARK 6.2. A-structures were introduced in [112]; a good introduction to the subject, with many examples, can be found in [10]. A comprehensive and accessible discussion of the results of [112] concerning actions of simple Lie groups can be found in [67].

Note that if N and N' are n-manifolds, and $h : N \to N'$ is a diffeomorphism, then h induces a bundle map $j^k(h) : F^k(N) \to F^k(N')$.

DEFINITION 6.3.

1) If $\phi : F^k(N) \to Z$, $\phi' : F^k(N') \to Z$ are A-structures, a diffeomorphism $h : N \to N'$ is an *isometry* from ϕ to ϕ' if $\phi' \circ j^k(h) = \phi$.

2) A *local isometry* of ϕ is a diffeomorphism $h : U_1 \to U_2$, for open sets $U_1, U_2 \subset N$, which is an isometry from $\phi|_{U_1}$ to $\phi|_{U_2}$.

For $p \in M$ denote by $Is_p^{loc}(\phi)$ the pseudogroup of local isometries of ϕ fixing p, and, for $l \geq k$, we denote by $Is_p^l(\phi)$ the set of elements $j_p^l(h) \in D_p^l$ such that $j_p^l(\phi \circ j_p^k(h)) = \phi^{l-k}$, where both sides are considered as maps $F^{k+l}(N) \to J^{l-k}(Z)$. Note that $Is_p^l(\phi)$ is a group, and there is a natural homomorphism $r_p^{l;m} : Is_p^l(\phi) \to Is_p^m(\phi)$ for $m < l$; in general, it is neither injective nor surjective.

DEFINITION 6.4. The structure ϕ is called *k-rigid* if for every point p, the map $r_p^{k+1;k}$ is injective.

A first remark worth making is that an affine connection is a rigid geometric structure and that any generalized quasi-affine action on a manifold of the form $K \backslash H / \Lambda \times M$ with Λ discrete preserves an affine connection and therefore also a torsion free affine connection. In order to provide some more examples, we recall the following lemma of Gromov.

LEMMA 6.5. *Let V be an algebraic variety and G a group acting algebraically on V. For every k, there is a tautological G invariant geometric structure of order k on V, given by $\omega : P^k(V) \to P^k(V)/G$. This structure is rigid if and only if the action of G on $P^k(V)$ is free and proper.*

The conclusion in the first sentence is obvious. The second sentence is proven in section 0.4, pages 69–70, of [112].

EXAMPLES

1) The action of $G = SL_n(\mathbb{R})$ on $\mathbb{R}^n$ is algebraic. So is the action of G on the manifold N_1 obtained by blowing up the origin. The reader can easily verify that the action of G on $P^2(N_1)$ is free and proper.

2) We can compactify N_1 by N_2 by viewing the complement of the blow up as a subset of the projective space P^n. Another description of the same action, which may make the rigid structure more visible to the naked eye, is as follows: $SL_{n+1}(\mathbb{R})$ acts on P^n. Let G be $SL_n(\mathbb{R}) < SL_{n+1}(\mathbb{R})$ as block diagonal matrices with blocks of size n and 1 and a 1×1 block equal to 1. Then G acts on P^n fixing a point p. We can obtain N_2 by blowing up the fixed point p. The G actions on both P^n and N_2 are algebraic and again the reader can verify that the action is free and proper on $P^2(N_2)$.

3) In the construction from (2) above, there is an action of a group H where $H = SL_n(\mathbb{R}) \ltimes \mathbb{R}^n$ and $G = SL_n(\mathbb{R}) < SL_n(\mathbb{R}) \ltimes \mathbb{R}^n < SL_{n+1}(\mathbb{R})$. The H action fixes the point p and so also acts on N_2 algebraically. However, over the exceptional divisor, the action is never free on any frame bundle, since the subgroup $\mathbb{R}^n$ acts trivially to all orders at the exceptional divisor.

The behavior in example (3) above illustrates the fact that existence of invariant rigid structures is more complicated for algebraic groups that are not semisimple; see the discussion in section 0.4.C. of [112].

DEFINITION 6.6. A rigid geometric structure is called a *finite type* geometric structure if V is a homogeneous D_n^k space.

This is by no means the original definition that is due to Cartan and pre-dates Gromov's notion of a rigid geometric structure by several decades. It is equivalent to Cartan's definition of a finite type structure by the work of Candel and Quiroga-Barranco [31, 32, 201]. Candel and Quiroga-Barranco also give a development of rigid geometric structures that more closely parallels the older notion of a structure of finite type as presented in, for example, [135].

All standard examples of rigid geometric structures are structures of finite type. However, example (2) above is not a structure of finite type. The notion of structure of finite type was first given in terms of prolongations of Lie algebras and so yields a criterion that is, in principle, computable. The work of Candel and Quiroga-Barranco extends this computable nature to general rigid geometric structures.

6.2. Rigid Structures and Representations of Fundamental Groups

In this subsection, we restrict our attention to actions of simple Lie groups G and lattices $\Gamma < G$. Many of the results of this section extend to semisimple G, though the formulations become more complicated.

A major impetus for Gromov's introduction of rigid geometric structures is the following theorem from [112].

THEOREM 6.7. *Let G be a simple noncompact Lie group and M a compact real analytic manifold. Assume G acts on M preserving an analytic rigid geometric structure ω and a volume form v. Further assume the action is ergodic. Then there is a linear representation $\rho : \pi_1(M) \to GL(n, \mathbb{R})$ such that the Zariski closure of $\rho(\pi_1(M))$ contains a group locally isomorphic to G.*

We remark that, by the Tits's alternative, the theorem immediately implies that actions of the type described cannot occur on manifolds with amenable fundamental group or even on manifolds whose fundamental groups do not contain a free group on two generators [229]. Expository accounts of the proof can be found in [67, 255, 258]. The representation ρ is actually defined on Killing fields of the lift $\tilde{\omega}$ of ω to $\tilde{M}$.

This is a worthwhile moment to indicate the weakness of Principle 2.6. If we start with a cocompact lattice $\Gamma < G$ and an action of Γ on a compact manifold M satisfying the assumptions of Theorem 6.7, we can induce the action to a G action on the compact manifold $N = (G \times M)/\Gamma$. However, in this setting, there is an obvious representation of $\pi_1(N)$ satisfying the conclusion of Theorem 6.7. This is because there is a surjection $\pi_1(N) \to \Gamma$. In fact for most cocompact lattices $\Gamma < G$ there are homomorphisms $\sigma : \Gamma \to K$ where K is compact and simply connected so that we can have Γ act on K by left translation and obtain examples where $\pi_1(N) = \Gamma$. For G of higher rank, the following theorem of Zimmer and the author shows that this is the only obstruction to a variant of Theorem 6.7 for lattices.

THEOREM 6.8. *Let $\Gamma < G$ be a lattice, where G is a simple group and $\mathbb{R} - rank(G) \geq 2$. Suppose Γ acts analytically and ergodically on a compact manifold M preserving a unimodular rigid geometric structure. Then either*

1) *the action is isometric and $M = K/C$ where K is a compact Lie group and the action is by right translation via $\rho : \Gamma \to K$, a dense image homomorphism, or*
2) *there exists an infinite image linear representation $\sigma : \pi_1(M) \to GL_n\mathbb{R}$, such that the algebraic automorphism group of the Zariski closure of $\sigma(\pi_1(M))$ contains a group locally isomorphic to G.*

The proof of this theorem makes fundamental use of a notion that does not occur in its statement: the entropy of the action. A key observation is the following formula for the entropy of the restriction of the induced action to Γ:

$$h_{(G \times M)/\Gamma}(\gamma) = h_{G/\Gamma}(\gamma) + h_M(\gamma).$$

We then use the fact that the last term on the right hand side is 0 if and only if the action preserves a measurable Riemannian metric as already explained in Proposition 5.1. In this setting, it then follows from Theorem 4.15 that the action is isometric and the other conclusions in (1) are simple consequences of ergodicity of the action.

When $h_M(\gamma) > 0$ for some Γ, we use relations between the entropy of the action and the Gromov representation discovered by Zimmer [257]. We discuss some of these ideas below in Section 8.

6.3. Affine Actions Close to the Critical Dimension

In this section we briefly describe a direction pursued by Feres, Goetze, Zeghib, and Zimmer that classify connection preserving actions of lattices in dimensions close to, but not less than, the critical dimension d. A representative result is the following:

THEOREM 6.9. *Let $n > 2$, $G = SL(n, \mathbb{R})$ and $\Gamma < G$ a lattice. Let M be a compact manifold with $\dim(M) \leq n + 1 = d(G) + 1$. Assume that Γ acts on M preserving a volume form and an affine connection. Then either*

1) $\dim(M) < n$ and the action is isometric,

2) $\dim(M) = n$ and either the action is isometric or, upon passing to finite covers and finite index subgroups, smoothly conjugate to the standard $SL(n, \mathbb{Z})$ action on $\mathbb{T}^n$, or

3) $\dim(M) = n + 1$ and either the action is isometric or, upon passing to finite covers and finite index subgroups, smoothly conjugate to the action of $SL(n, \mathbb{Z})$ on $\mathbb{T}^{n+1} = \mathbb{T}^n \times \mathbb{T}$ where the action on the second factor is trivial.

Variants of this theorem under more restrictive hypotheses were obtained by Feres [63], Goetze [104], and Zimmer [248]. The theorem as stated is due to Zeghib [243]. The cocycle superrigidity theorem is used in all of the proofs, mainly to force vanishing of curvature and torsion tensors of the associated connection.

One would expect similar results for a more general class of acting groups and also for a wider range of dimensions. Namely, if we let $d_2(G)$ be the dimension of the second non-trivial representation of G, we would expect a similar result to (3) for any affine, volume preserving action on a manifold M with $d(G) < d_2(G)$. With the further assumption that the connection is Riemannian (and still only for lattices in $SL(n, \mathbb{R})$) this is proven by Zeghib in [242].

Further related results are contained in other papers of Feres [60, 64]. No results of this kind are known for $\dim(M) \geq d_2(G)$. A major difficulty arises as soon as one has $\dim(M) \geq d_2(G)$, namely that more complicated examples can arise, including affine actions on nilmanifolds and left translation actions on spaces of the form G/Λ. All the results mentioned in this subsection depend on the particular fact that flat Riemannian manifolds have finite covers that are tori.

6.4. Zimmer's Borel Density Theorem and Generalizations

An obvious first question concerning G actions is the structure of the stabilizer subgroups. In this direction we have

THEOREM 6.10. *Let G be a simple Lie group acting essentially faithfully on a compact manifold M preserving a volume form, then the stabilizer of almost every point is discrete.*

As an application of this result, one can prove the following:

THEOREM 6.11. *Let G be a simple Lie group acting essentially faithfully on a compact manifold, preserving a volume form and a homogeneous geometric structure with structure group H. Then there is an inclusion $\mathfrak{g} \rightarrow \mathfrak{h}$.*

In particular, the theorem provides an obstruction to a higher rank simple Lie group having a volume preserving action on a compact Lorentz manifold. For more discussion of these theorems, see [258]. This observation led to a major series of works studying automorphism groups of Lorentz manifolds; see, for example, [1, 141, 240, 241]. We remark here that Theorem 6.11 does not require that the homogeneous structure be rigid.

More recently some closely related phenomena have been discovered in a joint work of Bader, Frances, and Melnick [5]. Their work uses yet another notion of a geometric structure, that of a *Cartan geometry.* We will not define this notion rigorously here; it suffices to note that any rigid homogeneous geometric structure defines a Cartan geometry, as discussed in [5, introduction]. The converse is not completely clear in general, but most classical examples of rigid geometric structures can also be realized as Cartan geometries. A Cartan geometry is essentially a way of saying that a manifold is *infinitesimally* modeled on some homogeneous space G/P. To recapture the notion of a Riemannian connection, $G = O(n) \ltimes \mathbb{R}^n$ and $P = O(n)$, to recapture the notion of an affine connection $G = Gl(n, \mathbb{R}) \ltimes \mathbb{R}^n$ and $P = Gl(n, \mathbb{R})$. Cartan connections come with naturally defined curvatures that vanish if and only if the manifold is *locally* modeled on G/P. We refer the reader to the book [217] for a general discussion of Cartan geometries and their use in differential geometry.

Given a connected linear group L, we define its real rank, $rk(L)$ as before to be the dimension of a maximal $\mathbb{R}$-diagonalizable subgroup of L, and let $n(L)$ denote the maximal nilpotence degree of a connected nilpotent subgroup. One of the main results of [5] is

THEOREM 6.12. *If a group L acts by automorphisms of a compact Cartan geometry modeled on G/P, then*

1) $rk(AdL^0) \leq rk(Ad_{\mathfrak{g}} P)$,
2) $n(AdL^0) \leq n(Ad_{\mathfrak{g}} P)$.

The main point is that L is not assumed either simple or connected. This theorem is deduced from an embedding theorem similar in flavor to Theorem 6.10.

In addition, when G is a simple group and P is a maximal parabolic subgroup, Bader, Frances, and Melnick prove rigidity results classifying all possible actions when the rank bound in Theorem 6.12 is achieved. This classification essentially says that all examples are algebraic; see [5] for a detailed discussion.

An earlier paper by Feres and Lampe also explored applications of Cartan geometries to rigidity and dynamical conditions for flatness of Cartan geometries [62].

6.5. Actions Preserving a Complex Structure

We mention here one recent result by Cantat and a few active related directions of research. In [250], Zimmer asked whether one had restrictions on low dimensional holomorphic actions of lattices. The answer was obtained in [36] and is

THEOREM 6.13. *Let G be a connected simple Lie group of real rank at least 2 and suppose $\Gamma < G$ is a lattice. If Γ admits a faithful action by automorphisms on a compact Kähler manifold M, then the rank of G is at least the complex dimension of M.*

The proof of the theorem depends primarily on results concerning Aut (M), particularly recent results on holomorphic actions of abelian groups by Dinh and Sibony [47]. It is worth noting that while the theorem does depend on Margulis's superrigidity theorem, it does not depend on Zimmer's cocycle superrigidity theorem. The condition that the action is faithful can be weakened considerably using Margulis's normal subgroups theorem.

More recently, Cantat [35], Déserti [46], and others have begun a program of studying large subgroups of automorphism groups of complex manifolds. In particular, Cantat has proven an analogue of Conjecture 4.7 in the context of birational actions on complex surfaces.

THEOREM 6.14. *Let S be a compact Kähler surface and G an infinite, countable group of birational transformations of S. If G has property (T), then there is a birational map $j : S \to \mathbb{P}^2(\mathbb{C})$ that conjugates G to a subgroup of Aut $(\mathbb{P}^2(\mathbb{C}))$.*

6.6. Conjectures and Questions

A major open problem, generalizing Conjecture 1.2, is the following:

CONJECTURE 6.15. (GROMOV-ZIMMER) *Let D be a semisimple Lie group with all factors having property (T) or a lattice in such a Lie group. Then any D action on a compact manifold preserving a rigid geometric structure and a volume form is generalized quasi-affine.*

A word is required on the attribution. Both Gromov [112] and Zimmer [249] made various less precise conjectures concerning the classification of actions as in the conjecture. The exact statement of the correct conjecture was muddy for several years while it was not known if the Katok-Lewis and Benveniste type examples admitted invariant rigid geometric structures. See Section 9 for more discussion of these examples. In the context of the results of [13], where Benveniste and the author prove that those actions do not preserve rigid geometric structures, this version of the classification seems quite plausible. It is perhaps more plausible if one assumes the action is ergodic or that the geometric structure is homogeneous.

A possibly easier question that is relevant is the following:

QUESTION 6.16. *Let M be a compact manifold equipped with a homogeneous rigid geometric structure ω. Assume* Aut (M, ω) *is ergodic or has a dense orbit. Is M locally homogeneous?*

Gromov's theorem on the open-dense implies that a dense open set in M is locally homogeneous even if ω is not homogeneous. The question is whether this homogeneous structure extends to all of M if ω is homogeneous. If ω is not homogeneous, the examples following Lemma 6.5 show that M need not be homogeneous.

It seems possible to approach the case of Conjecture 6.15 where ω is homogeneous and D acts ergodically by answering Question 6.16 positively. At this point, one is left with the problem of classifying actions of higher rank groups and lattices on locally homogeneous manifolds. Induction easily reduces one to considering G actions. The techniques Gromov uses to produce the locally homogeneous structure gives slightly more precise information in this setting: one is left trying to classify homogeneous manifolds modeled on H/L where G acts via an inclusion in H centralizing L. In fact this reduces one to problems about locally homogeneous manifolds studied by Zimmer and collaborators; see particularly [142, 143, 145, 256]. For a more general survey of locally homogeneous spaces, see also [136] and [137].

The following question concerns a possible connection between preserving a rigid geometric structure and having uniformly partially hyperbolic dynamics.

QUESTION 6.17. *Let D be any group acting on a compact manifold M preserving a volume form and a rigid geometric structure. Is it true that if some element d of D has positive Lyapunov exponents, then d is uniformly partially hyperbolic?*

This question is of particular interest for us when D is a semisimple Lie group or a lattice, but would be interesting to resolve in general. The main reason to believe that the answer might be yes is that the action of D on the space of frames of M is proper. Having a proper action on a tangent bundle minus the zero section implies that an action is Anosov; see Mañé's article for a proof [156].

7. Topological Super-rigidity: Regularity Results from Hyperbolic Dynamics and Applications

A major area of research in the Zimmer program has been the application of hyperbolic dynamics. This area might be described by the maxim: in the presence of hyperbolic dynamics, the straightening section from cocycle super-rigidity is often more regular. Some major successes in this direction are the work of Goetze-Spatzier and Feres-Labourie. In the context of hyperbolic dynamical approaches, there are two main settings. In the first, one considers actions of higher rank lattices on a particular class of compact manifolds and a second in which one makes no assumption on the topology of the manifold acted upon.

In this section we discuss a few such results, after first recalling some facts about the stability of hyperbolic dynamical systems in Subsection 7.1. We then discuss best-known rigidity results for actions on tori in Subsection 7.2 and after this we discuss the work of Goetze-Spatzier and Feres-Labourie in the more general context in Subsection 7.3.

The term "topological superrigidity" for this area of research was coined by Zimmer, whose early unpublished notes on the topic dramatically influenced research in the area [244].

This aspect of the Zimmer program has also given rise to a study of rigidity properties for other uniformly hyperbolic actions of large groups, most particularly higher rank abelian groups. See, for example, [126, 130, 209].

The following subsection recalls basic notions from hyperbolic dynamics that are needed in this section.

7.1. Stability in Hyperbolic Dynamics

A diffeomorphism f of a manifold X is said to be *Anosov* if there exists a continuous f invariant splitting of the tangent bundle $TX = E_f^u \oplus E_f^s$ and constants $a > 1$ and $C, C' > 0$ such that for every $x \in X$,

1) $\|Df^n(v^u)\| \geq Ca^n \|v^u\|$ for all $v^u \in E_f^u(x)$, and
2) $\|Df^n(v^s)\| \leq C'a^{-n} \|v^s\|$ for all $v^s \in E_f^s(x)$.

Note that the constants C and C' depend on the choice of metric, and that a metric can always be chosen so that $C = C' = 1$. There is an analogous notion for a flow f_t, where $TX = T\mathcal{O} \oplus E^u_{f_t} \oplus E^s_{f_t}$ where $T\mathcal{O}$ is the tangent space to the flow direction and vectors in $E^u_{f_t}$ (respectively, $E^s_{f_t}$) are uniformly expanded (respectively, uniformly contracted) by the flow.

This notion was introduced by Anosov [3] and named after Anosov by Smale [221], who popularized the notion in the United States. One of the earliest results in the subject is Anosov's proof that Anosov diffeomorphisms are *structurally stable*, that is, that any C^1 perturbation of an Anosov diffeomorphism is conjugate back to the original diffeomorphism by a homeomorphism. There is an analogous result for flows, though this requires that one introduce a notion of time change that we will not consider here. Since Anosov also showed that C^2 volume preserving Anosov flows and diffeomorphisms are ergodic, structural stability implies the existence of an open set of "chaotic" dynamical systems.

The notion of an Anosov diffeomorphism has had many interesting generalizations, for example: Axiom A diffeomorphisms, non-uniformly hyperbolic diffeomorphisms, and diffeomorphisms admitting a dominated splitting. A notion that has been particularly useful in the study of rigidity of group actions is the notion of a partially hyperbolic diffeomorphism as introduced by Hirsch, Pugh, and Shub [177]. Under strong enough hypotheses, these diffeomorphisms have a weaker stability property similar to structural stability. More or less, the diffeomorphisms are hyperbolic relative to some foliation, and any nearby action is hyperbolic to some nearby foliation. To describe more precisely the class of diffeomorphisms considered and the stability property they enjoy, we require some definitions.

The use of the word "foliation" varies with context. Here a *foliation by C^k leaves* will be a continuous foliation whose leaves are C^k injectively immersed submanifolds that vary continuously in the C^k topology in the transverse direction. To specify transverse regularity we will say that a foliation is transversely C^r. A foliation by C^k leaves that is tranversely C^k is called simply a C^k foliation. (Note our language does not agree with that in the reference [118].)

Given an automorphism f of a vector bundle $E \to X$ and constants $a > b \geq 1$, we say f is (a, b)-*partially hyperbolic* or simply *partially hyperbolic* if there is a metric on E, a constant and $C \geq 1$, and a continuous f invariant, nontrivial splitting $E = E^u_f \oplus E^c_f \oplus E^s_f$ such that for every x in X:

1) $\|f^n(v^u)\| \geq Ca^n \|v^u\|$ for all $v^u \in E^u_f(x)$,
2) $\|f^n(v^s)\| \leq C^{-1}a^{-n}\|v^s\|$ for all $v^s \in E^s_f(x)$, and
3) $C^{-1}b^{-n}\|v^0\| < \|f^n(v^0)\| \leq Cb^n\|v^0\|$ for all $v^0 \in E^c_f(x)$ and all integers n.

A C^1 diffeomorphism f of a manifold X is (a, b)-*partially hyperbolic* if the derivative action Df is (a, b)-partially hyperbolic on TX. We remark that for any partially hyperbolic diffeomorphism, there always exists an *adapted metric* for which $C = 1$. Note that E_f^c is called the *central distribution* of f, E_f^u is called the *unstable distribution* of f, and E_f^s the *stable distribution* of f.

Integrability of various distributions for partially hyperbolic dynamical systems is the subject of much research. The stable and unstable distributions are always tangent to invariant foliations that we call the stable and unstable foliations and denote by $\mathcal{W}_f^s$ and $\mathcal{W}_f^u$, respectively. If the central distribution is tangent to an f invariant foliation, we call that foliation a *central foliation* and denote it by $\mathcal{W}_f^c$. If there is a unique foliation tangent to the central distribution, we call the central distribution *uniquely integrable*. For smooth distributions unique integrability is a consequence of integrability, but the central distribution is usually not smooth. If the central distribution of an (a, b)-partially hyperbolic diffeomorphism f is tangent to an invariant foliation $\mathcal{W}_f^c$, then we say f is r-*normally hyperbolic* to $\mathcal{W}_f^c$ for any r such that $a > b^r$. This is a special case of the definition of r-normally hyperbolic given in [118].

Before stating a version of one of the main results of [118], we need one more definition. Given a group G, a manifold X, two foliations $\mathfrak{F}$ and $\mathfrak{F}'$ of X, and two actions ρ and ρ' of G on X, such that ρ preserves $\mathfrak{F}$ and ρ' preserves $\mathfrak{F}'$, following [118] we call ρ and ρ' *leaf conjugate* if there is a homeomorphism h of X such that

1) $h(\mathfrak{F}) = \mathfrak{F}'$, and
2) for every leaf $\mathcal{L}$ of $\mathfrak{F}$ and every $g \in G$, we have $h(\rho(g)\mathcal{L}) = \rho'(g)h(\mathcal{L})$.

The map h is then referred to as a *leaf conjugacy* between $(X, \mathfrak{F}, \rho)$ and $(X, \mathfrak{F}', \rho')$. This essentially means that the actions are conjugate modulo the central foliations.

We state a special case of some of the results of Hirsch-Pugh-Shub on perturbations of partially hyperbolic actions of $\mathbb{Z}$; see [118]. There are also analogous definitions and results for flows. As these are less important in the study of rigidity, we do not discuss them here.

THEOREM 7.1. *Let f be an (a, b)-partially hyperbolic C^k diffeomorphism of a compact manifold M that is k-normally hyperbolic to a C^k central foliation $\mathcal{W}_f^c$. Then for any $\delta > 0$, if f' is a C^k diffeomorphism of M that is sufficiently C^1 close to f, we have the following:*

1) f' is (a', b')-partially hyperbolic, where $|a - a'| < \delta$ and $|b - b'| < \delta$, and the splitting $TM = E^u_{f'} \oplus E^c_{f'} \oplus E^s_{f'}$, for f' is C^0 close to the splitting for f,

2) there exist f' invariant foliations by C^k leaves $\mathcal{W}^c_{f'}$, tangent to $E^c_{f'}$, which is close in the natural topology on foliations by C^k leaves to $\mathcal{W}^c_f$,

3) there exists a (nonunique) homeomorphism h of M with $h(\mathcal{W}^c_f) = \mathcal{W}^c_{f'}$, and h is C^k along leaves of $\mathcal{W}^c_f$, furthermore h can be chosen to be C^0 small and C^k small along leaves of $\mathcal{W}^c_f$, and

4) the homeomorphism h is a leaf conjugacy between the actions $(M, \mathcal{W}^c_f, f)$ and $(M, \mathcal{W}^c_{f'}, f')$.

Conclusion (1) is easy and probably older than [118]. One motivation for Theorem 7.1 is to study stability of dynamical properties of partially hyperbolic diffeomorphisms. See the survey [27] by Burns, Pugh, Shub, and Wilkinson for more discussion of that and related issues.

7.2. Uniformly Hyperbolic Actions on Tori

Many works have been written considering local rigidity of actions with some affine, quasi-affine, and generalized quasi-affine actions with hyperbolic behavior. For a discussion of this, we refer to [77]. Here we only discuss results that prove some sort of global rigidity of groups acting on manifolds. The first such results were contained in papers of Katok-Lewis [128] and Katok-Lewis-Zimmer [129]. As these are now special cases of later more general results, we do not discuss them in detail here.

In this section we discuss results that provide only continuous conjugacies to standard actions. This is primarily because these results are less technical and easier to state. In this context, one can improve regularity of the conjugacy given certain technical dynamical hypotheses on certain dynamical foliations.

DEFINITION 7.2. An action of a group Γ on a manifold M is weakly hyperbolic if there exist elements $\gamma_1, \ldots, \gamma_k$ each of which is partially hyperbolic such that the sum of the stable sub-bundles of the γ_i spans the tangent bundle to M at every point, that is, $\sum_i E_{\gamma_i} = TM$.

To discuss the relevant results we need a related topological notion that captures hyperbolicity at the level of fundamental group. This was introduced in [84] by Whyte and the author. If a group Γ acts on a manifold with torsion free nilpotent fundamental group and the action lifts to the universal cover, then the action of Γ on $\pi_1(M)$ gives rise to an action of Γ on the Malcev completion N of $\pi_1(M)$, which is a nilpotent Lie group. This yields a representation of Γ on the Lie algebra $\mathfrak{n}$.

DEFINITION 7.3. We say an action of Γ on a manifold M with nilpotent fundamental group is π_1-hyperbolic if for the resulting Γ representation on $\mathfrak{n}$, we have finitely many elements $\gamma_1, \ldots, \gamma_k$ such that the sum of their eigenspaces with eigenvalue of modulus less than 1 is all of $\mathfrak{n}$.

One can make this definition more general by considering M where $\pi_1(M)$ has a Γ equivariant nilpotent quotient; see [84]. We now discuss results that follow by combining the work of Margulis-Qian [165] with the later work of Schmidt [211] and Fisher-Hitchman [79].

THEOREM 7.4. *Let $M = N/\Lambda$ be a compact nilmanifold and let Γ be a lattice in a semisimple Lie group with property (T). Assume Γ acts on M such that the action lifts to the universal cover and is $\pi_1(M)$ hyperbolic, then the action is continuously semi-conjugate to an affine action.*

This theorem was proven by Margulis and Qian, who noted that if the Γ action contained an Anosov element, then the conjugacy could be taken to be a homeomorphism. In [84], the author and Whyte point out that this theorem extends easily to the case of any compact manifold with torsion free nilpotent fundamental group. In [84], we also discuss extensions to manifolds with fundamental group with a quotient that is nilpotent.

Margulis and Qian asked whether the assumption of π_1 hyperbolicity could be replaced by the assumption that the action on M was weakly hyperbolic. In the case of actions of Kazhdan groups on tori, Schmidt proved that weak hyperbolicity implies π_1 hyperbolicity, yielding:

THEOREM 7.5. *Let $M = \mathbb{T}^n$ be a compact torus and let Γ be a lattice in a semisimple Lie group with property (T). Any weakly hyperbolic Γ action M and that lifts to the universal cover is continuously semi-conjugate to an affine action.*

Remarks:

1) The contribution of Fisher-Hitchman in both theorems is just in extending cocycle superrigidity to a wider class of groups, as discussed above.

2) The assumption that the action lifts to the universal cover of M is often vacuous because of results concerning cohomology of higher rank lattices. In particular, it is vacuous for cocompact lattices in simple Lie groups of real rank at least 3.

It remains an interesting, open question to take this result and prove that the semiconjugacy is always a conjugacy and is also always a smooth diffeomorphism.

7.3. Rigidity Results for Uniformly Hyperbolic Actions

We begin by discussing some work of Goetze and Spatzier. To avoid technicalities we discuss only some of their results. We begin with the following definition:

DEFINITION 7.6. Let $\rho : \mathbb{Z}^k \times M \to M$ be an action and $\gamma_1, \ldots, \gamma_l$ be a collection of elements that generate for $\mathbb{Z}^k$. We call ρ a Cartan action if

1) each $\rho(\gamma_i)$ is an Anosov diffeomorphism,
2) each $\rho(\gamma_i)$ has 1 dimensional strongest stable foliation, and
3) the strongest stable foliations of the $\rho(\gamma_i)$ are pairwise transverse and span the tangent space to the manifold.

It is worth noting that Cartan actions are very special in three ways. First, we assume that a large number of elements in the acting group are Anosov diffeomorphisms; second, we assume that each of these has 1 dimensional strongest stable foliation; and lastly, we assume these 1 dimensional directions span the tangent space. All aspects of these assumptions are used in the following theorem. Reproving it even assuming 2 dimensional strongest stable foliations would require new ideas.

THEOREM 7.7. *Let G be a semisimple Lie group with all simple factors of real rank at least 2 and Γ in G a lattice. Then any volume preserving Cartan action of Γ is smoothly conjugate to an affine action on an infranilmanifold.*

This is slightly different than the statement in Goetze-Spatzier, where they pass to a finite cover and a finite index subgroup. It is not too hard to prove this statement from theirs. The proof spans the two papers [105] and [106]. The first paper [105] proves, in a somewhat more general context, that the π-simple section arising in the cocycle superrigidity theorem is in fact Hölder continuous. The second paper [105] makes use of the resulting Hölder

Riemannian metric in conjunction with ideas arising in other work of Katok and Spatzier to produce a smooth homogeneous structure on the manifold.

The work of Feres-Labourie differs from other work on rigidity of actions with hyperbolic properties in that it does not make any assumptions concerning the existence of invariant measures. Here we state only some consequences of their results, without giving the exact form of cocycle superrigidity that is their main result.

THEOREM 7.8. *Let* Γ *be a lattice in* $SL(n, \mathbb{R})$ *for* $n \geq 3$ *and assume* Γ *acts smoothly on a compact manifold* M *of dimension* n. *Further assume that for the induced action* $N = (G \times M)/\Gamma$ *we have*

1) *every* $\mathbb{R}$-*semisimple 1-parameter subgroup of* G *acts transitively on* N, *and*
2) *some element* g *in* G *is uniformly partially hyperbolic with* $E_s \oplus E_w$ *containing the tangent space to* M *at any point,*

then M *is a torus and the action on* M *is a standard affine action.*

These hypotheses are somewhat technical and essentially ensure that one can apply the topological version of cocycle superrigidity proven in [61]. The proof also uses a deep result of Benoist and Labourie classifying Anosov diffeomorphisms with smooth stable and unstable foliations [11].

The nature of the hypotheses of Theorem 7.8 make an earlier remark clear. Ideally one would only have hypotheses on the Γ action, but here we require hypotheses on the induced action instead. It is not clear how to reformulate these hypotheses on the induced action as hypotheses on the original Γ action.

Another consequence of the work of Feres and Labourie is a criterion for promoting invariance of rigid geometric structures. More precisely they give a criterion for a G action to preserve a rigid geometric structure on a dense open subset of a manifold M provided a certain type of subgroup preserves a rigid geometric structure on M.

7.4. Conjectures and Questions on Uniformly Hyperbolic Actions
We begin with a very general variant of Conjecture 1.3.

CONJECTURE 7.9. *Let* G *be a semisimple Lie group all of whose simple factors have property* (T), *let* $\Gamma < G$ *be a lattice. Assume* G *or* Γ *acts smoothly on a compact manifold* M *preserving volume such that some element* g *in the acting group is nontrivially uniformly partially hyperbolic. Then the action is generalized quasi-affine.*

There are several weaker variants on this conjecture, where, for example, one assumes the action is volume preserving or weakly hyperbolic. Even the following much weaker variant seems difficult:

CONJECTURE 7.10. *Let $M = \mathbb{T}^n$ and Γ as in Conjecture 7.9. Assume Γ acts on $\mathbb{T}^n$ weakly hyperbolicly and preserving a smooth measure. Then the action is affine.*

This conjecture amounts to conjecturing that the semiconjugacy in Theorem 7.5 is a diffeomorphism. In the special case where some element of Γ is Anosov, the semiconjugacy is at least a homeomorphism. If this is true and enough dynamical foliations are 1 and 2 dimensional and Γ has higher rank, one can then deduce smoothness of the conjugacy from the work of Rodriguez-Hertz on rigidity of actions of abelian groups [209]. Work in progress by the author, Kalinin, and Spatzier seems likely to provide a similar result when Γ contains many commuting Anosov diffeomorphisms without any assumptions on dimensions of foliations.

Finally, we recall an intriguing question from [84] that arises in this context.

QUESTION 7.11. *Let M be a compact manifold with $\pi_1(M) = \mathbb{Z}^n$ and assume $\Gamma < SL(n, \mathbb{Z})$ has finite index. Let Γ act on M fixing a point so that the resulting Γ action on $\pi_1(M)$ is given by the standard representation of $SL(n, \mathbb{Z})$ on $\mathbb{Z}^n$. Is it true that $\dim(M) \geq n$?*

The question is open even if the action on M is assumed to be smooth. The results of [84] imply that there is a continuous map from M to $\mathbb{T}^n$ that is equivariant for the standard Γ action on M. Since the image of M is closed and invariant, it is easy to check that it is all of $\mathbb{T}^n$. So the question amounts to one about the existence of equivariant "space filling curves," where curve is taken in the generalized sense of continuous map from a lower-dimensional manifold. In another context there are equivariant space filling curves, but they seem quite special. They arise as surface group equivariant maps from the circle to S^2 and come from three manifolds that fiber over the circle; see [34].

8. Representations of Fundamental Groups and Arithmetic Quotients

In this section we discuss some results and questions related to topological approaches to classifying actions; particularly, some related to Conjecture 1.4. In Subsection 8.2, we also discuss related results and questions concerning maximal generalized affine quotients of actions.

8.1. Linear Images of Fundamental Groups

This section is fundamentally concerned with the question:

QUESTION 8.1. *Let G be a semisimple Lie group all of whose factors are not compact and have property (T). Assume G acts by homeomorphisms on a manifold M preserving a measure. Can we classify linear representations of $\pi_1(M)$? Similarly for actions of $\Gamma < G$ a lattice on a manifold M.*

We remark that in this context, it is possible that $\pi_1(M)$ has no infinite image linear representations; see discussion in [84] and Section 9 below. In all known examples where this occurs there is an "obvious" infinite image linear representation on some finite index subgroup. Also, as first observed by Zimmer [251] for actions of Lie groups, under mild conditions on the action, representations of the fundamental group become severely restricted. Further work in this direction was done by Zimmer in conjunction with Spatzier [224] and later Lubotzky [152, 153]. Analogous results for lattices are surprisingly difficult in this context and constitute the authors dissertation [73, 74].

We recall a definition from [251]:

DEFINITION 8.2. Let D be a Lie group and assume that D acts on a compact manifold M preserving a finite measure μ and ergodically. We call the action *engaging* if the action of $\tilde{D}$ on any finite cover of M is ergodic.

There is a slightly more technical definition of engaging for non-ergodic actions that says there is no loss of ergodicity on passing to finite covers. That is that the ergodic decomposition of μ and its lifts to finite covers are canonically identified by the covering map. There are also two variants of this notion *totally engaging* and *topologically engaging*; see for example, [258] for more discussion.

It is worth noting that for ergodic D actions on M, the action on any finite cover has at most finitely many ergodic components, in fact at most the degree of the cover many ergodic components. We remark here that any generalized affine action of a Lie group is engaging if it is ergodic. The actions constructed below in Section 9 are not in general engaging.

THEOREM 8.3. *Let G be a simple Lie group of real rank at least 2. Assume G acts by homeomorphisms on a compact manifold M, preserving a finite measure μ and engaging. Assume $\sigma : \pi_1(M) \to GL(n, \mathbb{R})$ is an infinite image linear representation. Then $\sigma(\pi_1(M))$ contains an arithmetic group $H_{\mathbb{Z}}$ where $H_{\mathbb{R}}$ contains a group locally isomorphic to G.*

For an expository account of the proof of this theorem and a more detailed discussion of engaging conditions for Lie groups, we refer the reader to [258].

The extension of this theorem to lattice actions is non-trivial and is in fact the author's dissertation. Even the definition of engaging requires modification, since it is not at all clear that a discrete group action lifts to the universal cover.

DEFINITION 8.4. Let D be a discrete group and assume that D acts on a compact manifold M preserving a finite measure μ and ergodically. We call the action *engaging* if for every

1) finite index subgroup D' in D,
2) finite cover M' of M, and
3) lift of the D' action to M',

the action of D' of M' is ergodic.

The definition does immediately imply that every finite index subgroup D' of D acts ergodically on M. In [74], a definition is given that does not require the D action to be ergodic, but even in that context the ergodic decomposition for D' is assumed to be the same as that for D. Definition 8.4 is rigged to guarantee the following lemma:

LEMMA 8.5. *Let G be a Lie group and $\Gamma < G$ a lattice. Assume Γ acts on a manifold M preserving a finite measure and engaging, then the induced G action on $(G \times M)/\Gamma$ is engaging.*

We remark that with our definitions here, the lemma only makes sense for Γ cocompact, but this is not an essential difficulty. We can now state a first result for lattice actions. Let $\Lambda = \pi_1((G \times M)/\Gamma)$.

THEOREM 8.6. *Let G be a simple Lie group of real rank at least 2 and $\Gamma < G$ a lattice. Assume Γ acts by homeomorphisms on a compact manifold M, preserving a finite measure μ and engaging. Assume $\sigma : \Lambda \rightarrow GL(n, \mathbb{R})$ is a linear representation whose restriction to $\pi_1(M)$ has infinite image. Then $\sigma(\pi_1(M))$ contains an arithmetic group $H_{\mathbb{Z}}$ where $\mathrm{Aut}(H_{\mathbb{R}})$ contains a group locally isomorphic to G.*

This theorem is proven by inducing actions, applying Theorem 8.3, and analyzing the resulting output carefully. One would like to assume σ a priori only defined on $\pi_1(M)$, but there seems no obvious way to extend such a representation to the linear representation of Λ required by Theorem 8.3 without a priori information on Λ. This is yet another example of the difficulties in using induction to study lattice actions. As a consequence of Theorem 8.6, we discover that at least $\sigma(\Lambda)$ splits as a semidirect product of Γ and $\sigma(\pi_1(M))$. But this is not clear a priori and not clear a posteriori for Λ.

8.2. Arithmetic Quotients

In the context of Theorems 8.3 and 8.6, one can obtain much greater dynamical information concerning the relation of $H_{\mathbb{Z}}$ and the dynamics of the action on M. In particular, there is a compact subgroup $C < H_{\mathbb{R}}$ and a measurable equivariant map $\phi : M \to C\backslash H_{\mathbb{R}}/H_{\mathbb{Z}}$, which we refer to as a *measurable arithmetic quotient*. The papers [73] and [152] prove that there is always a canonical maximal quotient of this kind for any action of G or Γ on any compact manifold, essentially by using Ratner's theorem to prove that every pair of arithmetic quotients are dominated by a common, larger arithmetic quotient. Earlier results of Zimmer also obtained arithmetic quotients, but only under the assumption that there was an infinite image linear representation with discrete image [254]. The results we mention from [74] and [153] then show that there are "lower bounds" on the size of this arithmetic quotient, provided that the action is engaging in terms of the linear representations of the fundamental group of the manifold. In particular, one obtains arithmetic quotients where $H_{\mathbb{R}}$ and $H_{\mathbb{Z}}$ are essentially determined by $\sigma(\pi_1(M))$. In fact, $\sigma(\pi_1(M))$ contains $H_{\mathbb{Z}}$ and is contained in $H_{\mathbb{Q}}$.

The book [258] provides a good description of how to produce arithmetic quotients for G actions and the article [75] provides an exposition of the relevant constructions for Γ actions.

Under even stronger hypotheses, the papers [85] and [257] imply that the arithmetic quotient related to the "Gromov representation" discussed above in Subsection 6.2 has the same entropy as the original action. This means that, in a sense, the arithmetic quotient captures most of the dynamics of the original action.

8.3. Open Questions

As promised in the introduction, the analogue of Conjecture 1.4 for lattice actions follows.

CONJECTURE 8.7. *Let Γ be a lattice in a semisimple group G with all simple factors of real rank at least 2. Assume Γ acts faithfully, preserving volume on a compact manifold M. Further assume the action is not isometric. Then $\pi_1(M)$ has* *a finite index subgroup Λ such that Λ surjects onto an arithmetic lattice in a Lie group H where* Aut (H) *locally contains G.*

The following questions about arithmetic quotients are natural.

QUESTION 8.8. *Let G be a simple Lie group of real rank at least 2 and Γ < G a lattice. Assume that G or Γ act faithfully on a compact manifold M, preserving a smooth volume.*

1) *Is there a non-trivial measurable arithmetic quotient?*
2) *Can we take the quotient map φ smooth on an open dense set?*

Due to a construction in [84], one cannot expect that every M admitting a volume preserving G or Γ action has an arithmetic quotient that is even globally continuous or has $\pi_1(M)$ admitting an infinite image linear representation. However, the difficulties created in those examples all vanish on passage to a finite cover.

QUESTION 8.9. *Let G be a simple Lie group of real rank at least 2 and Γ < G a lattice. Assume that G or Γ act smoothly on a compact manifold M, preserving a smooth volume.*

1) *Is there a finite cover of M' of M such that $\pi_1(M')$ admits an infinite image linear representation?*
2) *Can we find a finite cover M' of M, a lift of the action to M' (on a subgroup of finite index) and an arithmetic quotient where the quotient map φ is continuous and smooth on an open dense set?*

The examples discussed in the next section imply that φ is at best Hölder continuous globally. It is not clear whether the questions and conjectures we have just discussed are any less reasonable for $SP(1, n)$, F_4^{-20} and their lattices.

9. Exotic Volume Preserving Actions

In this section, we discuss what is known about what are typically called "exotic actions." These are the only known smooth volume preserving actions of higher rank lattices and Lie groups that are not generalized affine algebraic. These actions make it clear that a clean classification of volume preserving actions is out of reach. In particular, these actions have continuous moduli and provide counterexamples to any naive conjectures of the form "the moduli

space of actions of some lattice Γ on any compact manifold M are countable."
In particular, I explain examples of actions of either Γ or G that have large
continuous moduli of deformations as well as manifolds where these moduli
have multiple connected components.

Essentially all of the examples given here derive from the simple construction of "blowing up" a point or a closed orbit, which was introduced to this
subject in [128]. The further developments after that result are all further
elaborations on one basic construction. The idea is to use the "blow up" construction to introduce distinguished closed invariant sets that can be varied
in some manner to produce deformations of the action. The "blow up" construction is a classical tool from algebraic geometry that takes a manifold N
and a point p and constructs from it a new manifold N' by replacing p by the
space of directions at p. Let $\mathbb{R}P^l$ be the l dimensional projective space. To blow
up a point, we take the product of $N \times \mathbb{R}P^{\dim(N)}$ and then find a submanifold
where the projection to N is a diffeomorphism off of p and the fiber of the
projection over p is $\mathbb{R}P^{\dim(N)}$. For detailed discussion of this construction we
refer the reader to any reasonable book on algebraic geometry.

The easiest example to consider is to take the action of $SL(n, \mathbb{Z})$, or any subgroup $\Gamma < SL(n, \mathbb{Z})$ on the torus $\mathbb{T}^n$ and blow up the fixed point, in this case the
equivalence class of the origin in $\mathbb{R}^n$. Call the resulting manifold M. Provided Γ
is large enough, for example, Zariski dense in $SL(n, \mathbb{R})$, this action of Γ does not
preserve the measure defined by any volume form on M. A clever construction
introduced in [128] shows that one can alter the standard blowing up procedure
in order to produce a one parameter family of $SL(n, \mathbb{Z})$ actions on M, only one
of which preserves a volume form. This immediately shows that this action
on M admits perturbations, since it cannot be conjugate to the nearby, non-
volume preserving actions. Essentially, one constructs different differentiable
structures on M that are diffeomorphic but not equivariantly diffeomorphic.

After noticing this construction, one can proceed to build more complicated examples by passing to a subgroup of finite index, and then blowing up
several fixed points. One can also glue together the "blown up" fixed points
to obtain an action on a manifold with more complicated topology. In particular, one can achieve a fundamental group that is an essentially arbitrary
free product with amalgamation or HNN extension of the fundamental group
of the original manifold over the fundamental group of (blown-up) orbits. In
the context described in more detail below, of blowing up along closed orbits
instead of points, it is not hard to do the blowing up and gluing in a way that
guarantees that there are no linear representations of the fundamental group
of the "exotic example." To prove nonexistence of linear representations, one

chooses examples where all groups involved are higher rank lattices and have very constrained linear representation theory. See [84, 128] for a discussion of the topological complications one can introduce.

While these actions do not preserve a rigid geometric structure, they do preserve a slightly more general object, an *almost* rigid structure introduced by Benveniste and the author in [13] and described below.

In [12] it is observed that a similar construction can be used for the action of a simple group G by left translations on a homogeneous space H/Λ where H is a Lie group containing G and $\Lambda < H$ is a cocompact lattice. Here we use a slightly more involved construction from algebraic geometry, and "blow up" the directions normal to a closed submanifold. That is, we replace some closed submanifold N in H/Λ by the projective normal bundle to N. In all cases we consider here, this normal bundle is trivial and so is just $N \times \mathbb{R}P^l$ where $l = \dim(H) - \dim(N)$.

Benveniste used his construction to produce more interesting perturbations of actions of higher rank simple Lie group G or a lattice Γ in G. In particular, he produced volume preserving actions that admit volume preserving perturbations. He does this by choosing $G < H$ such that not only are there closed G orbits but so that the centralizer $Z = Z_H(G)$ of G in H has nontrivial connected component. If we take a closed G orbit N, then any translate zN for z in Z is also closed and so we have a continuum of closed G orbits. Benveniste shows that if we choose two closed orbits N and zN to blow up and glue, and then vary z in a small open set, the resulting actions can only be conjugate for a countable set of choices of z.

This construction is further elaborated in [78]. Benveniste's construction is not optimal in several senses, nor is his proof of rigidity. In [78], I give a modification of the construction that produces non-conjugate actions for every choice of z in a small enough neighborhood. By blowing up and gluing more pairs of closed orbits, this allows me to produce actions where the space of deformations contains a submanifold of arbitrarily high, finite dimension. Further, Benveniste's proof that the deformation are nontrivial is quite involved and applies only to higher rank groups. In [78], I give a different proof of nontriviality of the deformations using consequences of Ratner's theorem [204] due to Shah [213] and Witte Morris [237]. This shows that the construction produces nontrivial perturbations for any semisimple G and any lattice Γ in G.

As mentioned above, in [13] we show that none of these actions preserve any rigid geometric structure in the sense of Gromov but that they do preserve a slightly more complicated object that we call an *almost rigid structure*. Both rigid and almost rigid structures are easiest to define in dimension 1. In this context a rigid structure is a non-vanishing vector field and an almost rigid structure is a vector field vanishing at isolated points and to finite degree.

We continue to use the notation of subsection 6.1 in order to give the precise definition of almost rigid geometric structure.

DEFINITION 9.1. An A-structure ϕ is called (j, k)-almost rigid (or just almost rigid) if for every point p, $r_p^{k,k-1}$ is injective on the subgroup $r^{k+j,k}(Is^{k+j}) \subset Is^k$.

Thus k-rigid structures are the $(0, k)$-almost rigid structures.

BASIC EXAMPLE. Let V be an n-dimensional manifold. Let $X_1, \ldots, X_n$ be a collection of vector fields on M. This defines an A-structure ψ of type $\mathbb{R}^{n^2}$ on M. If $X_1, \ldots, X_n$ span the tangent space of V at every point, then the structure is rigid in the sense of Gromov. Suppose instead that there exists a point p in V and $X_1 \wedge \ldots \wedge X_n$ vanishes to order $\leq j$ at p in V. Then ψ is a $(j, 1)$-almost rigid structure. Indeed, let $p \in M$, and let $(x_1, \ldots, x_n)$ be coordinates around p. Suppose that in terms of these coordinates, $X_l = a_l^m \frac{\partial}{\partial x_j}$. Suppose that $f \in Is_p^{j+1}$. We must show that $r_p^{j+1,1}(f)$ is trivial. Let $(f^1, \ldots, f^n)$ be the coordinate functions of f. Then $f \in Is_p^{j+1}$ implies that

9.1
$$a_k^l - a_k^m \frac{\partial f^l}{\partial x^m}$$

vanishes to order $j + 1$ at p for all k and l. Let (b_k^l) be the matrix so that $b_k^m a_m^l = \det(a_r^s)\delta_k^l$. Multiplying expression (9.1) by (b_k^l), we see that $\det(a_r^s)\left(\delta_k^l - \frac{\partial f^l}{\partial x^k}\right)$ vanishes to order $j + 1$. But since by assumption $\det(a_r^s)$ vanishes to order $\leq j$, this implies that $(\partial f^l/\partial x^k)(p) = \delta_k^l$, so $r_p^{j+1,1}(f)$ is the identity, as required.

If confused by the notation, the interested reader may find it enlightening to work out the basic example in the trivial case $n = 1$. Similar arguments can be given to show that frames that degenerate to subframes are also almost rigid, provided the order of vanishing of the form defining the frame is always finite.

QUESTION 9.2. *Does any smooth (or analytic) action of a higher rank lattice Γ admit a smooth (analytic) almost rigid structure in the sense of [13]? More generally does such an action admit a smooth (analytic) rigid geometric structure on an open dense set of full measure?*

This question is, in a sense, related to the discussion above about regularity of the straightening section in cocycle superrigidity. In essence, cocycle superrigidity provides one with a measurable invariant connection and what one wants to know is whether one can improve the measurable connection to a smooth geometric structure with some degeneracy on a small set. The examples described in this section show, among other things, that one cannot expect the straightening section to be smooth in general, though one might hope it is smooth in the complement of a closed submanifold of positive codimension or at least on an open dense set of full measure.

We remark that there are other possible notions of almost rigid structures. See the article in this volume by Dumitrescu for a detailed discussion of a different useful notion in the context of complex analytic manifolds [48]. Dumitrescu's notion is strictly weaker than the one presented here.

10. Non-Volume Preserving Actions

This section describes what is known for non-volume preserving actions. Subsections 10.1 and 10.2 describe examples which show that a classification in this setting is not in any sense possible. Subsection 10.3 describes some recent work of Nevo and Zimmer that proves surprisingly strong rigidity results in special settings.

10.1. Stuck's Examples

The following observations are from Stuck's paper [226]. Let G be any semisimple group. Let P be a minimal parabolic subgroup. Then there is a homomorphism $P \to \mathbb{R}$. As in the proof of Theorem 2.7, one can take any $\mathbb{R}$ action, view it is a P action, and induce to a G action. If we take an $\mathbb{R}$ action on a manifold M, then the induced action takes place on $(G \times M)/P$. We remark that the G action here is not volume preserving, simply because the G action on G/P is proximal. The same is true of the restriction to any Γ action when Γ is a lattice in G.

This construction shows that classifying G actions on all compact manifolds implicitly involves classifying all vector fields on compact manifolds. It is relatively easy to convince oneself that there is no reasonable sense in which the moduli space of vector fields can be understood up to smooth conjugacy.

10.2. Weinberger's Examples

This is a variant on the Katok-Lewis examples obtained by blowing up a point and is similar to a blowing up construction common in foliation theory. The idea is that one takes an action of a subgroup Γ of $SL(n, \mathbb{Z})$ on $M = \mathbb{T}^n$, removes a fixed or periodic point p, retracts onto a manifold with boundary $\bar{M}$, and then glues in a copy of the $\mathbb{R}^n$ compactified at infinity by the projective space of rays. It is relatively easy to check that the resulting space admits a continuous Γ action and even that there are many invariant measures for the action, but no invariant volume. One can also modify this construction by doing the same construction at multiple fixed or periodic points simultaneously and by doing more complicated gluings on the resulting $\bar{M}$.

This construction is discussed in [55] and a variant for abelian group actions is discussed in [125]. In the abelian case it is possible to smooth the action, but this does not seem to be the case for actions of higher rank lattices.

As far as I can tell, there is no obstruction to repeating this construction for closed orbits as in the case of the algebro-geometric blow up, but this does not seem to be written formally anywhere in the literature.

Also, a recent construction of Hurder shows that one can iterate this construction infinitely many times, taking retracts in smaller and smaller neighborhoods of periodic points of higher and higher orders. The resulting object is a kind of fractal admitting an $SL(n, \mathbb{Z})$ action [119].

10.3. Work of Nevo-Zimmer

In this subsection, we describe some work of Nevo and Zimmer from the sequence of papers [187, 188, 191, 186]. For more detailed discussion see the survey by Nevo and Zimmer [189] as well as the following two articles for related results [190, 192].

Given a group G acting on a space X and a measure μ on G, we call a measure ν on X *stationary* if $\mu * \nu = \nu$. We will only consider the case where the group generated by the support μ is G; such measures are often called *admissible*. This is a natural generalization of the notion of an invariant measure. If G is an amenable group, any action of G on a compact metric space admits an invariant measure. If G is not amenable, invariant measures need not exist, but stationary measures always do. We begin with the following cautionary example:

EXAMPLE 10.1. Let $G = SL(n, \mathbb{R})$ acting on $\mathbb{R}^{n+1}$ by the standard linear action on the first n coordinates and the trivial action on the last. Then the corresponding G action on $\mathbb{P}(\mathbb{R}^{n+1})$ has the property that any stationary measure is supported either on the subspace $\mathbb{P}(\mathbb{R}^n)$ given by the first n coordinates or on the subspace $\mathbb{P}(\mathbb{R})$ given by zeroing the first n coordinates.

The proof of this assertion is an easy exercise. The set $\mathbb{P}(\mathbb{R})$ is a collection of fixed points, so clearly admits invariant measures. The orbit of any point in $\mathbb{P}(\mathbb{R}^{n+1})\backslash(\mathbb{P}(\mathbb{R}^n) \bigsqcup \mathbb{P}(\mathbb{R}))$ is $SL(n, \mathbb{R})/(SL(n, \mathbb{R}) \ltimes \mathbb{R}^n)$. It is straightforward to check that no stationary measures can be supported on unions of sets of this kind. This fact should not be a surprise as it generalizes the fact that invariant measures for amenable group actions are often supported on minimal sets.

To state the results of Nevo and Zimmer, we need a slightly stronger notion of admissibility. We say a measure μ on a locally compact group G is *strongly admissible* if the support of μ generates G and μ^{*k} is absolutely continuous with respect to Haar measure on G for some positive k. In the papers of Nevo and Zimmer, this stronger notion is called admissible.

THEOREM 10.2. *Let X be a compact G space where G is a semisimple Lie group with all factors of real rank at least 2. Then for any admissible measure μ on G and any μ-stationary measure ν on X, we have either*

1) *ν is G invariant, or*
2) *there is a nontrivial measurable quotient of the G space (X, ν) that is of the form $(G/Q, Lebesgue)$ where $Q \subset G$ is a parabolic subgroup.*

The quotient space G/Q is called a *projective quotient* in the work of Nevo and Zimmer. Theorem 10.2 is most interesting for us for minimal actions, where the measure ν is necessarily supported on all of X and the quotient therefore reflects the Γ action on X and not some smaller set. Example 10.1 indicates the reason to be concerned, since there the action on the larger projective space is not detected by any stationary measure.

We remark here that Feres and Ronshausen have introduced some interesting ideas for studying group actions on sets not contained in the support of any stationary measure [69]. Similar ideas are developed in a somewhat different context by Deroin, Kleptsyn, and Navas [45].

In [186], Nevo and Zimmer show that for smooth non-measure preserving actions on compact manifolds that also preserve a rigid geometric structure, one can sometimes prove the existence of a projective factor where the factor map is smooth on an open dense set.

We also want to mention the main result of Stuck's paper [226], which we have already cited repeatedly for the fact that non-volume preserving actions of higher rank groups on compact manifolds cannot be classified.

THEOREM 10.3. *Let G be a semisimple Lie group with finite center. Assume G acts minimally by homeomorphisms on a compact manifold M. Then either the action* *is locally free or the action is induced from a minimal action by homeomorphisms of a proper parabolic subgroup of G on a manifold N.*

Stuck's theorem is proven by studying the Gauss map from M to the Grassman variety of subspaces of $\mathfrak{g}$ defined by taking a point to the Lie algebra of its stabilizer. This technique also plays an important role in the work of Nevo and Zimmer.

11. Groups That Do Act on Manifolds

Much of the work discussed so far begs an obvious question: "are there many interesting subgroups of Diff (M) for a general M?" So far we have only seen "large" subgroups of Diff (M) that arise in a geometric fashion, from the presence of a connected Lie group in either Diff (M) or Diff $(\tilde{M})$. In this section, we describe two classes of examples that make it clear that other phenomena exist. The following problem, however, seems open:

PROBLEM 11.1. *For a compact manifold M with a volume form ω construct a subgroup $\Gamma <$ Diff (M, ω) such that Γ has no linear representations.*

An example that is not often considered in this context is that Aut (F_n) acts on the space Hom (F_n, K) for K any compact Lie group. Since Hom $(F_n, K) \simeq K^n$ this defines an action of Aut (F_n) on a manifold. This action clearly preserves the Haar measure on K^n; see [107]. This action is not very well studied; we only know of [76, 97]. Similar constructions are possible and better known with mapping class groups, although in that case one obtains a representation variety that is not usually a manifold. Since Aut (F_n) has no faithful linear representations, this yields an example of truly "non-linear" action of a large group. This action is still very special and one expects many other examples of non-linear actions. We now describe two constructions that yield many examples if we drop the assumption that the action preserves volume.

11.1. Thompson's Groups

Richard Thompson introduced a remarkable family of groups, now referred to as Thompson's groups. These come in various flavors and have been studied from several points of view; see, for example, [33, 117]. For our purposes, the most important of these groups are the one's typically denoted T. One description of this group is the collection of piecewise linear diffeomorphisms of the circle where the break points and slopes are all dyadic rationals. (One can replace the implicit 2 here with other primes and obtain similar, but different, groups.) We record here two important facts about this group T of piecewise linear homeomorphisms of S^1.

THEOREM 11.2. (THOMPSON, SEF [33]) *The group T is simple.*

THEOREM 11.3. (GHYS-SERGIESCU [103]) *The defining piecewise linear action of the group T on S^1 is conjugate by a homeomorphism to a smooth action.*

These two facts together provide us with a rather remarkable class of examples of groups that act on manifolds. As finitely generated linear groups are always residually finite, the group T has no linear representations whatsoever. A simpler variant of Problem 11.1 is

PROBLEM 11.4. *Does T admit a volume preserving action on a compact manifold?*

It is easy to see that compactness is essential in this question. We can construct a smooth action of T on $S^1 \times \mathbb{R}$ simply by taking the Ghys-Sergiescu action on S^1 and acting on $\mathbb{R}$ by the inverse of the derivative cocycle. Replacing the derivative cocycle with the Jacobian cocycle, this procedure quite generally converts non-volume preserving actions on compact manifolds to volume preserving one's on non-compact manifolds, but we know of no real application of this in the present context. Another variant of Problem 11.4 is

PROBLEM 11.5. *Given a compact manifold M and a volume form ω does Diff (M, ω) contain a finitely generated, infinite discrete simple group?*

This question is reasonable for any degree of regularity on the diffeomorphisms.

In Ghys's survey on groups acting on the circle, he points out that Thompson's group can be realized as piecewise $SL(2, \mathbb{Z})$ homeomorphisms of the circle [102]. In [70], Whyte and the author point out that the group of piecewise $SL(n, \mathbb{Z})$ maps on either the torus or the real projective space is quite large. The following are natural questions; see [70] for more discussion.

QUESTION 11.6. *Are there interesting finitely generated or finitely presented subgroups of piecewise $SL(n, \mathbb{Z})$ maps on $\mathbb{T}^n$ or $\mathbb{P}^{n-1}$? Can any such group that is not a subgroup of $SL(n, \mathbb{Z})$ be made to act smoothly?*

11.2. Highly Distorted Subgroups of Diff (S^2)

In [30], Calegari and Freedman construct a very interesting class of subgroups of Diff^∞ (S^2). Very roughly, they prove:

THEOREM 11.7. *There is a finitely generated subgroup G of* Diff^∞ (S^2) *that contains a rotation r as an arbitrarily distorted element.*

Here by *arbitrarily distorted* we mean that we can choose the group G so that the function $f(n) = \|r^n\|_G$ grows more slowly than any function we choose. It is well known that for linear groups, the function $f(n)$ is at worst a logarithm, so this theorem immediately implies that we can find G with no faithful linear representations. This also answered a question raised by Franks and Handel in [88].

More recently, Avila has constructed similar examples in Diff^∞ (S^1) [4]. This answers a question raised in [30] where such subgroups were constructed in Diff^1 (S^1).

We are naturally led to the following questions. We say a diffeomorphism has *full support* if the complement of the fixed set is dense.

QUESTION 11.8. *For which compact manifolds M does* Diff^∞ *(M) contain arbitrarily distorted elements of full support? The same question for* Diff^ω *(M)? The same question for* Diff^∞ *(M, v) where v is a volume form on M? For the second two questions, we can drop the hypothesis of full support.*

The second and third questions here seem quite difficult and the answer could conceivably be "none." The only examples where anything is known in the volume preserving setting are compact surfaces of genus at least 1, where no element is more than quadratically distorted by a result of Polterovich [196]. However, this result depends heavily on the fact that in dimension 2, preserving a volume form is the same as preserving a symplectic structure.

11.3. Topological Construction of Actions on High Dimensional Spheres

In this subsection, we recall a construction due to Farrell and Lafont that allows one to construct actions of a large class of groups on closed disks, and so by doubling, to construct actions on spheres [59]. The construction yields actions on very high-dimensional disks and spheres and is only known to produce actions by homeomorphisms.

The class of groups involved is the set of groups that admit an *EZ*-structure. This notion is a modification of an earlier notion due to Bestvina of a *Z*-structure on a group [16]. We do not recall the precise definition here, but refer the reader to the introduction to [59] and remark here that both torsion free Gromov hyperbolic groups and CAT (0)-groups admit *EZ*-structures.

The result that concerns us is

THEOREM 11.9. *Given a group Γ with an EZ-structure, there is an action of Γ by homeomorphisms on a closed disk.*

In fact, Farrell and Lafont give a fair amount of information concerning their actions that are quite different from the actions we are usually concerned with here. In particular, the action is properly discontinuous off a closed subset Δ of the boundary of the disk. So from a dynamical viewpoint Δ carries the "interesting part" of the action, for example, it is the support of any Γ stationary measure. Farrell and Lafont point out an analogy between their construction and the action of a Kleinian group on the boundary of hyperbolic space. An interesting general question is

QUESTION 11.10. *When can the Farrell-Lafont action on a disk or sphere be chosen smooth?*

12. Rigidity for Other Classes of Acting Groups

In this section, we collect some results and questions concerning actions of other classes of groups. In almost all cases, little is known.

12.1. Lattices in Semi-direct Products

While it is not reasonable to expect classification results for arbitrary actions of *all* lattices in *all* Lie groups, there are natural broader classes to consider. To pick a reasonable class, a first guess is to try to exclude Lie groups whose lattices have homomorphisms onto $\mathbb{Z}$ or larger free groups. There are many such groups. For example, the groups $Sl(n, \mathbb{Z}) \ltimes \mathbb{Z}^n$ that are lattices in $Sl(n, \mathbb{R}) \ltimes \mathbb{R}^n$ have property (T) as soon as $n > 2$. As it turns out, a reasonable setting is to consider perfect Lie groups with no compact factors or factors locally isomorphic to $SO(1, n)$ or $SU(1, n)$. Any such Lie group will have property (T) and therefore so will its lattices. Many examples of such lattices are described in [230]. Some first rigidity results for these groups are contained in Zimmer's paper [252]. The relevant full generalization of the cocycle superrigidity theorem is contained in [238]. In this context it seems that there are probably many rigidity theorems concerning actions of these groups already implicit in the literature, following from the results in [238] and existing arguments.

12.2. Universal Lattices

As mentioned above in Subsection 4.7, the groups $SL(n, \mathbb{Z}[X])$ for $n > 2$ have property (T) by a result of Shalom [216]. His proof also works with larger collections of variables and some other arithmetic groups. The following is an interesting problem.

PROBLEM 12.1. *Prove cocycle superrigidity for universal lattices.*

For clarity, we indicate that we expect that all linear representations and cocycles (up to "compact noise") will be described in terms of representations of the ambient groups, for example, $SL(n, \mathbb{R})$, and will be determined by specifying a numerical value of X.

Some partial results toward superrigidity are known by Farb [52] and Shenfield [218]. The problem of completely classifying linear representations in this context does seem to be open.

12.3. SO(1,n) and SU(1,n) and Their Lattices

It is conjectured that all lattices in $SO(1, n)$ and $SU(1, n)$ admit finite-index subgroups with surjections to $\mathbb{Z}$; see [18]. The conjecture is usually attributed to Thurston and sometimes to Borel for the case of $SU(1, n)$. If this is true, it immediately implies that actions of those lattices can never be classified.

There are still some interesting results concerning actions of these lattices. In [214], Shalom places restrictions on the possible actions of $SO(1, n)$ and $SU(1, n)$ in terms of the fundamental group of the manifold acted upon. This work is similar in spirit to the work of Lubotzky and Zimmer described in Section 8, but requires more restrictive hypotheses.

In [78], the author exhibits large moduli spaces of ergodic affine algebraic actions that are constructed for certain lattices in $SO(1, n)$. These moduli spaces are, however, all finite dimensional. In [71], I construct an infinite-dimensional moduli of deformations of an isometric action of $SO(1, n)$. Both of these constructions rely on a notion of bending introduced by Johnson and Millson in the finite dimensional setting [121].

I do not formulate any precise questions or conjectures in this direction as I am not sure what phenomenon to expect.

12.4. Lattices in Other Locally Compact Groups

Much recent work has focused on developing a theory of lattices in locally compact groups other than Lie groups. This theory is fully developed for algebraic groups over other local fields. Though we did not mention it here, some of Zimmer's own conjectures were made in the context of S-arithmetic groups, that is, lattices in products of real and p-adic Lie groups. The following conjecture is natural in this context and does not seem to be stated in the literature.

CONJECTURE 12.2. *Let G be semisimple algebraic group defined over a field k of positive characteristic and $\Gamma < G$ a lattice. Further assume that all simple factors of G have k-rank at least 2. Then any Γ action on a compact manifold factors through a finite quotient.*

The existence of a measurable invariant metric in this context should be something one can deduce from the cocycle superrigidity theorems, though it is not clear that the correct form of these theorems is known or in the literature.

There is also a growing interest in lattices in locally compact groups that are not algebraic. We remark here that Kac-Moody lattices typically admit no nontrivial homomorphisms even to Homeo (M); see [83] for a discussion.

The only other interesting class of lattices known to the author is the lattices in the isometry group of a product of two trees constructed by Burger and Mozes [26, 25]. These groups are infinite simple finitely presented groups.

PROBLEM 12.3. *Do the Burger-Mozes lattices admit any nontrivial homomorphisms to Diff (M) when M is a compact manifold?*

That these groups do act by homeomorphisms on high dimensional spheres by the construction discussed in Subsection 11.3 was pointed out to the author by Lafont.

12.5. Automorphism Groups of Free and Surface Groups

We briefly mention a last set of natural questions. There is a long-standing analogy between higher rank lattices and two other classes of groups. These are mapping class groups of surfaces and the outer automorphism group of the free group. See [21] for a detailed discussion of this analogy. In the context of this article, this raises the following question. Here we denote the mapping class group of a surface by $MCG(\Sigma)$ and the outer automorphism group of F_n by Out (F_n).

QUESTION 12.4. *Assume Σ has genus at least 2 or that $n > 2$. Does $MCG(\Sigma)$ or $Out(F_n)$ admit a faithful action on a compact manifold? A faithful action by smooth, volume preserving diffeomorphisms?*

By not assuming that M is connected, we are implicitly asking the same question about all finite-index subgroups of $MCG(\Sigma)$ and $Out(F_n)$. Recall from Section 11 that $Aut(F_n)$ does admit a volume preserving action on a compact manifold. This makes the question above particularly intriguing.

13. Properties of Subgroups of Diff (M)

As remarked in Subsection 4.4, in the paper [100], Ghys attempts to reprove the classical Zassenhaus lemma for linear groups for groups of analytic diffeomorphisms. While the full strength of the Zassenhaus lemma does not hold in this setting, many interesting results do follow. This immediately leaves one wondering to what extent other properties of linear groups might hold for diffeomorphism groups or at least what analogues of many theorems might be true. This direction of research was initiated by Ghys and most of the questions below are due to him.

It is worth noting that some properties of finitely generated linear groups, like residual finiteness, do not appear to have reasonable analogues in the setting of diffeomorphism groups. For residual finiteness, the obvious example is the Thompson group discussed above, which is simple.

13.1. Jordan's Theorem

For linear groups, there is a classical (and not too difficult) result known as Jordan's theorem. This says that for any finite subgroup of $GL(n, \mathbb{C})$, there is a subgroup of index at most $c(n)$ that is abelian. One cannot expect better than this, as S^1 has finite subgroups of arbitrarily large order and is a subgroup of $GL(n, \mathbb{C})$. For proofs of Jordan's theorem as well as the theorems on linear groups mentioned in the next subsection, we refer the reader to, for example, [208].

QUESTION 13.1. *Given a compact manifold M and a finite subgroup F of Diff (M), is there a constant $c(M)$ such that F has an abelian subgroup of index $c(M)$?*

As above, one cannot expect more than this simply because one can construct actions of S^1 on M and, therefore, finite abelian subgroups of Diff (M) of arbitrarily large order.

For this question, it may be more natural to ask about finite groups of homeomorphisms and not assume any differentiability of the maps. Using the results in, for example, [157], one can show that at most finitely many simple finite groups act on a given compact manifold. To be clear, one can show this using the classification of finite simple groups. It would be most interesting to resolve Question 13.1 without reference to the classification.

A recent preprint of Mundet i Rieri provides some evidence for a positive answer to the question [180].

13.2. Burnside Problem

A group is called *periodic* if all of its elements have finite order. We say a periodic group G has *bounded exponent* if every element has order at most m. In 1905 Burnside proved that finitely generated linear groups of bounded exponent are finite and in 1911 Schur proved that finitely generated periodic linear groups are finite. For a general finitely generated group, this is not true, counterexamples to the *Burnside conjecture* were constructed in a sequence of works by many authors, including Novikov, Golod, Shafarevich, and Ol'shanskii. We refer the reader to the Website: $http : //www - groups.dcs.st - and.ac.uk/\ history/HistTopics/Burnside\_problem.html$ for a detailed discussion of the history. This page also discusses the *restricted Burnside conjecture* resolved by Zelmanov.

In this context, the following questions seem natural:

QUESTION 13.2. *Are there infinite, finitely generated periodic groups of diffeomorphisms of a compact manifold M? Are there infinite, finitely generated bounded exponent groups of diffeomorphisms of a compact manifold M?*

Some first results in this direction are contained in the paper of Rebelo and Silva [206].

13.3. Tits's alternative

In this section, we ask a sequence of questions related to a famous theorem of Tits' and more recent variants on it [229].

THEOREM 13.3. *Let Γ be a finitely generated linear group. Then either Γ contains a free subgroup on two generators or Γ is virtually solvable.*

The following conjecture of Ghys is a reasonable alternative to this for groups of diffeomorphisms.

CONJECTURE 13.4. *Let M be a compact manifold and Γ a finitely generated group of smooth diffeomorphisms of M. Then either Γ contains a free group on two generators or Γ preserves some measure on M.*

The best evidence for this conjecture to date is a theorem of Margulis' that proves the conjecture for $M = S^1$ [163]. Ghys has also asked if the more exact analogue of Tits's theorem might be true for analytic diffeomorphisms.

A closely related question is whether there exist groups of intermediate growth inside of diffeomorphism groups. The growth of a group is the rate of growth of the balls in the Cayley graph. For linear groups, this is either polynomial or exponential. In general, there are groups of intermediate growth, that is, subexponential but not polynomial. The first examples of these were discovered by Grigorchuk [109]. Some of these examples were shown by Grigorchuk and Maki to act by homeomorphisms on the interval [110]. More recently, it was shown by Navas that these examples do have an action by C^1 diffeomorphisms on the interval [185]. In the opposite direction, Navas showed that these phenomena disappear if one considers instead $C^{1+\alpha}$ diffeomorphisms for any positive α. In that setting, Navas shows that any finitely generated group of diffeomorphisms of the interval with subexponential growth must be nilpotent. It would be interesting to understand what can occur in higher dimensions.

A recent related line of research for linear groups concerns uniform exponential growth. A finitely generated group Γ has uniform exponential growth if the number of elements in a ball of radius r in a Cayley graph for Γ grows at least as λ^r for some $\lambda > 1$ that does not depend on the choice of generators. For linear groups, exponential growth implies uniform exponential growth by a theorem of Eskin, Mozes, and Oh [51]. There are examples of groups having exponential but not uniform exponential growth due to Wilson [235]. This raises the following question.

QUESTION 13.5. *Let M be a compact manifold and Γ a finitely generated subgroup of Diff (M). If Γ has exponential growth, does Γ have uniform exponential growth?*

The question seems most likely to have a positive answer if one further assumes that Γ is nonamenable.

References

[1]   Scot Adams, *Dynamics on Lorentz manifolds*, World Scientific Publishing Co., River Edge, NJ, 2001. MR1875067 (2003c:53101)

[2]   Jurgen Andersen, *Mapping class groups do not have Kazhdan's property (T)*, preprint 2007.

[3]   D. V. Anosov, *Geodesic flows on closed Riemann manifolds with negative curvature*, Proceedings of the Steklov Institute of Mathematics, No. 90 (1967). Translated from the Russian by S. Feder, American Mathematical Society, Providence, RI, 1969. MR0242194 (39 #3527)

[4]   Artur Avila, *Distortion elements in $Diff^\infty (R/Z)$*, preprint 2008.

[5]   Uri Bader, Charles Frances, and Karin Melnick, *An embedding theorem for automorphism groups of Cartan geometries*, (English summary). Geom. Funct. Anal. **19** (2009), no. 2, 333–355. MR254240

[6]   Uri Bader, Alex Furman, Tsachik Gelander, and Nicolas Monod, *Property (T) and rigidity for actions on Banach spaces*, Acta Math. **198** (2007), no. 1, 57–105. MR2316269 (2008g:22007)

[7]   Uri Bader and Yehuda Shalom, *Factor and normal subgroup theorems for lattices in products of groups*, Invent. Math. **163** (2006), no. 2, 415–454. MR2207022 (2006m:22017)

[8]   Augustin Banyaga, *Sur la structure du groupe des difféomorphismes qui préservent une forme symplectique*, Comment. Math. Helv. **53** (1978), no. 2, 174–227. MR490874 (80c:58005)

[9]   ———, *The structure of classical diffeomorphism groups*, Mathematics and its Applications, vol. 400, Kluwer Academic Publishers Group, Dordrecht, 1997. MR1445290 (98h:22024)

[10]   Yves Benoist, *Orbites des structures rigides (d'après M. Gromov)*, Integrable systems and foliations/Feuilletages et systèmes intégrables (Montpellier, 1995), Progr. Math., vol. 145, Birkhäuser, Boston, 1997, pp. 1–17. MR1432904 (98c:58126)

[11]   Yves Benoist and François Labourie, *Sur les difféomorphismes d'Anosov affines à feuilletages stable et instable différentiables*, Invent. Math. **111** (1993), no. 2, 285–308. MR11 98811 (94d:58114)

[12]   E. J. Benveniste, *University of Chicago, Ph.D. thesis*, unpublished, 1996.

[13]   E. Jerome Benveniste and David Fisher, *Nonexistence of invariant rigid structures and invariant almost rigid structures*, Comm. Anal. Geom. **13** (2005), no. 1, 89–111. MR2154667 (2006f:53056)

[14]   G. Besson, G. Courtois, and S. Gallot, *Entropies et rigidités des espaces localement symétriques de courbure strictement négative*, Geom. Funct. Anal. **5** (1995), no. 5, 731–799. MR1354289 (96i:58136)

[15]   Gérard Besson, Gilles Courtois, and Sylvestre Gallot, *Minimal entropy and Mostow's rigidity theorems*, Ergodic Theory Dynam. Systems **16** (1996), no. 4, 623–649. MR140 6425 (97e:58177)

[16]   Mladen Bestvina, *Local homology properties of boundaries of groups*, Michigan Math. J. **43** (1996), no. 1, 123–139. MR1381603 (97a:57022)

[17]   Mladen Bestvina and Koji Fujiwara, *Bounded cohomology of subgroups of mapping class groups*, Geom. Topol. **6** (2002), 69–89 (electronic). MR1914565 (2003f:57003)

[18]   A. Borel and N. Wallach, *Continuous cohomology, discrete subgroups, and representations of reductive groups*, second ed., Mathematical Surveys and Monographs, Vol. 67, American Mathematical Society, Providence, RI, 2000. MR1721403 (2000j:22015)

[19]   Armand Borel, *Compact Clifford-Klein forms of symmetric spaces*, Topology **2** (1963), 111–122. MR0146301 (26 #3823)

[20]   Martin Bridson and Karen Vogtmann, *Actions of automorphism groups of free groups on homology spheres and acyclic manifolds*, to appear Comm. Math. Helv.

[21]   Martin R. Bridson and Karen Vogtmann, *Automorphism groups of free groups, surface groups and free Abelian groups*, Problems on mapping class groups and related topics,

Proc. Sympos. Pure Math., Vol. 74, American Mathematical Society, Providence, RI, 2006, pp. 301–316. MR2264548 (2008g:20091)

[22] M. Burger and N. Monod, *Continuous bounded cohomology and applications to rigidity theory*, Geom. Funct. Anal. **12** (2002), no. 2, 219–280. MR1911660 (2003d:53065a)

[23] Marc Burger, *An extension criterion for lattice actions on the circle*, in this volume.

[24] ———, *Rigidity properties of group actions on CAT(0)-spaces*, Proceedings of the International Congress of Mathematicians, Vol. 1, 2 (Zürich, 1994) (Basel), Birkhäuser, 1995, pp. 761–769. MR1403976 (97j:20033)

[25] Marc Burger and Shahar Mozes, *Groups acting on trees: from local to global structure*, Inst. Hautes Études Sci. Publ. Math. (2000), no. 92, 113–150 (2001). MR1839488 (2002i:20041)

[26] ———, *Lattices in product of trees*, Inst. Hautes Études Sci. Publ. Math. (2000), no. 92, 151–194 (2001). MR1839489 (2002i:20042)

[27] Keith Burns, Charles Pugh, Michael Shub, and Amie Wilkinson, *Recent results about stable ergodicity*, Smooth ergodic theory and its applications (Seattle, WA, 1999), Proc. Sympos. Pure Math., Vol. 69, American Mathematical Society, Providence, RI, 2001, pp. 327–366. MR1858538 (2002m:37042)

[28] Eugenio Calabi, *On compact, Riemannian manifolds with constant curvature. I*, Proc. Sympos. Pure Math., Vol. III, American Mathematical Society, Providence, RI, 1961, pp. 155–180. MR0133787 (24 #A3612)

[29] Eugenio Calabi and Edoardo Vesentini, *On compact, locally symmetric Kähler manifolds*, Ann. of Math. (2) **71** (1960), 472–507. MR0111058 (22 #1922b)

[30] Danny Calegari and Michael H. Freedman, *Distortion in transformation groups*, Geom. Topol. **10** (2006), 267–293 (electronic), with an appendix by Yves de Cornulier. MR2207794 (2007b:37048)

[31] A. Candel and R. Quiroga-Barranco, *Gromov's centralizer theorem*, Geom. Dedicata **100** (2003), 123–155. MR2011119 (2004m:53075)

[32] Alberto Candel and Raul Quiroga-Barranco, *Parallelisms, prolongations of Lie algebras and rigid geometric structures*, Manuscripta Math. **114** (2004), no. 3, 335–350. MR2076451 (2005f:53056)

[33] J. W. Cannon, W. J. Floyd, and W. R. Parry, *Introductory notes on Richard Thompson's groups*, Enseign. Math. (2) **42** (1996), no. 3-4, 215–256. MR1426438 (98g:20058)

[34] James W. Cannon and William P. Thurston, *Group invariant Peano curves*, Geom. Topol. **11** (2007), 1315–1355. MR2326947 (2008i:57016)

[35] Serge Cantat, *Groupes de transformations birationnelles du plan*, preprint.

[36] ———, *Version kählérienne d'une conjecture de Robert J. Zimmer*, Ann. Sci. École Norm. Sup. (4) **37** (2004), no. 5, 759–768. MR2103473 (2006b:22010)

[37] Pierre-Emmanuel Caprace and Bertrand Rémy, *Simplicité abstraite des groupes de Kac-Moody non affines*, C. R. Math. Acad. Sci. Paris **342** (2006), no. 8, 539–544. MR2217912 (2006k:20102)

[38] Indira Chatterji and Martin Kassobov, *New examples of finitely presented groups with strong fixed point properties*, J. Topol. Anal. **1** (2009), no. 1, 1–12. MR2649346

[39] Christopher Connell and Benson Farb, *Some recent applications of the barycenter method in geometry*, Topology and geometry of manifolds (Athens, GA, 2001), Proc. Sympos. Pure Math., Vol. 71, American Mathematical Society, Providence, RI, 2003, pp. 19–50. MR2024628 (2005e:53058)

[40] Kevin Corlette, *Archimedean superrigidity and hyperbolic geometry*, Ann. of Math. (2) **135** (1992), no. 1, 165–182. MR1147961 (92m:57048)

[41]   Kevin Corlette and Robert J. Zimmer, *Superrigidity for cocycles and hyperbolic geometry*, Internat. J. Math. **5** (1994), no. 3, 273–290. MR1274120 (95g:58055)

[42]   S. G. Dani, *Continuous equivariant images of lattice-actions on boundaries*, Ann. of Math. (2) **119** (1984), no. 1, 111–119. MR736562 (85i:22009)

[43]   Pierre de la Harpe and Alain Valette, *La propriété (T) de Kazhdan pour les groupes localement compacts (avec un appendice de Marc Burger)*, Astérisque (1989), no. 175, 158, with an appendix by M. Burger. MR1023471 (90m:22001)

[44]   Patrick Delorme, *1-cohomologie des représentations unitaires des groupes de Lie semi-simples et résolubles. Produits tensoriels continus de représentations*, Bull. Soc. Math. France **105** (1977), no. 3, 281–336. MR0578893 (58 #28272)

[45]   Bertrand Deroin, Victor Kleptsyn and Andrés Navas, *Sur la dynamique unidimensionnelle en régularité intermédiaire*, Acta Math. **199** (2007), no. 2, 199–262. MR23 58052

[46]   Julie Déserti, *Groupe de Cremona et dynamique complexe: une approche de la conjecture de Zimmer*, Int. Math. Res. Not. (2006), Art. ID 71701, 27. MR2233717 (2007d:22013)

[47]   Tien-Cuong Dinh and Nessim Sibony, *Groupes commutatifs d'automorphismes d'une variété kählérienne compacte*, Duke Math. J. **123** (2004), no. 2, 311–328. MR2066940 (2005g:32020)

[48]   Sorin Dumitrescu, *Meromorphic almost rigid geometric structures*, in this volume.

[49]   Michael Entov and Leonid Polterovich, *Calabi quasimorphism and quantum homology*, Int. Math. Res. Not. (2003), no. 30, 1635–1676. MR1979584 (2004e:53131)

[50]   D. B. A. Epstein, *The simplicity of certain groups of homeomorphisms*, Compositio Math. **22** (1970), 165–173. MR0267589 (42 #2491)

[51]   Alex Eskin, Shahar Mozes, and Hee Oh, *On uniform exponential growth for linear groups*, Invent. Math. **160** (2005), no. 1, 1–30. MR2129706 (2006a:20081)

[52]   Benson Farb, *Group actions and Helly's theorem*, Adv. Math. **222** (2009), no. 5, 1574–1588. MR2555905

[53]   Benson Farb and Howard Masur, *Superrigidity and mapping class groups*, Topology **37** (1998), no. 6, 1169–1176. MR1632912 (99f:57017)

[54]   Benson Farb and Peter Shalen, *Real-analytic actions of lattices*, Invent. Math. **135** (1999), no. 2, 273–296. MR1666834 (2000c:22017)

[55]   ———, *Lattice actions, 3-manifolds and homology*, Topology **39** (2000), no. 3, 573–587. MR1746910 (2001b:57041)

[56]   ———, *Groups of real-analytic diffeomorphisms of the circle*, Ergodic Theory Dynam. Systems **22** (2002), no. 3, 835–844. MR1908556 (2003e:37030)

[57]   Benson Farb and Peter B. Shalen, *Real-analytic, volume-preserving actions of lattices on 4-manifolds*, C. R. Math. Acad. Sci. Paris **334** (2002), no. 11, 1011–1014. MR1913726 (2003e:57055)

[58]   Daniel S. Farley, *Proper isometric actions of Thompson's groups on Hilbert space*, Int. Math. Res. Not. (2003), no. 45, 2409–2414. MR2006480 (2004k:22005)

[59]   F. T. Farrell and J.-F. Lafont, *EZ-structures and topological applications*, Comment. Math. Helv. **80** (2005), no. 1, 103–121. MR2130569 (2006b:57022)

[60]   R. Feres, *Affine actions of higher rank lattices*, Geom. Funct. Anal. **3** (1993), no. 4, 370–394. MR1223436 (96d:22013)

[61]   R. Feres and F. Labourie, *Topological superrigidity and Anosov actions of lattices*, Ann. Sci. École Norm. Sup. (4) **31** (1998), no. 5, 599–629. MR1643954 (99k:58112)

[62]   R. Feres and P. Lampe, *Cartan geometries and dynamics*, Geom. Dedicata **80** (2000), no. 1–3, 29–41. MR1762497 (2001i:53059)

[63] Renato Feres, *Connection preserving actions of lattices in* SL_nR, Israel J. Math. **79** (1992), no. 1, 1–21. MR1195250 (94a:58039)

[64] ———, *Actions of discrete linear groups and Zimmer's conjecture*, J. Differential Geom. **42** (1995), no. 3, 554–576. MR1367402 (97a:22016)

[65] ———, *Dynamical systems and semisimple groups: an introduction*, Cambridge Tracts in Mathematics, Vol. 126, Cambridge University Press, Cambridge, 1998. MR1670703 (2001d:22001)

[66] ———, *An introduction to cocycle super-rigidity*, Rigidity in dynamics and geometry (Cambridge, 2000), Springer, Berlin, 2002, pp. 99–134. MR1919397 (2003g:37041)

[67] ———, *Rigid geometric structures and actions of semisimple Lie groups*, Rigidité, groupe fondamental et dynamique, Panor. Synthèses, Vol. 13, Soc. Math. France, Paris, 2002, pp. 121–167. MR1993149 (2004m:53076)

[68] Renato Feres and Anatole Katok, *Ergodic theory and dynamics of G-spaces (with special emphasis on rigidity phenomena)*, Handbook of dynamical systems, Vol. 1A, North-Holland, Amsterdam, 2002, pp. 665–763. MR1928526 (2003j:37005)

[69] Renato Feres and Emily Ronshausen, *Harmonic functions over group actions*, in this volume.

[70] D. Fisher and K. Whyte, *When is a group action determined by its orbit structure?* Geom. Funct. Anal. **13** (2003), no. 6, 1189–1200. MR2033836 (2004k:37045)

[71] David Fisher, *Cohomology of arithmetic groups and bending group actions*, in preparation.

[72] ———, *First cohomology and local rigidity of group actions*, in preparation.

[73] ———, *A canonical arithmetic quotient for actions of lattices in simple groups*, Israel J. Math. **124** (2001), 143–155. MR1856509 (2002f:22016)

[74] ———, *On the arithmetic structure of lattice actions on compact spaces*, Ergodic Theory Dynam. Systems **22** (2002), no. 4, 1141–1168. MR1926279 (2004j:37004)

[75] ———, *Rigid geometric structures and representations of fundamental groups*, Rigidity in dynamics and geometry (Cambridge, 2000), Springer, Berlin, 2002, pp. 135–147. MR1919398 (2003g:22007)

[76] ———, $Out(F_n)$ *and the spectral gap conjecture*, Int. Math. Res. Not. (2006), Art. ID 26028, 9. MR2250018 (2007f:22006)

[77] ———, *Local rigidity of group actions: past, present, future*, Dynamics, ergodic theory, and geometry, Math. Sci. Res. Inst. Publ., Vol. 54, Cambridge University Press, Cambridge, 2007, pp. 45–97. MR2369442

[78] ———, *Deformations of group actions*, Trans. Amer. Math. Soc. **360** (2008), no. 1, 491–505 (electronic). MR2342012

[79] David Fisher and Theron Hitchman, *Harmonic maps with infinite dimensional targets and cocycle superrigidity*, in preparation.

[80] ———, *Cocycle superrigidity and harmonic maps with infinite-dimensional targets*, Int. Math. Res. Not. (2006), 72405, 1–19. MR2211160

[81] David Fisher and G. A. Margulis, *Local rigidity for cocycles*, Surveys in differential geometry, Vol. VIII (Boston, MA, 2002), Surv. Differ. Geom., VIII, International Press, Somerville, MA, 2003, pp. 191–234. MR2039990 (2004m:22032)

[82] David Fisher and Gregory Margulis, *Almost isometric actions, property* (T)*, and local rigidity*, Invent. Math. **162** (2005), no. 1, 19–80. MR2198325

[83] David Fisher and Lior Silberman, *Groups not acting on manifolds*, Int. Math. Res. Not. (2008).

[84] David Fisher and Kevin Whyte, *Continuous quotients for lattice actions on compact spaces*, Geom. Dedicata **87** (2001), no. 1–3, 181–189. MR1866848 (2002j:57070)

[85] David Fisher and Robert J. Zimmer, *Geometric lattice actions, entropy and fundamental groups*, Comment. Math. Helv. **77** (2002), no. 2, 326–338. MR1915044 (2003e:57056)

[86] John Franks and Michael Handel, *Area preserving group actions on surfaces*, Geom. Topol. **7** (2003), 757–771 (electronic). MR2026546 (2004j:37042)

[87] ———, *Periodic points of Hamiltonian surface diffeomorphisms*, Geom. Topol. **7** (2003), 713–756 (electronic). MR2026545 (2004j:37101)

[88] ———, *Distortion elements in group actions on surfaces*, Duke Math. J. **131** (2006), no. 3, 441–468. MR2219247 (2007c:37018)

[89] Alex Furman, *A survey of measured group theory*, in this volume.

[90] ———, *Random walks on groups and random transformations*, Handbook of dynamical systems, Vol. 1A, North-Holland, Amsterdam, 2002, pp. 931–1014. MR1928529 (2003j:60065)

[91] Alex Furman and Nicolas Monod, *Product groups acting on manifolds*, preprint 2007.

[92] H. Furstenberg, *Recurrence in ergodic theory and combinatorial number theory*, Princeton University Press, Princeton, NJ, 1981, M. B. Porter Lectures. MR603625 (82j:28010)

[93] Harry Furstenberg, *A Poisson formula for semi-simple Lie groups*, Ann. of Math. (2) **77** (1963), 335–386. MR0146298 (26 #3820)

[94] ———, *Boundaries of Lie groups and discrete subgroups*, Actes du Congrès International des Mathématiciens (Nice, 1970), Tome 2, Gauthier-Villars, Paris, 1971, pp. 301–306. MR0430160 (55 #3167)

[95] ———, *Rigidity and cocycles for ergodic actions of semisimple Lie groups (after G. A. Margulis and R. Zimmer)*, Bourbaki Seminar, Vol. 1979/80, Lecture Notes in Math., vol. 842, Springer, Berlin, 1981, pp. 273–292. MR636529 (83j:22003)

[96] Jean-Marc Gambaudo and Étienne Ghys, *Commutators and diffeomorphisms of surfaces*, Ergodic Theory Dynam. Systems **24** (2004), no. 5, 1591–1617. MR2104597 (2006d:37071)

[97] Tsachik Gelander, *On deformtions of F_n in compact Lie groups*, Israel J. Math. **167** (2008), 15–26. MR2448015 (2009j:22011)

[98] Tsachik Gelander, Anders Karlsson, and Gregory A. Margulis, *Superrigidity, generalized harmonic maps and uniformly convex spaces*, Geom. Funct. Anal. **17** (2008), no. 5, 1524–1550. MR2377496

[99] Étienne Ghys, *Groupes d'homéomorphismes du cercle et cohomologie bornée*, The Lefschetz centennial conference, Part III (Mexico City, 1984), Contemp. Math., vol. 58, American Mathematical Society, Providence, RI, 1987, pp. 81–106. MR893858 (88m:58024)

[100] ———, *Sur les groupes engendrés par des difféomorphismes proches de l'identité*, Bol. Soc. Brasil. Mat. (N.S.) **24** (1993), no. 2, 137–178. MR1254981 (95f:58017)

[101] ———, *Actions de réseaux sur le cercle*, Invent. Math. **137** (1999), no. 1, 199–231. MR1703323 (2000j:22014)

[102] ———, *Groups acting on the circle*, Enseign. Math. (2) **47** (2001), no. 3–4, 329–407. MR1876932 (2003a:37032)

[103] Étienne Ghys and Vlad Sergiescu, *Sur un groupe remarquable de difféomorphismes du cercle*, Comment. Math. Helv. **62** (1987), no. 2, 185–239. MR896095 (90c:57035)

[104] Edward R. Goetze, *Connection preserving actions of connected and discrete Lie groups*, J. Differential Geom. **40** (1994), no. 3, 595–620. MR1305982 (95m:57052)

[105] Edward R. Goetze and Ralf J. Spatzier, *On Livšic's theorem, superrigidity, and Anosov actions of semisimple Lie groups*, Duke Math. J. **88** (1997), no. 1, 1–27. MR1448015 (98d:58134)

[106] ———, *Smooth classification of Cartan actions of higher rank semisimple Lie groups and their lattices*, Ann. of Math. (2) **150** (1999), no. 3, 743–773. MR1740993 (2001c:37029)

[107] William M. Goldman, *An ergodic action of the outer automorphism group of a free group*, Geom. Funct. Anal. **17** (2007), no. 3, 793–805. MR2346275 (2008g:57001)

[108] Alexander Gorodnik, *Open problems in dynamics and related fields*, J. Mod. Dyn. **1** (2007), no. 1, 1–35. MR2261070 (2007f:37001)

[109] R. I. Grigorchuk, *Degrees of growth of finitely generated groups and the theory of invariant means*, Izv. Akad. Nauk SSSR Ser. Mat. **48** (1984), no. 5, 939–985. MR764305 (86h:20041)

[110] R. I. Grigorchuk and A. Maki, *On a group of intermediate growth that acts on a line by homeomorphisms*, Mat. Zametki **53** (1993), no. 2, 46–63. MR1220809 (94c:20008)

[111] M. Gromov, *Hyperbolic groups*, Essays in group theory, Math. Sci. Res. Inst. Publ., Vol. 8, Springer, New York, 1987, pp. 75–263. MR919829 (89e:20070)

[112] Michael Gromov, *Rigid transformations groups*, Géométrie différentielle (Paris, 1986), Travaux en Cours, Vol. 33, Hermann, Paris, 1988, pp. 65–139. MR955852 (90d:58173)

[113] Mikhail Gromov and Richard Schoen, *Harmonic maps into singular spaces and p-adic superrigidity for lattices in groups of rank one*, Inst. Hautes Études Sci. Publ. Math. (1992), no. 76, 165–246. MR1215595 (94e:58032)

[114] Alain Guichardet, *Sur la cohomologie des groupes topologiques. II*, Bull. Sci. Math. (2) **96** (1972), 305–332. MR0340464 (49 #5219)

[115] Sigurdur Helgason, *Differential geometry, Lie groups, and symmetric spaces*, Graduate Studies in Mathematics, Vol. 34, American Mathematical Society, Providence, RI, 2001, Corrected reprint of the 1978 original. MR1834454 (2002b:53081)

[116] Michael-Robert Herman, *Simplicité du groupe des difféomorphismes de classe C^∞, isotopes à l'identité, du tore de dimension n*, C. R. Acad. Sci. Paris Sér. A-B **273** (1971), A232–A234. MR0287585 (44 #4788)

[117] Graham Higman, *Finitely presented infinite simple groups*, Department of Pure Mathematics, Department of Mathematics, I.A.S. Australian National University, Canberra, 1974, Notes on Pure Mathematics, no. 8 (1974). MR0376874 (51 #13049)

[118] M. W. Hirsch, C. C. Pugh, and M. Shub, *Invariant manifolds*, Lecture Notes in Mathematics, Vol. 583, Springer-Verlag, Berlin, 1977. MR0501173 (58 #18595)

[119] Steve Hurder, personal communication 2007.

[120] Steven Hurder, *A survey of rigidity theory for Anosov actions*, Differential topology, foliations, and group actions (Rio de Janeiro, 1992), Contemp. Math., Vol. 161, American Mathematical Society, Providence, RI, 1994, pp. 143–173. MR1271833 (95b:58112)

[121] Dennis Johnson and John J. Millson, *Deformation spaces associated to compact hyperbolic manifolds*, Discrete groups in geometry and analysis (New Haven, CT, 1984), Progr. Math., Vol. 67, Birkhäuser, Boston, 1987, pp. 48–106. MR900823 (88j:22010)

[122] Jürgen Jost and Shing-Tung Yau, *Harmonic maps and superrigidity*, Differential geometry: partial differential equations on manifolds (Los Angeles, CA, 1990), Proc. Sympos. Pure Math., Vol. 54, American Mathematical Society, Providence, RI, 1993, pp. 245–280. MR1216587 (94m:58060)

[123] Jürgen Jost and Kang Zuo, *Harmonic maps of infinite energy and rigidity results for representations of fundamental groups of quasiprojective varieties*, J. Differential Geom. **47** (1997), no. 3, 469–503. MR1617644 (99a:58046)

[124] Vadim A. Kaimanovich and Howard Masur, *The Poisson boundary of the mapping class group*, Invent. Math. **125** (1996), no. 2, 221–264. MR1395719 (97m:32033)

[125] Boris Kalinin, Anatole Katok, and Federico Rodriguez-Hertz, *Non-uniform measure rigidity*, preprint 2009.

[126] Boris Kalinin and Ralf Spatzier, *On the classification of Cartan actions*, Geom. Funct. Anal. **17** (2007), no. 2, 468–490. MR2322492 (2008i:37054)

[127] Martin Kassabov, personal communication, 2008.

[128] A. Katok and J. Lewis, *Global rigidity results for lattice actions on tori and new examples of volume-preserving actions*, Israel J. Math. **93** (1996), 253–280. MR1380646 (96k:22021)

[129] A. Katok, J. Lewis, and R. Zimmer, *Cocycle superrigidity and rigidity for lattice actions on tori*, Topology **35** (1996), no. 1, 27–38. MR1367273 (97e:22009)

[130] A. Katok and R. J. Spatzier, *Differential rigidity of Anosov actions of higher rank Abelian groups and algebraic lattice actions*, Tr. Mat. Inst. Steklova **216** (1997), no. Din. Sist. i Smezhnye Vopr., 292–319. MR1632177 (99i:58118)

[131] D. A. Každan, *On the connection of the dual space of a group with the structure of its closed subgroups*, Funkcional. Anal. i Priložen. **1** (1967), 71–74. MR0209390 (35 #288)

[132] David Kazhdan, *Some applications of the Weil representation*, J. Analyse Mat. **32** (1977), 235–248. MR0492089 (58 #11243)

[133] Bruce Kleiner, *The asymptotic geometry of negatively curved spaces: uniformization, geometrization and rigidity*, International Congress of Mathematicians. Vol. II, European Mathematical Society, Zürich, 2006, pp. 743–768. MR2275621 (2007k:53054)

[134] Anthony W. Knapp, *Lie groups beyond an introduction*, Progress in Mathematics, vol. 140, Birkhäuser, Boston, 1996. MR1399083 (98b:22002)

[135] Shoshichi Kobayashi, *Transformation groups in differential geometry*, Classics in Mathematics, Springer-Verlag, Berlin, 1995, Reprint of the 1972 edition. MR1336823 (96c:53040)

[136] Toshiyuki Kobayashi, *Discontinuous groups for non-Riemannian homogeneous spaces*, Mathematics unlimited—2001 and beyond, Springer, Berlin, 2001, pp. 723–747. MR1852186 (2002f:53086)

[137] Toshiyuki Kobayashi and Taro Yoshino, *Compact Clifford-Klein forms of symmetric spaces—revisited*, Pure Appl. Math. Q. **1** (2005), no. 3, part 2, 591–663. MR2201328 (2007h:22013)

[138] Nicholas Korevaar and Richard Schoen, *Global existence theorems for harmonic maps: finite rank spaces and approaches to rigidity of group actions*, preprint 1999.

[139] Nicholas J. Korevaar and Richard M. Schoen, *Sobolev spaces and harmonic maps for metric space targets*, Comm. Anal. Geom. **1** (1993), no. 3–4, 561–659. MR1266480 (95b:58043)

[140] ———, *Global existence theorems for harmonic maps to non-locally compact spaces*, Comm. Anal. Geom. **5** (1997), no. 2, 333–387. MR1483983 (99b:58061)

[141] Nadine Kowalsky, *Noncompact simple automorphism groups of Lorentz manifolds and other geometric manifolds*, Ann. of Math. (2) **144** (1996), no. 3, 611–640. MR1426887 (98g:57059)

[142] F. Labourie, *Quelques résultats récents sur les espaces localement homogènes compacts*, Manifolds and geometry (Pisa, 1993), Sympos. Math., XXXVI, Cambridge University Press, Cambridge, 1996, pp. 267–283. MR1410076 (97h:53055)

[143] F. Labourie, S. Mozes, and R. J. Zimmer, *On manifolds locally modelled on non-Riemannian homogeneous spaces*, Geom. Funct. Anal. **5** (1995), no. 6, 955–965. MR1361517 (97j:53053)

[144] François Labourie, *Large groups actions on manifolds*, Proceedings of the International Congress of Mathematicians, Vol. II (Berlin, 1998), no. Extra Vol. II, 1998, pp. 371–380 (electronic). MR1648087 (99k:53069)

[145] François Labourie and Robert J. Zimmer, *On the non-existence of cocompact lattices for* $SL(n)/SL(m)$, Math. Res. Lett. **2** (1995), no. 1, 75–77. MR1312978 (96d:22014)

[146] Lucy Lifschitz and Dave Witte Morris, *Bounded generation and lattices that cannot act on the line*, Pure Appl. Math. Q. **4** (2008), no. 1, part 2, 99–126. MR2405997 (2009m: 22106)

[147] Elon Lindenstrauss, *Rigidity of multiparameter actions*, Israel J. Math. **149** (2005), 199–226, Probability in mathematics. MR2191215 (2006j:37007)

[148] Ron Livne, *On certain covers of the universal elliptic curve*, unpublished Harvard PhD thesis.

[149] Alexander Lubotzky, *Discrete groups, expanding graphs and invariant measures*, Progress in Mathematics, Vol. 125, Birkhäuser Verlag, Basel, 1994, with an appendix by Jonathan D. Rogawski. MR1308046 (96g:22018)

[150] ———, *Free quotients and the first Betti number of some hyperbolic manifolds*, Transform. Groups **1** (1996), no. 1–2, 71–82. MR1390750 (97d:57016)

[151] Alexander Lubotzky, Shahar Mozes, and M. S. Raghunathan, *The word and Riemannian metrics on lattices of semisimple groups*, Inst. Hautes Études Sci. Publ. Math. (2000), no. 91, 5–53 (2001). MR1828742 (2002e:22011)

[152] Alexander Lubotzky and Robert J. Zimmer, *A canonical arithmetic quotient for simple Lie group actions*, Lie groups and ergodic theory (Mumbai, 1996), Tata Inst. Fund. Res. Stud. Math., Vol. 14, Tata Institute of Fundamental Research, Bombay, 1998, pp. 131–142. MR1699362 (2000m:22013)

[153] ———, *Arithmetic structure of fundamental groups and actions of semisimple Lie groups*, Topology **40** (2001), no. 4, 851–869. MR1851566 (2002f:22017)

[154] George W. Mackey, *Ergodic theory and virtual groups*, Math. Ann. **166** (1966), 187–207. MR0201562 (34 #1444)

[155] Iozhe Maleshich, *The Hilbert-Smith conjecture for Hölder actions*, Uspekhi Mat. Nauk **52** (1997), no. 2(314), 173–174. MR1480156 (99d:57026)

[156] Ricardo Mañé, *Quasi-Anosov diffeomorphisms and hyperbolic manifolds*, Trans. Amer. Math. Soc. **229** (1977), 351–370. MR0482849 (58 #2894)

[157] L. N. Mann and J. C. Su, *Actions of elementary p-groups on manifolds*, Trans. Amer. Math. Soc. **106** (1963), 115–126. MR0143840 (26 #1390)

[158] G. A. Margulis, *Arithmetic properties of discrete subgroups*, Uspehi Mat. Nauk **29** (1974), no. 1 (175), 49–98. MR0463353 (57 #3306a)

[159] ———, *Discrete groups of motions of manifolds of nonpositive curvature*, Proceedings of the International Congress of Mathematicians (Vancouver, BC, 1974), Vol. 2, Canadian Mathematical Congress, Montreal, Quetec, 1975, pp. 21–34. MR0492072 (58 #11226)

[160] ———, *Non-uniform lattices in semisimple algebraic groups*, Lie groups and their representations (Proc. Summer School on Group Representations of the Bolyai János Math. Soc., Budapest, 1971), Halsted, New York, 1975, pp. 371–553. MR0422499 (54 #10486)

[161] ———, *Arithmeticity of the irreducible lattices in the semisimple groups of rank greater than* 1, Invent. Math. **76** (1984), no. 1, 93–120. MR739627 (85j:22021)

[162] ———, *Discrete subgroups of semisimple Lie groups*, Ergebnisse der Mathematik und ihrer Grenzgebiete (3) [Results in Mathematics and Related Areas (3)], Vol. 17, Springer-Verlag, Berlin, 1991. MR1090825 (92h:22021)

[163] Gregory Margulis, *Free subgroups of the homeomorphism group of the circle*, C. R. Acad. Sci. Paris Sér. I Math. **331** (2000), no. 9, 669–674. MR1797749 (2002b:37034)

[164] ———, *Problems and conjectures in rigidity theory*, Mathematics: frontiers and perspectives, American Mathematical Society, Providence, RI, 2000, pp. 161–174. MR1754775 (2001d:22008)

[165] Gregory A. Margulis and Nantian Qian, *Rigidity of weakly hyperbolic actions of higher real rank semisimple Lie groups and their lattices*, Ergodic Theory Dynam. Systems **21** (2001), no. 1, 121–164. MR1826664 (2003a:22019)

[166] John N. Mather, *Commutators of diffeomorphisms*, Comment. Math. Helv. **49** (1974), 512–528. MR0356129 (50 #8600)

[167] ———, *Simplicity of certain groups of diffeomorphisms*, Bull. Amer. Math. Soc. **80** (1974), 271–273. MR0339268 (49 #4028)

[168] ———, *Commutators of diffeomorphisms. II*, Comment. Math. Helv. **50** (1975), 33–40. MR0375382 (51 #11576)

[169] ———, *Foliations and local homology of groups of diffeomorphisms*, Proceedings of the International Congress of Mathematicians (Vancouver, BC, 1974), Vol. 2, Canadian Mathematical Congress, Montreal, Quebec, 1975, pp. 35–37. MR0431203 (55 #4205)

[170] Yozô Matsushima and Shingo Murakami, *On vector bundle valued harmonic forms and automorphic forms on symmetric riemannian manifolds*, Ann. of Math. (2) **78** (1963), 365–416. MR0153028 (27 #2997)

[171] Ngaiming Mok, Yum Tong Siu, and Sai-Kee Yeung, *Geometric superrigidity*, Invent. Math. **113** (1993), no. 1, 57–83. MR1223224 (94h:53079)

[172] Nicolas Monod, *Superrigidity for irreducible lattices and geometric splitting*, J. Amer. Math. Soc. **19** (2006), no. 4, 781–814 (electronic). MR2219304 (2007b:22025)

[173] Dave Witte Morris, *Can lattices in SL(n, ℝ) act on the circle?* in this volume.

[174] ———, *Amenable groups that act on the line*, Algebr. Geom. Topol. **6** (2006), 2509–2518. MR2286034 (2008c:20078)

[175] ———, *An introduction to arithmetic groups*, book in preprint form 2007.

[176] Dave Witte Morris and Robert J. Zimmer, *Ergodic actions of semisimple Lie groups on compact principal bundles*, Geom. Dedicata **106** (2004), 11–27. MR2079831 (2005e:22016)

[177] G. D. Mostow, *Quasi-conformal mappings in n-space and the rigidity of hyperbolic space forms*, Inst. Hautes Études Sci. Publ. Math. (1968), no. 34, 53–104. MR0236383 (38 #4679)

[178] ———, *Strong rigidity of locally symmetric spaces*, Princeton University Press, Princeton, NJ, 1973, Annals of Mathematics Studies, No. 78. MR0385004 (52 #5874)

[179] ———, *personal communication*, 2000.

[180] I. Mundet Rieri, *Jordan's theorem for the diffeomorphism group of some manifolds*, preprint 2008.

[181] Andrés Navas, *Groups of circle diffeomorphisms*, book in preprint form, 2008.

[182] ———, *On the dynamics of (left) orderable groups*, preprint 2007.

[183] ———, *Actions de groupes de Kazhdan sur le cercle*, Ann. Sci. École Norm. Sup. (4) **35** (2002), no. 5, 749–758. MR1951442 (2003j:58013)

[184] ———, *Quelques nouveaux phénomènes de rang 1 pour les groupes de difféomorphismes du cercle*, Comment. Math. Helv. **80** (2005), no. 2, 355–375. MR2142246 (2006j:57003)

[185] ———, *Growth of groups and diffeomorphisms of the interval*, Geom. Funct. Anal. **18** (2008), no. 3, 988–1028. MR2439001

[186] Amos Nevo and Robert J. Zimmer, *Invariant rigid geometric structures and smooth projective factors*, Geom. Funct. Anal. **19** (2009), no. 2, 520–535. MR2545248 (2010h: 22029)

[187] ———, *Homogenous projective factors for actions of semi-simple Lie groups*, Invent. Math. **138** (1999), no. 2, 229–252. MR1720183 (2000h:22006)

[188] ———, *Rigidity of Furstenberg entropy for semisimple Lie group actions*, Ann. Sci. École Norm. Sup. (4) **33** (2000), no. 3, 321–343. MR1775184 (2001k:22009)

[189] ———, *Actions of semisimple Lie groups with stationary measure*, Rigidity in dynamics and geometry (Cambridge, 2000), Springer, Berlin, 2002, pp. 321–343. MR1919409 (2003j:22029)

[190] ———, *A generalization of the intermediate factors theorem*, J. Anal. Math. **86** (2002), 93–104. MR1894478 (2003f:22019)

[191] ———, *A structure theorem for actions of semisimple Lie groups*, Ann. of Math. (2) **156** (2002), no. 2, 565–594. MR1933077 (2003i:22024)

[192] ———, *Entropy of stationary measures and bounded tangential de-Rham cohomology of semisimple Lie group actions*, Geom. Dedicata **115** (2005), 181–199. MR2180047 (2006k: 22008)

[193] Yann Ollivier, *A January 2005 invitation to random groups*, Ensaios Matemáticos [Mathematical Surveys], Vol. 10, Sociedade Brasileira de Matemática, Rio de Janeiro, 2005. MR2205306

[194] Vladimir Platonov and Andrei Rapinchuk, *Algebraic groups and number theory*, Pure and Applied Mathematics, Vol. 139, Academic Press, Boston, 1994, Translated from the 1991 Russian original by Rachel Rowen. MR1278263 (95b:11039)

[195] Leonid Polterovich, *The geometry of the group of symplectic diffeomorphisms*, Lectures in Mathematics ETH Zürich, Birkhäuser Verlag, Basel, 2001. MR1826128 (2002g: 53157)

[196] ———, *Growth of maps, distortion in groups and symplectic geometry*, Invent. Math. **150** (2002), no. 3, 655–686. MR1946555 (2003i:53126)

[197] Sorin Popa, *Deformation and rigidity for group actions and von Neumann algebras*, International Congress of Mathematicians. Vol. I, European Mathematical Society, Zürich, 2007, pp. 445–477. MR2334200 (2008k:46186)

[198] Pierre Py, *Some remarks on area-preserving actions of lattices*, in this volume.

[199] ———, *Quasi-morphismes de Calabi et graphe de Reeb sur le tore*, C. R. Math. Acad. Sci. Paris **343** (2006), no. 5, 323–328. MR2253051 (2007e:53116)

[200] ———, *Quasi-morphismes et invariant de Calabi*, Ann. Sci. École Norm. Sup. (4) **39** (2006), no. 1, 177–195. MR2224660 (2007f:53116)

[201] R. Quiroga-Barranco and A. Candel, *Rigid and finite type geometric structures*, Geom. Dedicata **106** (2004), 123–143. MR2079838 (2005d:53068)

[202] M. Raghunathan, *Diskretnye podgruppy grupp Li*, Izdat. "Mir," Moscow, 1977, Translated from the English by O. V. Švarcman, Edited by È. B. Vinberg, with a supplement "Arithmeticity of irreducible lattices in semisimple groups of rank greater than 1" by G. A. Margulis. MR0507236 (58 #22394b)

[203] M. S. Raghunathan, *On the first cohomology of discrete subgroups of semisimple Lie groups*, Amer. J. Math. **87** (1965), 103–139. MR0173730 (30 #3940)

[204] Marina Ratner, *On Raghunathan's measure conjecture*, Ann. of Math. (2) **134** (1991), no. 3, 545–607. MR1135878 (93a:22009)

[205] Julio C. Rebelo, *On nilpotent groups of real analytic diffeomorphisms of the torus*, C. R. Acad. Sci. Paris Sér. I Math. **331** (2000), no. 4, 317–322. MR1787192 (2001m:22041)

[206] Julio C. Rebelo and Ana L. Silva, *On the Burnside problem in* Diff(*M*), Discrete Contin. Dyn. Syst. **17** (2007), no. 2, 423–439. MR2257443 (2007j:53107)

[207] Dušan Repovš and Evgenij Ščepin, *A proof of the Hilbert-Smith conjecture for actions by Lipschitz maps*, Math. Ann. **308** (1997), no. 2, 361–364. MR1464908 (99c:57066)

[208] Derek J. S. Robinson, *A course in the theory of groups*, second ed., Graduate Texts in Mathematics, Vol. 80, Springer-Verlag, New York, 1996. MR1357169 (96f:20001)

[209] Federico Rodriguez Hertz, *Global rigidity of certain Abelian actions by toral automorphisms*, J. Mod. Dyn. **1** (2007), no. 3, 425–442. MR2318497 (2008f:37063)

[210] Leslie Saper, *Tilings and finite energy retractions of locally symmetric spaces*, Comment. Math. Helv. **72** (1997), no. 2, 167–202. MR1470087 (99a:22019)

[211] B. Schmidt, *Weakly hyperbolic actions of Kazhdan groups on tori*, Geom. Funct. Anal. **16** (2006), no. 5, 1139–1156. MR2276535 (2008h:37045)

[212] Atle Selberg, *On discontinuous groups in higher-dimensional symmetric spaces*, Contributions to function theory (Internat. Colloq. Function Theory, Bombay, 1960), Tata Institute of Fundamental Research, Bombay, 1960, pp. 147–164. MR0130324 (24 #A188)

[213] Nimish A. Shah, *Invariant measures and orbit closures on homogeneous spaces for actions of subgroups generated by unipotent elements*, Lie groups and ergodic theory (Mumbai, 1996), Tata Inst. Fund. Res. Stud. Math., Vol. 14, Tata Institute of Fundamental Research, Bombay, 1998, pp. 229–271. MR1699367 (2001a:22012)

[214] Yehuda Shalom, *Rigidity, unitary representations of semisimple groups, and fundamental groups of manifolds with rank one transformation group*, Ann. of Math. (2) **152** (2000), no. 1, 113–182. MR1792293 (2001m:22022)

[215] ———, *Measurable group theory*, European Congress of Mathematics, European Mathematical Society, Zürich, 2005, pp. 391–423. MR2185757 (2006k:37007)

[216] ———, *The algebraization of Kazhdan's property (T)*, International Congress of Mathematicians. Vol. II, European Mathematical Society, Zürich, 2006, pp. 1283–1310. MR2275645 (2008a:22003)

[217] R. W. Sharpe, *Differential geometry*, Graduate Texts in Mathematics, Vol. 166, Springer-Verlag, New York, 1997, Cartan's generalization of Klein's Erlangen program, with a foreword by S. S. Chern. MR1453120 (98m:53033)

[218] Daniel Shenfield, *Semisimple representations of* $SL(n, Z[X_1, \ldots, X_k])$, to appear in Groups, Geometry and Dynamics.

[219] Y. T. Siu, *Remarks in talk at Margulis' 60th birthday conference*, February, 2006.

[220] Yum Tong Siu, *The complex-analyticity of harmonic maps and the strong rigidity of compact Kähler manifolds*, Ann. of Math. (2) **112** (1980), no. 1, 73–111. MR584075 (81j:53061)

[221] S. Smale, *Differentiable dynamical systems*, Bull. Amer. Math. Soc. **73** (1967), 747–817. MR0228014 (37 #3598)

[222] R. J. Spatzier, *Harmonic analysis in rigidity theory*, Ergodic theory and its connections with harmonic analysis (Alexandria, 1993), London Math. Soc. Lecture Note Ser., Vol. 205, Cambridge University Press, Cambridge, 1995, pp. 153–205. MR1325698 (96c:22019)

[223] ———, *An invitation to rigidity theory*, Modern dynamical systems and applications, Cambridge University Press, Cambridge, 2004, pp. 211–231. MR2090772 (2006a:53041)

[224] Ralf J. Spatzier and Robert J. Zimmer, *Fundamental groups of negatively curved manifolds and actions of semisimple groups*, Topology **30** (1991), no. 4, 591–601. MR1133874 (92m:57047)

[225] Melanie Stein, *Groups of piecewise linear homeomorphisms*, Trans. Amer. Math. Soc. **332** (1992), no. 2, 477–514. MR1094555 (92k:20075)

[226] Garrett Stuck, *Minimal actions of semisimple groups*, Ergodic Theory Dynam. Systems **16** (1996), no. 4, 821–831. MR1406436 (98a:57046)

[227] William Thurston, *Foliations and groups of diffeomorphisms*, Bull. Amer. Math. Soc. **80** (1974), 304–307. MR0339267 (49 #4027)

[228] William P. Thurston, *A generalization of the Reeb stability theorem*, Topology **13** (1974), 347–352. MR0356087 (50 #8558)

[229] J. Tits, *Free subgroups in linear groups*, J. Algebra **20** (1972), 250–270. MR0286898 (44 #4105)

[230] Alain Valette, *Group pairs with property (T), from arithmetic lattices*, Geom. Dedicata **112** (2005), 183–196. MR2163898 (2006d:22014)

[231] André Weil, *On discrete subgroups of Lie groups*, Ann. of Math. (2) **72** (1960), 369–384. MR0137792 (25 #1241)

[232] ———, *On discrete subgroups of Lie groups. II*, Ann. of Math. (2) **75** (1962), 578–602. MR0137793 (25 #1242)

[233] ———, *Remarks on the cohomology of groups*, Ann. of Math. (2) **80** (1964), 149–157. MR0169956 (30 #199)

[234] Shmuel Weinberger, *Some remarks inspired by the C^0 Zimmer program*, in this volume.

[235] John S. Wilson, *On exponential growth and uniformly exponential growth for groups*, Invent. Math. **155** (2004), no. 2, 287–303. MR2031429 (2004k:20085)

[236] Dave Witte, *Arithmetic groups of higher $\mathbf{Q}$-rank cannot act on 1-manifolds*, Proc. Amer. Math. Soc. **122** (1994), no. 2, 333–340. MR1198459 (95a:22014)

[237] ———, *Measurable quotients of unipotent translations on homogeneous spaces*, Trans. Amer. Math. Soc. **345** (1994), no. 2, 577–594. MR1181187 (95a:22005)

[238] ———, *Cocycle superrigidity for ergodic actions of non-semisimple Lie groups*, Lie groups and ergodic theory (Mumbai, 1996), Tata Inst. Fund. Res. Stud. Math., Vol. 14, Tata Institute of Fundamental Research, Bombay, 1998, pp. 367–386. MR1699372 (2000i:22008)

[239] Sai-Kee Yeung, *Representations of semisimple lattices in mapping class groups*, Int. Math. Res. Not. (2003), no. 31, 1677–1686. MR1981481 (2004d:53047)

[240] A. Zeghib, *Isometry groups and geodesic foliations of Lorentz manifolds. I. Foundations of Lorentz dynamics*, Geom. Funct. Anal. **9** (1999), no. 4, 775–822. MR1719606 (2001g:53059)

[241] ———, *Isometry groups and geodesic foliations of Lorentz manifolds. II. Geometry of analytic Lorentz manifolds with large isometry groups*, Geom. Funct. Anal. **9** (1999), no. 4, 823–854. MR1719610 (2001g:53060)

[242] Abdelghani Zeghib, *Le groupe affine d'une variété riemannienne compacte*, Comm. Anal. Geom. **5** (1997), no. 1, 199–211. MR1456311 (98g:53065)

[243] ———, *Sur les actions affines des groupes discrets*, Ann. Inst. Fourier (Grenoble) **47** (1997), no. 2, 641–685. MR1450429 (98d:57068)

[244] Robert J. Zimmer, *Topological superrigidity*, unpublished notes 1998.

[245] ———, *Ergodic theory and semisimple groups*, Monographs in Mathematics, Vol. 81, Birkhäuser Verlag, Basel, 1984. MR776417 (86j:22014)

[246] ———, *Actions of lattices in semisimple groups preserving a G-structure of finite type*, Ergodic Theory Dynam. Systems **5** (1985), no. 2, 301–306. MR796757 (87g:22011)

[247] ———, *Lattices in semisimple groups and distal geometric structures*, Invent. Math. **80** (1985), no. 1, 123–137. MR784532 (86i:57056)

[248] ——, *On connection-preserving actions of discrete linear groups*, Ergodic Theory Dynam. Systems **6** (1986), no. 4, 639–644. MR873437 (88g:57045)

[249] ——, *Actions of semisimple groups and discrete subgroups*, Proceedings of the International Congress of Mathematicians, Vol. 1, 2 (Berkeley, CA, 1986), American Mathematical Society, Providence, RI, 1987, pp. 1247–1258. MR934329 (89j:22024)

[250] ——, *Lattices in semisimple groups and invariant geometric structures on compact manifolds*, Discrete groups in geometry and analysis (New Haven, CT, 1984), Progr. Math., Vol. 67, Birkhäuser, Boston, 1987, pp. 152–210. MR900826 (88i:22025)

[251] ——, *Representations of fundamental groups of manifolds with a semisimple transformation group*, J. Amer. Math. Soc. **2** (1989), no. 2, 201–213. MR973308 (90i:22021)

[252] ——, *On the algebraic hull of an automorphism group of a principal bundle*, Comment. Math. Helv. **65** (1990), no. 3, 375–387. MR1069815 (92f:57050)

[253] ——, *Spectrum, entropy, and geometric structures for smooth actions of Kazhdan groups*, Israel J. Math. 75 (1991), no. 1, 65–80. MR1147291 (93i:22014)

[254] ——, *Superrigidity, Ratner's theorem, and fundamental groups*, Israel J. Math. **74** (1991), no. 2–3, 199–207. MR1135234 (93b:22019)

[255] ——, *Automorphism groups and fundamental groups of geometric manifolds*, Differential geometry: Riemannian geometry (Los Angeles, CA, 1990), Proc. Sympos. Pure Math., Vol. 54, American Mathematical Society, Providence, RI, 1993, pp. 693–710. MR1216656 (95a:58017)

[256] ——, *Discrete groups and non-Riemannian homogeneous spaces*, J. Amer. Math. Soc. **7** (1994), no. 1, 159–168. MR1207014 (94e:22021)

[257] ——, *Entropy and arithmetic quotients for simple automorphism groups of geometric manifolds*, Geom. Dedicata **107** (2004), 47–56. MR2110753 (2005i:22011)

[258] Robert J. Zimmer and Dave Witte Morris, *Ergodic theory, groups, and geometry*, CBMS Regional Conference Series in Mathematics, 109. Published for the Conference Board of the Mathematical Sciences, Washington, DC, by the American Mathematical Society, Providence, RI, 2008. MR2457556 (2009k:37059)

5

DAVE WITTE MORRIS

CAN LATTICES IN SL$(n, \mathbb{R})$ ACT ON THE CIRCLE?

TO ROBERT J. ZIMMER ON HIS 60TH BIRTHDAY

1. Overview

The following theorem is a simple example of the Zimmer program's principle that large groups should not be able to act on small manifolds. (For a description of the Zimmer program, see D. Fisher's survey article [Fis] elsewhere in this volume.) Unless stated otherwise, we assume actions are continuous, but we require no additional regularity.

THEOREM 1.1. (WITTE) *Let Γ be a finite-index subgroup of* SL$(n, \mathbb{Z})$, *with $n \geq 3$. Then Γ has no faithful action on the circle S^1.*

REMARK 1.2.

1) The proof of this theorem is not at all difficult, and will be given in Section 3, after the result is translated to a more algebraic form in Section 2. The proof illustrates the use of calculations with unipotent elements, which is a standard technique in the theory of arithmetic groups.
2) The assumption that $n \geq 3$ is essential. Indeed, some finite-index subgroup of SL$(2, \mathbb{Z})$ is a free group, which has countless actions on the circle (some faithful and some not).

A group Γ as in Theorem 1.1 is a lattice in SL$(n, \mathbb{R})$. It is an open question whether the theorem generalizes to the other lattices:

CONJECTURE 1.3. *Let Γ be a lattice in* SL$(n, \mathbb{R})$, *with $n \geq 3$. Then Γ has no faithful action on S^1.*

Note that if this conjecture is true, then every action of Γ on S^1 or $\mathbb{R}$ has a nontrivial kernel. Since the kernel is a normal subgroup, and the Margulis

158

normal subgroup theorem tells us that lattices in SL $(n, \mathbb{R})$ are "almost simple" (if $n \geq 3$), it follows that the kernel is a finite-index subgroup of Γ. Therefore, every Γ-orbit is finite. Such reasoning shows that the conjecture can be restated as follows:

CONJECTURE 1.3′. *Let Γ be a lattice in SL $(n, \mathbb{R})$, with $n \geq 3$. Whenever Γ acts on S^1, every orbit is finite.*

There is considerable evidence for the above conjectures, including the following important theorem:

THEOREM 1.4. (GHYS, BURGER-MONOD) *If Γ is any lattice in SL $(n, \mathbb{R})$, with $n \geq 3$, then every action of Γ on S^1 has at least one finite orbit.*

REMARK 1.5. The proof of É. Ghys utilizes amenability (or the Furstenberg boundary); see Sections 6 and 7. The proof of M. Burger and N. Monod is based on a vanishing theorem for bounded cohomology; see Section 8.

The above conjectures are about continuous actions, but they can be weakened by considering only differentiable actions. It will be explained in Section 4 that combining the Ghys-Burger-Monod theorem with the well-known Reeb-Thurston stability theorem establishes these weaker conjectures:

COROLLARY 1.6. (GHYS, BURGER-MONOD) *If Γ is any lattice in SL $(n, \mathbb{R})$, with $n \geq 3$, then:*

1) *Γ has no faithful C^1 action on S^1, and*
2) *whenever Γ acts on S^1 by C^1 diffeomorphisms, every orbit is finite.*

For $n \geq 3$, it is well known that every lattice in SL $(n, \mathbb{R})$ has Kazhdan's property (T) (see Definition 5.2), and the following result shows that if we strengthen the differentiability hypothesis slightly, then the conclusion is true for all groups that have that property. This is a very broad class of groups (including many groups that are not even linear), so one would expect a much stronger result to hold for the special case of lattices in SL $(n, \mathbb{R})$. Thus, this theorem constitutes good evidence for Conjecture 1.3. The proof will be presented in Section 5; it is both elegant and elementary.

THEOREM 1.7. (NAVAS) *If Γ is any infinite, discrete group with Kazhdan's property (T), then Γ has no faithful C^2 action on the circle.*

Bounded generation provides another approach to proving Theorem 1.1 and some other cases of Conjecture 1.3. This strategy is explained in Section 9.

In spite of the above results, Conjecture 1.3 remains completely open for cocompact lattices:

PROBLEM 1.8. Find a cocompact lattice Γ in SL $(n, \mathbb{R})$, for some n, such that no finite-index subgroup of Γ has a faithful action on S^1.

REMARK 1.9. A final section of the paper (§10) briefly discusses the generalization of Conjecture 1.3 to lattices in other semisimple Lie groups and two other topics: lattices that do act on the circle, and actions on trees.

NOTES FOR §1. The survey of É. Ghys [Gh3] and the forthcoming book of A. Navas [Na2] are excellent introductions to the study of group actions on the circle. Versions of Conjecture 1.3 were discussed in conversation as early as 1990, but the first published appearance may be in [Gh2, p. 200].

Expositions of the Margulis normal subgroup theorem can be found in [Ma2, chap. 4] and [Zi2, chap. 8].

Regarding the equivalence of Conjectures 1.3 and 1.3', see [Uom, thm. 1] for a proof that if every Γ-orbit on a connected manifold is finite, then the kernel of the action has finite index.

ACKNOWLEDGMENTS: Preparation of this paper was partially supported by research grants from the Natural Sciences and Engineering Research Council of Canada and the U. S. National Science Foundation. I am grateful to M. Burger, É. Ghys, N. Monod, A. Navas, and an anonymous referee, for helpful comments on a preliminary version of the manuscript.

2. Algebraic Reformulation of the Conjecture

In this section, we explain that Conjecture 1.3 can be reformulated in a purely algebraic form (see Proposition 2.8). The proof has two main parts: having an action on S^1 is almost the same as having an action on $\mathbb{R}$, and having an action on $\mathbb{R}$ is essentially the same as being left orderable.

DEFINITION 2.1. Γ is *left orderable* if there is a left-invariant total order on Γ. More precisely, there is an order relation $\prec$ on Γ, such that

1) $\prec$ is a total order (that is, for all $a, b \in \Gamma$, either $a \prec b$ or $a \succ b$, or $a = b$); and

2) $\prec$ is invariant under multiplication on the left (that is, for all $a, b, c \in \Gamma$, if $a \prec b$, then $ca \prec cb$).

EXERCISE 2.2. Show Γ is left orderable if and only if there exists a subset P of Γ, such that

1) $\Gamma = P \sqcup \{e\} \sqcup P^{-1}$ (disjoint union), where $P^{-1} = \{ a^{-1} \mid a \in P \}$; and
2) P is closed under multiplication (that is, $ab \in P$, for all $a, b \in P$).

Thus, being left orderable is a property of the internal algebraic structure of Γ.

[Hint: A subset P as above is the "positive cone" of an order $\prec$. Given P, define $a \prec b$ if $b^{-1}a \in P$. Given $\prec$, define $P = \{ a \in \Gamma \mid a \succ e \}$.]

REMARK 2.3. The group Γ is said to be *right orderable* if there is a *right*-invariant total order on Γ. The following exercise shows that the choice between "left orderable" and "right orderable" is entirely a matter of personal preference.

EXERCISE 2.4. *Show that Γ is left orderable if and only if Γ is right orderable.*

[Hint: Define $x \prec\!\!\prec y$ if $x^{-1} \prec y^{-1}$. Alternatively, the conclusion can be derived from Exercise 2.2 and its analogue for right-orderable groups.]

With this terminology, we can state the following algebraic conjecture, which will be seen to be equivalent to Conjecture 1.3.

CONJECTURE 2.5. *Let Γ be a lattice in SL$(n, \mathbb{R})$, with $n \geq 3$. Then Γ is not left orderable.*

As a tool for showing that Conjecture 2.5 is equivalent to Conjecture 1.3, let us give a geometric interpretation of being left orderable.

LEMMA 2.6. *A countable group Γ is left orderable if and only if there is an orientation-preserving, faithful action of Γ on $\mathbb{R}$.*

Proof. ($\Leftarrow$) For $a, b \in \Gamma$, define

$$a \prec b \quad \text{if} \quad a(0) < b(0).$$

It is easy to see that $\prec$ is transitive and antisymmetric, so it defines a partial order on Γ.

For any $c \in \Gamma$, the function $x \mapsto c(x)$ is a strictly increasing function on $\mathbb{R}$ (because the action is orientation preserving). Hence, if $a(0) < b(0)$, then $c(a(0)) < c(b(0))$. Therefore, $\prec$ is left-invariant (as required in Definition 2.1(2)).

However, the relation $\prec$ may not be a total order, because it could happen that $a(0) = b(0)$, even though $a \neq b$. It is not difficult to modify the definition to deal with this technical point (see Exercise 2.7).

($\Rightarrow$) We leave this as an exercise for the reader.[1] □

EXERCISE 2.7. *Complete the proof of Lemma 2.6($\Leftarrow$), by defining a left-invariant total order on Γ.*

[Hint: Make a list $q_1, q_2, \ldots$ of the rational numbers and define $a \prec b$ if $a(q_n) < b(q_n)$, where n is minimal with $a(q_n) \neq b(q_n)$.]

PROPOSITION 2.8. *Conjecture 1.3 and 2.5 are equivalent.*

Proof. We show there exists a counterexample to Conjecture 1.3 if and only if there exists a counterexample to Conjecture 2.5.

($\Leftarrow$) Suppose Γ is a counterexample to Conjecture 2.5, so Γ is left orderable. Then Lemma 2.6 tells us that Γ has a faithful action on $\mathbb{R}$. This implies that Γ acts faithfully on the one-point compactification of $\mathbb{R}$, which is homeomorphic to S^1. So Γ is a counterexample to Conjecture 1.3.

($\Rightarrow$) Suppose Γ is a counterexample to Conjecture 1.3, so Γ has a faithful action on S^1. From the Ghys-Burger-Monod theorem (1.4), we know that the action has a finite orbit. Therefore, some subgroup Γ' of finite index in Γ has a fixed point p. Then Γ' acts faithfully on the complement $S^1 \smallsetminus \{p\}$, which is homeomorphic to $\mathbb{R}$. Let Γ'' be the subgroup of Γ' consisting of the elements that act by orientation-preserving homeomorphisms (so Γ'' is either Γ', or a subgroup of index 2 in Γ'). Then Lemma 2.6 tells us that Γ'' is left orderable. Furthermore, Γ'' is a lattice in $\mathrm{SL}(n, \mathbb{R})$ (because it has finite index in Γ). Therefore, Γ'' is a counterexample to Conjecture 2.5. □

---

1. Hint: If the order relation $(\Gamma, \prec)$ is dense (that is, if $a \prec b \Rightarrow \exists c \in \Gamma, a \prec c \prec b$), then it is order-isomorphic to $(\mathbb{Q}, <)$. The action of Γ on $(\Gamma, \prec)$ by left multiplication extends to a continuous action of Γ on the Dedekind completion, which is homeomorphic to $\mathbb{R}$.

WARNING 2.9. The proof does not show that the assertions of the two conjectures are equivalent for each individual lattice Γ. Rather, if one of the conjectures is valid for all lattices Γ, then the other conjecture is also valid for all Γ. More precisely, if one of the conjectures is valid for every finite-index subgroup of Γ, then the other conjecture is also valid for Γ.

REMARK 2.10. In order to prove that Conjecture 2.5 implies Conjecture 1.3, we appealed to Theorem 1.4, which is quite deep. A stronger algebraic conjecture (that no central extension of Γ is left orderable) can easily be shown to imply Conjecture 1.3:

EXERCISE 2.11. A *central extension* of Γ is a group Λ, such that $\Lambda/Z \cong \Gamma$, for some subgroup Z of the center of Λ. Show that if Γ has an orientation-preserving, faithful action on S^1, then some central extension of Γ is left orderable.

[Hint: Because $\mathbb{R}$ is the universal cover of S^1, every homeomorphism of S^1 lifts to a homeomorphism of $\mathbb{R}$, and the lift is unique modulo an element of the fundamental group $\mathbb{Z}$. Let Λ be the subgroup of Homeo$_+$ $(\mathbb{R})$ consisting of all of the possible lifts of all of the elements of Γ, and show that Λ is a central extension that is left orderable.]

NOTES FOR §2. The material in this section is well known. See [KM] for a treatment of the algebraic theory of left-ordered groups. Informative discussions of orderings, actions on the line, and related topics appear in [Gh3, §6.5] and [Na2, §2.2.3–2.2.6].

3. SL$(n, \mathbb{Z})$ Has no Faithful Action on the Circle

In this section, we prove Theorem 1.1 by exploiting the interaction between some obvious nilpotent subgroups of Γ. The crucial ingredients are Lemma 3.6 and the fact that the theorem can be restated in the following algebraic form (cf. Proposition 2.8).

THEOREM 3.1. (WITTE) *If Γ is a finite-index subgroup of* SL $(n, \mathbb{Z})$, *with $n \geq 3$, then Γ is not left orderable.*

REMARK 3.2. It is easy to see that the group SL $(n, \mathbb{Z})$ itself is not left orderable, because it has elements of finite order. But there are subgroups of finite index that do not have any elements of finite order, and we will show that they, too, are not left orderable.

DEFINITION 3.3. Suppose $\prec$ is a left-invariant order relation on Γ. For elements a and b of Γ, we say a is *infinitely smaller* than b (denoted $a \ll b$) if either $a^k \prec b$ for all $k \in \mathbb{Z}$ or $a^k \prec b^{-1}$ for all $k \in \mathbb{Z}$. Notice that the relation $\ll$ is transitive.

NOTATION 3.4. The commutator $a^{-1}b^{-1}ab$ of elements a and b of a group Γ is denoted $[a, b]$. It is straightforward to check that a commutes with b if and only if $[a, b] = e$ (the identity element of Γ).

LEMMA 3.5. *Let a and b be elements of Γ. If $[a, b]$ commutes with both a and b, then $[b^k, a^m] = [a, b]^{-km}$ for all $k, m \in \mathbb{Z}$.*

Proof. Exercise (or see [Gor, Lems. 2.2.2(i) and 2.2.4(iii), pp. 19–20]). □

LEMMA 3.6. (AULT, RHEMTULLA) *Suppose a, b, z are nonidentity elements of a left-ordered group H, with $[a, b] = z^k$ for some nonzero $k \in \mathbb{Z}$, and $[a, z] = [b, z] = e$. Then either $z \ll a$ or $z \ll b$.*

Proof. Assume, for simplicity, that $a, b, z \geq e$ and $k > 0$. (All other cases can be reduced to this one by replacing some or all of a, b, z with their inverses, and/or interchanging a with b.)

We may assume $z \not\ll a$, so $a \prec z^p$ for some $p \in \mathbb{Z}^+$. Similarly, we may assume $b \prec z^q$ for some $q \in \mathbb{Z}^+$. Then (using the left-invariance of $\prec$) we have

$$e \prec a^{-1}z^p, \quad e \prec b^{-1}z^q, \quad e \prec a, \quad e \prec b, \quad e \prec z.$$

Hence, for all $m \in \mathbb{Z}^+$, we have

$$e \prec (b^{-1}z^q)^m (a^{-1}z^p)^m b^m a^m$$

$$= b^{-m}a^{-m}b^m a^m z^{(p+q)m} \qquad \text{(z commutes with a and b)}$$

$$= [b^m, a^m]z^{(p+q)m}$$

$$= z^{-km^2}z^{(p+q)m} \qquad \text{($[a, b] = z^k$ and see (lemma 3.5))}$$

$$= z^{-km^2+(p+q)m}$$

$$= z^{\text{negative}} \qquad \text{(if m is sufficiently large)}$$

$$\prec e.$$

This is a contradiction. □

Proof of Theorem 3.1. Suppose Γ is left orderable. (This will lead to a contradiction.) Because subgroups of left-orderable groups are left orderable, we may assume $n = 3$; that is, Γ has finite index in SL$(3, \mathbb{Z})$. Hence, there is some $k \in \mathbb{Z}^+$ such that the six matrices

3.7

$$a_1 = \begin{bmatrix} 1 & k & 0 \\ 0 & 1 & 0 \\ 0 & 0 & 1 \end{bmatrix}, \quad a_2 = \begin{bmatrix} 1 & 0 & k \\ 0 & 1 & 0 \\ 0 & 0 & 1 \end{bmatrix}, \quad a_3 = \begin{bmatrix} 1 & 0 & 0 \\ 0 & 1 & k \\ 0 & 0 & 1 \end{bmatrix},$$

$$a_4 = \begin{bmatrix} 1 & 0 & 0 \\ k & 1 & 0 \\ 0 & 0 & 1 \end{bmatrix}, \quad a_5 = \begin{bmatrix} 1 & 0 & 0 \\ 0 & 1 & 0 \\ k & 0 & 1 \end{bmatrix}, \quad a_6 = \begin{bmatrix} 1 & 0 & 0 \\ 0 & 1 & 0 \\ 0 & k & 1 \end{bmatrix}$$

all belong to Γ. A straightforward check shows that $[a_i, a_{i+1}] = e$ and $[a_{i-1}, a_{i+1}] = (a_i)^{\pm k}$ for $i = 1, \ldots, 6$, with subscripts read modulo 6. Thus, Lemma 3.6 asserts

3.8 $$\text{either } a_i \ll a_{i-1} \text{ or } a_i \ll a_{i+1}.$$

In particular, we must have either $a_1 \ll a_6$ or $a_1 \ll a_2$. Assume for definiteness that $a_1 \ll a_2$. (The other case is very similar.) For each i, Lemma 3.6 implies that if $a_{i-1} \ll a_i$, then $a_i \ll a_{i+1}$. Since $a_1 \ll a_2$, we conclude by induction that

$$a_1 \ll a_2 \ll a_3 \ll a_4 \ll a_5 \ll a_6 \ll a_1.$$

Thus $a_1 \ll a_1$, a contradiction. $\qquad\square$

For the reader acquainted with root systems, let us give a more conceptual presentation of the proof of Theorem 3.1.

Alternate Version of the Proof of Theorem 3.1. The root system of SL$(3, \mathbb{Z})$ is of type A_2, pictured in Figure 1. For $i = 1, \ldots, 6$, let U_i be the root space of Γ corresponding to the root α_i. (Note that $U_i \cong \mathbb{Z}$ is cyclic. In the notation of the above proof, the element a_i belongs to U_i.) Define

$$\alpha_i \ll \alpha_j \text{ if there exists } u \in U_j, \text{ such that } U_i \prec u$$

(that is, $v \prec u$, for all $v \in U_i$).

It is obvious, from Figure 1, that $\alpha_{i-1} + \alpha_{i+1} = \alpha_i$, and that $\alpha_i + \alpha_{i+1}$ is not a root, so

$$e \neq [U_{i-1}, U_{i+1}] \subset U_i \qquad \text{and} \qquad [U_i, U_{i+1}] = e.$$

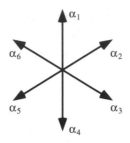

Fig. 1. The root system of SL $(3, \mathbb{Z})$ (type A_2).

Hence, Lemma 3.6 implies that either $\alpha_i \ll \alpha_{i-1}$ or $\alpha_i \ll \alpha_{i+1}$. Arguing as in the above proof, we conclude that

$$\alpha_1 \ll \alpha_2 \ll \alpha_3 \ll \alpha_4 \ll \alpha_5 \ll \alpha_6 \ll \alpha_1.$$

Thus $\alpha_1 \ll \alpha_1$, a contradiction. □

NOTES FOR §3. Theorem 3.1 is due to D. Witte [Wit]. (An exposition of the proof also appears in [Gh3, thm. 7.2].) Lemma 3.6 is a special case of a theorem of J. C. Ault [Aul] and A. H. Rhemtulla [Rhe]: if Λ is any nontrivial, finitely generated, left-ordered nilpotent group, then there is some $a \in \Lambda$, such that $[\Lambda, \Lambda] \prec a$.

An elementary proof that SL $(n, \mathbb{Z})$ has no nontrivial actions on the circle can be found in [BV], but the argument uses the existence of elements of finite order, so it does not apply to finite-index subgroups of SL $(n, \mathbb{Z})$.

4. The Reeb-Thurston Stability Theorem

The Ghys-Burger-Monod theorem (1.4) tells us that if Γ is a lattice in $G = $ SL $(3, \mathbb{R})$, then any action of Γ on S^1 has a finite orbit; in other words, some finite-index subgroup Γ' of Γ has a fixed point. In order to deduce Corollary 1.6 from this it suffices to show that if the action is by (orientation-preserving) C^1 diffeomorphisms, then Γ' acts trivially. This triviality of Γ' is immediate from the following result:

PROPOSITION 4.1. (REEB-THURSTON STABILITY THEOREM) *Suppose*

- *Λ is a finitely generated subgroup of $\mathrm{Diff}^1_+ (S^1)$,*
- *Λ has a fixed point, and*
- *the abelianization $\Lambda/[\Lambda, \Lambda]$ is finite.*

Then $\Lambda = \{e\}$ is trivial.

A differentiable action on S^1 that has a fixed point can be transformed into a differentiable action on the unit interval $[0, 1]$ (cf. pf. of Proposition 2.8($\Rightarrow$)). Thus, Proposition 4.1 can be reformulated as follows:

PROPOSITION 4.1′. (REEB-THURSTON STABILITY THEOREM) *Suppose*

- $I = [0, 1]$ *is the unit interval,*
- Λ *is a finitely generated subgroup of* $\mathrm{Diff}^1_+(I)$, *and*
- *the abelianization* $\Lambda/[\Lambda, \Lambda]$ *is finite.*

Then $\Lambda = \{e\}$ *is trivial.*

Proof in a special case. Define $\sigma : \Lambda \to \mathbb{R}^+$ by $\sigma(\lambda) = \lambda'(0)$. From the Chain Rule, we see that σ is a (multiplicative) homomorphism. Because $\Lambda/[\Lambda, \Lambda]$ is finite and $\mathbb{R}^+$ is abelian, this implies that $\sigma(\Lambda)$ is finite. However, $\mathbb{R}^+$ has no nontrivial, finite subgroups, so this implies that $\sigma(\Lambda)$ is trivial; therefore

$$\lambda'(0) = 1, \text{ for all } \lambda \in \Lambda.$$

For simplicity, let us assume, henceforth, that each element of Λ is real analytic, rather than merely C^1. (We remark that there is no need to assume Λ is finitely generated in the real-analytic case. The proof for C^1 actions is similar, but requires some additional effort, and does use the hypothesis that Λ is finitely generated.) Thus, each element λ of Λ can be expressed as a power series in a neighborhood of 0:

$$\lambda(x) = x + a_{\lambda,2}x^2 + a_{\lambda,3}x^3 + \cdots.$$

(There is no constant term, because $\lambda(0) = 0$; the coefficient of x is 1, because $\lambda'(0) = 1$.)

Now suppose Λ is nontrivial. (This will lead to a contradiction.) Then there exist n and λ, such that

4.2 $$a_{\lambda,n} \neq 0.$$

We may assume n is minimal; hence

$$\lambda(x) = x + a_{\lambda,n}x^n + a_{\lambda,n+1}x^{n+1} + \cdots \text{ for all } \lambda \in \Lambda.$$

Then, for $\lambda, \gamma \in \Lambda$, we have

$$x + a_{\lambda\gamma,n}x^n + O(x^{n+1}) = (\lambda\gamma)(x)$$

$$= \lambda(\gamma(x))$$

$$= \gamma(x) + a_{\lambda,n}(\gamma(x))^n + O(x^{n+1})$$

$$= (x + a_{\gamma,n}x^n + O(x^{n+1}))$$

$$+ a_{\lambda,n}(x + a_{\gamma,n}x^n + O(x^{n+1}))^n$$

$$+ O(x^{n+1})$$

$$= x + (a_{\gamma,n} + a_{\lambda,n})x^n + O(x^{n+1}).$$

Thus the map $\tau \colon \Lambda \to \mathbb{R}$, defined by $\tau(\lambda) = a_{\lambda,n}$, is an additive homomorphism. Because $\Lambda/[\Lambda,\Lambda]$ is finite, but $\mathbb{R}$ has no nontrivial finite subgroups, this implies that $\tau(\Lambda)$ is trivial; therefore $a_{\lambda,n} = 0$ for every $\lambda \in \Lambda$. This contradicts Equation (4.2). $\square$

REMARK 4.3 ON THE GENERAL CASE Suppose Λ is nontrivial. Then there is no harm in assuming that Λ acts nontrivially on every neighborhood of 0. Hence, letting Λ_0 be a finite generating set for Λ, we may choose $\lambda_0 \in \Lambda_0$ and a sequence $x_n \to 0^+$, such that, for all n, we have $\lambda_0(x_n) \neq x_n$ and

$$|\lambda(x_n) - x_n| \leq |\lambda_0(x_n) - x_n| \text{ for all } \lambda \in \Lambda_0.$$

By passing to a subsequence of $\{x_n\}$, we may assume

$$\tau(\lambda) = \lim_{n \to \infty} \frac{\lambda(x_n) - x_n}{\lambda_0(x_n) - x_n}$$

exists for all $\lambda \in \Lambda_0$. Because $\lambda'(0) = 1$ and λ is C^1, it can be shown that $\tau(\lambda\gamma) = \tau(\lambda) + \tau(\gamma)$. Therefore $\tau \colon \Lambda \to \mathbb{R}$ is a (nontrivial) homomorphism. This is a contradiction.

REMARK 4.4. (ZIMMER) Here is the outline of a nice proof of Proposition 4.1′, under the additional assumption that Λ has Kazhdan's property (T).

1) It suffices to show that the fixed points of Γ form a dense subset of I; thus, we may assume that 0 and 1 are the only points that are fixed by Γ.
2) We have $\gamma'(0) = 1$ for all $\gamma \in \Gamma$.

3) Define a unitary representation of Γ on $L^2(I)$ by

$$f^\gamma(t) = f(\gamma t) \, |\gamma'(t)|^{1/2}.$$

4) For any $\gamma \in \Gamma$ and any $\delta > 0$, if f is the characteristic function of a sufficiently small neighborhood of 0, then $\|f^\gamma - f\| < \delta \|f\|$. Hence, this unitary representation has almost-invariant vectors.

5) Because Λ has Kazhdan's property (T), this implies there are fixed vectors: there exists $f \in L^2(I) \smallsetminus \{0\}$, such that $f^\gamma = f$ for all $\gamma \in \Gamma$.

6) Every point in the essential support of f is a fixed point of Γ.

7) This is a contradiction.

NOTES FOR §4. Proposition 4.1 was proved by W. Thurston [Thu] in a more general form that also applies to actions on manifolds of higher dimension. (It generalizes a theorem of G. Reeb.)

See [RS] or [Sch] for details of the proof sketched in Remark 4.3.

The proof outlined in Remark 4.4 is due to R. J. Zimmer; details appear in [WZ, §5, pp. 108–109].

5. Smooth Actions of Kazhdan Groups on the Circle

In this section, we prove Theorem 1.7. First, let us recall one of the many equivalent definitions of Kazhdan's property (T).

NOTATION 5.1. For any real Hilbert space $\mathcal{H}$, we use Isom $(\mathcal{H})$ to denote the isometry group of $\mathcal{H}$. (We remark that each isometry of $\mathcal{H}$ is the composition of a translation with a norm-preserving linear transformation.)

DEFINITION 5.2. We say that a discrete group Γ has *Kazhdan's property* (T) if, for every homomorphism $\rho\colon \Gamma \to$ Isom $(\mathcal{H})$, where $\mathcal{H}$ is a real Hilbert space, there exists $v \in \mathcal{H}$, such that $\rho(g)v = v$, for every $g \in \Gamma$.

In short, to say that Γ has Kazhdan's property (T) means that every isometric action of Γ on any Hilbert space has a fixed point. The importance of this notion for our purposes stems from the following result, whose proof we omit.

THEOREM 5.3. (KAZHDAN) *If $n \geq 3$, then every lattice in* SL $(n, \mathbb{R})$ *has Kazhdan's property* (T).

EXERCISE 5.4. *Show that if Γ is an infinite group with Kazhdan's property (T), then Γ is not abelian.*

[Hint: Every group with Kazhdan's property (T) is finitely generated, and every finitely generated abelian group is either finite or has a quotient isomorphic to $\mathbb{Z}$.]

We now turn to the proof of Theorem 1.7. To simplify notation, we may think of S^1 as $[-\pi/2, \pi/2]$. In particular, for any diffeomorphism of S^1 and any $x \in S^1$, the derivative $g'(x)$ is a well-defined real number.

DEFINITION 5.5.

- Let $\mathcal{F}(S^1 \times S^1)$ be the vector space of measurable functions on $S^1 \times S^1$ (with two functions being identified if they are equal almost everywhere).
- Define an action of $\mathrm{Diff}^2(S^1)$ on $\mathcal{F}(S^1 \times S^1)$ by

$$F^g(x, y) = F(g(x), g(y)) |g'(x)|^{1/2} |g'(y)|^{1/2}$$

for $F \in \mathcal{F}(S^1 \times S^1)$ and $g \in \mathrm{Diff}^2(S^1)$.
- Let $\|F\|_2 = \left(\int_{S^1} \int_{S^1} F(x, y)^2 \, dx \, dy \right)^{1/2}$ be the L^2-norm of F; note that $\|F\|_2 = \infty$ if $F \notin L^2(S^1 \times S^1)$.

Note that

$$F^{gh} = (F^g)^h \quad \text{and} \quad \|F^g\|_2 = \|F\|_2 \quad \text{for } F \in \mathcal{F}(S^1 \times S^1) \text{ and } g, h \in \mathrm{Diff}^2(S^1).$$

NOTATION 5.6.

- Choose a positive function f on S^1, such that

—f has a $1/x$ singularity at the point 0 of S^1, and
—f is C^∞ everywhere else;

that is, identifying S^1 with $[-\pi/2, \pi/2]$, we have

5.7 $$f(x) - \frac{1}{|x|} \in C^\infty(S^1).$$

For example, one may take $f(x) = |\cot x|$.
- Now define
$$\Phi(x, y) = f(x - y) \text{ on } S^1 \times S^1.$$

Because of the $1/x$ singularity of f, it is easy to see that $\Phi \notin L^2(S^1 \times S^1)$.

For any $g \in \mathrm{Diff}^2(S^1)$, the following calculation shows that the singularity of Φ^g cancels the singularity of Φ.

LEMMA 5.8. *The difference $\Phi^g - \Phi$ is a bounded function on $S^1 \times S^1$, for any $g \in \mathrm{Diff}^2(S^1)$.*

Proof. From Equation (5.7), we see that there is no harm in working with the function $\Phi_0(x, y) = 1/|x - y|$, instead of Φ. Also, in order to reduce the number of absolute-value signs, let us assume $g' \geq 0$ everywhere.

$$|\Phi^g(x, y) - \Phi(x, y)|$$

$$\approx |\Phi_0^g(x, y) - \Phi_0(x, y)|$$

$$= \left| g'(x)^{1/2} g'(y)^{1/2} \Phi_0(g(x), g(y)) - \Phi_0(x, y) \right|$$

$$= \left| \frac{g'(x)^{1/2} g'(y)^{1/2}}{|g(x) - g(y)|} - \frac{1}{|x - y|} \right|$$

$$= \left| \frac{g'(x)^{1/2} g'(y)^{1/2}}{|g'(t)(x - y)|} - \frac{1}{|x - y|} \right| \qquad \left(\begin{array}{c} \exists t \in (x, y), \text{ by} \\ \text{Mean Value thm.} \end{array} \right)$$

$$= \frac{|g'(x)^{1/2} g'(y)^{1/2} - g'(t)|}{g'(t) |x - y|}$$

$$= \frac{|g'(x) g'(y) - g'(t)^2|}{(g'(x)^{1/2} g'(y)^{1/2} + g'(t)) g'(t) |x - y|} \qquad \left(\begin{array}{c} \text{multiply by conjugate} \\ \text{of numerator} \end{array} \right)$$

$$= O\left(\frac{|g'(x) g'(y) - g'(t)^2|}{|x - y|} \right) \qquad \left(\begin{array}{c} g \text{ diffeomorphism} \\ \Rightarrow g' \text{ is never } 0 \end{array} \right).$$

Now, we wish to show that the numerator is bounded by a constant multiple of the denominator.

$$|g'(x) g'(y) - g'(t)^2|$$

$$\leq g'(x) |g'(y) - g'(t)| + g'(t) |g'(x) - g'(t)| \qquad \text{(Triangle inequality)}$$

$$= g'(x) |g''(u)| |y - t| + g'(t) |g''(v)| |x - t| \qquad \left(\begin{array}{c} \text{Mean Value thm.} \\ \text{applied to } g' \end{array} \right)$$

$$\leq g'(x) |g''(u)| |x - y| + g'(t) |g''(v)| |x - y| \qquad \text{(because } t \in (x, y))$$

$$= O(|x - y|) \qquad \left(\begin{array}{c} g', g'' \text{ continuous,} \\ \text{so bounded} \end{array} \right).$$

$\square$

We will also use the following classical fact:

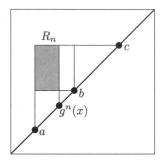

Fig. 2. A rectangle R_n (shaded) in $S^1 \times S^1$.

LEMMA 5.9. (HÖLDER, 1901) *Every nonabelian group of homeomorphisms of S^1 contains at least one nonidentity element that has a fixed point.*

Proof of Theorem 1.7. Suppose Γ is a discrete group with Kazhdan's property (T), and we have a faithful C^2 action of Γ on the circle S^1. From Lemma 5.8, we see that

5.10 $(\Phi + L^2(S^1 \times S^1))^g = \Phi + L^2(S^1 \times S^1)$ for all $g \in \mathrm{Diff}^2(S^1)$.

Thus, Γ acts (by isometries) on the affine Hilbert space $\Phi + L^2(S^1 \times S^1)$. Because Γ has Kazhdan's property, we conclude that Γ has a fixed point F in $\Phi + L^2(S^1 \times S^1)$:

$$F^g = F \text{ for all } g \in \Gamma.$$

Because $\Phi \notin L^2(S^1 \times S^1)$ and $F - \Phi \in L^2(S^1 \times S^1)$, it is obvious that $F \notin L^2(S^1 \times S^1)$.

Now, define a measure μ on $S^1 \times S^1$ by $\mu = F^2 \, dx \, dy$. Then

1) μ is a Γ-invariant measure on $S^1 \times S^1$ (because F is Γ-invariant); and
2) if $R = (a_1, b_1) \times (a_2, b_2)$ is a rectangle in $S^1 \times S^1$, then, from the $1/x$ singularity of Φ along the diagonal, we see that

$$\mu(R) = \begin{cases} \infty & \text{if } R \text{ intersects the diagonal;} \\ \text{finite} & \text{if } R \text{ is away from the diagonal.} \end{cases}$$

For example, in Figure 2, the rectangle $(a, b) \times (b, c)$ has infinite measure because it touches the diagonal at (b, b), but the shaded rectangle R_k has finite measure, because it does not touch the diagonal.

Because Γ is not abelian, Lemma 5.9 tells us we may choose $g \in \Gamma$, such that g has a fixed point. Then, by passing to a triple cover of S^1, we obtain an action of (a finite extension of) Γ in which g has at least three fixed points:

$$g(a) = a, \qquad g(b) = b, \qquad g(c) = c.$$

By perhaps replacing a and b with different fixed points, we may assume, for $x \in (a, b)$, that

$$\lim g^k(x) = \begin{cases} b & \text{as } k \to \infty \\ a & \text{as } k \to -\infty. \end{cases}$$

For each $k \in \mathbb{Z}$, define

$$R_k = (a, g^k(x)) \times (b, c)$$

(see Figure 2). Then

$$g(R_k) = R_{k+1},$$

so

$$\mu(R_k) = \mu(R_{k+1}).$$

Therefore

$$0 = \sum_{k=-\infty}^{\infty} (\mu(R_{k+1}) - \mu(R_k))$$

$$= \mu\left(\bigcup_{k=-\infty}^{\infty} (R_{k+1} \smallsetminus R_k) \right)$$

$$= \mu((a, b) \times (b, c)).$$

However, the rectangle $(a, b) \times (b, c)$ touches the diagonal at the point (b, b), so it has infinite measure. This is a contradiction. $\square$

REMARK 5.11. In the proof of Theorem 1.7, the assumption that the elements of Γ are C^2 was used only to establish Equation (5.10). For this, it is not necessary to show that $\Phi^g - \Phi$ is bounded (as in Lemma 5.8), but only that $\Phi^g - \Phi \in L^2(S^1 \times S^1)$. The calculations in the proof of Lemma 5.8 show that this holds under the weaker hypothesis that $g \in C^{3/2+\epsilon}$, for any $\epsilon > 0$. In fact, A. Navas observed that by using recent results that group with Kazhdan's property (T) also have fixed points in certain L^p spaces with $p \neq 2$, it can be shown that C^2 can be replaced with $C^{3/2}$ in the statement of Theorem 1.7.

This leads to the following well-known question:

PROBLEM 5.12. *Can an infinite (discrete) group with Kazhdan's property (T) have a faithful C^0 action on S^1?*

NOTES FOR §5. Theorem 1.7 was first proved in [Na1], but the idea to use Lemma 5.8 came from earlier work of A. Pressley and G. Segal [PS], and A. Reznikov [Rez, chap. 2]. Proofs also appear in [BHV, §2.9] and [Na2, §5.2].

The condition in Definition 5.2 was introduced by J.–P. Serre, and it is not at all obvious that it is equivalent to the original definition of property (T) that was given by D. Kazhdan. For a discussion of this, and much more, the standard reference on Kazhdan's property (T) is [BHV].

See [Gh3, thm. 6.10] or [Na2, §2.2.4] for a proof of Lemma 5.9.

Implications of Kazhdan's property (T) for fixed points of actions on L^p spaces (and other Banach spaces) are discussed in [BFGM].

6. Ghys's Proof That Actions Have a Finite Orbit

In this section, we present Ghys's proof of Theorem 1.4, modulo some facts that will be proved in Section 7. To get started, let us show that it suffices to find a Γ-invariant measure on the circle.

DEFINITION 6.1. A measure μ on a measure space X is a *probability measure* if $\mu(X) = 1$.

LEMMA 6.2. *If*

- Γ *is a discrete group, such that the abelianization $\Gamma/[\Gamma, \Gamma]$ is finite,*
- Γ *acts on S^1 by orientation-preserving homeomorphisms, and*
- *there is a Γ-invariant probability measure μ on S^1,*

then Γ has at least one finite orbit on S^1.

Proof. We consider two cases.

Case 1. Assume μ has at least one atom p. Let

$$\Gamma p = \{ gp \mid g \in \Gamma \}$$

be the orbit of p. Because μ is Γ-invariant, we have $\mu(gp) = \mu(p)$, for every $g \in \Gamma$. Therefore

$$\mu(\Gamma p) = \#(\Gamma p)\, \mu(p).$$

Since μ is a probability measure, we know that $\mu(\Gamma p) < \infty$, so we conclude that Γp is a finite set. That is, the orbit of p is finite.

Case 2. Assume μ has no atoms. To simplify the proof, let us assume that the support of μ is all of S^1. (In other words, every nonempty open subset of S^1 has positive measure.) For $x, y \in S^1$, define

6.3 $$d(x, y) = \mu([x, y]),$$

where $[x, y]$ is a path from x to y, and, for a given x and y, we choose the path $[x, y]$ to minimize $\mu([x, y])$. It is easy to see that d is a metric on S^1. Up to isometry, there is a unique metric on S^1, so we may assume that d is the usual arc-length metric.

Because μ is Γ-invariant, we know that d is Γ-invariant, so Γ acts by rotations of the circle. There is no harm in assuming that the action is faithful, so we conclude that Γ is abelian. But every abelian quotient of Γ is finite, so we conclude that Γ is finite. Hence, every orbit is finite. □

EXERCISE 6.4. Complete the proof of Lemma theorem 6.2, by eliminating the assumption that the support of μ is all of S^1 in Case 2.

[Hint: Modify the above proof to show that every orbit in the support of μ is finite.]

To simplify the notation, we assume henceforth that $n = 3$.

NOTATION 6.5. Let

- $G = \mathrm{SL}(3, \mathbb{R})$,
- Γ be a lattice in G, and

- $P = \begin{bmatrix} * & * & * \\ & * & * \\ & & * \end{bmatrix} \subset G$.

We remark that P is a minimal parabolic subgroup of G.

Ghys's proof is based on the following key fact that will be established in §7A by using the fact that the group P is "amenable."

NOTATION 6.6. Prob(X) denotes the set of all Radon probability measures on X (where X is any compact, Hausdorff space). This is a closed, convex subset of the unit ball in $C(X)^*$, with the weak* topology, so Prob(X) has a natural topology that makes it a compact Hausdorff space.

PROPOSITION 6.7. (FURSTENBERG) *If Γ acts continuously on any compact metric space X, then there is a Γ-equivariant, Borel measurable map $\psi: G/P \to$ Prob (X).*

If ψ were *invariant*, rather than *equivariant*, the following fundamental theorem would immediately imply that ψ is constant (a.e.). (This theorem will be proved in §7B.)

THEOREM 6.8. (MOORE ERGODICITY THEOREM) *If H is any closed, non-compact subgroup of G, then the action of Γ on G/H is ergodic: by definition, this means that every Γ-invariant, measurable function on G/H is constant (a.e.).*

The homogeneous space G/P plays a major role in Ghys's proof, because of its appearance in Proposition 6.7. We will now present a geometric interpretation of this space that is very helpful.

Recall that a *flag* in $\mathbb{R}^3$ consists of a pair (ℓ, π), where

- ℓ is a 1-dimensional vector subspace of $\mathbb{R}^3$ (a line), and
- π is a 2-dimensional vector subspace of $\mathbb{R}^3$ (a plane) that contains ℓ.

The group $G = \mathrm{SL}(3, \mathbb{R})$ acts transitively on the set of flags, and P is the stabilizer of the standard flag

$$F_0 = (\ell_0, \pi_0), \text{ where } \ell_0 = (*, 0, 0) \text{ and } \pi_0 = (*, *, 0).$$

Therefore:

PROPOSITION 6.9. *G/P can be identified with the space $\mathcal{F}$ of all flags, by identifying gP with the flag gF_0.*

Proof of Theorem 1.4. Suppose Γ acts by homeomorphisms on the circle S^1. From Proposition 6.7, we know there is a Γ-equivariant, measurable map

$$\psi: G/P \to \mathrm{Prob}\,(S^1).$$

It will suffice to show that ψ is constant (a.e.), for then

- the essential range of ψ consists of a single point $\mu \in \mathrm{Prob}\,(S^1)$, and
- the essential range of ψ is Γ-invariant, because ψ is Γ-equivariant.

So μ is Γ-invariant, and then Lemma 6.2 implies Γ has a finite orbit on S^1.

There are two basic cases to consider: either the measure $\psi(x)$ consists entirely of atoms, or $\psi(x)$ has no atoms. (Recall that an *atom* of a measure μ is a point p, such that $\mu(\{p\}) \neq 0$.) It is also possible that $\psi(x)$ consists partly of atoms, and partly of nonatoms, but Corollary 7.5 below tells us that we need only consider the two extreme cases.

Case 1. Assume $\psi(x)$ has no atoms, for a.e. $x \subset G/P$.

- Let $\mathrm{Prob}_0(S^1) = \{ \mu \in \mathrm{Prob}(S^1) \mid \mu \text{ has no atoms} \}$.
- By assumption, we know $\psi \colon G/P \to \mathrm{Prob}_0(S^1)$, so we may define

$$\psi^2 \colon (G/P)^2 \to (\mathrm{Prob}_0(S^1))^2 \quad \text{by} \quad \psi^2(x, y) = (\psi(x), \psi(y)).$$

Then ψ^2, like ψ, is measurable and Γ-equivariant.

- Define $d \colon (\mathrm{Prob}_0(S^1))^2 \to \mathbb{R}$ by

$$d(\mu_1, \mu_2) = \sup_J |\mu_1(J) - \mu_2(J)|,$$

where J ranges over all intervals (that is, over all connected subsets of S^1). Since Γ acts by homeomorphisms, and any homeomorphism of S^1 maps intervals to intervals, it is easy to see that d is Γ-invariant.

We claim that

6.10 $\qquad\qquad\qquad d$ is continuous,

with respect to the usual weak* topology that $\mathrm{Prob}_0(S^1)$ inherits from being a subset of $\mathrm{Prob}(S^1)$. To see this, we note that if $\mu \in \mathrm{Prob}_0(S^1)$ and $\epsilon > 0$, then, because μ has no atoms, we may partition S^1 into finitely many intervals $J_1, \ldots, J_k$, such that $\mu(J_i) < \epsilon$ for all i. By approximating the characteristic functions of these intervals from above and below, we may construct continuous functions $f_1, \ldots, f_{2k}$ and some $\delta > 0$, such that, for $\nu \in \mathrm{Prob}(S^1)$,

$$\text{if } |\nu(f_i) - \mu(f_i)| < \delta \text{ for every } i, \text{ then } |\nu(J_i) - \mu(J_i)| < \frac{\epsilon}{n} \text{ for every } i.$$

Then, for any interval J in S^1, we have $|\nu(J) - \mu(J)| < 2\epsilon$, so $d(\nu, \mu) < 2\epsilon$. This completes the proof of (6.10).

Because ψ^2 is measurable and d is continuous, we know that the composition $d \circ \psi^2$ is measurable. In addition, because ψ^2 is Γ-equivariant and d is Γ-invariant, we also know that $d \circ \psi^2$ is Γ-invariant. (*Note:* we are saying *invariant*, not just *equivariant*.) From the Moore ergodicity theorem (6.8), we know that Γ is ergodic on $(G/P)^2$ (see Corollary 7.6), so we conclude that $d \circ \psi^2$ is constant (a.e.); say

$$d(\psi^2(x, y)) = c, \quad \text{for a.e. } x, y \in G/P.$$

We wish to show that $c = 0$, for then it is clear that $\psi(x) = \psi(y)$ for a.e. $x, y \in G/P$, so ψ is constant (a.e.).

It is obvious that $d(\psi^2(x, x)) = 0$ for every $x \in G/P$. If ψ were constant everywhere, rather than merely *almost* everywhere, then it would follow immediately that $c = 0$. Unfortunately, the diagonal $\{(x, x)\}$ is a set of measure 0 in $(G/P)^2$, so a little bit of additional argument is required.

By Lusin's theorem, ψ is continuous on a set K of positive measure. Since the composition of continuous functions is continuous, we conclude that $d \circ \psi^2$ is continuous on $K \times K$. So, by continuity, it is constant on all of $K \times K$, not merely almost all. Since $d(\psi^2(x, x)) = 0$, this implies $d(\psi^2(x, y)) = 0$ for all $x, y \in K$. Since K is a set of positive measure, this implies $c = 0$, as desired.

Case 2. Assume $\psi(x)$ consists entirely of atoms, for a.e. $x \in G/P$. To simplify the notation, without losing the main ideas, let us assume that $\psi(x)$ consists of a single atom, for every $x \in G/P$. Thus, we may think of ψ as a Γ-equivariant, measurable map

$$\psi \colon G/P \to S^1.$$

Surprisingly, even with the simplifying assumption, the argument here seems to be more difficult than in Case 1. The idea is to obtain a contradiction from the Γ-equivariance of ψ, by contrasting two fundamental observations:

- Homeo$_+$ (S^1) is *not* triply transitive on S^1: if x, y, and z are distinct, then no orientation-preserving homeomorphism of S^1 can map the triple (x, y, z) to (y, x, z)—they have opposite orientations under the circular order on S^1.
- The action of GL $(2, \mathbb{R})$ on the projective line $\mathbb{R}P^1 = \mathbb{R} \cup \{\infty\}$ by linear-fractional transformations

$$g(x) = \frac{ax + b}{cx + d} \qquad \text{if } g = \begin{bmatrix} a & b \\ c & d \end{bmatrix}$$

  is triply transitive: if (x_1, y_1, z_1) and (x_2, y_2, z_2) are two ordered triples of distinct elements of $\mathbb{R} \cup \{\infty\}$, then there is some $g \in$ GL $(2, \mathbb{R})$ with $g(x_1) = x_2, g(y_1) = y_2,$ and $g(z_1) = z_2$.

To illustrate, let us give an easy proof that is not quite correct; the actual proof is a modified version of this. Define

$$\psi^3 \colon (G/P)^3 \to (S^1)^3 \qquad \text{by} \qquad \psi^3(x, y, z) = (\psi(x), \psi(y), \psi(z)).$$

Then ψ^3 is Γ-equivariant, so

$$X^+ = \{\, (x, y, z) \in (G/P)^3 \mid (\psi(x), \psi(y), \psi(z)) \text{ is positively oriented} \,\}$$

is a Γ-invariant, measurable subset of $(G/P)^3$. Let us assume that Γ is ergodic on $(G/P)^3$. (Unfortunately, this assumption is false, so it is where the proof breaks down.) Then X^+ must be (almost) all of $(G/P)^3$; thus, $(\psi(x), \psi(y), \psi(z))$ is positively oriented, for (almost) every $x, y, z \in G/P$. But this is nonsense: either $(\psi(x), \psi(y), \psi(z))$ or $(\psi(y), \psi(x), \psi(z))$ is negatively oriented, so there are many negatively oriented triples.

To salvage the above faulty proof, we replace $(G/P)^3$ with a subset X, on which Γ does act ergodically. Let

- $Q = \begin{bmatrix} * & * & * \\ * & * & * \\ 0 & 0 & * \end{bmatrix} \subset G$, and

- $X = \left\{\, (x_1, x_2, x_3) \in (G/P)^3 \,\middle|\, \begin{array}{c} x_1 Q = x_2 Q = x_3 Q, \\ x_1, x_2, x_3 \text{ distinct} \end{array} \,\right\}.$

Note that X is a submanifold of $(G/P)^3$. (If this is not obvious, it follows from the fact, proven below, that X is a single G-orbit in $(G/P)^3$.)

Assume, for the moment, that Γ is ergodic on X (with respect to any (hence, every) Lebesgue measure on the manifold X). Then the above proof, with X in the place of $(G/P)^3$, implies that

$$X^+ = \{\, (x, y, z) \in X \mid (\psi(x), \psi(y), \psi(z)) \text{ is positively oriented} \,\}$$

is a set of measure 0. Then it is not difficult to see that ψ is constant on X (a.e.). Hence ψ is right Q-invariant (a.e.): for each $q \in Q$, we have $\psi(xqP) = \psi(xP)$ for a.e. $x \in G/P$.

By a similar argument, we see that ψ is right Q'-invariant (a.e.), where

$$Q' = \begin{bmatrix} * & * & * \\ 0 & * & * \\ 0 & * & * \end{bmatrix} \subset G.$$

Because Q and Q', taken together, generate all of G, it then follows that ψ is right G-invariant (a.e.). Hence, ψ is constant (a.e.), as desired.

All that remains is to show that Γ is ergodic on X. By the Moore ergodicity theorem (6.8), we need only show that

1) G is transitive on X, and
2) the stabilizer of some element of X is not compact.

These facts are perhaps easiest to establish from a geometric perspective.

The (parabolic) subgroup Q is the stabilizer of the plane $\pi_0 = (*, *, 0)$. Hence, QF_0 is the set of all flags (ℓ, π) with $\pi = \pi_0$. Therefore, under the identification of G/P with $\mathcal{F}$, we have

6.11
$$X = \left\{ ((\ell_1, \pi_1), (\ell_2, \pi_2), (\ell_3, \pi_3)) \in \mathcal{F}^3 \;\middle|\; \begin{matrix} \pi_1 = \pi_2 = \pi_3, \\ \ell_1, \ell_2, \ell_3 \text{ distinct} \end{matrix} \right\}.$$

1) Let us show that G is transitive on X. Given

$$((\ell_1, \pi), (\ell_2, \pi), (\ell_3, \pi)), ((\ell_1', \pi'), (\ell_2', \pi'), (\ell_3', \pi')) \in X,$$

it suffices to show that there exists $g \in G$, such that

6.12
$$g((\ell_1, \pi), (\ell_2, \pi), (\ell_3, \pi)) = ((\ell_1', \pi'), (\ell_2', \pi'), (\ell_3', \pi')).$$

Because G is transitive on the set of 2-dimensional subspaces, we may assume

$$\pi = \pi' = \pi_0 = \mathbb{R}^2.$$

Then, because $\mathrm{GL}(2, \mathbb{R})$ is triply transitive on $\mathbb{R}P^1$, there exists $T \in \mathrm{GL}(2, \mathbb{R})$, such that

$$T(\ell_1, \ell_2, \ell_3) = (\ell_1', \ell_2', \ell_3').$$

Letting

$$g = \begin{bmatrix} T & & 0 \\ & & 0 \\ 0 & 0 & \frac{1}{\det T} \end{bmatrix}$$

yields Equation (6.12).

2) Let us show that the stabilizer of some element of X is not compact. Let

$$A = \left\{ \begin{bmatrix} a & 0 & 0 \\ 0 & a & 0 \\ 0 & 0 & 1/a^2 \end{bmatrix} \;\middle|\; a \in \mathbb{R}^\times \right\}.$$

Then A is a closed, noncompact subgroup of G. Furthermore, every element of A acts as a scalar on π_0, so every element of A fixes every 1-dimensional vector subspace of π_0. Thus, if

$$((\ell_1, \pi), (\ell_2, \pi), (\ell_3, \pi)) \in X$$

with $\pi = \pi_0$, then A is contained in the stabilizer of this element of X, so the stabilizer is not compact. $\qquad \square$

NOTES FOR §6. Ghys's proof first appeared in [Gh2]. See [Gh3, §7.3] for an exposition.

7. Additional Ingredients of Ghys's Proof

7A. Amenability and an Equivariant Map

A group is amenable if its action on every compact, convex set has a fixed point. More precisely:

DEFINITION 7.1   A Lie group G is amenable if, for every continuous action of G by linear operators on a locally convex topological vector space $\mathcal{V}$, and every nonempty, compact, convex G-invariant subset C of $\mathcal{V}$, the group G has a fixed point in C.

EXAMPLE 7.2.

1) If T is any continuous linear operator on $\mathcal{V}$ and v is any element of $\mathcal{V}$ such that $\{T^n v\}$ is bounded, then every accumulation point of the sequence

$$v_n = \frac{1}{n}(Tv + T^2 v + \cdots + T^n v)$$

is a fixed point for T. This implies that cyclic groups are amenable.

2) A generalization of this argument shows that all abelian groups are amenable; this statement is a version of the classical Kakutani-Markov fixed-point theorem.

3) It is not difficult to see that if N is a normal subgroup of G such that N and G/N are both amenable, then G is amenable.

4) Combining the preceding two observations implies that solvable groups are amenable.

5) In particular, the group P of Notation 6.5 is amenable.

Proof of Proposition 6.7. Since Prob (X) is a compact, convex set, a version of the Banach-Alaoglu theorem tells us that $L^\infty(G; \text{Prob}(X))$ is compact and convex in a natural weak* topology. Let

$$L^\infty_\Gamma(G; \text{Prob}(X)) = \left\{ \psi \in L^\infty(G; \text{Prob}(X)) \mid \psi \text{ is } \Gamma\text{-equivariant (a.e.)} \right\}.$$

This is a closed subset of $L^\infty(G; \text{Prob}(X))$, so it is compact. It is also convex and nonempty. To say ψ is Γ-equivariant (a.e.) means, for each $\gamma \in \Gamma$, that $\psi(\gamma x) = \gamma \cdot \psi(x)$ for a.e. $x \in G$; so G acts on $L^\infty_\Gamma(G; \text{Prob}(X))$ by translation on the right. Hence, the subgroup P acts on $L^\infty_\Gamma(G; \text{Prob}(X))$.

Since P is amenable (see Example 7.2(5)), it must have a fixed point ψ_0 in the compact, convex set $L_\Gamma^\infty(G; \text{Prob}(X))$. Then ψ_0 is invariant (a.e.) under translation on the right by elements of P, so it factors through (a.e.) to a well-defined map $\psi: G/P \to \text{Prob}(X)$. Because ψ_0 is Γ-equivariant, it is immediate that ψ is Γ-equivariant. $\qquad\square$

7B. Moore Ergodicity Theorem

We will obtain the Moore ergodicity theorem (6.8) as an easy consequence of the following result in representation theory:

THEOREM 7.3. (DECAY OF MATRIX COEFFICIENTS) *If*

- $G = \text{SL}(n, \mathbb{R})$,
- π *is a unitary representation of G on a Hilbert space $\mathcal{H}$, such that no non-zero vector is fixed by $\pi(G)$; and*
- $\{g_j\}$ *is a sequence of elements of G, such that $\|g_j\| \to \infty$,*

then $\langle \pi(g_j)\phi \mid \psi \rangle \to 0$, *for every* $\phi, \psi \in \mathcal{H}$.

Proof. By passing to a subsequence, we may assume $\pi(g_j)$ converges weakly to some operator E; that is,

$$\langle \pi(g_j)\phi \mid \psi \rangle \to \langle E\phi \mid \psi \rangle \text{ for every } \phi, \psi \in \mathcal{H}.$$

We wish to show $\ker E = \mathcal{H}$.

Let

$$U = \{ u \in G \mid g_j u g_j^{-1} \to e \}$$

and

$$U^- = \{ v \in G \mid g_j^{-1} v g_j \to e \}.$$

For $u \in U$, we have

$$
\begin{aligned}
\langle E\pi(u)\phi \mid \psi \rangle &= \lim \langle \pi(g_j u)\phi \mid \psi \rangle \\
&= \lim \langle \pi(g_j u g_j^{-1})\pi(g_j)\phi \mid \psi \rangle \\
&= \lim \langle \pi(g_j)\phi \mid \psi \rangle \\
&= \langle E\phi \mid \psi \rangle,
\end{aligned}
$$

so $E\pi(u) = E$. Therefore, letting $\mathcal{H}^U$ be the space of U-invariant vectors in $\mathcal{H}$, we have

$$(\mathcal{H}^U)^\perp \subset \ker E.$$

We have

$$\langle E^*\phi \mid \psi \rangle = \langle \phi \mid E\psi \rangle = \lim \langle \phi \mid \pi(g_j)\psi \rangle = \lim \langle \pi(g_j^{-1})\phi \mid \psi \rangle,$$

so the same argument, with E^* in the place of E and g_j^{-1} in the place of g_j, shows that

$$(\mathcal{H}^{U^-})^\perp \subset \ker E^*.$$

Assume, for simplicity, that each g_j is a positive-definite diagonal matrix:

$$g_j = \begin{bmatrix} a_j & & \\ & b_j & \\ & & c_j \end{bmatrix} \qquad \text{with } a_j, b_j, c_j > 0.$$

(It is not difficult to eliminate this hypothesis, by using the *Cartan decomposition* $G = KAK$, but that is not necessary for Ghys's proof.) Then the subgroup generated by $\{\pi(g_j)\}$ is commutative. Because π is unitary, this means that $\pi(g_j)$ commutes with both $\pi(g_k)$ and $\pi(g_k)^* = \pi(g_k^{-1})$ for every j and k. Therefore, the limit E commutes with its adjoint (that is, E is normal): we have $E^*E = EE^*$. Hence

$$\|E\phi\|^2 = \langle E\phi \mid E\phi \rangle = \langle (E^*E)\phi \mid \phi \rangle$$

$$= \langle (EE^*)\phi \mid \phi \rangle = \langle E^*\phi \mid E^*\phi \rangle = \|E^*\phi\|^2,$$

so $\ker E = \ker E^*$.

Thus,

$$\ker E = \ker E + \ker E^*$$

$$\supset (\mathcal{H}^U)^\perp + (\mathcal{H}^{U^-})^\perp$$

$$= (\mathcal{H}^U \cap \mathcal{H}^{U^-})^\perp$$

$$= (\mathcal{H}^{\langle U, U^- \rangle})^\perp.$$

By passing to a subsequence, and then permuting the basis vectors of $\mathbb{R}^3$, we may assume

$$a_j \geq b_j \geq c_j.$$

Since $\|g_j\| \to \infty$, we have

$$\lim_{j\to\infty} \max\left\{ \frac{a_j}{b_j}, \frac{b_j}{c_j} \right\} = \infty.$$

For definiteness, let us assume

$$\limsup_{j\to\infty} \frac{a_j}{b_j} < \infty \quad \text{and} \quad \lim_{j\to\infty} \frac{b_j}{c_j} = \infty,$$

so

$$U = \begin{bmatrix} 1 & & \\ & 1 & \\ * & * & 1 \end{bmatrix} \quad \text{and} \quad U^- = \begin{bmatrix} 1 & & * \\ & 1 & * \\ & & 1 \end{bmatrix}.$$

(Other cases are similar.) Then it is easy to see that $\langle U, U^- \rangle = G$, which means $\mathcal{H}^{\langle U, U^- \rangle} = \mathcal{H}^G = 0$, so

$$\ker E \supset (\mathcal{H}^{\langle U, U^- \rangle})^\perp = 0^\perp = \mathcal{H},$$

as desired. □

Proof of Theorem 6.8. Suppose there is a Γ-invariant, measurable function on G/H that is not constant (a.e.). Then:

∃ measurable function on $\Gamma \backslash G/H$ that is not constant (a.e.),

so ∃ measurable function on $H \backslash G/\Gamma$ that is not constant (a.e.),

so ∃ H-invariant, measurable function f on G/Γ that is not constant (a.e.).

There is no harm in assuming that f is bounded. Since Γ is a lattice in G, we know G/Γ has finite measure, so this implies $f \in L^2(G/\Gamma)$. Letting π be the natural unitary representation of G on $L^2(G/\Gamma)$, we know that f is $\pi(H)$-invariant.

Let $L^2(G/\Gamma)_0$ be the orthogonal complement of the constant functions. Since f is nonconstant, its projection $\overline{f}$ in $L^2(G/\Gamma)_0$ is nonzero. By normalizing, we may assume $\|\overline{f}\| = 1$. Since the orthogonal projection commutes with every unitary operator that preserves the space of constant functions, we know that $\overline{f}$, like f, is $\pi(H)$-invariant. So

$$\langle \pi(h_j)\overline{f} \mid \overline{f} \rangle = \langle \overline{f} \mid \overline{f} \rangle = 1, \quad \text{for all } h_j \in H.$$

On the other hand, since H is closed and noncompact, we may choose a sequence $\{h_j\}$ of elements of H, such that $\|h_j\| \to \infty$. Then, since no nonzero vector in $L^2(G/\Gamma)_0$ is fixed by $\pi(G)$, Theorem 7.3 tells us that

$$\langle \pi(h_j)\overline{f} \mid \overline{f} \rangle \to 0.$$

This is a contradiction. $\qquad\square$

REMARK 7.4. The assumption that H is not compact is necessary in the Moore ergodicity theorem (6.8): it is not difficult to see that if H is a *compact* subgroup of $G = \mathrm{SL}\,(3, \mathbb{R})$ and Γ is any lattice in G, then Γ is *not* ergodic on G/H.

COROLLARY 7.5. *In the situation of Proposition 6.7, one may assume that either:*

* *for a.e. $x \in G/P$, the measure $\psi_0(x)$ has no atoms, or*
* *for a.e. $x \in G/P$, the measure $\psi_0(x)$ consists entirely of atoms.*

Proof. For each $x \in G/P$, write $\psi(x) = \psi_{\mathrm{noatom}}(x) + \psi_{\mathrm{atom}}(x)$, where $\psi_{\mathrm{noatom}}(x)$ has no atoms, and $\psi_{\mathrm{atom}}(x)$ consists entirely of atoms. Since ψ_{noatom} and $\psi_{\mathrm{atom}}(x)$ are uniquely determined by ψ, it is not difficult to see that they, like ψ, are measurable and Γ-equivariant. One or the other must be nonzero on a set of positive measure, and then the ergodicity of Γ on G/P implies that this function is nonzero almost everywhere, so it can be normalized to define a (Γ-equivariant, measurable) map into $\mathrm{Prob}\,(S^1)$. $\qquad\square$

COROLLARY 7.6. *In the situation of Notation 6.5, Γ is ergodic on $(G/P)^2$.*

Proof. Two flags $F_1 = (\ell_1, \pi_1)$ and $F_2 = (\ell_2, \pi_2)$ are in *general position* if $\ell_1 \notin \pi_2$ and $\ell_2 \notin \pi_1$. It is not difficult to see that G is transitive on the set $\mathcal{F}_0^2$ of pairs of flags in general position, and that the complement of $\mathcal{F}_0^2$ has measure 0 in $\mathcal{F}^2$. Therefore, $(G/P)^2$ may be identified (a.e.) with G/H, where H is the stabilizer of some pair of flags in general position; we may take

$$H = \mathrm{Stab}_G\big(((*,0,0),(*,*,0)),((0,0,*),(0,*,*))\big) = \begin{bmatrix} * & & \\ & * & \\ & & * \end{bmatrix}.$$

Since H is not compact, the Moore Ergodicity theorem (6.8) tells us that Γ is ergodic on $G/H \approx (G/P)^2$. $\qquad\square$

NOTES FOR §7. Proposition 6.7 is due to Furstenberg [Fur]. It is a basic result in the theory of lattices, so proofs can be found in numerous references, including [Gh3, prop. 7.11] and [Zi2, prop. 4.3.9, p. 81].

The monograph [Pie] is a standard reference on amenability. See [Zi2, §4.1] for a brief treatment.

Theorem 6.8 is due to C. C. Moore [Moo]. The stronger Theorem 7.3 is due to R. Howe and C. C. Moore [HM, thm. 5.1] and (independently) R. J. Zimmer [Zi1, Thm. 5.2]. The elementary proof we give here was found by R. Ellis and M. Nerurkar [EN].

8. Bounded Cohomology and the Burger-Monod Proof

8A. Bounded Cohomology and Actions on the Circle

Suppose a discrete group Γ acts by orientation-preserving homeomorphisms on $S^1 = \mathbb{R}/\mathbb{Z}$. Since $\mathbb{R}$ is the universal cover of S^1, each element γ of Γ can be lifted to a homeomorphism $\widetilde{\gamma}$ of $\mathbb{R}$. The lift $\widetilde{\gamma}$ is not unique, but it is well defined if we require that $\widetilde{\gamma}(0) \in [0, 1)$.

DEFINITION 8.1. For $\gamma_1, \gamma_2 \in \Gamma$, the homeomorphisms $\widetilde{\gamma_1\gamma_2}$ and $\widetilde{\gamma_1}\widetilde{\gamma_2}$ are lifts of the same element $\gamma_1\gamma_2$ of Γ, so there exists $z = z(\gamma_1, \gamma_2) \in \mathbb{Z}$, such that

$$\widetilde{\gamma_1}\widetilde{\gamma_2} = \widetilde{\gamma_1\gamma_2} + z.$$

The map $z \colon \Gamma \times \Gamma \to \mathbb{Z}$ is called the *Euler cocycle* of the action of Γ on S^1.

It is easy to see that

- z is a bounded function (in fact, $z(\Gamma \times \Gamma) \subset \{0, 1\}$), and
- $z(\gamma_1, \gamma_2) + z(\gamma_1\gamma_2, \gamma_3) = z(\gamma_1, \gamma_2\gamma_3) + z(\gamma_2, \gamma_3)$, so z is an Eilenberg-MacLane 2-cocycle.

Therefore, the Euler cocycle determines a bounded cohomology class:

DEFINITION 8.2.

1) A *bounded k-cochain* is a bounded function $c \colon \Gamma^k \to \mathbb{Z}$.
2) The bounded cochains form a chain complex with respect to the differential

$$\delta \colon C^k_{\text{bdd}}(\Gamma; \mathbb{Z}) \to C^{k+1}_{\text{bdd}}(\Gamma; \mathbb{Z})$$

defined by

$$\delta c(\gamma_0, \gamma_1, \ldots, \gamma_k) = c(\gamma_1, \ldots, \gamma_k)$$

$$+ \sum_{i=1}^{k} (-1)^i c(\gamma_0, \ldots, \gamma_{i-1}\gamma_i, \ldots, \gamma_k)$$

$$+ (-1)^{k+1} c(\gamma_0, \gamma_1, \ldots, \gamma_{k-1}).$$

The cohomology of this complex is the *bounded cohomology* of Γ.

3) The *Euler class* of the action of Γ is the cohomology class $[z] \in H^2_{\text{bdd}}(\Gamma, \mathbb{Z})$ determined by the Euler cocycle.

REMARK 8.3. The Euler cocycle z depends on the choice of the covering map from $\mathbb{R}$ to S^1, but it is not difficult to see that the Euler class $[z]$ is well defined. Indeed, it is an invariant of the (orientation-preserving) homeomorphism class of the action.

The connection with Theorem 1.4 is provided by the following fundamental observation:

PROPOSITION 8.4. (GHYS) *The action of Γ on S^1 has a fixed point if and only if its Euler class is 0 in $H^2_{\text{bdd}}(\Gamma; \mathbb{Z})$.*

Proof. ($\Rightarrow$) We may assume the fixed point is the image of 0 under the covering map $\mathbb{R} \to S^1$. Then $\widetilde{\gamma}(0) = 0$ for every $\gamma \in \Gamma$, so it is clear that $z(\gamma_1, \gamma_2) = 0$ for all γ_1 and γ_2.

($\Leftarrow$) Assume $z = \delta\varphi$, where $\varphi \colon \Gamma \to \mathbb{Z}$ is bounded. If we set $\widehat{\gamma} = \widetilde{\gamma} - \varphi(\gamma)$, then the map $\gamma \to \widehat{\gamma}$ is a homomorphism; $\widehat{\Gamma}$ is a lift of Γ to a group of homeomorphisms of $\mathbb{R}$.

Since $\widetilde{\gamma}(0) \in [0, 1)$, and φ is bounded, it is clear that the $\widehat{\Gamma}$-orbit of 0 is bounded. Since the orbit is obviously a $\widehat{\Gamma}$-invariant set, its supremum is also $\widehat{\Gamma}$-invariant; in other words, the supremum is a fixed point for $\widehat{\Gamma}$ in $\mathbb{R}$. The image of this fixed point under the covering map is a fixed point for Γ in S^1. $\square$

The definition of $H^k_{\text{bdd}}(\Gamma; \mathbb{Z})$ can be generalized to allow any coefficient module in the place of $\mathbb{Z}$. For real coefficients, we have the following important fact:

COROLLARY 8.5. *If $H^2_{\text{bdd}}(\Gamma; \mathbb{R}) = 0$ and the abelianization of Γ is finite, then every action of Γ on S^1 has a finite orbit.*

Proof. The short exact sequence

$$0 \to \mathbb{Z} \to \mathbb{R} \to \mathbb{R}/\mathbb{Z} \to 0$$

of coefficient groups leads to a long exact sequence of bounded cohomology groups. A part of this sequence is

$$H^1_{\text{bdd}}(\Gamma; \mathbb{R}/\mathbb{Z}) \to H^2_{\text{bdd}}(\Gamma; \mathbb{Z}) \to H^2_{\text{bdd}}(\Gamma; \mathbb{R}).$$

The right end of this sequence is 0 by assumption. If we assume, for simplicity, that the abelianization of Γ is trivial (rather than merely finite), then the left end is also 0. Hence, the middle term must be 0. Then Proposition 8.4 implies that every (orientation-preserving) action of Γ on S^1 has a fixed point.

Without the simplifying assumption, one can obtain the weaker conclusion that the commutator subgroup $[\Gamma, \Gamma]$ has a fixed point. Since, by hypothesis, $\Gamma/[\Gamma, \Gamma]$ is finite, this implies that the action of Γ has a finite orbit. $\square$

We now need two observations:

1) Forgetting that a bounded k-cocycle is bounded yields a natural map from bounded cohomology to ordinary cohomology:

   8.6 $$\tau_\Gamma : H^k_{\text{bdd}}(\Gamma; \mathbb{R}) \to H^k(\Gamma; \mathbb{R}).$$

2) The cohomology of lattices in $\text{SL}(n, \mathbb{R})$ has been studied extensively; in particular, it is known that

   8.7 $\quad H^2(\Gamma; \mathbb{R}) = 0$ if Γ is any lattice in $\text{SL}(n, \mathbb{R})$, with $n \geq 6$.

Therefore, under the assumption that $n \geq 6$, the conclusion of Theorem 1.4 can be obtained by combining Corollary 8.5 with the following result.

THEOREM 8.8. (BURGER-MONOD) *If Γ is any lattice in $\text{SL}(n, \mathbb{R})$, with $n \geq 3$, then the comparison map in Equation (8.6) is injective for $k = 2$.*

COROLLARY 8.9. (BURGER-MONOD) *If Γ is any lattice in $\text{SL}(n, \mathbb{R})$, with $n \geq 3$, then $H^2_{\text{bdd}}(\Gamma; \mathbb{R}) = 0$.*

REMARK 8.10. See Remark 10.7 (3) for a brief mention of how to obtain Theorem 1.4 from Theorem 8.8, without needing to know that $H^2(\Gamma; \mathbb{R})$ vanishes.

8B. Outline of the Burger-Monod Proof of Injectivity

M. Burger and N. Monod developed an extensive and powerful theory for the study of bounded cohomology, but we will discuss only the parts that are used in the proof of Theorem 8.8, and even these will only be sketched.

ASSUMPTION 8.11. In the remainder of this section:

- $G = \text{SL}(n, \mathbb{R})$, with $n \geq 3$, and
- Γ is a lattice in G.

To avoid a serious technical complication, we will assume that G/Γ is compact.

Outline of the Proof of Theorem 8.8. We employ relations between the cohomology of Γ and the cohomology of G. (The bounded cohomology of G is introduced in Definition 8.16 below. When working with G, we always use *continuous* cochains.)

- We will see that there is a natural map $i_{\text{bdd}} \colon H^k_{\text{bdd}}(\Gamma; \mathbb{R}) \to H^k_{\text{bdd}}(G; L^2(G/\Gamma))$.
- It is a classical fact that if G/Γ is compact, then there is a natural map $i \colon H^k(\Gamma; \mathbb{R}) \to H^k(G; L^2(G/\Gamma))$.
- We have comparison maps τ_Γ and τ_G from bounded cohomology to ordinary cohomology.

Letting $k = 2$ yields the following commutative diagram:

$$
\begin{array}{ccc}
H^2_{\text{bdd}}(\Gamma; \mathbb{R}) & \xrightarrow{\ i_{\text{bdd}}\ } & H^2_{\text{bdd}}(G; L^2(G/\Gamma)) \\
{\scriptstyle \tau_\Gamma}\big\downarrow & & \big\downarrow{\scriptstyle \tau_G} \\
H^2(\Gamma; \mathbb{R}) & \xrightarrow{\ i\ } & H^2(G; L^2(G/\Gamma))
\end{array}
$$

We will show that i_{bdd} and τ_G are both injective (see Corollary 8.17 and Theorem 8.18). The commutativity of the diagram then implies that τ_Γ is also injective. $\qquad\square$

INJECTIVITY OF i_{bdd} Cohomology, whether bounded or not, can be described either in terms of inhomogeneous cocycles, or in terms of homogeneous cocycles. The Euler cocycle arose in Definition 8.1 as an inhomogeneous cocycle, but the injectivity of i_{bdd} is easier to explain in homogeneous terms. The following definition is written with real coefficients, because we no longer have any need for $\mathbb{Z}$-coefficients in our discussion.

DEFINITION 8.12.

1) A *homogeneous bounded k-cochain* on Γ is a bounded function $\dot{c}: \Gamma^{k+1} \to$ $\mathbb{R}$, such that

$$\dot{c}(\gamma\gamma_0, \gamma\gamma_1, \ldots, \gamma\gamma_k) = \dot{c}(\gamma_0, \gamma_1, \ldots, \gamma_k),$$

for all $\gamma, \gamma_0, \gamma_1, \ldots, \gamma_k \in \Gamma$.

2) The homogeneous bounded cochains form a chain complex with respect to the differential

$$\delta: \dot{C}^k_{\mathrm{bdd}}(\Gamma; \mathbb{R}) \to \dot{C}^{k+1}_{\mathrm{bdd}}(\Gamma; \mathbb{R})$$

defined by

$$\delta\dot{c}(\gamma_0, \gamma_1, \ldots, \gamma_{k+1}) = \sum_{i=0}^{k+1} (-1)^i \dot{c}(\gamma_0, \ldots, \widehat{\gamma_i}, \ldots, \gamma_{k+1}),$$

where $\widehat{\gamma_i}$ denotes that γ_i is omitted.

For any bounded k-cochain c, there is a corresponding homogeneous bounded k-cochain $\dot{c}$, defined by

$$\dot{c}(\gamma_0, \gamma_1, \ldots, \gamma_k) = c(\gamma_0^{-1}\gamma_1, \gamma_1^{-1}\gamma_2, \ldots, \gamma_{k-1}^{-1}\gamma_k).$$

Thus, the cohomology of the complex $\{\dot{C}^k_{\mathrm{bdd}}(\Gamma; \mathbb{R})\}$ is the bounded cohomology of Γ.

NOTATION 8.13.

1) P is the group of upper-triangular matrices in $G = \mathrm{SL}(n, \mathbb{R})$ (cf. Notation 6.5).

2) $ZL^\infty_{\mathrm{alt}}((G/P)^3; \mathbb{R})^\Gamma$ is the vector space of all $f \in L^\infty((G/P)^3; \mathbb{R})$, such that

a) f is *alternating*; that is,

$$f(x_{\sigma(1)}, x_{\sigma(2)}, x_{\sigma(3)}) = \mathrm{sgn}(\sigma)f(x_1, x_2, x_3)$$

for every permutation σ of $\{1, 2, 3\}$,

b) f is *Γ-invariant*; that is, for every $\gamma \in \Gamma$ we have

$$f(\gamma x_1, \gamma x_2, \gamma x_3) = f(x_1, x_2, x_3)$$

for a.e. x_1, x_2, x_3, and

c)f is a *cocycle*; that is, for a.e. x_0, x_1, x_2, x_3, we have

$$\sum_{i=0}^{3} (-1)^i f(x_0, \ldots, \widehat{x_i}, \ldots, x_3) = 0.$$

REMARK 8.14. If Γ were ergodic on $(G/P)^3$, then the following theorem would immediately imply that $H^2_{\text{bdd}}(\Gamma; \mathbb{R}) = 0$. This is because any Γ-invariant function on $(G/P)^3$ would have to be constant, so could not be alternating (unless it were identically 0).

THEOREM 8.15. (BURGER-MONOD) $H^2_{\text{bdd}}(\Gamma; \mathbb{R}) \cong ZL^\infty_{\text{alt}}((G/P)^3; \mathbb{R})^\Gamma$.

Proof. For each homogeneous bounded k-cocycle $\dot{c} \colon \Gamma^{k+1} \to \mathbb{R}$, we will show how to construct a corresponding $\check{c} \in ZL^\infty_{\text{alt}}((G/P)^{k+1}; \mathbb{R})^\Gamma$. The map $\dot{c} \mapsto \check{c}$ intertwines the differentials, so it induces a map from $H^k_{\text{bdd}}(\Gamma; \mathbb{R})$ to the cohomology of the chain complex

$$\left\{ ZL^\infty_{\text{alt}}((G/P)^{k+1}; \mathbb{R})^\Gamma \right\},$$

and, although we will not prove it, this map is an isomorphism on cohomology.

Since Γ is ergodic on $(G/P)^2$ (see Corollary 7.6), every Γ-invariant function on $(G/P)^2$ is constant, so 0 is the only such function that is alternating. Hence, there are no coboundaries in $ZL^\infty_{\text{alt}}((G/P)^3; \mathbb{R})^\Gamma$. This establishes the conclusion of the theorem.

To complete the proof, we now describe the construction of $\check{c}$. For simplicity, let us assume $k = 2$, so $\dot{c} \colon \Gamma^3 \to \mathbb{R}$ is a homogeneous bounded 2-cocycle. Also assume that $\dot{c}$ has been normalized (by subtracting a constant) so that $\dot{c}(e, e, e) = 0$. Then, by making use of the cocycle identity

$$\dot{c}(x_1, x_2, x_3) = \dot{c}(x_0, x_2, x_3) - \dot{c}(x_0, x_1, x_3) + \dot{c}(x_0, x_1, x_2),$$

one can show that $\dot{c}$ is alternating. Therefore, $\dot{c}$ can be extended to

$$\bar{c} \in ZL^\infty_{\text{alt}}(G^3; \mathbb{R})^\Gamma,$$

by choosing a fundamental domain $\mathcal{F}$ for Γ in G and making $\bar{c}$ constant on $\gamma_1 \mathcal{F} \times \gamma_2 \mathcal{F} \times \gamma_3 \mathcal{F}$, for all $\gamma_1, \gamma_2, \gamma_3 \in \Gamma$.

Now, because P is amenable (see Example 7.2(5)), there is a left-invariant mean μ on $L^\infty(P)$. Using μ to average on each left coset of P yields a map

$$\bar{\mu} \colon L^\infty(G) \to L^\infty(G/P).$$

Then a map

$$\overline{\mu}^3 \colon L^\infty(G^3) \to L^\infty((G/P)^3)$$

can be constructed by, roughly speaking, setting $\overline{\mu}^3 = \overline{\mu} \otimes \overline{\mu} \otimes \overline{\mu}$. Let $\check{c} = \overline{\mu}^3(\bar{c}) \in L^\infty((G/P)^3; \mathbb{R})$. □

DEFINITION 8.16.

1) The notion of a *homogeneous bounded k-cochain* on G is defined by replacing Γ with G in Definition 8.12 and requiring $\check{c}$ to be *continuous*.
2) The cohomology of the complex $\check{C}^k_{\mathrm{bdd}}(G; \mathbb{R})$ is $H_{\mathrm{bdd}}(G; \mathbb{R})$, the *(continuous) bounded cohomology* of G.

COROLLARY 8.17. (BURGER-MONOD) *There is a natural injection*

$$H^2_{\mathrm{bdd}}(\Gamma; \mathbb{R}) \hookrightarrow H^2_{\mathrm{bdd}}(G; L^2(G/\Gamma)).$$

Proof. For any $\check{c} \in ZL^\infty_{\mathrm{alt}}((G/P)^3; \mathbb{R})^\Gamma$ and $x \in (G/P)^3$, we can define

$$\check{c}_x \in L^\infty(G/\Gamma) \subset L^2(G/\Gamma) \quad \text{by} \quad \check{c}_x(g\Gamma) = \check{c}(gx).$$

The map $x \mapsto \check{c}_x$ is G-equivariant, so it is an element of $ZL^\infty_{\mathrm{alt}}((G/P)^3; L^2(G/\Gamma))^G$. Therefore, we have an injection

$$ZL^\infty_{\mathrm{alt}}((G/P)^3; \mathbb{R})^\Gamma \hookrightarrow ZL^\infty_{\mathrm{alt}}((G/P)^3; L^2(G/\Gamma))^G.$$

Theorem 8.15 identifies the domain of this injection with $H_{\mathrm{bdd}}(\Gamma; \mathbb{R})$, and the same argument identifies the target with $H_{\mathrm{bdd}}(G; L^2(G/\Gamma))$. □

THEOREM 8.18. (BURGER-MONOD) *The comparison map*

$$\tau_G \colon H^2_{\mathrm{bdd}}(G; L^2(G/\Gamma)) \to H^2(G; L^2(G/\Gamma))$$

is injective.

INJECTIVITY OF τ_G. The Hilbert space $L^2(G/\Gamma)$ decomposes as the direct sum of the constant functions $\mathbb{C}$ and the space $L^2_0(G/\Gamma)$ of functions with integral 0. The theorem is proved for the two summands individually (see Proposition 8.19 and Theorem 8.21). In both cases, we will argue with inhomogeneous cochains.

PROPOSITION 8.19. *The comparison map*

$$H^2_{bdd}(G; \mathbb{C}) \to H^2(G; \mathbb{C})$$

is injective.

Proof. Let c be an inhomogeneous cocycle that represents a class in the kernel of the comparison map. This implies that

- $c \colon G \times G \to \mathbb{C}$ is a bounded, continuous function; and
- there a a continuous function $\varphi \colon G \to \mathbb{C}$, such that $\delta\varphi = c$.

It suffices to show φ is bounded, for then c is the coboundary of a bounded cochain (namely, φ), so $[c] = 0$ in bounded cohomology.

Note that, for all $g, h \in G$, we have

8.20 $$|\varphi(gh) - \varphi(g) - \varphi(h)| = |\delta\varphi(g, h)| = |c(g, h)| \leq \|c\|_\infty.$$

Now assume, for concreteness, that $G = \mathrm{SL}(3, \mathbb{R})$, and let

$$U_{1,2} = \begin{bmatrix} 1 & * & \\ & 1 & \\ & & 1 \end{bmatrix}, \quad U_{1,3} = \begin{bmatrix} 1 & & * \\ & 1 & \\ & & 1 \end{bmatrix}, \quad U_{2,1} = \begin{bmatrix} 1 & & \\ * & 1 & \\ & & 1 \end{bmatrix},$$

$$U_{2,3} = \begin{bmatrix} 1 & & \\ & 1 & * \\ & & 1 \end{bmatrix}, \quad U_{3,1} = \begin{bmatrix} 1 & & \\ & 1 & \\ * & & 1 \end{bmatrix}, \quad U_{3,2} = \begin{bmatrix} 1 & & \\ & 1 & \\ & * & 1 \end{bmatrix}.$$

We will show that φ is bounded on $U_{1,2}$, and a similar argument shows that φ is bounded on each of the other elementary unipotent subgroups $U_{i,j}$. Then the desired conclusion that φ is bounded on all of G is obtained by combining these bounds with Equation (8.20) and the elementary observation that for some $N \in \mathbb{N}$, there exist $i_1, \cdots, i_N$ and $j_1, \ldots, j_N$, such that

$$G = U_{i_1, j_1} U_{i_2, j_2} U_{i_3, j_3} \cdots U_{i_N, j_N}.$$

To complete the proof, we now show that φ is bounded on $U_{1,2}$. To see this, note that for

$$a = \begin{bmatrix} 2 & & \\ & 1/2 & \\ & & 1 \end{bmatrix}$$

we have

$$\lim_{k \to \infty} a^{-k} u a^k = e \quad \text{for all } u \in U_{1,2}.$$

Therefore, for $u \in U_{1,2}$ and $k \in \mathbb{N}$, repeated application of Equation (8.20) yields

$$
\begin{aligned}
|h(u)| &\leq |h(a^k) + h(a^{-k}ua^k) + h(a^{-k})| + 2\|c\|_\infty \\
&\leq |h(a^{-k}ua^k)| + |h(a^k) + h(a^{-k})| + 2\|c\|_\infty \\
&\leq |h(a^{-k}ua^k)| + |h(e)| + 3\|c\|_\infty \\
&\to |h(e)| + |h(e)| + 3\|c\|_\infty \qquad\qquad \text{as } k \to \infty. \qquad \square
\end{aligned}
$$

THEOREM 8.21. (BURGER-MONOD) *The comparison map*

$$
H^2_{\mathrm{bdd}}(G; L^2_0(G/\Gamma)) \to H^2(G; L^2_0(G/\Gamma))
$$

is injective.

Proof. Let c be an inhomogeneous cocycle that represents a class in the kernel of the comparison map, so $c = \delta\varphi$ for some continuous $\varphi\colon G \to L^2_0(G/\Gamma)$. It suffices to show that φ is bounded.

Note that, letting π be the representation of G on $L^2_0(G/\Gamma)$, we have, for all $g, h \in G$,

8.22 $\qquad \|\varphi(gh) - \varphi(g) - \pi(g)\varphi(h)\| = |\delta\varphi(g, h)| = |c(g, h)\| \leq \|c\|_\infty.$

Let us assume $G = \mathrm{SL}(5, \mathbb{R})$. (The same argument works for all $n \geq 5$, but some modifications are needed when n is small.) Much as in the proof of Proposition 8.19, it suffices to show that φ is bounded on

$$
U_{1,2} = \begin{bmatrix} 1 & * & & & \\ & 1 & & & \\ & & 1 & & \\ & & & 1 & \\ & & & & 1 \end{bmatrix}.
$$

Let

$$
H = \begin{bmatrix} 1 & & & & \\ & 1 & & & \\ & & * & * & * \\ & & * & * & * \\ & & * & * & * \end{bmatrix} \cong \mathrm{SL}(3, \mathbb{R}).
$$

For all $u \in U_{1,2}$ and $h \in H$, we have

$$\| \varphi(u) - \varphi(h) - \pi(h)\varphi(u) - \pi(h)\pi(u)\,\varphi(h^{-1}) \|$$

$$= \| (\varphi(huh^{-1}) - \varphi(h) - \pi(h)\,\varphi(uh^{-1})) \qquad \begin{pmatrix} H \text{ commutes with } U_{1,2}, \\ \text{so } u = huh^{-1} \end{pmatrix}$$

$$+ \pi(h)\,(\varphi(uh^{-1}) - \varphi(u) - \pi(u)\,\varphi(h^{-1})) \|$$

$$\leq 2\|c\|_\infty. \qquad\qquad\qquad \text{(by Equation (8.22))}$$

So

$$\|(\mathrm{Id} - \pi(h))\varphi(u)\| \leq \|\varphi(h)\| + \|\varphi(h^{-1}\| + 2\|c\|_\infty.$$

Hence, for any function f on H whose integral is 1, we have

8.23 $\qquad \|(\mathrm{Id} - \pi(f))\,\varphi(u)\| \leq \sup_{h \in \mathrm{supp} f} \big(\|\varphi(h)\| + \|\varphi(h^{-1}\| + 2\|c\|_\infty\big).$

To complete the proof, we need to make a good choice of the function f. There are no G-invariant vectors in $L_0(G/\Gamma)$, so the Moore ergodicity theorem (6.8) implies there are no H-invariant vectors (see the proof on page 184). Since $H = \mathrm{SL}(3, \mathbb{R})$ has Kazhdan's property (T), we conclude that H has no almost-invariant vectors. Hence, it is not difficult to see that there is a continuous function f on H, such that

- f has compact support,
- $\int_H f \, d\mu_H = 1$ (where μ_H is the Haar measure on H), and
- $\|\pi(f)\| < 1$.

Then $\mathrm{Id} - \pi(f)$ is invertible, so Equation (8.23) implies $\|\varphi(u)\|$ is bounded (independent of u), as desired. $\qquad\qquad\qquad\qquad\qquad\qquad \square$

NOTES FOR §8. See [Mo2] for a recent introduction to bounded cohomology. Although our presentation of bounded cohomology takes a very naive approach, the work of Ghys and Burger-Monod is much more functorial.

Proposition 8.4 is due to É. Ghys [Gh1]. In fact, he proved that the Euler class determines the action up to semiconjugacy.

The Burger-Monod theory of bounded cohomology (including theorems 8.8 and 1.4) was developed in [BM1, BM2]. An exposition appears in [Mo1]. The improvement mentioned in Remark 8.10 appears in the paper of M. Burger [Bur] in this volume.

The cohomology vanishing result in Equation (8.7) is a special case of [Bo2, thm. 4.4(ii)].

9. Nonorderability from Bounded Orbits and Bounded Generation

Recall that Theorem 1.1 can be stated in any one of the following three equivalent forms (cf. Proposition 2.8 and Lemma 2.6):

1) If $n \geq 3$, then no finite-index subgroup of $SL(n, \mathbb{Z})$ has a faithful action on S^1.
2) If $n \geq 3$, then no finite-index subgroup of $SL(n, \mathbb{Z})$ has an orientation-preserving faithful action on $\mathbb{R}$.
3) If $n \geq 3$, then no finite-index subgroup of $SL(n, \mathbb{Z})$ is left orderable.

A short, elementary proof of this theorem was given in Section 3, but we will now describe a different approach that has the potential to work in a more general situation. It has two main ingredients:

- bounded orbits of unipotent elementary matrices (Proposition 9.2), and
- bounded generation by unipotent elementary matrices (Theorem 9.5).

To keep things simple, let us assume $n = 3$.

DEFINITION 9.1. A matrix u in $SL(3, \mathbb{R})$ is a *unipotent elementary matrix* if

- $u_{1,1} = u_{2,2} = u_{3,3} = 1$ (i.e., u has all 1's on the main diagonal), and
- u has only one nonzero entry off the main diagonal.

(In other words, a unipotent elementary matrix is one of the matrices $a_1, \ldots, a_6$ of Equation (3.7), for some $k \in \mathbb{R}$.)

PROPOSITION 9.2. *Suppose a finite-index subgroup Γ of $SL(3, \mathbb{Z})$ has an orientation-preserving faithful action on $\mathbb{R}$.*

If u is any unipotent elementary matrix in Γ, then the u-orbit of each point in $\mathbb{R}$ is a bounded set.

Proof. It suffices to show that the u-orbit of 0 is bounded above.

Define a left-invariant total order $\prec$ on Γ, as in Lemma 2.6($\Leftarrow$), so

$$a(0) < b(0) \quad \Rightarrow \quad a \prec b.$$

By permuting the standard basis vectors, we may assume

$$u = \begin{bmatrix} 1 & & * \\ & 1 & \\ & & 1 \end{bmatrix}.$$

Then Lemma 3.6 implies there is some $a \in \Gamma$, such that $u \ll a$ (see Equation (3.8) with $i = 2$); that is, $u^k \prec a$ for all $k \in \mathbb{Z}$. From the definition of $\prec$, this means that

$$u^k(0) < a(0) \quad \text{for all } k \in \mathbb{Z}.$$

So the u-orbit of 0 is bounded above (by $a(0)$). $\qquad \square$

It is a standard fact of undergraduate linear algebra that any invertible matrix is a product of elementary matrices. (This is a restatement of the fact that every invertible matrix can be reduced to the identity by elementary row operations.) Furthermore, it is not difficult to see that if the invertible matrix has determinant 1, then:

- the elementary matrices can be assumed to be unipotent, and
- there is a bound on the *number* of elementary matrices that is needed.

EXERCISE 9.3. Show that every matrix in SL$(3, \mathbb{R})$ is a product of < 10 unipotent elementary matrices.

It is a much deeper fact that the boundedness remains true when the field $\mathbb{R}$ is replaced with the ring $\mathbb{Z}$:

THEOREM 9.4. (CARTER-KELLER) *Every element of* SL$(3, \mathbb{Z})$ *is a product of* < 50 *unipotent elementary matrices in* SL$(3, \mathbb{Z})$.

There is also a bound for any finite-index subgroup Γ, but the bound may depend on Γ, and the unipotent elementary matrices may not generate quite all of Γ:

THEOREM 9.5. (CARTER-KELLER-PAIGE) *If* Γ *is a finite-index subgroup of* SL$(3, \mathbb{Z})$, *then there is a number* N, *and a finite-index subgroup* Γ' *of* Γ, *such that every element of* Γ' *is a product of* $\leq N$ *unipotent elementary matrices in* Γ'.

Combining Proposition 9.2 and Theorem 9.5 yields the following conclusion:

COROLLARY 9.6. *If* Γ *is a finite-index subgroup of* SL$(3, \mathbb{Z})$, *then every orientation-preserving action of* Γ *on* $\mathbb{R}$ *has a fixed point.*

Proof. It suffices to show that the orbit $\Gamma \cdot 0$ is bounded above, for then the supremum of this orbit is a fixed point.

For simplicity, let us ignore the difference between Γ' and Γ in Theorem 9.5, so there is a sequence $g_1, g_2, \ldots, g_M$ of unipotent elementary matrices in Γ, such that

$$\Gamma = \langle g_M \rangle \cdots \langle g_2 \rangle \langle g_1 \rangle.$$

We will show, for $k = 1, 2, \ldots, M$, that

$$(\langle g_k \rangle \cdots \langle g_2 \rangle \langle g_1 \rangle) \cdot 0 \text{ is bounded above.}$$

To this end, let x be the supremum of $(\langle g_{k-1} \rangle \cdots \langle g_2 \rangle \langle g_1 \rangle) \cdot 0$, and assume, by induction, that $x < \infty$. Then $x \in \mathbb{R}$, so Proposition 9.2 tells us that $\langle g_k \rangle \cdot x$ is bounded above by some $y \in \mathbb{R}$. Then all of $(\langle g_k \rangle \cdots \langle g_2 \rangle \langle g_1 \rangle) \cdot 0$ is bounded above by y, as desired. □

Proof of Theorem 1.1. Suppose Γ has a nontrivial, orientation-preserving action on $\mathbb{R}$. The set of fixed points is closed, so its complement is a disjoint union of open intervals; let I be one of those intervals.

Note that the interval I is Γ-invariant (because the endpoints I are fixed points), so Γ acts on I (by orientation-preserving homeomorphisms). Then, since the open interval I is homeomorphic to $\mathbb{R}$, Corollary 9.6 tells us that Γ has a fixed point in I. This contradicts the fact that, by definition, I is contained in the complement of the fixed point set. □

REMARK 9.7. It is hoped that the approach described in this section will (soon?) yield a proof of Conjecture 1.3 in the case where the lattice Γ is *not* cocompact.

More precisely, let Γ be a lattice in $G = \mathrm{SL}(3, \mathbb{R})$, such that G/Γ is not compact, and suppose we have an orientation-preserving action of Γ on $\mathbb{R}$. Then:

1) A matrix u in $\mathrm{SL}(3, \mathbb{R})$ is *unipotent* if it is conjugate to an element of

$$\begin{bmatrix} 1 & * & * \\ & 1 & * \\ & & 1 \end{bmatrix}.$$

2) Proposition 9.2 can be generalized to show that if u is any unipotent matrix in Γ, then the u-orbit of each point in $\mathbb{R}$ is a bounded set.

3) It is well known that some finite-index subgroup Γ' of Γ is generated by unipotent matrices, and it is conjectured that Theorem 9.5 generalizes: every element of Γ' should be the product of a *bounded* number of unipotent matrices in Γ'.

If the conjecture in (3) can be proved, then the above argument shows that Γ has no nontrivial, orientation-preserving action on $\mathbb{R}$.

NOTES FOR §9. This combination of bounded orbits and bounded generation was used in [LM1, LM2] to prove that some lattices cannot act on $\mathbb{R}$.

Theorem 9.4 is due to D. Carter and G. Keller [CK1]; an elementary proof is given in [CK2]. Theorem 9.5 is due to D. Carter, G. Keller, and E. Paige [CKP, Mor].

See the remarks leading up to theorem 10.5 for more discussion along the lines of Remark 9.7.

10. Complements

10A. Actions of Lattices in Other Semisimple Lie Groups

Conjecture 1.3 refers only to lattices in SL$(n, \mathbb{R})$. We can replace SL$(n, \mathbb{R})$ with any other (connected, linear) simple Lie group G whose real rank is at least 2, but some care is needed in stating a precise conjecture for groups that are semisimple, rather than simple. First of all, it should be assumed that the lattice Γ is *irreducible* (i.e., that no finite-index subgroup of Γ is a direct product $\Gamma_1 \times \Gamma_2$ of two infinite subgroups). But additional care is needed if SL$(2, \mathbb{R})$ is one of the simple factors of G:

EXAMPLE 10.1. Let $G = $ SL$(2, \mathbb{R}) \times$ SL$(2, \mathbb{R})$, so G is a semisimple Lie group, and $\mathbb{R}$-rank $G \geq 2$. Since SL$(2, \mathbb{R})$ acts on $\mathbb{R} \cup \{\infty\} \cong S^1$ (by linear-fractional transformations), the group G also acts on S^1, via projection to the first factor. It is then easy to see that any lattice Γ in G has an action on S^1 (by linear-fractional transformations) in which *every* orbit is infinite. Furthermore, the action is faithful if Γ is torsion free and irreducible.

CONJECTURE 10.2. *Let Γ be an irreducible lattice in a connected, semisimple Lie group G with finite center, such that $\mathbb{R}$-rank $G \geq 2$. Then*

1) *Γ has no nontrivial, orientation-preserving action on $\mathbb{R}$; and*
2) *Γ is not left orderable.*

Furthermore, if no simple factor of G is locally isomorphic to SL $(2, \mathbb{R})$, *then*

3) Γ *has no faithful action on* S^1, *and*
4) *whenever* Γ *acts on* S^1, *every orbit is finite.*

REMARK 10.3. The conjecture has been verified in some cases:

1) Theorem 1.1 verifies the conjecture in the special case where Γ is a finite-index subgroup of SL $(n, \mathbb{Z})$, with $n \geq 3$. A very similar argument applies when Γ is a finite-index subgroup of Sp $(2n, \mathbb{Z})$, with $n \geq 2$.
2) From the examples in (1), it follows that the conclusions of the conjecture hold whenever $\mathbb{Q}$-rank $\Gamma \geq 2$.
3) L. Lifschitz and D. W Morris verified the conjecture in the special case where

    a) some simple factor of G is locally isomorphic to either SL $(2, \mathbb{R})$ or SL $(2, \mathbb{C})$, and
    b) G/Γ is *not* compact.

It seems likely that the method of Section 9 will be able to prove the following cases of the conjecture:

CONJECTURE 10.4. *If* Γ *is any noncocompact lattice in either* SL $(3, \mathbb{R})$ *or* SL $(3, \mathbb{C})$, *then* Γ *has no nontrivial, orientation-preserving action on* $\mathbb{R}$.

If so, then we would have a proof of all of the noncocompact cases:

THEOREM 10.5. (CHERNOUSOV-LIFSCHITZ-MORRIS) *Assume Conjecture 10.4 is true. If G and* Γ *are as in Conjecture 10.2 and* G/Γ *is not compact, then the conclusions of Conjecture 10.2 are true.*

Other evidence for Conjecture 10.2 is provided by the Ghys-Burger-Monod theorem (1.4), which remains valid in this setting:

THEOREM 10.6. (GHYS, BURGER-MONOD) *If G and* Γ *are as in Conjecture 10.2, and no simple factor of G is locally isomorphic to* SL $(2, \mathbb{R})$, *then*

1) *Every action of* Γ *on* S^1 *has at least one finite orbit.*
2) Γ *has no faithful* C^1 *action on the circle* S^1.

REMARK 10.7.

1) The lattices that appear in Conjecture 10.2 are examples of arithmetic groups, and the conjecture could be extended to the class of S-arithmetic groups. The Ghys-Burger-Monod theorem has been generalized to this setting [WZ, Cor. 6.11], and the appropriate analogue of Conjecture 10.2 has been proved in the special case of S-arithmetic groups that are neither arithmetic nor cocompact [LM1].

2) The Burger-Monod proof of Theorem 10.6 applies to some cases where Γ is a lattice in a product group $G = G_1 \times G_2$ that is not assumed to be a Lie group. A generalization of this result on lattices in products has been proved by U. Bader, A. Furman, and A. Shaker [BFS].

3) The Burger-Monod injectivity theorem (cf. 8.8) does not immediately imply the conclusion of Theorem 10.6 in cases where $H^2(\Gamma; \mathbb{R}) \neq 0$. However, elsewhere in this volume, M. Burger [Bur] uses the injectivity to obtain a general theorem that includes both Theorem 8.8 and the results on lattices in products mentioned in the preceding paragraph. The rough idea is that if we have a Γ-action on S^1, such that the real Euler class of the action is in the image of $H^2_{\mathrm{bdd}}(G; \mathbb{R})$, then a certain quotient of the Γ-action must extend to a nontrivial, continuous action of G on the circle.

REMARK 10.8. Our discussion deals only with actions of lattices on the 1-dimensional manifolds S^1 and $\mathbb{R}$, but it is conjectured that large lattices also have no faithful actions on manifolds of other small dimensions. (For example, if $n \geq m + 2$, then no lattice in SL $(n, \mathbb{R})$ should have a faithful C^∞ action on any compact m-manifold.) Some discussion of this can be found in D. Fisher's survey paper [Fis] in this volume, or in Robert J. Zimmer's CBMS lectures [ZM].

NOTES FOR §10A. A weaker version of Conjecture 10.2 was suggested by D. Witte in 1990 (unpublished), but the definitive statement that deals correctly with SL $(2, \mathbb{R})$ factors is due to É. Ghys [Gh2, p. 200].

Parts (1) and (2) of Remark 10.3 are due to D. Witte [Wit]. See [LM1] or [LM2] for Part (3).

Theorem 10.5 is implicit in [LM2, §8]. The proof depends crucially on the main result of [CLM].

Theorem 10.6 is due to É. Ghys [Gh2] and (in slightly less generality) M. Burger and N. Monod [BM1, BM2]. Ghys's proof takes a geometric approach that relies on a case-by-case analysis of the possible Lie groups G; a modified (more algebraic) version of the proof that eliminates the case-by-case analysis was found by D. Witte and R. J. Zimmer [WZ].

10B. Some Lattices That Do Act on the Circle

The following two conjectures suggest that the conclusions of Conjecture 10.2 fail for lattices in SO $(1, n)$. Thus, the assumption that $\mathbb{R}$-rank $G \geq 2$ cannot be omitted, although it may be possible to weaken it.

CONJECTURE 10.9. (W. THURSTON) *If Γ is any lattice in SO $(1, n)$, then there is*

- *a finite-index subgroup Γ' of Γ, and*
- *a surjective homomorphism $\phi \colon \Gamma' \to \mathbb{Z}$.*

Because $\mathbb{Z}$ obviously has a C^∞ action on the circle with no finite orbits, this implies the following conjecture:

CONJECTURE 10.10. *If Γ is any lattice in SO $(1, n)$, then some finite-index subgroup of Γ has a C^∞ action on S^1 that has no finite orbits.*

These conjectures have been proved almost completely, under the additional assumption that Γ is arithmetic.

THEOREM 10.11. (LI, MILLSON, RAGHUNATHAN, VENKATARAMANA) *Suppose Γ is a lattice in SO $(1, n)$. If*

- *Γ is arithmetic, and*
- *$n \notin \{1, 3, 7\}$,*

then the conclusions of Conjectures 10.9 and 10.10 hold.

REMARK 10.12.

1) There exist lattices in SO $(1, 3)$ that act *faithfully* on S^1. (In fact, any torsion-free, cocompact lattice in SO $(1, 3)$ with infinite abelianization is left orderable.) It would be very interesting to know whether there exist such lattices in SO $(1, n)$ for $n \geq 4$, or in other groups of real rank 1.
2) The conclusions of Conjectures 10.9 and 10.10 hold for the nonarithmetic lattices constructed by M. Gromov and I. Piatetski-Shapiro[GP]. Thus,

a counterexample to these conjectures would have to be constructed by some other method.

REMARK 10.13. Generalizing Example 10.1, it is clear that if Γ is a torsion-free, irreducible lattice in G and G has a simple factor that is locally isomorphic to SL(2, $\mathbb{R}$), then Γ has a faithful action on S^1 by linear-fractional transformations. Conversely, under the additional assumption that $\mathbb{R}$-rank $G \geq 2$, É. Ghys [Gh2, thm. 1.2] proved that every action of Γ on S^1 either has a finite orbit or is semiconjugate to such an action by linear-fractional transformations.

NOTES FOR §10B. Conjecture 10.9 is attributed to W. Thurston (see [Bo1, p. 88]).

Theorem 10.11 combines work of several authors [Mil, Li, LiM, RV]. See [Rag] for a survey.

See [Gh3, §7.4] for a discussion of some lattices in SO (1, 3) that act on the circle. The general fact stated in Remark 10.12(1) is a theorem of S. Boyer, D. Rolfsen, and B. Wiest [BRW].

Remark 10.12(2) is due to A. Lubotzky [Lub].

10C. Actions on Trees

The real line and the circle are the only connected 1-manifolds, but there are many other 1-dimensional simplicial complexes. For short, a contractible, 1-dimensional simplicial complex is called a *tree*, and focusing our attention on the groups that act on trees leads to an interesting theory (the *Bass-Serre theory of group actions on trees*).

REMARK 10.14. Suppose, as usual, that Γ is a lattice in a (connected, linear) semisimple Lie group G. Then it is easy to construct a faithful action of Γ on some tree, even if we assume that the tree is locally finite. To do this,

1) Let $\Gamma = N_0 \supset N_1 \supset \cdots$ be a chain of finite-index, normal subgroups of Γ, such that $\bigcap_k N_k = \{e\}$.
2) Let the 0-skeleton T_0 be the disjoint union of all Γ/N_k.
3) Let the 1-skeleton T_1 have a 1-simplex (or "edge") joining γN_k and γN_{k+1}, for every $\gamma \in \Gamma$ and $k = 0, 1, \ldots$.

Then Γ has a natural action on T_0 by left translations, and this extends to an action on T (by isometries).

The following theorem states that under mild hypotheses, every action of Γ on a tree has a finite orbit. This can be thought of as an analogue of Theorem 1.4 for actions on trees.

THEOREM 10.15. *If*

- Γ *is as in Conjecture 10.2, or* Γ *has Kazhdan's property* (T),
- T *is a tree that is not homeomorphic to* $\mathbb{R}$, *and*
- Γ *acts on* T *by homeomorphisms,*

then Γ *has at least one finite orbit on* T.

REMARK 10.16.

1) More precisely, the finite orbit in the conclusion of Theorem 10.15 can be taken to consist of either a single vertex or two vertices of the tree.
2) In the Bass-Serre theory, it is usually assumed that the action is by isometries. In this case:

   a) there is no need to assume that T is not homeomorphic to $\mathbb{R}$, and
   b) the finite orbit can be taken to be a fixed point (and this fixed point is either a vertex or the midpoint of some edge).

REMARK 10.17. A fundamental conclusion of the Bass-Serre theory is that there is a finite orbit in every action of a countable group Λ on every tree (except possibly $\mathbb{R}$) if and only if

1) Λ is finitely generated,
2) $\Lambda/[\Lambda, \Lambda]$ is finite, and
3) Λ cannot be written in any nontrivial way as a free product with amalgamation $A *_C B$.

In the situation of Theorem 10.15, it is well known that Γ is finitely generated, and that $\Gamma/[\Gamma, \Gamma]$ is finite. Thus, in algebraic terms, Theorem 10.15 is the assertion that Γ is not a free product with amalgamation.

NOTES FOR §10C. J.–P. Serre's elegant book [Se2] is the standard introduction to the Bass-Serre theory of actions on trees.

In the special case where $\Gamma = \mathrm{SL}(3, \mathbb{Z})$, Theorem 10.15 is due to J.–P. Serre [Se1]. (See [Se2, thm. 16, p. 67] for an exposition.) The generalization to other lattices of higher rank is due to G. A. Margulis [Ma1, thm. 2]. The case of groups with Kazhdan's property (T) is due to R. Alperin [Alp] and Y. Watatani

[Wat], independently. (Proofs can also be found in [BHV, §2.3] and [Ma2, thm. 3.3.9 and ¶3.3.10].)

See [Se2, thm. 15, p. 58] for a proof of Remark 10.17.

References

[Alp]   R. C. Alperin: Locally compact groups acting on trees and property T *Mh. Math.* 93 (1982), 261–265.

[Aul]   J. C. Ault: Right-ordered locally nilpotent groups, *J. London Math. Soc.* (2) 4 (1972), 662–666.

[BFGM]   U. Bader, A. Furman, T. Gelander, and N. Monod: Property (1) and rigidity for actions on Banach spaces, *Acta Math.* 198 (2007), no. 1, 57–105.

[BFS]   U. Bader, A. Furman, and A. Shaker: Superrigidity, Weyl groups, and actions on the circle, (preprint).

[BHV]   B. Bekka, P. de la Harpe, and A. Valette: *Kazhdan's property (T)*, Cambridge University Press, Cambridge, 2008.

[Bo1]   A. Borel: Cohomologie de sous-groupes discrets et représentations de groupes semi-simples, *Astérisque* 32–33 (1976), 73–112.

[Bo2]   A. Borel: Stable real cohomology of arithmetic groups II, in: J. Hano et al., eds., *Manifolds and Lie groups (Notre Dame, Ind., 1980)*, Birkhäuser, Boston, 1981, pp. 21–55.

[BRW]   S. Boyer, D. Rolfsen, and B. Wiest: Orderable 3-manifold groups, *Ann. Inst. Fourier (Grenoble)* 55 (2005), no. 1, 243–288.

[BV]   M. Bridson and K. Vogtmann: Homomorphisms from automorphism groups of free groups, *Bull. London Math. Soc.* 35 (2003), no. 6, 785–792.

[Bur]   M. Burger: An extension criterion for lattice actions on the circle, (preprint).

[BM1]   M. Burger and N. Monod: Bounded cohomology of lattices in higher rank Lie groups, *J. Eur. Math. Soc.* 1 (1999), 199–235.

[BM2]   M. Burger and N. Monod: Continuous bounded cohomology and applications to rigidity theory, *Geom. Funct. Anal.* 12 (2002), no. 2, 219–280.

[CK1]   D. Carter and G. Keller: Bounded elementary generation of SL_n ($\mathcal{O}$), *Amer. J. Math.* 105 (1983), 673–687.

[CK2]   D. Carter and G. Keller: Elementary expressions for unimodular matrices, *Comm. Algebra* 12 (1984), 379–389.

[CKP]   D. Carter, G. Keller, and E. Paige: Bounded expressions in SL (n, A) (unpublished).

[CLM]   V. Chernousov, L. Lifschitz, and D. W. Morris: Almost-minimal nonuniform lattices of higher rank, *Michigan Math. J.* 56, no. 2, (2008), 453–478.

[EN]   R. Ellis and M. Nerurkar: Enveloping semigroup in ergodic theory and a proof of Moore's ergodicity theorem, in: J. C. Alexander., ed., *Dynamical systems (College Park, MD, 1986–87)*, Springer, New York, 1988, pp. 172–179.

[Fis]   D. Fisher: Groups acting on manifolds: around the Zimmer program, (preprint).

[Fur]   H. Furstenberg: Boundary theory and stochastic processes on homogeneous spaces, in: C. C. Moore, ed., *Harmonic Analysis on Homogeneous Spaces (Williamstown, MA., 1972)*. American Mathematical Society, Providence, RI, 1973, pp. 193–229.

[Gh1]   É. Ghys: Groupes d'homéomorphismes du cercle et cohomologie bornée, in: *The Lefschetz Centennial Conference, Part III (Mexico City, 1984)*, Contemp. Math., vol. 58, Part III, American Mathematical Society, Providence, RI, 1987, pp. 81–106.

[Gh2]   É. Ghys: Actions de réseaux sur le cercle, *Invent. Math.* 137 (1999), 199–231.

[Gh3]   É. Ghys: Groups acting on the circle, *L'Enseign. Math.* 47 (2001), 329–407.

[Gor]   D. Gorenstein: *Finite Groups*, Chelsea, New York, 1980.

[GP]   M. Gromov and I. Piatetski-Shapiro: Nonarithmetic groups in Lobachevsky spaces, *Inst. Hautes Études Sci. Publ. Math.* 66 (1988), 93–103.

[HM]   R. E. Howe and C. C. Moore: Asymptotic properties of unitary representations, *J. Funct. Anal.* 32 (1979), no. 1, 72–96.

[KM]   V. M. Kopytov and N. Ya. Medvedev: *Right-Ordered Groups*, Consultants Bureau, New York, 1996.

[Li]   J. S. Li: Non-vanishing theorems for the cohomology of certain arithmetic quotients, *J. reine angew. Math.* 428 (1992), 177–217.

[LiM]   J. S. Li and J. Millson: On the first Betti number of a hyperbolic manifold with an arithmetic fundamental group, *Duke Math. J.* 71 (1993), 365–401.

[LM1]   L. Lifschitz and D. Morris: Isotropic nonarchimedean S-arithmetic groups are not left orderable, *C. R. Math. Acad. Sci. Paris* 339 (2004), no. 6, 417–420.

[LM2]   L. Lifschitz and D. Morris: Bounded generation and lattices that cannot act on the line, *Pure Appl. Math. Q.* 4 (2008), no. 1, 99–126.

[Lub]   A. Lubotzky: Free quotients and the first Betti number of some hyperbolic manifolds, *Transformation Groups* 1 (1996), 71–82.

[Ma1]   G. A. Margulis: On the decomposition of discrete subgroups into amalgams, *Selecta Math. Sovietica* 1 (1981), 197–213.

[Ma2]   G. A. Margulis: *Discrete Subgroups of Semisimple Lie Groups*, Springer, New York, 1991.

[Mil]   J. Millson: On the first Betti number of a constant negatively curved manifold, *Ann. Math.* 104 (1976), 235–247.

[Mo1]   N. Monod: *Continuous bounded cohomology of locally compact groups*, Springer, Berlin, 2001.

[Mo2]   N. Monod: An invitation to bounded cohomology, in: M. Sanz-Solé et al., eds., *Proc. International Congress of Mathematicians, Vol. II*, European Mathematical Society, Zürich, 2006, pp. 1183–1211.

[Moo]   C. C. Moore: Ergodicity of flows on homogeneous spaces, *Amer. J. Math.* 88 (1966), 154–178.

[Mor]   D. W. Morris: Bounded generation of SL (n, A) (after D. Carter, G. Keller, and E. Paige), *New York J. Math.* 13 (2007), 383–421.

[Na1]   A. Navas: Actions de groupes de Kazhdan sur le cercle, *Ann. Sci. École Norm. Sup.* (4) 35 (2002), no. 5, 749–758.

[Na2]   A. Navas: *Groups of Circle Diffeomorphisms*, (preprint, 2008).

[Pie]   J.-P. Pier: *Amenable Locally Compact Groups*, Wiley, New York, 1984.

[PS]   A. Pressley and G. Segal: *Loop Groups*, Oxford University Press, Oxford, 1986.

[Rag]   M. S. Raghunathan: The first Betti number of compact locally symmetric spaces, in: S. D. Adhikari, ed., *Current Trends in Mathematics and Physics—A Tribute to Harish-Chandra*, Narosa Publishing House, New Delhi, 1995, pp. 116–137.

[RV]   M. S. Raghunathan and T. N. Venkataramana: The first Betti number of arithmetic groups and the congruence subgroup problem, *Contemp. Math.* 153 (1993), 95–107.

[RS]   G. Reeb and P. Schweitzer: Un théorème de Thurston établi au moyen de l'analyse
         non standard, in: P. A. Schweitzer, ed., *Differential Topology, Foliations and Gelfand-
         Fuks Cohomology* (Proc. Sympos., Pontificia Univ. Católica, Rio de Janeiro, January
         1976), Lecture Notes in Math, Vol. 652, Springer, New York, 1978, p. 138.

[Rez]   A. Reznikov: Analytic topology of groups, actions, strings and varieties, in:
         M. Kapranov et al., eds. *Geometry and dynamics of groups and spaces*, Birkhäuser,
         Basel, 2008, pp. 3–93, http://arxiv.org/abs/math/0001135

[Rhe]   A. H. Rhemtulla: Right-ordered groups, *Canad. J. Math.* 24 (1972), 891–895.

[Sch]   W. Schachermayer: Une modification standard de la demonstration non standard
         de Reeb et Schweitzer, in: P. A. Schweitzer, ed., *Differential Topology, Foliations
         and Gelfand-Fuks Cohomology* (Proc. Sympos., Pontificia Univ. Católica, Rio de
         Janeiro, January 1976), Lecture Notes in Math., vol. 652, Springer, New York,
         1978, pp. 139–140.

[Se1]   J.–P. Serre: Amalgames et points fixes, in: *Proceedings of the Second Inter-
         national Conference on the Theory of Groups (Australian Nat. Univ., Canberra, 1973)*,
         Springer, Berlin, 1974, pp. 633–640.

[Se2]   J.–P. Serre: *Trees*, Springer, New York, 1980.

[Thu]   W. Thurston: A generalization of the Reeb stability theorem, *Topology* 13 (1974),
         347–352.

[Uom]   R. Uomini: A proof of the compact leaf conjecture for foliated bundles, *Proc. Amer.
         Math. Soc.* 59 (1976), no. 2, 381–382.

[Wat]   Y. Watatani, Property (*T*) of Kazhdan implies property (*FA*) of Serre, *Math. Japon.*
         27 (1981), 97–103.

[Wit]   D. Witte: Arithmetic groups of higher $\mathbb{Q}$-rank cannot act on 1-manifolds, *Proc.
         Amer. Math. Soc.* 122 (1994), 333–340.

[WZ]   D. Witte and R. J. Zimmer: Actions of semisimple Lie groups on circle bundles,
         Geometriae Dedicata 87 (2001), 91–121.

[Zi1]   R. J. Zimmer: Orbit spaces of unitary representations, ergodic theory, and simple
         Lie groups, *Ann. Math.* (2) 106 (1977), no. 3, 573–588.

[Zi2]   R. J. Zimmer: *Ergodic Theory and Semisimple Groups*, Birkhäuser, Basel, 1984.

[ZM]   R. J. Zimmer and D. W. Morris: *Ergodic Theory, Groups, and Geometry*, American
         Mathematical Society, Providence, RI, 2008.

6

SOME REMARKS ON AREA-PRESERVING ACTIONS OF LATTICES

TO ROBERT ZIMMER

Abstract

In the spirit of Zimmer's study of large group actions on closed manifolds, we discuss the existence of area-preserving actions of higher-rank lattices on surfaces. We explain why certain vanishing results in bounded cohomology, due to Burger and Monod, combined with some constructions by Gambaudo and Ghys, give some constraints on the measure theoretical properties of such actions.

1. Introduction

The aim of this note is to discuss the existence of area-preserving actions of higher-rank lattices on surfaces. All the diffeomorphisms and vector fields we will consider in the text will be of class C^∞. Although much of our discussion would be valid with a lower regularity, we will not be concerned with this issue here.

We consider a lattice Γ in a connected simple Lie group G, with finite center and of real rank greater than 1. See [5] for some classical constructions of lattices. Let Σ be a closed oriented surface endowed with an area form ω. According to a well-known conjecture of Zimmer [48], any homomorphism $\rho : \Gamma \to \mathrm{Diff}(\Sigma, \omega)$ from Γ to the group of area-preserving diffeomorphisms of Σ should have finite image. Note that this conjecture can be seen as part of a more general program proposed by Zimmer [48] for the study of (not necessarily volume preserving) actions of higher-rank lattices on closed manifolds. For recent advances on this program, the reader might consult [7, 10, 11, 16, 22, 23, 26, 27, 37, 38, 46, 47, 49].

From now on, we will assume that a homomorphism $\rho : \Gamma \to \mathrm{Diff}(\Sigma, \omega)$ is given. We will say that the associated action of Γ on Σ is trivial if $\rho(\Gamma)$ is a finite group. We will also denote by μ the measure associated to the 2-form ω on Σ, and will suppose that $\int_\Sigma d\mu = 1$.

Although the conjecture is still open, a certain number of results already show that such an action of Γ have poor dynamical properties. For instance,

208

Zimmer [47, 49] showed that in this situation Γ must preserve a measurable Riemannian metric on Σ. As a consequence, all the elements in the group $\rho(\Gamma)$ have zero metric entropy. Another consequence is that the action has discrete spectrum [49]: the space $L^2(\Sigma, \mu)$ breaks down as a direct sum of finite-dimensional subrepresentations for the natural Γ-action. One approach to the conjecture (explained in [18], for instance) would be to study the regularity of the invariant Riemannian metric provided by Zimmer's theorem. Here we will follow a different route and give a few more evidences that such an action of Γ, if not trivial, has simple dynamical properties (at least from the measure theoretical point of view).

As we have just said, a consequence of Zimmer's results is that the group $\rho(\Gamma)$ does not contain any area-preserving diffeomorphism with positive metric entropy for the measure μ. On the other side of the spectrum of possible dynamical behaviors are area-preserving diffeomorphisms that are contained in a Hamiltonian flow. A consequence of our discussion will be that this kind of diffeomorphism cannot appear in the group $\rho(\Gamma)$ (see Section 3).

Observe first that we can compose ρ with the projection π from $\mathrm{Diff}(\Sigma, \omega)$ to the mapping class group of Σ, denoted by $\Lambda(\Sigma)$, to get a homomorphism $\pi \circ \rho : \Gamma \to \Lambda(\Sigma)$. According to a result of Farb and Masur [15], such a homomorphism has finite image. Hence, up to replacing Γ by a subgroup of finite index, we can assume that the image of ρ lies in the group $\mathrm{Diff}_0(\Sigma, \omega)$ of area-preserving diffeomorphisms of Σ that are isotopic to the identity. In fact, we can do one more reduction and assume that the image of ρ lies in the group of *Hamiltonian diffeomorphisms* of Σ. We recall its definition now.

Rotation Vectors and Hamiltonian Diffeomorphisms

There is a canonical homomorphism a from the group $\mathrm{Diff}_0(\Sigma, \omega)$ to an abelian group A (that depends on Σ). Its kernel is the group $\mathrm{Ham}(\Sigma, \omega)$ of Hamiltonian diffeomorphisms of Σ. When Σ is the 2-sphere, the group A is trivial and the groups $\mathrm{Diff}_0(S^2, \omega)$ and $\mathrm{Ham}(S^2, \omega)$ coincide (in that case, these two groups also coincide with the whole group of area-preserving diffeomorphisms of the sphere, which is connected [43]). When $\Sigma = T^2$ is the torus, the group A equals R^2/Z^2 and the homomorphism a is defined as follows. For each diffeomorphism $f \in \mathrm{Diff}_0(T^2, \omega)$, choose a lift $F : R^2 \to R^2$ of f. Since f is isotopic to the identity, the map F commutes with integral translations. Hence, the map $x \mapsto F(x) - x \in R^2$ is invariant under integral translations and defines a map from T^2 to R^2. One defines:

$$a(f) = \int_{T^2} (F(x) - x) d\mu(x) \bmod Z^2.$$

As suggested by the notation, $a(f) \in \mathbf{T}^2$ depends only on f: any lift of f differs from F by a translation by an element of $\mathbf{Z}^2$, hence the integral above is well-defined modulo $\mathbf{Z}^2$. It is often called the *rotation vector* of the diffeomorphism f. The map

$$a : \mathrm{Diff}_0(\mathbf{T}^2, \omega) \to \mathbf{R}^2/\mathbf{Z}^2$$

is a homomorphism. When Σ has genus greater than 1, the group A equals the first homology group $H_1(\Sigma, \mathbf{R})$ of Σ and one can construct the homomorphism a in the same spirit as above, see [20, 33]. Note that, as opposed to the case of the torus, we do not need to divide by the group $H_1(\Sigma, \mathbf{Z})$ since the group $\mathrm{Diff}_0(\Sigma, \omega)$ is simply connected [12]. The homomorphism $a : \mathrm{Diff}_0(\Sigma, \omega) \to H_1(\Sigma, \mathbf{R})$ is dual to the classical *flux homomorphism* from $\mathrm{Diff}_0(\Sigma, \omega)$ to $H^1(\Sigma, \mathbf{R})$; see for instance [35] and section 2 of [21].

It can be shown that the group $\mathrm{Ham}(\Sigma, \omega)$ is exactly the group of diffeomorphisms that are time 1 maps of a *Hamiltonian isotopy* $(f_t)_{t\in[0,1]}$. Recall that this means that the isotopy $(f_t)_{t\in[0,1]}$ satisfies the following differential equation

$$\begin{cases} f_0(x) = x \\ \frac{d}{dt}\left(f_t(x)\right) = X_t(f_t(x)), \end{cases}$$

for all $x \in \Sigma$, for a time-dependent vector field X_t, which is the *symplectic gradient* of a smooth time-dependent function H_t on Σ. This means by definition that X_t satisfies the relation:

$$dH_t(u) = \omega(X_t, u)$$

for any tangent vector $u \in T\Sigma$.

Coming back to our lattice, we see that the homomorphism $a \circ \rho$ has finite image, Γ having Kazhdan's property (T) (see [31]). Hence, up to considering once again a subgroup of finite index, one can assume that the homomorphism ρ has its image contained in the group of Hamiltonian diffeomorphisms of Σ. *The study of area-preserving actions of higher-rank lattices on closed surfaces is thus reduced to the study of actions by Hamiltonian diffeomorphisms.*

Let me mention that when Γ is a *nonuniform* lattice, and Σ has positive genus, Zimmer's conjecture has been proved by Polterovich [38], using methods from symplectic topology. His results have been reproved by Franks and Handel [22, 23] (and extended to the case where $\Sigma = \mathbf{S}^2$ for a particular class of nonuniform lattices) using methods from 2-dimensional dynamics. Note that one of the central theorems established by Polterovich in [38] to study actions of lattices is the following (in the statement below, we assume that Σ has genus at least 2).

If Λ *is a finitely generated subgroup of* $\mathrm{Ham}(\Sigma, \omega)$ *endowed with any word metric* $|\cdot|$, *and if* $\gamma \in \Lambda$ *is different from the identity, there exists* $\varepsilon > 0$ *such that* $|\gamma^n| \geq \varepsilon n$ $(n \in \mathbf{N})$.

This allows one to exclude the existence of nontrivial actions of nonuniform lattices thanks to the presence of unipotent elements; see [34]. But it also has different applications: for instance, any finitely generated nilpotent subgroup of $\mathrm{Ham}(\Sigma, \omega)$ is in fact abelian.

Quasimorphisms and Bounded Cohomology

Recall that a *quasi morphism* on a group Λ is a map $\phi : \Lambda \to \mathbf{R}$ for which there exists a constant $C > 0$ such that

$$|\phi(xy) - \phi(x) - \phi(y)| \leq C$$

for any $x, y \in \Lambda$. The quasi morphism ϕ is called *homogeneous* if moreover it satisfies $\phi(x^n) = n\phi(x)$ $(n \in \mathbf{Z}, x \in \Lambda)$; that is, ϕ is a true homomorphism when restricted to cyclic subgroups of Λ. For any quasimorphism ϕ, the limit

$$\phi_h(x) = \lim_{n \to \infty} \frac{\phi(x^n)}{n}$$

exists. The map ϕ_h is the unique homogeneous quasimorphism such that $\phi - \phi_h$ is bounded. We will denote by $\mathrm{QM}(\Lambda, \mathbf{R})$ the vector space of all quasimorphisms on Λ, and by $\mathrm{QM}_h(\Lambda, \mathbf{R})$ the subspace of homogeneous quasimorphisms. For more details on quasi morphisms and bounded cohomology see [4, 6] as well as Gromov's seminal paper [30]. We will return to the link between quasi morphisms and bounded cohomology in the next section.

According to a result by Burger and Monod (see [7] or [8]), any homogeneous quasimorphism on a higher-rank lattice Γ as above is a homomorphism, and hence is identically zero since Γ is a Kazhdan group. On the other hand, the group of Hamiltonian diffeomorphisms of a closed surface (which is simple and hence admits no nontrivial homomorphism to $\mathbf{R}$ [3]) admits many quasimorphisms: Gambaudo and Ghys [25] proved that the vector space $\mathrm{QM}_h(\mathrm{Ham}(\Sigma, \omega), \mathbf{R})$ is infinite-dimensional. Their proof consists in constructing explicitly a family of linearly independent homogeneous quasimorphisms. More constructions of quasimorphisms on groups of Hamiltonian diffeomorphisms can be found in the work of Entov and Polterovich [13] and of the author [40]. It is now well known (see [29]) that one could try to use this contrast between the lattice Γ and groups of Hamiltonian diffeomorphisms of surfaces to attack the problem of the existence of nontrivial homomorphisms $\rho : \Gamma \to \mathrm{Ham}(\Sigma, \omega)$. According to the discussion above, for any homogeneous quasimorphism $\phi : \text{Ham}(\Sigma, \omega) \to \mathbf{R}$ the invariant $\phi \circ \rho : \Gamma \to \mathbf{K}$ vanishes identically. This should give some constraints on the homomorphism ρ. This circle of ideas is nicely discussed in Ghys's survey [29].

Here, we will explain how one can use more subtly the vanishing results of Burger and Monod in bounded cohomology to obtain much more precise constraints on the homomorphism ρ. Namely, we will use the vanishing of certain bounded cohomology groups with unitary coefficients.

Note that the use of bounded cohomology for the study of group actions is not new. Bounded cohomology already appears in the study of groups acting on the circle; see [7, 28].

2. Vanishing Results and Consequences

We begin with a brief reminder on the *second bounded cohomology group* of a discrete group Λ, with unitary coefficients (see for instance [36] for more details).

We consider a unitary representation π of Λ, that is, a homomorphism from Λ to the group of unitary operators of a Hilbert space $\mathcal{H}$. We will write $\|v\|$ for the norm of a vector $v \in \mathcal{H}$. All the unitary representations we will consider are obtained in the following way: the group Λ is acting by measure-preserving transformations on a probability space (X, μ) and we consider the representation π of Λ on the Hilbert space $\mathcal{H} = L^2(X, \mu)$ of square integrable functions on X, defined by $\pi(\gamma)(f) = f \circ \gamma^{-1}$ $(f \in L^2(X, \mu))$. We will say that a map $c : \Lambda^j \to \mathcal{H}$ is bounded if the quantity

$$|c|_{\infty, \Lambda^j} := \sup_{(\gamma_1, \dots, \gamma_j) \in \Lambda^j} \|c(\gamma_1, \dots, \gamma_j)\|$$

is finite. The space $Z^2(\Lambda, \pi)$ of 2-cocycles on Λ with values in $\mathcal{H}$ is the space of maps $c : \Lambda^2 \to \mathcal{H}$ such that:

$$\pi(\gamma_1)(c(\gamma_2, \gamma_3)) - c(\gamma_1 \gamma_2, \gamma_3) + c(\gamma_1, \gamma_2 \gamma_3) - c(\gamma_1, \gamma_2) = 0.$$

We will denote by $Z_b^2(\Lambda, \pi) \subset Z^2(\Lambda, \pi)$ the subspace of bounded 2-cocycles. In the same way, the space $B^1(\Lambda, \pi) \subset Z^2(\Lambda, \pi)$ of coboundaries consists of the maps $c : \Lambda^2 \to \mathcal{H}$ that satisfy the equation $c(\gamma_1, \gamma_2) = \pi(\gamma_1)(v(\gamma_2)) + v(\gamma_1) - v(\gamma_1 \gamma_2)$ for some map $v : \Lambda \to \mathcal{H}$. The subspace of coboundaries c for which the map v in the equation above can be chosen bounded will be denoted by $B_b^1(\Lambda, \pi)$. The second bounded cohomology group of Λ with coefficients in $\mathcal{H}$ is the quotient

$$H_b^2(\Lambda, \pi) = Z_b^2(\Lambda, \pi)/B_b^1(\Lambda, \pi),$$

and the second usual cohomology group of Λ with coefficients in $\mathcal{H}$ is the quotient $H^2(\Lambda, \pi) = Z^2(\Lambda, \pi)/B^1(\Lambda, \pi)$. Of course, there is a natural map from $H_b^2(\Lambda, \pi)$ to $H^2(\Lambda, \pi)$ that sends the class of a bounded cocycle to its usual cohomology class. We will denote by $EH_b^2(\Lambda, \pi)$ the kernel of this map.

In the following proposition $\mathscr{C}(\Lambda, \mathscr{H})$ is the space of all maps from Λ to $\mathscr{H}$, $\mathscr{C}_b(\Lambda, \mathscr{H})$ the subspace of those maps that are bounded and $\partial^1 u$ is the coboundary of a map $u \in \mathscr{C}(\Lambda, \mathscr{H})$:

$$\partial^1 u(\gamma_1, \gamma_2) = \pi(\gamma_1)(u(\gamma_2)) + u(\gamma_1) - u(\gamma_1\gamma_2).$$

PROPOSITION 2.1. *The space $EH_b^2(\Lambda, \pi)$ is isomorphic to*

$$\{u \in \mathscr{C}(\Lambda, \mathscr{H}), |\partial^1 u|_{\infty, \Lambda^2} < \infty\} / \left(Z^1(\Lambda, \pi) + \mathscr{C}_b(\Lambda, \mathscr{H})\right).$$

Proof. If $u \in \mathscr{C}(\Lambda, \mathscr{H})$ is such that $|\partial^1 u|_{\infty, \Lambda^2} < \infty$, the coboundary $\partial^1 u$ of u is a bounded 2-cocycle, hence defines a class $[\partial^1 u] \in H_b^2(\Lambda, \pi)$ that is obviously trivial in usual cohomology. We have thus constructed a map $u \mapsto [\partial^1 u]$ from $\{u \in \mathscr{C}(\Lambda, \mathscr{H}), |\partial^1 u|_{\infty, \Lambda^2} < \infty\}$ to $EH_b^2(\Lambda, \pi)$ that is (by definition) surjective. Let us examine its kernel. The class $[\partial^1 u]$ vanishes in bounded cohomology precisely if there exists a bounded map $v : \Lambda \to \mathscr{H}$ such that $\partial^1 u = \partial^1 v$. In that case $w = u - v$ is a 1-cocycle and we have $u = w + v \in Z^1(\Lambda, \pi) + \mathscr{C}_b(\Lambda, \mathscr{H})$. □

When $\mathscr{H} = \mathbf{R}$, with the trivial representation π_0 of Λ, the previous proposition simply proves the following classical fact [4, 6]: the group $EH_b^2(\Lambda, \pi_0)$ is isomorphic to the quotient

$$\mathrm{QM}(\Lambda, \mathbf{R}) / \left(\mathrm{Hom}(\Lambda, \mathbf{R}) \oplus \mathscr{C}_b(\Lambda, \mathbf{R})\right).$$

Note that this last quotient is also isomorphic to the space $\mathrm{QM}_h(\Lambda, \mathbf{R})/\mathrm{Hom}(\Lambda, \mathbf{R})$. Let us come back to the case of a higher-rank lattice Γ, as in the introduction. In that case, we have already said that the group $EH_b^2(\Gamma, \pi_0)$ is trivial, according to a result of Burger and Monod [7, 8]. In fact, their result is much more general: they prove that the group $EH_b^2(\Gamma, \pi)$ vanishes, for any unitary representation π of Γ on a Hilbert space $\mathscr{H}$ (the result even holds for more general Banach spaces, but we will not really discuss this here). This can be seen as a strengthening of property (T): the group Γ has the property that for any unitary representation π, the group $H^1(\Gamma, \pi)$ as well as the group $EH_b^2(\Gamma, \pi)$ vanishes. Following Monod [36], we will say that a discrete group with this property has property (TT). We now discuss the dynamical consequences of property (TT).

Suppose that $\Lambda \curvearrowright (X, \mu)$ is a measure-preserving action of Λ on a probability space. We will say that a map $u : \Lambda \to L^2(X, \mu)$ is a *quasicocycle* if there exists a constant $C > 0$ such that

$$|u(\gamma_1 \gamma_2) - \pi(\gamma_1)(u(\gamma_2)) - u(\gamma_1)| \leq C$$

almost everywhere, for all $\gamma_1, \gamma_2 \in \Lambda$. We will give some examples of quasi-cocycles defined on groups of Hamiltonian diffeomorphisms of surfaces in the next section. According to the subadditive ergodic theorem (see [32] for instance), if u is a quasicocycle and $\gamma \in \Lambda$, the sequence of functions

$$\frac{u(\gamma^n)}{n}$$

converges almost everywhere to a function $\widehat{u}(\gamma)$. The reader should observe that since the action of Λ on X is measure preserving, the map

$$\gamma \mapsto \phi(\gamma) = \int_X u(\gamma) d\mu$$

is a quasimorphism on Λ. It is not difficult to check that the map $\gamma \mapsto \phi_u(\gamma) = \int_X \widehat{u}(\gamma) d\mu$ is the unique homogeneous quasimorphism at a bounded distance from ϕ. Note that all the quasimorphisms defined by Gambaudo and Ghys [25] and by the author [40] are obtained in this way: by integration of a quasicocycle.

REMARK 1 The quasimorphisms constructed by Entov and Polterovich [13] are defined in a completely different way; see [39] for a survey. It is not a priori clear how to relate them (in a nontrivial way) to some bounded cohomology class with nontrivial coefficients.

REMARK 2 A weaker definition of quasicocycle already appears in the littera-ture; see [44]. There, a map $u : \Lambda \to L^2(X, \mu)$ is called a quasicocycle if

$$|\partial^1 u|_{\infty, \Lambda^2} = \sup_{(\gamma_1, \gamma_2) \in \Lambda^2} |\partial^1 u(\gamma_1, \gamma_2)|_{L^2}$$

is finite. In our definition, we require the stronger condition that

$$\sup_{(\gamma_1, \gamma_2) \in \Lambda^2} |\partial^1 u(\gamma_1, \gamma_2)|_{L^\infty}$$

is finite, to ensure that $(\frac{u(\gamma^n)}{n})_{n \geq 0}$ converges almost everywhere for all γ. The notion of quasicocycle is also related to the notion of *rough action* (see [36], chapter V).

We now see the advantage of using the full result of Burger and Monod: if Λ is a group with property (TT) acting on X and u is a quasicocycle, the vanishing result with trivial coefficients tells us that for any $\gamma \in \Lambda$ the integral

$$\phi_u(\gamma) = \int_X \widehat{u}(\gamma) d\mu$$

is zero, while the full vanishing result tells us that the function $\widehat{u}(\gamma)$ is zero almost everywhere, as the following proposition shows.

PROPOSITION 2.2. *Suppose Λ has property (TT). Then, for any measure-preserving action $\Lambda \curvearrowright (X, \mu)$ on a probability space and for any quasicocycle $u : \Lambda \to L^2(X, \mu)$ we have*

$$\widehat{u}(\gamma) = 0,$$

almost everywhere (for all $\gamma \in \Lambda$).

Proof. If u is a quasicocycle, the map $\partial^1 u$ is a bounded 2-cocycle. Since the group $EH_b^2(\Lambda, \pi)$ is trivial, there exists, according to the previous proposition, a bounded map $v : \Lambda \to L^2(X, \mu)$ and a 1-cocycle $w : \Lambda \to L^2(X, \mu)$ such that $u = v + w$. We write $D := \sup_{\gamma \in \Lambda} ||v(\gamma)||$. Since Λ has property (T), there exists a map $\varphi \in L^2(X, \mu)$ such that $w(\gamma) = \varphi \circ \gamma - \varphi$. We obtain:

$$\frac{u(\gamma^n)}{n} = \frac{\varphi \circ \gamma^n - \varphi}{n} + \frac{v(\gamma^n)}{n}.$$

We already know that the sequence $\frac{u(\gamma^n)}{n}$ converges almost everywhere to $\widehat{u}(\gamma)$. The function

$$\frac{\varphi \circ \gamma^n - \varphi}{n} = \frac{1}{n} \sum_{j=0}^{n-1} (\varphi \circ \gamma - \varphi) \circ \gamma^j$$

converges almost everywhere by Birkhoff's theorem; it is a classical lemma in ergodic theory that its limit is 0 almost everywhere. Finally, $\frac{v(\gamma^n)}{n}$ converges almost everywhere to $\widehat{u}(\gamma)$. Since the norm of $\frac{v(\gamma^n)}{n}$ in $L^2(X, \mu)$ is less than $\frac{D}{n}$, there exists a subsequence $\frac{v(\gamma^{n_k})}{n_k}$ that converges almost everywhere to 0. Hence $\widehat{u}(\gamma) = 0$ almost everywhere. □

3. Some Examples

We now describe some examples of concrete quasicocycles on groups of Hamiltonian diffeomorphisms of surfaces, taken from [17, 20, 24, 25, 33, 40]. These functions are constructed in the spirit of Schwartzman's classical asymptotic cycle [42] or of Arnold's asymptotic Hopf invariant; see [1]. The interested reader will find many more examples in the work of Gambaudo and Ghys [25].

Since we are dealing with diffeomorphisms that are isotopic to the identity, we will always make use of isotopies $(f_t)_{t\in[0,1]}$ from the identity $\mathbf{1} = f_0$ to a given diffeomorphism $f = f_1$ and will consider various (kinds of) rotation numbers associated to the orbit $(f_t(x))$ of a point x or to a pair of orbits $(f_t(x))$, $(f_t(y))$ ($x \neq y$). To ensure that the number we define does not depend on the choice of the isotopy $(f_t)_{t\in[0,1]}$ but only on f we will often appeal to some results on the topology of the groups $\mathrm{Ham}(\Sigma, \omega)$ or $\mathrm{Diff}_0(\Sigma, \omega)$; see [12].

EXAMPLE 1 [17, 24] We first give an example of a 1-cocycle defined on the group $\mathrm{Diff}_c(\mathbf{D}^2, \omega)$ of area-preserving diffeomorphisms of the disc $\mathbf{D}^2 = \{(x, y) \in \mathbf{R}^2, |x|^2 + |y|^2 \leq 1\}$, which coincide with the identity near the boundary. This group is torsion free (see [39]). Although there is no nontrivial homomorphism from a Kazhdan group to $\mathrm{Diff}_c(\mathbf{D}^2, \omega)$ (thanks to Thurston's stability theorem), it is worth studying this first example. Consider a diffeomorphism $f \in \mathrm{Diff}_c(\mathbf{D}^2, \omega)$ and choose an isotopy (f_t) from $f_0 = \mathbf{1}$ to $f_1 = f$. For any two distinct points x, y in the disc, consider the nonzero vector $f_t(x) - f_t(y) \in \mathbf{R}^2$. When t goes from 0 to 1 the argument $e^{2i\pi u(t)}$ of this vector varies of a quantity $\mathrm{angle}_f(x, y) := u(1) - u(0)$. This defines a continuous function

$$\mathrm{angle}_f : \mathbf{D}^2 \times \mathbf{D}^2 - \Delta \to \mathbf{R},$$

where $\Delta = \{(x, x), x \in \mathbf{D}^2\}$ is the diagonal. One easily establishes that this function does not depend on the choice of the isotopy $(f_t)_{t\in[0,1]}$ but only on f, and that it is bounded; see [24]. If $f, g \in \mathrm{Diff}_c(\mathbf{D}^2, \omega)$ and (f_t) and (g_t) are two isotopies from the identity to f and g, respectively, we can consider the isotopy $(h_t) = (g_t) * (f_t g)$ from the identity to fg. It allows us to establish the relation:

$$\mathrm{angle}_{fg}(x, y) = \mathrm{angle}_g(x, y) + \mathrm{angle}_f(g(x), g(y)).$$

Hence the map $f \mapsto \mathrm{angle}_{f^{-1}}$ defines a 1-cocycle on the group $\mathrm{Diff}_c(\mathbf{D}^2, \omega)$ with values in the space $\mathrm{L}^2(\mathbf{D}^2 \times \mathbf{D}^2, \mathbf{R})$.

Problem. Suppose $\Lambda \subset \mathrm{Diff}_c(\mathbf{D}^2, \omega)$ is a finitely generated group. Can we give a condition on Λ that ensures that the class of the cocycle angle is nonzero in $H^1(\Lambda, \mathrm{L}^2(\mathbf{D}^2 \times \mathbf{D}^2, \mu^2))$?

The following proposition shows that the class of the cocycle $f \mapsto \mathrm{angle}_{f^{-1}}$ is always nonzero in the space $H^1(\Lambda, \mathscr{C}^0(\mathbf{D}^2 \times \mathbf{D}^2 - \Delta))$, unless Λ is trivial (here $\mathscr{C}^0(\mathbf{D}^2 \times \mathbf{D}^2 - \Delta)$ is the space of continuous functions on $\mathbf{D}^2 \times \mathbf{D}^2 - \Delta$).

PROPOSITION 3.1. *Fix a diffeomorphism $f \in \mathrm{Diff}_c(\mathbf{D}^2, \omega)$. Assume that there exists a continuous function $\varphi : \mathbf{D}^2 \times \mathbf{D}^2 - \Delta \to \mathbf{R}$ such that $\mathrm{angle}_f = \varphi \circ f - \varphi$. Then f is the identity.*

Proof. We assume that f is not the identity. Then, it is enough to find an integer n and two fixed points x, y of f^n such that $\text{angle}_{f^n}(x, y) \neq 0$. Indeed the relation $\text{angle}_f = \varphi \circ f - \varphi$ implies $\text{angle}_{f^n} = \varphi \circ f^n - \varphi$. If φ is continuous, this forces the equality $\text{angle}_{f^n}(x, y) = 0$ for any pair (x, y) of distinct fixed points of f^n.

We will see at the end of this section (see Lemma 3.2 and the paragraph following it) that if f is distinct from the identity there exists a fixed point x_0 and a point $y_0 \neq x_0$ such that the number

$$\widetilde{\text{angle}}_f(x_0, y_0) := \lim_{n \to \infty} \frac{1}{n} \text{angle}_{f^n}(x_0, y_0)$$

exists and is nonzero. Consider now the compact annulus $\mathbf{A}$ obtained from the disc by blowing up the fixed point x_0 thanks to the action of the differential df_{x_0}. The diffeomorphism f naturally extends to a homeomorphism of $\mathbf{A}$, still denoted f. Let F be the lift of f to the universal cover $\widetilde{\mathbf{A}}$ of the annulus that pointwise fixes the component of the boundary of $\widetilde{\mathbf{A}}$ corresponding to the boundary of the disc. If z is in the interior of $\mathbf{A}$ (identified with the interior of the disc, minus the point x_0), the limit

$$\lim_{n \to \infty} \frac{1}{n} \text{angle}_{f^n}(x_0, z),$$

if it exists, is the rotation number $\rho_F(z)$ (determined by the lift F) of the point z in the annulus. We know that

- the point y_0 has a nonzero rotation number, and
- points close to the boundary have zero rotation number.

Now, let us recall a result of Franks (this is corollary 2.4 of [19]). Let $g : \mathbf{A} \to \mathbf{A}$ be a homeomorphism isotopic to the identity and G be a lift of g to $\widetilde{\mathbf{A}}$. If x is a point of $\mathbf{A}$ we will denote by $\rho_G(x)$ the rotation number of x determined by G, if it exists. Let $K \subset \mathbf{A}$ be a chain transitive compact invariant set for g. See [19] for the notion of chain transitivity; this is a very weak form of recurrence. Assume that there exist two points x_1 and x_2 in K whose rotation numbers $\rho_G(x_1)$ and $\rho_G(x_2)$ exist and satisfy $\rho_G(x_1) < \rho_G(x_2)$. Then, for any rational number $\frac{p}{q} \in]\rho_G(x_1), \rho_G(x_2)[$, g has a periodic point with rotation number $\frac{p}{q}$.

Let us apply the above result to f. Since f is area preserving, the chain transitivity hypothesis will be easily satisfied. Almost every point of the interior of $\mathbf{A}$ is recurrent, hence every point of $\mathbf{A}$ is nonwandering. In particular every point of $\mathbf{A}$ is chain recurrent. The compact set $K = \mathbf{A}$ is connected, f-invariant, and all its points are chain recurrent, hence it is chain transitive (this is proposition 1.2 in [19]). Franks's result applies with $K = \mathbf{A}$. The existence of the point y_0 and of points with zero rotation number implies that there exists a periodic point γ_1 with nonzero rotation number. If the period of γ_1 is n, this exactly means that the quantity angle$_{f^n}(x_0, \gamma_1)$ is nonzero. □

EXAMPLE 2 [20, 33] If f is a Hamiltonian diffeomorphism of the torus, we have seen in the introduction that for any lift $F : \mathbf{R}^2 \to \mathbf{R}^2$ of f, one has $\int_{\mathbf{T}^2} (F(x) - x)d\mu(x) \in \mathbf{Z}^2$. We can therefore choose a particular lift f_* for f: the unique one for which $\int_{\mathbf{T}^2} (f_*(x) - x)d\mu(x) = 0$. Since the map

$$\begin{aligned} \mathbf{R}^2 &\to \mathbf{R}^2 \\ x &\mapsto f_*(x) - x \end{aligned}$$

is invariant under integral translations, there exists a map $v_f : \mathbf{T}^2 \to \mathbf{R}^2$ such that $v_f(p(x)) = f_*(x) - x$ (where $p : \mathbf{R}^2 \to \mathbf{T}^2$ is the natural projection). From the relation $(f \circ g)_* = f_* \circ g_*$, one deduces the cocycle relation $v_{f \circ g} = v_g + v_f \circ g$. The map $f \mapsto v_{f^{-1}} \subset L^2(\mathbf{T}^2, \mathbf{R}^2)$ is therefore a 1-cocycle.

Suppose now that $\Lambda \subset \mathrm{Ham}(\mathbf{T}^2, \omega)$ is a finitely generated subgroup on which the previous cocycle is cohomologically trivial: there exists a measurable map $\varphi : \mathbf{T}^2 \to \mathbf{R}^2$ such that $v_f = \varphi \circ f - \varphi$ ($f \in \Lambda$). Recall first that for almost every point x in the torus, we have

$$\liminf_n \left| \varphi(f^n(x)) - \varphi(x) \right| = 0.$$

This is an easy consequence of Poincaré's recurrence theorem and Lusin's theorem. But we also have the following relation: $f_*^n(x) - x = \varphi(f^n(p(x))) - \varphi(p(x))$ ($x \in \mathbf{R}^2$). Hence, we get that almost every point of the plane is recurrent for the dynamics of the diffeomorphism f_* (for all $f \in \Lambda$). In fact the function

$$\begin{aligned} \psi : \mathbf{R}^2 &\to \mathbf{R}^2 \\ x &\mapsto x - \varphi(p(x)) \end{aligned}$$

is invariant under the lifted action of Λ on the plane (through the diffeomorphisms f_*). The domain $\psi^{-1}([0, 1[^2) \subset \mathbf{R}^2$, as well as its translates by integral vectors, has finite measure and is invariant under Λ. Therefore the vanishing of the cohomology class of the cocycle $f \mapsto v_{f^{-1}}$ gives some constraints on the (measurable) dynamics of Λ.

EXAMPLE 3 [25] We assume here that the genus of Σ is greater than 1. Endow Σ with a hyperbolic metric and identify its universal cover with the Poincaré disc Δ. For each 1-form η on Σ, we will define a quasicocycle u_η with values in the space of continuous functions on Σ. This quasicocycle is a true cocycle if η is closed, and a coboundary if η is exact.

If $f : \Sigma \to \Sigma$ is a Hamiltonian diffeomorphism, we will name $f_* : \Delta \to \Delta$ the unique lift of f, which commutes with the action of the fundamental group of Σ on Δ. If $\widetilde{x} \in \Delta$, one can consider the unique geodesic arc $\alpha(\widetilde{x})$ between $\widetilde{x}$ and $f_*(\widetilde{x})$. Let $\widetilde{\eta}$ denote the lift of the 1-form η to Δ. The map

$$\begin{aligned} \Delta &\to \mathbf{R} \\ \widetilde{x} &\mapsto \int_{\alpha(\widetilde{x})} \widetilde{\eta} \end{aligned}$$

descends to a function $u_\eta(f, \cdot)$ on Σ. If $x \in \Sigma$ and $\widetilde{x} \in \Delta$ is any lift of x, the quantity

$$u_\eta(fg, x) \quad u_\eta(g, x) - u_\eta(f, g(x))$$

equals the integral of the 2-form $d\widetilde{\eta}$ on the geodesic triangle of Δ with vertices $\widetilde{x}, g_*(\widetilde{x}), f_* \circ g_*(\widetilde{x})$. Since the 2-form $d\widetilde{\eta}$ is bounded (it is invariant by the action of the fundamental group of Σ) this quantity is bounded by the norm of $d\widetilde{\eta}$ times the area of a hyperbolic triangle. We obtain

$$\left| u_\eta(fg, x) - u_\eta(g, x) - u_\eta(f, g(x)) \right| \leq \pi \cdot |d\eta|_\infty.$$

Hence the map $f \mapsto u_\eta(f^{-1}, \cdot)$ is a quasicocycle. The reader will find an alternative description of this quasicocycle in [39].

EXAMPLE 4 [25] We now give an example of a quasicocycle defined on the group $\mathrm{Diff}_0(\mathbf{T}^2, \omega)$. We will make use of the group structure of the torus.

We fix a point x_* in $\mathbf{T}^2 - \{0\}$ and choose a homogeneous quasimorphism

$$\phi : \pi_1(\mathbf{T}^2 - \{0\}, x_*) \to \mathbf{R}$$

(there are many since $\pi_1(\mathbf{T}^2 - \{0\}, x_*)$ is a free group; see [6] as well as [14] for generalizations). For each point $v \in \mathbf{T}^2 - \{0\}$ we fix a path $(\alpha_v(t))_{t \in [0,1]}$ from x_* to v in $\mathbf{T}^2 - \{0\}$ whose length is bounded for a Riemannian metric defined on the compact surface with boundary $\overline{\mathbf{T}^2 - \{0\}}$ obtained by blowing up the origin on the torus. If $(f_t)_{t \in [0,1]}$ is an area-preserving isotopy on $\mathbf{T}^2$ and x and y two distinct points on $\mathbf{T}^2$, we can consider the curve

$$f_t(x) - f_t(y) \in \mathbf{T}^2 - \{0\}.$$

We close it to form a loop $\alpha(f, x, y) = \alpha_{x-y} * (f_t(x) - f_t(y)) * \overline{\alpha_{f_1(x) - f_1(y)}}$. Then, we define a function $u(\phi, f, \cdot, \cdot)$ on $\mathbf{T}^2 \times \mathbf{T}^2 - \Delta$ by $u(\phi, f, x, y) = \phi(\alpha(f, x, y))$. One can easily check that this function does not depend on the choice of the isotopy $(f_t)_{t \in [0,1]}$ but only on f, is measurable, and bounded; see [25]. From the relation

$$\alpha(fg, x, y) = \alpha(g, x, y) * \alpha(f, g(x), g(y))$$

and the fact that ϕ is a quasimorphism, we deduce

$$|u(\phi, fg, x, y) - u(\phi, g, x, y) - u(\phi, f, g(x), g(y))| \leq C.$$

This means precisely that the map $f \mapsto u(\phi, f^{-1}, \cdot, \cdot) \in L^2(\mathbf{T}^2 \times \mathbf{T}^2, \mu \otimes \mu)$ is a quasicocycle.

EXAMPLE 5 [40] Consider the 2-sphere, with an area form ω of total area 2. Let $\Lambda(\mathbf{S}^2)$ be the space of oriented tangent lines to $\mathbf{S}^2$ and $p : \Lambda(\mathbf{S}^2) \to \mathbf{S}^2$ the canonical projection. There is a natural action of the circle $\mathbf{R}/\mathbf{Z}$ on $\Lambda(\mathbf{S}^2)$: if $\ell \in \Lambda(\mathbf{S}^2)$ is a tangent line at $x \in \mathbf{S}^2$, $e^{2i\pi s} \cdot \ell$ is the tangent line at x whose angle with ℓ is $2\pi s$. Let X be the (periodic) vector field that generates this action of $\mathbf{R}/\mathbf{Z}$ on $\Lambda(\mathbf{S}^2)$. We can find a 1-form α on $\Lambda(\mathbf{S}^2)$, invariant by rotations, such that

$$\alpha(X) = 1 \quad \text{and} \quad d\alpha = p^*\omega.$$

This implies that α is a contact form whose Reeb vector field is X. Let f be an area-preserving diffeomorphism of $\mathbf{S}^2$ and (f_t) a Hamiltonian isotopy from the identity to f, generated by the Hamiltonian (H_t). Let X_t be the symplectic gradient of H_t. The isotopy (f_t) induces an isotopy on $\Lambda(\mathbf{S}^2)$ through the action of the differential $df_t : \Lambda(\mathbf{S}^2) \to \Lambda(\mathbf{S}^2)$ of f_t. The key point of our last example is that there exists a second isotopy on $\Lambda(\mathbf{S}^2)$ that lifts (f_t). It is constructed as follows. Let $\widehat{X}_t$ be the horizontal lift of X_t to $\Lambda(\mathbf{S}^2)$, that is, the vector field on $\Lambda(\mathbf{S}^2)$ defined by the equations

$$\alpha(\widehat{X}_t) = 0 \quad \text{and} \quad p_*(\widehat{X}_t) = X_t.$$

The time-dependent vector field $\widehat{X}_t - (H_t \circ p)X$ generates an isotopy $\theta(f_t) : \Lambda(\mathbf{S}^2) \to \Lambda(\mathbf{S}^2)$ that preserves α and whose isotopy class depends only on the isotopy class of (f_t); see [2]. We can now measure a kind of rotation number associated to any tangent line $\ell \in \Lambda(\mathbf{S}^2)$. Indeed for all t, $\theta(f_t)(\ell)$ and $df_t(\ell)$ are tangent lines at the point $f_t(p(\ell)) \in \mathbf{S}^2$. Hence, we can write $df_t(\ell) = e^{2i\pi u(t)} \cdot \theta(f_t)(\ell)$ where $e^{2i\pi u(t)}$ is the angle between $\theta(f_t)(\ell)$ and $df_t(\ell)$. The quantity $w(f, \ell) = u(1) - u(0)$ depends only on f and ℓ. Then $w(f, \cdot) : \Lambda(\mathbf{S}^2) \to \mathbf{R}$ is a continuous map that satisfies

$$w(fg, \ell) = w(g, \ell) + w(f, dg(\ell)).$$

This is not yet exactly what we need: since the action $(f, \ell) \mapsto df(\ell)$ of the group $\mathrm{Ham}(\mathbf{S}^2, \omega)$ on $\Lambda(\mathbf{S}^2)$ does not have an invariant measure we cannot consider the function $w(f, \cdot)$ as a vector in some unitary representation of $\mathrm{Ham}(\Sigma, \omega)$. Nevertheless, we can prove the following inequality; see [40]: if ℓ_1

and ℓ_2 are two tangent lines at the same point $x = p(\ell_1) = p(\ell_2) \in \mathbf{S}^2$, then

$$|w(f, \ell_1) - w(f, \ell_2)| \le 10.$$

Thus, if we define $\underline{w}(f, x) = \max_{p(\ell)=x} w(f, \ell)$, $\underline{w}(f, \cdot)$ is a bounded measurable function on $\mathbf{S}^2$ and we have

$$|\underline{w}(fg, x) - \underline{w}(g, x) - \underline{w}(f, g(x))| \le 30.$$

The map $f \mapsto \underline{w}(f^{-1}, \cdot)$ is therefore a quasicocycle.

Motivated by these examples and by the vanishing results of the previous section, we can raise the following question.

Question 1

If $f : \Sigma \to \Sigma$ is a Hamiltonian diffeomorphism of a closed surface, distinct from the identity, can we find a quasicocycle u defined on the group $\mathrm{Ham}(\Sigma, \omega)$, with values in the space $L^2(\Sigma, \mu)$, or $L^2(\Sigma \times \Sigma, \mu \otimes \mu)$, or even $L^2(\Sigma^n, \mu^n)$ for some integer n, such that the function $\hat{u}(f)$ is not zero almost everywhere? In other words, can we find a set of positive measure (in Σ or $\Sigma \times \Sigma \ldots$) of points for which some kind of rotation number or asymptotic linking number is nonzero?

A positive answer to this question would yield a positive answer to Zimmer's conjecture. Note, however, that we should certainly add some hypothesis on f to get a positive answer. For instance, if f is a rotation of the sphere, $\hat{u}(f)$ vanishes for all known quasicocycles. It is probably possible to construct more subtle counterexamples. Assume for instance that we can construct a Hamiltonian diffeomorphism f with the following two properties: f is weakly mixing, and for some algebraic reason, $\phi(f) = 0$ for any homogeneous quasimorphism defined on the group of Hamiltonian diffeomorphisms. In that case, for any quasicocycle u with values in $L^2(\Sigma^n, \mu^n)$ one has $\hat{u}(f) = 0$ almost everywhere. Indeed, since the diagonal action of f on Σ^n is ergodic for any n, one has $\hat{u}(f) = \phi_u(f) = 0$ almost everywhere on Σ^n. Note, however, that this kind of counterexample cannot appear inside the image $\rho(\Gamma)$ of a higher-rank lattice, since all the diffeomorphisms in the group $\rho(\Gamma)$ have discrete spectrum.

We now give two small hints that tend to show that a nontrivial Hamiltonian diffeomorphism of a compact surface contains somewhere a set of positive measure of points that "rotate around each other."

Consider first a diffeomorphism $f \in \mathrm{Diff}_c(\mathbf{D}^2, \omega)$, distinct from the identity. We write

$$\widetilde{\mathrm{angle}}_f(x, y) = \lim_{n \to \infty} \frac{1}{n} \mathrm{angle}_{f^n}(x, y)$$

when this limit exists. Note that, according to Birkhoff's theorem, $\widetilde{\text{angle}}_f(x, y)$ exists for $\mu \otimes \mu$-almost every point $(x, y) \in \mathbf{D}^2 \times \mathbf{D}^2$, and if x_0 is a fixed point of f, $\widetilde{\text{angle}}_f(y, x_0)$ exists for μ-almost every point $y \in \mathbf{D}^2$. A result of Viterbo [45] implies that there exists a fixed point $x_0 \in \mathbf{D}^2$ for f and a Borel set $B \subset \mathbf{D}^2$ of positive measure such that

$$\widetilde{\text{angle}}_f(y, x_0) \neq 0,$$

for all $y \in B$. We now explain this fact. Recall first that one can define the *symplectic action* $\mathcal{A}(x_0)$ of a fixed point x_0 of f in the following way. If $(f_t)_{t \in [0,1]}$ is a Hamiltonian isotopy from the identity to f, generated by a Hamiltonian (H_t) one defines

$$\mathcal{A}(x_0) = \int_\gamma \lambda + \int_0^1 H_t(f_t(x))dt,$$

where γ is the loop $(f_t(x_0))_{t \in [0,1]}$ and λ is a primitive of the area form ω on $\mathbf{D}^2$. This number does not depend on the choice of the isotopy (f_t) from the identity to f. The next lemma, which I learned from Patrice Le Calvez, relates the symplectic action of a fixed point to the cocycle angle_f.

LEMMA 3.2. *For any fixed point x_0 of f one has*

$$\mathcal{A}(x_0) = \int_{\mathbf{D}^2} \text{angle}_f(y, x_0)d\mu(y) = \int_{\mathbf{D}^2} \widetilde{\text{angle}}_f(y, x_0)d\mu(y).$$

Proof. To compute the action $\mathcal{A}(x_0)$, we can choose an isotopy $(f_t)_{t \in [0,1]}$ such that $f_t(x_0) = x_0$ for all t. We denote by X_t the vector field that generates the isotopy and H_t the corresponding compactly supported Hamiltonian. Let $d\theta$ be a closed 1-form on the annulus $\mathbf{D}^2 - \{x_0\}$ whose integral over a generator of the first homology of the annulus equals 1. The integral of the function $\text{angle}_f(\cdot, x_0)$ over $\mathbf{D}^2$ equals $\int_{\mathbf{D}^2} \int_0^1 d\theta(X_t)dt\,\omega$. Using the fact that the 3-form $d\theta \wedge \omega$ is 0 we get

$$d\theta(X_t)\,\omega = d\theta \wedge \omega(X_t, \cdot)$$
$$= -d(H_t d\theta).$$

Choose now a small disc D_ε around x_0. We have

$$\int_{\mathbf{D}^2} \text{angle}_f(y, x_0)d\mu(y) = \lim_{\varepsilon \to 0} \int_{\mathbf{D}^2 - D_\varepsilon} \int_0^1 d\theta(X_t)dt\,\omega$$
$$= \lim_{\varepsilon \to 0} \int_{\partial D_\varepsilon} \int_0^1 H_t dt\,d\theta.$$

Here we have used Stokes's theorem and the fact that H_t vanishes near the boundary of $\mathbf{D}^2$. The last limit in the equality above coincides with the action

$\mathcal{A}(x_0) = \int_0^1 H_t(x_0)dt$. We have therefore proved the first equality of the lemma. The second equality follows from Birkhoff's theorem. □

Now, using symplectic methods, Viterbo shows that if f is distinct from the identity, there exists a fixed point x_0 with nonzero action. Thanks to the interpretation of the action of x_0 as the average linking number around x_0, we deduce the existence of the set B. We get that the map angle_f is nonzero on the set $B \times \{x_0\} \subset \mathbf{D}^2 \times \mathbf{D}^2$. However, from our point of view, it would be more natural to obtain a set of positive measure in $\mathbf{D}^2 \times \mathbf{D}^2$ for the *product measure*.

Second, we consider the simplest examples of Hamiltonian diffeomorphisms: integrable ones. For instance, let us take a smooth function $H : \mathbf{T}^2 \to \mathbf{R}$ and consider the associated Hamiltonian flow $\varphi_H^t : \mathbf{T}^2 \to \mathbf{T}^2$. Assume that H is nonconstant. Then, we will show that there is a quasicocycle u on $\mathrm{Ham}(\mathbf{T}^2, \omega)$ with values in $L^2(\mathbf{T}^2, \mathbf{R}^2)$ or in $L^2(\mathbf{T}^2 \times \mathbf{T}^2, \mathbf{R})$ such that $\widehat{u}(\varphi_H^1)$ is not zero almost everywhere (for the measure μ on $\mathbf{T}^2$ in the first case, for the measure $\mu \otimes \mu$ in the second one). As H is nonconstant there is an embedding of the annulus

$$i : A = \mathbf{S}^1 \times [0, 1] \to \mathbf{T}^2$$

on which the flow φ_H^t reads $\varphi_H^t(i(\theta, s)) = i(\theta + t\vartheta(s), s)$ for a nonvanishing function $\vartheta : [0, 1] \to \mathbf{R}^*$. To prove this fact, one just considers a connected component $\mathcal{C}$ of a regular level of H. In a neighborhood of $\mathcal{C}$ all orbits of φ_H^t are periodic; hence $\mathcal{C}$ is contained in an annulus that is invariant under the flow φ_H^t and in which each orbit is periodic. From this one easily deduces the existence of the embedding i.

We write $\hat{i} : \mathbf{R} \times [0, 1] \to \mathbf{R}^2$ for a lift of the map i. Assume first that the annulus A is essential in $\mathbf{T}^2$. In that case there exists a nonzero vector $w \in \mathbf{Z}^2$ such that $\hat{i}(\theta + k, s) = \hat{i}(\theta, s) + kw$. Consequently, if $v_{\varphi_H^1}$ is the cocycle from Example 2, we have

$$\frac{v_{\varphi_H^n}(i(\theta, s))}{n} \xrightarrow[n \to \infty]{} \vartheta(s)w \neq 0.$$

Assume now that the annulus A is inessential. We will make use of the quasicocycle from Example 4. We can assume that the curve $i(\mathbf{S}^1 \times \{0\})$ bounds an embedded disc D that is disjoint from $i(\mathbf{S}^1 \times]0, 1])$ (up to changing the parameter s by $1 - s$). Assume that $s_2 > s_1$ and consider the trajectories of the two points $i(\theta_1, s_1)$ and $i(\theta_2, s_2)$. The family of curves $(\varphi_H^t(i(\theta_2, s_2)) - \varphi_H^t(i(\theta_1, s_1)))_{t \in [0, n]}$ winds around 0 in the disc $D \cup i(A)$ (see Figure 1). The loops $\alpha(\varphi_H^n, i(\theta_2, s_2), i(\theta_1, s_1))$ are therefore almost equal to a power of the commutator $[a, b]$, where a and b are the standard generators of the group $\pi_1(\mathbf{T}^2 - \{0\}, x_*)$.

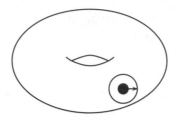

Fig. 1. Linking of points in an inessential annulus on the torus.

More precisely, we can write

$$\alpha(\varphi_H^n, i(\theta_2, s_2), i(\theta_1, s_1)) = \alpha_n * [a, b]^{[n\vartheta(s_2)]} * \beta_n$$

where α_n and β_n stay in a fixed finite subset of $\pi_1(\mathbf{T}^2 - \{0\}, x_*)$ and $[n\vartheta(s_2)]$ is the integer part of $n\vartheta(s_2)$. From this we deduce that $u(\phi, \varphi_H^n, i(\theta_2, s_2), i(\theta_1, s_1)) - \phi([a, b]^{[n\vartheta(s_2)]})$ is bounded independently of n. Therefore, for any homogeneous quasimorphism ϕ on $\pi_1(\mathbf{T}^2 - \{0\}, x_*)$ we have

$$\frac{u(\phi, \varphi_H^n, i(\theta_2, s_2), i(\theta_1, s_1))}{n} \to \vartheta(s_2)\phi([a, b]).$$

To conclude we now choose a quasimorphism ϕ such that $\phi([a, b])$ is nonzero and observe that the set $\{(i(\theta_1, s_1), i(\theta_2, s_2)), s_2 > s_1\} \subset \mathbf{T}^2 \times \mathbf{T}^2$ has positive $\mu \otimes \mu$-measure. Hence we have proved:

THEOREM *Let $\rho : \Gamma \to \mathrm{Ham}(\mathbf{T}^2, \omega)$ be a homomorphism, where Γ is a higher-rank lattice, as above. If a diffeomorphism $f \in \rho(\Gamma)$ is contained in a Hamiltonian flow, then f is the identity.*

In fact, this result could have been deduced directly from the existence of Zimmer's invariant Riemannian metric. Indeed, in the situation above, one can always find an embedding i such that the function ϑ is not constant. This implies that for an open set of points x the sequence of differentials $(d\varphi_H^n(x))_{n \geq 0}$ tends to infinity. This contradicts the existence of Zimmer's metric. Yet, the above proof was presented in order to illustrate the possible use of the quasicocycles that we described.

Also, it is not difficult to establish a similar result on a surface of genus greater than 1, or on the sphere if we assume that the flow is not conjugated to a 1-parameter group of rotations; see [41].

4. Final Remarks

We have already mentioned that, according to a result of Zimmer, the group $\rho(\Gamma)$ preserves (almost everywhere) a measurable Riemannian metric. In the case where the surface is the torus $\mathbf{T}^2$, we can slightly improve this fact: the group $\rho(\Gamma)$ preserves a pair (X^1, X^2) of linearly independent measurable vector fields. If g is a Hamiltonian diffeomorphism of $\mathbf{T}^2$, let us consider a Hamiltonian isotopy (g_t) from the identity to g. We identify the differential $dg_t(x)$ of the diffeomorphism g_t at a point x with a 2×2 matrix. Hence, the path of matrices $dg_t(x)$ defines an element $\widetilde{dg}(x) \in \widetilde{SL}_2(\mathbf{R})$ in the universal cover of $SL_2(\mathbf{R})$, which depends only on g and x. The map

$$
\begin{aligned}
\mathrm{Ham}(\mathbf{T}^2) \times \mathbf{T}^2 &\to \widetilde{SL}_2(\mathbf{R}) \\
(g, x) &\mapsto \widetilde{dg}(x)
\end{aligned}
$$

is a cocycle. If p denotes the projection from $\widetilde{SL}_2(\mathbf{R})$ to $SL_2(\mathbf{R})$, the map $(g, x) \mapsto dg(x) = p(\widetilde{dg}(x))$ is the usual derivative cocycle. Suppose $\Gamma \hookrightarrow \mathrm{Ham}(\mathbf{T}^2, \omega)$ is a Kazhdan group and consider the restriction of the previous cocycle to Γ. According to Zimmer [47, 49], there exists a measurable map $\varphi : \mathbf{T}^2 \to SL_2(\mathbf{R})$ such that

$$
\varphi(\gamma \cdot x)^{-1} \circ d\gamma(x) \circ \varphi(x) \in SO(2),
$$

almost everywhere (for $\gamma \in \Gamma$). We denote by $\tau : \widetilde{SL}_2(\mathbf{R}) \to \mathbf{R}$ the translation number associated to the action of $\widetilde{SL}_2(\mathbf{R})$ on the line (see [28]) and by T the generator of the center of $\widetilde{SL}_2(\mathbf{R})$. Let $\widetilde{\varphi} : \mathbf{T}^2 \to \widetilde{SL}_2(\mathbf{R})$ be a measurable lift of φ chosen in such a way that the map $x \mapsto \tau(\widetilde{\varphi}(x))$ is bounded. This is always possible since one has $\tau(a \cdot T^k) = \tau(a) + k$ ($a \in \widetilde{SL}_2(\mathbf{R})$, $k \in \mathbf{Z}$). The map $(\gamma, x) \mapsto \widetilde{\varphi}(\gamma \cdot x)^{-1} \circ \widetilde{d\gamma}(x) \circ \widetilde{\varphi}(x)$ is a cocycle whose image is contained in the inverse image of $SO(2)$ in $\widetilde{SL}_2(\mathbf{R})$, which is canonically isomorphic to $\mathbf{R}$.

LEMMA 4.1. *For each $\gamma \in \Gamma$ the map $x \mapsto s_\gamma(x) := \widetilde{\varphi}(\gamma \cdot x)^{-1} \circ \widetilde{d\gamma}(x) \circ \widetilde{\varphi}(x) \in \mathbf{R}$ is bounded. Hence the cocycle $\gamma \mapsto s_\gamma$ determines a class in $H^1(\Gamma, L^2(\mathbf{T}^2, \mathbf{R}))$.*

Proof. Since the restriction of τ to the inverse image of $SO(2)$ in $\widetilde{SL}_2(\mathbf{R})$ is the identity, it is enough to show that the map $x \mapsto \tau(s_\gamma(x))$ is bounded. Since the two maps $x \mapsto \tau(\widetilde{\varphi}(x))$ and $x \mapsto \tau(\widetilde{\varphi}(\gamma \cdot x)^{-1})$ are bounded thanks to the choice of the lift $\widetilde{\varphi}$ and since τ is a quasimorphism, it is enough to check that the map $x \mapsto \tau(\widetilde{d\gamma}(x))$ is bounded. But this is obvious since this is a continuous map from $\mathbf{T}^2$ to $\mathbf{R}$. $\qquad\square$

From the lemma, we deduce that there exists a measurable (in fact L^2) map $\psi : \mathbf{T}^2 \to \mathbf{R}$ such that $s_\gamma(x) = \psi(\gamma \cdot x) - \psi(x)$. Coming back to $SL_2(\mathbf{R})$, we obtain

$$\varphi(\gamma \cdot x)^{-1} \circ d\gamma(x) \circ \varphi(x) = e^{2i\pi(\psi(\gamma(x)) - \psi(x))},$$

almost everywhere, where $e^{2i\pi u}$ stands for the rotation of angle $2\pi u$ in $SL_2(\mathbf{R})$. This tells us that the derivative cocycle is cohomologous to the constant cocycle (equal to the identity), which is equivalent to the existence of the vector fields X^1 and X^2.

Although the results we have discussed here give more evidence toward Zimmer's conjecture, it seems difficult to exploit them. In particular, the quest for a set of positive measure of points that "rotates" under the action of a diffeomorphism seems delicate. Note also that we have simply used global cohomological properties of the lattice Γ to give some constraints on the dynamics of each single diffeomorphism in the group $\rho(\Gamma)$ but I was not able to investigate the dynamics of the whole group Γ. Rather than studying measure theoretical properties, it might be useful to study the topological dynamics of the action. In this direction, the work of Calegari [9] contains many promising ideas.

ACKNOWLEDGMENTS. This work is part of my thesis realized at École Normale Supérieure de Lyon. I would like to thank Étienne Ghys for all the discussions we had around the problems discussed in this text and for his warm encouragement. I would also like to thank François Béguin, Patrice Le Calvez, Frédéric Le Roux, and Leonid Polterovich for the numerous conversations we had around Zimmer's conjecture. This text was completed while I was enjoying the hospitality of Tel-Aviv University. This work was also partly supported by the French Agence Nationale de la Recherche.

Finally, I would like to thank the referee for his or her constructive comments on this text.

References

[1] V. I. Arnold and B. A. Khesin, *Topological methods in hydrodynamics*, Applied Mathematical Sciences **125**, Springer-Verlag, New York, (1998).

[2] A. Banyaga, *The group of diffeomorphisms preserving a regular contact form*, Topology and algebra (Proc. Colloq., Eidgenöss. Tech. Hochsch., Zurich, 1977), 47–53, Monograph. Enseign. Math. **26**, Univ. Genève (1978).

[3] A. Banyaga, *The structure of classical diffeomorphism groups*, Mathematics and its applications **400**, Kluwer Academic Publishers Group, Dordrecht, (1997).

[4] C. Bavard, *Longueur stable des commutateurs*, Enseign. Math. (2) **37**, No. 1–2 (1991), 109–150.

[5] A. Borel, *Compact Clifford-Klein forms of symmetric spaces*, Topology **2** (1963), 111–122.

[6]   R. Brooks, *Some remarks on bounded cohomology*, Riemann Surfaces and Related Topics: Proceedings of the 1978 Stony Brook Conference (State Univ. New York, Stony Brook, NY, 1978), 53–63, Ann. of Math. Stud. **97**, Princeton University Press, Princeton, NJ, (1981).

[7]   M. Burger and N. Monod, *Bounded cohomology of lattices in higher rank Lie groups*, J. Eur. Math. Soc. (JEMS) **1**, No. 2 (1999), 199–235.

[8]   M. Burger and N. Monod, *Continuous bounded cohomology and applications to rigidity theory*, Geom. Funct. Anal. **12**, No. 2 (2002), 219–280.

[9]   D. Calegari, *Circular groups, planar groups, and the Euler class*, Proceedings of the Casson Fest, Geom. Topol. Monogr. **7**, Geom. Topol. Publ., Coventry (2004), 431–491.

[10] S. Cantat, *Version kählérienne d'une conjecture de Robert J. Zimmer*, Ann. Sci. École Norm. Sup. (4) **37**, No. 5 (2004), 759–768.

[11] J. Déserti, *Groupe de Cremona et dynamique complexe: une approche de la conjecture de Zimmer*, Int. Math. Res. Not. (2006).

[12] C. J. Earle and J. Eells, *A fibre bundle description of Teichmüller theory*, J. Diff. Geometry **3**, (1969), 19–43.

[13] M. Entov and L. Polterovich, *Calabi quasimorphism and quantum homology*, Int. Math. Res. Not., No. 30 (2003), 1635–1676.

[14] D. Epstein and K. Fujiwara, *The second bounded cohomology of word-hyperbolic groups*, Topology **36**, No. 6 (1997), 1275–1289.

[15] B. Farb and H. Masur, *Superrigidity and mapping class groups*, Topology **37**, No. 6 (1998), 1169–1176.

[16] B. Farb and P. Shalen, *Real-analytic actions of lattices*, Invent. Math. **135**, No. 2 (1999), 273–296.

[17] A. ÊFathi, *Transformations et homéomorphismes préservant la mesure. Systèmes dynamiques minimaux*, Thèse, Orsay, (1980).

[18] D. Fisher, *Talk at the conference "Geometry, Rigidity, and Group Actions," in honor of R. J. Zimmer's 60th birthday*, University of Chicago (September 2007).

[19] J. Franks, *Recurrence and fixed points of surface homeomorphisms*, Ergodic Theory Dynam. Systems **8***, Charles Conley Memorial Issue (1988), 99–107.

[20] J. Franks, *Rotation vectors for surface diffeomorphisms*, Proceedings of the International Congress of Mathematicians, Vol.1, 2, (Zürich, 1994), Birkhäuser, Basel (1995), 1179–1186.

[21] J. Franks and M. Handel, *Periodic points of Hamiltonian surface diffeomorphisms*, Geom. Topol. **7**, (2003), 713–756.

[22] J. Franks and M. Handel, *Area preserving group actions on surfaces*, Geom. Topol. **7**, (2003), 757–771.

[23] J. Franks and M. Handel, *Distortion elements in group actions on surfaces*, Duke Math. J. **131**, No. 3 (2006), 441–468.

[24] J.-M. Gambaudo and É. Ghys, *Enlacements asymptotiques*, Topology **36**, No. 6 (1997), 1355–1379.

[25] J.-M. Gambaudo and É. Ghys, *Commutators and diffeomorphisms of surfaces*, Ergodic Theory Dynam. Systems **24**, No. 5 (2004), 1591–1617.

[26] É. Ghys, *Sur les groupes engendrés par des difféomorphismes proches de l'identité*, Bol. Soc. Brasil. Mat. (N.S.) **24**, No. 2 (1993), 137–178.

[27] É. Ghys, *Actions de réseaux sur le cercle*, Invent. Math. **137**, No. 1 (1999), 199–231.

[28] É. Ghys, *Groups acting on the circle*, Enseign. Math. (2) **47**, No. 3–4 (2001), 329–407.

[29] É. Ghys, *Knots and dynamics*, International Congress of Mathematicians, Vol. I, Eur. Math. Soc., Zürich (2007), 247–277.

[30] M. Gromov, *Volume and bounded cohomology*, Inst. Hautes Études Sci. Publ. Math. **56** (1983), 5–99.

[31] P. de la Harpe and A. Valette, *La propriété (T) de Kazhdan pour les groupes localement compacts (avec un appendice de Marc Burger)*, Astérisque No. 175 (1989).

[32] J. F. C. Kingman, *Subadditive processes*, École d'Été de Probabilités de Saint-Flour, V 1975, Lecture Notes in Math. **539**, Springer, Berlin, (1976), 167–223.

[33] P. Le Calvez, *Identity isotopies on surfaces*, Dynamique des difféomorphismes conservatifs des surfaces: un point de vue topologique, Panor. Synthèse **21**, Soc. Math. France, Paris (2006), 105–143.

[34] A. Lubotzky, S. Mozes, and M. S. Raghunathan, *The word and Riemannian metrics on lattices of semisimple groups*, Inst. Hautes Études Sci. Publ. Math., No. 91 (2000), 5–53.

[35] D. McDuff and D. Salamon, *Introduction to symplectic topology*, Second edition, Oxford Mathematical Monographs, Clarendon Press, Oxford University Press, New York, (1998).

[36] N. Monod, *Continuous bounded cohomology of locally compact groups*, Lecture Notes in Mathematics **1758**, Springer-Verlag, Berlin, (2001).

[37] A. Navas, *Actions de groupes de Kazhdan sur le cercle*, Ann. Sci. École Norm. Sup. (4) **35**, No. 5 (2002), 749–758.

[38] L. Polterovich, *Growth of maps, distortion in groups and symplectic geometry*, Invent. Math. **150**, No. 3 (2002), 655–686.

[39] L. Polterovich, *Floer homology, dynamics and groups*, Morse theoretic methods in nonlinear analysis and in symplectic topology, NATO Sci. Ser. II Math. Phys. Chem. **217**, Springer, Dordrecht (2006), 417–438.

[40] P. Py, *Quasi-morphismes et invariant de Calabi*, Ann. Sci. École Norm. Sup. (4) **39**, No. 1 (2006), 177–195.

[41] P. Py, *Quasi-morphismes et difféomorphismes hamiltoniens*, Ph.D thesis, École Normale Supérieure de Lyon (2008), available at http://tel.archives-ouvertes.fr/tel-00263607/fr/.

[42] S. Schwartzman, *Asymptotic cycles*, Ann. of Maths (2) **66**, (1957), 270–284.

[43] S. Smale, *Diffeomorphisms of the 2-sphere*, Proc. Amer. Math. Soc. **10** (1959), 621–626.

[44] A. Thom, *Low degree bounded cohomology and L^2-invariants for negatively curved groups*, Groups, Geometry and Dynamics **3** (2009), 343–358.

[45] C. Viterbo, *Symplectic topology as the geometry of generating functions*, Math. Ann. **292**, No. 4 (1992), 685–710.

[46] D. Witte, *Arithmetic groups of higher Q-rank cannot act on 1-manifolds*, Proc. Amer. Math. Soc. **122**, No. 2 (1994), 333–340.

[47] R. J. Zimmer, *Kazhdan groups acting on compact manifolds*, Invent. Math. **75**, No. 3 (1984), 425–436.

[48] R. J. Zimmer, *Actions of semisimple groups and discrete subgroups*, Proceedings of the International Congress of Mathematicians, Vol. 1, 2 (Berkeley, CA., 1986), 1247–1258, Amer. Math. Soc., Providence, RI (1987).

[49] R. J. Zimmer, *Spectrum, entropy, and geometric structures for smooth actions of Kazhdan groups*, Israel J. Math. **75**, No. 1 (1991), 65–80.

7

ISOMETRIC ACTIONS OF SIMPLE GROUPS AND TRANSVERSE STRUCTURES: THE INTEGRABLE NORMAL CASE

WITH GRATITUDE TO ROBERT J. ZIMMER ON THE OCCASION
OF HIS 60TH BIRTHDAY

Abstract

For actions with a dense orbit of a connected noncompact simple Lie group G, we obtain some global rigidity results when the actions preserve certain geometric structures. In particular, we prove that for a G-action to be equivalent to 1 on a space of the form $(G \times K\backslash H)/\Gamma$, it is necessary and sufficient for the G-action to preserve a pseudo-Riemannian metric and a transverse Riemannian metric to the orbits. A similar result proves that the G-actions on spaces of the form $(G \times H)/\Gamma$ are characterized by preserving transverse parallelisms. By relating these techniques to the notion of the algebraic hull of an action, we obtain infinitesimal Lie algebra structures on certain geometric manifolds acted upon by G.

1. Introduction

In the rest of this work, let G be a connected noncompact simple Lie group with Lie algebra $\mathfrak{g}$ and M a smooth connected manifold acted upon smoothly by G. There are several examples of such actions that preserve a finite volume. Some of the most interesting are obtained from Lie group homomorphisms $G \hookrightarrow L$ where L is a connected Lie group that admits a lattice Γ. The relevant G-action is then given on the double coset $K\backslash L/\Gamma$, where K is some compact subgroup that centralizes G. Robert Zimmer proposed in [15] to study the finite measure-preserving G-actions on M and determine to what extent these are given by such double cosets.

To understand how to tackle Zimmer's program, it has been proved very useful to consider G-actions preserving a suitable geometric structure (see [2] and [18]). This is a natural condition since the above double coset examples

This work was supported by SNI-Mexico and the grants Conacyt 44620 and Concyteg 07-02-K662-091.

carry a G-invariant pseudo-Riemannian metric when L is semisimple. Such metric comes from the bi-invariant metric on L obtained from the Killing form of its Lie algebra.

The properties of the G-orbits of finite measure-preserving G-actions are reasonably well known. For example, such G-actions are known to be everywhere locally free when they preserve suitable pseudo-Riemannian metrics (see [2], [13], and [14]). Also in such case, the metric restricted to the orbits can be described precisely (see Lemma 2.6). However, there is still a lack of knowledge of the properties of a manifold, acted upon by G, in the transverse direction to the orbits.

In this work, we propose to study the G-actions on M by emphasizing the need to understand the properties of the transverse to the G-orbits. For this, we will be dealing with G-actions that have a dense orbit and preserve a finite-volume pseudo-Riemannian metric. As observed above, in this case the results from [13] show that the G-orbits define a foliation, which from now on we will denote with $\mathcal{O}$. By assuming that the orbits are nondegenerate for the pseudo-Riemannian metric on M, we can consider the normal bundle $T\mathcal{O}^\perp$ to the orbits as realizing the transverse direction in M. In this respect, we obtain the following structure theorem for G-actions on M when this normal bundle is integrable.

THEOREM 1.1. *Suppose that the G-action on M has a dense orbit and preserves a finite-volume complete pseudo-Riemannian metric. If the G-orbits in M are nondegenerate and the normal bundle to the orbits $T\mathcal{O}^\perp$ is integrable, then there exist:*

1) *an isometric finite covering map $\widehat{M} \to M$ to which the G-action lifts,*
2) *a simply connected complete pseudo-Riemannian manifold $\widetilde{N}$, and*
3) *a discrete subgroup $\Gamma \subset G \times \mathrm{Iso}(\widetilde{N})$,*

such that $\widehat{M}$ is G-equivariantly isometric to $(G \times \widetilde{N})/\Gamma$.

We observe that in the double coset examples $K \backslash L / \Gamma$ as above with L semisimple and with the metric coming from the Killing form, the G-orbits are always nondegenerate. Besides that, we prove that for a general G-action on M, the orbits are always nondegenerate when $\dim(M) < 2\dim(G)$ (see Lemma 2.7). Also, by developing criteria for the normal bundle $T\mathcal{O}^\perp$ to be integrable, we obtain some results where the conclusion of Theorem 1.1 holds. For example, Corollary 3.1 ensures such conclusion when G has high enough real rank. Also, Corollary 3.2 does the same for manifolds M whose dimensions have a suitable bound in terms of G. In this last result we can even dispense with the assumption of having nondegenerate orbits by applying Lemma 2.7.

Without assuming in Theorem 1.1 that the G-action has a dense orbit it is still possible to draw some conclusions. More specifically, if we assume the rest of the hypotheses, a description of the universal covering space of the manifold as a warped product is obtained. This sort of result already appears in [3].

For a G-action on M preserving a pseudo-Riemannian metric, in [10] we considered a certain geometric condition between the metrics on G and M (the former for a bi-invariant metric) from which we concluded that M is, up to a finite covering space, a double coset of the form $(G \times K\backslash H)/\Gamma$. In other words, a double coset as above where G appears as a factor in L. One of the steps used in [10] to achieve this was to prove that the normal bundle $T\mathcal{O}^\perp$ is Riemannian. Considering the relevance we are giving to the transverse to the orbits, it is natural to determine the properties of the G-actions that preserve a transverse Riemannian structure. In this context, we obtain the following result, which proves that, up to a finite covering, the double cosets of the form $(G \times K\backslash H)/\Gamma$ are characterized as those isometric G-actions that preserve a transverse Riemannian structure on the foliation $\mathcal{O}$. We recall that a semisimple Lie group is isotypic if the complexification of its Lie algebra is isomorphic to a sum of identical simple complex ideals.

THEOREM 1.2. *If G is a connected noncompact simple Lie group acting faithfully on a compact manifold M, then the following conditions are equivalent:*

1) *There is a G-equivariant finite covering map $(G \times K\backslash H)/\Gamma \to M$ where H is a connected Lie group with a compact subgroup K and $\Gamma \subset G \times H$ is a discrete cocompact subgroup such that $G\Gamma$ is dense in $G \times H$.*
2) *There is a finite covering map $\widehat{M} \to M$ for which the G-action on M lifts to a G-action on $\widehat{M}$ with a dense orbit and preserving*
   - *a pseudo-Riemannian metric for which the orbits are nondegenerate, and*
   - *a transverse Riemannian structure for the foliation $\mathcal{O}$ by G-orbits.*

Furthermore, if G has finite center and real rank at least 2, then we can assume in (1) of the above equivalence that $G \times H$ is semisimple isotypic with finite center and that Γ is an irreducible lattice.

Based on the previous result, we prove in Theorem 5.1 that the G-actions on M preserving a Lorentzian metric are, up to a finite covering, those given by double cosets $(G \times K\backslash H)/\Gamma$ with G locally isomorphic to $SL(2, \mathbb{R})$. We observe that this result improves a similar one found in [2] (see also the introduction of [4]).

Continuing with our study of transverse structures, we next consider isometric G-actions preserving a transverse parallelism for the foliation $\mathcal{O}$. We prove that, up to a finite covering, such actions characterize the double cosets of the form $(G \times H)/\Gamma$.

THEOREM 1.3. *If G is a connected noncompact simple Lie group acting faithfully on a compact manifold M, then the following conditions are equivalent:*

1) *There is a G-equivariant finite covering map $(G \times H)/\Gamma \to M$ where H is a connected Lie group and $\Gamma \subset G \times H$ is a discrete cocompact subgroup such that $G\Gamma$ is dense in $G \times H$.*

2) *There is a finite covering map $\widehat{M} \to M$ for which the G-action on M lifts to a G-action on $\widehat{M}$ with a dense orbit and preserving*
   - *a pseudo-Riemannian metric for which the orbits are nondegenerate, and*
   - *a transverse parallelism for the foliation $\mathcal{O}$ by G-orbits.*

Furthermore, if G has finite center and real rank at least 2, then we can assume in (1) of the above equivalence that $G \times H$ is semisimple isotypic with finite center and that Γ is an irreducible lattice.

The notion of the algebraic hull of an action on a bundle, introduced by Zimmer, is a fundamental tool to understand G-actions as they relate to geometric structures. We recall that the algebraic hull is the smallest algebraic subgroup for which there is a measurable G-invariant reduction of the bundle being acted upon (see [17]). For our setup, where we are interested in the transverse to the orbits, it is then natural to consider the algebraic hull of the G-action on the bundle $L(T\mathcal{O}^\perp)$ for an isometric G-action with nondegenerate orbits. Since a G-action of this sort preserves a pseudo-Riemannian metric on $T\mathcal{O}^\perp$, the algebraic hull for $L(T\mathcal{O}^\perp)$ is in this case a subgroup of $O(p, q)$, for some (p, q). The next result shows that for weakly irreducible manifolds and when the algebraic hull for $L(T\mathcal{O}^\perp)$ is the largest possible for this setup, the manifold being acted upon has infinitesimally at some point the structure of a specific Lie algebra that contains $\mathfrak{g}$. By the last claim in the statement, such Lie algebra structure is nontrivially linked to the geometry of the manifold. For the explicit description of the Lie algebra structure obtained in this result we refer to Lemma 7.1 and Theorem 7.2.

We recall that a pseudo-Riemannian manifold is weakly irreducible if the tangent space at some (and hence any) point has no proper nondegenerate subspaces invariant under the restricted holonomy group at that point. Also recall that every simple Lie group with a bi-invariant metric is weakly irreducible.

THEOREM 1.4. *Suppose that G has finite center and real rank at least 2, and that the G-action on M preserves a finite-volume complete pseudo-Riemannian metric. Also assume that G acts ergodically on M and that the foliation by G-orbits is nondegenerate. Denote with L the algebraic hull for the G-action on the bundle $L(TO^\perp)$ and with $\mathfrak{l}$ its Lie algebra. In particular, there is an embedding of Lie algebras $\mathfrak{l} \hookrightarrow \mathfrak{so}(p,q)$, where (p,q) is the signature of the metric of M restricted to $TO^\perp$. If this embedding is surjective and M is weakly irreducible, then the following holds:*

1) *G is locally isomorphic to $SO_0(p,q)$ and* $\dim(M) = (p+q)(p+q+1)/2$, *and*
2) *for some $x \in M$ the tangent space $T_x M$ admits a Lie algebra structure isomorphic to either $\mathfrak{so}(p, q+1)$ or $\mathfrak{so}(p+1,q)$ such that $T_x O$ is a Lie subalgebra isomorphic to $\mathfrak{g}$.*

Furthermore, with respect to the representation of Lemma 2.1, there is a Lie algebra of local Killing fields vanishing at x, which is isomorphic to $\mathfrak{g}$ and acts nontrivially on $T_x M$ by derivations of the Lie algebra structure given in (2).

With the results developed in this work, we try to show the importance of considering transverse geometric structures to understand the actions of noncompact simple Lie groups. In fact, we believe that some form of the following conjecture could provide a geometric characterization of the double coset examples of G-actions mentioned above.

CONJECTURE 1.5. *Consider the double coset G-spaces of the form $K \backslash L / \Gamma$, where L is a semisimple Lie group with an irreducible lattice Γ and a compact subgroup K; the G-action is then induced by a nontrivial homomorphism $G \to L$ whose image centralizes K. Then, a G-action on a manifold M is equivalent to such a double coset G-action for some L, Γ, K if and only if the G-action on M:*

- *has a dense orbit,*
- *preserves a pseudo-Riemannian metric, and*
- *preserves a transverse geometric structure to the orbits suitably related to the geometry of $GK \backslash L$.*

Note that for $L = G \times H$ and K a compact subgroup of H, we obtain quotients $GK \backslash L = K \backslash H$ and $G \backslash L = H$, which naturally carry a Riemannian metric and a parallelism, respectively. Thus Theorems 1.2 and 1.3 verify the conjecture in these cases.

One of the main tools used to obtain our results is Gromov's machinery on geometric G-actions (see [1], [2], [10], and [18]). Such machinery ensures the existence of large families of local Killing fields. We develop these techniques in Section 2 emphasizing the fact that the Killing fields thus obtained yield g-module structures on the tangent spaces to M; in this respect, our main result is Proposition 2.3, which is due to Gromov [2] in the analytic/compact case (see also [18]) and was extended to the smooth/finite volume case in [1] and [10]. In Section 3, we prove that the latter impose restrictions strong enough to guarantee the integrability of the normal bundle under suitable conditions (Corollaries 3.1 to 3.3). We observe that Theorem 1.1 and its consequences obtained in Section 3 are in fact extensions of results obtained in [2]: theorem 5.3.E on page 129 of [2] states, under the assumptions of our Corollary 3.1, that M has a covering G-equivariantly diffeomorphic to $G \times N$ for some N. Note, however, that the result in [2] yields only a topological covering and does not further describe the covering group.

Theorem 1.2 is obtained in Section 4 by applying the main results and arguments from [10]. As mentioned above, this yields in Section 5 our characterization of Lorentzian manifolds acted upon with a dense orbit by a simple noncompact Lie group. In Section 6 we establish the characterization of G-spaces of the form $(G \times H)/\Gamma$ provided by Theorem 1.3; for this, one of the main ingredients is given by Theorem 1.1. We observe that the arguments used in Section 6 are based on those found in [11].

To obtain Theorem 1.4 in Section 7, an application of Theorem 1.1 is used again. But now, the notion of the algebraic hull and the deep result about it found in [17] are also fundamental. Ultimately, this is somewhat to be expected, since a computation of the algebraic hull for frame bundles over M is in fact an important step to build Gromov's machinery for geometric G-actions on M. This is evident in our proof of Proposition 2.3.

We want to observe that the conclusion of Theorem 1.4 can be paraphrased by saying that the manifold M has infinitesimally the structure of either $SO(p, q+1)$ or $SO(p+1, q)$. It remains the question as to whether or not M can be related more precisely to either of these groups. This problem will be pursued elsewhere (see [7]) by requiring some additional conditions.

The author wishes to thank Uri Bader and Shmuel Weinberger for some fruitful conversations.

2. Killing Fields on Manifolds with a Simple Group of Isometries

By theorem 4.17 from [13], if the G-action on M has a dense orbit and preserves a finite-volume pseudo-Riemannian metric, then the action is locally free and so the orbits define a foliation that we have agreed to denote with $\mathcal{O}$. In this case, it is well known that the bundle $T\mathcal{O}$ tangent to the foliation $\mathcal{O}$ is a trivial vector bundle isomorphic to $M \times \mathfrak{g}$, under the isomorphism given by

$$M \times \mathfrak{g} \to T\mathcal{O}$$

$$(x, X) \mapsto X_x^*.$$

For every $x \in M$, this induces an isomorphism between the fiber $T_x\mathcal{O}$ and $\mathfrak{g}$, which we will refer to as the natural isomorphism. Furthermore, if we consider on $M \times \mathfrak{g}$ the product G-action, where the G-action on $\mathfrak{g}$ is the adjoint one, then such isomorphism is G-equivariant. More precisely, we have

$$dg(X^*) = \mathrm{Ad}(g)(X)^*$$

for every $g \in G$ and $X \in \mathfrak{g}$. Note that for X in the Lie algebra of a group acting on a manifold, we denote with X^* the vector field on the manifold whose 1-parameter group of diffeomorphisms is given by $(\exp(tX))_t$ through the action on the manifold.

For any given pseudo-Riemannian manifold N, we will denote with $\mathrm{Kill}(N, x)$ the Lie algebra of germs at x of local Killing vector fields defined in a neighborhood of x, and with $\mathrm{Kill}_0(N, x)$ we will denote the Lie subalgebra consisting of those germs that vanish at x. The following result is a consequence of the Jacobi identity and the fact that the Lie derivative of a metric with respect to its Killing fields vanishes. In the rest of this work, for a vector space W with a nondegenerate symmetric bilinear form, we will denote with $\mathfrak{so}(W)$ the Lie algebra of linear maps on W that are skew-symmetric with respect to the bilinear form.

LEMMA 2.1. *Let N be a pseudo-Riemannian manifold and $x \in N$. Then, the map*

$$\lambda_x : \mathrm{Kill}_0(N, x) \to \mathfrak{so}(T_x N)$$

$$\lambda_x(Z)(v) = [Z, V]_x,$$

where V is any vector field such that $V_x = v$, is a well-defined homomorphism of Lie algebras.

From now on, for a given point x of a pseudo-Riemannian manifold, the map λ_x will denote the homomorphism from the previous lemma.

For the proof of our next result we will present some facts about infinitesimal automorphisms and Killing fields, and we refer to [1] for further details. The tangent bundle of order k of a manifold N is the smooth bundle $T^{(k)}N$ whose fiber at x is the space $T_x^{(k)}N$ of $(k-1)$-jets at x of vector fields of N. For every (local) diffeomorphism $\varphi : N_1 \to N_2$ mapping x_1 to x_2 we have a linear isomorphism:

$$T_{x_1}^{(k)} N_1 \to T_{x_2}^{(k)} N_2$$

$$j_{x_1}^{k-1}(X) \mapsto j_{x_2}^{k-1}(d\varphi(X))$$

that depends only on the jet $j_{x_1}^k(\varphi)$. For $N_1 = N_2 = N$, $x_1 = x_2 = x$ this yields the group $D_x^{(k)}(N)$ of k-jets at x of local diffeomorphisms fixing x, whose group structure is induced by the composition of maps. In the case when $N = \mathbb{R}^n$ and $x = 0$, we will denote this group with $\mathrm{GL}^{(k)}(n)$. We also recall that the k-th order frame bundle over N is the smooth bundle $L^{(k)}(N)$ that consists of the k-jets at 0 of local diffeomorphisms $\mathbb{R}^n \to N$; any such a jet $j_0^k(\varphi)$ thus defines a linear isomorphism $T_0^{(k)} \mathbb{R}^n \to T_{\varphi(0)}^{(k)} N$. The structure group of $L^{(k)}(N)$ is $\mathrm{GL}^{(k)}(n)$. With respect to vector fields, we denote with $\mathcal{D}_x^{(k)}(N)$ the space of k-jets at x of vector fields vanishing at x, and we use the special notation $\mathfrak{gl}^{(k)}(n)$ when $N = \mathbb{R}^n$ and $x = 0$. If N carries a pseudo-Riemannian metric, for every $x \in N$, we will denote with $\mathrm{Aut}^k(N, x)$ the subgroup of $D_x^{(k)}(N)$ consisting of those k-jets that preserve the metric up to order k at x. Correspondingly, for vector fields we denote with $\mathrm{Kill}^k(N, x)$ the space of k-jets at x of vector fields on N that preserve the metric up to order k at x; the subspace of those k-jets whose 0-jet vanishes is denoted with $\mathrm{Kill}_0^k(N, x)$. The next result provides a natural representation of $D_x^{(k)}(N)$ from which the Lie algebras of this group and of $\mathrm{Aut}^k(N, x)$ are described in terms of $\mathcal{D}_x^{(k)}(N)$ and $\mathrm{Kill}_0^k(N, x)$, respectively; the proof is elementary, but it is detailed in sections 2 and 4 of [1].

LEMMA 2.2. *For a smooth manifold N and any given point $x \in N$ the following properties hold for every $k \geq 1$:*

1) The map

$$\Theta_x : D_x^{(k)}(N) \to \mathrm{GL}(T_x^{(k)} N)$$

$$\Theta_x(j_x^k(\varphi))(j_x^{k-1}(X)) = j_x^{k-1}(d\varphi(X))$$

is a Lie group monomorphism.

2) The assignment $[j_x^k(X), j_x^k(Y)]^k = -j_x^k([X, Y])$ yields a well-defined Lie algebra structure on $\mathcal{D}_x^{(k)}(N)$.

3) The map

$$\theta_x : \mathcal{D}_x^{(k)}(N) \to \mathfrak{gl}(T_x^{(k)} N)$$

$$\theta_x(j_x^k(X))(j_x^{k-1}(Y)) = -j_x^{k-1}([X, Y])$$

is a Lie algebra monomorphism for the Lie algebra structure on $\mathcal{D}_x^{(k)}(N)$ given by $[\cdot, \cdot]^k$. Furthermore, $\theta_x(\mathcal{D}_x^{(k)}(N)) = \mathrm{Lie}(\Theta_x(D_x^{(k)}(N)))$.

4) If N has a pseudo-Riemannian metric, then we have $\theta_x(\mathrm{Kill}_0^k(N, x)) = \mathrm{Lie}(\Theta_x(\mathrm{Aut}^k(N, x)))$.

In particular, with respect to the homomorphisms Θ_x and θ_x, the Lie algebra of $\mathrm{Aut}^k(N, x)$ is realized by $\mathrm{Kill}_0^k(N, x)$ with the Lie algebra structure given by $[\cdot, \cdot]^k$.

The following result is due to Gromov in the analytic case (see [2]) and it was extended to the smooth case in [10]. We present here a fairly detailed proof based on the results from [1]. In congruence with our notation for M, in the rest of this work we will use $\mathcal{O}$ to denote the foliation by $\widetilde{G}$-orbits in $\widehat{M}$ for the lifted $\widetilde{G}$-action on every covering space $\widehat{M}$ of M; this will be so for any covering whether finite or not.

PROPOSITION 2.3. *Suppose that the G-action on M has a dense orbit and preserves a finite-volume pseudo-Riemannian metric. Then, there exists a dense subset $S \subset \widetilde{M}$ such that for every $x \in S$ the following properties are satisfied.*

1) *There is a homomorphism of Lie algebras $\rho_x : \mathfrak{g} \to \mathrm{Kill}(\widetilde{M}, x)$, which is an isomorphism onto its image $\rho_x(\mathfrak{g}) = \mathfrak{g}(x)$.*
2) *$\mathfrak{g}(x) \subset \mathrm{Kill}_0(\widetilde{M}, x)$; that is, every element of $\mathfrak{g}(x)$ vanishes at x.*
3) *For every $X, Y \in \mathfrak{g}$ we have*

$$[\rho_x(X), Y^*] = [X, Y]^* = -[X^*, Y^*]$$

in a neighborhood of x. In particular, the elements in $\mathfrak{g}(x)$ and their corresponding local flows preserve both $\mathcal{O}$ and $T\mathcal{O}^\perp$ in a neighborhood of x.
4) *The homomorphism of Lie algebras $\lambda_x \circ \rho_x : \mathfrak{g} \to \mathfrak{so}(T_x\widetilde{M})$ induces a $\mathfrak{g}$-module structure on $T_x\widetilde{M}$ for which the subspaces $T_x\mathcal{O}$ and $T_x\mathcal{O}^\perp$ are $\mathfrak{g}$-submodules.*

Proof. Since the proof builds on the notions and results found in [1], we will mostly follow its notation. We will be careful to define the objects considered but we refer to [1] for further details.

For every $k \geq 1$, let us denote with $\sigma^k : L^{(k)}(M) \to Q_k$ the $\mathrm{GL}^{(k)}(n)$-equivariant map that defines the k-th order extension of the geometric structure defined by the pseudo-Riemannian metric on M.

Consider the set

$$\mathcal{G}^k = \{j_x^{k-1}(X^*) : X \in \mathfrak{g}, x \in M\},$$

which, by the local freeness of the G-action, is a smooth subbundle of $T^{(k)}M$. In fact, we have $\mathcal{G} = T\mathcal{O}$, and as with this bundle there is a trivialization given by

$$M \times \mathfrak{g} \to \mathcal{G}^k$$

$$(x, X) \mapsto j_x^{k-1}(X^*).$$

The corresponding trivialization of the frame bundle of $\mathcal{G}^k$ is given by

$$M \times \mathrm{GL}(\mathfrak{g}) \to L(\mathcal{G}^k)$$

$$(x, A) \mapsto A_x,$$

where $A_x(X) = j_x^{k-1}((AX)^*)$. Note that we have taken $\mathfrak{g}$ as the standard fiber of the bundle $\mathcal{G}^k$.

Choose a subspace $\mathcal{G}_0$ of $T_0^{(k)}\mathbb{R}^n$ isomorphic to $\mathfrak{g}$. We will fix such subspace as well as an isomorphism with $\mathfrak{g}$ through which we will identify these two spaces. Let us now consider

$$L^{(k)}(M, \mathcal{G}^k) = \{\alpha \in L^{(k)}(M) : \alpha(\mathcal{G}_0) = \mathcal{G}_x^k \text{ if } \alpha \in L^{(k)}(M)_x\},$$

which is a smooth reduction of $L^{(k)}(M)$ to the subgroup of $\mathrm{GL}^{(k)}(n)$ that preserves the subspace $\mathcal{G}_0$; we will denote such subgroup with $\mathrm{GL}^{(k)}(n, \mathcal{G}_0)$. Recall from the remarks preceding Lemma 2.2 that for every $j_0^k(\varphi) \in L^{(k)}(M)$ we obtain a linear isomorphism

$$T_0^{(k)}\mathbb{R}^n \to T_{\varphi(0)}^{(k)}M$$

$$j_0^{k-1}(X) \mapsto j_{\varphi(0)}^{k-1}(d\varphi(X)).$$

In particular, if we let

$$f_k : L^{(k)}(M, \mathcal{G}^k) \to L(\mathcal{G}^k)$$

$$j_0^k(\varphi) \mapsto j_0^k(\varphi)|_{\mathcal{G}_0},$$

then, by the identification between $\mathcal{G}_0$ and $\mathfrak{g}$, we can consider f_k as a well-defined smooth principal bundle morphism that covers the identity. The associated homomorphism of structure groups for f_k is given by

$$\pi_k : \mathrm{GL}^{(k)}(n, \mathcal{G}_0) \to \mathrm{GL}(\mathfrak{g})$$

$$j_0^k(\varphi) \mapsto j_0^k(\varphi)|_{\mathcal{G}_0},$$

which is clearly surjective. Note that we have used again our identification between $\mathfrak{g}$ and $\mathcal{G}_0$.

Fix μ an arbitrary ergodic component for the G-action on M for the pseudo-Riemannian volume. Then, there is a measurable reduction P of $L^{(k)}(M, \mathcal{G}^k)$ so that $\sigma^k(P)$ is (μ-a.e. over M) a single point $q_0 \in Q_k$. Furthermore, the structure group of P is the subgroup of $GL^{(k)}(n, \mathcal{G}_0)$ that stabilizes q_0, and in particular it is algebraic. This claim is a consequence of the fact that the $GL^{(k)}(n, \mathcal{G}_0)$-action on Q_k is algebraic, which in turn follows from the fact that a pseudo-Riemannian metric is a geometric structure of algebraic type; we refer to section 4 and the proof of proposition 8.4 of [1] for further details.

On the other hand, since π_k is a surjection and f_k is G-equivariant, by proposition 8.2 of [1], there exist reductions Q_1 and Q_2 of $L^{(k)}(M, \mathcal{G}^k)$ and $L(\mathcal{G}^k)$, respectively, to subgroups $L_1 \subset GL^{(k)}(n, \mathcal{G}_0)$ and $\overline{Ad(G)}^Z \subset GL(\mathfrak{g})$, such that $f_k(Q_1) \subset Q_2$ (μ-a.e. over M) and such that $\pi_k(L_1)$ is a finite-index subgroup of $\overline{Ad(G)}^Z$. Here L_1 can be chosen to be the algebraic hull of $L^{(k)}(M, \mathcal{G}^k)$ for the G-action on M with respect to the ergodic measure μ. This claim uses the well-known fact that $\overline{Ad(G)}^Z$ is the algebraic hull of $M \times GL(\mathfrak{g})$ for the product action. It is not difficult to see that this can be chosen so that $Q_2 = M \times \overline{Ad(G)}^Z$ (μ-a.e. over M) with respect to the above identification $M \times GL(\mathfrak{g}) \cong L(\mathcal{G}^k)$. We can also assume that $Q_1 \subset P$, μ-a.e. over M, because the reduction Q_1 is the smallest one to an algebraic subgroup.

The above discussion ensures that for μ-a.e. $x \in M$, we have the following relations:

$$f_k((Q_1)_x) \subset (Q_2)_x = \{x\} \times \overline{Ad(G)}^Z$$

$$(Q_1)_x \subset (P)_x \subset L^{(k)}(M, \mathcal{G}^k)_x$$

$$\sigma^k((P)_x) = \{q_0\}.$$

Let us now fix a point x such that these conditions hold. Choose $\alpha_x \in (Q_1)_x$ and let $f_k(\alpha_x) = (x, k_x)$ where $k_x \in \overline{Ad(G)}^Z$. Since π_k is surjective, there exists $\widehat{k}_x \in GL^{(k)}(n, \mathcal{G}_0)$ such that $\pi_k(\widehat{k}_x) = k_x$. In particular, by the π_k-equivariance of f_k we have $f_k(\alpha_x \widehat{k}_x^{-1}) = (x, e)$. We also have by the same reason

$$f_k(\alpha_x g \widehat{k}_x^{-1}) = f_k(\alpha_x \widehat{k}_x^{-1} \widehat{k}_x g \widehat{k}_x^{-1}) = (x, k_x \pi_k(g) k_x^{-1}),$$

for every $g \in L_1$. Also, the inclusion $(Q_1)_x \subset L^{(k)}(M, \mathcal{G}^k)_x$ implies that, for every $g \in L_1$, the k-jets of diffeomorphisms $\alpha_x \widehat{k}_x^{-1}, \alpha_x g \widehat{k}_x^{-1}$ considered as linear isomorphisms $T_0^{(k)} \mathbb{R}^n \to T_x^{(k)} M$ map $\mathcal{G}_0$ onto $\mathcal{G}_x^k$. Hence, from the definition of f_k it follows that $\alpha_x g \alpha_x^{-1} = (\alpha_x g \widehat{k}_x^{-1})(\alpha_x \widehat{k}_x^{-1})^{-1}$ is a k-jet of local diffeomorphism of M at x whose associated isomorphism $T_x^{(k)} M \to T_x^{(k)} M$ maps $\mathcal{G}_x^k$ onto itself by the assignment

$$j_x^{k-1}(X^*) \mapsto j_x^{k-1}((k_x \pi_k(g) k_x^{-1} X)^*),$$

for which we have used the above trivialization $M \times \mathrm{GL}(\mathfrak{g}) \cong L(\mathcal{G}^k)$. Since $\pi_k(L_1)$ has finite index in $\overline{\mathrm{Ad}(G)}^Z$ it contains the identity component $\mathrm{Ad}(G)$, and because $k_x \in \overline{\mathrm{Ad}(G)}^Z$ the group $k_x \pi_k(L_1) k_x^{-1}$ also contains $\mathrm{Ad}(G)$. It follows that $\alpha_x L_1 \alpha_x^{-1}$ is a subgroup of $D_x^{(k)}(M)$ for which the homomorphism from Lemma 2.2(1) induces a homomorphism

$$H_x : \alpha_x L_1 \alpha_x^{-1} \to \mathrm{GL}(\mathcal{G}_x^k)$$

$$\alpha_x g \alpha_x^{-1} \mapsto \Theta_x(\alpha_x g \alpha_x^{-1})|_{\mathcal{G}_x^k},$$

whose image contains $\mathrm{Ad}(G) \subset \mathrm{GL}(\mathfrak{g})$ with respect to the identification between $\mathfrak{g}$ and $\mathcal{G}_x^k$ given by the above isomorphism $M \times \mathfrak{g} \cong \mathcal{G}^k$. This implies that the induced Lie algebra homomorphism

$$h_x : \mathrm{Lie}(\alpha_x L_1 \alpha_x^{-1}) \to \mathfrak{gl}(\mathcal{G}_x^k)$$

has image $\mathrm{ad}(\mathfrak{g})$, again with respect to the referred identification between $\mathfrak{g}$ and $\mathcal{G}_x^k$.

On the other hand, we have for every $g \in L_1$

$$\sigma^k((\alpha_x g \alpha_x^{-1}) \alpha_x) = \sigma^k(\alpha_x g) = \sigma^k(\alpha_x)$$

because $\sigma^k((Q_1)_x) \subset \sigma^k((P)_x) = \{q_0\}$ is a single point. But this identity proves that every such k-jet $\alpha_x g \alpha_x^{-1}$ preserves the pseudo-Riemannian metric up order k (see [1]). In other words, $\alpha_x L_1 \alpha_x^{-1}$ is a subgroup of $\mathrm{Aut}^k(M, x)$ and by Lemma 2.2 we also have that $\mathrm{Lie}(\alpha_x L_1 \alpha_x^{-1})$ is a Lie subalgebra of $\mathrm{Kill}_0^k(M, x)$.

From the above remarks, it follows that there is a Lie algebra homomorphism $\widehat{\rho}_x^k : \mathfrak{g} \to \mathrm{Kill}_0^k(M, x)$ such that

$$(*) \qquad \theta_x(\widehat{\rho}_x^k(X))(j_x^{k-1}(Y^*)) = j_x^{k-1}([X, Y]^*) \quad \text{for every } X, Y \in \mathfrak{g}.$$

For k fixed, the existence of the homomorphism $\widehat{\rho}_x^k$ has been established for μ-a.e. $x \in M$, where μ is an arbitrary ergodic component of the pseudo-Riemannian volume of M. Thus, for k fixed, it follows that the homomorphism $\widehat{\rho}_x^k$ exists for every $x \in S_k$, where S_k is some subset of M, which is conull with respect to the pseudo-Riemannian volume of M. Finally, if we let $S_0 = \cap_{k=1}^{\infty} S_k$, then S_0 is conull with respect to the pseudo-Riemannian volume and for every $x \in S_0$ and every $k \geq 1$ there exists a homomorphism $\widehat{\rho}_x^k : \mathfrak{g} \to \mathrm{Kill}_0^k(M, x)$ satisfying $(*)$.

In [6] the notion of ℓ-regular point for a metric in a manifold is introduced. Such regular points satisfy two key properties relevant to our discussion. First, the set of regular points U of M is an open dense subset. Second, for $x \in U$ there is some integer $k(x) \geq 1$ so that, for $k \geq k(x)$, every element of $\mathrm{Kill}_0^k(M, x)$ extends uniquely to an element of $\mathrm{Kill}_0(M, x)$. The first property is found in [6] and the second one is proved in [1], both just using smoothness. Note that the results in [6] are stated for Riemannian manifolds but, as remarked in [1], the ones we consider here apply without change to general pseudo-Riemannian manifolds. The upshot of these remarks is that for every $x \in U$, there is some $k(x) \geq 1$ so that the map

$$J_x^k : \mathrm{Kill}_0(M, x) \to \mathrm{Kill}_0^k(M, x)$$

$$X \mapsto j_x^k(X),$$

is a linear isomorphism for every $k \geq k(x)$. Note that in this case, for the usual brackets in $\mathrm{Kill}_0(M, x)$ and the brackets $[\cdot, \cdot]^k$ in $\mathrm{Kill}_0^k(M, x)$ considered above, the map J_x^k is a Lie algebra anti-isomorphism.

For S_0 and U as above, consider the dense subset $S = S_0 \cap U \subset M$. Next choose $x \in S$ and $k \geq \max(k(x), 2)$. Then, the map J_x^k is a Lie algebra anti-isomorphism, and there exists a Lie algebra homomorphism $\widehat{\rho}_x^k : \mathfrak{g} \to \mathrm{Kill}_0^k(M, x)$ satisfying (*). If we let $\rho_x = -(J_x^k)^{-1} \circ \widehat{\rho}_x^k : \mathfrak{g} \to \mathrm{Kill}_0(M, x)$, then ρ_x defines a Lie algebra homomorphism such that

$$j_x^{k-1}([\rho_x(X), Y^*]) = j_x^{k-1}([X, Y]^*),$$

for every $X, Y \in \mathfrak{g}$. For this, we have used (*) and the definition of θ_x from Lemma 2.2. Since $k - 1 \geq 1$ and because germs of Killing fields are determined by any jet of order at least 1, we conclude that, at our chosen point x, ρ_x satisfies (from our statement) (1), (2), and the identity in (3) in a neighborhood of x with $\widetilde{M}$ replaced with M. The identity in (3) now proves that every element of $\mathfrak{g}(x)$ preserves the tangent bundle to $\mathcal{O}$ in a neighborhood of x: that is, the corresponding Lie derivatives map sections of $T\mathcal{O}$ into sections of $T\mathcal{O}$. By proposition 2.2 of [5] we conclude that the local flows of the elements of $\mathfrak{g}(x)$ preserve $\mathcal{O}$ as well in a neighborhood of x. Since the elements of $\mathfrak{g}(x)$ are Killing fields, we conclude that they (and their local flows) also preserve the normal bundle $T\mathcal{O}^\perp$ in a neighborhood of x. This completes the proof of our statement for the dense subset $S \subset M$ and for $\widetilde{M}$ replaced with M in (1)–(4). Finally, this yields the statement for $\widetilde{M}$ for the dense subset, which is the inverse image of S under the covering map since such map is a local isometry. $\qquad\square$

REMARK 2.4. The conclusions of Proposition 2.3 hold without change for some dense subset $S \subset M$ by replacing $\widetilde{M}$ with M in (1)–(4). In fact, our proof first obtains the required Killing fields on M, which are then translated into corresponding ones on $\widetilde{M}$.

In this work, we will be dealing with and interested in the case where the G-orbits in M are nondegenerate submanifolds with respect to the pseudo-Riemannian metric. In this case, the $\widetilde{G}$-orbits on $\widetilde{M}$ are nondegenerate as well and we have a direct sum decomposition $T\widetilde{M} = T\mathcal{O} \oplus T\mathcal{O}^{\perp}$. When this holds, we can consider the $\mathfrak{g}$-valued 1-form ω on $\widetilde{M}$ that is given, at every $x \in \widetilde{M}$, by the composition $T_x\widetilde{M} \to T_x\mathcal{O} \cong \mathfrak{g}$, where the first map is the natural projection and the second map is the natural isomorphism. From this, we then define the $\mathfrak{g}$-valued 2-form given by $\Omega = d\omega|_{\wedge^2 T\mathcal{O}^{\perp}}$. The following result will provide us with a criterion, in terms of Ω, for the normal bundle $T\mathcal{O}^{\perp}$ to be integrable. At the same time, we relate Ω with the $\mathfrak{g}$-module structures from Proposition 2.3. This result is very well known and it is essentially contained in [2], [3], and [10], but we include its proof here for the sake of completeness.

LEMMA 2.5. *Let G, M, and S be as in Proposition 2.3. If we assume that the G-orbits are nondegenerate, then*

1) *for every $x \in S$, the maps $\omega_x : T_x\widetilde{M} \to \mathfrak{g}$ and $\Omega_x : \wedge^2 T_x\mathcal{O}^{\perp} \to \mathfrak{g}$ are both homomorphisms of $\mathfrak{g}$-modules, for the $\mathfrak{g}$-module structures from Proposition 2.3, and*
2) *the normal bundle $T\mathcal{O}^{\perp}$ is integrable if and only if $\Omega = 0$.*

Proof. In the rest of the proof, let $X \in \mathfrak{g}$ and $x \in S$ be fixed but arbitrarily given.

For Z a vector field over $\widetilde{M}$, let $Z^{\top}, Z^{\perp}$ be its $T\mathcal{O}$ and $T\mathcal{O}^{\perp}$ components, respectively. Since $\rho_x(X)$ is a Killing field preserving $\mathcal{O}$ and $T\mathcal{O}^{\perp}$ it follows that

$$[\rho_x(X), Z]^{\top} = [\rho_x(X), Z^{\top}],$$

$$[\rho_x(X), Z]^{\perp} = [\rho_x(X), Z^{\perp}].$$

Denote with $\alpha : T_x\mathcal{O} \to \mathfrak{g}$ the inverse map of $X \mapsto X_x^*$. Then we have

$$\omega_x(X \cdot Z_x) = \omega_x([\rho_x(X), Z]_x)$$

$$= \alpha([\rho_x(X), Z^{\top}]_x)$$

$$= \alpha([\rho_x(X), \omega(Z)^*]_x)$$

$$= \alpha([X, \omega(Z)]_x^*)$$

$$= [X, \omega_x(Z)]$$

$$= X \cdot \omega_x(Z_x),$$

thus showing that ω_x is a homomorphism of $\mathfrak{g}$-modules. Here we used in the second and third identities the definition of ω, and in the fourth identity the formula from Proposition 2.3(3); the rest follows from the definition of the $\mathfrak{g}$-module structures involved.

Next, observe that for every pair of sections Z_1, Z_2 of $T\mathcal{O}^\perp$ we have

$$\Omega(Z_1 \wedge Z_2) = Z_1(\omega(Z_2)) - Z_2(\omega(Z_1)) - \omega([Z_1, Z_2])$$

$$= -\omega([Z_1, Z_2]),$$

which clearly implies (2).

Now let $u, v \in T_x\mathcal{O}^\perp$ be given and choose U, V sections of $T\mathcal{O}^\perp$ extending them, respectively. Hence, using that ω is a homomorphism of $\mathfrak{g}$-modules, the Jacobi identity and the above expression relating Ω and ω, we obtain

$$\Omega_x(X \cdot (u \wedge v)) = \Omega_x((X \cdot u) \wedge v) + \Omega_x(u \wedge (X \cdot v))$$

$$= \Omega_x([\rho_x(X), U] \wedge V) + \Omega_x(U \wedge [\rho_x(X), V])$$

$$= -\omega_x([[\rho_x(X), U], V]) - \omega_x([U, [\rho_x(X), V]])$$

$$= -\omega_x([\rho_x(X), [U, V]])$$

$$= -\omega_x(X \cdot [U, V]_x)$$

$$= -[X, \omega_x([U, V])]$$

$$= [X, \Omega_x(U \wedge V)]$$

$$= X \cdot \Omega_x(u \wedge v),$$

thus showing that Ω_x is a homomorphism of $\mathfrak{g}$-modules. Note that we have used that both $[\rho_x(X), U], [\rho_x(X), V]$ are sections of $T\mathcal{O}^\perp$. $\square$

The following result allows us to relate the metric $T\mathcal{O}$ coming from M to suitable metrics on G. The proof presented here is due to Gromov (see [2]) and provides our first application of Proposition 2.3.

LEMMA 2.6. *Suppose that the G-action on M has a dense orbit and preserves a finite-volume pseudo-Riemannian metric. Then, for every $x \in M$ and with respect to the natural isomorphism $\mathfrak{g} \cong T_x\mathcal{O}$, the metric of M restricted to $T_x\mathcal{O}$ defines an Ad(G)-invariant symmetric bilinear form on $\mathfrak{g}$ independent of the point x.*

Proof. With the above-mentioned trivialization of $T\mathcal{O}$, the metric h on M restricted to the orbits and pulled back to $M \times \mathfrak{g}$ yields a map

$$\psi : M \to \mathfrak{g}^* \otimes \mathfrak{g}^*$$

$$x \mapsto B_x$$

where $B_x(X, Y) = h_x(X_x^*, Y_x^*)$.

By Remark 2.4, there is a dense subset $S \subset M$ so that the conclusions of Proposition 2.3 are satisfied for every $x \in S$ with the tangent spaces and Killing fields of $\widetilde{M}$ replaced by those of M. Hence, for every $x \in S$ the inner product h_x is preserved, in the sense of Proposition 2.3(4), by the Killing vector fields that belong to $\mathfrak{g}(x)$. In particular, for every $x \in S$ and $X, Y, Z \in \mathfrak{g}$ we have

$$h_x([\rho_x(X), Y^*]_x, Z_x^*) = -h_x(Y_x^*, [\rho_x(X), Z^*]_x),$$

which, by Proposition 2.3(3) yields

$$h_x([X, Y]_x^*, Z_x^*) = -h_x(Y_x^*, [X, Z]_x^*).$$

In other words, for every $x \in S$ we have

$$B_x([X, Y], Z) = -B_x(Y, [X, Z]),$$

for all $X, Y, Z \in \mathfrak{g}$. This implies that $\psi(x) = B_x$ is an $\mathrm{Ad}(G)$-invariant form on $\mathfrak{g}$ for every $x \in S$. By the density of S in M, we conclude that the image of ψ lies in the set of $\mathrm{Ad}(G)$-invariant forms.

On the other hand, at every $x \in M$ and for $g \in G$, $X, Y \in \mathfrak{g}$ we have

$$\psi(gx)(X, Y) = h_{gx}(X_{gx}^*, Y_{gx}^*)$$

$$= h_x(dg_{gx}^{-1}(X_{gx}^*), dg_{gx}^{-1}(Y_{gx}^*))$$

$$= h_x(\mathrm{Ad}(g^{-1})(X)_x^*, \mathrm{Ad}(g^{-1})(Y)_x^*)$$

$$= \psi(x)(\mathrm{Ad}(g^{-1})(X), \mathrm{Ad}(g^{-1})(Y)),$$

which shows that ψ is G-equivariant. Note that we used in the second identity that G preserves the metric, and in the third identity we used the remarks at the beginning of this section.

The G-equivariance of ψ and the fact that its image lies in G-fixed points implies that ψ is G-invariant. Then, the result follows from the existence of a dense G-orbit. $\qquad\square$

As a simple application of the previous result, we prove the nondegeneracy of the orbits when the manifold acted upon has a suitably bounded dimension.

LEMMA 2.7. *Suppose that the G-action on M has a dense orbit and preserves a finite-volume pseudo-Riemannian metric. If* $\dim(M) < 2 \dim(G)$, *then the G-orbits are nondegenerate with respect to the metric on M.*

Proof. By Lemma 2.6, for every $x \in M$ the metric restricted to $T_x \mathcal{O}$ corresponds to an Ad(G)-invariant form in $\mathfrak{g}$. The kernel of such a form is an ideal and so the metric h_x restricted to $T_x \mathcal{O}$ is either nondegenerate or zero.

Suppose that h_x is zero when restricted to $T_x \mathcal{O}$ for some $x \in M$. Then, $T_x \mathcal{O}$ lies in the null cone of $T_x M$ for the metric h_x. Hence, for (m_1, m_2) the signature of M, we have $\dim(G) = \dim(T_x \mathcal{O}) \leq \min(m_1, m_2)$. And this implies $2 \dim(G) \leq m_1 + m_2 = \dim(M)$, which is impossible. $\square$

3. Proof of Theorem 1.1 and Some Consequences

We start this section by proving Theorem 1.1.

Proof of Theorem 1.1. Assuming that $T\mathcal{O}^\perp$ is integrable, let $\mathcal{F}$ be the induced foliation. We will first prove that $\mathcal{F}$ is totally geodesic, that is, its leaves are totally geodesic submanifolds of M. We will denote with h the metric on M preserved by G.

First note that, if Y, Z are local sections of $T\mathcal{O}^\perp$ that preserve the foliation, then we have for every $X \in \mathfrak{g}$:

$$X^*(h(Y,Z)) = h([X^*, Y], Z) + h(Y, [X^*, Z]) = 0,$$

because our choices imply that $[X^*, Y]$ and $[X^*, Z]$ are sections of $T\mathcal{O}$. In particular, for every Y, Z as above the function $h(Y, Z)$ is constant along the G-orbits. In the notation of [5], we conclude that h is a bundle-like metric for the foliation $\mathcal{O}$. Hence, by the remarks on page 79 of [5] it follows that h induces a transverse metric to the foliation $\mathcal{O}$. By the construction of such transverse metric from h and the arguments in the proof of proposition 3.2 of [5], it is easy to conclude that the foliation $\mathcal{O}$ is given by pseudo-Riemannian submersions that define the transverse metric. More precisely, at every point in M there is an open subset U of M and a pseudo-Riemannian submersion $\pi : U \to B$ such that the fibers of π define the foliation $\mathcal{O}$ restricted to U. We observe that the results of [5] are stated for Riemannian metrics, but those that we use here extend without change to arbitrary pseudo-Riemannian metrics.

We will now use the properties of the structural equations from [8] for a pseudo-Riemannian submersion $\pi : U \to B$ as above. Again, the results in [8] are stated for Riemannian submersions, but the ones that we will use are easily seen to hold for pseudo-Riemannian submersions as well. For π as above, let A be the associated fundamental tensor defined in [8]. In particular, by the definition of A, the second fundamental form for the leaves of $\mathcal{F}$ is given by $A_X Y$, for X, Y tangent to $\mathcal{F}$. But by lemma 2 of [8] we have for X, Y tangent to $\mathcal{F}$ the identity $A_X Y = \frac{1}{2}[X, Y]^\top$, where $Z^\top$ denotes the projection of Z onto $T\mathcal{O}$. Hence, the integrability of $T\mathcal{O}^\perp$ to $\mathcal{F}$ shows that A vanishes on vector fields tangent to $\mathcal{F}$, thus showing that the leaves of $\mathcal{F}$ are totally geodesic.

Choose a leaf N of $\mathcal{F}$. Then, one can prove fairly easily that every geodesic in M that is tangent at some point to N remains in N for every value of the parameter of the geodesic; this uses the fact that N is a maximal integral submanifold of $T\mathcal{O}^\perp$ and that the leaves of $\mathcal{F}$ are totally geodesic. Hence, the completenes of M implies that of N.

For our chosen leaf N of $\mathcal{F}$, consider the G-action map restricted to $G \times N$. This defines a smooth map $\varphi : G \times N \to M$, which is G-equivariant. By Lemma 2.6, it follows easily that φ is a local isometry for $G \times N$ endowed with the product metric where G carries a suitable bi-invariant metric. In particular, $G \times N$ is complete and so we conclude from corollary 29 on page 202 from [9] that φ is an isometric covering map. Hence, the universal covering map of M is given by $\widetilde{\varphi} : \widetilde{G} \times \widetilde{N} \to M$ and the $\widetilde{G}$-action on M lifted to $\widetilde{G} \times \widetilde{N}$ is the left action on the first factor.

We now claim that $\pi_1(M) \subset Iso(\widetilde{G}) \times Iso(\widetilde{N})$, that is, that every element in $\pi_1(M)$ preserves the factors in the product $\widetilde{G} \times \widetilde{N}$. To see this, let $\gamma \in \pi_1(M)$ be given with $\gamma = (\gamma_1, \gamma_2)$ its component decomposition. Observe that in $\widetilde{G} \times \widetilde{N}$ we have

$$T_{(g,x)}\mathcal{O} = T_g\widetilde{G}, \quad T_{(g,x)}\mathcal{O}^\perp = T_x\widetilde{N},$$

for every $(g, x) \in \widetilde{G} \times \widetilde{N}$. Since γ commutes with the $\widetilde{G}$-action, it preserves both $T\mathcal{O}$ and $T\mathcal{O}^\perp$ and so

$$d\gamma(u) = d\gamma_1(u) + d\gamma_2(u) \in T\mathcal{O}$$

$$d\gamma(v) = d\gamma_1(v) + d\gamma_2(v) \in T\mathcal{O}^\perp,$$

for every $u \in T\widetilde{G}$ and $v \in T\widetilde{N}$. We conclude that $d\gamma_2(T\mathcal{O}) = 0$ and $d\gamma_1(T\mathcal{O}^\perp) = 0$, which implies that γ_1 is independent of $\widetilde{N}$ and γ_2 is independent of $\widetilde{G}$. This yields our claim about $\pi_1(M)$.

On the other hand, since $\widetilde{G}$ carries a bi-invariant metric, by the results from section 4 of [10] we know that the connected component of the identity of $Iso(\widetilde{G})$ is given by $Iso_0(\widetilde{G}) = L(\widetilde{G})R(\widetilde{G})$ (the left and right translations) and that it is a finite index subgroup of $Iso(\widetilde{G})$. Hence, the group $\Lambda = \pi_1(M) \cap (Iso_0(\widetilde{G}) \times Iso(\widetilde{N}))$ has finite index in $\pi_1(M)$, and so the induced map $(\widetilde{G} \times$

$\widetilde{N})/\Lambda \to M$ is a finite covering. Moreover, every $\gamma \in \Lambda$ can be written as $\gamma = (L_{g_1} R_{g_2}, \gamma_2)$, where $g_1, g_2 \in \widetilde{G}$ and $\gamma_2 \in Iso(\widetilde{N})$, and since such γ commutes with the $\widetilde{G}$-action we conclude that

$$(g_1 g g_2, \gamma_2(x)) = (L_{g_1} R_{g_2}, \gamma_2)(g(e, x)) = g((L_{g_1} R_{g_2}, \gamma_2)(e, x)) = (g g_1 g_2, \gamma_2(x))$$

for every $g \in \widetilde{G}$ and $x \in \widetilde{N}$, which implies $g_1 \in Z(\widetilde{G})$. Hence, $L_{g_1} = R_{g_1}$ and then $\gamma \in R(\widetilde{G}) \times Iso(\widetilde{N})$, thus showing that $\Lambda \subset R(\widetilde{G}) \times Iso(\widetilde{N})$.

Also note that the covering map $G \times \widetilde{N} \to M$ realizes $\pi_1(G) \subset \Lambda$, which induces a covering map $G \times \widetilde{N} \to (\widetilde{G} \times \widetilde{N})/\Lambda$. We claim that $G \times \widetilde{N} \to (\widetilde{G} \times \widetilde{N})/\Lambda$ is a normal covering map. For this we need to check that $\pi_1(G)$ is a normal subgroup of Λ under the inclusion $z \mapsto (R_z, e)$. But by the above remarks, every $\gamma \in \Lambda$ can be written as $\gamma = (R_{g_1}, \gamma_2)$, where $g_1 \in \widetilde{G}$ and $\gamma_2 \in Iso(\widetilde{N})$, from which we obtain

$$\gamma(R_z, e)\gamma^{-1} = (R_{g_1} R_z R_{g_1^{-1}}, \gamma_2 \gamma_2^{-1}) = (R_z, e)$$

since $\pi_1(G)$ is central in $\widetilde{G}$. It follows that the group of deck transformations for $G \times \widetilde{N} \to (\widetilde{G} \times \widetilde{N})/\Lambda$ is given by the group $\Gamma = \Lambda/\pi_1(G)$ and that we also have $(G \times \widetilde{N})/\Gamma = (\widetilde{G} \times \widetilde{N})/\Lambda$.

From the above, we conclude that $\Gamma \subset R(G) \times Iso(\widetilde{N}) = G \times Iso(\widetilde{N})$ is a group of deck transformations of $G \times \widetilde{N} \to M$ that induces a G-equivariant finite covering map $(G \times \widetilde{N})/\Gamma \to M$ that satisfies the properties required to obtain Theorem 1.1. $\qquad \square$

The integrability of the normal bundle can be ensured for suitable relations between the Lie group G and the geometry of the manifold on which it acts, thus providing the following results. In what follows, the signature of G, as a pseudo-Riemannian manifold, is always considered with respect to a bi-invariant metric. Note that if (n_1, n_2) is the signature of some bi-invariant metric on G, then the signature of any other bi-invariant metric is either (n_1, n_2) or (n_2, n_1).

COROLLARY 3.1. *Suppose that the G-action on M has a dense orbit and pre-serves a finite-volume complete pseudo-Riemannian metric. If the G-orbits are nondegenerate and either one of the following holds,*

1) *there is no Lie algebra embedding of $\mathfrak{g}$ into $\mathfrak{so}(T_x \mathcal{O}^{\perp})$ for every $x \in M$, or*
2) *for $n_0 = \min(n_1, n_2)$ and $m_0 = \min(m_1, m_2)$, where (n_1, n_2) and (m_1, m_2) are the signatures of G and M, respectively, we have $\operatorname{rank}_{\mathbb{R}}(\mathfrak{g}) > m_0 - n_0$,*

then the conclusion of Theorem 1.1 holds.

Proof. Let us consider a subset $S \subset \widetilde{M}$ given as in Proposition 2.3.

First suppose that condition (1) is satisfied. Then, for every $x \in S$, the $\mathfrak{g}$-module structure on $T_x \mathcal{O}^\perp$ given by Proposition 2.3(4) is trivial. By Lemma 2.5(1), being a homomorphism of $\mathfrak{g}$-modules, the map Ω_x is trivial for every $x \in S$. Since S is dense, we conclude that $\Omega = 0$ and so $T \mathcal{O}^\perp$ is integrable by Lemma 2.5(2). Hence, Theorem 1.1 can be applied.

Let us now assume that (2) holds. Note that by Lemma 2.6, the signature of $T \mathcal{O}$ is either (n_1, n_2) or (n_2, n_1). If we let (k_1, k_2) be the signature of $T \mathcal{O}^\perp$, then it is easily seen that

$$\min (k_1, k_2) \leq m_0 - n_0.$$

Since the real rank of $\mathfrak{so}(T_x \mathcal{O}^\perp)$ is precisely $\min (k_1, k_2)$ the result in this case follows from the first part. $\qquad \square$

COROLLARY 3.2. *Suppose that the G-action on M has a dense orbit and preserves a finite-volume complete pseudo-Riemannian metric. Let n be the dimension of the smallest $\mathfrak{g}$-module V such that $\wedge^2 V$ contains a $\mathfrak{g}$-submodule isomorphic to $\mathfrak{g}$. If $\dim (M) < \dim (G) + n$, then the conclusion of Theorem 1.1 holds.*

Proof. First observe that the Lie brackets in $\mathfrak{g}$ define a surjective homomorphism of $\mathfrak{g}$-modules from $\wedge^2 \mathfrak{g}$ onto $\mathfrak{g}$. Hence, $\wedge^2 \mathfrak{g}$ contains a submodule isomorphic to $\mathfrak{g}$, thus implying that $n \leq \dim (\mathfrak{g})$. Then, Lemma 2.7 shows that the G-orbits in M are nondegenerate with respect to the metric of M.

Let $S \subset \widetilde{M}$ be a subset given as in Proposition 2.3. From our hypotheses and since the G-orbits are nondegenerate, we have $\dim (T_x \mathcal{O}^\perp) < n$. In particular, for every $x \in S$ and for the $\mathfrak{g}$-module structure from Proposition 2.3(4), $\wedge^2 T_x \mathcal{O}^\perp$ does not contain a $\mathfrak{g}$-submodule isomorphic to $\mathfrak{g}$. Hence, Lemma 2.5(1) implies that $\Omega_x = 0$ for every $x \in S$ and so that $\Omega = 0$. By Lemma 2.5(2) the bundle $T \mathcal{O}^\perp$ is integrable and so Theorem 1.1 can be applied. $\qquad \square$

On a compact manifold with Riemannian normal bundle, we can obtain the conclusions of Theorem 1.1 without having to a priori assume that the manifold is complete and that the normal bundle is integrable.

COROLLARY 3.3. *Suppose that the G-action on M has a dense orbit and preserves a pseudo-Riemannian metric. If M is compact, the G-orbits are nondegenerate and the normal bundle $T \mathcal{O}^\perp$ is Riemannian, then the conclusions of Theorem 1.1 hold. Moreover, we can assume that $\widetilde{N}$ is Riemannian homogeneous and that $\Gamma \subset G \times Iso_0(\widetilde{N})$.*

Proof. The proof is a refinement of that of Theorem 1.1, so we will follow the notation of the latter.

First observe that the integrability of $T\mathcal{O}^\perp$ follows from the proof of Corollary 3.1(1) since $\mathfrak{so}(T_x\mathcal{O}^\perp)$ is compact for every $x \in \widetilde{M}$. In particular, we have the hypotheses of Theorem 1.1 except for the completeness of M. In this respect, it is easy to check that the compactness of M and the fact that $T\mathcal{O}^\perp$ is Riemannian imply that the geodesics in M perpendicular to the G-orbits are complete. This completeness is enough for the rest of the arguments in the proof of Theorem 1.1 to apply.

Finally, we observe that the existence of a dense G-orbit implies that $\widetilde{N}$ has a dense orbit by its local isometries and, being Riemannian, we conclude that it is homogeneous. The latter follows from the infinitesimal characterization of homogeneous Riemannian manifolds obtained in [12], and the fact that the orthogonal group is compact for definite metrics; we refer to [10] for further details. But for a homogeneous Riemannian manifold the group of isometries has finitely many connected components, and so we can intersect Γ with $G \times Iso_0(\widetilde{N})$ to obtain the last claim after passing to a finite covering. $\quad\square$

4. Manifolds with a Transverse Riemannian Structure: Proof of Theorem 1.2

In this section we will characterize those actions that preserve a Riemannian structure transverse to the G-orbits. We start with the following basic result relating transverse geometric structures for the foliation by G-orbits with geometric structures on the normal bundle to such orbits.

LEMMA 4.1. *Suppose that the G-action on M has a dense orbit and preserves a finite-volume pseudo-Riemannian metric. Also assume that the G-orbits are nondegenerate. If H is a subgroup of $GL(k, \mathbb{R})$, where $k = \dim(M) - \dim(G)$, then there is a one-to-one correspondence between the G-invariant H-reductions of $L(TM/T\mathcal{O})$ and the G-invariant H-reductions of $L(T\mathcal{O}^\perp)$. In particular, every transverse H-structure for the foliation $\mathcal{O}$ by G-orbits induces a G-invariant H-reduction of $L(T\mathcal{O}^\perp)$.*

Proof. From the decomposition $TM = T\mathcal{O} \oplus T\mathcal{O}^\perp$ we obtain a natural G-equivariant isomorphism $TM/T\mathcal{O} \to T\mathcal{O}^\perp$, which clearly yields the first claim.

Next, we recall that a transverse H-structure to the foliation $\mathcal{O}$ is given by a reduction P of $L(TM/T\mathcal{O})$, which is invariant under the local flows of vectors fields tangent to the foliation $\mathcal{O}$. In particular, P is invariant under the G-action on $L(TM/T\mathcal{O})$ thus showing the last claim. $\qquad\qquad\square$

The following result is obtained by applying the main theorems from [10].

PROPOSITION 4.2. *Suppose that the G-action on M has a dense orbit and preserves a pseudo-Riemannian metric. Also assume that M is compact. If the normal bundle to the orbits $T\mathcal{O}^{\perp}$ is Riemannian, then there exist*

 1) *a finite covering map $\widehat{M} \to M$,*
 2) *a connected Lie group H with a compact subgroup K, and*
 3) *a discrete cocompact subgroup $\Gamma \subset G \times H$ such that $G\Gamma$ is dense in $G \times H$,*

for which the G-action on M lifts to $\widehat{M}$ so that $\widehat{M}$ is G-equivariantly diffeomorphic to $(G \times K \backslash H)/\Gamma$. Furthermore, if G has finite center and real rank at least 2, then we can assume that $G \times H$ is a finite-center isotypic semisimple Lie group and Γ is an irreducible lattice.

Proof. Let us denote with (m_1, m_2) and (n_1, n_2) the signatures of M and G, respectively. Also, let us denote $m_0 = \min(m_1, m_2)$ and $n_0 = \min(n_1, n_2)$. Observe that since $T\mathcal{O}^{\perp}$ is Riemannian, the bundle $T\mathcal{O}$ is nondegenerate. Hence, by Lemma 2.6 the signature of $T\mathcal{O}$ is either (n_1, n_2) or (n_2, n_1); this is because the signature of every $\mathrm{Ad}(G)$-invariant metric on $\mathfrak{g}$ is the signature of either the Killing form or its negative. Without loss of generality, we can assume that the signature of $T\mathcal{O}$ is precisely (n_1, n_2).

Theorems A and B from [10] provide precisely our required conclusions when $n_0 = m_0$ holds, except for explicitly stating the property that $G\Gamma$ is dense in $G \times H$. The latter is obtained as follows. The conclusion of theorem A from [10] ensures the existence of a G-invariant ergodic smooth measure, which in turn implies the existence of a dense G-orbit in $(G \times H)/\Gamma$. From this, it is easily seen that we necessarily have the density of the G-orbit of $(e, e)\Gamma$ in $(G \times H)/\Gamma$. Since the map $G \times H \to (G \times H)/\Gamma$ is a covering, we conclude that the inverse image under this covering of such orbit, which is $G\Gamma$, is dense in $G \times H$.

We now consider the possibilities for n_0. We will now assume that the signatures are given in the form $(+, -)$. If $n_0 = n_2$, then the fact that $T\mathcal{O}^{\perp}$ is Riemannian implies that $(m_1, m_2) = (n_1 + k, n_2)$, where k is the rank of $T\mathcal{O}^{\perp}$. In particular, $n_0 = m_0$ in this case and the result follows from [10].

In case $n_0 = n_1$, we can replace the metric on $T\mathcal{O}^{\perp}$ by its negative to obtain a new G-invariant metric for which the signature of M is $(m_1, m_2) = (n_1, n_2 + k)$, where k is again the rank of $T\mathcal{O}^{\perp}$. Hence, for this new metric $n_0 = m_0$ and so the result follows also from [10]. □

We now proceed to prove Theorem 1.2.

Proof of Theorem 1.2. Assume that (1) holds. Let us take $\widehat{M} = (G \times K\backslash H)/\Gamma$ and verify that it satisfies (2).

Note that, with respect to the quotient map $G \times H \to (G \times H)/\Gamma$, the set $G\Gamma$ projects onto the G-orbit of the class of the identity. Hence, there is a dense G-orbit in $\widehat{M}$.

Next, endow $G \times H$ with the product metric given by a bi-invariant metric on G and a Riemannian metric on H, which is left K-invariant and right H-invariant. The latter exists because K is compact. This induces a pseudo-Riemannian metric on $G \times K\backslash H$, which is left G-invariant and right Γ-invariant. Note that the G-orbits are nondegenerate with normal bundle given by the tangent bundle to the factor $K\backslash H$. Furthermore, by construction the projection $G \times K\backslash H \to K\backslash H$ is a pseudo-Riemannian submersion and so it defines a transverse Riemannian structure for the foliation by G-orbits. Since the left G-action and the Γ-action commute with each other, the transverse Riemannian structure on $G \times K\backslash H$ induces a corresponding one on the double coset $(G \times K\backslash H)/\Gamma$, which is G-invariant. This provides the geometric structures required by (2).

Let us now assume that (2) holds, so that in particular $\widehat{M}$ has a dense G-orbit and carries the indicated geometric structures. Since the G-orbits in $\widehat{M}$ are nondegenerate then we have an orthogonal decomposition $T\widehat{M} = T\mathcal{O} \oplus T\mathcal{O}^{\perp}$ which is invariant under G. The existence of a transverse Riemannian structure for the foliation by G-orbits in $\widehat{M}$ and Lemma 4.1 imply the existence of a G-invariant Riemannian metric on $T\mathcal{O}^{\perp}$. Hence, if we replace with such Riemannian metric the metric on $T\mathcal{O}^{\perp}$ induced from $\widehat{M}$, we obtain a new G-invariant pseudo-Riemannian metric on $\widehat{M}$ for which $T\mathcal{O}^{\perp}$ is Riemannian. Hence, the G-action on $\widehat{M}$ satisfies the hypotheses of Proposition 4.2, and the latter provides a finite covering of $\widehat{M}$, and thus of M, with the properties required by (1).

Finally, let us assume that G has finite center and real rank at least 2. First observe that (1) replaced by the conditions in the last part of the statement still implies (2). The only nontrivial property to check is the existence of a dense G-orbit in $(G \times K\backslash H)/\Gamma$ for Γ an irreducible lattice. But in this case, $G\Gamma$ is dense in $G \times H$ (see for example lemma 6.3 from [10]), which yields the required dense orbit. On the other hand, that (2) implies (1) with the properties from the last part of the statement is a consequence of Proposition 4.2. □

5. Actions on Lorentzian Manifolds

In this section, we present a characterization of the G-actions on M preserving a Lorentzian metric.

THEOREM 5.1. *Let G be a connected noncompact simple Lie group acting faithfully on a compact manifold M. Then the following conditions are equivalent:*

1) *The group G is locally isomorphic to $\mathrm{SL}(2, \mathbb{R})$ and there is a G-equivariant finite covering map $(G \times K\backslash H)/\Gamma \to M$ where H is a connected Lie group with a compact subgroup K and $\Gamma \subset G \times H$ is a discrete cocompact subgroup such that $G\Gamma$ is dense in $G \times H$.*
2) *There is a finite covering map $\widehat{M} \to M$ for which the G-action on M lifts to a G-action on $\widehat{M}$ with a dense orbit and preserving a Lorentzian metric.*

Proof. Let us assume (1) and consider the finite covering $\widehat{M} = (G \times K\backslash H)/$ $\Gamma \to M$. Endow $G \times H$ with the product metric given by the bi-invariant metric coming from the Killing form in $\mathfrak{g}$ and a left K-invariant and right H-invariant Riemannian metric on H. Then consider the induced G-invariant metric on $\widehat{M}$. Since G is locally isomorphic to $\mathrm{SL}(2, \mathbb{R})$, such metric on $\widehat{M}$ is Lorentzian. As in the proof of Theorem 1.2, the density of $G\Gamma$ implies the existence of a dense G-orbit in $\widehat{M}$. This proves (2).

Let us now assume that (2) holds. By Lemma 2.6, for every $x \in M$, the metric in $T_x \mathcal{O}$ corresponds to an $\mathrm{Ad}(G)$-invariant symmetric bilinear form on $\mathfrak{g}$. Such form is either 0 or nondegenerate. The former case implies the existence of a null tangent subspace of dimension at least 3, which is impossible; in particular, the G-orbits are nondegenerate. Since G is noncompact and simple, we conclude the existence of an $\mathrm{Ad}(G)$-invariant form on $\mathfrak{g}$, which is Lorentzian. But $\mathfrak{sl}(2, \mathbb{R})$ is the only simple Lie algebra admitting such a form, which implies that G is locally isomorphic to $\mathrm{SL}(2, \mathbb{R})$. We also conclude that $T\mathcal{O}^\perp$ is Riemannian and so the rest of the claims in (1) follow from Proposition 4.2. □

6. Manifolds with a Transverse Parallelism: Proof of Theorem 1.3

The following result is an immediate consequence of the properties of Lie foliations. A proof can be found in [10].

LEMMA 6.1. *Let X be a compact manifold carrying a foliation with a transverse Lie structure. If the foliation has a dense leaf, then the lifted foliation to any finite covering space of X has a dense leaf as well.*

We now obtain the next result, which describes actions preserving a metric and a transverse parallelism. Its proof is based on some of the arguments found in [11].

PROPOSITION 6.2. *Suppose that the G-action on M has a dense orbit and preserves a pseudo-Riemannian metric. Also assume that M is compact. If the foliation by G-orbits is nondegenerate (with respect to the pseudo-Riemannian metric) and carries a transverse parallelism, then there exist*

1) *a finite covering map $\widehat{M} \to M$,*
2) *a connected Lie group H, and*
3) *a discrete cocompact subgroup $\Gamma \subset G \times H$ such that $G\Gamma$ is dense in $G \times H$,*

for which the G-action on M lifts to $\widehat{M}$ so that $\widehat{M}$ is G-equivariantly diffeomorphic to $(G \times H)/\Gamma$. Furthermore, if G has finite center and real rank at least 2, then we can assume that $G \times H$ is a finite center isotypic semisimple Lie group and Γ is an irreducible lattice.

Proof. By Lemma 4.1, the transverse parallelism to the G-orbits yields a G-invariant trivialization of $L(T\mathcal{O}^{\perp})$. Hence, there is a family of G-invariant sections of $T\mathcal{O}^{\perp}$, say $X_1, \dots, X_k$, that defines a basis of $T\mathcal{O}^{\perp}$ on every fiber. Let us consider the G-invariant Riemannian metric on $T\mathcal{O}^{\perp}$ for which these vector fields are orthonormal at every point. Because of the orthogonal decomposition $TM = T\mathcal{O} \oplus T\mathcal{O}^{\perp}$, if we replace the metric on $T\mathcal{O}^{\perp}$ induced from M with the G-invariant Riemannian metric thus defined from the parallelism, then we obtain a pseudo-Riemannian metric h on M, which is G-invariant, defines the same orthogonal complement $T\mathcal{O}^{\perp}$ to the orbits and such that this orthogonal complement is Riemannian. In the rest of this proof we will consider M endowed with this new metric h. Hence, by Corollary 3.3 there is a simply connected homogeneous Riemannian manifold $\widetilde{N}$ and a discrete cocompact subgroup $\Gamma \subset G \times Iso_0(\widetilde{N})$ for which there is a G-equivariant finite covering map

$$\widehat{M} = (G \times \widetilde{N})/\Gamma \to M.$$

Note that by the proof of Corollary 3.3, the normal bundle $T\mathcal{O}^{\perp}$ is integrable. In particular, the vector fields X_i given above satisfy:

$$[X_i, X_j] = \sum_{r=1}^{k} f_{ij}^r X_r$$

for some smooth functions f_{ij}^r defined on M. The G-invariance of the parallelism then implies that the functions f_{ij}^r are G-invariant and so constant because of the existence of a dense G-orbit. We conclude that the linear span over $\mathbb{R}$ of the vector fields $X_1, \ldots, X_k$ is a Lie algebra.

On the other hand, the fact that the vector fields X_i are preserved by the G-action is easily seen to imply that such fields are foliate in the notation of [5]. Hence, the parallelism $X_1, \ldots, X_k$ defines a transverse Lie structure for the foliation by G-orbits in M. Clearly, this transverse Lie structure induces a corresponding one on $\widehat{M}$. Let H be a simply connected Lie group that models the transverse Lie structure on $\widehat{M}$ and consider a corresponding development $D : \widetilde{G} \times \widetilde{N} \to H$. In particular, D is a submersion whose fibers have connected components given precisely by subsets of the form $\widetilde{G} \times \{x\}$. Hence, we conclude that $D_{\{e\} \times \widetilde{N}} : \widetilde{N} \to H$ defines a local diffeomorphism.

By the proofs of Corollary 3.3 and Theorem 1.1, the manifold $\widetilde{N}$ is the (isometric) universal covering space of a leaf N for the foliation in M defined by $T\mathcal{O}^\perp$. Hence, if we let $\widetilde{X}_1, \ldots, \widetilde{X}_k$ be the pull-backs to $\widetilde{N}$ of the restrictions of the fields $X_1, \ldots, X_k$ to N, then the induced fields on $\widetilde{G} \times \widetilde{N}$ (which we will denote with the same symbols) define the transverse Lie structure on $\widetilde{G} \times \widetilde{N}$. Furthermore, from the previous construction of the metric h, the metric on $\widetilde{N}$ is given by the condition of the fields $\widetilde{X}_1, \ldots, \widetilde{X}_k$ being orthonormal at every point. In particular, if we let $\mathfrak{h}$ be the Lie algebra of H, then there is a basis $v_1, \ldots, v_k$ of $\mathfrak{h}$ such that the development D maps the vector field $\widetilde{X}_i$ into v_i, for every $i = 1, \ldots, k$. Without loss of generality we will assume that the transverse H-structures are modeled by taking H with its right translations, and so transverse Lie parallelisms are modeled on right-invariant vector fields. With this convention, each v_i is considered as a right-invariant vector field on H.

From the above remarks, if we endow H with the right-invariant Riemannian metric for which the vector fields $v_1, \ldots, v_k$ are orthonormal at every point, then $D_{\{e\} \times \widetilde{N}} : \widetilde{N} \to H$ is a local isometry. By corollary 29 on page 202 of [9] and since $\widetilde{N}$ is complete, we conclude that $D_{\{e\} \times \widetilde{N}} : \widetilde{N} \to H$ is an isometry and so it induces in $\widetilde{N}$ a Lie group structure with respect to which it is an isomorphism. Hence, we can replace the Riemannian manifold $\widetilde{N}$ with H carrying the above right invariant Riemannian metric and assume that the natural projection $\widetilde{G} \times H \to H$ is a development for the transverse Lie structure on $\widehat{M}$. Moreover, by the definition of $\widehat{M}$, the space $G \times \widetilde{N}$ covers $\widehat{M}$, and so we can assume that the natural projection $\pi : G \times H \to H$ is also a development for the transverse Lie structure on $\widehat{M}$ with a corresponding holonomy representation $\widehat{\rho} : \Gamma \to H$.

Since $\Gamma \subset G \times Iso_0(\widetilde{N}) = G \times Iso_0(H)$, if we choose $\gamma \in \Gamma$, then we can write $\gamma = (R_{\gamma_1}, \gamma_2)$, where $\gamma_1 \in G$ and $\gamma_2 \in Iso_0(H)$. And so, the $\widehat{\rho}$-equivariance of π yields for every $(g, x) \in G \times H$:

$$\gamma_2(x) = \pi(g\gamma_1, \gamma_2(x)) = \pi((g, x)\gamma) = \pi(g, x)\widehat{\rho}(\gamma) = xy$$

where $y \in H$. It follows that γ_2 is given by a right translation by an element in H, and so we have $\Gamma \subset G \times R(H) = G \times H$. Since $\widehat{M} = (G \times H)/\Gamma \to M$ is a finite covering, by Lemma 6.1 we conclude that $\widehat{M}$ has a dense G orbit and so $G\Gamma$ is dense in $G \times H$ as in the proof of Proposition 4.2. This concludes the proof of the first part of the statement.

For the last claim, if G has finite center and real rank at least 2, then H is semisimple by the main results from [16]. Also, by modding out by a suitable central group we can assume that H has finite center (see [10]). The arguments at the end of Section 6 from [10] also prove that Γ is an irreducible lattice. Finally, recall that an irreducible lattice in a semisimple Lie group can only exist if the group is isotypic. $\qquad\square$

We can now prove our characterization of actions with a transverse parallelism.

Proof of Theorem 1.3. First, let us assume that (1) holds and consider the finite covering $\widehat{M} = (G \times H)/\Gamma \to M$. As in the proof of Theorem 1.2 we conclude that $\widehat{M}$ has a dense G-orbit. Note that the right-invariant vector fields on H define both a transverse Lie structure on $G \times H$ for the foliation given by the factor G, and a right H-invariant Riemannian metric on H. If we consider the product pseudo-Riemannian metric on $G \times H$ using a bi-invariant metric on G, then we also obtain a pseudo-Riemannian metric for which the foliation given by the factor G is nondegenerate. Both of these geometric structures on $G \times H$ are left G-invariant and right Γ-invariant and so descend to corresponding G-invariant geometric structures on $\widehat{M}$ thus establishing (2).

If we now assume (2), then Proposition 6.2 applied to $\widehat{M}$ yields (1). The last claim is also a consequence of Proposition 6.2. $\qquad\square$

7. Proof of Theorem 1.4

Let us consider G and M as in Proposition 2.3 with $S \subset M$ a dense subset provided by this result and Remark 2.4. With the notation of Proposition 2.3, the representation $\lambda_x \circ \rho_x$ leaves invariant the subspace $T_x \mathcal{O}^\perp$ thus defining its

$\mathfrak{g}$-module structure. This induces a homomorphism $\mathfrak{g} \to \mathfrak{so}(T_x\mathcal{O}^\perp)$ obtained by restricting $\lambda_x \circ \rho_x$ to the submodule $T_x\mathcal{O}^\perp$. By composition with such homomorphism, we can consider Ω_x obtained from Lemma 2.5 as a map $\wedge^2 T_x\mathcal{O}^\perp \to \mathfrak{so}(T_x\mathcal{O}^\perp)$.

For a vector space W with inner product $\langle \cdot, \cdot \rangle$, we will say that a bilinear map $T : W \times W \to \mathfrak{gl}(W)$ is of curvature type if it satisfies the following conditions for every $x, y, z, v, w \in W$:

1) $T(x, y) = -T(x, y)$,
2) $\langle T(x, y)v, w \rangle = -\langle v, T(x, y)w \rangle$,
3) $T(x, y)z + T(y, z)x + T(z, x)y = 0$, and
4) $\langle T(x, y)v, w \rangle = \langle T(v, w)x, y \rangle$.

Note that (1) and (2) together are equivalent to T inducing a map $\wedge^2 W \to \mathfrak{so}(W)$. Also, by the proof of proposition 36 on page 75 of [9] we know that (1), (2), and (3) together imply (4).

With the above notation, we have the following result.

LEMMA 7.1. *Let G, M and $S \subset M$ be as in Proposition 2.3 and Remark 2.4. For every $x \in S$, define the bilinear operation $[\cdot, \cdot]_0$ in T_xM by the assignments:*

- $[X_x^*, Y_x^*]_0 = [X, Y]_x^*$, *for every $X_x^*, Y_x^* \in T_x\mathcal{O}$, $(X, Y \in \mathfrak{g})$,*
- $[X_x^*, v]_0 = -[v, X_x^*]_0 = X(v)$ *for every $X_x^* \in T_x\mathcal{O}, v \in T_x\mathcal{O}^\perp$, $(X \in \mathfrak{g})$, and*
- $[v_1, v_2]_0 = \Omega_x(v_1, v_2)_x^*$ *for every $v_1, v_2 \in T_x\mathcal{O}^\perp$.*

Then, $[\cdot, \cdot]_0$ yields a Lie algebra structure on T_xM if and only if Ω_x is of curvature type when considered as a map $\wedge^2 T_x\mathcal{O}^\perp \to \mathfrak{so}(T_x\mathcal{O}^\perp)$. In this case, the representation of $\mathfrak{g}$ in T_xM, given by Proposition 2.3(4), preserves the Lie algebra structure of T_xM.

Proof. Note that by the above remarks, Ω_x is of curvature type if and only if it satisfies the above condition (3). Also observe that $[\cdot, \cdot]_0$ always defines a skew-symmetric bilinear form in T_xM. Hence, for the first claim, we need to show that $[\cdot, \cdot]_0$ satisfies the Jacobi identity if and only if Ω_x satisfies (3).

For the Jacobi identity to hold we only need to verify the following cases.

- $u_1, u_2, u_3 \in T_x\mathcal{O}$. Note that $[\cdot, \cdot]_0$ maps $T_x\mathcal{O} \times T_x\mathcal{O}$ into $T_x\mathcal{O}$ in such a way that it defines a skew-symmetric operation that corresponds to the Lie brackets of $\mathfrak{g}$ under the natural isomorphism $\mathfrak{g} \to T_x\mathcal{O}$ given by $X \mapsto X_x^*$. Hence, the Jacobi identity always holds in this case.
- $u_1, u_2 \in T_x\mathcal{O}$ and $u_3 \in T_x\mathcal{O}^\perp$. In this case we can write $u_1 = X_x^*$ and $u_2 = Y_x^*$ for some $X, Y \in \mathfrak{g}$. Then, by the definition of $[\cdot, \cdot]_0$

$$[[X_x^*, Y_x^*]_0, u_3]_0 = [[X, Y]_x^*, u_3]_0 = [X, Y](u_3)$$

$$= X(Y(u_3)) - Y(X(u_3))$$

$$= X([Y_x^*, u_3]_0) - Y([X_x^*, u_3]_0)$$

$$= [X_x^*, [Y_x^*, u_3]_0]_0 - [Y_x^*, [X_x^*, u_3]_0]_0,$$

which proves that the Jacobi identity holds in this case in general.

- $u_1 \in T_x\mathcal{O}$ and $u_2, u_3 \in T_x\mathcal{O}^\perp$. We can now choose $X \in \mathfrak{g}$ such that $u_1 = X_x^*$. Hence, using from Lemma 2.5 the fact that Ω_x is a homomorphism of $\mathfrak{g}$-modules, we obtain

$$[X_x^*, [u_2, u_3]_0]_0 = [X_x^*, \Omega_x(u_2, u_3)_x^*]_0 = [X, \Omega_x(u_2, u_3)]_x^*$$

$$= \Omega_x(X(u_2), u_3)_x^* + \Omega_x(u_2, X(u_3))_x^*$$

$$= \Omega_x([X_x^*, u_2]_0, u_3)_x^* + \Omega_x(u_2, [X_x^*, u_3]_0)_x^*$$

$$= [[X_x^*, u_2]_0, u_3]_0 + [u_2, [X_x^*, u_3]_0]_0,$$

which yields again the Jacobi identity without extra conditions.

- $u_1, u_2, u_3 \in T_x\mathcal{O}^\perp$. The definition of $[\cdot, \cdot]_0$ yields now:

$$[[u_1, u_2]_0, u_3]_0 = [\Omega_x(u_1, u_2)_x^*, u_3]_0 = \Omega_x(u_1, u_2)(u_3),$$

and so the Jacobi identity is satisfied in this case exactly when Ω_x satisfies condition (3).

The above proves the equivalence in the statement, and so it remains to obtain the last claim. For this we need to show that

$$X([u_1, u_2]_0) = [X(u_1), u_2]_0 + [u_1, X(u_2)]_0,$$

for every $X \in \mathfrak{g}$ and $u_1, u_2 \in T_x M$. This is now dealt with through the following cases, which just basically apply the definitions involved and properties already considered.

- $u_1, u_2 \in T_x\mathcal{O}$. Then, we can write $u_1 = Y_x^*, u_2 = Z_x^*$ for some $Y, Z \in \mathfrak{g}$ and

$$X([Y_x^*, Z_x^*]_0) = X([Y, Z]_x^*) = [\rho_x(X), [Y, Z]^*]_x = [X, [Y, Z]]_x^*$$

$$= [[X, Y], Z]_x^* + [Y, [X, Z]]_x^*$$

$$= [[X, Y]_x^*, Z_x^*]_0 + [Y_x^*, [X, Z]_x^*]_0$$

$$= [[\rho_x(X), Y^*]_x, Z_x^*]_0 + [Y_x^*, [\rho_x(X), Z^*]_x]_0$$

$$= [X(Y_x^*), Z_x^*]_0 + [Y_x^*, X(Z_x^*)]_0.$$

- $u_1 \in T_x\mathcal{O}$ and $u_2 \in T_x\mathcal{O}^\perp$. Now we can write $u_1 = Y_x^*$ for $Y \in \mathfrak{g}$ and

$$X([Y_x^*, u_2]_0) = X(Y(u_2)) = [X, Y](u_2) + Y(X(u_2))$$
$$= [[X, Y]_x^*, u_2]_0 + [Y_x^*, X(u_2)]_0$$
$$= [[\rho_x(X), Y^*]_x, u_2]_0 + [Y_x^*, X(u_2)]_0$$
$$= [X(Y_x^*), u_2]_0 + [Y_x^*, X(u_2)]_0.$$

- $u_1, u_2 \in T_x\mathcal{O}^\perp$. We now have

$$X([u_1, u_2]_0) = X(\Omega_x(u_1, u_2)_x^*) = [\rho_x(X), \Omega_x(u_1, u_2)^*]_x$$
$$= [X, \Omega_x(u_1, u_2)]_x^*$$
$$= \Omega_x(X(u_1), u_2)_x^* + \Omega_x(u_1, X(u_2))_x^*$$
$$= [X(u_1), u_2]_0 + [u_1, X(u_2)]_0. \qquad \square$$

The following result will allow us to prove Theorem 1.4. It also provides an explicit description of the Lie algebra structure considered in Theorem 1.4.

THEOREM 7.2. *Suppose that G has finite center and real rank at least 2, and that the G-action on M preserves a finite-volume complete pseudo-Riemannian metric. Also assume that G acts ergodically on M and that the foliation by G-orbits is nondegenerate. Denote with L the algebraic hull for the G-action on the bundle $L(T\mathcal{O}^\perp)$ and with $\mathfrak{l}$ its Lie algebra. In particular, there is an embedding of Lie algebras $\mathfrak{l} \hookrightarrow \mathfrak{so}(p, q)$, where (p, q) is the signature of the metric of M restricted to $T\mathcal{O}^\perp$. If this embedding is surjective, then one of the following occurs:*

1) *The conclusion of Theorem 1.1 holds, or*
2) *G is locally isomorphic to $SO_0(p, q)$, $\dim(M) = (p + q)(p + q + 1)/2$, for some $x \in M$ the bilinear map Ω_x is nonzero and of curvature type and the Lie algebra structure on $T_x M$ obtained from Lemma 7.1 is isomorphic to either $\mathfrak{so}(p, q + 1)$ or $\mathfrak{so}(p + 1, q)$.*

Proof. Since the G-orbits are nondegenerate, G preserves a pseudo-Riemannian metric on $T\mathcal{O}^\perp$. Hence, the algebraic hull of $L(T\mathcal{O}^\perp)$ for the G-action can be embedded into the structure group $O(p, q)$ for such metric, where (p, q) is the signature of $T\mathcal{O}^\perp$. This yields the first claim.

Let us now assume that the induced embedding $\mathfrak{l} \hookrightarrow \mathfrak{so}(p, q)$ is surjective. This implies that L is a finite-index subgroup of $O(p, q)$. By section 3 of [17], we can write $L = ZS$ where S is semisimple without compact factors, Z is compact and centralizes S, and the product is almost direct. In particular, we have a direct product $\mathfrak{l} = \mathfrak{z} \times \mathfrak{s}$. Note that for $p + q \leq 2$ the conclusion of Theorem 1.1 holds by Corollary 3.1(1). Hence, we can assume from now on that $p + q \geq 3$. Then, the only cases in which $\mathfrak{so}(p, q)$ is not simple is for $\mathfrak{so}(4) \cong \mathfrak{so}(3) \times \mathfrak{so}(3)$ and $\mathfrak{so}(2, 2) \cong \mathfrak{sl}(2, \mathbb{R}) \times \mathfrak{sl}(2, \mathbb{R})$. And so, one of the following holds:

- $\mathfrak{s} = 0$ and $\mathfrak{z} \cong \mathfrak{so}(p, q)$ is compact, or
- $\mathfrak{z} = 0$ and $\mathfrak{s} \cong \mathfrak{so}(p, q)$ is noncompact.

In the first case, we conclude that $\mathfrak{so}(T_x\mathcal{O}^\perp)$ is compact for every $x \in M$, thus implying that the conclusion of Theorem 1.1 follows by Corollary 3.1(1). In particular, we can assume that $\mathfrak{l} = \mathfrak{s} \cong \mathfrak{so}(p, q)$. Also, by the arguments in section 3 of [17], there is a surjection $\mathfrak{g} \to \mathfrak{s}$ of Lie algebras, which implies that $\mathfrak{g} \cong \mathfrak{so}(p, q)$. Hence, G is locally isomorphic to $SO_0(p, q)$ and because of the decomposition $TM = T\mathcal{O} \oplus T\mathcal{O}^\perp$ we also have $\dim(M) = \dim(SO_0(p, q)) + \dim(T_x\mathcal{O}^\perp) = \dim(SO_0(p, q)) + p + q = (p + q)(p + q + 1)/2$.

On the other hand, for $S \subset M$ given as in Proposition 2.3, if $\Omega|_S = 0$, then the conclusion of Theorem 1.1 holds by Lemma 2.5(2). Then, we can choose $x \in S$ such that $\Omega_x \neq 0$. In particular, by Lemma 2.5(1) the $\mathfrak{g}$-module $T_x\mathcal{O}^\perp$ is a nontrivial one. If we consider the representation $\mathfrak{g} \to \mathfrak{so}(T_x\mathcal{O}^\perp)$ that defines the $\mathfrak{g}$-module structure of $T_x\mathcal{O}^\perp$ from Proposition 2.3, then the above remarks show that this representation is an isomorphism. We conclude that $T_x\mathcal{O}^\perp$ is a $\mathfrak{g}$-module isomorphic to $\mathbb{R}^{p,q}$ with respect to the isomorphism $\mathfrak{g} \to \mathfrak{so}(T_x\mathcal{O}^\perp)$ thus obtained.

For $\langle \cdot, \cdot \rangle_{p,q}$ the metric in $\mathbb{R}^{p,q}$ preserved by $\mathfrak{so}(p, q)$ we have the following:

Claim: For every $c \in \mathbb{R}$, the map $T_c : \wedge^2 \mathbb{R}^{p,q} \to \mathfrak{so}(p, q)$ given by

$$T_c(u \wedge v) = c \langle v, \cdot \rangle_{p,q} u - c \langle u, \cdot \rangle_{p,q} v,$$

where $u, v \in \mathbb{R}^{p,q}$ is a well-defined homomorphism of $\mathfrak{so}(p, q)$-modules. Moreover, these maps exhaust all the $\mathfrak{so}(p, q)$-module homomorphisms $\wedge^2 \mathbb{R}^{p,q} \to \mathfrak{so}(p, q)$.

The only nontrivial part of the claim is the last statement, which we will now prove. Let $T : \wedge^2 \mathbb{R}^{p,q} \to \mathfrak{so}(p, q)$ be a homomorphism of $\mathfrak{so}(p, q)$-modules. Consider the map $T \circ T_1^{-1}$, which is a homomorphism of $\mathfrak{so}(p, q)$-modules $\mathfrak{so}(p, q) \to \mathfrak{so}(p, q)$. Hence, for $n = p + q$ and by complexifying, the map $T \circ T_1^{-1}$ yields a homomorphism $\widehat{T} : \mathfrak{so}(n, \mathbb{C}) \to \mathfrak{so}(n, \mathbb{C})$ of $\mathfrak{so}(n, \mathbb{C})$-modules such that $\widehat{T}|_{\mathfrak{so}(p,q)} = T \circ T_1^{-1}$. In particular, $\widehat{T}(\mathfrak{so}(p, q)) \subset \mathfrak{so}(p, q)$ and so $\widehat{T}$ commutes with the conjugation σ of $\mathfrak{so}(n, \mathbb{C})$ whose fixed point set is $\mathfrak{so}(p, q)$, that is, $\widehat{T} \circ \sigma = \sigma \circ \widehat{T}$. Note that since $\mathfrak{g} \cong \mathfrak{so}(p, q)$ is simple with real rank at least 2, the Lie algebra $\mathfrak{so}(n, \mathbb{C})$ is simple as well. Hence, the irreducibility of

$\mathfrak{so}(n, \mathbb{C})$ as $\mathfrak{so}(n, \mathbb{C})$-module implies, by Schur's Lemma, that there is a complex number c such that $\widehat{T} = c\mathrm{Id}_{\mathfrak{so}(n,\mathbb{C})}$. But then, the condition $\widehat{T} \circ \sigma = \sigma \circ \widehat{T}$ implies that c is real and so $T = T_c$.

By the Claim and Lemma 2.5(1), with respect to the above-established isomorphisms $\mathfrak{g} \to \mathfrak{so}(T_x\mathcal{O}^\perp)$ and $T_x\mathcal{O}^\perp \cong \mathbb{R}^{p,q}$, we conclude that the map Ω_x corresponds to a homomorphism T_c for some real $c \neq 0$. It is straightforward to check that T_c is of curvature type; in fact, it yields the curvature of the pseudo-Riemannian manifolds of constant sectional curvature c and signature (p, q) (see corollary 43 on page 80 of [9]). Hence, Lemma 7.1 applies and it provides a Lie algebra structure on T_xM. It remains to show that such structure is isomorphic to either $\mathfrak{so}(p, q+1)$ or $\mathfrak{so}(p+1, q)$.

By our identifications, the Lie algebra structure on T_xM is isomorphic to the one obtained on $\mathfrak{so}(p, q) \oplus \mathbb{R}^{p,q}$ from the map T_c with the Lie brackets $[\cdot, \cdot]_c$ defined with formulas similar to those in Lemma 7.1 replacing Ω_x with T_c. Also, it is straightforward to show that the map given by

$$(\mathfrak{so}(p, q) \oplus \mathbb{R}^{p,q}, [\cdot, \cdot]_c) \to \mathfrak{so}(\mathbb{R}^{p+q+1}, I_{p,q}(c))$$

$$X + u \to \begin{pmatrix} X & cu \\ u^t I_{p,q} & 0 \end{pmatrix},$$

is an isomorphism of Lie algebras, where

$$I_{p,q}(c) = \begin{pmatrix} I_{p,q} & 0 \\ 0 & c \end{pmatrix}.$$

Since $\mathfrak{so}(\mathbb{R}^{p+q+1}, I_{p,q}(c))$ is clearly isomorphic to either $\mathfrak{so}(p, q+1)$ or $\mathfrak{so}(p+1, q)$ this yields (2) from our statement. $\qquad\square$

We can now complete the following proof.

Proof of Theorem 1.4. Since the hypotheses of Theorem 7.2 are a subset of those of Theorem 1.4, the conclusions of the former hold. Since M is weakly irreducible its universal covering space cannot split isometrically and so part (2) of Theorem 7.2 necessarily holds at some $x \in M$. By the definition of the Lie algebra structure in T_xM from Lemma 7.1 we conclude that (1) and (2) of Theorem 1.4 are satisfied.

Finally we observe that, by the proof of Theorem 7.2, we can assume that $x \in S$ where $S \subset M$ is given by Proposition 2.3. Consider the Lie algebra $\mathfrak{g}(x)$ of local Killing fields provided by Proposition 2.3. Then, the last claim of Lemma 7.1 and the definition of the representations involved implies the last claim of Theorem 1.4 for the Lie algebra $\mathfrak{g}(x)$. $\qquad\square$

References

[1] A. Candel and R. Quiroga-Barranco, Gromov's centralizer theorem, Geom. Dedicata **100** (2003), 123–155.

[2] M. Gromov, Rigid transformations groups, in Géométrie différentielle, Colloque Géométrie et Physique de 1986 en l'honneur de André Lichnerowicz (D. Bernard and Y. Choquet-Bruhat, eds.), Hermann, Paris, 1988, 65–139.

[3] F. G. Hernández-Zamora, Isometric splitting for actions of simple Lie groups on pseudo-Riemannian manifolds, Geom. Dedicata **109** (2004), 147–163.

[4] N. Kowalsky, Noncompact simple automorphism groups of Lorentz manifolds and other geometric manifolds, Ann. of Math. (2) **144** (1996), no. 3, 611–640.

[5] P. Molino, Riemannian foliations, translated from the French by Grant Cairns, with appendices by Cairns, Y. Carrire, É. Ghys, E. Salem, and V. Sergiescu, Progress in Mathematics, 73, Birkhäuser Boston, Boston, MA, 1988.

[6] K. Nomizu, On local and global existence of Killing vector fields, Ann. of Math. **72** (1960), 105–120.

[7] G. Olafsson and R. Quiroga-Barranco, Isometric actions of $SO_0(p, q)$ on low dimensional weakly irreducible pseudo-Riemannian manifolds, in preparation.

[8] B. O'Neill, The fundamental equations of a submersion, Michigan Math. J. **13** (1966), 459–469.

[9] B. O'Neill, Semi-Riemannian geometry. With applications to relativity, Pure and Applied Mathematics, 103. Academic Press, New York, 1983.

[10] R. Quiroga-Barranco, Isometric actions of simple Lie groups on pseudo-Riemannian manifolds, Ann. of Math. (2) **164** (2006), no. 6, 941–969.

[11] R. Quiroga-Barranco, Arithmeticity of totally geodesic Lie foliations with locally symmetric leaves, Asian J. Math. 12 (2008), no. 3, 289–297.

[12] I. M. Singer, Infinitesimally homogeneous spaces, Comm. Pure Appl. Math. 13 (1960), 685–697.

[13] J. Szaro, Isotropy of semisimple group actions on manifolds with geometric structure, Amer. J. Math. **120** (1998), 129–158.

[14] A. Zeghib, On affine actions of Lie groups, Math. Z. **227** (1998), no. 2, 245–262.

[15] R. J. Zimmer, Actions of semisimple groups and discrete subgroups, Proceedings of the International Congress of Mathematicians, Vol. 1, 2 (Berkeley, CA, 1986), 1247–1258, American Mathematical Society, Providence, RI, 1987.

[16] R. J. Zimmer, Arithmeticity of holonomy groups of Lie foliations, J. Amer. Math. Soc. **1** (1988), no. 1, 35–58.

[17] R. J. Zimmer, On the algebraic hull of an automorphism group of a principal bundle, Comment. Math. Helv. **65** (1990), no. 3, 375–387.

[18] R. J. Zimmer, Automorphism groups and fundamental groups of geometric manifolds, in Differential Geometry: Riemannian Geometry (Los Angeles, CA, 1990), 693–710, Proc. Sympos. Pure Math. 54, Part 3, American Mathematical Society, Providence, RI, 1993.

8

SHMUEL WEINBERGER

SOME REMARKS INSPIRED BY THE C⁰ ZIMMER PROGRAM

TO BOB ZIMMER, WITH FRIENDSHIP, GRATITUDE, AND ADMIRATION

In the mid-1980s, when I had just arrived in Chicago, Bob Zimmer was lecturing on his recent work showing how perturbations of isometric actions of Kazhdan groups were dynamically simple (preserve a measurable Riemannian metric) [Zi1]. The heady mixture of the beauty of the results, the audacity of the vision (this is really 0% of what should be true.. . .[1]), and Bob's charisma and charm attracted me so that I still cannot but help think about the problem of what large discrete groups of Cat-isomorphisms of a manifold can conceivably look like.

Alas, my ardor has not been well rewarded in this pursuit. In this paper I would like to make some comments and raise some questions about the case of Cat = Top, that is, the group of homeomorphisms of manifolds. Some of these remarks are rather old, but I will give some details if I am not aware of a discussion elsewhere—while a few are new (at least to me). I hope that this paper spurs more exploration of the fascinating C⁰ aspect of this nexus of problems.

Although this note is short, it has a bit of structure. The beginning is about groups that don't act, the middle is about lattices and hyperbolic groups and quite tame actions, and the end has an ergodic aspect and relates directly to a paper of Lashof and Zimmer [LZ]. There is also an appendix about deforming homotopy actions to actions at the cost of stabilization.

Besides Bob Zimmer, who is the obvious inspiration for this paper, I would also like to thank Kevin Whyte, and the 3Fs: Benson Farb, Steve Ferry, and David Fisher for many conversations that have strongly influenced my perspective. Courtney Thatcher's thesis on cyclic group actions on products of spheres has some overlap with the perspective in Section 4.

---

1. Maybe now we are above the 0% mark, at least in this direction, thanks to the wonderful papers [Be] and [FiM].

1. Some Groups That Don't Act

Proposition 1. The infinite special linear group SL(Z) does not act topologically, nontrivially, on any compact manifold, or indeed on any manifold whose homology with coefficients in a field of positive characteristic is finitely generated.

[Zi2] only mentions the compact case. The more general case is proven simultaneously (and is necessary for the proof, because one must analyze complements of fixed sets as well as fixed sets in the coming argument), and is motivated by the recent paper [ABJLMS], as is Proposition 2 below (see also [FiS] that was written very recently and has some quite deep further observations about the volume-preserving case).

Proof. The proof is based on Smith theory considerations. A p-group acting on a mod p-homology manifold has fixed set that is also a mod p-homology manifold of lower dimension. Moreover, the sum of the Betti numbers of the fixed sets can be at most as large for the fixed set as for the original manifold. As a formal consequence, by considerations of the open strata, if $(Z/p)^k$ is acting effectively, for k large enough, there must be a very large $(Z/p)^l$ acting freely on some mod p-homology manifold with bounded Betti numbers. However, if l is large enough, this action will automatically have a large kernel for its action on homology (GL_r doesn't have that much p-torsion). Consequently, one can compute the homology of the quotient space by means of the usual Serre spectral sequence. Now if the rank of this kernel is greater than the homology of the space, one sees that the quotient is infinite-dimensional. Putting this all together, we get elements of order p that are acting trivially on M. Since SL(Z) is the normal closure of such elements, we see that the action is actually trivial. □

Proposition 2. There are finitely presented groups that do not act nontrivially on any compact manifold.

Proof. The idea is to produce finitely presented groups that contain high-rank finite subgroups and that are normally generated by enough elements of order p. Many examples are constructed in [ABJLMS]; it might be worth noting that old ideas related to the unsolvability of the triviality problem (a la Rabin) can be used for this. We follow the exposition in [We1]. Suppose that π is an effectively acting group with arbitrarily large elementary p-groups generated by specific words we know. We can make infinitely many HNN extensions, each one conjugating one of these elements to another, so that they are now all conjugate in a larger group, called π'. We denote by g one of these elements of order p. Now, Higman embed π' into a group π''. By taking an amalgamated free product $(\pi'' * \mathbf{Z})^*\mathbf{Z}^*\mathbf{Z}A$ where A is a knot group (say the trefoil group for concreteness), $\mathbf{Z}^*\mathbf{Z}$ is a free subgroup of A that contains a meridian as its first generator, and $\mathbf{Z}^*\mathbf{Z}$ is embedded into $(\pi'' * \mathbf{Z})$ so that the first generator goes to $[g, t]$ in the obvious notation. We thus have arranged that any homomorphism that sends any of the elements of order p in our list to e kills the whole group. $\qquad\square$

Whether one prefers Proposition 1 or 2 is a matter of taste. Of course these both beg the question of whether there is a torsion-free group that does not act continuously on any manifold. We note:

Proposition 3. Any countable group of finite homological dimension does have a continuous action on some sphere.

Proof. If $B\pi$ is a finite-dimensional complex, then it can be thickened to a manifold. The universal cover of this manifold is contractible, and if not a Euclidean space, its product with $\mathbf{R}$ is (by a well-known theorem of Stallings). This manifold has a free π action. Its one point compactification is a sphere, and it has a π action with a single fixed point. $\qquad\square$

This action is rather different than the usual ones we think of coming from symmetric spaces and lattices. It has no dynamics!

A smooth version of Proposition 3 seems unlikely (but I have no ideas about how to prove such a thing). On the other hand, the hypothesis is far from necessary. Every manifold (of dimension >0) has an action of a free abelian group of infinite rank. On the other hand, presumably the obvious versions of congruence subgroups of SL(Z) should not act, and we know nothing about this.

Problem: What groups are discrete subgroups of $\mathrm{Homeo}(D^n, \mathrm{rel}\ \partial)$? Needless to say, they are torsion free (and they are subgroups of $\mathrm{Homeo}(M)$ for any m-manifold for $m \geq n$).

2. Actions of Lattices

Of course, Zimmer actually suggests that $SL_n(Z)$ shouldn't act on any manifold of dimension $< n{-}1$ (except via finite quotients). In dimension n, the action on T^n should be rigid. Indeed, the beautiful work of Hurder shows that, with enough smoothness, that action is infinitesimally rigid.

It seems worth pointing out that in C^0 there are many deformations possible. One deformed action is explained in [FS1], but the process is one that can be repeated infinitely often producing uncountably many new and different actions.

The technique here is "insertion" a la [KL]. Whenever one has a C^1 action with a fixed point (or indeed, finite orbits, or indeed, an invariant submanifold) one can glue in a tubular nighborhood acted on via the differential, and extend the action to the complement. Essentially one uses the diffeomorphism of the complement of a disk with the complement of a point inside $M - D = M - p$. In the neighborhood one can alternatively glue in the cone of the action on the sphere at infinity. These actions are quite different from each other or the original action, as one can see by examining orbits of points.

Now, if one wishes, one can insert many times, producing many rings around the fixed point. Also, one can make some of these rings "thick" by coning for a little while before doing a more standard linear insertion. Clearly, this builds uncountably many actions that differ only around a fixed point. (They are distinguished by one another by looking at the limit points of orbits. Note that for compact groups, the action given by insertion is topologically equivalent to the original action.)

But these "bull's-eye" structures can be inserted at will at smaller and smaller scales around any of the finite orbits associated to the rational points on the torus. As a result, we have beautiful moduli of "leopard spots," each of which is itself variable according to the bull's-eye patterns. Thus we have seen a much more beautiful structure than is asserted in the following:

Proposition 4. There are uncountably many nontopologically-conjugate $SL_n(Z)$ actions on T^n. They can all be deformed to the usual action.

These deformations are 1-parameter families of actions that start at the "exotic" action at time 0, and end at the standard action at time 1.

And all of these actions map equivariantly to the standard affine action.

These actions are just "modifications" of arithmetic actions. In its most vague form, the Zimmer program asserts that, in some sense,[2] all actions of high-rank lattices are built up out of such actions. Thus, Proposition 4 is just par for the course.

---

2. It is part of the program to figure out the correct sense.

There are other deformations that arise from decomposition space theory (aka "Bing topology"); I will give just one example and then move on.

Proposition 5. Let π be a countable group with a dense embedding in $SO(n)$. Then $SO(n)$ acts on S^{n+k} with quotient D^{1+k}. If one considers the induced action over $D^{1+k}/\sim$ where $\sim$ is a cell-like semicontinuous decomposition (these exist in profusion if $k > 1$), the total space is still S^{n+k}, but the new action of π is never equivalent to any of the other ones.

The equivalence of total spaces follows readily from Edwards's characterization of manifolds and the CE approximation theorem (of Siebenmann, which is a consequence of Edwards's theorem, as well), see [Da] and [We2] for such deformations. The actions are thus distinguished by the maximal Hausdorff quotients of their orbit spaces, which will be nonmanifold ANR homology manifolds.

Part of the Zimmer program, although it is far from the usual part where the actions are ergodic, seems to me to include considerations of free, and more generally, proper discontinuous actions. Of course for these, in the geometric setting, the basic theorem is Mostow rigidity—and superrigidity can be easily viewed as a description of the infinite covolume geometric actions that follow from the action. The C^0 part of this portion of the Zimmer program is then an extension of the well-known Borel conjecture.

On the topological side there is the fabulous and celebrated work of Farrell and Jones [FJ] that tells one that in the cocompact case, there are only the most obvious actions. For uniform lattices, it then follows that if one is interested in actions on Euclidean spaces, the dimension must be strictly larger than the homogeneous space, but there are interesting remarks that should be made about the proper discontinuous actions that exist in (1) the nonuniform case, (2) in dimensions larger than the original action, and (3) when there is torsion.

Regarding the first of these, it is clear that spirit of the Zimmer program demands that for any aspherical manifold with fundamental group Γ the dimension should be at least that of G/K where G is the Lie group in which Γ sits. It is the **Q-rank** that governs the difference between this dimension and the cohomological dimension of Γ $\dim(G/K)-cd(\Gamma)=Q-rk(\Gamma)$. When this is 1, it is not hard to give an ad hoc proof that there isn't a geometric realization (the fundamental group of the end is a subgroup, which is a Poincaré duality group of $\dim=\dim(G/K)-1$ and of infinite index). However, for higher Q-rank, I had been unable to settle it for a number of years, and was very happy when the papers [BKK] and [BF] came out and beautifully confirmed the Zimmer predictions that lower-dimensional examples do not exist.

Problem: In the minimal dimension (i.e., dim(G/K)), is there any useful way to parametrize the proper homotopy types of aspherical manifolds with fundamental group Γ?

First of all, there is an issue of the fundamental groups at infinity. It is very easy to build in any nonuniform case (of dim > 2) uncountably many modifications of the standard actions that are distinguished by the fundamental group system of the complements of compact sets (viewed as a prosystem of groups). This is a straightforward modification of constructions of contractible manifolds. However, such actions have proper comparison maps to the standard action, and are concordant to it (i.e., there is a degeneration of each of these to the standard action).

Much more interesting is the observation that Kevin Whyte made (when he was my student) that the manifolds with fundamental group $(F2)^k$ obtained by taking various products of thrice-punctured spheres and punctured tori are all different from each other even up to proper type. It would be quite interesting to get a systematic understanding in some sort of homological fashion. Here the problem should be much more doable when the Q-rank is large. Perhaps the answer is quadratic in part of the group homology when Q-rank > $\frac{1}{2}$ dim(G/K), cubic when the fraction is around $\frac{1}{3}$ dim(G/K), and so on by analogy to the work of Goodwillie, Klein, and Weiss [GKW]. Nevertheless, this is quite unclear to me at the moment and the discussion below of case (2) suggests that it could be much simpler than that.

If we assume that we have the correct proper homotopy type, then one might have a hope that the proper homotopy equivalence is properly homotopic to a homeomorphism, but this is rarely the case. As Stanley Chang and I noticed, when Q-rank > 2, there is always a finite cover where this fails [ChW].[3] Our obstruction is of exponent 2 and our examples are all virtually standard. On the other hand, whenever $H^{4i}(\Gamma; \mathbf{R}) \neq 0$ for some i, surgery arguments (e.g., like the ones in that paper that make use, of course, of the Borel-Serre picture of arithmetic manifolds [BoS][4]) produce (infinitely many) manifolds in the given proper homotopy type distinguished by pi, the i-th Pontrjagin class. These examples will not go away by passing to a finite-index subgroup. (In characteristic 0, cohomology can only get bigger on passing to a finite cover.) And, many examples then follow from known results (e.g., see Borel [Bo2]).

---

3. There is a good heuristic for proper rigidity when Q-rank=2. It is true when Q-rank<2 by the work of [FJ].

4. Borel-Serre theory implies that Q-rank>2 boils down to the fundamental group condition at infinity. (In [ChW] this is interpreted and verified using Margulis arithmeticity for all, even nonarithmetic, lattices.)

The situation of being above the cd in the uniform case (or the nonuniform case) is very similar to the issues involved in the proper nonuniform discussion we just had. However, in the uniform case we can be much more explicit about what occurs if, say, the dimension supercedes the cd by > 2 and we assume that the fundamental group system is trivial (i.e., equivalent to the constant system Γ). In that case, it is not hard to show that the manifold is the interior of a manifold with boundary, which by the theorem of Browder-Casson-Haefliger-Sullivan-Wall (see [We3]) will automatically be a topological block bundle over $K\backslash G/\Gamma$. The proper homotopy types are then parametrized by the homotopy classes of maps $[K\backslash G/\Gamma : \mathrm{BAut}(S^{c-1})]$ where c is the difference of dimensions. Rationally this is isomorphic to a group cohomology. For c even, the invariant is simply the Euler class of this bundle. When c is odd, a relatively straightforward calculation shows that it lies in $H^{2(c-1)}(K\backslash G/\Gamma ; \mathbf{Q})$ and all elements are realized.[5]

Note: as here c is the analogue of the **Q**-rank in the previous discussion, I currently hope that the answers might be simpler than one thinks in the nonuniform case. Including the results of classification[6] one gets:

Summary Proposition 6: The actions of a uniform lattice Γ on $K\backslash G \times \mathbf{R}^c$ that have appropriate fundamental group systems at ∞ (in their quotient) $\rightarrow$ $[K\backslash G/\Gamma:\mathrm{BTop}_c]$ if c > 2. Rationally[7] this can be calculated as a sum of group cohomology groups $H^t(K\backslash G/\Gamma; \mathbf{Q}) \oplus H^{4i}(K\backslash G/\Gamma ; \mathbf{Q})$ where $t = c$, if c is even, and $t = 2\,(c{-}1)$ if c is odd, and the sum is overall positive i.

When there is torsion in Γ the situation is much more complicated in several respects. Even in the situation of **cocompact** proper discontinuous actions there can be different failures of rigidity. Some examples appear in [CK]—and are based on failures of excision in algebraic and hermitian K-theory, that is, the functors Nil and UNil. When nonzero, these groups tend to be infinitely generated torsion groups.

---

5. The interpretation of this, not as well known as it should be, invariant is as follows: every odd-dimensional bundle (rationally) has a section that splits off a lower-dimensional bundle. While the Euler class of this complementary bundle is not uniquely determined, its square is.

6. See the discussion after Proposition 7 below.

7. Since there are no group structures here at least on the left, it is worth clarifying here that the map from left to right is finite to 1 and has image that contains a lattice in the target vector space.

Rather different examples, based on a connection to embedding theory and somewhat technical arguments involving stratified space theory, can be found in [We3, We4][8] for actions of crystallographic groups. These were extended in [Shi] to many other lattices. They can be infinitely generated nontorsion, for example, detected by analogues of the Alexander polynomial. However, all of these examples (e.g., the algebraic and the embedding theoretic) are virtually trivial in the sense that they become standard on passing to a sufficiently large cover that keeps all of the torsion. (When one passes to a cover, only some of the coefficients of the polynomial survive, so any particular example will be killed on passing to a suitable finite cover.) This can be proved by a geometric argument related to [SW] (the equivariant version of the α-approximation theorem [ChF]) when there are no fixed sets of subgroups that are included in one another in codimension < 3.[9]

It is worth noting that the above cohomological discussion about the proper analysis for nonuniform lattices can be married to the discussion of the role of the singularities (and embeddings) that come from torsion. In any case, suffice it to say that the analogue of the $\oplus H^{4i}(K\backslash G/\Gamma; \mathbf{Q})$ part of the previous theorem is a sum of similar terms, one for each stratum that has $\mathbf{G}$-rank > 2, and the form of the term is an equivariant K-group. As a result, nonuniform lattices of Q-rank > 2 with torsion almost never have proper rigidity.

In any case, all of these examples certainly can be viewed as "obtained from some procedure applied to the arithmetic examples"—although the exact procedures involved are somewhat involved and might be hard to visualize.

3. Leaving the Aspherical Setting

One of the audacious aspects of the Zimmer program is that the manifolds studied are general compact manifolds. Here is a theorem of this sort from [FW].

Theorem: If $M = X/\pi$ is a compact manifold so that $\pi = \pi^1(M)$ is torsion free and has no normal abelian subgroups, and the isometry group Iso(X) is not

---

8. We are just now getting to the point where we can give complete classifications of some proper cocompact actions that are not rigid. However, it is my feeling that the picture of this subject is currently too complicated to be able to give a survey here.

9. When there are codimension 2 situations, one can use counterexamples to the "Smith conjecture" on unknottedness of fixed sets of cyclic group actions on the sphere, (e.g., [Gi]) to build knot theoretic failures of rigidity that do not die in any finite cover.

a compact extension of π, then π is a lattice in a semisimple Lie group and there is a Riemannian fiber bundle M $\to$ K\G/π.

If π has torsion, one has to allow the possibility of "orbifibering" over an orbifold.

Note that all the fibers are isometric here, and that therefore the structural group of the bundle is compact. Moreover, the representations $\pi \to$ Iso(F) control the possible M's. (Taking a representation with infinite image gives an example where no intermediate covers are Riemannian products. Of course many of them are differentiable products.)

The following result, sketched in chapter 12 of [We3], frames our discussion.

Proposition 7. Suppose that M= X/π is a compact manifold whose fundamental group is a torsion-free lattice in a Lie group and that H*(X) is finitely generated. Then, unless dimM = 4, there is a map M $\to$ K\G/π so that the inverse image of every open ball is homeomorphic to X.

So from a slightly blurred perspective this map is like a fiber bundle map: rather than controlling the inverse image of each point, we control the inverse image of each little open ball. The relevant M's that occur here, though, are much richer than the ones that occur in the previous theorem. Rather than homomorphisms from π to a Lie group, the objects that occur in this proposition are much more algebraic topological. For example, when $X = S^k \times \mathbf{R}^n$, if one restricts attention just to the homotopically trivial fibrations, then the relevant classification is (independent of $k > 1^{10}$) the homotopy classes of maps [K\G/π : G/Top], where the space G/Top was analyzed completely by Sullivan-Kirby-Siebenmann; rationally it is $\oplus H^{4i}(\text{K}\backslash\text{G}/\pi \ ; \mathbf{Q})$. Incidentally, even when π has nontrivial homomorphisms to O($k + 1$) they never give rise to nontrivial elements in this cohomology.[11] These actions of π on $S^k \times \mathbf{R}^n$ are never "Lie theoretic."

Conjecture 7': If M is a manifold with torsion-free fundamental group M = $X/\pi (\pi = \pi_1 M)$ so that $H_*(X)$ is finitely generated, then there is an aspherical

---

10. This is due to G_c/Top_c stability as explained in [We3].

11. These cohomology classes are essentially Pontrjagin classes (that can be very high-dimensional in Top despite the low dimensionality of the "fiber"); the vanishing for representations is immediate from the Chern-Weil description of characteristic classes.

homology manifold N with fundamental group π, and for any such N, there is a map f: M $\to$ N so that for any small open set O $\subset$ N, $f^{-1}(O) < O \times X$.

There are several things that need to be explained about this conjecture. The first is that the maps allowed here, when N is a manifold, are exactly the maps produced in the previous theorem; moreover, despite allowing more general N's than K\G/π in the conjecture, by Quinn's work on resolution of homology manifolds [Q], the conjecture is a consequence of the theorem.

Proposition 8. The conjecture is true for $\pi \subset GL_n(\mathbf{R})$ discretely embedded and of finite type[12] (but not necessarily a lattice) and for hyperbolic groups, assuming that $cd(\pi) > 4$.

This follows the same lines as the previous proposition, aside from a few points. The first is that one must exclude, say, nonuniform lattices. More precisely, if π is the fundamental group of an aspherical manifold, it must satisfy Poincaré duality. That Poincaré duality follows, assuming finiteness of $K(\pi, 1)$, from a manifold's having a finite universal cover as shown in [BW1] (as is the coarseness of satisfying Poincaré duality among such groups)[13]. Now, the following is an immediate consequence of [BFMW1]:

Proposition 9. If π is a Poincaré duality group of dimension > 4 and satisfies the Borel conjecture (see e.g., [FRR][14]), then there is an aspherical homology manifold with fundamental group π.

---

12. I believe that if B π has some infinite skeleton, then there cannot be a manifold such as our M, but I haven't excluded this yet.

13. Nonuniform lattices never satisfy Poincaré duality. They satisfy Bieri-Eckmann duality with an infinitely generated dualizing module.

14. The Borel conjecture required here is the following standard generalization of the most commonly stated one: (ignoring orientation issues for simplicity) suppose that M is an aspherical manifold and f:W $\to$ M is a proper homotopy equivalence that is a homeomorphism outside of a compact set, then f is homotopic through such maps to a homeomorphism. For a group, we demand that this hold for all aspherical M with that fundamental group. This extends the usual Borel conjecture from fundamental groups of compact aspherical manifolds to all countable groups of finite cohomological dimensional, and in the case of fundamental groups of compact aspherical manifolds is equivalent to it if one considers M and M $\times$ tori simultaneously. (In other words a counterexample to the extended Borel for some bundle over M will translate into a counterexample to ordinary Borel for M $\times$ some torus.)

Thus the existence of the aspherical homology manifold follows from the results of [FJ] and [BL]. I will soon outline some evidence for not believing that it is possible to improve the existence result to be a manifold. In any case:

Proposition 10. Among torsion-free hyperbolic groups, being the fundamental group of a closed aspherical manifold is a coarse condition; it follows from the Gromov boundary being a sphere.

The idea for this appears in a discussion of the case of homotopy tori in [BFMW1]. One glues the boundary sphere onto the universal cover to obtain a homology manifold with boundary, and then relates the local index of the interior to that of the boundary. Details will appear in a forthcoming paper with Barthels and Lueck [BLW] that discusses more generally the role of the boundary.

It is also worth noting that the conjecture (affirmed in Proposition 8) includes the Borel conjecture in its usual formulation. If M and N are both aspherical manifolds with fundamental group π, there is then a CE map (see [Da]) M $\to$ N, which according to [Si] is automatically a uniform limit of homeomorphisms. The conjecture thus accomplishes the liberation of the Borel conjecture from the setting of aspherical manifolds.

Finally, the map comes directly from the argument in [BFMW2]. The underlying homotopy theoretic assumption enables one to set up a controlled surgery problem, which is solved by that technology. More precisely, it produces a DDP[15] homology manifold controlled homotopy equivalent to the original manifold that "approximately fibers over N." However, Edwards's characterization theorem [Da], Quinn's resolution theory [Q], and the α-approximation theorem [ChF] then apply to show that the homology manifold is homeomorphic to the original manifold M.

The situation for groups with torsion is not nearly as pleasant: there are many more sources of obstruction—however, it is consideration of these that leads me to believe that the homology manifolds arising in the conjecture cannot be replaced by manifolds.

The naive suggestion would be that π should act proper discontinuously on a contractible (homology) manifold C and that there should be an equivariant

---

15. DDP stands for "Disjoint Disks Property"; it demands of a space Z that any pair of maps of $D^2 \to Z$ can be approximated by a disjoint pair. According to Edwards [Da] a resolvable homology manifold of dimension > 4 with this property is a manifold.

map $X \rightarrow$ C that has the properties of the conjecture. This is completely deflated by the failure of the Nielsen conjecture [BW2].

In any case, one would suspect that C actually should have more properties, for example, that the fixed set of every finite subgroup is also contractible, like K\G in the lattice case. However, even then there are some additional obstructions related to Nil and UNil that would have to vanish. If one is willing to cross M with a manifold with zero Euler characteristic, then one can get rid of the Nil obstructions and then, if π is a lattice with only odd torsion, in rank 1, then one can affirm this conjecture—presumably it is only a matter of time—until the general case will follow.

However, for general groups π, even when C exists, nothing implies that the action is locally smoothable. It is very possible for action near the fixed set to be modeled on nonlinearizable (say homotopy linear) actions of a finite group on the sphere. (These are constructed in [BW1].) This seems to be entirely parallel to having homology manifolds that are locally homotopy spheres (=DDP) but not actually being "locally linear," that is, resolvable.

Remark: In [BFMW1] there were a number of conjectures made about the geometric topology of DDP homology manifolds. Despite the years that have passed almost no progress has been made. The reason for this is probably because they have never been "seen": they are constructed as Gromov-Hausdorff limits of polyhedra that are themselves constructed using fairly high-power machinery. If one of these arose as a boundary of a hyperbolic group, it would naturally have many self-homeomorphisms, and other additional structures that could lead an optimist to hope that it would be an importat step toward understanding the nonresolvable homology manifolds.

Appendix: Actions on $M \times R^n$

This brief appendix gives a bit more information about the possible monodromies possible for the manifolds just discussed (in the torsion-free case). But we shall phrase the result in terms of the Zimmer program. Here Cat is any of the geometric categories Top, PL, or Diff and Aut(?) is the space of self-homotopy equivalences of the space?.

Proposition A1. Suppose that Γ is a countable group of finite cohomological dimension and we have a homomorphism $\rho : \Gamma \rightarrow \pi_0 \text{Aut}(M)$. Then there is a Cat action of Γ on $M \times R^n$ for some n iff there is a lift of $B\rho$ to BAut(M, t) where (M, t) is the component of the space of maps M $\rightarrow$ BCat that contains the stable tangent bundle of M, and Aut denotes the automorphism group. Moreover, the action can be taken free.

Note: Note that if the original $\rho : \Gamma \to \pi_0 \mathrm{Aut}(M)$ does not preserve the stable tangent bundle, then the automorphism space is empty and the proposition is vacuous. If M is stably parallelizable, then the condition is equivalent to just solving the lifting problem:

$$\mathrm{BAut}(M)$$

$$\downarrow$$

$$B\rho : B\Gamma \to B\pi_0 \mathrm{Aut}(M)$$

This propositon is the (un)natural marriage of Cooke's obstruction theory for lifting homotopy actions to actions [Co] combined with Mazur's theory of stable differential topology [Ma] where one uses finite dimensionality of $B\Gamma$ to get a manifold structure when "assembling." The proof is quite straightforward and is left to the reader. (In the parallelizable case, one simply argues that the action produced by Cooke gives rise to a finite-dimensional CW complex with a map $\pi_1 \to \Gamma$ whose induced cover is homotopy equivalent to M. "Thickening" this complex to be a parallelizable manifold gives one whose Γ cover is $M \times R^n$ according to Mazur.)

Remark: On the other hand, for finite groups this approach seems doomed to failure. The theorem *does* give rise to obstructions: for example, the Nielson realization problem was disproved for nilmanifolds via this obstruction; see [RS]. However, for trivial ρ clearly it is impossible to ever accomplish the task achieved by the proposition, that is, constructing a free action. There is some literature on this problem for finite groups, but it is hemmed in by natural and strong hypotheses.

Note that according to Sullivan [Su] and Wilkerson [Wi] when M is a compact (or even finite-type) simply connected manifold, $\pi_0 \mathrm{Aut}(M)$ is an arithmetic group. In practice the lifting problem has its subtle aspects, and I hope to devote a future paper to some examples of it.

The actions produced by the method are always free and proper discontinuous. As a result, the n here can be rather larger than the *cd*. (Think about the case of a product of free groups [BKK] and $M = $ a point.)

Much more in the spirit of the Zimmer program would be to produce actions with small values of n. With some trepidation, I would like to suggest:

Conjecture: If M is a compact manifold of dimension less than that of the smallest representation of an irreducible lattice Γ in a Lie group G of rank > 1, and $n << \dim(G/K)$, then all actions of Γ on $M \times R^n$ factor through a finite group.

The condition on n should force a certain amount of recurrence. If n is (rather) less than $\dim(G/K)$, one can suspect that the amount is enough to impose the features of rigidity.

4. Conjugacy of Translations

Suppose G is a connected Lie group and g is an element. Left translation by g defines a dynamical system on G. In this section, I would like to discuss in some simple case when these dynamical systems are topologically conjugate in what is just an experimental exploration. My hope is that this will serve as a first setting for which surgical ideas have dynamical applications. In any case, it gives some more examples in the style of [LZ] of lattices that act ergodically[16] on a compact manifold in infinitely many distinct ways.

Proposition 11. Suppose g is an element of a torus, then translation by g is topologically conjugate to tranlation by h iff there is an element of $GL_n(\mathbf{Z})$ taking g to h. If $<g>$ is dense in the torus, then this element is unique and is the unique continuous conjugating map.

Note that g can be recovered by its eigenvalues on $L^2(T)$.

Corollary. If G is a compact Lie group, then any g can be conjugated to an element of a maximal torus. Translation by g is conjugate measure theoretically to h iff there is an element of $GL_n(\mathbf{Z})$ taking g to h (for the usual action of $GL_n(\mathbf{Z})$ as automorphisms of the torus).

Much more interesting is the situation of continuous conjugacy.

Corollary. (G, g) is topologically conjugate to (G, h) iff $G/c1(<g>) \approx G/cl(<h>)$ by an isomorphism that pulls back the principle $cl <g> = cl <h>$ bundles.

---

16. Of course, many nonergodic actions—even with continuous deformation—appear in earlier sections.

As a special case, consider $G = SU(n)$ and g generic, so that $<g>$ is dense in T the maximal torus. Then the maps associated to g and h, $SU(n)/T \to BT$ classify the principle bundles differ by the automorphism of $GL_n(\mathbf{Z})$; in other words, these are the same bundles, but their structure as principle bundles is different. So we want to know what kinds of automorphisms of G/T are possible. The automorphisms of the cohomology algebra of this space was much studied and for $G = SU(n)$ the only possibility is the Weyl group combined with complex conjugation (see, e.g., [EL]). Thus we obtain and generically there is rigidity:

Corollary. For generic $g \in SU(n)$, translation by g is only topologically conjugate to translation by elements conguate in $SU(n)$ to g or its complex conjugate.

Much more interesting is the situation for nongeneric elements. We will see that these are frequently topologically conjugate. We shall focus on the extreme cases: where $cl <g>$ is finite or a circle.[17]

First of all I want to show some conjugacies that are "soft," that is, follow from general principles with no calculations.[18] We will consider Z_k where k is a product of a large number of primes. For each prime separately, we have the action of the Weyl group moving the generator to another element of the maximal torus. However, if there are r prime factors, one has something like $(\#\text{Weyl})^{r-1}$ associated representations of Z_k modulo action of the Weyl element on the whole maximal torus.

Proposition 12.[19] These "Weyl-mixed" translations are all topologically conjugate for $G = SU(n)$, $n > 2$.

Proof. For simplicity of notation let us consider the case of two factors. Thus we are considering $G/(P \times Q)$ versus $G/(P' \times Q')$ where P and Q have relative

---

17. In general is a product $Z_r \times T$ for some torus.

18. On the other hand, they have the usual shocking feeling that follows from applying obviously discontinuous p-adic constructions on standardly manifold theoretic objects: in other words, they are reminiscent of the use of Frobenius to get actions on etale homotopy types that arose in the proof of the Adams conjecture.

19. This proposition is true for all simply connected compact Lie groups other than $SU(2)$; Indeed, I suspect it is true smoothly and for arbitrary compact Lie groups of rank > 1 and perhaps even more general abelian subgroups of G.

orders, and P and P' are conjugate in G, as are Q and Q'. First of the quotients are homotopy equivalent[20] by localization theory [HMR]. To build an equivalence between (simple) spaces,[21] it suffices to build rationally compatible equivalences between their localizations at each prime. At P, G/P → G/(P×Q) is an equivalence, and G/P ≈ G/P' by a conjugacy that is homotopic to the identity as a map of G (it is induced by an element of G, which is connected). At Q, the argument is the same. At other primes, it's even easier; both quotients are equivalent to G. These are all visibly rationally compatible.

In fact they are simple homotopy equivalent. As $n > 2$, the Lie group we are considering has rank > 1. Therefore one can use an extra circle from the maximal torus to compute that the Reidemeister torsion of these spaces vanish. As the Whitehead groups of cyclic groups are torsion free [BMS], the Whitehead torsion is detected by Reidemeister torsion and we get its vanishing as well.

Thus we can consider G/(P'×Q') ∈ S(G/(P×Q)), where S denotes the structure set of surgery theory. We shall analyze this in terms of the "old-fashioned surgery exact sequence" of Sullivan [SU]. The first obstruction we have is the normal invariant of this homotopy equivalence C∈ [G/(P×Q) : F/Cat]. However, by passing to P and Q covers, this map clearly vanishes (on those covers, it's homotopic to a diffeomorphism), so the map is nullhomotopic. As a result G/(P×Q) and G/(P'×Q') are normally cobordant. To complete the proof, we have to analyze a final surgery obstruction. If dim(G) is even, the relevant surgery group vanishes.[22]

However, if dim(G) is odd, then there are additional obstructions, the most prominent being the ρ–invariant. This is the G-signature of any free G-manifold bounding our manifold. If S is an S^1 in T-disjoint from P×Q and P'×Q', we can use it compute the ρ–invariants of G/(P×Q) and G/(P'×Q') simultaneously. Note that G bounds the mapping cylinder of G → G/S, and that both finite groups act homologically trivially on this manifold with boundary. Thus for both, the ρ–invariant is the signature of this manifold

20. All homotopy equivalences are assumed to preserve identification of fundamental groups.

21. A space is simple if its fundamental group is abelian and it acts trivially on higher homotopy. Localization theory works well for nilpotent, and hence, simple spaces (see [HMR]).

22. If P and Q are odd. Otherwise there can be a Z_2 that is a codimension 1 arf invariant. This element always acts trivially on topological structure sets as is well known (but can be nontrivial smoothly). In many cases this obstruction can be seen to be trivial even smoothly, e.g., if there is a fixed circle for all of the elements of the Weyl group used at P and Q, using a variant of the trick used above and the "numerical Levine formula" of [CW, Dv, Pa].

with boundary (almost always 0) × the trivial representation. In all cases, these manifolds cannot be distinguished by their ρ–invariants, completing the argument.   □

Remark: It is clear that one can apply the same method to some positive dimensional topologically cyclic subgroups of the maximal torus—although one will be compelled to only mix using subgroups of the Weyl group. This, though, would give an amusing example of the use of "soft" topological methods from homotopy theory together with surgery giving rise to topological conjugacies of dynamical systems that do have some recurrence.

In order to make further progress (i.e., to show that the quotient of G under the relation of topological conjugacy has trivial hausdorfification) we have to study what happens for individual primes. We shall facilitate matters greatly by assuming that **p is a large prime** compared to the dimension. Moreover, the calculations are suggestive of what occurs for the case of a single circle.

Consider now $Z_p \subset T$ as a diagonal matrix in $SU(n)$, denoted by $(a_1, a_2, \ldots, a_n)$ where the a's are integers mod p. We have $\Sigma a_i = 0$ since our torus is in $SU(n)$. The Chern classes are, of course, the symmetric functions of the a_i's.

Proposition 13. For $G = SU(n)$, two G/Z_p's have the same homotopy type (for p large w.r.t. n) iff their first nonzero Chern classes agree. Moreover, they then have the same homeomoprhism (=diffeomorphism) type.

There is a fibration $G/Z_p \to BZ_p \to BSU(n)$. The map $BZ_p \to BSU(n)$ factors through $BZ_p \to BT$, where T is the maximal torus and it is easy enough to understand. $BSU(n)$ through dimension $= \dim(G)$ at a large prime can be thought of as a product of Eilenberg-MacLane spaces determined by the Chern classes. So $BSU(n) \to \prod K(Z, 2i)$ is an isomorphism. The Chern classes thus determine the Postnikov decomposition of G/Z_p. The r-th k-invariant is obtained by pulling back the $r + 1$st Chern class from BZ_p. However, once a Chern class is nonzero, the cohomology from BZ_p pulls back trivially in higher dimensions, and G/Z_p looks like that Postnikov piece (at p) × with a product of the remaining odd-dimensional spheres. This proves the proposition by the classification of spaces via Postnikov towers (and the fact that two finite-dimensional spaces are homotopy equvalent iff their Postnikov towers agree through their dimensions).

A similar method computes their Pontrjagin classes in terms of the Chern class of the representation ρ pulled back to BZ_p. As a result, only one nonzero Pontrjagin class enters. If our prime is sufficiently large, F/Cat can be caught in terms of Pontrjagin classes and we've computed the normal invariant. The rest of the surgery exact sequence is computed as before. Also, if p is sufficiently large, Top/O has no homotopy groups in the relevant range, so all of these manifolds are diffeomophic because they are homeomorphic and their universal covers are diffeomorphic.

Corollary. By choosing p large it is possible to get topological conjugacy classes of elements as dense in G as we would like. So the largest Hausdorff quotient of $G/\sim$ is a point.

Now let us consider the same analysis applied to BS^1 in place of BZ_p. Again we have $\Sigma a_i = 0$, but now as an equation in $\mathbf{Z}$. Now consider the Second Chern class $= \frac{1}{2}(0^2 - \Sigma a_i^2)$. So there are only finitely many circles in the maximal torus that give rise to bundles with the same Second Chern class.

I have not checked the following but it seems reasonable in light of the previous discussion to believe the following:

Problem: Is it true that if G is irreducible, then the map from G-conjugacy classes to topological conjugacy of the translations is finite to 1?

The first case where one sees infinite indeterminacy is $SU(2) \times SU(2)$. Here the maximal torus is of dimension 2. The quotient by any circle is topologically $S^2 \times S^3$. This follows very easily from old work of Smale [Sm]. Smale showed that any simply connected spin 5-manifold with H^2 E $\mathbf{Z}$ is diffeomorphic to $S^2 \times S^3$. That the quotient has these properties is easy enough to see from the Gysin sequence associated to the cover $G/S \to G/T$. (S is the circle, and T is the maximal 2-torus).

The classifying map for the circle bundle $G \to G/S$ is a generator of H^2, and these are equivalent under diffeomorphism. So all of these circle actions are the same, and thus we have an analysis of the topological conjugacy in this way.

Remark: As Lashof and Zimmer [LZ] have observed, whenever the compact group G has lattice subgroup, examples where G/A E G/B for nonconjugate subgroups A and B give rise to different ergodic lattice actions on the same manifold. Thus, the observations and methods of this section shed a bit of light on these phenomena.

References

[ABJLMS] G. Arzhantseva, M. Bridson, T. Janszkiewicz, I. Leary, A. Minyasin, J. Swiatkowski, Infinite groups with fixed point properties (preprint).

[BL] A. Barthels and W. Lueck, The Borel conjecture for Cat(0) and hyperbolic groups (preprint).

[BLW] A. Barthels, W. Lueck, and S. Weinberger, Aspherical manifolds with hyperbolic fundamental groups (to appear).

[BMS] H. Bass, J. Milnor, and J.-P. Serre, Solution of the congruence subgroup problem for SL_n, $n \geq 3$ and Sp_{2n}, $n \geq 2$. Inst. Hautes Études Sci. Publ. Math. no. 33 (1967), 59–137.

[Be] J. Benveniste, Rigidity of isometric lattice actions on compact Riemannian manifolds. Geom. Funct. Anal. 10 (2000), no. 3, 516–542.

[BF] M. Bestvina and M. Feign, Proper actions of lattices on contractible manifolds. Invent. Math. 150 (2002), no. 2, 237–256.

[BKK] M. Bestvina, M. Kapovich, and B. Kleiner, Van Kampen's embedding obstruction for discrete groups. Invent. Math. 150 (2002), no. 2, 219–235.

[BM] M. Bestvina and G. Mess, The boundary of negatively curved groups. J. Amer. Math. Soc. 4 (1991), no. 3, 469–481.

[BW1] J. Block and S. Weinberger, Large scale homology theories and geometry. Geometric topology (Athens, GA, 1993), 522–569, AMS/IP Stud. Adv. Math., 2.1, American Mathematical Society, Providence, RI, 1997.

[BW2] J. Block and S. Weinberger, On the generalized Nielsen realization problem. Comm. Math. Helv. 83 (2008), 21–33.

[Bo1] A. Borel, Seminar on transformation groups. Princeton University Press, Princeton, NJ, 1960.

[Bo2] A. Borel, Introduction to the cohomology of arithmetic groups. Lie groups and automorphic forms, 51–86, AMS/IP Stud. Adv. Math., 37, American Mathematical Society, Providence, RI, 2006.

[BoS] A. Borel and J.-P. Serre. Corners and arithmetic groups. Avec un appendice: Arrondissement des variétés à coins, par A. Douady et L. Hérault. Comment. Math. Helv. 48 (1973), 436–491.

[BFMW1] J. Bryant, S. Ferry, W. Mio, and S. Weinberger, Topology of homology manifolds. Ann. of Math. (2) 143 (1996), no. 3, 435–467.

[BFMW2] J. Bryant, S. Ferry, W. Mio, and S. Weinberger, Desingularizing homology manifolds. Geom. Topol. 11 (2007), 1289–1314.

[CW] S. Cappell and S. Weinberger, Which H-spaces are manifolds? I. Topology 27 (1988), no. 4, 377–386.

[ChF] T. Chapman and S. Ferry, Approximating homotopy equivalences by homeomorphisms. Amer. J. Math. 101 (1979), no. 3, 583–607.

[ChW] S. Chang and S. Weinberger, Topological nonrigidity of nonuniform lattices. Comm. Pure Appl. Math. 60 (2007), no. 2, 282–290.

[CK]   F. Connolly and T. Kosniewski, Examples of lack of rigidity in crystallographic groups. Algebraic topology Poznan 1989, 139–145, Lecture Notes in Math., 1474, Springer, Berlin, 1991.

[Co]   G. Cooke, Replacing homotopy actions by topological actions. Trans. Amer. Math. Soc. 237 (1978), 391–406.

[Da]   R. Daverman, Decompositions of manifolds, Reprint of the 1986 original. AMS Chelsea Publishing, Providence, RI, 2007.

[Dv]   J. Davis, Evaluation of odd-dimensional surgery obstructions with finite fundamental group. Topology 27 (1988), no. 2, 179–204.

[EL]   J. Ewing and A. Liulevicius, Homotopy rigidity of linear actions on homogeneous spaces. J. Pure Appl. Algebra 18 (1980), no. 3, 259–267.

[FS1]   B. Farb and P. Shalen, Lattice actions, 3 manifolds and homology. Topology 39 (2000), no. 3, 573–587.

[FS2]   B. Farb and P. Shalen, Real-analytic actions of lattices. Invent. Math. 135 (1999), no. 2, 273–296.

[FW]   B. Farb and S. Weinberger, Isometries, rigidity, and universal covers, Ann. of Math. 168 (2008), 915–940.

[FJ]   T. Farrell and L. Jones, Rigidity for aspherical manifolds with $\pi_1 \subset GL_m(\mathbb{R})$. Asian J. Math. 2 (1998), no. 2, 215–262.

[Fe]   S. Ferry, Homotoping ε–maps to homeomorphisms, Amer. J. Math. 101 (1979), no. 3, 567–582.

[FRR]   S. Ferry, A. Ranicki, and J. Rosenberg, A history and survey of the Novikov conjecture. Novikov conjectures, index theorems and rigidity, Vol. 1 (Oberwolfach, 1993), 7–66, London Math. Soc. Lecture Note Ser., 226, Cambridge University Press, Cambridge, 1995.

[FiM]   D. Fisher and G. Margulis, Almost isometric actions, property (T), and local rigidity. Invent. Math. 162 (2005), no. 1, 19–80.

[FiS]   D. Fisher and L. Silberman, Groups not acting on manifolds, IMRN 2008 (to appear).

[Gi]   C. Giffen, The generalized Smith conjecture. Amer. J. Math. 88 (1966), 187–198.

[GKW]   T. Goodwillie, J. Klein, and M. Weiss, Spaces of smooth embeddings, disjunction and surgery. Surveys on surgery theory, Vol. 2, 221–284, Ann. of Math. Stud., 149, Princeton University Press, Princeton, NJ, 2001.

[HMR]   P. Hilton, G. Mislin, and J. Roitberg, Localization of nilpotent groups and spaces. North-Holland Mathematics Studies, No. 15. Notas de Matemática, No. 55. North-Holland Publishing Co., Amsterdam-Oxford; American Elsevier Publishing Co., New York, 1975.

[H]   S. Hurder, Rigidity for Anosov actions of higher rank lattices. Ann. of Math. (2) 135 (1992), no. 2, 361–410.

[KL]   A. Katok and J. Lewis, Global rigidity results for lattice actions on tori and new examples of volume-preserving actions. Israel J. Math. 93 (1996), 253–280.

[LZ]   R. Lashof and R. Zimmer, Manifolds with infinitely many actions of an arithmetic group. Illinois J. Math. 34 (1990), no. 4, 765–768.

[Ma]   B. Mazur, Differential topology from the point of view of simple homotopy theory. Inst. Hautes Études Sci. Publ. Math. no. 15 (1963), 93 pp.

[Mi]   J. Milnor, Whitehead torsion, Bull. Amer. Math. Soc. 72 (1966), 358–426.

[Pa]   W. Pardon, The exact sequence of a localization for Witt groups. II. Numerical invariants of odd-dimensional surgery obstructions. Pacific J. Math. 102 (1982), no. 1, 123–170.

[Q]   F. Quinn, An obstruction to the resolution of homology manifolds. Michigan Math. J. 34 (1987), no. 2, 285–291.

[RS]   F. Raymond and L. Scott, Failure of Nielsen's theorem in higher dimensions. Arch. Math. (Basel) 29 (1977), no. 6, 643–654.

[Shi]   N. Shirokova, thesis (University of Chicago, 1998).

[Si]   L. Siebenmann, Approximating cellular maps by homeomorphisms. Topology 11 (1972), 271–294.

[Sm]   S. Smale, On the structure of 5-manifolds, Ann. of Math. (2) 75 (1962), 38–46.

[SW]   M. Steinberger and J. West, Approximation by equivariant homeomorphisms. Trans. Amer. Math. Soc. 302 (1987), no. 1, 297–317.

[Su]   D. Sullivan, Infinitesimal computations in topology. Publ. Math. d'IHES 47 (1977), 269–331.

[We1]   S. Weinberger, Computers, rigidity, and moduli: The large scale fractal geometry of Riemannian moduli space. Princeton University Press, Princeton, NJ, 2004.

[We2]   S. Weinberger, Continuous versus discrete symmetry, Geometry and topology (Athens, GA, 1985), 319–323, Lecture Notes in Pure and Appl. Math., 105. Dekker, New York, 1987.

[We3]   S. Weinberger, The topological classification of stratified spaces. Chicago Lectures in Mathematics. University of Chicago Press, Chicago, 1994.

[We4]   S. Weinberger, Nonlinear averaging, embeddings, and group actions. Tel Aviv Topology Conference: Rothenberg Festschrift (1998), 307–314, Contemp. Math., 231. American Mathematical Society, Providence, RI, 1999.

[Wi]   C. W. Wilkerson, Applications of minimal simplicial groups. Topology 15 (1976), 111–130.

[Zi1]   R. Zimmer, Lattices in semisimple groups and distal geometric structures. Invent. Math. 80 (1985), no. 1, 123–137.

[Zi2]   R. Zimmer, Lattices in semisimple groups and invariant geometric structures on compact manifolds. Discrete groups in geometry and analysis (New Haven, CT, 1984), 152–210, Progr. Math., 67. Birkhäuser Boston, Boston, 1987.

PART 2

Analytic, Ergodic, and Measurable Group Theory

9

MICHAEL G. COWLING

CALCULUS ON NILPOTENT LIE GROUPS

TO MY FRIEND BOB

Abstract

Mostow's rigidity theorem states that if two locally symmetric spaces $\Gamma\backslash G/K$ (where G/K is a symmetric space of the noncompact type, and Γ is a uniform lattice in the semisimple group G) are homeomorphic, then they are diffeomorphic; the case where $G = \mathrm{SL}(2, \mathbb{R})$ has to be excluded. Mostow's proof treats the rank 1 case and the higher-rank case differently: the former analyzes quasiconformal mappings, while the latter appeals to a theorem of Tits on mappings of spherical buildings. This note discusses a more analytical approach to the higher-rank case, which aims to unify the rank 1 and higher-rank cases, and links to ideas of Gromov and Pansu. Only the case of $\mathrm{SL}(n, \mathbb{R})$ is discussed here.

1. Introduction

A key step in the proof of Mostow's rigidity theorem [6] is to identify maps of the boundary space G/P that arise as Γ-equivariant boundary maps of quasi-isometries, and to show that all these maps are of the form $xP \mapsto gxP$, for some g in G (here P is a minimal parabolic subgroup of G). In the rank 1 case, the boundary maps to be considered are all "contact" mappings, and in the cases of the groups $\mathrm{SO}(1, n)$ and $\mathrm{SU}(1, n)$, the ergodicity of Γ is also needed, but for the groups $\mathrm{Sp}(1, n)$ and the exceptional group $F_{4,20}$, P. Pansu [8] showed that the contact condition alone implies that the maps all arise from the action of G. Later, the differential geometer K. Yamaguchi [10] showed that, with a few exceptions, sufficiently smooth "contact" mappings of the boundary G/P all come from the action of the group G. However, the smoothness results needed to apply the machinery of differential geometry are not natural in the context of Mostow's theorem. We show how the weaker condition of "Pansu differentiability" (defined later), similar to the condition considered by Mostow in the rank 1 case, implies that contact mappings come from the action of G

It is a pleasure to thank the anonymous referee of this chapter for his or her useful and constructive criticism.

(with a few exceptions); for simplicity, here we consider only the case where G is SL$(n, \mathbb{R})$ and n is at least 4. We intend to use this theory elsewhere to give another proof of Mostow's theorem.

If a mapping of a stratified nilpotent Lie group N is "Pansu differentiable," then its derivative is a strata-preserving homomorphism of the Lie group or of the Lie algebra (depending on the definition used); if the mapping is invertible, then the mapping is an automorphism. We characterize the strata-preserving automorphisms of the Lie algebra $\mathfrak{utn}(n, \mathbb{R})$ of the group of upper-triangular unipotent matrices UTU$(n, \mathbb{R})$, which is the Iwasawa N group for the group SL$(n, \mathbb{R})$. We first of all consider the case where $n = 4$, and then treat the case of larger n by induction. Using this characterization, we show that mappings that are differentiable everywhere, or almost everywhere and satisfying a simple ergodicity condition, arise out of the group action.

To state our theorems, we need to define the "reflection in the ant-diagonal" $\phi \colon \text{SL}(n, \mathbb{R}) \to \text{SL}(n, \mathbb{R})$, by the formula

$$\phi(g)_{ij} = g_{n+1-j, n+1-i} \qquad \forall i, j \in \{1, \ldots, n\}.$$

It is clear that this is an automorphism of SL$(n, \mathbb{R})$ that preserves UTU$(n, \mathbb{R})$. Indeed, it is the automorphism that arises from the obvious nontrivial automorphism of the Dynkin diagram of the root system A_{n-1}. Other automorphisms of SL$(n, \mathbb{R})$ that preserve UTU$(n, \mathbb{R})$ include conjugations by diagonal matrices. We write the group of these automorphisms as Ad (MA).

We will prove the following results.

THEOREM 1. *Suppose that N is the group of $n \times n$ unipotent upper-triangular real matrices, where $n \geq 4$, and that $f \colon N \to N$ is Pansu differentiable at p, with a Pansu differentiable local inverse at $f(p)$. Then the Pansu derivative $df(p)$ lies in Ad (MA) or ϕ Ad (MA).*

THEOREM 2. *Suppose that N is the group of $n \times n$ unipotent upper-triangular real matrices, where $n \geq 3$, that the continuous mapping $f \colon N \to N$ is Pansu differentiable almost everywhere, and that df is in Ad (MA) almost everywhere. Then f is an affine map, that is, the composition of a group automorphism (that commutes with dilations) and a translation.*

Theorem 2 has a number of interesting features. First, only pointwise differentiability is assumed. Most theorems in analysis require more than this. For instance, the Fundamental Theorem of Calculus, that is, for a function $u \colon [a, b] \to \mathbb{R}$,

$$u(b) - u(a) = \int_a^b u'(x)\, dx,$$

requires u' to be integrable. The Cauchy-Goursat theorem in complex analysis is an exception. There are interesting parallels between the Cauchy-Goursat theorem and Theorem 2: in both cases differentiability and algebraic structure of the derivative leads to strong conclusions.

Next, local hypotheses give global conclusions. In complex analysis, the exponential map is differentiable and has a differentiable local inverse, but it is not one to one.

Finally, this theorem also has a local version, when f is defined in an open set. In this case, f comes from the action of $GL(n, \mathbb{R})$ (augmented by the map ϕ).

2. The Pansu Derivative

A Lie algebra $\mathfrak{n}$ is said to be *stratified of step s* if there are nontrivial subspaces $\mathfrak{n}_1, \mathfrak{n}_2, \ldots, \mathfrak{n}_s$ such that

$$\mathfrak{n} = \mathfrak{n}_1 \oplus \mathfrak{n}_2 \oplus \cdots \oplus \mathfrak{n}_s$$

1

$$[\mathfrak{n}_1, \mathfrak{n}_j] = \mathfrak{n}_{j+1},$$

where $\mathfrak{n}_{s+1} := \{0\}$. A stratified algebra is graded, that is, $[\mathfrak{n}_k, \mathfrak{n}_j] \subseteq \mathfrak{n}_{k+j}$, and nilpotent, because $[\mathfrak{n}, [\mathfrak{n}, \ldots [\mathfrak{n}, \mathfrak{n}] \ldots]] = \{0\}$, where there are $s+1$ Lie brackets. This follows by induction.

Stratified Lie algebras admit *dilations* δ_t, for t in $\mathbb{R}^+$, defined by requiring that $\delta_t X := t^j X$ for all X in $\mathfrak{n}_j$ and extending δ_t to $\mathfrak{n}$ by linearity. Clearly

$$\delta_t[X, Y] = [\delta_t X, \delta_t Y] \qquad \forall X \in \mathfrak{n}_j, \quad \forall Y \in \mathfrak{n}_k$$

because $\mathfrak{n}$ is graded; hence δ_t is a Lie algebra automorphism.

An example of a stratified Lie algebra is the Lie algebra $\mathfrak{utn}(n, \mathbb{R})$ of $n \times n$ upper-triangular unipotent matrices. Denote by $E_{i,j}$ the $n \times n$ matrix whose entries are all 0 except for the i, jth entry, which is 1. Define $\mathfrak{n}_k$ to be span$\{E_{i,i+k} : 1 \le i \le n - k\}$ (in other words, the kth layer of $\mathfrak{n}$ is the sum of all the root spaces where the root is the sum of exactly k simple roots). It is easy to check that the stratification conditions 1 hold, and that $\mathfrak{utn}(n, \mathbb{R})$ is of step $n - 1$.

A stratified Lie group is a connected, simply connected, necessarily nilpotent Lie group N whose Lie algebra $\mathfrak{n}$ is stratified. These occur in areas of mathematics including sub-Riemannian geometry, subelliptic differential operators, and nonholonomic mechanics.

The exponential map exp is a bijection from a stratified Lie algebra n to the corresponding stratified Lie group N, and the dilations δ_t may be extended to N by conjugating by the exponential map:

$$\delta_t p = \exp(\delta_t \log(p)) \qquad \forall p \in N.$$

A map $f \colon \mathbb{R}^m \to \mathbb{R}^n$ is (totally) differentiable at a point p in $\mathbb{R}^m$ if and only if

$$\lim_{t \to 0+} \frac{f(p + th) - f(p)}{t}$$

exists, uniformly on compacta in h; if the limit exists, then we write it as $Df(p)(h)$. The map $Df(p) \colon \mathbb{R}^m \to \mathbb{R}^n$ is linear. It "is" the matrix of partial derivatives of the components of f.

Suppose that N^1 and N^2 are stratified Lie groups. A map $f \colon N^1 \to N^2$ is *Pansu differentiable* at a point p in N^1 if and only if

$$\lim_{t \to 0+} \delta_{t^{-1}}[f(p)^{-1} f(p\delta_t h)]$$

exists, uniformly on compacta in h; if the limit exists, then we write it as $Df(p)(h)$. The map $Df(p) \colon N^1 \to N^2$ is a group homomorphism that commutes with dilations.

It is often more convenient to work with a linearized version of the derivative; we define

$$df(p)X := \log(DF(p) \exp(X)) \qquad \forall X \in n^1.$$

The map $df(p)$ is a Lie algebra homomorphism from n^1 to n^2 that commutes with dilations, or equivalently, respects the stratifications of the algebras.

3. Proof of Theorem 1

In this section, we prove Theorem 1, that is, we characterize the strata-preserving automorphisms of $\mathfrak{utn}_n(\mathbb{R})$.

Write N for the group $\mathrm{UTU}_n(\mathbb{R})$ of all upper-triangular unipotent $n \times n$ real matrices, whose Lie algebra n is the set $\mathfrak{utn}_n(\mathbb{R})$ of all upper-triangular nilpotent $n \times n$ real matrices.

3.1. The Case When n = 3

It is easy to check that there are automorphisms of the algebra $\mathfrak{utn}(3, \mathbb{R})$ that preserve both strata, but act as arbitrary elements of $\mathrm{GL}(2, \mathbb{R})$ on the first stratum. Thus Pansu differentiability imposes no restrictions on the derivative, other than the strata-preserving condition and an evident determinant condition, and Theorem 1 fails.

A. Korànyi and H.M. Reimann [3, 4] developed a theory of quasiconformal mappings on the Heisenberg group $\mathrm{UTU}_3(\mathbb{R})$; these quasiconformal mappings are Pansu differentiable, and the space thereof is infinite-dimensional. Thus Pansu differentiability in the case of $\mathrm{UTU}_3(\mathbb{R})$ is not a very restrictive condition.

However, Theorem 2 does hold in this case, as we shall see; indeed, the restriction that the Pansu derivative lie in Ad (MA) almost everywhere is very strong, not only for groups of upper-triangular matrices, but also for more general Iwasawa N groups. For instance, for $\mathrm{SO}(n, 1)$, where $n \geq 3$, the hypotheses of Theorem 2 lead us to consider 1-quasiregular mappings, which form quite a limited class!

3.2. The Case When n = 4

When we take n to be 4, elements of $\mathfrak{n}$ and of N are of the form

$$
\mathbf{2} \qquad
\begin{pmatrix}
0 & u & x & z \\
0 & 0 & v & y \\
0 & 0 & 0 & w \\
0 & 0 & 0 & 0
\end{pmatrix}
\quad \text{and} \quad
\begin{pmatrix}
1 & u & x & z \\
0 & 1 & v & y \\
0 & 0 & 1 & w \\
0 & 0 & 0 & 1
\end{pmatrix},
$$

respectively, where $u, v, \dots, z$ are real numbers. For this example, the dilations are given by

$$
\delta_t
\begin{pmatrix}
1 & u & x & z \\
0 & 1 & v & y \\
0 & 0 & 1 & w \\
0 & 0 & 0 & 1
\end{pmatrix}
=
\begin{pmatrix}
1 & tu & t^2 x & t^3 z \\
0 & 1 & tv & t^2 y \\
0 & 0 & 1 & tw \\
0 & 0 & 0 & 1
\end{pmatrix}.
$$

Let $\{U, V, W, X, Y, Z\}$ be the basis of $\mathfrak{utn}_4(\mathbb{R})$ such that the left-hand matrix above is equal to

$$
\mathbf{3} \qquad\qquad uU + vV + wW + xX + yY + zZ.
$$

In formula **2**, the group element shown is not the exponential of the algebra element shown. It is easy to check that

$$
[U, V] = X, \quad [V, W] = Y, \quad [U, Y] = Z, \quad \text{and} \quad [X, W] = Z,
$$

and all other (undetermined) commutators are zero.

Take a map f of $\mathrm{UTU}_4(\mathbb{R})$, locally invertible near p. Its Pansu derivative is a strata-respecting automorphism of $\mathfrak{utn}_4(\mathbb{R})$. We may therefore represent $df(p)$ in the basis $\{U, V, W, X, Y, Z\}$ introduced in formula **3** by a 6×6 matrix:

$$\begin{pmatrix} a & b & c & 0 & 0 & 0 \\ d & e & f & 0 & 0 & 0 \\ g & h & i & 0 & 0 & 0 \\ 0 & 0 & 0 & j & k & 0 \\ 0 & 0 & 0 & l & m & 0 \\ 0 & 0 & 0 & 0 & 0 & n \end{pmatrix}.$$

It is easy to check that span$\{U, W\}$ is the unique 2-dimensional abelian subalgebra of n_1, and that span$\{V\}$ is the centralizer of n_2 in n_1. Since $df(p)$ preserves these subalgebras, we may represent $df(p)$ by

$$\begin{pmatrix} a & 0 & c & 0 & 0 & 0 \\ 0 & e & 0 & 0 & 0 & 0 \\ g & 0 & i & 0 & 0 & 0 \\ 0 & 0 & 0 & j & k & 0 \\ 0 & 0 & 0 & l & m & 0 \\ 0 & 0 & 0 & 0 & 0 & n \end{pmatrix}.$$

The commutation relation $[U, V] = X$ implies that $[TU, TV] = TX$, whence (using other relations as needed)

$$aeX - egY = [aU + gW, eV] = jX + lY,$$

and hence $j = ae$ and $l = -ge$. The relation $[V, W] = Y$ implies that $k = -ce$ and $m = ie$ analogously. Similarly, the relations $[U, X] = 0$, $[U, Y] = Z$, $[W, Y] = 0$, and $[W, X] = Z$ yield the equations

$$0 = al - gj \qquad n = am - gk$$
$$0 = ik - cm \qquad n = ij - cl.$$

All these equations together also imply that

$$aeg = 0 \qquad cei = 0 \qquad n = aei + cge.$$

It follows that we may represent $df(p)$ by a matrix, as shown below. All the entries indicated by letters have to be nonzero for the matrix to be invertible; they may depend on the point p.

$$\begin{pmatrix} a & 0 & 0 & 0 & 0 & 0 \\ 0 & e & 0 & 0 & 0 & 0 \\ 0 & 0 & i & 0 & 0 & 0 \\ 0 & 0 & 0 & ae & 0 & 0 \\ 0 & 0 & 0 & 0 & ei & 0 \\ 0 & 0 & 0 & 0 & 0 & aei \end{pmatrix} \quad \text{or} \quad \begin{pmatrix} 0 & 0 & c & 0 & 0 & 0 \\ 0 & e & 0 & 0 & 0 & 0 \\ g & 0 & 0 & 0 & 0 & 0 \\ 0 & 0 & 0 & 0 & -ce & 0 \\ 0 & 0 & 0 & -eg & 0 & 0 \\ 0 & 0 & 0 & 0 & 0 & cge \end{pmatrix}.$$

3.3. The Case When n > 4

Suppose that T is a strata-preserving automorphism of $\mathfrak{n}$. It is easy to verify that X in $\mathfrak{n}_1$ centralizes $\mathfrak{n}_{n-2}$ if and only if X is in $\mathfrak{s}$, the span of the $E_{i,i+1}$ where $2 \leq i \leq n - 2$. Consequently, T maps $\mathfrak{s}$ to $\mathfrak{s}$. Suppose that

4 $$TE_{1,2} = aE_{1,2} + S + bE_{n-1,n},$$

where a and b are real numbers and $S \in \mathfrak{s}$. We write $\mathrm{alg}\,(X, Y, \dots,)$ for the Lie algebra generated by X, Y, ... If any argument of alg is a set, then of course we mean the Lie algebra generated by the other arguments and the elements of this set. It is straightforward to check that $\dim \mathrm{alg}\,(E_{1,2}, \mathfrak{s}) = (n - 1)(n - 2)/2$, while $\dim \mathrm{alg}\,(aE_{1,2} + S + bE_{n-1,n}, \mathfrak{s})$ is equal to

$$\begin{cases} n(n-1)/2 - 1 & \text{if neither of } a \text{ or } b \text{ is } 0 \\ (n-1)(n-2)/2 & \text{if exactly one of } a \text{ or } b \text{ is } 0 \\ (n-2)(n-3)/2 & \text{if both of } a \text{ or } b \text{ is } 0. \end{cases}$$

A strata-preserving automorphism preserves the dimensions of algebras generated by sets, and so exactly one of a and b is nonzero. By composing with ϕ, if necessary we may suppose that b is zero and a is not. Now T preserves $\mathbb{R}E_{1,2} + \mathfrak{s}$, and hence preserves the algebra that this generates, which is isomorphic to $\mathfrak{utn}(n - 1, \mathbb{R})$. By induction, we may suppose that this restricted version of T lies in $\mathrm{Ad}\,(MA)$, and in particular that $S = 0$ in formula 4. A similar argument then shows that T also maps $E_{n-1,n}$ to a multiple of itself, and from this it follows that T lies in $\mathrm{Ad}\,(MA)$, as enunciated.

4. The Proof of Theorem 2

Theorem 2 includes the hypothesis that $df(p)$ is in $\mathrm{Ad}\,(MA)$ almost everywhere. This hypothesis is necessary, because we cannot say much about a map of N if its derivative changes from being in $\mathrm{Ad}\,(MA)$ to being in $\phi\,\mathrm{Ad}\,(MA)$ in some possibly uncontrolled way.

Suppose that f is Pansu differentiable everywhere. A theorem of Darboux says that the derivative of a function f that is differentiable in an interval has the intermediate value property: if $f'(x) = a$ and $f'(y) = b$, and if c lies between a and b, then there exists z between x and y such that $f'(z) = c$. Alternatively, we may say that f' maps connected sets to connected sets. In this formulation, Darboux's theorem has been generalized to $\mathbb{R}^n$ and beyond—see, for instance, J. Malý [5]. In our case, the set of possible derivatives $df(p)$ is the union of two disjoint open sets $\mathrm{Ad}\,(MA)$ and $\phi\,\mathrm{Ad}\,(MA)$. This means that $df(p)$ is always of one type or always of the other. By composing with the group automorphism ϕ if necessary, we may suppose that $df(p)$ is always diagonal.

Alternatively, if we knew that f was differentiable almost everywhere and that a group Γ acted ergodically, preserving the set of points p where $df(p) \in$ Ad (MA) and the set of points p where $df(p) \in \phi MA$, then we could deduce that df was of one type almost everywhere.

In any case, we have stated Theorem 2 with the hypotheses that f is continuous and differentiable almost everywhere, and that $df(p)$ is in Ad (MA). For any small neighborhood Ω of a point p in N, the set $\Omega \exp(\mathbb{R}E_{i,i+1})$ is sent into a set $\Upsilon \exp(\mathbb{R}E_{i,i+1})$, where Υ is a neighborhood of $f(p)$; by shrinking Ω, we may deduce that cosets of $\exp(\mathbb{R}E_{i,i+1})$ are mapped into cosets of $\exp(\mathbb{R}E_{i,i+1})$.

Such a map is affine, that is, it is composed of a (left) translation and a group automorphism. This is proved in the forthcoming PhD thesis of Rupert McCallum; a special case is considered in the article [1]. We give an analytic proof of the result here that conveys less geometric information than McCallum's proof. It should be pointed out that McCallum also considers maps defined on connected open subsets of a stratified group N, and the proof given here does not (obviously) extend to this case.

We first recall that it is shown in [1] that a map f of the group $\mathrm{UTU}_3(\mathbb{R})$ that sends cosets of $\exp(\mathbb{R}E_{1,2})$ to cosets of $\exp(\mathbb{R}E_{1,2})$ and cosets of $\exp(\mathbb{R}E_{2,3})$ to cosets of $\exp(\mathbb{R}E_{2,3})$ is affine.

Now we consider the case where $n = 4$. For notational convenience, we again use the basis $\{U, V, W, X, Y, Z\}$ of $\mathfrak{utn}_4(\mathbb{R})$ such that

$$\begin{pmatrix} 0 & u & x & z \\ 0 & 0 & v & y \\ 0 & 0 & 0 & w \\ 0 & 0 & 0 & 0 \end{pmatrix} = uU + vV + wW + xX + yY + zZ.$$

We consider four copies of $\mathrm{UTU}_3(\mathbb{R})$ in $\mathrm{UTU}_4(\mathbb{R})$; their Lie algebras are spanned by $\{U, V, X\}$, by $\{V, W, Y\}$, by $\{X, W, Z\}$ and $\{U, Y, Z\}$. We know already that f preserves cosets of $\exp(\mathbb{R}U)$, $\exp(\mathbb{R}V)$, and $\exp(\mathbb{R}W)$, and show successively that f preserves cosets of $\exp(\mathbb{R}X)$, $\exp(\mathbb{R}Y)$, and $\exp(\mathbb{R}Z)$. In addition, we show that various partial derivatives are constant on larger and larger sets, in order to conclude that the derivative of f itself is constant.

First, f sends cosets of $\exp(\mathbb{R}U)$ into cosets of $\exp(\mathbb{R}U)$, and cosets of $\exp(\mathbb{R}V)$ into cosets of $\exp(\mathbb{R}V)$, and $\exp(\mathbb{R}U)$ and $\exp(\mathbb{R}V)$ generate $\exp(\mathbb{R}U + \mathbb{R}V + \mathbb{R}X)$, which is isomorphic to $\mathrm{UTU}(3, \mathbb{R})$, so f sends the coset $p \exp(\mathbb{R}U + \mathbb{R}V + \mathbb{R}X)$ into the coset $f(p) \exp(\mathbb{R}U + \mathbb{R}V + \mathbb{R}X)$. By translating by p and $f(p)$, we may temporarily suppose that p and $f(p)$ are both the identity. By the result for the special case of SL(3, $\mathbb{R}$) in [1], we deduce that f is affine on $\exp(\mathbb{R}U + \mathbb{R}V + \mathbb{R}X)$; by translating back, we deduce that f is affine on $p\exp(\mathbb{R}U + \mathbb{R}V + \mathbb{R}X)$; in particular, the derivative of f, restricted to this coset, is constant, and f maps cosets of $\exp(\mathbb{R}X)$ into cosets of $\exp(\mathbb{R}X)$.

Next, we repeat this argument starting with cosets of $\exp(\mathbb{R}V)$ and of $\exp(\mathbb{R}W)$, to show that f is affine on $p\exp(\mathbb{R}V + \mathbb{R}W + \mathbb{R}Y)$; in particular, the derivative of f, restricted to this coset, is constant, and f maps cosets of $\exp(\mathbb{R}Y)$ into cosets of $\exp(\mathbb{R}Y)$.

Third, we repeat the argument starting with cosets of $\exp(\mathbb{R}X)$ and of $\exp(\mathbb{R}W)$, to show that f is affine on $p\exp(\mathbb{R}X + \mathbb{R}W + \mathbb{R}Z)$; in particular, the derivative of f, restricted to this coset, is constant.

Finally, we repeat the argument, starting with cosets of $\exp(\mathbb{R}U)$ and of $\exp(\mathbb{R}Y)$, to show that f is affine on $p\exp(\mathbb{R}U + \mathbb{R}Y + \mathbb{R}Z)$; in particular, the derivative of f, restricted to this coset, is constant.

Recall that the derivative $df(p)$ of f at p may be represented as the matrix

$$\begin{pmatrix} a & 0 & 0 & 0 & 0 & 0 \\ 0 & e & 0 & 0 & 0 & 0 \\ 0 & 0 & i & 0 & 0 & 0 \\ 0 & 0 & 0 & ae & 0 & 0 \\ 0 & 0 & 0 & 0 & ei & 0 \\ 0 & 0 & 0 & 0 & 0 & aei \end{pmatrix},$$

where the entries $a, e, \ldots$, are functions of p. Our four arguments above have shown that these entries are constant along certain cosets of certain subgroups, and more precisely they show that

$$5 \qquad\qquad \tilde{U}a = \tilde{V}a = \tilde{U}e = \tilde{V}e = 0,$$

$$6 \qquad\qquad \tilde{V}e = \tilde{W}e = \tilde{V}i = \tilde{W}i = 0,$$

$$7 \qquad \tilde{X}(ae) = \tilde{W}(ae) = \tilde{X}i = \tilde{W}i = 0, and$$

$$8 \qquad\qquad \tilde{U}a = \tilde{Y}(ei) = \tilde{U}a = \tilde{Y}(ai) = 0,$$

where $\tilde{T}$ stands for the left-invariant vector field associated to the lie algebra element T (T being any of U, V, and so on).

From Equations 5 and 6, we see that $\tilde{U}e = \tilde{V}e = \tilde{W}e = 0$. This implies that e is constant on the whole group. Further, from Equation 7, we see that $e\tilde{W}a + a\tilde{W}e = 0$, whence from Equation 6, $\tilde{W}a = 0$, and combined with Equation 5, this implies that $\tilde{U}a = \tilde{V}a = \tilde{W}a = 0$, whence a is constant on the whole group. Similarly, i is constant on the whole group, and this implies that f is affine.

The case when $n \geq 5$ is proved similarly. There are many more copies of UTU$_3$ inside UTU$_n$, but we can still apply the result of [1] to each of these, and build up to the desired conclusion in much the same way as in the case when $n = 4$.

5. Rigidity for Stratified Groups

One of the key ideas of Pansu [8] is that most stratified groups are rigid, in the sense that the dimension of the space of differentiable maps thereon is finite (even when we consider maps defined on an open subset). As mentioned above, most Iwasawa N groups are rigid in this sense. The exceptions are the N groups coming from SO$(n, 1)$, SU$(n, 1)$, SL$(3, \mathbb{F})$, and Sp$(2, \mathbb{F})$, where $\mathbb{F}$ is either $\mathbb{R}$ or $\mathbb{C}$. At the present time, the published proofs of this fact known to this author involve prolongation and use of the Borel-Weil theorem in one case [10] and a careful study of contact flows in the other [2], and require more smoothness than just differentiability.

In a paper in preparation, P. Ciatti and the author of this paper classify the strata-preserving automorphisms of all the Iwasawa N groups, by simpler, hands-on methods. The groups listed above are precisely those where the automorphism group is not a finite extension of Ad (MA) (where KAN and MAN are an Iwasawa decomposition of the associated semisimple group G and a minimal parabolic subgroup of G). If we include the stronger hypothesis that the automorphisms preserve the root spaces, then the two rank 1 families are the only exceptions. I believe that this result, combined with the methods of this chapter, will lead to another proof of rigidity, which treats the rank 1 and higher-rank cases similarly.

There is interesting related work going on at the moment on differentiable maps of more general stratified groups. For instance, it has been shown recently that differentiable maps of free nilpotent Lie groups that are of step at least 4, or of step 3 with at least three generators, are all affine [9]. General criteria for rigidity are also appearing [7].

References

[1]   A. Čap, M.G. Cowling, F. De Mari, M.G. Eastwood, and R. McCallum, The Heisenberg group, SL(3, $\mathbb{R}$) and rigidity, pp. 41–52 in: *Harmonic analysis, group representations, automorphic forms and invariant theory in honor of Roger E. Howe*, Lecture Notes Series, Institute for Mathematical Sciences, National University of Singapore, Vol. 12. World Scientific, Singapore, 2007.

[2]   M.G. Cowling, F. De Mari, A. Koráányi, and H.M. Reimann, Contact and conformal maps in parabolic geometry. I, *Geom. Dedic.* 111 (2005), 65–86.

[3]   A. Korányi and H.M. Reimann, Quasiconformal mappings on the Heisenberg group, *Invent. Math.* 80 (1985), 309–338.

[4]   A. Korányi and H.M. Reimann, Foundations of the theory of quasiconformal mappings on the Heisenberg group, *Adv. Math.* 111 (1995), 1–87.

[5]   J. Malý, "The Darboux property for gradients", *Real Anal. Exchange* 22 (1996–1997), 167–173.

[6]   G.D. Mostow, *Strong rigidity of locally symmetric spaces*, Annals of Math. Studies, Vol. 78. Princeton University Press, Princeton, NJ 1973.

[7]   A. Ottazzi, A sufficient condition for nonrigidity of Carnot groups, *Math. Zeits.* 259 (2008), 617–629.

[8]   P. Pansu, Métriques de Carnot-Caratheodory et quasiisométries des espaces symétriques de rang un, *Annals of Math.* 129 (1989), 1–60.

[9]   B. Warhurst, Tanaka prolongation of free Lie algebras, *Geom. Dedic.* 130 (2007), 59–69.

[10] K. Yamaguchi, Differential systems associated with simple graded Lie algebras, *Adv. Studies in Pure Math.* 22 (1993), 413–494.

A SURVEY OF MEASURED GROUP THEORY

TO ROBERT ZIMMER ON THE OCCASION OF HIS 60TH BIRTHDAY

Abstract

Measured group theory is an area of research that studies infinite groups using measure-theoretic tools, and studies the restrictions that group structure imposes on ergodic-theoretic properties of their actions. This paper is a survey of recent developments focused on the notion of measure equivalence between groups, and orbit equivalence between group actions.

Contents

Supported by NSF Grant DMS 0604611, and BSF Grant 2004345

1. Introduction

This survey concerns an area of mathematics that studies infinite countable groups using measure-theoretic tools, and studies ergodic theory of group actions, emphasizing the impact of group structure on the actions. *Measured group theory* is a particularly fitting title as it suggests an analogy with *geometric group theory*. The origins of measured group theory go back to the seminal paper of Robert Zimmer [139], which established a deep connection between questions on orbit equivalence in ergodic theory to Margulis's celebrated superrigidity theorem for lattices in semisimple groups. The notion of amenable actions, introduced by Zimmer in an earlier work [138], became an indispensable tool in the field. Zimmer continued to study orbit structures of actions of large groups in [32, 41, 140–144, 146, 147] and [135]. The monograph [146] had a particularly big impact on both ergodic theorists and people studying big groups, as well as researchers in other areas, such as operator algebras and descriptive set theory.[1]

In recent years several new layers of results have been added to what we called measured group theory, and this paper aims to give an overview of the current state of the subject. Such a goal is unattainable—any survey is doomed to be partial, biased, and outdated before it appears. Nevertheless, we shall try our best, hoping to encourage further interest in this topic. The reader is also referred to Gaboriau's paper [58], which contains a very nice overview of some of the results discussed here, and to Shalom's survey [133], which is even closer to the present paper (hence the similarity of the titles). The monographs by Kechris and Miller [81] and the forthcoming one [80] by Kechris include topics

---

1. Zimmer's cocycle superrigidity proved in [139] plays a central role in another area of research, vigorously pursued by Zimmer and others, concerning actions of large groups on manifolds. David Fisher surveys this direction in [42] in this volume.

in descriptive set theory related to measured group theory. Readers interested in connections to von Neumann algebras are referred to Vaes's [136], Popa's [114], and references therein.

The scope of this paper is restricted to interaction of infinite groups with ergodic theory, leaving out the connections to the theory of von Neumann algebras and descriptive set theory. When possible, we try to indicate proofs or ideas of proofs for the stated results. In particular, we chose to include a proof of one cocycle superrigidity theorem (Theorem 5.21), which enables a self-contained presentation of a number of important results: a very rigid equivalence relation (Theorem 4.19) with trivial fundamental group and outer automorphism group (Theorem 4.15), an equivalence relation that cannot be generated by an essentially free action of any group (§4.3.1).

DISCLAIMER. As usual, the quoted results are often presented not in the full possible generality, so the reader should consult the original papers for full details. The responsibility for inaccuracies, misquotes, and other flaws lies solely with the author of these notes.

ACKNOWLEDGMENTS: I would like to express my deep appreciation to Bob Zimmer for his singular contribution to the subject. I would also like to thank Miklos Abert, Aurélien Alvarez, Uri Bader, Damien Gaboriau, Alexander Kechris, Sorin Popa, Yehuda Shalom, and the referee for the corrections and comments on the earlier drafts of the paper.

ORGANIZATION OF THE PAPER. The paper is organized as follows: the next section is devoted to a general introduction that emphasizes the relations among measure equivalence, quasi-isometry, and orbit equivalence in ergodic theory. One may choose to skip most of this, but read Definition 2.1 and the following remarks. Section 3 concerns groups considered up to measure equivalence. Section 4 focuses on the notion of equivalence relations with orbit relations as a prime (but not only) example. In both of these sections we consider separately the *invariants* of the studied objects (groups and relations) and *rigidity* results, which pertain to questions of classification. Section 5 describes the main techniques used in these theories (mostly for rigidity): a discussion of superrigidity phenomena and some of the ad hoc tools used in the subject; generalities on cocycles appear in Appendix A.

2. Preliminary Discussion and Remarks

This section contains an introduction to measure equivalence and related topics and contains a discussion of this framework. Readers familiar with the subject (especially Definition 2.1 and the following remarks) may skip to the next section in the first reading.

There are two natural entry points to Measured group theory, corresponding to the ergodic-theoretic and group-theoretic perspectives. Let us start from the latter.

2.1. Lattices and Other Countable Groups

When should two infinite discrete groups be viewed as closely related? Isomorphism of abstract groups is an obvious, maybe trivial, answer. The next degree of closeness would be **commensurability**: two groups are commensurable if they contain isomorphic subgroups of finite index. This relation might be relaxed a bit further, by allowing to pass to a quotient modulo finite normal subgroups. The algebraic notion of being commensurable, modulo finite kernels, can be vastly generalized in two directions: measure equivalence (measured group theory) and quasi-isometry (geometric group theory).

The key notion discussed in this paper is that of **measure equivalence** of groups, introduced by Gromov in [66, 0.5.E].

DEFINITION 2.1. Two infinite discrete countable groups Γ, Λ are **measure equivalent** (abbreviated as ME, and denoted $\Gamma \overset{\text{ME}}{\sim} \Lambda$) if there exists an infinite measure space (Ω, m) with a measurable, measure-preserving action of $\Gamma \times \Lambda$, so that both actions $\Gamma \curvearrowright (\Omega, m)$ and $\Lambda \curvearrowright (\Omega, m)$ admit finite-measure fundamental domains $Y, X \subset \Omega$:

$$\Omega = \bigsqcup_{\gamma \in \Gamma} \gamma Y = \bigsqcup_{\lambda \in \Lambda} \lambda X.$$

The space (Ω, m) is called a (Γ, Λ)-**coupling** or ME-coupling. The **index** of Γ to Λ in Ω is the ratio of the measures of the fundamental domains

$$[\Gamma : \Lambda]_\Omega = \frac{m(X)}{m(Y)} \quad \left(= \frac{\text{meas}(\Omega/\Lambda)}{\text{meas}(\Omega/\Gamma)} \right).$$

We shall motivate this definition after making a few immediate comments.

a) The index $[\Gamma : \Lambda]_\Omega$ is well defined—it does not depend on the choice of the fundamental domains X, Y for Ω/Λ, Ω/Γ respectively, because their measures are determined by the group actions on (Ω, m). However, a given pair (Γ, Λ) might have ME-couplings with different indices (the set $\{[\Gamma : \Lambda]_\Omega\}$ is a coset of a subgroup of $\mathbf{R}_+^*$ corresponding to possible indices $[\Gamma : \Gamma]_\Omega$ of self-Γ-couplings. Here it makes sense to focus on *ergodic* couplings only).

b) Any ME-coupling can be decomposed into an integral over a probability space of *ergodic* ME-couplings, that is, ones for which the $\Gamma \times \Lambda$-action is ergodic.

c) measure equivalence is indeed an *equivalence relation* between groups: for any countable Γ the action of $\Gamma \times \Gamma$ on $\Omega = \Gamma$ with the counting measure m_Γ by $(\gamma_1, \gamma_2) : \gamma \mapsto \gamma_1 \gamma \gamma_2^{-1}$ provides the **trivial** self-ME-coupling, giving reflexivity; symmetry is obvious from the definition;[2] while transitivity follows from the following construction of **composition**, or **fusion**, of ME-couplings. If (Ω, m) is a (Γ_1, Γ_2) coupling and (Ω', m') is a (Γ_2, Γ_3) coupling, then the quotient $\Omega'' = \Omega \times_{\Gamma_2} \Omega'$ of $\Omega \times \Omega'$ under the diagonal action $\gamma_2 : (\omega, \omega') \mapsto (\gamma_2 \omega, \gamma_2^{-1} \omega')$ inherits a measure $m'' = m \times_{\Gamma_2} m'$ so that (Ω'', m'') becomes a (Γ_1, Γ_3) coupling structure. The indices satisfy:

$$[\Gamma_1 : \Gamma_3]_{\Omega''} = [\Gamma_1 : \Gamma_2]_\Omega \cdot [\Gamma_2 : \Gamma_3]_{\Omega'}.$$

d) The notion of ME can be extended to the broader class of all *unimodular locally compact second countable* groups: a ME-coupling of G and H is a measure space (Ω, m) with measure space isomorphisms

$$i : (G, m_G) \times (Y, \nu) \cong (\Omega, m), \qquad j : (H, m_H) \times (X, \mu) \cong (\Omega, m)$$

with (X, μ), (Y, ν) being *finite* measure spaces, so that the actions $G \curvearrowright (\Omega, m)$, $H \curvearrowright (\Omega, m)$ given by $g : i(g', y) \mapsto i(gg', y)$, $h : j(h', x) \mapsto j(hh', x)$ commute. The index is defined by $[G : H]_\Omega = \mu(X)/\nu(Y)$.

e) Measure equivalence between countable groups can be viewed as a *category*, whose *objects* are countable groups and *morphisms* between, say Γ and Λ, are possible (Γ, Λ) couplings. Composition of morphisms is the operation of composition of ME-couplings as in (c). The trivial ME-coupling (Γ, m_Γ) is nothing but the *identity* of the object Γ. It is also useful to consider **quotient maps** $\Phi : (\Omega_1, m_1) \to (\Omega_2, m_2)$ between (Γ, Λ)-couplings (these are 2-morphisms in the category), which are assumed to be $\Gamma \times \Lambda$ nonsingular maps, that is, $\Phi_*[m_1] \sim m_2$. Since

---

2. One should formally distinguish between (Ω, m) as a (Γ, Λ) coupling, and the same space with the same actions as a (Λ, Γ) coupling; hereafter we shall denote the latter by $(\check{\Omega}, \check{m})$. Example 2.2 illustrates the need to do so.

preimage of a fundamental domain is a fundamental domain, it follows (under the ergodicity assumption) that $m_1(\Phi^{-1}(E)) = c \cdot m_2(E)$, $E \subset \Omega_2$, where $0 < c < \infty$. ME self-couplings of Γ that have the trivial Γ-coupling are especially useful, their cocycles are conjugate to isomorphisms. Similarly, (Γ, Λ)-couplings, which have a discrete coupling as a quotient, correspond to virtual isomorphisms (see Lemma 4.18).

f) Finally, one might relax the definition of quotients by considering equivariant maps $\Phi : \Omega_1 \to \Omega_2$ between (Γ_i, Λ_i)-couplings (Ω_i, m_i) with respect to homomorphisms $\Gamma_1 \to \Gamma_2$, $\Lambda_1 \to \Lambda_2$ with finite kernels and cokernels.

Gromov's motivation for ME comes from the theory of lattices. Recall that a subgroup Γ of a locally compact second countable (**lcsc** for short) group G is a **lattice** if Γ is discrete in G and the quotient space G/Γ carries a finite G-invariant Borel regular measure (necessarily unique up to normalization); equivalently, if the Γ-action on G by left (equivalently, right) translations admits a Borel fundamental domain of finite positive Haar measure. A discrete subgroup $\Gamma < G$ with G/Γ being compact is automatically a lattice. Such lattices are called **uniform** or **cocompact**; others are **nonuniform**. The standard example of a nonuniform lattice is $\Gamma = \mathrm{SL}_n(\mathbf{Z})$ in $G = \mathrm{SL}_n(\mathbf{R})$. Recall that a lcsc group that admits a lattice is necessarily unimodular.

A common theme in the study of lattices (say in Lie, or algebraic groups over local fields) is that certain properties of the ambient group are inherited by its lattices. From this perspective it is desirable to have a general framework in which lattices in the same group are considered equivalent. Measure equivalence provides such a framework.

EXAMPLE 2.2. If Γ and Λ are lattices in the same lcsc group G, then $\Gamma \overset{\mathrm{ME}}{\sim} \Lambda$; the group G with the Haar measure m_G is a (Γ, Λ) coupling where

$$(\gamma, \lambda) : g \mapsto \gamma g \lambda^{-1}.$$

(In fact, $\Gamma \overset{\mathrm{ME}}{\sim} G \overset{\mathrm{ME}}{\sim} \Lambda$ if ME is considered in the broader context of unimodular lcsc groups: $G \times \{pt\} \cong \Gamma \times G/\Gamma$). This example also illustrates the fact that the *dual* (Λ, Γ)-coupling $\check{G}$ is better related to the original (Γ, Λ)-coupling G via $g \mapsto g^{-1}$ rather than the identity map.

In **geometric group theory** the basic notion of equivalence is **quasi-isometry** (QI). Two metric spaces (X_i, d_i), $i = 1, 2$ are quasi-isometric (notation: $X_1 \overset{\mathrm{QI}}{\sim} X_2$) if there exist maps $f : X_1 \to X_2, g : X_2 \to X_1$, and constants M, A so that

$$d_2(f(x), f(x')) < M \cdot d_1(x, x') + A \qquad\qquad (x, x' \in X_1)$$

$$d_2(g(y), g(y')) < M \cdot d_2(y, y') + A \qquad\qquad (y, y' \in X_2)$$

$$d_1(g \circ f(x), x) < A \qquad\qquad (x \in X_1)$$

$$d_2(f \circ g(y), y) < A \qquad\qquad (y \in X_2).$$

Two finitely generated groups are QI if their Cayley graphs (with respect to some/any finite sets of generators) are QI as metric spaces. It is easy to see that finitely generated groups commensurable modulo finite groups are QI.

Gromov observes that QI between groups can be characterized as **topological equivalence** (TE) defined in the following statement.

THEOREM 2.3. (GROMOV [66, THEOREM 0.2.C$'_2$]) *Two finitely generated groups* Γ *and* Λ *are quasi-isometric iff there exists a locally compact space* Σ *with a continuous action of* $\Gamma \times \Lambda$, *where both actions* $\Gamma \curvearrowright \Sigma$ *and* $\Lambda \curvearrowright \Sigma$ *are properly discontinuous and cocompact.*

The space X in the above statement is called a **TE-coupling**. Here is an idea for the proof. Given a TE-coupling Σ one obtains a quasi-isometry from any point $p \in \Sigma$ by choosing $f : \Gamma \to \Lambda$, $g : \Lambda \to \Gamma$ so that $yp \in f(y)X$ and $\lambda p \in g(\lambda)Y$, where $X, Y \subset \Sigma$ are open sets with compact closures and $\Sigma = \bigcup_{y \in \Gamma} yY = \bigcup_{\lambda \in \Lambda} \lambda X$. To construct a TE-coupling Σ from a quasi-isometry $f : \Gamma \to \Lambda$, consider the pointwise closure of the $\Gamma \times \Lambda$-orbit of f in the space of all maps $\Gamma \to \Lambda$ where Γ acts by precomposition on the domain and Λ by postcomposition on the image. For more details see the guided exercise in [67, p. 98].

A nice instance of QI between groups is a situation where the groups admit a common **geometric model**. Here a **geometric model** for a finitely generated group Γ is a (complete) separable metric space (X, d) with a properly discontinuous and cocompact action of Γ on X by *isometries*. If X is a common geometric model for Γ_1 and Γ_2, then $\Gamma_1 \overset{QI}{\sim} X \overset{QI}{\sim} \Gamma_2$. For example, fundamental groups $\Gamma_i = \pi_1(M_i)$ of compact locally symmetric manifolds M_1 and M_2 with the same universal cover $\tilde{M}_1 \cong \tilde{M}_2 = X$ have X as a common geometric model. Notice that the common geometric model X itself does not serve as a TE-coupling because the actions of the two groups do not commute. However, a TE-coupling can be explicitly constructed from the group $G = \text{Isom}(X, d)$, which is a locally compact (in fact, compactly generated due to finite generation assumption on Γ_i) second countable group. Indeed, the isometric actions

$\Gamma_i \curvearrowright (X, d)$ define homomorphisms $\Gamma_i \to G$ with finite kernels and images being uniform lattices. Moreover, the converse is also true: if Γ_1, Γ_2 admit homomorphisms with finite kernels and images being uniform lattices in the same compactly generated second countable group G, then they have a common geometric mode—take G with a (pseudo-)metric arising from an analogue of a word metric using compact sets.

Hence all uniform lattices in the same group G are QI to each other. Yet, typically, nonuniform lattices in G are not QI to uniform ones—see Farb's survey [37] for the QI classification for lattices in semisimple Lie groups.

To summarize this discussion: the notion of measure equivalence is an equivalence relation between countable groups, an important instance of which is given by groups that can be embedded as lattices (uniform or not) in the same lcsc group. It can be viewed as a measure-theoretic analogue of the equivalence relation of being quasi-isometric (for finitely generated groups) by taking Gromov's topological equivalence point of view. An important instance of QI/TE is given by groups that can be embedded as uniform lattices in the same lcsc group. In this situation one has both ME and QI. However, we should emphasize that this is merely an *analogy*: the notions of QI and ME do not imply each other.

2.2. Orbit Equivalence in Ergodic Theory

Ergodic theory investigates dynamical systems from a measure-theoretic point of view. Hereafter we shall be interested in measurable, measure-preserving group actions on a standard nonatomic probability measure space, and will refer to such actions as *probability measure preserving* (**p.m.p.**). It is often convenient to assume the action to be **ergodic**, that is, to require all measurable Γ-invariant sets to be **null** or **conull** (that is, $\mu(E) = 0$ or $\mu(X \setminus E) = 0$).

A basic question in this context concerns possible *orbit structures* of actions. Equivalence of orbit structures is captured by the following notions of orbit equivalence (the notion of an orbit structure itself is discussed in §4.1).

DEFINITION 2.4. Two p.m.p. actions $\Gamma \curvearrowright (X, \mu)$ and $\Lambda \curvearrowright (Y, \nu)$ are **orbit equivalent** (abbreviated OE, denoted $\Gamma \curvearrowright (X, \mu) \overset{\text{OE}}{\sim} \Lambda \curvearrowright (Y, \nu)$) if there exists a measure space isomorphism $T : (X, \mu) \cong (Y, \nu)$ which takes Γ-orbits onto Λ-orbits. More precisely, an orbit equivalence is a Borel isomorphism $T : X' \cong Y'$ between conull subsets $X' \subset X$ and $Y' \subset Y$ with $T_*\mu(E) = \mu(T^{-1}E) = \nu(E)$, $E \subset Y'$ and $T(\Gamma.x \cap X') = \Lambda.T(x) \cap Y'$ for $x \in X'$.

A **weak OE**, or **stable OE** (SOE) is a Borel isomorphism $T : X' \cong Y'$ between *positive measure* subsets $X' \subset X$ and $Y' \subset Y$ with $T_*\mu_{X'} = \nu_{Y'}$, where $\mu_{X'} = \mu(X')^{-1} \cdot \mu|_{X'}$, $\nu_{Y'} = \nu(Y')^{-1} \cdot \nu|_{Y'}$, so that $T(\Gamma.x \cap X') = \Lambda.T(x) \cap Y'$ for all $x \in X'$. The **index** of such an SOE-map T is $\mu(Y')/\nu(X')$.

In the study of orbit structure of dynamical systems in the topological or smooth category, one often looks at such objects as fixed or periodic points/orbits. Despite the important role these notions play in the underlying dynamical system, periodic orbits have zero measure and therefore are invisible from the purely measure-theoretic standpoint. Hence OE in ergodic theory is a study of the global orbit structure. This point of view is consistent with the general philosophy of "noncommutative measure theory," that is, von Neumann algebras. Specifically, OE in ergodic theory is closely related to the theory of II_1 factors as follows.

In the 1940s Murray and von Neumann introduced the so-called "group-measure space" construction to provide interesting examples of von Neumann factors:[3] given a probability measure preserving (or more generally, nonsingular) group action $\Gamma \curvearrowright (X, \mu)$ the associated von Neumann algebra $M_{\Gamma \curvearrowright X}$ is a cross-product of Γ with the abelian algebra $L^\infty(X, \mu)$, namely the weak closure in bounded operators on $L^2(\Gamma \times X)$ of the algebra generated by the operators $\{f(g, x) \mapsto f(\gamma g, \gamma.x) : \gamma \in \Gamma\}$ and $\{f(g, x) \mapsto \phi(x)f(g, x) : \phi \in L^\infty(X, \mu)\}$. Ergodicity of $\Gamma \curvearrowright (X, \mu)$ is equivalent to $M_{\Gamma \curvearrowright X}$ being a factor. It turns out that (for *essentially free*) OE actions $\Gamma \curvearrowright X \overset{OE}{\sim} \Lambda \curvearrowright Y$ the associated algebras are isomorphic $M_{\Gamma \curvearrowright X} \cong M_{\Lambda \curvearrowright Y}$, with the isomorphism identifying the abelian subalgebras $L^\infty(X)$ and $L^\infty(Y)$. The converse is also true (one has to specify, in addition, an element in $H^1(\Gamma \curvearrowright X, \mathbf{T})$)—see Feldman-Moore [39, 40]. So orbit equivalence of (essentially free p.m.p. group actions) fits into the study of II_1 factors $M_{\Gamma \curvearrowright X}$ with a special focus on the so-called *Cartan subalgebra* given by $L^\infty(X, \mu)$. We refer the reader to Popa's 2006 ICM lecture [114] and Vaes's Seminar Bourbaki paper [136] for some more recent reports on this rapidly developing area.

The above-mentioned assumption of **essential freeness** of an action $\Gamma \curvearrowright (X, \mu)$ means that, up to a null set, the action is free; equivalently, for μ-a.e. $x \in X$ the stabilizer $\{\gamma \in \Gamma : \gamma.x = x\}$ is trivial. This is a natural assumption, when one wants the acting group Γ to "fully reveal itself" in a.e. orbit of the action. Let us now link the notions of OE and ME.

---

3. von Neumann algebras whose center consists only of scalars.

THEOREM 2.5. *Two countable groups Γ and Λ are measure equivalent iff they admit essentially free (ergodic) probability measure-preserving actions $\Gamma \curvearrowright (X, \mu)$ and $\Lambda \curvearrowright (Y, \nu)$ that are stably orbit equivalent.*

(SOE) $\Longrightarrow$ (ME) direction is more transparent in the special case of orbit equivalence, that is, index 1. Let $\alpha : \Gamma \times X \to \Lambda$ be the cocycle associated to an orbit equivalence $T : (X, \mu) \to (Y, \nu)$ defined by $T(g.x) = \alpha(g, x).T(x)$ (here freeness of $\Lambda \curvearrowright Y$ is used). Consider $(\Omega, m) = (X \times \Lambda, \mu \times m_\Lambda)$ with the actions

$$2.1 \qquad g : (x, h) \mapsto (gx, \alpha(g, x)h), \qquad h : (x, k) \mapsto (x, hk^{-1}) \qquad (g \in \Gamma, h \in \Lambda).$$

Then $X \times \{1\}$ is a common fundamental domain for both actions (note that here freeness of $\Gamma \curvearrowright X$ is used). Of course, the same coupling (Ω, m) can be viewed as $(Y \times \Gamma, \nu \times m_\Gamma)$ with the Λ-action defined using $\beta : \Lambda \times Y \to \Gamma$ given by $T^{-1}(h.y) = \beta(h, y).T^{-1}(y)$. In the more general setting of *stable* OE one needs to adjust the definition for the cocycles (see [45]) to carry out a similar construction.

Alternative packaging for the (OE) $\Longrightarrow$ (ME) argument uses the language of equivalence relations (see §4.1). Identifying Y with X via T^{-1}, one views $\mathscr{R}_{\Lambda \curvearrowright Y}$ and $\mathscr{R}_{\Gamma \curvearrowright X}$ as a single relation $\mathscr{R}$. Taking $\Omega = \mathscr{R}$ equipped with the measure $\tilde{\mu}$ (§4.1) consider the actions

$$g : (x, y) \mapsto (g.x, y), \qquad h : (x, y) \mapsto (x, h.y) \qquad (g \in \Gamma, h \in \Lambda).$$

Here the diagonal embedding $X \mapsto \mathscr{R}$, $x \mapsto (x, x)$, gives the fundamental domain for both actions.

(ME) $\Longrightarrow$ (SOE). Given an ergodic (Γ, Λ) coupling (Ω, m), let $X, Y \subset \Omega$ be fundamental domains for the Λ, Γ actions; these may be chosen so that $m(X \cap Y) > 0$. The finite measure-preserving actions

$$2.2 \qquad \Gamma \curvearrowright X \cong \Omega/\Lambda, \qquad \Lambda \curvearrowright Y \cong \Omega/\Gamma$$

have weakly isomorphic orbit relations, since they appear as the restrictions to X and Y of the relation $\mathscr{R}_{\Gamma \times \Lambda \curvearrowright \Omega}$ (of type II_∞); these restrictions coincide on $X \cap Y$. The index of this SOE coincides with the ME-index $[\Gamma : \Lambda]_\Omega$ (if $[\Gamma : \Lambda]_\Omega = 1$ one can find a common fundamental domain $X = Y$). The only remaining issue is that the actions $\Gamma \curvearrowright X \cong \Omega/\Lambda$, $\Lambda \curvearrowright Y \cong \Omega/\Gamma$ may not be essential free. This can be fixed (see [56]) by passing to an extension $\Phi :$ $(\bar{\Omega}, \bar{m}) \to (\Omega, m)$ where $\Gamma \curvearrowright \bar{\Omega}/\Lambda$ and $\Lambda \curvearrowright \bar{\Omega}/\Gamma$ are essentially free. Indeed, take $\bar{\Omega} = \Omega \times Z \times W$, where $\Lambda \curvearrowright Z$ and $\Lambda \curvearrowright W$ are *free probability measure-preserving actions* and let

$$g : (\omega, z, w) \mapsto (g\omega, gz, w), \qquad h : (\omega, z, w) \mapsto (h\omega, z, hw) \qquad (g \in \Gamma, h \in \Lambda).$$

REMARK 2.6. *Freeness of actions is mostly used in order to define the rearrangement cocycles for a (stable) orbit equivalence between actions. However, if SOE comes from a ME-coupling, the well-defined ME-cocycles satisfy the desired rearrangement property (such as $T(g.x) = \alpha(g, x).T(x)$) and freeness becomes superfluous.*

If $\Phi : \bar{\Omega} \to \Omega$ is as above and $\bar{X}, \bar{Y}$ denote the preimages of X, Y, then $\bar{X}, \bar{Y}$ are Λ, Γ fundamental domains and the OE-cocycles $\Gamma \curvearrowright \bar{X} \overset{SOE}{\sim} \Lambda \curvearrowright \bar{Y}$ coincide with the ME-cocycles associated with $X, Y \subset \Omega$.

Another, essentially equivalent, point of view is that ME-coupling defines a weak isomorphism between the **groupoids** *$\Gamma \curvearrowright \Omega/\Lambda$ and $\Lambda \curvearrowright \Omega/\Gamma$. In case of free actions these groupoids reduce to their relations groupoids, but in general the information about stabilizers is carried by the ME-cocycles.*

2.3. Further Comments on QI, ME, and Related Topics

Let Σ be Gromov's topological equivalence between Γ and Λ. Then any point $x \in \Sigma$ defines a quasi-isometry $q_x : \Gamma \to \Lambda$ (see the sketch of proof of Theorem 2.3). In ME the maps $\alpha(-, x) : \Gamma \to \Lambda$ defined for a.e. $x \in X$ play a similar role. However, due to their measure-theoretic nature, such maps are insignificant taken individually, and are studied as a measured family with the additional structure given by the cocycle equation.

Topological and measure equivalences are related to the following interesting notion, introduced by Nicolas Monod in [98] under the appealing term "**randomorphisms.**" Consider the Polish space Λ^Γ of all maps $f : \Gamma \to \Lambda$ with the product uniform topology, and let

$$[\Gamma, \Lambda] = \left\{ f : \Gamma \to \Lambda : f(e_\Gamma) = e_\Lambda \right\}.$$

Then Γ acts on $[\Gamma, \Lambda]$ by $g : f(x) \mapsto f(xg)f(g)^{-1}$, $x \in \Gamma$. The basic observation is that homomorphisms $\Gamma \to \Lambda$ are precisely Γ-fixed points of this action.

DEFINITION 2.7. A **randomorphism** is a Γ-invariant probability measure on $[\Gamma, \Lambda]$.

A measurable cocycle $c : \Gamma \times X \to \Lambda$ over a p.m.p. action $\Gamma \curvearrowright (X, \mu)$ defines a randomorphism by pushing forward the measure μ by the cocycle $x \mapsto c(-, x)$. Thus orbit equivalence cocycles (see Appendix A.2) correspond to randomorphisms supported on *bijections* in $[\Gamma, \Lambda]$. Also note that the natural composition operation for randomorphisms, given by the push-forward of the measures under the natural map

$$[\Gamma_1, \Gamma_2] \times [\Gamma_2, \Gamma_3] \to [\Gamma_1, \Gamma_3], \qquad (f, g) \mapsto g \circ f$$

corresponds to the composition of couplings. The viewpoint of topological dynamics of the Γ-action on $[\Gamma, \Lambda]$ may be related to quasi-isometries and topological equivalence. For example, points in $[\Gamma, \Lambda]$ with precompact Γ-orbits correspond to Lipschitz embeddings $\Gamma \to \Lambda$.

2.3.1. USING ME FOR QI Although measure equivalence and quasi-isometry are parallel in many ways, these concepts are different and neither one implies the other. Yet, Yehuda Shalom has shown [132] how one can use ME ideas to study QI of *amenable* groups. The basic observation is that a topological coupling Σ of amenable groups Γ and Λ carries a $\Gamma \times \Lambda$-invariant measure m (coming from a Γ-invariant *probability measure* on Σ/Λ), which gives a measure equivalence. It can be thought of as an invariant distribution on quasi-isometries $\Gamma \to \Lambda$, and can be used to induce unitary representations, cohomology with unitary coefficients, and so on from Λ to Γ. Using such constructions, Shalom [132] was able to obtain a list of new QI invariants in the class of amenable groups, such as (co)homology over **Q**, ordinary Betti numbers $\beta_i(\Gamma)$ among nilpotent groups, and others. Shalom also studied the notion of **uniform embedding** (UE) between groups and obtained group invariants, which are *monotonic* with respect to UE.

In [125] Roman Sauer obtains further QI-invariants and UE-monotonic invariants using a combination of QI, ME, and homological methods.

In another work [126] Sauer used ME point of view to attack problems of purely topological nature related to the work of Gromov.

2.3.2. ℓ^p-MEASURE EQUIVALENCE Let Γ and Λ be finitely generated groups, equipped with some word metrics $|\cdot|_\Gamma$, $|\cdot|_\Lambda$. We say that a (Γ, Λ) coupling (Ω, m) is ℓ^p for some $1 \leq p \leq \infty$ if there exist fundamental domains $X, Y \subset \Omega$ so that the associated ME-cocycles (see Appendix A.3) $\alpha : \Gamma \times X \to \Lambda$ and $\beta : \Lambda \times Y \to \Gamma$ satisfy

$$\forall g \in \Gamma : \quad |\alpha(g, -)|_\Lambda \in L^p(X, \mu), \qquad \forall h \in \Lambda : \quad |\beta(h, -)|_\Gamma \in L^p(Y, \nu).$$

If an ℓ^p-ME-coupling exists, say that Γ and Λ are ℓ^p-ME. Clearly any ℓ^p-ME-coupling is ℓ^q for all $q \leq p$. So ℓ^1-ME is the weakest and ℓ^∞-ME is the most stringent among these relations. One can check that ℓ^p-ME is an equivalence relation on groups (the ℓ^p condition is preserved under composition of couplings), so we obtain a hierarchy of ℓ^p-ME categories with ℓ^1-ME being the weakest (largest classes) and at $p = \infty$ one arrives at ME + QI. Thus ℓ^p-ME amounts to measure equivalence with some geometric flavor.

The setting of ℓ^1-ME is considered in [13, 14] by Uri Bader, Roman Sauer, and the author to analyze rigidity of the least rigid family of lattices—lattices in $SO_{n,1}(\mathbf{R}) \simeq \text{Isom}(\mathbf{H}_{\mathbf{R}}^n)$, $n \geq 3$, and fundamental groups of general negatively curved manifolds. It should be noted that examples of non amenable ME groups that are not ℓ^1-ME seem to be rare (surface groups and free groups seem to be the main culprits). In particular, it follows from Shalom's computations in [131] that for $n \geq 3$ all lattices in $SO_{n,1}(\mathbf{R})$ are mutually ℓ^1-ME. We shall return to invariants and rigidity in the ℓ^1-ME framework in Sections 3.1.8 and 3.2.4.

3. Measure Equivalence Between Groups

This section is concerned with the notion of measure equivalence between countable groups $\Gamma \overset{\text{ME}}{\sim} \Lambda$ (Definition 2.1). First recall the following deep result (extending previous work of Dye [33, 34] on some amenable groups, and followed by Connes-Feldman-Weiss [27] concerning all nonsingular actions of all amenable groups).

THEOREM 3.1. (ORNSTEIN-WEISS [105]) *Any two ergodic probability measure-preserving actions of any two infinite countable amenable groups are orbit equivalent.*

This result implies that all infinite countable amenable groups are ME; moreover, for any two infinite amenable groups Γ and Λ there exists an ergodic ME-coupling Ω with index $[\Gamma : \Lambda]_{\Omega} = 1$ (hereafter we shall denote this situation by $\Gamma \overset{\text{OE}}{\sim} \Lambda$). Measure equivalence of all amenable groups shows that many QI-invariants are not ME-invariants; these include growth type, being virtually nilpotent, (virtual) cohomological dimension, finite generations/presentation, and so on.

The following are basic constructions and examples of measure equivalent groups:

1) If Γ and Λ can be embedded as lattices in the same lcsc group, then $\Gamma \overset{\text{ME}}{\sim} \Lambda$.

2) If $\Gamma_i \overset{\mathrm{ME}}{\sim} \Lambda_i$ for $i = 1, \ldots, n$, then $\Gamma_1 \times \cdots \times \Gamma_n \overset{\mathrm{ME}}{\sim} \Lambda_1 \times \cdots \times \Lambda_n$.

3) If $\Gamma_i \overset{\mathrm{OE}}{\sim} \Lambda_i$ for $i \in I$ (i.e. the groups admit an ergodic ME-coupling with index 1), then $(*_{i \in I} \Gamma_i) \overset{\mathrm{OE}}{\sim} (*_{i \in I} \Lambda_i)$.[4]

For $2 \neq n, m < \infty$ the free groups $\mathbf{F}_n$ and $\mathbf{F}_m$ are commensurable, and therefore are ME (however, $\mathbf{F}_\infty \overset{\mathrm{ME}}{\not\sim} \mathbf{F}_2$). The measure equivalence class $\mathrm{ME}(\mathbf{F}_{2 \leq n < \infty})$ is very rich and remains mysterious (see [58]). For example, it includes surface groups $\pi_1(\Sigma_g)$, $g \geq 2$, nonuniform (infinitely generated) lattices in $\mathrm{SL}_2 (F_p[[X]])$, the automorphism group of a regular tree, free products $*_{i=1}^n A_i$ of arbitrary infinite amenable groups, more complicated free products such as $\mathbf{F}_2 * \pi_1(\Sigma_g) * \mathbf{Q}$, and so on. In the aforementioned paper by Gaboriau he constructs interesting geometric examples of the form $*_c^n \mathbf{F}_{2g}$, which are fundamental groups of certain "branched surfaces." Bridson, Tweedale, and Wilton [19] prove that a large class of *limit groups*, namely all *elementarily free* groups, are ME to $\mathbf{F}_2$. Notice that $\mathrm{ME}(\mathbf{F}_{2 \leq n < \infty})$ contains uncountably many groups.

The fact that some ME classes are so rich and complicated should emphasize the impressive list of ME invariants and rigidity results below.

3.1. Measure Equivalence Invariants

By ME-**invariants** we mean properties of groups that are preserved under measure equivalence, and numerical invariants that are preserved or predictably transformed as a function of the ME index.

3.1.1. AMENABILITY, KAZHDAN'S PROPERTY (T), AND A-T-MENABILITY These

properties are defined using the language of unitary representations. Let $\pi : \Gamma \to U(\mathscr{H})$ be a unitary representation of a (topological) group. Given a finite (respectively compact) subset $K \subset G$ and $\epsilon > 0$, we say that a unit vector $v \in \mathscr{H}$ is (K, ϵ)-**almost invariant** if $\|v - \pi(g)v\| < \epsilon$ for all $g \in K$. A unitary Γ-representation π that has (K, ϵ)-almost invariant vectors for all $K \subset \Gamma$ and $\epsilon > 0$ is said to **weakly contain** the trivial representation 1_Γ, denoted $1_\Gamma \prec \pi$. The trivial representation 1_Γ is **(strongly) contained** in π, denoted $1_\Gamma < \pi$, if there exist non-zero $\pi(G)$-invariant vectors, that is, $\mathscr{H}^{\pi(\Gamma)} \neq \{0\}$. Of course $1_\Gamma < \pi$ trivially implies $1_\Gamma \prec \pi$. We recall:

AMENABILITY: Γ is **amenable** if the trivial representation is weakly contained in the regular representation $\rho : \Gamma \to U(\ell^2(\Gamma))$, $\rho(g)f(x) = f(g^{-1}x)$.

---

4. The appearance of the sharper condition $\overset{\mathrm{OE}}{\sim}$ in (2) is analogous to the one in the QI context: if groups Γ_i and Λ_i are *bi-Lipschitz*, then $*_{i \in I} \Gamma_i \overset{\mathrm{QI}}{\sim} *_{i \in I} \Lambda_i$.

PROPERTY (T):Γ has **property (T)** (Kazhdan [79]) if for every unitary Γ-representation π: $1_\Gamma \prec \pi$ implies $1_\Gamma < \pi$. This is equivalent to an existence of a compact $K \subset \Gamma$ and $\epsilon > 0$ so that any unitary Γ-representation π with (K, ϵ)-almost invariant vectors has nontrivial invariant vectors. For compactly generated groups, another equivalent characterization (Delorme and Guichardet) is that any affine isometric Γ-action on a Hilbert space has a fixed point, that is, if $H^1(\Gamma, \pi) = \{0\}$ for any (orthogonal) Γ-representation π. We refer to [17] for the details.

(HAP):Γ is **a-T-menable** (or has **Haagerup approximation property**) if the following equivalent conditions hold: (i) Γ has a mixing Γ-representation weakly containing the trivial one, or (ii) Γ has a proper affine isometric action on a (real) Hilbert space. The class of infinite a-T-menable groups contains amenable groups, free groups but is disjoint from infinite groups with property (T). See [24] as a reference.

Measure equivalence allows us to relate unitary representations of one group to another. More concretely, let (Ω, m) be a (Γ, Λ) coupling, and $\pi : \Lambda \to U(\mathcal{H})$ be a unitary Λ-representation. Denote by $\tilde{\mathcal{H}}$ the Hilbert space consisting of equivalence classes (mod null sets) of all measurable, Λ-equivariant maps $\Omega \to \mathcal{H}$ with square-integrable norm over a Λ-fundamental domain:

$$\tilde{\mathcal{H}} = \left\{ f : \Omega \to \mathcal{H} : f(\lambda x) = \pi(\lambda) f(x), \int_{\Omega/\Lambda} \|f\|^2 < \infty \right\} \qquad \text{mod null sets.}$$

The action of Γ on such functions by translation of the argument defines a unitary Γ-representation $\tilde{\pi} : \Gamma \to U(\tilde{\mathcal{H}})$. This representation is said to be **induced** from $\pi : \Lambda \to U(\mathcal{H})$ via Ω. (In Example 2.2 this is precisely the usual Mackey induction of a unitary representations, of a lattice to the ambient group, followed by a restriction to another lattice).

The ME invariance of the properties above (amenability, property (T), Haagerup approximation property) can be deduced from the following observations. Let (Ω, m) be a (Γ, Λ) ME-coupling, $\pi : \Lambda \to U(\mathcal{H})$ a unitary representation, and $\tilde{\pi} : \Gamma \to U(\tilde{\mathcal{H}})$ the corresponding induced representation. Then

1) If π is the regular Λ-representation on $\mathcal{H} = \ell^2(\Lambda)$, then $\tilde{\pi}$ on $\tilde{\mathcal{H}}$ can be identified with the Γ-representation on $L^2(\Omega, m) \cong n \cdot \ell^2(\Gamma)$, where $n = \dim L^2(\Omega/\Lambda) \in \{1, 2, \ldots, \infty\}$.

2) If $1_\Lambda \prec \pi$, then $1_\Gamma \prec \tilde{\pi}$.

3) If (Ω, m) is $\Gamma \times \Lambda$ ergodic and π is *weakly mixing* (i.e., $1_\Lambda \not\prec \pi \otimes \pi^*$), then $1_\Gamma \not\prec \tilde{\pi}$.

4) If (Ω, m) is $\Gamma \times \Lambda$ ergodic and π is *mixing* (i.e., for all $v \in \mathcal{H}$: $\langle \pi(h)v, v \rangle \to$ 0 as $h \to \infty$ in Λ), then $\tilde{\pi}$ is a mixing Γ-representation.

Combining (1) and (2) we obtain that being amenable is an ME-invariant. The deep result of Ornstein-Weiss [105] and Theorem 2.5 imply that any two infinite countable amenable groups are ME. This gives

COROLLARY 3.2. *The measure equivalence class of* $\mathbf{Z}$ *is the class of all infinite countable amenable groups*

$$\mathrm{ME}(\mathbf{Z}) = \mathrm{Amen}.$$

Bachir Bekka and Alain Valette [16] showed that if Λ does not have property (T), then it admits a *weakly mixing* representation π weakly containing the trivial one. By (2) and (3) this implies that property (T) is an ME-invariant (this is the argument in [44, corollary 1.4]; see also Zimmer [146, theorem 9.1.7(b)]). The ME-invariance of amenability and Kazhdan's property for groups indicates that it should be possible to define these properties for *equivalence relations* and then relate them to groups. This was done by Zimmer in [138, 141] and was recently further studied in the context of measured groupoids in [8, 9]. We return to this discussion in §4.2.1. The ME-invariance of a-T-menability follows from (2) and (4); see [24, 76].

3.1.2. COST OF GROUPS The notion of the cost of an action/relation was introduced by Levitt [90] and developed by Damien Gaboriau [53, 54, 56]; the monographs [81] and [80] also contain an extensive discussion of this topic.

The cost of an essentially free p.m.p. action $\Gamma \curvearrowright (X, \mu)$, denoted cost $(\Gamma \curvearrowright X)$, is the cost of the corresponding orbit relations cost $(\mathcal{R}_{\Gamma \curvearrowright X})$ as defined in §4.2.4 (it is the infimum of the weights of generating systems for the groupoid where the "weight" is the sum of the measures of the domain/image sets of the generating system). The cost of an action can be turned into a group invariant/(s) by setting

$$\mathscr{C}_*(\Gamma) = \inf_X \ \mathrm{cost}\,(\Gamma \curvearrowright X), \qquad \mathscr{C}^*(\Gamma) = \sup_X \ \mathrm{cost}\,(\Gamma \curvearrowright X)$$

where the infimum/supremum are taken over all essentially free p.m.p. actions of Γ (we drop ergodicity assumption here; in the definition of $\mathscr{C}_*(\Gamma)$ essential freeness is also superfluous). Groups Γ for which $\mathscr{C}_*(\Gamma) = \mathscr{C}^*(\Gamma)$ are said to have fixed price, or **prix fixe** (abbreviated P.F.). For general groups,

Gaboriau defined **the cost of a group** to be the lower one:

$$\mathscr{C}(\Gamma) = \mathscr{C}_*(\Gamma).$$

To avoid confusion, we shall use here the notation $\mathscr{C}_*(\Gamma)$ for general groups, and reserve $\mathscr{C}(\Gamma)$ for P.F. groups only.

QUESTION 3.3. *Do all countable groups have property P.F.?*

The importance of this question will be illustrated in §4.2.4; for example, a positive answer to an apparently weaker Question 4.11 would have applications to groups theory and 3-manifold (Abert-Nikolov [1]).

The properties $\mathscr{C}_* = 1$, $1 < \mathscr{C}_* < \infty$, and $\mathscr{C}_* = \infty$ are ME-invariants. More precisely:

THEOREM 3.4. *If* $\Gamma \overset{ME}{\sim} \Lambda$, *then* $\mathscr{C}_*(\Lambda) - 1 = [\Gamma : \Lambda]_\Omega \cdot (\mathscr{C}_*(\Gamma) - 1)$ *for some/any* (Γ, Λ)-*coupling* Ω.

We do not know whether the same holds for $\mathscr{C}^*$. Note that in [54] this ME-invariance is stated for P.F. groups only.

Proof. Let Ω be a (Γ, Λ)-coupling with $\Gamma \curvearrowright X = \Omega/\Lambda$ and $\Lambda \curvearrowright Y = \Omega/\Gamma$ being free, where $X, Y \subset \Omega$ are Λ-, Γ-fundamental domains. Given any essentially free p.m.p. action $\Lambda \curvearrowright Z$, consider the (Γ, Λ)-coupling $\bar{\Omega} = \Omega \times Z$ with the actions

$$g : (\omega, z) \mapsto (g\omega, z), \qquad h : (\omega, z) \mapsto (h\omega, hz) \qquad (g \in \Gamma, \, h \in \Lambda).$$

The actions $\Gamma \curvearrowright \bar{X} = \bar{\Omega}/\Lambda$ and $\Lambda \curvearrowright \bar{Y} = \bar{\Omega}/\Gamma$ are stably orbit equivalent with index $[\Gamma : \Lambda]_{\bar{\Omega}} = [\Gamma : \Lambda]_\Omega = c$. Hence (using Theorem 4.7 below) we have

$$c \cdot (\mathrm{cost}\,(\mathscr{R}_{\Gamma \curvearrowright \bar{X}}) - 1) = \mathrm{cost}\,(\mathscr{R}_{\Lambda \curvearrowright \bar{Y}}) - 1.$$

While $\Gamma \curvearrowright \bar{X}$ is a skew product over $\Gamma \curvearrowright X$, the action $\Lambda \curvearrowright \bar{Y}$ is the diagonal action on $\bar{Y} = Y \times Z$. Since $\bar{Y} = Y \times Z$ has Z as a Λ-equivariant quotient, it follows (by considering preimages of any "graphing system") that

$$\mathrm{cost}\,(\Lambda \curvearrowright \bar{Y}) \leq \mathrm{cost}\,(\Lambda \curvearrowright Z).$$

Since $\Lambda \curvearrowright Z$ was arbitrary, we deduce $\mathscr{C}_*(\Lambda) - 1 \geq c \cdot (\mathscr{C}_*(\Gamma) - 1)$. A symmetric argument completes the proof. $\square$

THEOREM 3.5. (GABORIAU [53,54,56]) *The following classes of groups have P.F.:*

1) *Any finite group Γ has $\mathscr{C}_*(\Gamma) = \mathscr{C}^*(\Gamma) = 1 - \frac{1}{|\Gamma|}$.*
2) *Infinite amenable groups have $\mathscr{C}_*(\Gamma) = \mathscr{C}^*(\Gamma) = 1$.*
3) *Free group $\mathbf{F}_n$, $1 \leq n \leq \infty$, have $\mathscr{C}_*(\mathbf{F}_n) = \mathscr{C}^*(\mathbf{F}_n) = n$.*
4) *Surface groups $\Gamma = \pi_1(\Sigma_g)$ where Σ_g is a closed orientable surface of genus $g \geq 2$ have $\mathscr{C}_*(\Gamma) = \mathscr{C}^*(\Gamma) = 2g - 1$.*
5) *Amalgamated products $\Gamma = A *_C B$ of finite groups have P.F. with*

$$\mathscr{C}_*(\Gamma) = \mathscr{C}^*(\Gamma) = 1 - \left(\frac{1}{|A|} + \frac{1}{|B|} - \frac{1}{|C|}\right).$$

In particular, $\mathscr{C}_*(\mathrm{SL}_2(\mathbf{Z})) = \mathscr{C}^*(\mathrm{SL}_2(\mathbf{Z})) = 1 + \frac{1}{12}$.
6) *Assume Γ_1, Γ_2 have P.F., then the free product $\Gamma_1 * \Gamma_2$ and more general amalgamated free products $\Lambda = \Gamma_1 *_A \Gamma_2$ over an amenable group A, has P.F. with*

$$\mathscr{C}(\Gamma_1 * \Gamma_2) = \mathscr{C}(\Gamma_1) + \mathscr{C}(\Gamma_2), \quad \mathscr{C}(\Gamma_1 *_A \Gamma_2) = \mathscr{C}(\Gamma_1) + \mathscr{C}(\Gamma_2) - \mathscr{C}(A).$$

7) *Products $\Gamma = \Gamma_1 \times \Gamma_2$ of infinite nontorsion groups have $\mathscr{C}_*(\Gamma) = \mathscr{C}^*(\Gamma) = 1$.*
8) *Finitely generated groups Γ containing an infinite amenable normal subgroup have $\mathscr{C}_*(\Gamma) = \mathscr{C}^*(\Gamma) = 1$.*
9) *Arithmetic lattices Γ of higher $\mathbf{Q}$-rank (e.g., $\mathrm{SL}_{n \geq 3}(\mathbf{Z})$) have $\mathscr{C}_*(\Gamma) = \mathscr{C}^*(\Gamma) = 1$.*

Note that for an infinite group $\mathscr{C}^*(\Gamma) = 1$ iff Γ has P.F. of cost 1. So the content of cases (2), (7), (8), and (9) is that $\mathscr{C}^*(\Gamma) = 1$.

QUESTION 3.6. *Is it true that for all (irreducible) lattices Γ in a (semi)simple Lie group G of higher rank have P.F. of $\mathscr{C}^*(\Gamma) = 1$? Is it true that any infinite group Γ with Kazhdan's property (T) has P.F. with $\mathscr{C}^*(\Gamma) = 1$?*

Item (9) in Theorem 3.5 provides a positive answer to the first question for some *nonuniform lattices* in higher-rank Lie groups, but the proof relies on the internal structure of such lattices (chains of pairwise commuting elements), rather than on its relation to the ambient Lie group G (which also has a lot of commuting elements). Note also that Theorem 3.4 implies that $\mathscr{C}_*(\Gamma) = 1$ for all higher-rank lattices. The motivation for the second question is that property (T) implies vanishing of the first ℓ^2-Betti number, $\beta_1^{(2)}(\Gamma) = 0$; while for infinite groups it was shown by Gaboriau that

3.1 $$\beta_1^{(2)}(\Gamma) = \beta_1^{(2)}(\mathscr{R}_{\Gamma \curvearrowright X}) \leq \mathrm{cost}(\mathscr{R}_{\Gamma \curvearrowright X}) - 1.$$

Furthermore, there are no known examples of strict inequality. Lattices Γ in higher-rank semisimple Lie groups without property (T) still satisfy $\beta_1^{(2)}(\Gamma) = 0$ (an argument in the spirit of the current discussion is $\beta_1^{(2)}$ for ME groups are positively proportional by Gaboriau's theorem 3.8, an irreducible lattice in a product is ME to a product of lattices and products of infinite groups have $\beta_1^{(2)} = 0$ by the Küneth formula. Shalom's [130] provides a completely geometric explanation).

To give the flavor of the proofs let us indicate the argument for (8) in Theorem 3.5. Let Γ be a group generated by a finite set $\{g_1, \ldots, g_n\}$ and containing an infinite normal amenable subgroup A and $\Gamma \curvearrowright (X, \mu)$ be an essentially free (ergodic) p.m.p. action. Since A is amenable, there is a $\mathbf{Z}$-action on X with $\mathcal{R}_{A \curvearrowright X} = \mathcal{R}_{\mathbf{Z} \curvearrowright X}$ (mod null sets), and we let $\phi_0 : X \to X$ denote the action of the generator of $\mathbf{Z}$. Given $\epsilon > 0$ one can find a subset $E \subset X$ with $0 < \mu(E) < \epsilon$ so that $\bigcup_{a \in A} aE - \bigcup \psi_0^n E = X$ mod null sets (if A-action is ergodic, any positive measure set works; in general, one uses the ergodic decomposition). For $i = 1, \ldots, n$ let ϕ_i be the restriction of g_i to E. Now one easily checks that the normality assumption implies that $\Phi = \{\phi_0, \phi_1, \ldots, \phi_n\}$ generates $\mathcal{R}_{\Gamma \curvearrowright X}$, while $\mathrm{cost}\,(\Phi) = 1 + n\epsilon$.

For general (not necessarily P.F.) groups Γ_i a version of (6) still holds:

$$\mathscr{C}_* (\Gamma_1 * \Gamma_2) = \mathscr{C}_* (\Gamma_1) + \mathscr{C}_* (\Gamma_2),$$

$$\mathscr{C}_* (\Gamma_1 *_A \Gamma_2) = \mathscr{C}_* (\Gamma_1) + \mathscr{C}_* (\Gamma_2) - \mathscr{C} (A)$$

where A is finite or, more generally, amenable.

Very recently Miklos Abert and Benjamin Weiss [2] showed:

THEOREM 3.7. (ABERT-WEISS [2]) *For any discrete countable group Γ, the highest cost $\mathscr{C}^* (\Gamma)$ is attained by nontrivial Bernoulli actions $\Gamma \curvearrowright (X_0, \mu_0)^{\Gamma}$ and their essentially free quotients.*

Some comments are in order. Kechris [80] introduced the following notion: for probability measure-preserving actions of a fixed group Γ say that $\Gamma \curvearrowright (X, \mu)$ **weakly contains** $\Gamma \curvearrowright Y$ if given any finite measurable partition $Y = \bigsqcup_{i=1}^{n} Y_i$, a finite set $F \subset \Gamma$ and an $\epsilon > 0$, there is a finite measurable partition $X = \bigsqcup_{i=1}^{n} X_i$ so that

$$\left| \mu(gX_i \cap X_j) - \nu(gY_i \cap Y_j) \right| < \epsilon \qquad (1 \leq i, j \leq n, \; g \in F).$$

The motivation for the terminology is the fact that weak containment of actions implies (but not equivalent to) weak containment of the corresponding unitary representations: $L^2(Y) \preceq L^2(X)$. It is clear that a quotient is (weakly) contained in the larger action. It is also easy to see that the cost of a quotient action is no less than that of the original (because one can lift any graphing from a quotient to the larger action maintaining the cost of the graphing). Kechris [80] proves that this (anti)monotonicity still holds in the context of weak containment of essentially free actions of finitely generated groups, namely:

$$\Gamma \curvearrowright Y \preceq \Gamma \curvearrowright X \quad \Longrightarrow \quad \text{cost}\,(\Gamma \curvearrowright Y) \geq \text{cost}\,(\Gamma \curvearrowright X).$$

In fact, it follows from the more general fact that cost is upper semicontinuous in the topology of actions. Abert and Weiss prove that Bernoulli actions (and their quotients) are *weakly contained* in any essentially free action of a group. Thus Theorem 3.7 follows from the monotonicity of the cost.

3.1.3. ℓ^2-BETTI NUMBERS

The ℓ^2-Betti numbers of (coverings of) manifolds were introduced by Atiyah in [11]. Cheeger and Gromov [23] defined ℓ^2-Betti numbers $\beta_i^{(2)}(\Gamma) \in [0, \infty]$, $i \in \mathbf{N}$, for arbitrary countable group Γ as dimensions (in the sense of Murray von Neumann) of certain homology groups (which are Hilbert Γ-modules). For reference we suggest [35], [92]. Here let us just point out the following facts:

1) If Γ is infinite amenable, then $\beta_i^{(2)}(\Gamma) = 0$, $i \in \mathbf{N}$.
2) For free groups $\beta_1^{(2)}(\mathbf{F}_n) = n - 1$ and $\beta_i^{(2)}(\mathbf{F}_n) = 0$ for $i > 1$.
3) For groups with property (T), $\beta_1^{(2)}(\Gamma) = 0$.
4) Küneth formula: $\beta_k^{(2)}(\Gamma_1 \times \Gamma_2) = \sum_{i+j=k} \beta_i^{(2)}(\Gamma_1) \cdot \beta_j^{(2)}(\Gamma_2)$.
5) Kazhdan's conjecture, proved by Lück, states that for residually finite groups satisfying appropriate finiteness properties (e.g., finite $K(pi, 1)$) the ℓ^2-Betti numbers are the stable limit of Betti numbers of finite-index subgroups normalized by the index: $\beta_i^{(2)}(\Gamma) = \lim \frac{\beta_i(\Gamma_n)}{[\Gamma:\Gamma_n]}$ where $\Gamma > \Gamma_1 > \ldots$ is a chain of normal subgroups of finite index.
6) The ℓ^2 Euler characteristic $\chi^{(2)}(\Gamma) = \sum (-1)^i \cdot \beta_i^{(2)}(\Gamma)$ coincides with the usual Euler characteristic $\chi(\Gamma) = \sum (-1)^i \cdot \beta_i(\Gamma)$, provided both are defined, as is the case for fundamental group $\Gamma = \pi_1(M)$ of a compact aspherical manifold.
7) According to the Hopf-Singer conjecture the ℓ^2-Betti numbers for a fundamental group $\Gamma = \pi_1(M)$, of a compact aspherical manifold M vanish except, possibly, in the middle dimension n. Atiyah's conjecture states that ℓ^2-Betti numbers are integers.

The following remarkable result of Damien Gaboriau states that these intricate numeric invariants of groups are preserved under measure equivalence, after a rescaling by the coupling index.

THEOREM 3.8. (GABORIAU [55], [57]) *Let $\Gamma \overset{ME}{\sim} \Lambda$ be ME-countable groups. Then*

$$\beta_i^{(2)}(\Lambda) = c \cdot \beta_i^{(2)}(\Gamma) \qquad (i \in \mathbf{N})$$

where $c = [\Gamma : \Lambda]_\Omega$ is an/the index of some/any (Γ, Λ)-coupling.

In fact, Gaboriau introduced the notion of ℓ^2-Betti numbers for II_1-relations and related them to ℓ^2-Betti numbers of groups in case of the orbit relation for an essentially free p.m.p. action—see more comments in §4.2.5 below.

Thus the geometric information encoded in the ℓ^2-Betti numbers for fundamental groups of aspherical manifolds, such as Euler characteristic and sometimes the dimension, pass through measure equivalence. In particular, if lattices Γ_i ($i = 1, 2$) (uniform or not) in $\mathrm{SU}_{n_i,1}(\mathbf{R})$ are ME, then $n_1 = n_2$; the same applies to $\mathrm{Sp}_{n_i,1}(\mathbf{R})$ and $\mathrm{SO}_{2n_i,1}(\mathbf{R})$. (The higher-rank lattices are covered by stronger rigidity statements—see §3.2.1 below). Furthermore, it follows from Gaboriau's result that in general the set

$$D^{(2)}(\Gamma) = \left\{ i \in \mathbf{N} \ : \ 0 < \beta_i^{(2)}(\Gamma) < \infty \right\}$$

is an ME-invariant. Conjecture (7) relates this to the dimension of a manifold M in the case of $\Gamma = \pi_1(M)$. One shouldn't expect dim (M) to be an ME-invariant of $\pi_1(M)$, as the examples of tori show; note also that for any manifold M one has $\pi_1(M \times \mathbf{T}^n) \overset{ME}{\sim} \pi_1(M \times \mathbf{T}^k)$. However, among negatively curved manifolds Theorem 3.13 below shows that dim (M) is an invariant of ℓ^1-ME.

For closed aspherical manifolds M the dimension dim (M) is a QI-invariant of $\pi_1(M)$. Pansu proved that the whole set $D^{(2)}(\Gamma)$ is a QI-invariant of Γ. However, positive proportionality of ℓ^2-Betti numbers for ME fails under QI; in fact, there are QI groups whose Euler characteristics have opposite signs. Yet

COROLLARY 3.9. *For ME groups Γ and Λ with well-defined Euler characteristic, say fundamental groups of compact manifolds, one has*

$$\chi(\Lambda) = c \cdot \chi(\Gamma), \qquad \text{where} \qquad c = [\Gamma : \Lambda]_\Omega \in (0, \infty).$$

In particular, the sign (positive, zero, negative) of the Euler characteristic is an ME-invariant.

3.1.4. COWLING-HAAGERUP Λ-INVARIANT This numeric invariant Λ_G, taking values in $[1, \infty]$, is defined for any lcsc group G in terms of norm bounds on unit approximation in the Fourier algebra $A(G)$ (see Cowling and Haagerup [31]). The Λ-invariant coincides for a lcsc group and its lattices. Moreover, Cowling and Zimmer [32] proved that $\Gamma_1 \overset{OE}{\sim} \Gamma_2$ implies $\Lambda_{\Gamma_1} = \Lambda_{\Gamma_2}$. In fact, their proof implies the invariance under measure equivalence (see [76]). So, Λ_Γ is an ME-invariant.

Cowling and Haagerup [31] computed the Λ-invariant for simple Lie groups and their lattices: in particular, proving that $\Lambda_G = 1$ for $G \simeq SO_{n,1}(\mathbf{R})$ and $SU_{n,1}(\mathbf{R})$, $\Lambda_G = 2n - 1$ for $G \simeq Sp_{n,1}(\mathbf{R})$, and $\Lambda_G = 21$ for the exceptional rank-1 group $G = F_{4(-20)}$.

One may observe that simple Lie groups split into two classes: (1) $SO_{n,1}(\mathbf{R})$ and $SU_{n,1}(\mathbf{R})$ family, and (2) $G \simeq Sp_{n,1}(\mathbf{R})$, $F_{4(-20)}$ and higher rank. Groups in the first class have haagerup approximation property (HAP, a.k.a. a-T-menability) and $\Lambda_G = 1$, while groups in the second class have Kazhdan's property (T) and $\Lambda_G > 1$. Cowling conjectured that $\Lambda_G = 1$ and (HAP) might be equivalent. Recently one implication of this conjecture has been disproved: Cornulier, Stalder, and Valette [30] proved that the wreath product $H \wr F_2$ of a finite group H by the free group F_2 has (HAP), while Ozawa and Popa [108] prove that $\Lambda_{H \wr F_2} > 1$. The question whether $\Lambda_\Gamma = 1$ implies (HAP) is still open.

One may deduce now that if Γ is a lattice in $G \simeq Sp_{n,1}(\mathbf{R})$ or in $F_{4(-20)}$ and Λ is a lattice in a simple Lie group H, then $\Gamma \overset{ME}{\sim} \Lambda$ iff $G \simeq H$. Indeed, higher-rank H are ruled out by Zimmer's theorem 3.15; H cannot be in the families $SO_{n,1}(\mathbf{R})$ and $SU_{n,1}(\mathbf{R})$ by property (T) or Haagerup property; and within the family of $Sp_{n,1}(\mathbf{R})$ and $F_{4(-20)}$ the Λ-invariant detects G (ℓ^2-Betti numbers can also be used for this purpose).

3.1.5. TREEABILITY, ANTITREEABILITY, AND ERGODIC DIMENSION In [4] Scott Adams introduced the notion of **treeable** equivalence relations (see §4.2.3). Following [81], a group Γ is

Treeable: if there *exists* an essentially free p.m.p. Γ-action with a treeable orbit relation.

Strongly treeable: if *every* essentially free p.m.p. Γ-action gives a treeable orbit relation.

Antitreeable: if there *are no* essentially free p.m.p. Γ-actions with a treeable orbit relation.

Amenable groups and free groups are strongly treeable. It seems to be still unknown whether there exist treeable but not strongly treeable groups; in particular, it is not clear whether surface groups (that are treeable) are strongly treeable.

The properties of being treeable or antitreeable are ME-invariants. Moreover, Γ is treeable iff Γ is amenable (i.e., $\overset{ME}{\sim} F_1 = \mathbf{Z}$), or is ME to either F_2 or F_∞ (this fact uses Hjorth's [69]; see [81, theorems 28.2 and 28.5]). Groups with Kazhdan's property (T) are antitreeable [6]. More generally, it follows from the recent work of Alvarez and Gaboriau [7] that a nonamenable group Γ with $\beta_1^{(2)}(\Gamma) = 0$ is antitreeable (in view of Equation (3.1) this also strengthens [54, corollaire VI.22], where Gaboriau showed that a nonamenable Γ with $\mathscr{C}_*(\Gamma) = 1$ is antitreeable).

A treeing of a relation can be seen as a Γ-invariant assignment of pointed trees with Γ as the set of vertices. One may view the relation *acting* on this measurable family of pointed trees by moving the marked point. More generally, one might define actions by relations, or measured groupoids, on fields of simplicial complexes. Gaboriau defines (see [57]) the **geometric dimension** of a relation $\mathscr{R}$ to be the smallest possible dimension of such a field of *contractible* simplicial complexes; the **ergodic dimension** of a group Γ will be the minimal geometric dimension over orbit relations $\mathscr{R}_{\Gamma \curvearrowright X}$ of all essentially free p.m.p. Γ-actions. In this terminology $\mathscr{R}$ is treeable iff it has geometric dimension 1, and a group Γ is treeable if its ergodic dimension is 1. There is also a notion of an **approximate geometric/ergodic dimension** [57] describing the dimensions of a sequence of subrelations approximating a given orbit relation.

THEOREM 3.10. (GABORIAU [57]) *Ergodic dimension and approximate ergodic dimension are ME-invariants.*

This notion can be used to obtain some information about ME of lattices in the family of rank-1 groups $SO_{n,1}(\mathbf{R})$. If $\Gamma_i < SO_{n_i,1}(\mathbf{R})$, $i = 1, 2$ are lattices and $\Gamma_1 \overset{ME}{\sim} \Gamma_2$, then Gaboriau's result on ℓ^2-Betti numbers shows that if one of n_i is even, then $n_1 = n_2$. However, for $n_i = 2k_i + 1$ all $\beta_i^{(2)}$ vanish. In this case Gaboriau shows, using the above ergodic dimension, that $k_1 \leq k_2 \leq 2k_1$ or $k_2 \leq k_1 \leq 2k_2$.

3.1.6. FREE PRODUCTS It was mentioned above that if $\Gamma_i \overset{OE}{\sim} \Lambda$, then $*_{i \in I} \Gamma_i \overset{OE}{\sim} *_{i \in I} \Lambda_i$ (here $\Gamma \overset{OE}{\sim} \Lambda$ means that the two groups admit an ergodic ME-coupling with index 1, and equivalently admit essentially free actions that are *orbit equivalent*). To what extent does the converse hold? Namely, when can one recognize the free factors on the level of measure equivalence?

This problem was extensively studied by Ioana, Peterson, and Popa in [75] where strong rigidity results were obtained for orbit relations under certain assumptions on the actions (see §4.2.8). Here let us formulate a recent result from Alvarez and Gaboriau [7] that can be stated in purely group theoretic terms. In [7] a notion of **measurably freely indecomposable** groups (MFI) is introduced, and it is shown that this class includes all nonamenable groups with $\beta_1^{(2)} = 0$. Thus, infinite property (T) groups, nonamenable direct products, are examples of MFI groups.

THEOREM 3.11. (ALVAREZ-GABORIAU [7]) *Suppose that* $*_{i=1}^n \Gamma_i \overset{ME}{\sim} *_{j=1}^m \Lambda_j$, *where* $\{\Gamma_i\}_{i=1}^n$ *and* $\{\Lambda_j\}_{j=1}^m$ *are two sets of MFI groups with* $\Gamma_i \overset{ME}{\not\sim} \Gamma_{i'}$ *for* $1 \leq i \neq i' \leq n$, *and* $\Lambda_j \overset{ME}{\not\sim} \Lambda j'$ *for* $1 \leq j \neq j' \leq m$. *Then* $n = m$ *and, up to a permutation of indices,* $\Gamma_i \overset{ME}{\sim} \Lambda_i$.

Another result from [7] concerning decompositions of equivalence relations as free products of subrelations is discussed in §4.2.8.

Let us also mention recent works of Kida [88] and Popa and Vaes [120] that describe extremely strong rigidity properties for certain amalgamated products of various rigid groups.

3.1.7. THE CLASSES $\mathcal{C}_{reg}$ AND $\mathcal{C}$ In §3.2.2 below we shall discuss rigidity results obtained by Nicolas Monod and Yehuda Shalom in [101] (see also [99, 100] and jointly with Mineyev [96]). These results involve **second-bounded cohomology** with unitary coefficients: $H_b^2(\Gamma, \pi)$—a certain vector space associated to a countable group Γ and a unitary representation $\pi : \Gamma \rightarrow U(\mathcal{H}_\pi)$. (Some background on bounded cohomology can be found in [101, §3] or [98]; for more details see [21, 97]). Monod and Shalom define the class $\mathcal{C}_{reg}$ of groups characterized by the property that

$$H_b^2(\Gamma, \ell^2(\Gamma)) \neq \{0\}$$

and (potentially larger) class $\mathcal{C}$ of groups Γ with nonvanishing $H_b^2(\Gamma, \pi)$ for some *mixing* Γ-representation π. Known examples of groups in $\mathcal{C}_{reg} \subset \mathcal{C}$ include groups admitting "hyperboliclike" actions of the following types (see [100], [96]):

   i) nonelementary simplicial action on some simplicial tree, proper on the set of edges;

   ii) nonelementary proper isometric action on some proper CAT(-1) space; and

   iii) nonelementary proper isometric action on some Gromov-hyperbolic graph of bounded valency.

Hence C_{reg} includes free groups, free products of arbitrary countable groups, and free products amalgamated over a finite group (with the usual exceptions of order 2), fundamental groups of negatively curved manifolds, Gromov hyperbolic groups, and nonelementary subgroups of the above families. Examples of groups not in C include amenable groups, products of at least two infinite groups, lattices in higher-rank simple Lie groups (over any local field), and irreducible lattices in products of general compactly generated nonamenable groups (see [101, §7]).

THEOREM 3.12. (MONOD-SHALOM [101])

1) *Membership in C_{reg} or C is an ME-invariant.*
2) *For direct products $\Gamma = \Gamma_1 \times \cdots \times \Gamma_n$ where $\Gamma_i \in C_{reg}$ are torsion free, the number of factors and their ME types are ME-invariants.*
3) *For Γ as above, if $\Lambda \overset{ME}{\sim} \Gamma$, then Λ cannot be written as a product of $m > n$ infinite torsion-free factors.*

3.1.8. DIMENSION AND SIMPLICIAL VOLUME (ℓ^1-ME)

Geometric properties are hard to capture with the notion of measure equivalence. The ℓ^2-Betti numbers is an exception, but this invariant benefits from its Hilbert space nature. In [13, 14] Uri Bader, Roman Sauer, and the author consider a restricted version of measure equivalence, namely, ℓ^1-ME (see §2.3.2 for a definition). Being ℓ^1-ME is an equivalence relation between finitely generated groups, in which any two **integrable** lattices in the same lcsc group are ℓ^1-ME. All uniform lattices are integrable, and so are all lattices in $SO_{n,1}(\mathbf{R}) \simeq \mathrm{Isom}\,(\mathbb{H}^n_{\mathbf{R}})$ (see §3.2.4).

THEOREM 3.13. (BADER-FURMAN-SAUER [14])

Let $\Gamma_i = \pi_1(M_i)$ where M_i are closed manifolds that admit a Riemannian metric of negative sectional curvature. Assume that Γ_1 and Γ_2 admit an ℓ^1-ME-coupling Ω. Then

$$\dim (M_1) = \dim (M_2) \qquad and \qquad \|M_1\| = [\Gamma_2 : \Gamma_1]_\Omega \cdot \|M_2\|,$$

where $\|M_i\|$ denotes the simplicial volume of M_i.

The simplicial volume $\|M\|$ of a closed manifold M, introduced by Gromov in [65], is the norm of the image of the fundamental class under the comparison map $H_n(M) \to H_n^{\ell^1}(M)$ into the ℓ^1-homology, which is an ℓ^1-completion of the usual homology. This is a homotopy invariant of manifolds. Manifolds carrying a Riemannian metric of negative curvature have $\|M\| > 0$ (Gromov [65]).

3.2. Orbit/Measure Equivalence Rigidity

Let us now turn to measure equivalence rigidity results, that is, classification results in the ME category. In the introduction to this section we mentioned that the ME class ME(**Z**) is precisely all infinite amenable groups. The (distinct) classes ME($\mathbf{F}_{2 \le n < \infty}$) and ME($\mathbf{F}_\infty$) are very rich and resist precise description. However, much is known about more rigid families of groups.

3.2.1. HIGHER-RANK-LATTICES

THEOREM 3.14. (ZIMMER [139]) *Let G and G' be center-free simple Lie groups with $\mathrm{rk}_\mathbf{R}(G) \ge 2$, let $\Gamma < G$, $\Gamma' < G'$ be lattices, and $\Gamma \curvearrowright (X, \mu) \overset{\mathrm{OE}}{\sim} \Gamma' \curvearrowright (X', \mu')$ be orbit equivalence between essentially free probability measure-preserving actions. Then $G \cong G'$. Moreover, the induced actions $G \curvearrowright (G \times_\Gamma X)$, $G' \curvearrowright (G' \times_{\Gamma'} Y)$ are isomorphic up to a choice of the isomorphism $G \cong G'$.*

In other words ergodic (infinite) p.m.p. actions of lattices in distinct higher-rank semisimple Lie groups always have distinct orbit structures,[5] for example,

$$2 \le n < m \quad \Longrightarrow \quad SL_n(\mathbf{Z}) \curvearrowright \mathbf{T}^n \overset{\mathrm{OE}}{\not\sim} SL_m(\mathbf{Z}) \curvearrowright \mathbf{T}^m.$$

This remarkable result (a contemporary of Ornstein-Weiss Theorem 3.1) not only showed that the variety of orbit structures of *nonamenable* groups is very rich, but more importantly established a link between OE in ergodic theory and the theory of algebraic groups and their lattices; in particular, introducing Margulis's *superrigidity* phenomena into ergodic theory. This seminal result can be considered as the birth of the subject discussed in this survey. Let us record an ME conclusion of the above.

---

5. There is no need here to assume that the actions are essentially free. Stuck and Zimmer [135] showed that all non-atomic ergodic p.m.p. actions of higher-rank lattices are essentially free; this is based on and generalizes the famous factor theorem of Margulis [94]; see [95].

COROLLARY 3.15. (ZIMMER) *Let G, G' be connected center-free simple Lie groups with $\mathrm{rk}_R(G) \geq 2$, $\Gamma < G$ and $\Gamma' < G'$ lattices. Then $\Gamma \overset{ME}{\sim} \Gamma'$ iff $G \cong G'$.*

The picture of ME classes of lattices in higher-rank simple Lie groups can be sharpened as follows.

THEOREM 3.16. ([44]) *Let G be a center-free simple Lie group with $\mathrm{rk}_R(G) \geq 2$, $\Gamma < G$ a lattice, Λ some group measure equivalent to Γ.*

Then Λ is commensurable up to finite kernels to a lattice in G. Moreover, any ergodic (Γ, Λ)-coupling has a quotient that is either an atomic coupling (in which case Γ and Λ are commensurable), or G, or $\mathrm{Aut}(G)$ with the Haar measure.

(Recall that $\mathrm{Aut}(G)$ contains $\mathrm{Ad}(G) \cong G$ as a finite-index subgroup). The main point of this result is a construction of a representation $\rho : \Lambda \to \mathrm{Aut}(G)$ for the *unknown* group Λ using ME to a higher-rank lattice Γ. It uses Zimmer's cocycle superrigidity theorem and a construction involving a bi-Γ-equivariant measurable map $\Omega \times_\Lambda \check{\Omega} \to \mathrm{Aut}(G)$. An updated version of this construction is stated in §5.5. The by-product of this construction is a map $\Phi : \Omega \to \mathrm{Aut}(G)$ satisfying

$$\Phi(\gamma\omega) = \gamma\,\Phi(\omega), \qquad \Phi(\lambda\omega) = \Phi(\omega)\rho(\lambda)^{-1}.$$

It defines the above quotients (the push-forward measure $\Phi_* m$ is identified as either atomic or Haar measure on $G \cong \mathrm{Ad}(G)$ or on all of $\mathrm{Aut}(G)$, using Ratner's theorem [121]). This additional information is useful to derive OE rigidity results (see Theorem 4.19).

3.2.2. PRODUCTS OF HYPERBOLICLIKE GROUPS The results above use in an essential way the cocycle superrigidity theorem of Zimmer, which exploits *higher-rank phenomena* as in Margulis's superrigidity. A particular situation where such phenomena take place are *irreducible* lattices in products of (semi)simple groups, starting from $SL_2(R) \times SL_2(R)$; or cocycles over **irreducible actions** of a product of $n \geq 2$ simple groups. Here irreducibility of an action $G_1 \times \cdots \times G_n \curvearrowright (X, \mu)$ means ergodicity of $G_i \curvearrowright (X, \mu)$ for each $1 \leq i \leq n$.[6] It recently became clear that higher-rank phenomena occur also for irreducible lattices in products of $n \geq 2$ of rather general lcsc groups; and in the cocycle setting, for cocycles over irreducible actions of products of $n \geq 2$ of rather general groups (see the introduction to [101]). This is to say that the

---

6. Sometimes this can be relaxed to ergodicity of $G'_i \curvearrowright (X, \mu)$ where $G'_i = \prod_{j \neq i} G_j$.

product structure alone seems to provide sufficient *"higher-rank thrust"* to the situation. The following breakthrough results of Nicolas Monod and Yehuda Shalom is an excellent illustration of this fact (see §5.2). Similar phenomena were independently discovered by Greg Hjorth and Alexander Kechris in [70].

THEOREM 3.17. (MONOD-SHALOM [101, THEOREM 1.16]) *Let* $\Gamma = \Gamma_1 \times \cdots \times \Gamma_n$ *and* $\Lambda = \Lambda_1 \times \cdots \times \Lambda_m$ *be products of torsion-free countable groups, where* $\Gamma_i \in C_{\text{reg}}$. *Assume that* $\Gamma \overset{\text{ME}}{\sim} \Lambda$.

Then $n \geq m$. *If* $n = m$, *then, after a permutation of the indices,* $\Gamma_i \overset{\text{ME}}{\sim} \Lambda_i$. *In the latter case* ($n = m$) *any ergodic ME-coupling of* $\Gamma \cong \Lambda$ *has the trivial coupling as a quotient.*

THEOREM 3.18. (MONOD-SHALOM [101]) *Let* $\Gamma = \Gamma_1 \times \cdots \times \Gamma_n$ *where* $n \geq 2$ *and* Γ_i *are torsion-free groups in class* C, *and* $\Gamma \curvearrowright (X, \mu)$ *be an irreducible action (i.e., every* $\Gamma_i \curvearrowright (X, \mu)$ *is ergodic); let* Λ *be a torsion free countable group and* $\Lambda \curvearrowright (Y, \nu)$ *be a mildly mixing action. If* $\Gamma \curvearrowright X \overset{\text{SOE}}{\sim} \Lambda \curvearrowright Y$, *then this SOE has index 1, and* $\Lambda \cong \Gamma$ *and the actions are isomorphic.*

THEOREM 3.19. (MONOD-SHALOM [101]) *For* $i = 1, 2$ *let* $1 \to A_i \to \bar{\Gamma}_i \to \Gamma_i \to 1$ *be a short exact sequence of groups with* A_i *amenable and* Γ_i *are in* C_{reg} *and are torsion free. Then* $\bar{\Gamma}_1 \overset{\text{ME}}{\sim} \bar{\Gamma}_2$ *implies* $\Gamma_1 \overset{\text{ME}}{\sim} \Gamma_2$.

A key tool in the proofs of these results is a cocycle superrigidity Theorem 5.5, which involves *second-bounded cohomology* H_b^2 of groups. In [12] (see also [15]) Uri Bader and the author develop a different approach to higher-rank phenomena, in particular showing an analogue of Monod-Shalom Theorem 5.5, as stated in Theorem 5.6. This result concerns a class of groups that admit *convergence action* on a compact metrizable space (i.e., a continuous action $H \curvearrowright M$ where the action $H \curvearrowright M^3 \setminus \text{Diag}$ on the locally compact space of distinct triples is proper). Following Furstenberg [50] we denote this class as $\mathcal{D}$, and distinguish a subclass $\mathcal{D}_{\text{ea}}$ of groups admitting convergent action $H \curvearrowright M$ with amenable stabilizers. As a consequence of this superrigidity theorem it follows that Theorems 3.17–3.19 remain valid if class C_{reg} is replaced by $\mathcal{D}_{\text{ea}}$. Recently Hiroki Sako [122, 123] has obtained similar results for groups in Ozawa's class $\mathcal{S}$ (see [107]).

Let us point out that each of the classes C_{reg}, $\mathcal{D}_{\text{ea}}$, $\mathcal{S}$ include all Gromov hyperbolic groups (and many relatively hyperbolic ones), are closed undertaking subgroups, and exclude direct products of two infinite groups. These are key features of what one would like to call a "hyperboliclike" group.

3.2.3. MAPPING CLASS GROUPS The following remarkable result of Yoshikata Kida concerns mapping class groups of surfaces. Given a compact orientable surface $\Sigma_{g,p}$ of genus g with p boundary components the **extended mapping class group** $\Gamma(\Sigma_{g,p})^\diamond$ is the group of isotopy components of diffeomorphisms of $\Sigma_{g,p}$ (the **mapping class group** itself is the index 2 subgroup of isotopy classes of orientation-preserving diffeomorphisms). In the following assume $3g + p > 0$, that is, rule out the torus $\Sigma_{1,0}$, once-punctured torus $\Sigma_{1,1}$, and spheres $\Sigma_{0,p}$ with $p \le 4$ punctures.

THEOREM 3.20. (KIDA [86]) *Let Γ be a finite-index subgroup in $\Gamma(\Sigma_{g,p})^\diamond$ with $3g + p - 4 > 0$, or in a finite product of such mapping class groups $\prod_{i=1}^n \Gamma(\Sigma_{g,p})^\diamond$.*

Then any group $\Lambda \overset{ME}{\sim} \Gamma$ is commensurable up to finite kernels to Γ, and ergodic ME-coupling has a discrete (Γ, Λ)-coupling as a quotient.

This work (spanning [83, 85, 86]) is a real tour de force. Mapping class groups $\Gamma(\Sigma)$ are often compared to a lattice in a semisimple Lie group G: the Teichmüller space $\mathcal{T}(\Sigma)$ is analogous to the symmetric space G/K, Thurston boundary $\mathcal{PML}(\Sigma)$ analogous to Furstenberg boundary $B(G) = G/P$, and the curve complex $C(\Sigma)$ to the spherical Tits's building of G. The MCG has been extensively studied as a geometric object, while Kida's work provides a new ergodic-theoretic perspective. For example, Kida proves that Thurston boundary $\mathcal{PML}(\Sigma)$ with the Lebesgue measure class is Γ-**boundary** in the sense of Burger-Monod for the mapping class group, that is, the action of the latter is amenable and doubly ergodic with unitary coefficients. Properties of the MCG action on $\mathcal{PML}(\Sigma)$ allow Kida to characterize certain subrelations/subgroupoids arising in self-measure equivalence of a MCG; leading to the proof of a cocycle (strong) rigidity Theorem 5.7, which can be viewed as a groupoid version of Ivanov's rigidity theorem. This strong rigidity theorem can be used with §5.5 to recognize arbitrary groups ME to an MCG.

Note that a mapping class group behaves like a "lattice without ambient Lie group"—all its ME-couplings have discrete quotients. Moreover, Kida's ME rigidity results extend to products of MCGs without any irreducibility assumptions. From this point of view MCGs are more ME rigid than higher-rank lattices, despite the fact that they lack many other rigidity attributes, such as property (T) (see Andersen [10]).

Added in proof. Very recently additional extremely strong ME rigidity results were obtained in Kida [88] and Popa and Vaes [120] for certain amalgamated products of higher-rank lattices and also mapping class groups. The latter paper also establishes W^*-rigidity.

3.2.4. HYPERBOLIC LATTICES AND ℓ^1-ME Measure equivalence is motivated by the theory of lattices, with ME-couplings generalizing the situation of groups embedded as lattices in the same ambient lcsc group. Thus, in the context of semisimple groups, one wonders whether ME rigidity results would parallel Mostow rigidity; and in particular would apply to (lattices in) all simple groups with the usual exception of $SL_2(\mathbf{R}) \simeq SO_{2,1}(\mathbf{R}) \simeq SU_{1,1}(\mathbf{R})$. The higher-rank situation (at least that of simple groups) is well understood (§3.2.1). In the rank 1 case (consisting of the families $SO_{n,1}(\mathbf{R})$, $SU_{m,1}(\mathbf{R})$, $Sp_{k,1}(\mathbf{R})$, and $F_{4(-20)}$) known ME-invariants discussed above (namely: property (T), ℓ^2-Betti numbers, Λ-invariant, ergodic dimension) allow to distinguish lattices among most rank 1 groups. This refers to statements of the form: if $\Gamma_i < G_i$ are lattices, then $\Gamma_1 \overset{ME}{\sim} \Gamma_2$ iff $G_1 \simeq G_2$. However, ME classification such as in Theorems 3.16, 3.17, and 3.20 are not known for rank 1 cases. The ingredient that is missing in the existing approach is an appropriate cocycle superrigidity theorem.[7]

In a joint work with Uri Bader and Roman Sauer a cocycle strong rigidity theorem is proved for ME-cocycles for lattices in $SO_{n,1}(\mathbf{R}) \simeq \mathrm{Isom}(\mathbf{H}_{\mathbf{R}}^n)$, $n \geq 3$, under a certain ℓ^1-assumption (see §2.3.2). It is used to obtain the following:

THEOREM 3.21. (BADER-FURMAN-SAUER [13]) *Let Γ is a lattice in $G = \mathrm{Isom}(\mathbb{H}^n)$, $n \geq 3$, and Λ is some finitely generated group ℓ^1-ME to Γ then Λ is a lattice in G modulo a finite normal subgroup. Moreover, any ergodic (Γ, Λ)-coupling has a quotient, which is either discrete, or $G = \mathrm{Aut}(G)$, or G^0 with the Haar measure.*

Recently Sorin Popa has introduced a new set of ideas for studying orbit equivalence. These results, rather than relying on rigidity of the acting groups alone, exploit rigidity aspects of *groups actions* of certain type. We shall discuss them in §Sections 5.4, and 5.6, and 5.7.

3.3. How Many Orbit Structures Does a Given Group Have?

Theorem 3.1 of Ornstein and Weiss [105] implies that for an infinite amenable countable group Γ all ergodic probability measure preserving actions $\Gamma \curvearrowright (X, \mu)$ define the same orbit structure, namely, $\mathscr{R}_{amen}$. What happens for non-amenable groups Γ?

---

7. For $Sp_{n,1}(\mathbf{R})$ and $F_{4(-20)}$ a cocycle superrigidity theorem was proved by Corlette and Zimmer [29] (see also Fisher and Hitchman [43]), but these results require boundness assumptions that preclude them from being used for ME-cocycles.

THEOREM 3.22. (EPSTEIN [36], AFTER IOANA [73] AND GABOTIAU-LYONS [60]) *Any nonamenable countable group* Γ *has a continuum of essentially free ergodic probability measure-preserving actions* $\Gamma \curvearrowright (X, \mu)$, *no two of which are stably orbit equivalent.*

Let us briefly discuss the problem and its solution. Since Card(Aut $(X, \mu)^{\Gamma}) = \aleph = 2^{\aleph_0}$ there are at most continuum many *actions* for a fixed countable group Γ. In fact, this upper bound on the cardinality of isomorphism classes of actions is achieved, using the corresponding fact about unitary representations and the Gaussian construction. Hence one might expect at most $\aleph$-many non-OE actions for any given Γ. OE rigidity results showed that some specific classes of groups indeed have many mutually non-OE actions; this includes higher-rank lattices [64], products of hyperboliclike groups [101, theorem 1.7], and some other classes of groups [110, 113]). But the general question, regarding an arbitrary nonamenable Γ, remained open.

Most invariants of equivalence relations depend on the acting group rather than the action, and thus could not be used to distinguish actions of the fixed group Γ. The notable exception to this metamathematical statement appears for nonamenable groups that do not have property (T). For such groups *two* non-SOE actions can easily be constructed: (1) a *strongly ergodic* action (using Schmidt's [129]), and (2) an ergodic action that is *not strongly ergodic* (using Connes-Weiss [28]). Taking a product with an essentially free weakly mixing strongly ergodic Γ-actions (e.g., the Bernoulli action $(X_0, \mu_0)^{\Gamma}$) makes the above two actions essentially free and distinct.

In [68] Greg Hjorth showed that if Γ has property (T), the set of isomorphism classes of orbit structures for essentially free Γ-actions has cardinality $\aleph$, by proving that the natural map from the isomorphism classes of essentially free ergodic Γ-actions to the isomorphism classes of Γ-orbit structures is at most countable-to-one. More precisely, the space of Γ-actions producing a fixed-orbit structure is equipped with a structure of a Polish space (separability) where any two nearby actions are shown to be conjugate. This is an example of proving *rigidity up to countable classes* combining *separability* of the ambient space with a *local rigidity* phenomenon (stemming from property (T); see §5.6 below). These ideas can be traced back to Connes [26] and Popa [106], and play a central role in the most recent developments—see [114] §4].

The challenge now became to show that other nonamenable groups have infinitely, or even $\aleph$-many, non-OE essentially free ergodic actions. Damien Gaboriau and Sorin Popa [59] achieved this goal for the quintessential representative of a nonamenable group without property (T), namely for the free group F_2. Using a sophisticated *rigidity vs. separability* argument they showed that within a certain rich family of F_2-actions the map from isomorphism classes of actions to orbit structures is countable-to-one. The *rigidity* component of the argument was this time provided by Popa's notion of *w-rigid* actions such as $SL_2(Z) \curvearrowright T^2$, with the rigidity related to the *relative property (T)* for the semidirect product $SL_2(Z) \ltimes Z^2$ viewing Z^2 as the Pontryagin dual of T^2.

In [73] Adrian Ioana obtained a sweeping result showing that any Γ containing a copy of F_2 has $\aleph$-many mutually non-SOE essentially free actions. The basic idea of the construction being to use a family of non-SOE of F_2-actions $F_2 \curvearrowright X_t$ to construct co-induced Γ-actions $\Gamma \curvearrowright X_t^{\Gamma/F_2}$ and pushing the solution of F_2-problem to the analysis of the co-induced actions. The class of groups containing F_2 covers "most of" the class of nonamenable groups with few, very hard to obtain, exceptions. The ultimate solution to the problem, covering all nonamenable groups, was shortly obtained by Inessa Epstein [36] using a result by Damien Gaboriau and Russel Lyons [60], who proved that any nonamenable Γ contains an F_2 in a sort of measure-theoretical sense. Epstein was able to show that this sort of containment suffices to carry out an analogue of Ioana's co-induction argument [73] to prove Theorem 3.22.

Furthermore, in [74] Ioana, Kechris, and Tsankov, jointly with Epstein, show that for any nonamenable Γ the space of all ergodic-free p.m.p. actions taken up to OE not only has cardinality of the continuum, but is also impossible to classify in a very strong sense. One may also add that most of the general results mentioned above show that within certain families of actions the grouping into SOE-ones has countable classes, therefore giving only implicit families of non-SOE actions. In [71] Ioana provided an explicit list of a continuum of mutually non-SOE actions of F_2.

4. Measured Equivalence Relations

4.1. Basic Definitions

We start with the notion of **countable equivalence relations** in the Borel setting. It consists of a standard Borel space $(X, \mathcal{X})$ (cf. [39] for definitions) and a Borel subset $\mathscr{R} \subset X \times X$, an equivalence relation whose equivalence classes $\mathscr{R}[x] = \{y \in X : (x, y) \in \mathscr{R}\}$ are all countable.

To construct such relations choose a countable collection $\Phi = \{\phi_i\}_{i \in I}$ of Borel bijections $\phi_i : A_i \to B_i$ between Borel subsets $A_i, B_i \in \mathcal{X}$, $i \in I$; and let $\mathscr{R}_\Phi$ be the smallest equivalence relation including the graphs of all ϕ_i, $i \in I$. More precisely, $(x, y) \in \mathscr{R}_\Phi$ iff there exists a finite sequence $i_1, \ldots, i_k \in I$ and

$\epsilon_1, \ldots, \epsilon_k \in \{-1, 1\}$ so that

$$y = \phi_{i_k}^{\epsilon_k} \circ \cdots \circ \phi_{i_2}^{\epsilon_2} \circ \phi_{i_1}^{\epsilon_1}(x).$$

We shall say that the family Φ **generates** the relation $\mathscr{R}_\Phi$. The particular case of a collection $\Phi = \{\phi_i\}$ of Borel isomorphisms of the whole space X generates a countable group $\Gamma = \langle \Phi \rangle$ and

$$\mathscr{R}_\Phi = \mathscr{R}_{\Gamma \curvearrowright X} = \{(x, y) \; : \; \Gamma x = \Gamma y\} = \{(x, y.x) \; : \; x \in X, \; y \in \Gamma\}.$$

Feldman and Moore [39] proved that any countable Borel equivalence relation admits a generating set whose elements are defined on all of X; in other words, any equivalence relation appears as the orbit relation $\mathscr{R}_{\Gamma \curvearrowright X}$ of a Borel action $\Gamma \curvearrowright X$ of some countable group Γ (see §4.3.1).

Given a countable Borel equivalence relation $\mathscr{R}$ the **full group** $[\mathscr{R}]$ is defined by

$$[\mathscr{R}] = \{\phi \in \text{Aut}(X, \mathcal{X}) \; : \; \forall x \in X : (x, \phi(x)) \in \mathscr{R}\}.$$

The **full pseudogroup** $[[\mathscr{R}]]$ consists of partially defined Borel isomorphisms

$$\psi : \text{Dom}(\psi) \to \text{Im}(\psi), \quad \text{so that} \quad \text{Graph}(\psi) = \{(x, \psi(x)) \; : \; x \in \text{Dom}(\psi)\} \subset \mathscr{R}.$$

If $\mathscr{R}$ is the orbit relation $\mathscr{R}_{\Gamma \curvearrowright X}$ of a group action $\Gamma \curvearrowright (X, \mathcal{X})$, then any $\phi \in [\mathscr{R}]$ has the following "piecewise Γ-structure": there exist countable partitions $\bigsqcup A_i = X = \bigsqcup B_i$ into Borel sets and elements $y_i \in \Gamma$ with $y_i(A_i) = B_i$ so that $\phi(x) = y_i x$ for $x \in A_i$. Elements ψ of the full pseudogroup $[[\mathscr{R}_\Gamma]]$ have a similar "piecewise Γ-structure" with $\bigsqcup A_i = \text{Dom}(\psi)$ and $\bigsqcup B_i = \text{Im}(\psi)$.

Let $\mathscr{R}$ be a countable Borel equivalence relation on a standard Borel space $(X, \mathcal{X})$. A measure μ on $(X, \mathcal{X})$ is $\mathscr{R}$-**invariant** (respectively, $\mathscr{R}$-**quasi-invariant**) if for all $\phi \in [\mathscr{R}]$, $\phi_* \mu = \mu$ (respectively, $\phi_* \mu \sim \mu$). Note that if $\Phi = \{\phi_i : A_i \to B_i\}$ is a generating set for $\mathscr{R}$, then μ is $\mathscr{R}$-invariant iff μ is invariant under each ϕ_i, that is, $\mu(\phi_i^{-1}(E) \cap A_i) = \mu(E \cap B_i)$ for all $E \in \mathcal{X}$. Similarly, quasi-invariance of a measure can be tested on a generating set. The $\mathscr{R}$-**saturation** of $E \in \mathcal{X}$ is $\mathscr{R}[E] = \{x \in X \; : \; \exists y \in E, (x, y) \in \mathscr{R}\}$. A $\mathscr{R}$ (quasi-)invariant measure μ is **ergodic** if $\mathscr{R}[E]$ is either μ-null or μ-conull for any $E \in \mathcal{X}$. In this section we shall focus on countable Borel equivalence relations $\mathscr{R}$ on $(X, \mathcal{X})$ equipped with an ergodic, invariant, nonatomic, probability measure μ on $(X, \mathcal{X})$. Such a quadruple $(X, \mathcal{X}, \mu, \mathscr{R})$ is called **type II$_1$-relation**. These are precisely the orbit relations of ergodic measure-preserving actions of countable groups on nonatomic standard probability measure spaces (the nontrivial implication follows from the above-mentioned theorem of Feldman and Moore).

Given a countable Borel relation $\mathscr{R}$ on $(X, \mathcal{X})$ and an $\mathscr{R}$-quasi-invariant probability measure μ, define infinite measures $\tilde{\mu}_L$, $\tilde{\mu}_R$ on $\mathscr{R}$ by

$$\tilde{\mu}_L(E) = \int_X \# \{ y \ : \ (x, y) \in E \cap \mathscr{R} \} \, d\mu(x),$$

$$\tilde{\mu}_R(E) = \int_X \# \{ x \ : \ (x, y) \in E \cap \mathscr{R} \} \, d\mu(y).$$

These measures are equivalent, and coincide if μ is $\mathscr{R}$-invariant, which is our main focus, In this case we shall denote

4.1 $$\tilde{\mu} = \tilde{\mu}_L = \tilde{\mu}_R$$

Hereafter, saying that some property holds a.e. on $\mathscr{R}$ would refer to $\tilde{\mu}$-a.e. (this makes sense even if μ is only $\mathscr{R}$-quasi-invariant).

REMARK 4.1. *In some situations a Borel equivalence relation $\mathscr{R}$ on $(X, \mathcal{X})$ has only one (nonatomic) invariant probability measure. For example, this is the case for the orbit relation of the standard action of a finite-index subgroup[8] $\Gamma < \mathrm{SL}_n(\mathbf{Z})$ on the torus $\mathbf{T}^n = \mathbf{R}^n/\mathbf{Z}^n$, or for a lattice Γ in a simple center-free Lie group G acting on H/Λ, where H is a simple Lie group, $\Lambda < H$ is a lattice, and Γ acts by left translations via an embedding $j : G \to H$ with $j(G)$ having trivial centralizer in H. In such situations one may gain understanding of the countable Borel equivalence relation $\mathscr{R}$ via the study of the II_1-relation corresponding to the unique $\mathscr{R}$-invariant probability measure.*

As always in the *measure-theoretic* setting null sets should be considered negligible. So an isomorphism T between (complete) measure spaces $(X_i, \mathcal{X}_i, \mu_i)$, $i = 1, 2$, is a Borel isomorphism between μ_i-conull sets $T : X_1' \to X_2'$ with $T_*(\mu_1) = \mu_2$. In the context of II_1-relations, we declare two relations $(X_i, \mathcal{X}_i, \mu_i, \mathscr{R}_i)$, $i = 1, 2$ to be **isomorphic**, if there exists a measure space isomorphism $T : (X_1, \mu_1) \cong (X_2, \mu_2)$ so that $T \times T : (\mathscr{R}_1, \tilde{\mu}_1) \to (\mathscr{R}_2, \tilde{\mu}_2)$ is an isomorphism. In other words, after a restriction to conull sets, T satisfies

$$(x, y) \in \mathscr{R}_1 \quad \Longleftrightarrow \quad (T(x), T(y)) \in \mathscr{R}_2.$$

Let us also adapt the notions of the full group and the full pseudogroup to the measure-theoretic setting, by passing to a quotient $\mathrm{Aut}(X, \mathcal{X}) \to \mathrm{Aut}(X, \mathcal{X}, \mu)$ where two Borel isomorphisms ϕ and ϕ' that agree μ-a.e. are identified. This allows us to focus on the essential measure-theoretic issues. The following easy but useful Lemma illustrates the advantage of this framework.

8. Or just Zariski dense subgroup; see [18].

LEMMA 4.2. *Let* $(X, \mathcal{X}, \mu, \mathcal{R})$ *be a* II_1*-relation. Then for* $A, B \in \mathcal{X}$ *one has* $\mu(\phi(A) \triangle B) = 0$ *for some* $\phi \in [\mathcal{R}]$ *iff* $\mu(A) = \mu(B)$.

4.1.1. RESTRICTION AND WEAK ISOMORPHISMS Equivalence relations admit a natural operation of **restriction**, sometimes called **induction**, to a subset: given a relation $\mathcal{R}$ on X and a measurable subset $A \subset X$ the restriction $\mathcal{R}_A$ to A is

$$\mathcal{R}_A = \mathcal{R} \cap (A \times A).$$

In the presence of, say, $\mathcal{R}$-invariant, measure μ on $(X, \mathcal{X})$ the restriction to a subset $A \subset X$ with $\mu(A) > 0$ preserves the restricted measure $\mu|_A$, defined by $\mu|_A(E) = \mu(A \cap E)$. If μ is a probability measure, we shall denote by μ_A the normalized restriction $\mu_A = \mu(A)^{-1} \cdot \mu|_A$. It is easy to see that ergodicity is preserved, so a restriction of a II_1-relation $(X, \mu, \mathcal{R})$ to a positive measure subset $A \subset X$ is a II_1-relation $(A, \mu_A, \mathcal{R}_A)$.

REMARK 4.3. *Note that it follows from Lemma 4.2 that the isomorphism class of* $\mathcal{R}_A$ *depends only on* $\mathcal{R}$ *and on the size* $\mu(A)$*, so* $\mathcal{R}_A$ *may be denoted* $\mathcal{R}^t$ *where* $t = \mu(A)$ *is* $0 < t \leq 1$*. One may also define* $\mathcal{R}^t$ *for* $t > 1$*. For an integer* $k > 1$ *let* $\mathcal{R}^k$ *denote the product of* $\mathcal{R}$ *with the full relation on the finite set* $\{1, \ldots, k\}$*, namely the relation on* $X \times \{1, \ldots, k\}$ *with* $((x, i), (y, j)) \in \mathcal{R}^k$ *iff* $(x, y) \in \mathcal{R}$*. So* $(\mathcal{R}^k)^{1/k} \cong \mathcal{R}^1 \cong \mathcal{R}$*. The definition of* $\mathcal{R}^t$ *can now be extended to all* $0 < t < \infty$ *using an easily verified formula* $(\mathcal{R}^t)^s \cong \mathcal{R}^{ts}$*. This construction is closely related to the notion of an* **amplification** *in von Neumann algebras: the Murray von Neumann group-measure space construction* $M_{\mathcal{R}}$ *satisfies* $M_{\mathcal{R}^t} = (M_{\mathcal{R}})^t$*.*

The operation of restriction/induction allows one to relax the notion of isomorphism of II_1-relations as follows:

DEFINITION 4.4. Two II_1-relations $\mathcal{R}_1$ and $\mathcal{R}_2$ are *weakly isomorphic* if $\mathcal{R}_1 \cong \mathcal{R}_2^t$ for some $t \in \mathbf{R}_+^\times$. Equivalently, there exist positive measurable subsets $A_i \subset X_i$ with $\mu_2(A_2) = t \cdot \mu_1(A_1)$ and an isomorphism between the restrictions of $\mathcal{R}_i$ to A_i.

Observe that two ergodic probability measure-preserving actions $\Gamma_i \curvearrowright (X_i, \mathcal{X}_i, \mu_i)$ of countable groups are orbit equivalent iff the corresponding orbit relations $\mathcal{R}_{\Gamma_i \curvearrowright X_i}$ are isomorphic.

4.2. Invariants of Equivalence Relations

Let us now discuss in some detail several qualitative and numerical properties of II$_1$ equivalence relations that are preserved under isomorphisms and often preserved or rescaled by the index under weak isomorphisms. We refer to such properties as **invariants** of equivalence relations. Many of these properties are motivated by properties of groups, and often an orbit relation $\mathscr{R}_{\Gamma \curvearrowright X}$ of an essentially free action of a countable group would be a reflection of the corresponding property of Γ.

4.2.1. AMENABILITY, STRONG ERGODICITY, AND PROPERTY (T)

Amenability of an equivalence relation can be defined in a number of ways. In [138] Zimmer introduced the notion of **amenability** for a **group action** on a space with quasi-invariant measure. This notion plays a central role in the theory. This definition is parallel to the fixed-point characterization of amenability for groups. For equivalence relation $\mathscr{R}$ on $(X, \mathcal{X})$ with a *quasi-invariant* measure μ it reads as follows.

Let E be a separable Banach space, and $c : \mathscr{R} \to$ Isom (E) be a measurable 1-**cocycle**, that is, a measurable (with respect to the weak topology on E) map, satisfying $\tilde{\mu}$-a.e.:

$$c(x, z) = c(x, y) \circ c(y, z).$$

Let $X \ni x \mapsto Q_x \subset E^*$ be a measurable family of nonempty convex compact subsets of the dual space E^* taken with the $*$-topology, so that $c(x, y)^*(Q_x) = Q_y$. The relation $\mathscr{R}$ is **amenable** if any such family contains a measurable invariant section, that is, a measurable assignment $X \ni x \mapsto p(x) \in Q_x$, so that a.e.:

$$c(x, y)^* p(x) = p(y).$$

The (original) definition of amenability for group actions concerned general cocycles $c : G \times X \to$ Isom (E) rather than the ones depending only on the orbit relation $\mathscr{R}_{\Gamma \curvearrowright X}$. The language of measured groupoids provides a common framework for both settings (see [9]).

Any nonsingular action of an amenable group is amenable, because any cocycle $c : \Gamma \times X \to$ Isom (E) can be used to define an affine Γ-action on the closed convex subset of $L^\infty(X, E^*) = L^1(X, E)^*$ consisting of all measurable sections $x \to p(x) \in Q_x$; the fixed-point property of Γ provides the desired c^*-invariant section. The converse is not true: any (countable or lcsc) group admits essentially free amenable action with a quasi-invariant measure—this is the main use of the notion of amenable actions. However, for essentially *free*, *probability measure-preserving* actions, amenability of the II_1-relation $\mathscr{R}_{\Gamma \curvearrowright X}$ implies (hence is equivalent to) amenability of Γ. Indeed, given an affine Γ action α on a convex compact $Q \subset E^*$, one can take $Q_x = Q$ for all $x \in X$ and set $c(gx, x) = \alpha(g)$; amenability of $\mathscr{R}_{\Gamma \curvearrowright X}$ provides an invariant section $p : X \to Q$ whose barycenter $q = \int_X p(x) \, d\mu(x)$ would be an $\alpha(\Gamma)$-fixed point in Q.

Connes, Feldman, and Weiss [27] proved that amenable relations are **hyperfinite** in the sense that they can be viewed as an increasing union of finite subrelations; they also showed that such a relation can be generated by an action of $\mathbf{Z}$ (see also [78] by Kaimanovich for a nice exposition and several other nice characterizations of amenability). It follows that there is only one amenable II_1-relation, which we denote hereafter by

$$\mathscr{R}_{amen}.$$

In [141] Zimmer introduced the notion of **property (T)** for group actions on measure spaces with quasi-invariant measure. The equivalence relation version can be stated as follows. Let $\mathscr{H}$ be a separable Hilbert space and let $c : \mathscr{R} \to U(\mathscr{H})$ be a measurable 1-*cocycle*, that is, c satisfies

$$c(x, z) = c(x, y) \circ c(y, z).$$

Then $\mathscr{R}$ has property (T) if any such cocycle for which there exists a sequence $v_n : X \to S(\mathscr{H})$ of measurable maps into the unit sphere $S(\mathscr{H})$ with

$$\|v_n(y) - c(x, y)v_n\| \to 0 \qquad [\tilde{\mu}]\text{-a.e.}$$

admits a measurable map $u : X \to S(\mathscr{H})$ with $u(y) = c(x, y)u(x)$ for $\tilde{\mu}$-a.e. $(x, y) \in \mathscr{R}$. For an essentially free probability measure-preserving action $\Gamma \curvearrowright (X, \mu)$ the orbit relation $\mathscr{R}_{\Gamma \curvearrowright X}$ has property (T) if and only if the group Γ has Kazhdan's property (T) (in [141] weak mixing of the action was assumed for the "only if" implication, but this can be removed as in §3.1.1 relying on Bekka - Valette [16]). In [8] Anantharaman-Delaroche studied the notion of property (T) in the context of general measured groupoids.

Let $\mathscr{R}$ be a II_1-equivalence relation on (X, μ). A sequence $\{A_n\}$ of measurable subsets of X is **asymptotically $\mathscr{R}$-invariant**, if $\mu(\phi(A_n) \triangle A_n) \to 0$ for every $\phi \in [\mathscr{R}]$. This is satisfied trivially if $\mu(A_n) \cdot (1 - \mu(A_n)) \to 0$. Relation $\mathscr{R}$ is **strongly ergodic** if any asymptotically $\mathscr{R}$-invariant sequence of sets is trivial in the above sense. (Note that the condition of asymptotic invariance may be checked on elements ϕ_i of any generating system Φ of $\mathscr{R}$.)

The amenable relation $\mathscr{R}_{amen}$ is not strongly ergodic. If an action $\Gamma \curvearrowright (X, \mu)$ has a **spectral gap** (i.e., does not have almost-invariant vectors) in the Koopman representation on $L^2(X,\mu) \ominus \mathbf{C}$, then $\mathscr{R}_{\Gamma \curvearrowright X}$ is strongly ergodic. Using the fact that the Koopman representation of a Bernoulli action $\Gamma \curvearrowright (X_0, \mu_0)^\Gamma$ is contained in a multiple of the regular representation $\infty \cdot \ell^2(\Gamma)$, Schmidt [128] characterized nonamenable groups by the property that they admit p.m.p. actions with strongly ergodic orbit relation. If $\mathscr{R}$ is not strongly ergodic, then it has an amenable relation as a nonsingular quotient (Jones and Schmidt [77]). Connes and Weiss [28] showed that all p.m.p. actions of a group Γ have strongly ergodic orbit relations if and *only if* Γ has Kazhdan's property (T). In this short elegant paper they introduced the idea of Gaussian actions as a way of constructing a *p.m.p. action* from a given unitary *representation*.

In general, strong ergodicity of the orbit relation $\mathscr{R}_{\Gamma \curvearrowright X}$ does not imply a spectral gap for the action $\Gamma \curvearrowright (X,\mu)$ ([128], [70]). However, this implication does hold for generalized Bernoulli actions (Kechris and Tsankov [82]), and when the action has an ergodic centralizer (Chifan and Ioana [25, lemma 10]).

4.2.2. FUNDAMENTAL GROUP-INDEX VALUES OF SELF-SIMILARITY The term "fundamental group" of a II_1-relation $\mathscr{R}$ refers to a subgroup of $\mathbf{R}_+^\times$ defined by

$$\mathscr{F}(\mathscr{R}) = \left\{ t \in \mathbf{R}_+^\times : \mathscr{R} \cong \mathscr{R}^t \right\}.$$

Equivalently, for $\mathscr{R}$ on (X,μ), the fundamental group $\mathscr{F}(\mathscr{R})$ consists of all ratios $\mu(A)/\mu(B)$ where $A, B \subset X$ are positive measure subsets with $\mathscr{R}_A \cong \mathscr{R}_B$ (here one can take one of the sets to be X without loss of generality). The notion is borrowed from a similarly defined concept of the fundamental group of a von Neumann algebra, introduced by Murray and von Neumann [104]: $\mathscr{F}(M) = \left\{ t \in \mathbf{R}_+^\times : M^t \cong M \right\}$. However, the connection is not direct: even for group space construction $M = \Gamma \ltimes L^\infty(X)$ isomorphisms $M \cong M^t$ (or even automorphisms of M) need not respect the Cartan subalgebra $L^\infty(X)$ in general.

Since the restriction of the amenable relation $\mathscr{R}_{amen}$ to any positive measure subset $A \subset X$ is amenable, it follows

$$\mathscr{F}(\mathscr{R}_{amen}) = \mathbf{R}_+^\times.$$

The same obviously applies to the product of any relation with an amenable one.

On another extreme are orbit relations $\mathscr{R}_{\Gamma \curvearrowright X}$ of essentially free ergodic action of ICC groups Γ with property (T): for such relations the fundamental group $\mathscr{F}(\mathscr{R}_{\Gamma \curvearrowright X})$ is at most countable (Gefter and Golodets [64, Corollary 1.8]).

Many relations have a trivial fundamental group. This includes all II_1 relations with a nontrivial numeric invariant that scales under restriction:

1) Relations with $1 < \text{cost}(\mathscr{R}) < \infty$; in particular, orbit relation $\mathscr{R}_{\Gamma \curvearrowright X}$ for essentially free actions of $\mathbf{F}_n$, $1 < n < \infty$, or surface groups.
2) Relations with some nontrivial ℓ^2-Betti number $0 < \beta_i^{(2)}(\mathscr{R}) < \infty$ for some $i \in \mathbf{N}$; in particular, orbit relation $\mathscr{R}_{\Gamma \curvearrowright X}$ for essentially free actions of a group Γ with $0 < \beta_i^{(2)}(\Gamma) < \infty$ for some $i \in \mathbf{N}$, such as lattices in $SO_{2n,1}(\mathbf{R})$, $SU_{m,1}(\mathbf{R})$, and $Sp_{k,1}(\mathbf{R})$.

Triviality of the fundamental group often appears as a by-product of rigidity of groups and group actions. For example, $\mathscr{F}(\mathscr{R}_{\Gamma \curvearrowright X}) = \{1\}$ in the following situations:

1) Any (essentially free) action of a lattice Γ in a simple Lie group of higher rank ([64]);
2) Any essentially free action of (finite-index subgroups of products) mapping class groups ([86]);
3) Actions of $\Gamma = \Gamma_1 \times \cdots \times \Gamma_n$, $n \geq 2$, of hyperboliclike groups Γ_i where each of them acts ergodically ([101]); and
4) $\mathscr{G}_{dsc}$-cocycle superrigid actions $\Gamma \curvearrowright X$ such as Bernoulli actions of groups with property (T) ([111, 112, 117]).

What are other possibilities for the fundamental group beyond the two extreme cases $\mathscr{F}(\mathscr{R}) = \mathbf{R}_+^\times$ and $\mathscr{F}(\mathscr{R}) = \{1\}$? The most comprehensive answer (to date) to this question is contained in the following result of S. Popa and S. Vaes (see [118] for further references):

THEOREM 4.5. (POPA-VAES, [118, THM 1.1]) *There exists a family S of additive subgroups of $\mathbf{R}$ that contains all countable groups, and (uncountable) groups of arbitrary Hausdorff dimension in $(0, 1)$, so that for any $F \in S$ and any totally disconnected locally compact unimodular group G there exists uncountably many mutually non-SOE essentially free p.m.p. actions of $\mathbf{F}_\infty$ whose orbit relations $\mathscr{R} = \mathscr{R}_{\mathbf{F}_\infty \curvearrowright X}$ have $\mathscr{F}(\mathscr{R}) \cong \exp(F)$ and $\mathrm{Out}(\mathscr{R}) \cong G$.*

Moreover, in these examples the Murray von Neumann group space factor $M = \Gamma \ltimes L^\infty(X)$ has $\mathscr{F}(M) \cong \mathscr{F}(\mathscr{R}) \cong \exp(F)$ and $\mathrm{Out}(M) \cong \mathrm{Out}(\mathscr{R}) \ltimes H^1(\mathscr{R}, \mathbf{T})$, where $H^1(\mathscr{R}, \mathbf{T})$ is the first cohomology with coefficients in the 1-torus.

4.2.3. TREEABILITY An equivalence relation $\mathscr{R}$ is said **treeable** (Adams [4]) if it admits a generating set $\Phi = \{\phi_i\}$ so that the corresponding (nonoriented) graph on a.e. $\mathscr{R}$-class is a tree. Basic examples of treeable relations include: $\mathscr{R}_{amen}$ viewing the amenable II_1-relation as the orbit relation of some/any action of $\mathbf{Z} = \mathbf{F}_1$, and more generally, $\mathscr{R}_{\mathbf{F}_n \curvearrowright X}$ where $\mathbf{F}_n \curvearrowright X$ is an essentially free action of the free group $\mathbf{F}_n$, $1 \le n \le \infty$. Any restriction of a treeable relation is treeable, and $\mathscr{R}$ is treeable iff $\mathscr{R}^t$ is.

If $\mathscr{R}_1 \to \mathscr{R}_2$ is a (weak) *injective relation morphism* and $\mathscr{R}_2$ is treeable, then so is $\mathscr{R}_1$—the idea is to lift a treeing graphing from $\mathscr{R}_2$ to $\mathscr{R}_1$ piece by piece. This way, one shows that if a group Λ admits an essentially free action $\Lambda \curvearrowright Z$ with treeable $\mathscr{R}_{\Lambda \curvearrowright Z}$, and Γ and Λ admit (S)OE essentially free actions $\Gamma \curvearrowright X$ and $\Lambda \curvearrowright Y$, then the Γ-action on $X \times Z$, $g : (x,z) \mapsto (gx, \alpha(g,x)z)$ via the (S)OE cocycle $\alpha : \Gamma \times X \to \Lambda$ has a treeable orbit structure $\mathscr{R}_{\Gamma \curvearrowright X \times Z}$. Since surface groups $\Gamma = \pi_1(\Sigma_g)$, $g \ge 2$, and $\mathbf{F}_2$ are lattices in $PSL_2(\mathbf{R})$, hence ME, the former groups have free actions with treeable orbit relations. Are all orbit relations of free actions of a surface group treeable?

4.2.4. COST The notion of **cost** for II_1-relations corresponds to the notion of **rank** for discrete countable groups. The notion of cost was introduced by G. Levitt [90] and extensively studied by D. Gaboriau [53, 54, 60].

DEFINITION 4.6. Given a generating system $\Phi = \{\phi_i : A_i \to B_i\}_{i \in \mathbf{N}}$ for a II_1-equivalence relation $\mathscr{R}$ on (X, μ) the **cost of the graphing** Φ is

$$\text{cost}\,(\Phi) = \sum_i \mu(A_i) = \sum_i \mu(B_i)$$

and the **cost of the relation** is

$$\text{cost}\,(\mathscr{R}) = \inf\left\{\text{cost}\,(\Phi)\ :\ \Phi \text{ generates } \mathscr{R}\right\}.$$

A generating system Φ defines a graph structure on every $\mathscr{R}$-class and cost (Φ) is half of the average valency of this graph over the space (X, μ).

The cost of a II_1-relation takes values in $[1, \infty]$. In the definition of the cost of a relation it is important that the relation is probability measure preserving, but ergodicity is not essential. The broader context includes relations with finite classes; such relations can have values less than 1. For instance, from the orbit relation of a (nonergodic) probability measure-preserving action of a finite group $\Gamma \curvearrowright (X, \mu)$ one gets

$$\text{cost}\,(\mathscr{R}_{\Gamma \curvearrowright X}) = 1 - \frac{1}{|\Gamma|}.$$

If $\mathscr{R}$ is the orbit relation of some (not necessarily free) action $\Gamma \curvearrowright (X, \mu)$, then $\text{cost}(\mathscr{R}) \leq \text{rank}(\Gamma)$, where the latter stands for the minimal number of generators for Γ. Indeed, any generating set $\{g_1, \ldots, g_k\}$ for Γ gives a generating system $\Phi = \{\gamma_i : X \to X\}_{i=1}^k$ for $\mathscr{R}_{\Gamma \curvearrowright X}$. Recall that the amenable II_1-relation $\mathscr{R}_{amen}$ can be generated by (any) action of $\mathbf{Z}$. Hence

$$\text{cost}(\mathscr{R}_{amen}) = 1.$$

The cost behaves nicely with respect to restriction:

THEOREM 4.7. (GABORIAU [54]) *For a II_1-relation $\mathscr{R}$:*

$$t \cdot (\text{cost}(\mathscr{R}^t) - 1) = \text{cost}(\mathscr{R}) - 1 \qquad (t \in \mathbf{R}_+^{\times}).$$

The following is a key tool for computations of the cost:

THEOREM 4.8. (GABORIAU [54]) *Let $\mathscr{R}$ be a treeable equivalence relation, and Φ be a graphing of $\mathscr{R}$ giving a tree structure to $\mathscr{R}$-classes. Then*

$$\text{cost}(\mathscr{R}) = \text{cost}(\Phi).$$

Conversely, for a relation $\mathscr{R}$ with $\text{cost}(\mathscr{R}) < \infty$, if the cost is attained by some graphing Ψ, then Ψ is a treeing of $\mathscr{R}$.

The above result (the first part) implies that for any essentially free action $\mathbf{F}_n \curvearrowright (X, \mu)$ one has $\text{cost}(\mathscr{R}_{\mathbf{F}_n \curvearrowright X}) = n$. This allowed Gaboriau to prove the following fact, answering a long-standing question:

COROLLARY 4.9. (GABORIAU [53], [54]) *If essentially free probability measure-preserving actions of $\mathbf{F}_n$ and $\mathbf{F}_m$ are orbit equivalent, then $n = m$.*

Note that $\mathbf{F}_n$ and $\mathbf{F}_m$ are commensurable for $2 \leq n, m < \infty$, hence they have essentially free actions that are *weakly isomorphic*. The index of such weak isomorphism will necessarily be $\frac{n-1}{m-1}$, or $\frac{m-1}{n-1}$ (these free groups have P.F.-fixed price). It should be pointed out that one of the major open problems in the theory of von Neumann algebras is whether it is possible for the factors $L(\mathbf{F}_n)$ and $L(\mathbf{F}_m)$ to be isomorphic for $n \neq m$ (it is known that either all $L(\mathbf{F}_n)$, $2 \leq n < \infty$, are isomorphic, or all distinct).

The following powerful result of Greg Hjorth provides a link from treeable relations back to actions of free groups:

THEOREM 4.10. (HJORTH [69]) *Let $\mathscr{R}$ be a treeable equivalence relation with $n = \text{cost}(\mathscr{R})$ in $\{1, 2, \ldots, \infty\}$. Then $\mathscr{R}$ can be generated by an essentially free action of $\mathbf{F}_n$.*

The point of the proof is to show that a relation $\mathscr{R}$ that has a treeing graphing with *average valency* $2n$ admits a (treeing) graphing with a.e. constant valency $2n$.

The behavior of the cost under passing to a subrelation of finite index is quite subtle—the following question is still open (to the best of the author's knowledge).

QUESTION 4.11. (GABORIAU) *Let Γ' be a subgroup of finite index in Γ, and $\Gamma \curvearrowright (X, \mu)$ be an essentially free p.m.p. action. Is it true that the costs of the orbit relations of Γ and Γ' are related by the index $[\Gamma : \Gamma']$*

$$\text{cost}(\mathscr{R}_{\Gamma' \curvearrowright X}) - 1 = [\Gamma : \Gamma'] \cdot (\text{cost}(\mathscr{R}_{\Gamma \curvearrowright X}) - 1)?$$

In general Γ' has at most $[\Gamma : \Gamma']$-many ergodic components. The extreme case where the number of Γ'-ergodic components is maximal: $[\Gamma : \Gamma']$ corresponds to $\Gamma \curvearrowright (X, \mu)$ being a co-induction from an ergodic Γ'-action. In this case the above formula easily holds. The real question lies in the other extreme where Γ' is ergodic.

Recall that the notion of the cost is analogous to the notion of *rank* for groups, where $\text{rank}(\Gamma) = \inf\{n \in \mathbf{N} : \exists \text{ epimorphism } \mathbf{F}_n \to \Gamma\}$. Schreier's theorem states that for $n \in \mathbf{N}$ any subgroup $F < \mathbf{F}_n$ of finite-index $[\mathbf{F}_n : F] = k$ is itself free: $F \cong \mathbf{F}_{k(n-1)+1}$. This implies that for any finitely generated Γ and any finite-index subgroup of $\Gamma' < \Gamma$ one has

$$\text{rank}(\Gamma') - 1 \leq [\Gamma : \Gamma'] \cdot (\text{rank}(\Gamma) - 1)$$

with equality in the case of free groups. Let $\Gamma > \Gamma_1 > \ldots$ be a chain of subgroups of finite index. One defines the **rank gradient** (Lackenby [89]) of the chain $\{\Gamma_n\}$ as the limit of the monotonic (!) sequence:

$$\text{RG}(\Gamma, \{\Gamma_n\}) = \lim_{n \to \infty} \frac{\text{rank}(\Gamma_n) - 1}{[\Gamma : \Gamma_n]}.$$

It is an intriguing question whether (or when) is it true that $\text{RG}(\Gamma, \{\Gamma_n\})$ depends only on Γ and not on a particular chain of finite-index subgroups. One should, of course, assume that the chains in question have trivial intersection, and one might require the chains to consist of normal subgroups in the original group. In the case of free groups RG is indeed independent of the chain.

In [1] Abert and Nikolov prove that the rank gradient of a chain of finite-index subgroups of Γ is given by the cost of a certain associated ergodic p.m.p. Γ-action. Let us formulate a special case of this relation where the chain $\{\Gamma_n\}$ consists of *normal* subgroups Γ_n with $\bigcap \Gamma_n = \{1\}$. Let $K = \varprojlim \Gamma / \Gamma_n$ denote the profinite completion corresponding to the chain. The Γ-action by left translations on the compact totally disconnected group K preserves the Haar measure m_K and $\Gamma \curvearrowright (K, m_K)$ is a free ergodic p.m.p. action. (Let us point out in passing that this action has a spectral gap, implying strong ergodicity, iff the chain has property (τ) introduced by Lubotzky and Zimmer [91]).

THEOREM 4.12. (ABERT-NIKOLOV [1]) *With the above notations:*

$$\mathrm{RG}\,(\Gamma, \{\Gamma_n\}) = \mathrm{cost}\,(\mathscr{R}_{\Gamma \curvearrowright K}) - 1.$$

One direction ($\geq$) is easy to explain. Let K_n be the closure of Γ_n in K. Then K_n is an open normal subgroup of K of index $m = [\Gamma : \Gamma_n]$. Let $1 = g_1, g_2, \ldots, g_n \in \Gamma$ be representatives of Γ_n-cosets, and $h_1, \ldots, h_k$ generators of Γ_n with $k = \mathrm{rank}\,(\Gamma_n)$. Consider the graphing $\Phi = \{\phi_2, \ldots, \phi_m, \psi_1, \ldots, \psi_k\}$, where $\phi_i : K_n \to g_i K_n$ are restrictions of g_i ($2 \leq i \leq m$), and $\psi_j : K_n \to K_n$ are restrictions of h_j ($1 \leq j \leq k$). These maps are easily seen to generate $\mathscr{R}_{\Gamma \curvearrowright K}$, with the cost of

$$\mathrm{cost}\,(\Phi) = k \cdot m_K(K_n) + (m-1) \cdot m_K(K_n) = \frac{k-1}{m} + 1 = 1 + \frac{\mathrm{rank}\,(\Gamma_n) - 1}{[\Gamma : \Gamma_n]}.$$

Abert and Nikolov observed that a positive answer to Question 4.11 combined with the above result shows that RG (Γ) is independent of the choice of a (normal) chain, and therefore is a numeric invariant associated to any residually finite finitely generated group. Surprisingly, this turns out to be related to a problem in the theory of compact hyperbolic 3-manifolds concerning rank versus Heegard genus [89]—see [1] for the connection and further discussions.

The above result has an application, independent of Question 4.11. Since amenable groups have P.F. with $\mathscr{C} = 1$, it follows that a finitely generated, residually finite amenable group Γ has sublinear rank growth for finite-index normal subgroups with trivial intersection, that is, RG (Γ) = 0 for any such chain.

4.2.5. ℓ^2-BETTI NUMBERS I have already mentioned the ℓ^2-Betti numbers $\beta_i^{(2)}(\Gamma)$ associated with a discrete group Γ and Gaboriau's proportionality result in Theorem 3.8 for measure equivalence between groups. In fact, rather than relating the ℓ^2-Betti numbers of groups via ME, in [57] Gaboriau

- *defines* the notion of ℓ^2-Betti numbers $\beta_i^{(2)}(\mathscr{R})$ for a II_1-equivalence relation $\mathscr{R}$;
- proves that $\beta_i^{(2)}(\Gamma) = \beta_i^{(2)}(\mathscr{R}_{\Gamma \curvearrowright X})$ for essentially free ergodic action $\Gamma \curvearrowright (X, \mu)$; and
  observes that $\beta_i^{(2)}(\mathscr{R}^t) = t \cdot \beta_i^{(2)}(\mathscr{R})$ for any II_1-relation.

The definition of $\beta_i^{(2)}(\mathscr{R})$ is inspired by the definition of $\beta_i^{(2)}(\Gamma)$ by Cheeger and Gromov [23]: it uses $\mathscr{R}$-action (or groupoid action) on pointed contractible simplicial complexes, corresponding complexes of Hilbert modules with $\mathscr{R}$-action, and von Neumann dimension with respect to the algebra $M_{\mathscr{R}}$.

In the late 1990s Wolgang Lück developed an algebraic notion of dimension for arbitrary modules over von Neumann algebras, in particular giving an alternative approach to ℓ^2-Betti numbers for groups (see [92]). In [124] Roman Sauer used Lück's notion of dimension to define ℓ^2-Betti numbers of equivalence relations, and more general measured groupoids, providing an alternative approach to Gaboriau's results. In [127] Sauer and Thom develop further homological tools (including a spectral sequence associated to strongly normal subrelations) to study ℓ^2-Betti numbers for groups, relations, and measured groupoids.

4.2.6. OUTER AUTOMORPHISM GROUP

Given an equivalence relation $\mathscr{R}$ on (X, μ) define the corresponding **automorphism group** as the group of self-isomorphisms:

$$\mathrm{Aut}\,(\mathscr{R}) = \left\{ T \in \mathrm{Aut}\,(X, \mu) \,:\, T \times T(\mathscr{R}) = \mathscr{R} \quad (\text{modulo null sets}) \right\}.$$

The subgroup $\mathrm{Inn}\,(\mathscr{R})$ of inner automorphisms is

$$\mathrm{Inn}\,(\mathscr{R}) = \left\{ T \in \mathrm{Aut}\,(X, \mu) \,:\, (x, T(x)) \in \mathscr{R} \text{ for a.e. } x \in X \right\}.$$

This is just the full group $[\mathscr{R}]$, but the above notation emphasizes the fact that it is normal in $\mathrm{Aut}\,(\mathscr{R})$ and suggests to consider the **outer automorphism group**

$$\mathrm{Out}\,(\mathscr{R}) = \mathrm{Aut}\,(\mathscr{R})/\,\mathrm{Inn}\,(\mathscr{R}).$$

One might think of $\mathrm{Out}\,(\mathscr{R})$ as the group of all measurable permutations of the $\mathscr{R}$-classes on X. Recall (Lemma 4.2) that $\mathrm{Inn}\,(\mathscr{R})$ is a huge group as it acts transitively on (classes mod null sets of) measurable subsets of any given size in X. Yet the quotient Out $(\mathscr{R})$ might be small (even finite or trivial), and can sometimes be explicitly computed.

REMARK 4.13. *As an abstract group $H = $ Inn $(\mathscr{R})$ is simple, and its auto-morphisms come from automorphisms of $\mathscr{R}$; in particular Out $(H) = $ Out $(\mathscr{R})$. Moreover, Dye's reconstruction theorem states that (the isomorphism type of) $\mathscr{R}$ is determined by the structure of Inn $(\mathscr{R})$ as an abstract group (see [80, §I.4] for proofs and further facts).*

Let us also note that the operation of restriction/amplification of the relation does not alter the outer automorphism group (cf. [47, lemma 2.2]):

$$\text{Out}(\mathscr{R}^t) \cong \text{Out}(\mathscr{R}) \qquad (t \in \mathbf{R}^\times_+).$$

The group Aut $(\mathscr{R})$ has a natural structure of a Polish group [62, 64]. First, recall that if (Y, ν) is a finite- or infinite-measure Lebesgue space, then Aut (Y, ν) is a Polish group with respect to the **weak topology** induced from the weak (=strong) operator topology of the unitary group of $L^2(Y, \nu)$. This defines a Polish topology on Aut $(\mathscr{R})$ when the latter is viewed as acting on the infinite-measure space $(\mathscr{R}, \tilde{\mu})$. [9] However, Inn $(\mathscr{R})$ is not always closed in Aut $(\mathscr{R})$, so the topology on Out $(\mathscr{R})$ might be complicated. Alexander Kechris recently found the following surprising connection:

THEOREM 4.14. (KECHRIS [80, THEOREM 8.1]) *If* Out $(\mathscr{R})$ *fails to be a Polish group, then* cost $(\mathscr{R}) = 1$.

Now assume that $\mathscr{R}$ can be presented as the orbit relation of an essentially free action $\Gamma \curvearrowright (X, \mu)$, so Aut $(\mathscr{R})$ is the group of self-orbit equivalences of $\Gamma \curvearrowright X$. The centralizer Aut$_\Gamma (X, \mu)$ of Γ in Aut (X, μ) embeds in Aut $(\mathscr{R})$ and if Γ is ICC (i.e., has infinite conjugacy classes), then the quotient map Aut $(\mathscr{R}) \overset{\text{out}}{\longrightarrow}$ Out $(\mathscr{R})$ is *injective* on Aut$_\Gamma (X, \mu)$ (cf. [62, lemma 2.6]). So Out $(\mathscr{R})$ has a copy of Aut$_\Gamma (X, \mu)$, and the latter might be very big. For example, in the Bernoulli action $\Gamma \curvearrowright (X, \mu) = (X_0, \mu_0)^\Gamma$, it contains Aut (X_0, μ_0) acting diagonally on the factors. Yet, if Γ has property (T), then Aut$_\Gamma (X, \mu) \cdot$ Inn $(\mathscr{R}_{\Gamma \curvearrowright X})$ is open in the Polish group Aut $(\mathscr{R}_{\Gamma \curvearrowright X})$. In this case the image of Aut$_\Gamma (X, \mu)$ has finite or countable index in Out $(\mathscr{R}_{\Gamma \curvearrowright X})$. This fact was observed by Gefter and Golodets in [64, §2], and can be deduced from Proposition 5.14.

---

9. This topology coincides with the restriction to Aut $(\mathscr{R})$ of the **uniform** topology on Aut (X, μ) given by the metric $d(T, S) = \mu \{x \in X : T(x) \neq S(x)\}$. On all of Aut (X, μ) the uniform topology is complete but not separable; but its restriction to Aut $(\mathscr{R})$ is separable.

To get a handle on Out $(\mathscr{R}_{\Gamma \curvearrowright X})$ one looks at OE-cocycles $c_T : \Gamma \times X \to \Gamma$ corresponding to elements $T \in \mathrm{Aut}\,(\mathscr{R}_{\Gamma \curvearrowright X})$. It is not difficult to see that c_T is conjugate in Γ to the identity (i.e., $c_T(g,x) = f(gx)^{-1} g f(x)$ for $f : X \to \Gamma$) iff T is in $\mathrm{Aut}_\Gamma\,(X, \mu) \cdot \mathrm{Inn}\,(\mathscr{R})$. Thus, starting from a group Γ or action $\Gamma \curvearrowright X$ with strong rigidity properties for cocycles, one controls Out $(\mathscr{R}_{\Gamma \curvearrowright X})$ via $\mathrm{Aut}_\Gamma\,(X, \mu)$. This general scheme (somewhat implicitly) is behind the first example of an equivalence relation with trivial Out $(\mathscr{R})$ constructed by Gefter [61, 62]. Here is a variant of this construction:

THEOREM **4.15**. *Let* Γ *be a torsion-free group with property (T),* K *a compact connected Lie group without outer automorphisms, and* $\tau : \Gamma \to K$ *a dense embedding. Let* $L < K$ *be a closed subgroup and consider the ergodic actions* $\Gamma \curvearrowright (K, m_K)$ *and* $\Gamma \curvearrowright (K/L, m_{K/L})$ *by left translations. Then*

$$\mathrm{Out}\,(\mathscr{R}_{\Gamma \curvearrowright K}) \cong K, \qquad \mathrm{Out}\,(\mathscr{R}_{\Gamma \curvearrowright K/L}) \cong N_K(L)/L.$$

In particular, taking $K = \mathrm{PO}_n(\mathbf{R})$ and $L \cong \mathrm{PO}_{n-1}(\mathbf{R}) < K$ to be the stabilizer of a line in $\mathbf{R}^n$, the space K/L is the projective space P^{n-1}, and we get

$$\mathrm{Out}\,(\mathscr{R}_{\Gamma \curvearrowright P^{n-1}}) = \{1\}$$

for any property (T) dense subgroup $\Gamma < \mathrm{PO}_n(\mathbf{R})$. Such a group Γ exists iff $n \geq 5$, Zimmer [145, theorem 7]. The preceding discussion, combined with the cocycles superrigidity Theorem 5.21 below, and an easy observation that $\mathrm{Aut}_\Gamma\,(K/L, m_{K/L})$ is naturally isomorphic to $N_K(L)/L$, provide a self contained sketch of the proof of the theorem.

In the above statement Out (K) is assumed to be trivial and Γ to be torsion free just to simplify the statement. However, the assumption that K is connected is essential. Indeed, the dense embedding of $\Gamma = \mathrm{PSL}_n(\mathbf{Z})$ in the compact profinite group $K = \mathrm{PSL}_n(\mathbf{Z}_p)$ where p is a prime, gives

$$\mathrm{Out}\,(\mathscr{R}_{\mathrm{PSL}_n(\mathbf{Z}) \curvearrowright \mathrm{PSL}_n(\mathbf{Z}_p)}) \cong \mathrm{PSL}_n(\mathbf{Q}_p) \rtimes \mathbf{Z}/2\mathbf{Z}$$

where the $\mathbf{Z}/2$-extension is given by the transpose $g \mapsto g^{tr}$. The inclusion $\supset$ was found in [62], and the equality is proved in [47, theorem 1.6], where many other computations of Out $(\mathscr{R}_{\Gamma \curvearrowright X})$ are carried out for actions of lattices in higher-rank Lie groups.

Finally, we recall that the recent preprint [117] of Popa and Vaes quoted above (Theorem 4.5) shows that an arbitrary totally disconnected lcsc group G can arise as Out $(\mathscr{R}_{\Gamma \curvearrowright X})$ for an essentially free action of a *free* group $\mathbf{F}_\infty$.

4.2.7. COHOMOLOGY Equivalence relations have groups of cohomology associated to them similar to cohomology of groups. These were introduced by Singer [134] and largely emphasized by Feldman and Moore [40]. Given, say, a type II_1 equivalence relation $\mathcal{R}$ on (X, μ) consider

$$\mathcal{R}^{(n)} = \left\{ (x_0, \ldots, x_n) \in X^{n+1} : (x_i, x_{i+1}) \in \mathcal{R} \right\}$$

equipped with the infinite Lebesgue measure $\tilde{\mu}^{(n)}$ defined by

$$\tilde{\mu}^{(n)}(A) = \int_X \# \left\{ (x_1, \ldots, x_n) : (x_0, \ldots, x_n) \in \mathcal{R}^{(n)} \right\} d\mu(x_0).$$

Take $(\mathcal{R}^{(0)}, \tilde{\mu}^0)$ to be (X, μ). Note that $(\mathcal{R}^{(1)}, \mu^{(1)})$ is just $(\mathcal{R}, \tilde{\mu})$ from §4.1. Since μ is assumed to be $\mathcal{R}$-invariant, the above formula is invariant under permutations of $x_0, \ldots, x_n$.

Fix a Polish abelian group A written multiplicatively (usually $A = \mathbf{T}$). The space $C^n(\mathcal{R}, A)$ of n-cochains consists of equivalence classes (modulo $\tilde{\mu}^{(n)}$-null sets) of measurable maps $\mathcal{R}^{(n)} \to A$, linked by the operators $d_n : C^n(\mathcal{R}, A) \to C^{n+1}(\mathcal{R}, A)$

$$d_n(f)(x_0, \ldots, x_{n+1}) = \prod_{i=0}^{n+1} f(x_0, \ldots, \hat{x}_i, \ldots, x_0)^{(-1)^i}.$$

Call $Z^n(\mathcal{R}) = \text{Ker}(d_n)$ the n-cocycles, and $B^n(\mathcal{R}) = \text{Im}(d_{n-1})$ the n-coboundaries; the cohomology groups are defined by $H^n(\mathcal{R}) = Z^n(\mathcal{R})/B^n(\mathcal{R})$. In degree $n = 1$ the 1-cocycles are measurable maps $c : (\mathcal{R}, \mu) \to A$ such that

$$c(x, y)c(y, x) = c(x, z)$$

and 1-coboundaries have the form $b(x, y) = f(x)/f(y)$ for some measurable $f : X \to A$.

If A is a compact abelian group, such as $\mathbf{T}$, then $C^1(\mathcal{R}, A)$ is a Polish group (with respect to convergence in measure). Being a closed subgroup in $C^1(\mathcal{R}, A)$, the 1-cocycles $Z^1(\mathcal{R}, A)$ form a Polish group. Schmidt [129] showed that $B^1(\mathcal{R}, A)$ is closed in $Z^1(\mathcal{R}, A)$ iff $\mathcal{R}$ is strongly ergodic.

There are only few cases where $H^1(\mathcal{R}, \mathbf{T})$ were computed: C.C. Moore [102] constructed a relation with trivial $H^1(\mathcal{R}, \mathbf{T})$. Gefter [63] considered $H^1(\mathcal{R}_{\Gamma \curvearrowright G}, \mathbf{T})$ for actions of property (T) group Γ densely embedded in a semisimple Lie group G. More recently Popa and Sasyk [116] studied $H^1(\mathcal{R}_{\Gamma \curvearrowright X}, \mathbf{T})$ for property (T) groups Γ with Bernoulli actions $(X, \mu) = (X_0, \mu_0)^\Gamma$. In both cases $H^1(\mathcal{R}_{\Gamma \curvearrowright X}, \mathbf{T})$ is shown to coincide with the group of characters $\text{Hom}(\Gamma, \mathbf{T})$. Higher cohomology groups remain mysterious.

The fact that A is abelian is essential to the definition of $H^n(\mathscr{R}, A)$ for $n > 1$. However, in degree $n = 1$ one can define $H^1(\mathscr{R}, \Lambda)$ as a *set* for a general target group Λ[10]. In fact, this notion is commonly used in this theory under the name of **measurable cocycles** (see Appendix A and §5.1 below). For the definition in terms of equivalence relations let $Z^1(\mathscr{R}, \Lambda)$ denote the *set* of all measurable maps (mod $\tilde{\mu}$-null sets)

$$c : (\mathscr{R}, \tilde{\mu}) \to \Lambda \qquad s.t. \qquad c(x, z) = c(x, y)c(y, z)$$

and let $H^1(\mathscr{R}, \Lambda) = Z^1(\mathscr{R}, \Lambda)/ \sim$ where the equivalence $\sim$ between $c, c' \in Z^1(\mathscr{R}, \Lambda)$ is declared if $c(x, y) = f(x)^{-1}c'(x, y)f(y)$ for some measurable $f : (X, \mu) \to \Lambda$.

If $\mathscr{R} = \mathscr{R}_{\Gamma \curvearrowright X}$ is the orbit relation of an essentially free action, then $Z^1(\mathscr{R}_{\Gamma \curvearrowright X}, \Lambda)$ coincides with the set of measurable cocycles $\alpha : \Gamma \times X \to \Lambda$ by $\alpha(g, x) = c(x, gx)$. Note that $\mathrm{Hom}(\Gamma, \Lambda)/\Lambda$ maps into $H^1(\mathscr{R}, \Lambda)$, via $c_\pi(x, y) = \pi(g)$ for the unique $g \in \Gamma$ with $x = gy$. The point of cocycles superrigidity theorems is to show that under favorable conditions this map is surjective.

4.2.8. FREE DECOMPOSITIONS Group theoretic notions such as free products, amalgamated products, and HNN-extensions can be defined in the context of equivalence relations—see Gaboriau [54, section IV]. For example, a II_1-relation $\mathscr{R}$ is said to split as a free product of subrelations $\{\mathscr{R}_i\}_{i \in I}$, denoted $\mathscr{R} = *_{i \in I}\mathscr{R}_i$, if

1) $\mathscr{R}$ is generated by $\{\mathscr{R}_i\}_{i \in I}$, that is, $\mathscr{R}$ is the smallest equivalence relation containing the latter family; and
2) almost every chain $x = x_0, \ldots, x_n = y$, where $x_{j-1} \neq x_j$, $(x_{j-1}, x_j) \in \mathscr{R}_{i(j)}$ and $i(j+1) \neq i(j)$, has $x \neq y$.

If $\mathscr{S}$ is yet another subrelation, one says that $\mathscr{R}$ splits as a **free product** of $\mathscr{R}_i$ **amalgamated over** $\mathscr{S}$, $\mathscr{R} = *_{\mathscr{S}}\mathscr{R}_i$, if in condition (2) one replaces $x_{j-1} \neq x_j$ by $(x_{j-1}, x_j) \notin \mathscr{S}$.

The obvious example of the situation above is an essentially free action of a free product of groups $\Gamma_3 = \Gamma_1 * \Gamma_2$ (respectively amalgamated product $\Gamma_5 = \Gamma_1 *_{\Gamma_4} \Gamma_2$) on a probability space (X, μ); in this case the orbit relations $\mathscr{R}_i = \mathscr{R}_{\Gamma_i \curvearrowright X}$ satisfy $\mathscr{R}_3 = \mathscr{R}_1 * \mathscr{R}_2$ (respectively $\mathscr{R}_5 = \mathscr{R}_1 *_{\mathscr{R}_4} \mathscr{R}_2$).

Another useful construction (see Ioana, Peterson, Popa [75]) is as follows. Given measure preserving (possibly ergodic) relations $\mathscr{R}_1, \mathscr{R}_2$ on a probability

---

10. In order to define the notion of measurability Λ should have a Borel structure, and better be a Polish group; often it is a discrete countable group, or a Lie group.

space (X, μ) for $T \in \text{Aut}(X, \mu)$ consider the relation generated by $\mathcal{R}_1$ and $T(\mathcal{R}_2)$. It can be shown that for a residual set of $T \in \text{Aut}(X, \mu)$ the resulting relation is a free product of $\mathcal{R}_1$ and $T(\mathcal{R}_2)$. A similar construction can be carried out for amalgamated products. Note that in contrast with the category of groups the isomorphism type of the free product is not determined by the free factors alone.

Ioana, Peterson, and Popa [75] obtained strong rigidity results for free and amalgamated products of ergodic measured equivalence relations and II_1-factors with certain rigidity properties. Here let us describe some results obtained by Alvarez and Gaboriau [7], which are easier to state; they may be viewed as an analogue of Bass-Serre theory in the context of equivalence relations. Say that a II_1-relation $\mathcal{R}$ is **freely indecomposable** (FI) if $\mathcal{R}$ is not a free product of its subrelations. A group Γ is said to be **measurably freely indecomposable** (MFI) if all its essentially free action give freely indecomposable orbit relations. A group may fail to be MFI even if it is freely indecomposable in the group theoretic sense (surface groups provide an example). Not surprisingly, groups with property (T) are MFI (cf. Adams Spatzier [6]); but more generally

THEOREM 4.16. (ALVAREZ-GABORIAU [7]) *If Γ is nonamenable and $\beta_1^{(2)}(\Gamma) = 0$, then Γ is MFI.*

THEOREM 4.17. (ALVAREZ-GABORIAU [7]) *Let I, J be two finite or countable-index sets, $\{\Gamma_i\}_{i \in I}$ and $\{\Lambda_j\}_{j \in J}$ be two families of MFI groups, $\Gamma = *_{i \in I} \Gamma_i$, $\Lambda = *_{j \in J} \Lambda_j$, and $\Gamma \curvearrowright (X, \mu)$, $\Lambda \curvearrowright (Y, \nu)$ be essentially free p.m.p. actions where each $\Gamma_i \curvearrowright (X, \mu)$ and $\Lambda_j \curvearrowright (Y, \nu)$ are ergodic. Assume that $\Gamma \curvearrowright X \overset{\text{SOE}}{\sim} \Lambda \curvearrowright Y$.*

Then $|I| = |J|$ and there is a bijection $\theta : I \to J$ so that $\Gamma_i \curvearrowright X \overset{\text{SOE}}{\sim} \Lambda_{\theta(i)} \curvearrowright Y$.

The assumption that each free factor is ergodic is important here; Alvarez and Gaboriau also give an analysis of the general situation (where this assumption is dropped).

4.3. Rigidity of Equivalence Relations

The close relation between ME and SOE allows us to deduce that certain orbit relations $\mathcal{R}_{\Gamma \curvearrowright X}$ remember the acting group Γ and the action $\Gamma \curvearrowright (X, \mu)$ up to isomorphism, or up to a **virtual isomorphism**. This slightly technical concept is described in the following:

LEMMA 4.18. *Suppose an ergodic ME-coupling (Ω, m) of Γ with Λ corresponds to an SOE between ergodic actions* $T : \Gamma \curvearrowright (X, \mu) \overset{SOE}{\sim} \Lambda \curvearrowright (Y, \nu)$. *Then the following are equivalent:*

1) *There exist short exact sequences*

$$1 \to \Gamma_0 \to \Gamma \to \Gamma_1 \to 1, \qquad 1 \to \Lambda_0 \to \Lambda \to \Lambda_1 \to 1$$

where Γ_0 and Λ_0 are finite, a discrete (Γ_1, Λ_1)-coupling (Ω_1, m_1), and an equivariant map $\Phi : (\Omega, m) \to (\Omega_1, m_1)$; and

2) *There exist isomorphism between finite-index subgroups*

$$\Gamma_1 > \Gamma_2 \cong \Lambda_2 < \Lambda_1,$$

so that $\Gamma_1 \curvearrowright X_1 = X/\Gamma_0$ and $\Lambda_1 \curvearrowright Y_1 = Y/\Lambda_0$ are induced from some isomorphic ergodic actions $\Gamma_2 \curvearrowright X_2 \cong \Lambda_2 \curvearrowright Y_2$.

3) *The SOE (or ME) cocycle $\Gamma \times X \to \Lambda$ is conjugate in Λ to a cocycle whose restriction to some finite-index subgroup Γ_1 is a homomorphism $\Gamma_1 \to \Lambda$ (the image is necessarily of finite index).*

Let us now state two general forms of relation rigidity. Here is one form

THEOREM 4.19. *Let $\Gamma \curvearrowright (X, \mu)$ be an ergodic essentially free action of one of the types below, Λ an arbitrary group, and $\Lambda \curvearrowright (Y, \nu)$ as essentially free p.m.p. action whose orbit relation $\mathscr{R}_{\Lambda \curvearrowright Y}$ is weakly isomorphic to $\mathscr{R}_{\Gamma \curvearrowright X}$.*

Then Λ is commensurable up to finite kernels to Γ and the actions $\Gamma \curvearrowright X$ and $\Lambda \curvearrowright Y$ are virtually isomorphic; in particular, the SEO-index is necessarily rational.

The list of actions $\Gamma \curvearrowright X$ with this SOE-rigidity property includes:

1) *Γ is a lattice in a connected, center-free, simple Lie group G of higher rank, and $\Gamma \curvearrowright X$ has no equivariant quotients of the form $\Gamma \curvearrowright G/\Gamma'$ where $\Gamma' < G$ is a lattice ([45, theorem A]);*

2) *$\Gamma = \Gamma_1 \times \cdots \times \Gamma_n$ where $n \geq 2$, $\Gamma_i \in C_{\text{reg}}$, and $\Gamma_i \curvearrowright (X, \mu)$ are ergodic; in addition assume that $\Lambda \curvearrowright (Y, \nu)$ is mildly mixing (Monod-Shalom [101]); and*

3) *Γ is a finite-index subgroup in a (product of) mapping class groups as in Theorem 3.20 (Kida [85]).*

a) For a concrete example for (1)–(3) one might take Bernoulli actions $\Gamma \curvearrowright (X_0, \mu_0)^\Gamma$. In (1) one might also take $SL_n(\mathbf{Z}) \curvearrowright \mathbf{T}^n$ or $SL_n(\mathbf{Z}) \curvearrowright SL_n(\mathbf{Z}_p)$ with $n \geq 3$. In (2) one might look at $\mathbf{F}_n \times \mathbf{F}_m$ acting on a compact Lie group K, for example $SO_3(\mathbf{R})$, by $(g, h) : k \mapsto gkh^{-1}$ where $\mathbf{F}_n$, $\mathbf{F}_m$ are embedded densely in K.

b) In (1) the assumption that there are no Γ-equivariant quotient maps $X \to G/\Gamma'$ is necessary, since given such a quotient there is a Γ'-action on some (X', μ') with $\Gamma' \curvearrowright X' \overset{\text{SOE}}{\sim} \Gamma \curvearrowright X$. The rigidity statement in this case is that this is a complete list of groups and their essentially free actions up to virtual isomorphism ([45, theorem C]). The appearance of these factors has to do with (G, m_G) appearing as a quotient of a (Γ, Λ)-coupling.

c) The basic technique for establishing the stated rigidity in cases (1)–(3) is to establish condition (1) in Lemma 4.18. This is done by analyzing a self-Γ-coupling of the form $\Omega \times_\Lambda \check{\Omega}$ (where $X = \Omega/\Lambda$ and $Y = \Omega/\Gamma$) and invoking an analogue of the construction in §5.5.

d) In all cases one can sharpen the results (eliminate the "virtual") by imposing some benign additional assumptions: rule out torsion in the acting groups, and impose ergodicity for actions of finite-index subgroups.

The second *stronger* form of relation rigidity refers to **rigidity of relation morphisms** that are obtained from $\mathscr{G}_{\text{dsc}}$-**cocycle superrigid actions** discovered by Sorin Popa (see §5.4). We illustrate this framework by the following particular statement (see [113, theorem 0.4] and [48, theorem 1.8]).

THEOREM 4.20. *Let $\Gamma \curvearrowright (X, \mu)$ be a mixing $\mathscr{G}_{\text{dsc}}$-cocycle superrigid action, such as:*

1. *A Bernoulli Γ-action on $(X_0, \mu_0)^\Gamma$, where Γ has property (T), or $\Gamma = \Gamma_1 \times \Gamma_2$ with Γ_1 nonamenable and Γ_2 being infinite; and*
2. *$\Gamma \curvearrowright K/L$ where $\Gamma \to K$ is a homomorphism with dense image in a simple compact Lie group K with trivial $\pi_1(K)$, $L < K$ is a closed subgroup, and Γ has (T).*

Let Λ be some group with an ergodic essentially free measure-preserving action $\Lambda \curvearrowright (Y, \nu)$,[11] $X' \subset X$ a positive measure subset, and $T : X' \to Y$ a measurable map with $T_* \mu \prec \nu$ and

$$(x_1, x_2) \in \mathscr{R}_{\Gamma \curvearrowright X} \cap (X' \times X') \qquad \Longrightarrow \qquad (T(x_1), T(x_2)) \in \mathscr{R}_{\Lambda \curvearrowright Y}.$$

Then there exists

- *an exact sequence $\Gamma_0 \longrightarrow \Gamma \overset{\rho}{\longrightarrow} \Lambda_1$ with finite Γ_0 and $\Lambda_1 < \Lambda$; and*
- *a Λ_1-ergodic subset $Y_1 \subset Y$ with $0 < \nu(Y_1) < \infty$; and*

---

11. The space (Y, ν) might be finite- or infinite-measure Lebesgue space.

- *denoting $(X_1, \mu_1) = (X, \mu)/\Gamma_0$ and $\nu_1 = \nu(Y_1)^{-1} \cdot \nu|_{Y_1}$, there is an isomorphism $T_1 : (X_1, \mu_1) \cong (Y_1, \nu_1)$ of Λ_1-actions.*

Moreover, μ-a.e. $T(x)$ and $T_1(\Gamma_0 x)$ are in the same Λ-orbit.

4.3.1. A QUESTION OF FELDMAN AND MOORE Feldman and Moore showed
[39] that any countable Borel equivalence relation $\mathscr{R}$ can be generated by a Borel action of a countable group. They asked whether one can find a *free* action of some group, so that $\mathscr{R}$-classes would be in one-to-one correspondence with the acting group. This question was answered in the negative by Adams [3]. In the context of measured relations, say of type II_1, the question is whether it is possible to generate $\mathscr{R}$ (up to null sets) by an *essentially free* action of some group. This question was also settled in the negative in [45, theorem D], using the following basic constructions:

1) Start with an essentially free action $\Gamma \curvearrowright (X, \mu)$ that is rigid as in Theorem 4.19 or 4.20, and let $\mathscr{R} = (\mathscr{R}_{\Gamma \curvearrowright X})^t$ with an *irrational* t.
2) Consider a proper embedding $G \hookrightarrow H$ of higher-rank simple Lie groups and choose a lattice $\Gamma < H$, say $G = \mathrm{SL}_3(\mathbf{R}) \subset H = \mathrm{SL}_4(\mathbf{R})$ with $\Gamma = \mathrm{SL}_4(\mathbf{Z})$. Such actions always admit a Borel cross-section $X \subset H/\Gamma$ for the G-action, equipped with a holonomy-invariant probability measure μ. Take $\mathscr{R}$ on (X, μ) to be the relation of being in the same G-orbit.

In case (1) one argues as follows: if some group Λ has an essentially free action $\Lambda \curvearrowright (Y, \nu)$ with $(\mathscr{R}_{\Gamma \curvearrowright X})^t = \mathscr{R} \cong \mathscr{R}_{\Lambda \curvearrowright Y}$, then the rigidity implies that Γ and Λ are commensurable up to finite kernel, and $\Gamma \curvearrowright X$ is virtually isomorphic to $\Lambda \curvearrowright Y$. But this would imply that the index t is rational, contrary to the assumption. This strategy can be carried out in other cases of very rigid actions as in [72, 85, 101, 113]. Theorem 5.21 provides an example of this type $\mathscr{R} = (\mathscr{R}_{\Gamma \curvearrowright K})^t$ where Γ is a Kazhdan group densely embedded in a compact connected Lie group K. So the reader has a sketch of the full proof for a II_1-relation that cannot be generated by an essentially free action of any group.

Example of type (2) was introduced by Zimmer in [147], where it was proved that the relation $\mathscr{R}$ on such a cross-section cannot be essentially freely generated by a group Λ, which admits a linear representation with an infinite image. The linearity assumption was removed in [45]. This example is particularly interesting since it cannot be "repaired" by restriction/amplification; as any $\mathscr{R}^t$ can be realized as a cross-section of the same G-flow on H/Γ.

QUESTION 4.21. (VERSHIK) *Let $\mathscr{R}$ on (X, μ) be a II_1-relation that cannot be generated by an essentially free action of a group; and let $\Gamma \curvearrowright (X, \mu)$ be some action* *producing $\mathscr{R}$. One may assume that the action is faithful, that is, $\Gamma \to \mathrm{Aut}(X, \mu)$ is an embedding. What can be said about Γ and the structure of the measurable family $\{\Gamma_x\}_{x \in X}$ of the stabilizers of points in X?*

In [119] S. Popa and S. Vaes give an example of a II_1-relation $\mathscr{R}$ (which is a restriction of the II_∞-relation $\mathscr{R}_{\mathrm{SL}_5(\mathbf{Z}) \curvearrowright \mathbf{R}^5}$ to a subset $A \subset \mathbf{R}^5$ of positive finite measure), which has property (T) but cannot be generated by an action (not necessarily free) of *any* group with property (T).

5. Techniques

5.1. Superrigidity in Semisimple Lie Groups

The term "*superrigidity*" refers to a number of phenomena originated and inspired by the following celebrated discovery of G. A. Margulis.

THEOREM 5.1. (MARGULIS [93]) *Let G and G' be (semi)simple connected real center-free Lie groups without compact factors with $\mathrm{rk}(G) \geq 2$, $\Gamma < G$ be an irreducible lattice, and $\pi : \Gamma \to G'$ a homomorphism with $\pi(\Gamma)$ being Zariski dense in G' and not precompact. Then π extends to a (rational) epimorphism $\bar{\pi} : G \to G'$.*

The actual result is more general than stated, as it applies to products of semisimple algebraic groups over general local fields. We refer the reader to the comprehensive monograph (Margulis [95]) for the general statements, proofs, and further results and applications, including the famous arithmeticity theorem.

The core of (some of the available) proofs of Margulis's superrigidity theorem is a combination of the theory of algebraic groups and purely ergodic-theoretic arguments. The result applies to uniform and nonuniform lattices alike, and it also covers irreducible lattices in higher-rank Lie groups such as $\mathrm{SL}_2(\mathbf{R}) \times \mathrm{SL}_2(\mathbf{R})$. Let us also note that the assumption that $\pi(\Gamma)$ is not precompact in G' is redundant if $\pi(\Gamma)$ is Zariski dense in a real Lie group G' (since compact groups over $\mathbf{R}$ are algebraic), but is important in general (cf. $\mathrm{SL}_n(\mathbf{Z}) < \mathrm{SL}_n(\mathbf{Q}_p)$ is Zariski dense but precompact).

In [139] R. J. Zimmer has obtained a far-reaching generalization of Margulis's superrigidity, passing from the context of representations of lattices to the framework of measurable cocycles over probability measure-preserving actions (representations of "virtual subgroups" in Mackey's terminology). The connection can be briefly summarized as follows: given a transitive action $G \curvearrowright X = G/\Gamma$ and some topological group H, there is a bijection between measurable cocycles $G \times G/\Gamma \to H$ modulo cocycle conjugation and homomorphisms $\Gamma \to H$ modulo conjugation in H

$$H^1(G \curvearrowright G/\Gamma, H) \cong \mathrm{Hom}(\Gamma, H)/H$$

(see §A.1, and [52, 146]). In this correspondence, a representation $\pi : \Gamma \to H$ extends to a homomorphism $G \to H$ iff the corresponding cocycle $\pi \circ c :$ $G \times G/\Gamma \to H$ is conjugate to a homomorphism $G \to H$. Zimmer's cocycle superrigidity theorem states that under appropriate nondegeneracy assumptions a measurable cocycle over an *arbitrary* p.m.p. ergodic action $G \curvearrowright (X, \mu)$ is conjugate to a homomorphism.

THEOREM 5.2. (ZIMMER [139], SEE ALSO [146]) *Let G, G' be a semisimple Lie group as in Theorem 5.1, in particular $\mathrm{rk}_{\mathbf{R}}(G) \geq 2$, let $G \curvearrowright (X, \mu)$ be an irreducible probability measure-preserving action and $c : G \times X \to G'$ be a measurable cocycle that is Zariski dense and not compact. Then there exist a (rational) epimorphism $\pi : G \to G'$ and a measurable $f : X \to G'$ so that $c(g, x) = f(gx)^{-1}\pi(g)f(x)$.*

In the above statement **irreducibility** of $G \curvearrowright (X, \mu)$ means mere ergodicity if G is a simple group, and ergodicity of the action $G_i \curvearrowright (X, \mu)$ for each factor G_i in the case of a semisimple group $G = \prod_{i=1}^{n} G_i$ with $n \geq 2$ factors. For a lattice $\Gamma < G = \prod G_i$ in a semisimple group the transitive action $G \curvearrowright G/\Gamma$ is irreducible precisely iff Γ is an irreducible lattice in G. The notion of being **Zariski dense** (respectively **not compact**) for a cocycle $c : G \times X \to H$ means that c is not conjugate to a cocycle c^f taking values in a proper algebraic (respectively compact) subgroup of H.

The setting of cocycles over p.m.p. actions adds a great deal of generality to the superrigidity phenomena. The first illustration of this is the fact that once cocycle superrigidity is known for actions of G it passes to actions of lattices in G: given an action $\Gamma \curvearrowright (X, \mu)$ of a lattice $\Gamma < G$ one obtains a G-action on $\bar{X} = G \times_\Gamma X$ by acting on the first coordinate (just like the composition operation of ME-coupling in §2.1). A cocycle $c : \Gamma \times X \to H$ has a natural lift to $\bar{c} : G \times \bar{X} \to H$ and its cohomology is directly related to that of the original cocycle. So cocycle superrigidity theorems have an almost automatic bootstrap from lcsc groups to their lattices. The induced action $G \curvearrowright \bar{X}$ is ergodic iff $\Gamma \curvearrowright X$ is ergodic; however, irreducibility is more subtle. Yet, if $\Gamma \curvearrowright (X, \mu)$ is mixing, then $G \curvearrowright \bar{X}$ is mixing and therefore is irreducible.

Theorem 3.14 was the first application of Zimmer's cocycle superrigidity 5.2 (Theorem; see [139]). Indeed, if $\alpha : \Gamma \times X \to \Gamma'$ is the rearrangement cocycle associated to an Orbit Equivalence $T : \Gamma \curvearrowright (X, \mu) \overset{\text{OE}}{\sim} \Gamma' \curvearrowright (X', \mu')$ where $\Gamma <$ $G, \Gamma' < G'$ are lattices, then, viewing α as taking values in G', Zimmer observes that α is Zariski dense using a form of Borel's density theorem and deduces that $G \cong G'$ (here for simplicity the ambient groups are assumed to be simple, connected, center free, and $\text{rk}_{\mathbf{R}}(G) \geq 2$). Moreover, there is a homomorphism $\pi : \Gamma \to G'$ and $f : X \to G'$ so that $\alpha(\gamma, x) = f(\gamma x)\pi(\gamma)f(x)^{-1}$ with $\pi : \Gamma \to$ $\pi(\Gamma) < G'$ being an isomorphism of lattices.

REMARK 5.3. *At this point it is not clear whether $\pi(\Gamma)$ should be (conjugate to) Γ', and even assuming $\pi(\Gamma) = \Gamma'$ whether f takes values in Γ'. In fact, the self orbit equivalence of the Γ action on G/Γ given by $g\Gamma \mapsto g^{-1}\Gamma$ gives a rearrangement cocycle $c : \Gamma \times G/\Gamma \to \Gamma$, which is conjugate to the identity $\Gamma \to \Gamma$ by a unique map $f : G/\Gamma \to G$ with $f_*(m_{G/\Gamma}) \prec m_G$. However, if $\pi(\Gamma) = \Gamma'$ and f takes values in Γ', it follows that the original actions $\Gamma \curvearrowright (X, \mu)$ and $\Gamma' \curvearrowright (X', \mu')$ are isomorphic via the identification $\pi : \Gamma \cong \Gamma'$. We return to this point below.*

5.1.1. SUPERRIGIDITY AND ME-COUPLINGS Zimmer's cocycle superrigidity theorem applied to OE- or ME-cocycles (see SectionsA.2 and A.3) has a natural interpretation in terms of ME-couplings. Let G be a higher-rank simple Lie group (hereafter implicitly, connected, and center free) and, denote by $i : G \to$ Aut (G) the adjoint homomorphism (which is an embedding since G is center free).

THEOREM 5.4. ([44, THEOREM 4.1]) *Let G be a higher-rank simple Lie group, $\Gamma_1, \Gamma_2 < G$ lattices, and (Ω, m) an ergodic (Γ_1, Γ_2)-coupling. Then there exists a unique measurable map $\Phi : \Omega \to$ Aut (G) so that m-a.e. on Ω*

$$\Phi(\gamma_1 \omega) = i(\gamma_1)\Phi(\omega), \qquad \Phi(\gamma_2 \omega) = \Phi(\omega)i(\gamma_2)^{-1} \qquad (\gamma_i \in \Gamma_i).$$

Moreover, $\Phi_* m$ is either the Haar measure on a group $G \cong$ Ad $(G) \leq G' \leq$ Aut (G) or is atomic, in which case Γ_1 and Γ_2 are commensurable.

Sketch of the proof. To construct such a Φ, choose a fundamental domain $X \subset$ Ω for the Γ_2-action and look at the ME-cocycle $c : \Gamma_1 \times X \to \Gamma_2 < G$. Apply Zimmer's cocycle superrigidity theorem to find $\pi : \Gamma_1 \to G$ and $\phi : X \to G$. Viewing G as a subgroup in Aut (G), one may adjust π and $\phi : X \to$ Aut (G) by some $\alpha \in$ Aut (G), so that π is the isomorphism $i : \Gamma_1 \to \Gamma_1$ to get

$$c(\gamma_1, x) = \phi(\gamma_1.x)^{-1}i(\gamma_1)\phi(x).$$

Define $\Phi : \Omega \to \mathrm{Aut}\,(G)$ by $\Phi(\gamma_2 x) = \phi(x)i(\gamma_2)^{-1}$ and check that it satisfies the required relation. To identify the measure $\Phi_* m$ on $\mathrm{Aut}\,(G)$ one uses Ratner's theorem, which provides the classification of Γ_1-ergodic *finite* measures on $\bar{G}/\Gamma_2$. $\hfill \square$

Theorem 3.16 is then proved using this fact with $\Gamma_1 = \Gamma_2$ plugged into the construction in §5.5 that describes an unknown group Λ essentially as a lattice in G.

Note that there are two distinct cases in Theorem 5.4: either $\Phi_* m$ is atomic, in which case (Ω, m) has a discrete ME coupling as a quotient, or $\Phi_* m$ is a Haar measure on a Lie group. The former case leads to a virtual isomorphism between the groups and the actions (this is case (1) in Theorem 4.19); in the latter, $\Gamma_1 \curvearrowright X \cong \Omega/\Gamma_2$ has a quotient of the form $\Gamma_1 \curvearrowright \bar{G}/\Gamma_2$ (which is [45, theorem C]). This dichotomy clarifies the situation in Remark 5.3 above.

5.2. Superrigidity for Product Groups

Let us now turn to a brief discussion of Monod-Shalom rigidity (see Sections 3.1.7 and 3.2.2). Consider a special case of Margulis-Zimmer superrigidity results where the target group G' has rank 1 (say $G' = \mathrm{PSL}_2(\mathbf{R})$), while G has higher rank. The conclusion of the superrigidity Theorem 5.1 (respectively Theorem 5.2) is that either a representation (respectively cocycle) is degenerate,[12] or there is an epimorphism $\pi : G \to G'$. The latter case occurs *if and only if* G is semisimple $G = \prod G_i$, with one of the factors $G_i \simeq G'$, and $\pi : G \to G'$ factoring through the projection $\pi : G \xrightarrow{\mathrm{pr}_i} G_i \simeq G'$. In this case the given representation of the lattice extends to π (respectively the cocycle is conjugate to the epimorphism π).

This special case of Margulis-Zimmer superrigidity, that is, from higher-rank G to rank 1 G', was generalized by a number of authors [5, 6, 22] replacing the assumption that the target group G' has rank 1 by more geometric notions, such as $G' = \mathrm{Isom}\,(X)$ where X is a proper CAT(-1) space. In the setting considered by Monod and Shalom the target group is "hyperboliclike" in a very general way, while the source group G rather than being a higher-rank semisimple Lie group, is just a product $G = G_1 \times \cdots \times G_n$ of $n \geq 2$ *arbitrary* compactly generated (in fact, just lcsc) groups. The philosophy is that the number $n \geq 2$ of direct factors provides enough *higher-rank* properties for such statements.

---

12. Here precompact, or contained in a parabilic.

THEOREM 5.5. (MONOD-SHALOM [100]) *Let $G = G_1 \times \cdots \times G_n$ be a product of $n \geq 2$ lcsc groups, $G \curvearrowright (X, \mu)$ an irreducible p.m.p. action, H is a hyperboliclike group, and $c : G \times X \to H$ is a nonelementary measurable cocycle.*

Then there is a nonelementary closed subgroup $H_1 < H$, a compact normal subgroup $K \lhd H_1$, a measurable $f : X \to H$, and a homomorphism $\rho : G_i \to H_1/K$ from one of the factors G_i of G, so that the conjugate cocycle c^f takes values in H_1 and $G \times X \to H_1 \to H_1/K$ is the homomorphism $\pi : G \xrightarrow{pr_i} G_i \xrightarrow{\rho} H_1/K$.

This beautiful theorem is proved using the technology of *second-bounded cohomology* (developed in [20, 21, 97] and applied in this setting in [96, 100]) with the notions of *hyperboliclike* and *nonelementary* interpreted in the context of the class C_{reg}.

Suppose $\Gamma = \Gamma_1 \times \cdots \times \Gamma_n$, $n \geq 2$, is a product of "hyperboliclike" groups. Let (Ω, m) be a self-ME-coupling of Γ. Consider an ME-cocycle $\Gamma \times X \to \Gamma$, which can be viewed as a combination of n cocycles

$$c_i : \Gamma \times X \xrightarrow{c} \Gamma \xrightarrow{pr_i} \Gamma_i \qquad (i = 1, \ldots, n)$$

and assume that $\Gamma \curvearrowright \Omega/\Gamma$ is an *irreducible* action. Viewing the source group Γ as a product of $n \geq 2$ factors acting irreducibly, and recalling that the target groups Γ_i are "hyperboliclike," Monod and Shalom apply Theorem 5.5. The cocycles arising from ME-coupling turn out to be nonelementary, leading to the conclusion that each cocycle c_i is conjugate to a homomorphism $\rho_i : \Gamma_{j(i)} \to \Gamma'_i$, modulo some reductions and finite kernels. Since Γ_i commute, the conjugations can be performed independently and *simultaneously* on all the cocycles c_i. After some intricate analysis of the map $i \to j(i)$, kernels and cokernels of ρ_i, Monod and Shalom show that in the setting of ME-couplings as above the map, $i \to j(i)$ is a permutation and ρ_i are isomorphisms. Thus the original cocycle c can be conjugate to an *automorphism* of Γ.

This ME-cocycle superrigidity can now be plugged into an analogue of Theorem 5.4 to give a measurable bi-Γ-equivariant map $\Omega \to \Gamma$, which can be used as an input to a construction like Theorem 5.13. This allows us to identify *unknown* groups Λ measure equivalent to $\Gamma = \Gamma_1 \times \cdots \times \Gamma_n$. The only delicate point is that starting from a $\Gamma \curvearrowright X \overset{\text{SOE}}{\sim} \Lambda \curvearrowright Y$ and the corresponding (Γ, Λ)-coupling Ω one needs to look at the self-Γ-coupling $\Sigma = \Omega \times_\Lambda \check{\Omega}$ and apply the cocycle superrigidity result to $\Gamma \curvearrowright \Sigma/\Gamma$. In order to guarantee that the latter action is irreducible, Monod and Shalom require $\Gamma \curvearrowright X$ to be irreducible and $\Lambda \curvearrowright Y$ to be *mildly mixing*. They also show that the assumption on mild mixing is necessary for the result.

In [12] Uri Bader and the author proposed to study higher-rank superrigidity phenomena using a notion of a (generalized) Weyl group, which works well for higher-rank simple Lie groups, arbitrary products $G = G_1 \times \cdots \times G_n$ of $n \geq 2$ factors, and exotic $\tilde{A}_2$ groups, which are close relatives to lattices in $SL_3(\mathbf{Q}_p)$. In particular:

THEOREM 5.6. (BADER-FURMAN [12]) *Theorem 5.5 holds for target groups from class $\mathcal{D}_{\mathrm{ea}}$.*

Here $\mathcal{D}_{\mathrm{ea}}$ is a class of *hyperboliclike* groups that include many of the examples in $\mathcal{C}_{\mathrm{reg}}$. Plugging this into the Monod-Shalom machine one obtains the same results of products of groups in class $\mathcal{D}_{\mathrm{ea}}$.

5.3. Strong Rigidity for Cocycles

In the proof of Theorem 5.4 Zimmer's cocycle superrigidity was applied to a measure equivalence cocycle. This is a rather special class of cocycles (see §A.3). If cocycles are analogous to representations of lattices, then ME-cocycles are analogous to isomorphisms between lattices; in particular, they have an "inverse." Kida's work on ME for mapping class groups focuses on rigidity results for such cocycles. We shall not attempt to explain the ingredients used in this work, but will just formulate the main technical result analogous to Theorem 5.4. Let Γ be a subgroup of finite index in $\Gamma(\Sigma_{g,p})^\diamond$ with $3g + p - 4 > 0$, $C = C(\Sigma_{g,p})$ denoting its curve complex, and $\mathrm{Aut}(C)$ the group of its automorphisms; this is a countable group commensurable to Γ.

THEOREM 5.7. (KIDA [86]) *Let (Ω, m) be a self-ME-coupling of Γ. Then there exists a measurable map $\Gamma \times \Gamma$-equivariant map $\Phi : \Omega \to \mathrm{Aut}(C)$.*

Returning to the point that ME-cocycles are analogous to isomorphism between lattices, one might wonder whether Theorem 5.4 holds in cases where Mostow rigidity applies, specifically for G of rank 1 with $PSL_2(\mathbf{R})$ excluded. In [13] this is proved for $G \simeq \mathrm{Isom}(\mathbb{H}^n_{\mathbf{R}})$, $n \geq 3$, and a restricted ME.

THEOREM 5.8. (BADER-FURMAN-SAUER [13]) *Theorem 5.4 applies to ℓ^1-ME-couplings of lattices in $G = SO_{n,1}(\mathbf{R})$, $n \geq 3$.*

The proof of this result uses homological methods (ℓ^1 and other completions of the usual homology) combined with a version of the Gromov-Thurston proof of Mostow rigidity (for $\mathrm{Isom}(\mathbb{H}^n_{\mathbf{R}})$, $n \geq 3$) adapted to this setting.

5.4. Cocycle Superrigid Actions

In all the previous examples the structure of the acting group was the sole source for (super)rigidity. Recently Sorin Popa has developed a number of remarkable cocycle superrigidity results of a completely different nature [109–115]. These results exhibit an extreme form of cocycle superrigidity and rather than relying only on the properties of the acting group Γ, take advantage of the *action* $\Gamma \curvearrowright (X, \mu)$.

DEFINITION 5.9. An action $\Gamma \curvearrowright (X, \mu)$ is $\mathscr{C}$-**cocycle superrigid**, where $\mathscr{C}$ is some class of topological groups, if for every $\Lambda \in \mathscr{C}$ every measurable cocycle $c : \Gamma \curvearrowright X \to \Lambda$ has the form $c(g, x) = f(gx)^{-1} \rho(g) f(x)$ for some homomorphism $\rho : \Gamma \to \Lambda$ and some measurable $f : X \to \Lambda$.

Here we shall focus on the class $\mathscr{G}_{dsc}$ of all countable groups; however, the following results hold for all cocycles taking values in a broader class $\mathscr{U}_{fin}$, which contains $\mathscr{G}_{dsc}$ and $\mathscr{G}_{cpt}$—separable compact groups. Note that the concept of $\mathscr{G}_{dsc}$-cocycle superrigidity is unprecedentedly strong: there is no assumption on the cocycle, the assumption on the target group is extremely weak, and the "untwisting" takes place in the same target group.

THEOREM 5.10. (POPA [113]) *Let Γ be a group with property (T) and $\Gamma \curvearrowright (X, \mu) = (X_0, \mu_0)^\Gamma$ be the Bernoulli action. Then $\Gamma \curvearrowright (X, \mu)$ is $\mathscr{G}_{dsc}$-cocycle superrigid.*

In fact, the result is stronger: it suffices to assume that Γ has **relative property (T)** with respect to a **w-normal** subgroup Γ_0, and $\Gamma \curvearrowright (X, \mu)$ has a **relatively weakly mixing** extension $\Gamma \curvearrowright (\bar{X}, \bar{\mu})$ that is **s-malleable**, while $\Gamma_0 \curvearrowright (\bar{X}, \bar{\mu})$ is weakly mixing. Under these conditions $\Gamma \curvearrowright (X, \mu)$ is $\mathscr{U}_{fin}$-cocycle superrigid. See [113] and [48] for the relevant definitions and more details. We indicate the proof (of the special case above) in §5.8. Vaguely speaking, Popa's approach exploits the tension between certain (local) *rigidity* provided by the acting group and *deformations* supplied by the action. In the following remarkable result, Popa further relaxed the property (T) assumption.

THEOREM 5.11. (POPA [115]) *Let Γ be a group containing a product $\Gamma_1 \times \Gamma_2$ where Γ_1 is nonamenable, Γ_2 is infinite, and $\Gamma_1 \times \Gamma_2$ is w-normal in Γ. Then any Bernoulli action $\Gamma \curvearrowright (X, \mu)$ is $\mathscr{U}_{fin}$-cocycle superrigid.*

The deformations alluded to above take place for the diagonal Γ-action on the square $(X \times X, \mu \times \mu)$. This action is supposed to be ergodic; equivalently, the original action should be weakly *mixing* and satisfy addition properties. *Isometric actions*, or staying within the ergodic-theoretic terminology, actions with *discrete spectrum* provide the opposite type of dynamics. These actions have the form $\Gamma \curvearrowright K/L$ where $L < K$ are compact groups, $\Gamma \to K$ a homomorphism with dense image, and Γ acts by left translations. Totally disconnected K corresponds to profinite completion $\varprojlim \Gamma/\Gamma_n$ with respect to a chain of normal subgroups of finite index. Isometric actions $\Gamma \curvearrowright K/L$ with profinite K, can be called *profinite ergodic* actions of Γ—these are precisely inverse limits $X = \varprojlim X_n$ of transitive Γ-actions on finite spaces. Adrian Ioana found the following "virtually $\mathscr{G}_{\mathrm{dsc}}$-cocycle superrigidity" phenomenon for profinite actions of Kazhdan groups.

THEOREM 5.12. (IOANA [72]) *Let $\Gamma \curvearrowright X = K/L$ be an ergodic profinite action. Assume that Γ has property (T), or a relative property (T) with respect to a normal subgroup Γ_0, which acts ergodically on X. Then any measurable cocycle $c : \Gamma \curvearrowright X \to \Lambda$ into a discrete group, is conjugate to a cocycle coming from a finite quotient $X \to X_n$, that is, c is conjugate to a cocycle induced from a homomorphism $\Gamma_n \to \Lambda$ of a finite-index subgroup.*

In §5.8.2 a similar result is proven for all discrete spectrum actions (not necessarily profinite ones).

5.5. Constructing Representations

In Geometric group theory many QI-rigidity results are proved using the following trick. Given a metric space X one declares self-quasi-isometries $f, g : X \to X$ to be equivalent if

$$\sup_{x \in X} d(f(x), g(x)) < \infty.$$

Then equivalence classes of quasi-isometries (hereafter q.i.) form a *group*, denoted $\mathrm{QI}(X)$. This group contains (a quotient of) $\mathrm{Isom}\,(X)$, which can sometimes be identified within $\mathrm{QI}(X)$ in coarse geometric terms. If Γ is a group with well-understood $\mathrm{QI}(\Gamma)$ and Λ is an *unknown* group q.i. to Γ, then one gets a *homomorphism*

$$\rho : \Lambda \to \mathrm{Isom}\,(\Lambda) \to \mathrm{QI}(\Lambda) \cong \mathrm{QI}(\Gamma)$$

whose kernel and image can then be analyzed.

Facing a similar problem in the measure equivalence category, there is a difficulty in defining an analogue for QI(Γ). Let us describe a construction that allows us to analyze the class of all groups ME to a given group Γ from an information about self-ME-couplings of Γ.

Let G be a lcsc unimodular group. Let us assume that G has the **strong ICC** property, by which we mean that the only regular Borel conjugation-invariant probability measure on G is the trivial one, namely the Dirac mass δ_e at the origin. For countable groups this is equivalent to the condition that all nontrivial conjugacy classes are infinite, that is, the usual ICC property. Connected, (semi) simple Lie groups with trivial center and no compact factors provide other examples of strongly ICC groups.

Theorems 5.4, 5.7 and 5.8 are instances where a strongly ICC group G has the property that for any[13] ME-self-coupling (Ω, m) of a lattice Γ in G there exists a bi-Γ-equivariant measurable map to G, that is, a Borel map $\Phi : \Omega \to G$ satisfying m-a.e.:

$$\Phi((\gamma_1, \gamma_2)\omega) = \gamma_1 \Phi(\omega)\gamma_2^{-1} \qquad (\gamma_1, \gamma_2 \in \Gamma).$$

It is not difficult to see that the strong ICC property implies that such a map is also unique. (It should also be pointed out that the existence of such maps for self-couplings of lattices is equivalent to the same property for self-couplings of the lcsc group G itself; but here we shall stay in the framework of countable groups). The following general tool shows how these properties of G can be used to classify all groups ME to a lattice $\Gamma < G$; up to finite kernels these turn out to be lattices in G.

THEOREM 5.13. (BADER-FURMAN-SAUER [13]) *Let G be a strongly ICC lcsc unimodular group, $\Gamma < G$ a lattice, and Λ some group ME to Γ and (Ω, m) be a (Γ, Λ)-coupling. Assume that the self-ME-coupling $\Sigma = \Omega \times_\Lambda \check{\Omega}$ of Γ admits a Borel map $\Phi : \Sigma \to G$, satisfying a.e.:*

$$\Phi([\gamma_1 x, \gamma_2 y]) = \gamma_1 \cdot \Phi([x, y]) \cdot \gamma_2^{-1} \qquad (\gamma_1, \gamma_2 \in \Gamma).$$

Then there exists a short exact sequence $K \longrightarrow \Lambda \longrightarrow \bar{\Lambda}$ with K finite and $\bar{\Lambda}$ being a lattice in G, and a Borel map $\Psi : \Omega \to G$ so that a.e.:

$$\Phi([x, y]) = \Psi(x) \cdot \Psi(y)^{-1}, \quad \Psi(\gamma z) = \gamma \cdot \Psi(z), \quad \Psi(\lambda z) = \Psi(z) \cdot \bar{\lambda}^{-1}.$$

---

13. In the case of $G = \text{Isom}\,(\mathbf{H}^n)$ we restrict to all ℓ^1-ME-couplings.

Moreover, the push-forward of $\Psi_* m$ is a Radon measure on G-invariant under the maps

$$g \mapsto \gamma g \bar{\lambda}, \qquad (\gamma \in \Gamma, \ \bar{\lambda} \in \bar{\Lambda}).$$

If G is a (semi)simple Lie group, the last condition on the push-forward measure can be analyzed using Ratner's theorem (as in Theorem 5.4) to deduce that assuming ergodicity $\Psi_* m$ is either a Haar measure on G, or on a coset of its finite-index subgroup, or it is (proportional to) counting measure on a coset of a lattice Γ' containing Γ and a conjugate of $\bar{\Lambda}$ as finite-index subgroups.

Theorem 5.13 is a streamlined and improved version of similar statements obtained in [44] for higher-rank lattices, in [101] for products, and in [86] for mapping class groups.

5.6. Local Rigidity for Measurable Cocycles

The *rigidity vs. deformations* approach to rigidity results developed by Sorin Popa led to a number of striking results in von Neumann algebras and in Ergodic theory (some have been mentioned in §5.4). Let us illustrate the *rigidity* side of this approach by the following simple purely ergodic-theoretic statement, which is a variant of Hjorth's [68, lemma 2.5].

Recall that one of the several equivalent forms of property (T) is the following statement: a lcsc group G has (T) if there exists a compact $K \subset G$ and $\epsilon > 0$ so that for any unitary G-representation π and any (K, ϵ)-almost invariant unit vector v there exists a G-invariant unit vector w with $\|v - w\| < \frac{1}{4}$.

PROPOSITION 5.14. *Let G be a group with property (T) and (K, ϵ), as above. Then for any ergodic probability measure-preserving action $G \curvearrowright (X, \mu)$, any countable group Λ and any pair of cocycles $\alpha, \beta : G \times X \to \Lambda$ with*

$$\mu \left\{ x \in X \ : \ \alpha(g, x) = \beta(g, x) \right\} > 1 - \frac{\epsilon^2}{2} \qquad (\forall g \in K)$$

there exists a measurable map $f : X \to \Lambda$ so that $\beta = \alpha^f$. Moreover, one can assume that

$$\mu \left\{ x \ : \ f(x) = e \right\} > \frac{3}{4}.$$

Proof. Let $\tilde{X} = X \times \Lambda$ be equipped with the infinite measure $\tilde{\mu} = \mu \times m_\Lambda$ where m_Λ stands for the counting measure on Λ. Then G acts on $(\tilde{X}, \tilde{\mu})$ by

$$g : (x, \lambda) \mapsto (g.x, \alpha(g, x) \lambda \beta(g, x)^{-1}).$$

This action preserves $\tilde{\mu}$ and we denote by π the corresponding unitary G-representation on $L^2(\tilde{X}, \tilde{\mu})$. The characteristic function $v = 1_{X \times \{e\}}$ satisfies

$$\|v - \pi(g)v\|^2 = 2 - 2\mathrm{Re}\langle\pi(g)v, v\rangle < 2 - 2\left(1 - \frac{\epsilon^2}{2}\right) = \epsilon^2 \qquad (g \in \Gamma)$$

and therefore there exists a $\pi(G)$-invariant unit vector $w \in L^2(\tilde{X}, \tilde{\Lambda})$ with $\|v - w\| < \frac{1}{4}$. Since $1 = \|w\|^2 = \int_X \sum_\lambda |w(x, \lambda)|^2$ we may define

$$p(x) = \max_\lambda |w(x, \lambda)|, \qquad \Lambda(x) = \{\lambda \ : \ |w(x, \lambda)| = p(x)\}$$

and observe that $p(x)$ and the cardinality $k(x)$ of the finite set $\Lambda(x)$ are measurable Γ-invariant functions on (X, μ); hence are a.e. constants $p(x) = p \in (0, 1]$, $k(x) = k \in \{1, 2, \dots\}$. Since $\frac{1}{16} > \|v - w\|^2 \geq (1 - p)^2$ we have $p > \frac{3}{4}$. It follows that $k = 1$ because $1 = \|w\|^2 \geq kp^2$. Therefore $\Lambda(x) = \{f(x)\}$ for some measurable map $f : X \to \Lambda$. The $\pi(G)$-invariance of w gives $\pi(G)$-invariance of the characteristic function of $\{(x, f(x)) \in \tilde{X} \ : \ x \in X\}$, which is equivalent to

5.1 $\qquad f(gx) = \alpha(g, x)f(x)\beta(g, x)^{-1} \qquad$ and $\qquad \beta = \alpha^f.$

Let $A = f^{-1}(\{e\})$ and $a = \mu(A)$. Since $\sum_\lambda |w(x, \lambda)|^2$ is a G-invariant function it is a.e. constant $\|w\|^2 = 1$. Hence for $x \notin A$ we have $|w(x, e)|^2 \leq 1 - |w(x, f(x))|^2 = 1 - p^2$, and

$$\frac{1}{16} > \|v - w\|^2 \geq a \cdot (1 - p^2) + (1 - a) \cdot (1 - (1 - p^2)) \geq (1 - a) \cdot p^2 > \frac{9(1 - a)}{16}.$$

Thus $a = \mu\{x \in X \ : \ f(x) = e\} > \frac{8}{9} > \frac{3}{4}$, as required. $\qquad \square$

5.7. Cohomology of Cocycles

Let us fix two groups Γ and Λ. There is no real assumption on Γ, it may be any lcsc group, but we shall impose an assumption on Λ. One might focus on the case where Λ is a countable group (class $\mathscr{G}_{\mathrm{dsc}}$), but versions of the statements below would apply also to separable compact groups, or groups in a larger class $\mathscr{U}_{\mathrm{fin}}$ of all Polish groups, which embed in the unitary group of a von Neumann algebra with finite faithful trace,[14] or a potentially even larger class $\mathscr{G}_{\mathrm{binv}}$ of groups with a bi-invariant metric, and the class $\mathscr{G}_{\mathrm{alg}}$ of connected algebraic groups over local fields, say of zero characteristic.

---

14. This class, introduced by Popa, contains both discrete countable groups and separable compact ones.

Given a (not necessarily free) p.m.p. action $\Gamma \curvearrowright (X, \mu)$ let $Z^1(X, \Lambda)$ or $Z^1(\Gamma \curvearrowright X, \Lambda)$ denote the space of all measurable cocycles $c : \Gamma \times X \to \Lambda$ and by $H^1(X, \Lambda)$, or $H^1(\Gamma \curvearrowright X, \Lambda)$, the space of equivalence classes of cocycles up to conjugation by measurable maps $f : X \to \Lambda$. If $\Lambda \in \mathscr{G}_{\text{alg}}$, we shall focus on a subset $H^1_{ss}(X, \Lambda)$ of (classes of) cocycles whose algebraic hull is connected, semisimple, center free, and has no compact factors.

Any Γ-equivariant quotient map $\pi : X \to Y$ defines a pull-back $Z^1(Y, \Lambda) \to Z^1(X, \Lambda)$ by $c^\pi (g, x) = c(g, \pi(x))$, which descends to

$$H^1(Y, \Lambda) \xrightarrow{\pi^*} H^1(X, \Lambda).$$

Group inclusions $i : \Lambda < \bar{\Lambda}$, and $j : \Gamma' < \Gamma$ give rise to push-forward maps

$$H^1(X, \Lambda) \xrightarrow{i_*} H^1(X, \bar{\Lambda}), \qquad H^1(\Gamma \curvearrowright X, \Lambda) \xrightarrow{j_*} H^1(\Gamma' \curvearrowright X, \Lambda).$$

QUESTION 5.15. *What can be said about these maps of the cohomology ?*

The discussion here is inspired and informed by Popa's [113]. In particular, the following statements Propositions 5.16(2), 5.17, and 5.18(1), and Corollary 5.20 are variations on Popa's original [113, lemma 2.11, proposition 3.5, lemma 3.6, theorem 3.1]. Working with class $\mathscr{G}_{\text{binv}}$ makes the proofs more transparent than in $\mathscr{U}_{\text{fin}}$—this was done in [48, §3]. Proposition 5.16(3) for semisimple target is implicit in [45, lemma 3.5]. The full treatment of the statements below, including Theorem 5.19, will appear in [49].

PROPOSITION 5.16. *Let $\pi : X \to Y$ be a Γ-equivariant quotient map. Then*

$$H^1(Y, \Lambda) \xrightarrow{\pi^*} H^1(X, \Lambda)$$

is injective in the following cases:

1) Λ *is discrete and torsion free.*
2) $\Lambda \in \mathscr{G}_{\text{binv}}$ *and* $\pi : X \to Y$ *is relatively weakly mixing.*
3) $\Lambda \in \mathscr{G}_{\text{alg}}$ *and* $H^1(-, \Lambda)$ *is replaced by* $H^1_{ss}(-, \Lambda)$.

The notion of **relative weakly mixing** was introduced independently by Zimmer [137] and Furstenberg [51]: a Γ-equivariant map $\pi : X \to Y$ is relatively weakly mixing if the Γ-action on the fibered product $X \times_Y X$ is ergodic (or ergodic relatively to Y); this turns out to be equivalent to the condition that $\Gamma \curvearrowright X$ contains no intermediate isometric extensions of $\Gamma \curvearrowright Y$.

PROPOSITION 5.17. *Let* $i : \Lambda < \bar{\Lambda} \in \mathcal{G}_{\text{binv}}$ *be a closed subgroup, and* $\Gamma \curvearrowright$ (X, μ) *some p.m.p. action. Then*

$$H^1(X, \Lambda) \xrightarrow{i_*} H^1(X, \bar{\Lambda})$$

is injective.

This useful property fails in the $\mathcal{G}_{\text{alg}}$ setting: if $\Gamma < G$ is a lattice in a (semi) simple Lie group and $c : \Gamma \times G/\Gamma \to \Gamma$ in the canonical class, then viewed as a cocycle into $G > \Gamma$, c is conjugate to the identity embedding $\Gamma \cong \Gamma < G$, but as a Γ-valued cocycle it cannot be "untwisted."

PROPOSITION 5.18. *Let* $\pi : X \to Y$ *be a quotient map of ergodic actions, and* $j :$ $\Gamma' < \Gamma$ *be a normal, or sub-normal, or w-normal closed subgroup acting ergodically on X. Assume that either*

1) $\Lambda \in \mathcal{G}_{\text{binv}}$ *and* π *is relatively weakly mixing, or*
2) $\Lambda \in \mathcal{G}_{\text{alg}}$ *and one considers* $H^1_{\text{ss}}(-, \Lambda)$.

Then $H^1(\Gamma \curvearrowright Y, \Lambda)$ *is the push-out of the rest of the following diagram:*

$$
\begin{array}{ccc}
H^1(\Gamma \curvearrowright X, \Lambda) & \xrightarrow{j_*} & H^1(\Gamma' \curvearrowright X, \Lambda) \\
\pi^* \uparrow & & \pi^* \uparrow \\
H^1(\Gamma \curvearrowright Y, \Lambda) & \xrightarrow{j_*} & H^1(\Gamma' \curvearrowright Y, \Lambda).
\end{array}
$$

In other words, if the restriction to $\Gamma' \curvearrowright X$ *of a cocycle* $c : \Gamma \times X \to \Lambda$ *is conjugate to one descending to* $\Gamma' \times Y \to \Lambda$, *then c has a conjugate that descends to* $\Gamma \times X \to \Lambda$.

The condition $\Gamma' < \Gamma$ is **w-normal** (weakly normal), which means that there exists a well-ordered chain Γ_i of subgroups starting from Γ' and ending with Γ, so that $\Gamma_i \lhd \Gamma_{i+1}$ and for limit ordinals $\Gamma_j = \bigcup_{i<j} \Gamma_i$ (Popa).

Let $\pi_i : X \to Y_i$ is a collection of Γ-equivariant quotient maps. Then X has a unique Γ-equivariant quotient $p : X \to Z = \bigwedge Y_i$, which is maximal among all common quotients $p_i : Y_i \to Z$. Identifying Γ-equivariant quotients with Γ-equivariant complete sub-σ-algebras of $\mathcal{X}$, one has $p^{-1}(\mathcal{Z}) = \bigcap_i \pi_i^{-1}(\mathcal{Y}_i)$; or in the operator algebra formalism $p^{-1}(L^\infty(Z)) = \bigcap_i \pi_i^{-1}(L^\infty(Y_i))$.

THEOREM 5.19. *Let* $\pi_i : X \to Y_i$, $1 \leq i \leq n$, *be a finite collection of* Γ-*equivariant quotients, and* $Z = \bigwedge_{i=1}^{n} Y_i$. *Then* $H^1(Z, \Lambda)$ *is the push-out of* $H^1(Y_i, \Lambda)$ *under conditions (1)–(3) of Proposition 5.16:*

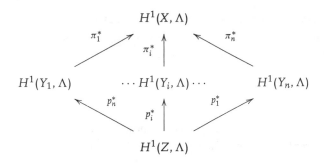

More precisely, if $c_i : \Gamma \times X \to \Lambda$ *are cocycles (in case (3) assume* $[c_i] \in H^1_{ss}(Y_i, \Lambda))$, *whose pullbacks* $c_i(g, \pi_i(x))$ *are conjugate over* X, *then there exists a unique class* $[c] \in H^1(Z, \Lambda)$, *so that* $c(g, p_i(y)) \sim c_i(g, y)$ *in* $Z^1(Y_i, \Lambda)$ *for all* $1 \leq i \leq n$.

The proof of this Theorem relies on Proposition 5.16 and contains it as a special case $n = 1$.

This result can be useful to push cocycles to deeper and deeper quotients; if $\pi : X \to Y$ is *a minimal* quotient to which a cocycle or a family of cocycles can descend up to conjugacy, then it is **the minimal** or **characteristic** quotient for these cocycles: if they descend to any quotient $X \to Y'$, then necessarily $X \to Y' \to Y$. For example if $\Gamma < G$ is a higher-rank lattice, Λ is a discrete group and $c : \Gamma \times X \to \Lambda$ is an OE (or ME) cocycle, then either c descends to a Γ-action on a finite set (virtual isomorphism case), or to $X \xrightarrow{\pi} G/\Lambda'$ with $\Lambda \simeq \Lambda'$ lattice in G, where π is *uniquely defined* by c (initial OE or ME).

An important special (and motivating) case of Theorem 5.19 is that of $X = Y \times Y$ where $\Gamma \curvearrowright Y$ is a weakly mixing action. Then the projections $\pi_i : X \to Y_i = Y$, $i = 1, 2$, give $Z = Y_1 \wedge Y_2 = \{pt\}$ and $H^1(\Gamma \curvearrowright \{pt\}, \Lambda) = \mathrm{Hom}(\Gamma, \Lambda)$. So

COROLLARY 5.20. (POPA [113, THEOREM 3.1], SEE ALSO [48, THEOREM 3.4]) *Let* $\Gamma \curvearrowright Y$ *be a weakly mixing action and* $c : \Gamma \times Y \to \Lambda$ *a cocycle into* $\Lambda \in \mathcal{G}_{\mathrm{binv}}$. *Let* $X = Y \times Y$ *with the diagonal* Γ-*action,* $c_1, c_2 : \Gamma \times X \to \Lambda$ *the cocycles* $c_i(g, (y_1, y_2)) = c(g, y_i)$. *If* $c_1 \sim c_2$ *over* X, *then there exists homomorphism* $\rho : \Gamma \to \Lambda$ *and a measurable* $f : Y \to \Lambda$, *so that* $c(g, y) = f(gy)^{-1} \rho(g) f(y)$.

5.8. Proofs of Some Results

In this section we shall give relatively self-contained proofs of some of the results mentioned above.

5.8.1. SKETCH OF A PROOF FOR POPA'S COCYCLE SUPERRIGIDITY THEOREM

5.10 First note that without loss of generality the base space (X_0, μ_0) of the Bernoulli action may be assumed to be nonatomic. Indeed, Proposition 5.16(2) implies that for each of the classes $\mathscr{G}_{dsc} \subset \mathscr{U}_{fin} \subseteq \mathscr{G}_{binv}$ the corresponding cocycle superrigidity descends through relatively weakly mixing quotients, and $([0, 1], dx)^{\Gamma} \to (X_0, \mu_0)^{\Gamma}$ is such.

Given any action $\Gamma \curvearrowright (X, \mu)$ consider the diagonal Γ-action on $(X \times X, \mu \times \mu)$ and its centralizer $\mathrm{Aut}_{\Gamma}(X \times X)$ in the Polish group $\mathrm{Aut}(X \times X, \mu \times \mu)$. It always contain the flip $F : (x, y) \mapsto (y, x)$. Bernoulli actions $\Gamma \curvearrowright X - [0, 1]^{\Gamma}$ have the special property (called **s-malleability** by Popa) that there is a path

$$p : [1, 2] \to \mathrm{Aut}_{\Gamma}(X \times X), \qquad \text{with} \qquad p_1 = \mathrm{Id}, \qquad p_2 = F.$$

Indeed, the diagonal component-wise action of $\mathrm{Aut}([0, 1] \times [0, 1])$ on $X \times X = ([0, 1] \times [0, 1])^{\Gamma}$ embeds into $\mathrm{Aut}_{\Gamma}(X \times X)$ and can be used to connect Id to F.

Fix a cocycle $c : \Gamma \curvearrowright X \to \Lambda$. Consider the two lifts to $X \times X \to X$:

$$c_i : \Gamma \curvearrowright X \times X \to \Lambda, \qquad c_i(g, (x_1, x_2)) = c(g, x_i), \qquad (i = 1, 2).$$

Observe that they are connected by the continuous path of cocycles $c_t(g, (x, y)) = c_1(g, p_t(x, y))$, $1 \leq t \leq 2$. Local rigidity in Proposition 5.14 implies that c_1 and c_2 are conjugate over $X \times X$, and the proof is completed invoking Corollary 5.20. Under the weaker assumption of relative property (T) with respect to a w-normal subgroup, Popa uses Proposition 5.18.

5.8.2. A COCYCLE SUPERRIGIDITY THEOREM

We state and prove a cocycle superrigidity theorem, inspired and generalizing Adrian Ioana's Theorem 5.12. Thus a number of statements (Theorems 4.15 and 4.20(2) and §4.3.1) in this survey get a relatively full treatment. The proof is a good illustration of Popa's *deformation vs. rigidity* approach.

Recall that an ergodic p.m.p. action $\Gamma \curvearrowright (X, \mu)$ is said to have a **discrete spectrum** if the Koopman Γ-representation on $L^2(X, \mu)$ is a Hilbert sum of finite-dimensional subrepresentations. Mackey proved (generalizing the Halmos–von Neumann theorem for $\mathbf{Z}$, and using Peter-Weyl ideas) that discrete spectrum action is measurably isomorphic to the isometric Γ-action on

$(K/L, m_{K/L})$, $g : kL \mapsto \tau(g)kL$, where $L < K$ are compact separable groups and $\tau : \Gamma \to K$ is a homomorphism with dense image.

THEOREM 5.21. (AFTER IOANA'S THEOREM 5.12, [72]) *Let $\Gamma \curvearrowright (X, \mu)$ be an ergodic p.m.p. action with discrete spectrum. Assume that Γ has property (T), or contains a w-normal subgroup Γ_0 with property (T) acting ergodically on (X, μ). Let Λ be an arbitrary torsion-free discrete countable group and $c : \Gamma \times X \to \Lambda$ be a measurable cocycle.*

Then there is a finite-index subgroup $\Gamma_1 < \Gamma$, a Γ_1-ergodic component $X_1 \subset X$ (of measure $\mu(X_1) = [\Gamma : \Gamma_1]^{-1}$), a homomorphism $\rho : \Gamma_1 \to \Lambda$, and a measurable map $\phi : X \to \Lambda$, so that the conjugate cocycle c^ϕ restricted to $\Gamma_1 \curvearrowright X_1 \to \Lambda$, is the homomorphism $\rho : \Gamma_1 \to \Lambda$. The cocycle $c^\phi : \Gamma \times X \to \Lambda$ is induced from ρ.

The assumption that Λ is torsion free is not essential; in general, one might need to lift the action to a finite cover $\hat{X}_1 \to X_1$ via a finite group that embeds in Λ. If K is a connected Lie group, then $\Gamma_1 = \Gamma$ and $X_1 = X = K/L$. The stated result is deduced from the case where L is trivial, that is, $X = K$, using Proposition 5.16(1). We shall make this simplification and assume Γ has property (T) (the modification for the more general case uses an appropriate version of Propositions 5.14 and 5.17). An appropriate modification of the result handles compact groups as a possible target group Λ for the cocycle.

Proof. The K-action by right translations: $t : x \mapsto xt^{-1}$ commutes with the Γ-action on K; in fact, K is precisely the centralizer of Γ in $\text{Aut}\,(K, m_K)$. This allows us to deform the initial cocycle $c : \Gamma \times X \to \Lambda$, setting

$$c_t(g, x) = c(g, xt^{-1}) \qquad (t \in K).$$

Let $F \subset \Gamma$ and $\epsilon > 0$ be as in the "local rigidity" Proposition 5.14. Then for some open neighborhood U of $e \in K$ for every $t \in U$ there is a unique measurable $f_t : K \to \Lambda$ with

$$c_t(g, x) = c(g, xt^{-1}) = f_t(gx)c(g, x)f_t(x)^{-1} \qquad \mu\left\{x \,:\, f_t(x) = e\right\} > \frac{3}{4}.$$

Suppose $t, s \in U$, and $ts \in U$. Then

$$f_{ts}(gx)\, c(g, x)\, f_{ts}(x)^{-1} = c_{ts}(g, x) = c(g, xs^{-1}t^{-1})$$

$$= f_t(gxs^{-1})\, c(g, xs^{-1})\, f_t(xs^{-1})^{-1}$$

$$= f_t(gxs^{-1})f_s(gx)\, c(g, x)\, [f_t(xs^{-1})f_s(x)^{-1}]^{-1}.$$

This can be rewritten as

$$F(gx) = c(g, x) F(x) c(g, x)^{-1}, \qquad \text{where} \qquad F(x) = f_{ts}(x)^{-1} f_t(xs^{-1}) f_s(x).$$

Since f_t, f_s, f_{ts} takes value e with probability $> 3/4$, it follows that $A = F^{-1}(\{e\})$ has $\mu(A) > 0$. The equation implies Γ-invariance of A. Thus $\mu(A) = 1$ by ergodicity. Hence, whenever $t, s, ts \in U$

5.2 $$f_{ts}(x) = f_t(xs^{-1}) f_s(x).$$

If K is a totally disconnected group, that is, a profinite completion of Γ as in Ioana's Theorem 5.12, then U contains an open subgroup $K_1 < K$. In this case one can skip the following paragraph.

In general, let V be a symmetric neighborhood of $e \in K$ so that $V^2 \subset U$, and let $K_1 = \bigcup_{n=1}^{\infty} V^n$. Then K_1 is an open (hence, also closed) subgroup of K; in the connected case $K_1 = K$. We shall extend the family $\{f_t : K \to \Lambda\}_{t \in V}$ to be defined for all $t \in K_1$ while satisfying Equation (5.2), using a "cocycle continuation" procedure akin to analytic continuation. For $t, t' \in K_1$ a V-**quasipath** $p_{t \to t'}$ from t to t' is a sequence $t = t_0, t_1, \ldots, t_n = t'$ where $t_i \in t_{i-1} V$. Two V-quasipaths from t to t' are V-**close** if they are within V-neighborhoods from each other. Two V-quasipaths $p_{t \to t'}$ and $q_{t \to t'}$ are V-**homotopic** if there is a chain $p_{t \to t'} = p_{t \to t'}^{(0)}, \ldots, p_{t \to t'}^{(k)} = q_{t \to t'}$ of V-quasipaths where $p^{(i-1)}$ and $p^{(i)}$ are V-close, $1 \le i \le k$. Iterating Equation (5.2) one may continue the definition of f from t to t' along a V-quasipath; the continuation being the same for V-close quasipaths, and therefore for V-homotopic quasipaths as well (all from t to t'). The possible ambiguity of this cocycle continuation procedure is encoded in the **homotopy group** $\pi_1^{(V)}(K_1)$ consisting of equivalence classes of V-quasipaths from $e \to e$ modulo V-homotopy. We claim that this group is finite. In the case of a connected Lie group K_1, $\pi_1^{(V)}(K_1)$ is a quotient of $\pi_1(K_1)$ that is finite since K_1, contains a dense property (T) group and cannot have torus factors. This covers the general case as well since $\pi_1^{(V)}(K_1)$ "feels" only finitely many factors when K_1 is written as an inverse limit of connected Lie groups and finite groups. Considering the continuations of f along V-quasipaths $e \to e$ we get a homomorphism $\pi_1^{(V)}(K_1) \to \Lambda$, which must be trivial since Λ was assumed to be torsion free. Therefore, we obtain a family of measurable maps $f_t : K_1 \to \Lambda$ indexed by $t \in K_1$ and still satisfying Equation (5.2).

Let $\Gamma_1 = \tau^{-1}(K_1)$. Then the index $[\Gamma : \Gamma_1] = [K : K_1]$ is finite. We shall focus on the restriction c_1 of c to $\Gamma_1 \curvearrowright K_1$. Note that Equation (5.2) is a cocycle equation for the *simply transitive action* K_1 on itself. It follows by a standard argument that it is a coboundary. Indeed, for a.e. $x_0 \in K_1$ Equation (5.2) holds for a.e. $t, s \in K_1$. In particular, for a.e. $t, x \in K_1$, using $s = x^{-1}x_0$, one obtains $f_{tx^{-1}x_0}(x_0) = f_t(x)f_{x^{-1}x_0}(x_0)$. This gives

$$f_t(x) = \phi(xt^{-1})\phi(x)^{-1}, \qquad \text{where} \qquad \phi(x) = f_{x^{-1}x_0}(x_0).$$

Equation $c_t = c^{f_t}$ translates into the fact that the cocycle $c^\phi(g, x) = \phi(gx)^{-1} c(g, x)\phi(x)$ satisfies for a.e. x, t

$$c^\phi(g, xt^{-1}) = c^\phi(g, x).$$

Thus $c(g, x)$ does not depend on the space variable. Hence, it is a homomorphism

$$c^\phi(g, x) = \rho(g).$$

Finally, the fact that c^ϕ is induced from c_1^ϕ is straightforward. □

Appendix A. Cocycles

Let $G \curvearrowright (X, \mu)$ be a measurable, measure-preserving (sometimes just measure class preserving) action of a topological group G on a standard Lebesgue space (X, μ), and H be a topological group. A Borel measurable map $c : G \times X \to H$ forms a **cocycle** if for every $g_1, g_2 \in G$ for μ-a.e. $x \in X$ one has

$$c(g_2g_1, x) = c(g_2, g_1.x) \cdot c(g_1, x).$$

If $f : X \to H$ is a measurable map and $c : G \times X \to H$ is a measurable cocycle, define the f-conjugate c^f of c to be

$$c^f(g, x) = f(g.x)^{-1} c(g, x)f(x).$$

It is straightforward to see that c^f is also a cocycle. One says that c and c^f are (measurably) **conjugate**, or **cohomologous** cocycles. The space of all measurable cocycles $\Gamma \times X \to \Lambda$ is denoted by $Z^1(\Gamma \curvearrowright X, \Lambda)$ and the space of equivalence classes by $H^1(\Gamma \curvearrowright X, \Lambda)$.

Cocycles that do not depend on the space variable $c(g, x) = c(g)$ are precisely homomorphisms $c : G \to H$. So cocycles may be viewed as generalized homomorphisms. In fact, any group action $G \curvearrowright (X, \mu)$ defines a **measured groupoid** $\mathcal{G}$ with $\mathcal{G}^{(0)} = X$, and $\mathcal{G}^{(1)} = \{(x, gx) : x \in X, g \in G\}$ (see [9] for the background). In this context cocycles can be viewed as homomorphisms $\mathcal{G} \to H$.

If $\pi : (X, \mu) \to (Y, \nu)$ is an equivariant quotient map between Γ-actions (so $\pi_*\mu = \nu$ and $\pi \circ \gamma = \gamma$ for $\gamma \in \Gamma$), then for any target group Λ any cocycle $c : \Gamma \times Y \to \Lambda$ lifts to $\bar{c} : \Gamma \times X \to \Lambda$ by

$$\bar{c}(g, x) = c(g, \pi(x)).$$

Moreover, if $c' = c^f \sim c$ in $Z^1(\Gamma \curvearrowright Y, \Lambda)$, then the lifts $\bar{c}' = \bar{c}^{f \circ \pi} \sim \bar{c}$ in $Z^1(\Gamma \curvearrowright X, \Lambda)$; so $X \xrightarrow{\pi} Y$ induces

$$H^1(\Gamma \curvearrowright X, \Lambda) \xleftarrow{\pi^\circ} H^1(\Gamma \curvearrowright Y, \Lambda).$$

Note that $\mathrm{Hom}(\Gamma, \Lambda)$ is $Z^1(\Gamma \curvearrowright \{pt\}, \Lambda)$ and classes of cocycles on $\Gamma \times X \to \Lambda$ cohomologous to homomorphisms is precisely the pull-back of $H^1(\Gamma \curvearrowright \{pt\}, \Lambda)$.

A.1. The Canonical Class of a Lattice, (Co)induction

Let $\Gamma < G$ be a lattice in a lcsc group. By definition the transitive G-action on $X = G/\Gamma$ has an invariant Borel regular probability measure μ. Let $\mathcal{F} \subset G$ be a Borel fundamental domain for the right Γ-action on G (i.e., $\mathcal{F}$ is a Borel subset of G set that meets every coset $g\Gamma$ precisely once). Fundamental domains correspond to Borel cross-section $\sigma : G/\Gamma \to G$ of the projection $G \to G/\Gamma$. Define

$$c_\sigma : G \times G/\Gamma \to \Gamma \qquad \text{by} \qquad c_\sigma(g, h\Gamma) = \sigma(gh\Gamma)^{-1} g \, \sigma(h\Gamma).$$

Clearly, this is a cocycle (a conjugate of the identity homomorphism $G \to G$); however, c_σ takes values in the subgroup Γ of G. This cocycle is associated to a choice of the cross-section σ (equivalently, the choice of the fundamental domain); starting from another Borel cross-section $\sigma' : G/\Gamma \to G$ results in a cohomologous cocycle:

$$c_{\sigma'} = c_\sigma^f \quad \text{where} \quad f : G/\Gamma \to \Gamma \quad \text{is defined by} \quad \sigma(x) = f(x)\sigma'(x).$$

Let Γ be a lattice in G. Then any action $\Gamma \curvearrowright (X, \mu)$ gives rise to the **induced** G-action (a.k.a. **suspension**) on $\bar{X} = G \times_\Gamma X$ where G acts on the first coordinate. Equivalently, $\bar{X} = G/\Gamma \times X$ and $g : (g'\Gamma, x) \mapsto (gg'\Gamma, c(g, g'\Gamma)x)$ where $c : G \times G/\Gamma \to \Gamma$ is in the canonical class. Here the G-invariant finite measure $\bar{\mu} = m_{G/\Gamma} \times \mu$ is ergodic iff μ is Γ-ergodic. If $\alpha : \Gamma \times X \to H$ is a cocycle, the **induced cocycle** $\bar{\alpha} : G \times \bar{X} \to H$ is given by $\bar{\alpha}(g, (g'\Gamma, x)) = \alpha(c(g, g'\Gamma), x)$. The cohomology of $\bar{\alpha}$ is the same as that of α (one relates maps $F : \bar{X} \to H$ to $f : X \to H$ by $f(x) = F(e\Gamma, x)$ taking instead of $e\Gamma$ a generic point in G/Γ). In particular, $\bar{\alpha}$ is cohomologous to a homomorpism $\bar{\pi} : G \to H$ iff α is cohomologous to a homomorphism $\Gamma \to H$; see [146] for details.

Cocycles appear quite naturally in a number of situations such as (volume-preserving) smooth actions on manifolds, where choosing a measurable trivialization of the tangent bundle, the derivative becomes a matrix-valued cocycle. We refer the reader to David Fisher's survey [42] where this type of cocycle is extensively discussed in the context of Zimmer's program. Here we shall be interested in a different type of cocycles: "rearrangement" cocycles associated to orbit equivalence, measure equivalence, and so on. as follows.

A.2. OE-Cocycles

Let $\Gamma \curvearrowright (X, \mu)$ and $\Lambda \curvearrowright (Y, \nu)$ be two measurable, measure-preserving ergodic actions on probability spaces, and $T : (X, \mu) \to (Y, \nu)$ be an Orbit Equivalence. Assume that the Λ-action is *essentially free*, that is, for ν-.a.e $y \in Y$, the stabilizer $\Lambda_y = \{h \in \Lambda : h.y = y\}$ is trivial. Then for every $g \in \Gamma$ and μ-a.e. $x \in X$, the points $T(g.x), T(x) \in Y$ lie on the same Λ-orbit. Let $\alpha(g, x) \in \Lambda$ denote the (a.e. unique) element of Λ with

$$T(g.x) = \alpha(g, x).T(x).$$

Considering $x, g.x, g'g.x$ one checks that α is actually a cocycle $\alpha : \Gamma \times X \to \Lambda$. We shall refer to such α as the **OE-cocycle**, or the **rearrangement** cocycle, corresponding to T.

Note that for μ-a.e. x, the map $\alpha(-, x) : \Gamma \to \Lambda$ is a bijection; it describes how the Γ-names of points $x' \in \Gamma.x$ translate into the Λ-names of $y' \in \Lambda.T(x)$ under the map T. The inverse map $T^{-1} : (Y, \nu) \to (X, \mu)$ defines an OE-cocycle $\beta : \Lambda \times Y \to \Gamma$, which serves as an "inverse" to α in the sense that a.e.:

$$\beta(\alpha(g, x), T(X)) = g \qquad (g \in \Gamma).$$

A.3. ME-Cocycles

Let (Ω, m) be an ME-coupling of two groups Γ and Λ and let $Y, X \subset \Omega$ be fundamental domains for Γ, Λ actions, respectively. The natural identification $\Omega/\Lambda \cong X$, $\Lambda\omega \mapsto \Lambda\omega \cap X$, translates the Γ-action on Ω/Λ to $\Gamma \curvearrowright X$ by

$$\gamma : X \ni x \mapsto g\alpha(g, x)x \in X$$

where $\alpha(\gamma, x)$ is the unique element in Λ taking $\gamma x \in \Omega$ into $X \subset \Omega$. It is easy to see that $\alpha : \Gamma \times X \to \Lambda$ is a cocycle with respect to the above Γ-action on X, which we denote by a dot $\gamma \cdot x$ to distinguish it from the Γ-action on Ω. (If Γ and Λ are lattices in G, then $\alpha : \Gamma \times G/\Lambda \to \Lambda$ is the restriction of the canonical cocycle $G \times G/\Lambda \to \Lambda$). Similarly we get a cocycle $\beta : \Lambda \times Y \to \Gamma$. So the (Γ, Λ) ME-coupling Ω and a choice of fundamental domains $Y \cong \Omega/\Gamma$, $X \cong \Omega/\Lambda$ define a pair of cocycles

A.1 $\alpha : \Gamma \times \Omega / \Lambda \to \Lambda, \qquad \beta : \Lambda \times \Omega / \Gamma \to \Gamma.$

Changing the fundamental domains amounts to conjugating the cocycles and vice versa.

REMARK 5.22. *One can characterize ME-cocycles among all measurable cocycles* $\alpha : \Gamma \times X \to \Lambda$ *as* **discrete** *ones with* **finite covolume**. *These concepts refer to the following construction: let* $(\tilde{X}, \tilde{\mu}) = (X \times \Lambda, \mu \times m_\Lambda)$ *and let* Γ *act by* $g : (x, h) \mapsto$ $(g.x, \alpha(g, x)h)$. *Say that the cocycle is discrete and has finite covolume if the action* $\Gamma \curvearrowright (\tilde{X}, \tilde{\mu})$ *admits a finite-measure fundamental domain.*

References

[1]   M. Abert and N. Nikolov, *Rank gradient, cost of groups and the rank versus Heegaard genus problem* (2007), available at arXiv:0701.361.

[2]   M. Abert and B. Weiss, *Bernoulli actions are weakly contained in any free action*, in preparation.

[3]   S. Adams, *An equivalence relation that is not freely generated*, Proc. Amer. Math. Soc. **102** (1988), no. 3, 565–566.

[4]   ———, *Trees and amenable equivalence relations*, Ergod Th. & Dynam. Sys. **10** (1990), no. 1, 1–14.

[5]   ———, *Reduction of cocycles with hyperbolic targets*, Ergod. Th. & Dynam. Sys. **16** (1996), no. 6, 1111–1145.

[6]   S. R. Adams and R. J. Spatzier, *Kazhdan groups, cocycles and trees*, Amer. J. Math. **112** (1990), no. 2, 271–287.

[7]   A. Alvarez and D. Gaboriau, *Free products, orbit equivalence and measure equivalence rigidity* (2008), available at arXiv:0806.2788.

[8]   C. Anantharaman-Delaroche, *Cohomology of property T groupoids and applications*, Ergod. Th. & Dynam. Sys. **25** (2005), no. 4, 977–1013.

[9]   C. Anantharaman-Delaroche and J. Renault,, *Amenable groupoids*, Monographies de L'Enseignement Mathématique [Monographs of L'Enseignement Mathématique], vol. 36, L'Enseignement Mathématique, Geneva, 2000. With a foreword by Georges Skandalis and Appendix B by E. Germain.

[10]   J. E. Andersen, *Mapping class groups do not have Kazhdan's property (T)* (2007), available at arXiv:0706.2184.

[11]   M. F. Atiyah, *Elliptic operators, discrete groups and von Neumann algebras*, Colloque "Analyse et Topologie" en l'Honneur de Henri Cartan (Orsay, 1974),Soc. Math. France, Paris, 1976, pp. 43–72. Astérisque, No. 32–33.

[12]   U. Bader and A. Furman, *Superrigidity via Weyl groups: hyperbolic-like targets* (preprint, 2010).

[13]   U. Bader, A. Furman, and R. Sauer, *Intergrable measure equivalence and rigidity of hyperbolic lattices* (2010), available at arXiv:1006.5193.

[14]   ———, *Efficient subdivision in hyperbolic groups and applications* (2010), available at arXiv:1003.1562.

[15] U. Bader, A. Furman, and A. Shaker, *Superrigidity, Weyl groups, and actions on the circle* (2006), available at arXiv:math/0605276.

[16] M. E. B. Bekka and A. Valette, *Kazhdan's property* (T) *and amenable representations*, Math. Z. **212** (1993), no. 2, 293–299.

[17] B. Bekka, P. de la Harpe, and A. Valette, *Kazhdan's property (T)*, New Mathematical Monographs, Vol. 11, Cambridge University Press, Cambridge, 2008.

[18] J. Bourgain, A. Furman, E. Lindenstrauss, and S. Mozes, *Invariant measures and stiffness for non-abelian groups of toral automorphisms*, C. R. Math. Acad. Sci. Paris **344** (2007), no. 12, 737–742.

[19] M. R. Bridson, M. Tweedale, and H. Wilton, *Limit groups, positive-genus towers and measure-equivalence*, Ergod. Th. & Dynam. Sys. 2) (2007), no. 3, 703–710.

[20] M. Burger and N. Monod, *Bounded cohomology of lattices in higher rank Lie groups*, J. Eur. Math. Soc. (JEMS), **1**, (1999), no. 2, 199–235.

[21] ———, *Continuous bounded cohomology and applications to rigidity theory*, Geom. Funct. Anal. **12** (2002), no. 2, 219–280.

[22] M. Burger and S. Mozes, CAT(−1)-*spaces, divergence groups and their commensurators*, J. Amer. Math. Soc. **9** (1996), no. 1, 57–93.

[23] J. Cheeger and M. Gromov, L_2-*cohomology and group cohomology*, Topology **25** (1986), no. 2, 189–215.

[24] P.-A. Cherix, M. Cowling, P. Jolissaint, P. Julg, and A. Valette, *Groups with the Haagerup property*, Progress in Mathematics, vol. 197 Birkhäuser Verlag, Basel, 2001. Gromov's a-T-menability.

[25] I. Chifan and A. Ioana, *Ergodic subequivalence relations induced by a Bernoulli action* (2008), available at arXiv:0802.2353.

[26] A. Connes, *A factor of type II$_1$ with countable fundamental group*, J. Operator Theory **4** (1980), no. 1, 151–153.

[27] A. Connes, J. Feldman, and B. Weiss, *An amenable equivalence relation is generated by a single transformation*, Ergod. Th. & Dynam. Sys. **1** (1981), no. 4, 431–450 (1982).

[28] A. Connes and B. Weiss, *Property T and asymptotically invariant sequences*, Israel J. Math. **37** (1980), no. 3, 209–210.

[29] K. Corlette and R. J. Zimmer, *Superrigidity for cocycles and hyperbolic geometry*, Internat. J. Math. **5** (1994), no. 3, 273–290.

[30] Y. de Cornulier, Y. Stalder, and A. Valette, *Proper actions of lamplighter groups associated with free groups*, C. R. Math. Acad. Sci. Paris **346** (2008), no. 3–4, 173–176 (English, with English and French summaries).

[31] M. Cowling and U. Haagerup, *Completely bounded multipliers of the Fourier algebra of a simple Lie group of real rank one*, Invent. Math. **96** (1989), no. 3, 507–549.

[32] M. Cowling and R. J. Zimmer, *Actions of lattices in Sp(1,n)*, Ergod. Th. & Dynam. Sys. **9** (1989), no. 2, 221–237.

[33] H. A. Dye, *On groups of measure preserving transformation. I*, Amer. J. Math. **81** (1959), 119–159.

[34] ———, *On groups of measure preserving transformations. II*, Amer. J. Math. **85** (1963), 551–576.

[35] B. Eckmann, *Introduction to l_2-methods in topology: reduced l_2-homology, harmonic chains, l_2-Betti numbers*, Israel J. Math. **117** (2000), 183–219. Notes prepared by Guido Mislin.

[36] I. Epstein, *Orbit inequivalent actions of non-amenable groups* (2007), available at arXiv:0707.4215.

[37] B. Farb, *The quasi-isometry classification of lattices in semisimple Lie groups*, Math. Res. Lett. **4** (1997), no. 5, 705–717.

[38] B. Farb and S. Weinberger, *The intrinsic asymmetry and inhomogeneity of Teichmuller space* (2008), available at arXiv:0804.4428.

[39] J. Feldman and C. C. Moore, *Ergodic equivalence relations, cohomology, and von Neumann algebras. I*, Trans. Amer. Math. Soc. **234** (1977), no. 2, 289–324.

[40] ———, *Ergodic equivalence relations, cohomology, and von Neumann algebras. II*, Trans. Amer. Math. Soc. **234** (1977), no. 2, 325–359.

[41] J. Feldman, C. E. Sutherland, and R. J. Zimmer, *Subrelations of ergodic equivalence relations*, Ergod. Th. & Dynam. Sys. **9** (1989), no. 2, 239–269.

[42] D. Fisher, *Groups acting on manifolds: around the Zimmer program*, in this volume.

[43] D. Fisher and T. Hitchman, *Cocycle superrigidity and harmonic maps with infinite dimensional targets* (2005), available at arXiv:math/0511666v3.

[44] A. Furman, *Gromov's measure equivalence and rigidity of higher rank lattices*, Ann. of Math. (2), **150** (1999), no. 3, 1059–1081.

[45] ———, *Orbit equivalence rigidity*, Ann. of Math. (2) **150** (1999), no. 3, 1083–1108.

[46] ———, *Mostow-margulis rigidity with locally compact targets*, Geom. Funct. Anal. **11** (2001), no. 1, 30–59.

[47] ———, *Outer automorphism groups of some ergodic equivalence relations*, Comment. Math. Helv. **80** (2005), no. 1, 157–196.

[48] ———, *On Popa's cocycle superrigidity theorem*, Int. Math. Res. Not. IMRN **19** (2007), Art. ID rnm073, 46.

[49] ———, *Cohomology of measurable cocycles*, in preparation.

[50] H. Furstenberg, *Poisson boundaries and envelopes of discrete groups*, Bull. Amer. Math. Soc. **73** (1967), 350–356.

[51] ———, *Ergodic behavior of diagonal measures and a theorem of Szemerédi on arithmetic progressions*, J. Analyse Math. **31** 1977, 204–256.

[52] ———, *Rigidity and cocycles for ergodic actions of semisimple Lie groups (after G. A. Margulis and R. Zimmer)*, Bourbaki Seminar, Vol. 1979/80, Lecture Notes in Math., vol. 842, Springer, Berlin, 1981, pp. 273–292.

[53] D. Gaboriau, *Mercuriale de groupes et de relations*, C. R. Acad. Sci. Paris Sér. I Math. **326** (1998), no. 2, 219–222.

[54] ———, *Coût des relations d'équivalence et des groupes*, Invent. Math. **139** (2000), no. 1, 41–98.

[55] ———, *Sur la (co-)homologie L^2 des actions préservant une mesure*, C. R. Acad. Sci. Paris Sér. I Math. **330** (2000), no. 5, 365–370.

[56] ———, *On orbit equivalence of measure preserving actions*, Rigidity in dynamics and geometry (Cambridge, 2000), Springer, Berlin, 2002, pp. 167–186.

[57] ———, *Invariants l^2 de relations d'équivalence et de groupes*, Publ. Math. Inst. Hautes Etudes Sci. **95**, (2002), 93–150.

[58] ———, *Examples of groups that are measure equivalent to the free group*, Ergod. Th. & Dynam. Sys. **25** (2005), no. 6, 1809–1827.

[59] D. Gaboriau and S. Popa, *An uncountable family of non-orbit equivalent actions of F_n*, J. Amer. Math. Soc. **18** (2005), no. 3, 547–559 (electronic).

[60] D. Gaboriau and R. Lyons, *A measurable-group-theoretic solution to von Neumann's problem* (2007), available at arXiv:0711.1643.

[61] S. L. Gefter, *Ergodic equivalence relation without outer automorphisms*, Dopov./Dokl. Akad. Nauk Ukraïni, no. 11 (1993), 25–27.

[62] ———, *Outer automorphism group of the ergodic equivalence relation generated by translations of dense subgroup of compact group on its homogeneous space*, Publ. Res. Inst. Math. Sci. **32** (1996), no. 3, 517–538.

[63] ———, *On cohomologies of ergodic actions of a T-group on homogeneous spaces of a compact Lie group (Russian)*, Operators in functional spaces and questions of function theory, Collect. Sci. Works, 1987, pp. 77–83.

[64] S. L. Gefter and V. Ya. Golodets, *Fundamental groups for ergodic actions and actions with unit fundamental groups*, Publ. Res. Inst. Math. Sci. **24** (1988), no. 6, 821–847 (1989).

[65] M. Gromov, *Volume and bounded cohomology*, Inst. Hautes Études Sci. Publ. Math. **56** (1982), 5–99 (1983).

[66] ———, *Asymptotic invariants of infinite groups*, Geometric group theory, Vol. 2 (Sussex, 1991), 1993, pp. 1–295.

[67] P. de la Harpe, *Topics in geometric group theory*, Chicago Lectures in Mathematics, University of Chicago Press, Chicago, 2000.

[68] G. Hjorth, *A converse to Dye's theorem*, Trans. Amer. Math. Soc. **357** (2005), no. 8, 3083–3103 (electronic).

[69] ———, *A lemma for cost attained*, Ann. Pure Appl. Logic **143** (2006), no. 1-3, 87–102.

[70] G. Hjorth and A. S. Kechris, *Rigidity theorems for actions of product groups and countable Borel equivalence relations*, Mem. Amer. Math. Soc. **177** (2005), 833, i–109.

[71] A. Ioana, *A construction of non-orbit equivalent actions of F_2* (2006), available at arXiv:math/0610452.

[72] ———, *Cocycle superrigidity for profinite actions of property (T) groups* (2008), available at arXiv:0805.2998.

[73] ———, *Orbit inequivalent actions for groups containing a copy of F_2* (2007), available at arXiv:0701.5027.

[74] A. Ioana, A. S. Kechris, and T. Tsankov, *Subequivalence relations and positive-definite functions* (2008), available at arXiv:0806.0430.

[75] A. Ioana, J. Peterson, and S. Popa, *Amalgamated free products of weakly rigid factors and calculation of their symmetry groups*, Acta Math. **200**, (2008), no. 1, 85–153.

[76] P. Jolissaint, *Approximation properties for measure equivalent groups*, 2001. preprint.

[77] V. F. R. Jones and K. Schmidt, *Asymptotically invariant sequences and approximate finiteness*, Amer. J. Math. **109** (1987), no. 1, 91–114.

[78] V. A. Kaimanovich, *Amenability, hyperfiniteness, and isoperimetric inequalities*, C. R. Acad. Sci. Paris Sér. I Math. **325** (1997), no. 9, 999–1004.

[79] D. A. Každan, *On the connection of the dual space of a group with the structure of its closed subgroups*, Funkcional. Anal. i Priložen. **1** (1967), 71–74 Russian.

[80] A. S. Kechris, *Global aspects of ergodic group actions*, Mathematical surveys and monographs, Vol. 160, American Mathematical Society, 2010. to appear.

[81] A. S. Kechris, B. D. Miller, *Topics in orbit equivalence*, Lecture Notes in Mathematics, Vol. 1852, Springer-Verlag, Berlin, 2004.

[82] A. S. Kechris and T. Tsankov, *Amenable actions and almost invariant sets*, Proc. Amer. Math. Soc. **136** (2008), no. 2, 687–697 (electronic).

[83] Y. Kida, *The mapping class group from the viewpoint of measure equivalence theory*, Mem. Amer. Math. Soc. **196** (2008), no. 916, i–190.

[84] ———, *Classification of the mapping class groups up to measure equivalence*, Proc. Japan Acad. Ser. A Math. Sci. **82** (2006), no. 1, 4–7.

[85] ———, *Orbit equivalence rigidity for ergodic actions of the mapping class group*, Geom Dedicata **131** (2008 February), no. 1, 99–109.

[86] ———, *Measure equivalence rigidity of the mapping class group*, arXiv **math.GR** (2006 July), available at arXiv:0607600.

[87] ———, *Outer automorphism groups of equivalence relations for mapping class group actions*, J. Lond. Math. Soc. (2) **78** (2008), no. 3, 622–638.

[88] ———, *Rigidity of amalgamated free products in measure equivalence theory* (2009), available at arXiv:0902.2888.

[89] M. Lackenby, *Heegaard splittings, the virtually Haken conjecture and property τ* (2002), available at arXiv:0205327.

[90] G. Levitt, *On the cost of generating an equivalence relation*, Ergod. Th. & Dynam. Sys. **15** (1995), no. 6, 1173–1181.

[91] A. Lubotzky and R. J. Zimmer, *Variants of Kazhdan's property for subgroups of semisimple groups*, Israel J. Math. **66** (1989), no. 1-3, 289–299.

[92] W. Lück, *L^2-invariants: theory and applications to geometry and K-theory*, Ergebnisse der Mathematik und ihrer Grenzgebiete. 3. Folge. A series of modern surveys in mathematics [Results in Mathematics and Related Areas. 3rd Series. A Series of Modern Surveys in Mathematics], Vol. 44, Springer-Verlag, Berlin, 2002.

[93] G. A. Margulis, *Discrete groups of motions of manifolds of nonpositive curvature*, Proceedings of the International Congress of Mathematicians (Vancouver, BC, 1974), Vol. 2, Canad. Math. Congress, Montreal, Que. bec, 1975, pp. 21–34 (Russian).

[94] ———, *Factor groups of discrete subgroups and measure theory*, Funktsional. Anal. i Prilozhen. **12** (1978), no. 4, 64–76 (Russian).

[95] ———, *Discrete subgroups of semisimple Lie groups*, Ergebnisse der Mathematik und ihrer Grenzgebiete (3) [Results in Mathematics and Related Areas (3)], Vol. 17 Springer-Verlag, Berlin, 1991.

[96] I. Mineyev, N. Monod, and Y. Shalom, *Ideal bicombings for hyperbolic groups and applications*, Topology **43** (2004), no. 6, 1319–1344.

[97] N. Monod, *Continuous bounded cohomology of locally compact groups*, Lecture Notes in Mathematics, Vol. 1758, Springer-Verlag, Berlin, 2001.

[98] ———, *An invitation to bounded cohomology*, International Congress of Mathematicians. Vol. II, Eur. Math. Soc., Zürich, 2006, pp. 1183–1211.

[99] N. Monod and Y. Shalom, *Negative curvature from a cohomological viewpoint and cocycle superrigidity*, C. R. Math. Acad. Sci. Paris **337** (2003), no. 10, 635–638.

[100] ———, *Cocycle superrigidity and bounded cohomology for negatively curved spaces*, J. Differential Geom. **67** (2004), no. 3, 395–455.

[101] ———, *Orbit equivalence rigidity and bounded cohomology*, Ann. of Math. (2) **164** (2006), no. 3, 825–878.

[102] C. C. Moore, *Ergodic theory and von Neumann algebras*, Operator algebras and applications, Part 2, Kingston, Ontario, 1980), Proc. Sympos. Pure Math., Vol. 38, Amer. Math. Soc., Providence, RI, 1982, pp. 179–226.

[103] G. D. Mostow, *Strong rigidity of locally symmetric spaces*, Princeton University Press, Princeton, NJ, 1973. Annals of Mathematics Studies, no. 78.

[104] F. J. Murray and J. von Neumann, *On rings of operators. IV*, Ann. of Math. (2) **44** (1943), 716–808.

[105] D. S. Ornstein and B. Weiss, *Ergodic theory of amenable group actions. I. The Rohlin lemma*, Bull. Amer. Math. Soc. (N.S.) **2** (1980), no. 1, 161–164.

[106] S. Popa, *Correspondences*, Prépublication Institul Natinoal Pentru Creatie Stiintica si Tehnica (1986).

[107] N. Ozawa, *A Kurosh-type theorem for type* II_1 *factors*, Int. Math. Res. Not., (2006), Art. ID 97560, 21.

[108] N. Ozawa and S. Popa, *On a class of* II_1 *factors with at most one Cartan subalgebra II* (2008), available at arXiv:0807.4270v2.

[109] S. Popa, *On a class of type* II_1 *factors with Betti numbers invariants*, Ann. of Math. (2) **163** (2006), no. 3, 809–899.

[110] ———, *Some computations of 1-cohomology groups and construction of non-orbit-equivalent actions*, J. Inst. Math. Jussieu **5** (2006), no. 2, 309–332.

[111] ———, *Strong rigidity of* II_1 *factors arising from malleable actions of w-rigid groups. I*, Invent. Math. **165** (2006), no. 2, 369–408.

[112] ———, *Strong rigidity of* II_1 *factors arising from malleable actions of w-rigid groups. II*, Invent. Math. **165** (2006), no. 2, 409–451.

[113] ———, *Cocycle and orbit equivalence superrigidity for malleable actions of w-rigid groups*, Invent. Math. **170** (2007), no. 2, 243–295.

[114] ———, *Deformation and rigidity for group actions and von Neumann algebras*, International Congress of Mathematicians. Vol. I, Eur. Math. Soc., Zürich, 2007, pp. 445–477.

[115] ———, *On the superrigidity of malleable actions with spectral gap*, J. Amer. Math. Soc. **21** (2008), no. 4, 981–1000.

[116] S. Popa and R. Sasyk, *On the cohomology of Bernoulli actions*, Ergod. Th. & Dynam. Sys. **27** (2007), no. 1, 241–251.

[117] S. Popa and S. Vaes, *Strong rigidity of generalized Bernoulli actions and computations of their symmetry groups*, Adv. Math. **217** (2008), no. 2, 833–872.

[118] ———, *Actions of* F_∞ *whose* II_1 *factors and orbit equivalence relations have prescribed fundamental group* (2008), available at arXiv:0803.3351.

[119] ———, *Cocycle and orbit superrigidity for lattices in* SL (n, R) *acting on homogeneous spaces*, in this volume.

[120] ———, *Group measure space decomposition of* II_1 *factors and* W^**-superrigidity*, available at arXiv:0906.2765.

[121] M. Ratner, *Interactions between ergodic theory, Lie groups, and number theory*, 2, (Zürich, 1994), Birkhäuser, Basel, 1995, pp. 157–182.

[122] H. Sako, *The class* S *is an ME invariant*, Int. Math. Res. Not. (2009), available at arXiv:0901.3374. rnp025.

[123] ———, *Measure equivalence rigidity and bi-exactness of groups*, J. Funct. Anal. (2009), available at arXiv:0901.3376. to appear.

[124] R. Sauer, L^2*-Betti numbers of discrete measured groupoids*, Internat. J. Algebra Comput. **15** (2005), 5-6, 1169–1188.

[125] ———, *Homological invariants and quasi-isometry*, Geom. Funct. Anal. **16** (2006), no. 2, 476–515.

[126] ———, *Amenable covers, volume and* l^2*-Betti numbers of aspherical manifolds* (2006), available at arXiv:math/0605627.

[127] R. Sauer and A. Thom, *A Hochschild-Serre spectral sequence for extensions of discrete measured groupoids* (2007), available at arXiv:0707.0906.

[128] K. Schmidt, *Asymptotically invariant sequences and an action of* SL(2, Z) *on the 2-sphere*, Israel J. Math. **37** (1980), no. 3, 193–208.

[129] ———, *Amenability, Kazhdan's property T, strong ergodicity and invariant means for ergodic group-actions*, Ergod. Th. & Dynam. Sys. **1** (1981), no. 2, 223–236.

[130] Y. Shalom, *Rigidity of commensurators and irreducible lattices*, Invent. Math. **141** (2000), no. 1, 1–54.

[131] ———, *Rigidity, unitary representations of semisimple groups, and fundamental groups of manifolds with rank one transformation group*, Ann. of Math. (2) **152** (2000), no. 1, 113–182.

[132] ———, *Harmonic analysis, cohomology, and the large-scale geometry of amenable groups*, Acta Math. **192** (2004), no. 2, 119–185.

[133] ———, *Measurable group theory*, European Congress of Mathematics, Eur. Math. Soc., Zürich, 2005, pp. 391–423.

[134] I. M. Singer, *Automorphisms of finite factors*, Amer. J. Math. **77** (1955), 117–133.

[135] G. Stuck and R. J. Zimmer, *Stabilizers for ergodic actions of higher rank semisimple groups*, Ann. of Math. (2) **139** (1994), no. 3, 723–747.

[136] S. Vaes, *Rigidity results for Bernoulli actions and their von Neumann algebras (after Sorin Popa)*, Astérisque **311** (2007), Exp. No. 961, viii, 237–294. Séminaire Bourbaki. Vol. 2005/2006.

[137] R. J. Zimmer, *Extensions of ergodic group actions*, Illinois J. Math. **20** (1976), no. 3, 373–409.

[138] ———, *Amenable ergodic group actions and an application to Poisson boundaries of random walks*, J. Functional Analysis **27** (1978), no. 3, 350–372.

[139] ———, *Strong rigidity for ergodic actions of semisimple Lie groups*, Ann. of Math. (2) **112** (1980), no. 3, 511–529.

[140] ———, *Orbit equivalence and rigidity of ergodic actions of Lie groups*, Ergod. Th. & Dynam. Sys. **1** (1981), no. 2, 237–253.

[141] ———, *On the cohomology of ergodic actions of semisimple Lie groups and discrete subgroups*, Amer. J. Math. **103** (1981), no. 5, 937–951.

[142] ———, *Ergodic theory, semisimple Lie groups, and foliations by manifolds of negative curvature*, Inst. Hautes Études Sci. Publ. Math. **55**, (1982), 37–62.

[143] ———, *Ergodic theory, group representations, and rigidity*, Bull. Amer. Math. Soc. (N.S.) **6** (1982), no. 3, 383–416.

[144] ———, *Ergodic actions of semisimple groups and product relations*, Ann. of Math. (2) **118** (1983), no. 1, 9–19.

[145] ———, *Kazhdan groups acting on compact manifolds*, Invent. Math. **75** (1984), no. 3, 425–436.

[146] ———, *Ergodic theory and semisimple groups*, Monographs in Mathematics, Vol. 81, Birkhäuser Verlag, Basel, 1984.

[147] ———, *Groups generating transversals to semisimple Lie group actions*, Israel J. Math. **73** 1991, no. 2, 151–159.

11

ALESSANDRA IOZZI

ON RELATIVE PROPERTY (T)

TO BOB ZIMMER ON HIS 60TH BIRTHDAY

Abstract

We present families of pairs $(H \ltimes A, A)$ with relative property (T), where H is a locally compact group acting continuously by automorphisms on a locally compact abelian group A. The paper is completely self-contained.

1. Introduction

Property (T) for groups and for pairs (also referred to as relative property (T)) was first introduced by D. Kazhdan in his thesis [14] to show that lattices in semisimple Lie groups are finitely generated. His proof that $SL(n, \mathbb{R})$ has property (T) for $n \geq 3$ follows from the fact that the pair $(SL(2, \mathbb{R}) \ltimes \mathbb{R}^2, \mathbb{R}^2)$ has the relative property (T).

Recall that a unitary representation $\pi : G \to \mathcal{U}(\mathcal{H})$ of a locally compact group G almost has invariant vectors if for every compact subset $K \subset G$ and for every $\epsilon > 0$ there exists $\xi \in \mathcal{H}$ with $\|\xi\| = 1$ such that

$$\sup_{g \in K} \|\pi(g)\xi - \xi\| < \epsilon.$$

Equivalently, one says that the representation π weakly contains the trivial representation $\mathbf{1}_G$. Then a group G has property (T) if any representation that almost has invariant vectors actually has invariant vectors. A typical example of a group with property (T) is a connected semisimple real Lie group all of whose factors have real rank at least 2, for example $SL(n, \mathbb{R})$ for $n \geq 3$, or any lattice therein.

DEFINITION 1.1. Let $L < G$ be a subgroup of a locally compact group G. The pair (G, L) has *relative property (T)* if whenever $(\pi, \mathcal{H})$ is a continuous unitary representation of G, which almost has invariant vectors; there are nonzero L-invariant vectors.

Partially supported by FNS grant PP002-102765.

Property (T) has several diverse applications in representation theory, ergodic theory, operator algebras, lattices in algebraic groups over local fields, geometric group theory, and the theory of networks. Aside the implicit use of relative property (T) in [14], among its first applications there is the construction of an infinite family of expanders [15], or the solution of Ruziewicz problem for $\mathbb{R}^n$, [16], both due to Margulis. Further applications include Gaboriau and Popa's construction [11] of a noncountable family of free ergodic measure-preserving non-orbit equivalent actions of the free group $\mathbb{F}_r$ in $r \geq 2$ generators on a standard probability space (see also [21] or [10] for generalizations) or Popa's construction [20, 22, 23] of a factor of type II_1 with trivial fundamental group. Finally, although the fact that a pair (G, L) has relative property (T) is a weaker condition than requiring that the group G itself has property (T), in some situations relative property (T) will suffice; for example, Navas [19] has proven that for the nonexistence of "interesting" C^2-actions on the circle, the relative property (T) of a pair (Γ, Γ_0), where Γ_0 is normal in Γ, will suffice.

Just to give a flavor of how relative property (T) differs from property (T), although we will not use these facts here, observe that if (G, L) has relative property (T) this does not imply that either G or L have property (T), and moreover, $(SL(2, \mathbb{Z}) \ltimes \mathbb{Z}^2, \mathbb{Z}^2)$ has relative property (T) but neither $SL(2, \mathbb{Z})$ nor $\mathbb{Z}^2$ have property (T). Furthermore, the fact that (G, L) has relative property (T) does not imply that either G or L is compactly generated, as the example of $(SL(3, \mathbb{Z}) \times \Lambda, \mathbb{F}_\infty \times \{e\})$, where Λ is any group that is not finitely generated shows.[1]

Just as for property (T), several different characterizations of relative property (T) are available, for example, in terms of strongly ergodic actions, in terms of Von Neumann algebras, of positive definite functions, of isometric actions of Hilbert spaces, and so on; moreover, it is possible to define relative property (T) for pairs (G, X) where X is any subset [6]. We refer the reader to [13] and [6] and to the references therein.

It goes without saying that if either G or L is a group with property (T), then the pair (G, L) has the relative property (T); nontrivial examples of pairs with relative property (T) have been constructed in [5], in [8] (see Theorem 1.3 and the comments thereafter) or in [25] (see the last paragraph of this introduction). The scope of this note is to present families of pairs (G, L) with relative property (T) whose general form is the semidirect product $G = H \ltimes A$ of a

---

1. Note, however, that if (G, L) has relative property (T), there exists a compactly generated subgroup H containing L such that (H, L) has relative property (T); see [6].

locally compact group H acting continuously by automorphisms on a locally compact abelian group A and $L = A$.

1.1.

Let Γ be a discrete group, S a finite set of prime numbers, $\mathbb{Z}[S]$ the ring obtained inverting the primes in S and $\rho : \Gamma \to GL_N(\mathbb{Z}[S])$ a homomorphism. We will denote by $\rho_p : \Gamma \to GL_N(\mathbb{Q}_p)$ the representation obtained by injecting $\mathbb{Z}[S]$ into $\mathbb{Q}_p$, where $\mathbb{Q}_\infty := \mathbb{R}$.

THEOREM 1.2. *Assume that the image $\rho(\Gamma)$ of Γ is Zariski dense in SL_N. Then the following are equivalent:*

i) *the pair $(\Gamma \ltimes \mathbb{Z}[S]^N, \mathbb{Z}[S]^N)$ has relative property (T);*
ii) *the $\mathbb{Z}[\Gamma]$-module $\mathbb{Z}[S]^N$ is finitely generated;*
iii) *for every $p \in S \cup \{\infty\}$ the image $\rho_p(\Gamma) < SL_N(\mathbb{Q}_p)$ is not bounded; and*
iv) *there is no $\rho_p^*(\Gamma)$-invariant probability measure on $\mathbb{P}((\mathbb{Q}_p^N)^*)$, for $p \in S \cup \{\infty\}$.*

We remark that (ii) is in contrast with the fact that, unless $S = \emptyset$, the ring $\mathbb{Z}[S]$ is not finitely generated as a $\mathbb{Z}$-module; moreover, it is remarkable that although the relative property (T) is analytic in nature, the equivalent property in (ii) above is purely algebraic. As it will be clear from the proof (see Proposition 4.5) Zariski density is not needed for the equivalence of the assertions (i) and (iv).

The motivation to establish results of the type of Theorem 1.2 comes from [8], where the following is shown:

THEOREM 1.3. (FERNÓS, [8]) *Let Γ be a finitely generated group. Then the following are equivalent:*

i) *there exists a representation $\rho : \Gamma \to SL_n(\mathbb{R})$ such that the real points $\overline{\rho(\Gamma)}^Z(\mathbb{R})$ of the Zariski closure of the image of ρ is not an amenable group; and*
ii) *there exists a finite set of primes S, an integer $N \in \mathbb{N}$, and a representation $\Gamma \to SL_N(\mathbb{Z}[S])$ such that $(\Gamma \ltimes \mathbb{Z}[S]^N, \mathbb{Z}[S]^N)$ has relative property (T).*

EXAMPLE 1.4.

1) According to the Tits alternative, if k is any local field of zero characteristic and $\Gamma < GL(n, k)$ is a finitely generated subgroup, then either Γ is

virtually solvable or it contains a free subgroup $\mathbb{F}_2$ in two generators that is Zariski dense in Γ. Since $\Gamma = \mathrm{SL}_N(\mathbb{Z})$ is not virtually solvable, then there is a Zariski-dense $\mathbb{F}_2 < \mathrm{SL}_N(\mathbb{Z})$ and thus $(\mathbb{F}_2 \ltimes \mathbb{Z}^N, \mathbb{Z}^N)$ has relative property (T).

2) A topological version of the Tits alternative shown by Breuillard and Gelander [4] asserts that if k_i are local fields of zero characteristic and $\Gamma < \prod \mathrm{GL}(n, k_i)$ is a finitely generated subgroup, then either Γ contains an open solvable subgroup, or a finitely generated free subgroup that is dense in Γ.

Using this result we show the following:

COROLLARY 1.5. *For every natural number $N \in \mathbb{N}$ and every nonempty finite set of primes S, there is a finitely generated free group $\Gamma < \mathrm{SL}_N(\mathbb{Z}[S])$ such that the pair $(\Gamma \ltimes \mathbb{Z}[S]^N, \mathbb{Z}[S]^N)$ has relative property (T).*

1.2.

Now let $\rho : G \to \mathrm{GL}(V)$ be a rational representation defined over $\mathbb{Q}$, where G is a connected, semisimple $\mathbb{Q}$-group. Let $\rho_\mathbb{R} : G_\mathbb{R} \to \mathrm{GL}(V_\mathbb{R})$ be the representation on the level of real points. We say that $\rho_\mathbb{R}$ is *totally unbounded* if for any $G_\mathbb{R}$-invariant subspace $W_\mathbb{R} \subset V_\mathbb{R}$ of positive dimension, the group $\rho(G_\mathbb{R})|_{W_\mathbb{R}} < \mathrm{GL}(W_\mathbb{R})$ is unbounded.

THEOREM 1.6. *Let $\Lambda \subset V_\mathbb{Q}$ be a $\mathbb{Z}$-module of maximal rank invariant under $G_\mathbb{Z}$. Let $\Gamma < G_\mathbb{Z}$ and assume that Γ is Zariski dense in G. The following are equivalent:*

i) *the pair $(\Gamma \ltimes \Lambda, \Lambda)$ has relative property (T); and*
ii) *the representation $\rho_\mathbb{R} : G_\mathbb{R} \to \mathrm{GL}(V_\mathbb{R})$ is totally unbounded.*

A remark is in order regarding the hypothesis of Zariski density of Γ in G. There is a result of Borel [3, theorem 1] to the extent that if G contains no connected normal $\mathbb{Q}$-subgroups $N \neq \{e\}$ such that $N_\mathbb{R}$ is compact, then $G_\mathbb{Z}$ is Zariski dense in G; this avoids the situation in which if $G = G_1 \times G_2$ with G_i $\mathbb{Q}$-groups, then $G_{1,\mathbb{Z}} \times G_{2,\mathbb{Z}} = G_\mathbb{Z} \subset G_\mathbb{R} = G_{1,\mathbb{R}} \times G_{2,\mathbb{R}}$ with $G_{2,\mathbb{R}}$ compact, which would prevent the Zariski density of $G_\mathbb{Z}$. So Theorem 1.6 applies to any finite-index subgroup Γ of $G_\mathbb{Z}$, provided $G_\mathbb{R}$ has no compact factors.

If $\rho_\mathbb{R}$ is irreducible with unbounded image, it is trivially totally unbounded and hence the pair $(\Gamma \ltimes \Lambda, \Lambda)$ has relative property (T). The construction of such a representation for a group defined over $\mathbb{Q}$, which is either adjoint or simply connected will be given in Lemma 4.7. As an application we have the following:

COROLLARY 1.7. *If G is a connected real algebraic semisimple Lie group without compact factors and $\Gamma < G$ is an arithmetic lattice, then there is a linear representation $\Gamma' \to SL_N(\mathbb{Z})$ of a finite-index subgroup $\Gamma' < \Gamma$ such that $(\Gamma' \ltimes \mathbb{Z}^N, \mathbb{Z}^N)$ has relative property (T).*

Theorem 1.6 is motivated by [25, theorems 1 and 4], where the same assertion as in Corollary 1.7 is proven, under the additional hypothesis that the group G is absolutely simple. Corollary 1.7 can also be deduced from [8, theorems 3 and 7.1] by observing that such an arithmetic lattice Γ satisfies property (F_∞) in [8]. Just like in [25] the value of the integer N is explicitly given (see the proof of Lemma 4.7). Moreover, if $\Gamma' < \Gamma$ is of finite-index $k := [\Gamma, \Gamma']$ and $(\Gamma' \ltimes \mathbb{Z}^N, \mathbb{Z}^N)$ has relative property (T), then $(\Gamma \ltimes \mathbb{Z}^{kN}, \mathbb{Z}^{kN})$ has relative property (T), [8, step B in the proof of theorem 7.1].

Acknowledgments: The author thanks Marc Burger for useful conversations during the preparation of this paper and the referee for helpful comments.

2. Generalities

We collect here, with proofs, few classical facts about relative property (T). We start by recording here the following observation:

LEMMA 2.1. *Let H_i, B_i be locally compact groups such that H_i acts on B_i via a continuous action $H_i \to \mathrm{Aut}(B_i)$, $i = 1, 2$, and let $q : H_1 \ltimes B_1 \to H_2 \ltimes B_2$ be a continuous homomorphism such that $q(B_1) = B_2$. If $(H_1 \ltimes B_1, B_1)$ has relative property (T), then also $(H_2 \ltimes B_2, B_2)$ has relative property (T).*

LEMMA 2.2. *Let $H, A_1, A_2, \ldots, A_n$ be locally compact groups such that H acts on A_j via a continuous action $H \to \mathrm{Aut}(A_j)$. The following are equivalent:*

   i) *the pair $(H \ltimes (A_1 \times \cdots \times A_n), A_1 \times \cdots \times A_n)$ has relative property (T); and*
   ii) *the pair $(H \ltimes A_j, A_j)$ has relative property (T) for $j = 1, \ldots, n$.*

Proof. The implication (i)$\Rightarrow$(ii) follows from Lemma 2.1 by considering the homomorphisms $H \ltimes (A_1 \times \cdots \times A_n) \to H \ltimes A_j$, while (ii)$\Rightarrow$(i) is in [8, lemma 5.2], where it is used that relative property (T) is a property closed under certain extensions. $\qquad\square$

The following corollary is essential in the proof of Proposition 4.5.

COROLLARY 2.3. *Let H and M be locally compact groups, $\Lambda < M$ a cocompact lattice and let $\rho : H \to \mathrm{Aut}\,(M)$ be a continuous action by automorphisms preserving Λ. Then the pair $(H \ltimes \Lambda, \Lambda)$ has relative property (T) if and only if the same holds for the pair $(H \ltimes M, M)$.*

Although the above result is all we need, we are going to prove the following more general result from which the corollary can be obtained at once by setting $L := H \ltimes \Lambda$ and $A := M \trianglelefteq H \ltimes M =: G$.

PROPOSITION 2.4. *Let G be a locally compact group, $A \trianglelefteq G$ a normal subgroup, and $L < G$ a closed subgroup.*

 1) *If (G, A) has relative property (T) and $L \backslash G$ has a finite G-invariant measure, then the pair $(L, L \cap A)$ also has relative property (T).*
 2) *If $A/L \cap A$ is compact with finite A-invariant measure and $(L, L \cap A)$ has relative property (T), then (G, A) has relative property (T).*

In order to prove this, we will use a strengthening of relative property (T). We recall first that if $\pi : G \to \mathcal{U}(\mathcal{H})$ is a unitary representation of a locally compact group G, we say that a sequence $\{\xi_n\} \subset \mathcal{H}$ of vectors is *asymptotically $\pi(G)$-invariant* if $\|\xi_n\| = 1$ for all n and for every compact subset $K \subset G$ we have that

$$2.1 \qquad \lim_{n \to \infty} \sup_{k \in K} \|\pi(k)\xi_n - \xi_n\| = 0\,.$$

If a group is σ-compact, a representation almost has invariant vectors if and only if it has a sequence of asymptotically invariant vectors. Denote by $\mathcal{H}^G$ the subspace of $\mathcal{H}$ consisting of $\pi(G)$-invariant vectors. Then it is well known that if G is a group with property (T) and $\pi : G \to \mathcal{U}(\mathcal{H})$ is a unitary representation of G with a sequence $\{\xi_n\}$ of asymptotically $\pi(G)$-invariant vectors, then

$$\lim_{n \to \infty} d(\xi_n, \mathcal{H}^G) = 0,$$

Where d denotes the distance to a subspace.

Similarly one has the following:

LEMMA 2.5. *Let $A \trianglelefteq G$ be a normal subgroup of a locally compact group G such that the pair (G, A) has relative property (T). If $\pi : G \to \mathcal{U}(\mathcal{H})$ is a continuous* *unitary representation and* $\{\xi_n\} \subset \mathcal{H}$ *is a sequence of asymptotically* $\pi(G)$-*invariant vectors, then*

$$\lim_{n \to \infty} d(\xi_n, \mathcal{H}^A) = 0.$$

Proof. Since A is normal in G, the Hilbert subspace $\mathcal{H}^A$ is $\pi(G)$-invariant, so that

$$\mathcal{H} = \mathcal{H}^A \oplus (\mathcal{H}^A)^\perp$$

is an orthogonal decomposition into $\pi(G)$-invariant subspaces. Let $\{\xi_n\} \subset \mathcal{H}$ be a sequence of asymptotically invariant vectors and let $\xi_n = \zeta_n + \zeta_n'$ be the corresponding orthogonal decomposition, so that showing that $\lim_{n \to \infty} d(\xi_n, \mathcal{H}^A) = 0$ is equivalent to showing that $\lim_{n \to \infty} \|\zeta_n'\| = 0$.

Let us assume by contradiction that $\lim \sup_{n \to \infty} \|\zeta_n'\| \neq 0$, that is, for some $\epsilon > 0$, let $\{\zeta_{n_k}'\} \in (\mathcal{H}^A)^\perp$ be a subsequence such that

2.2 for all $k \in \mathbb{N}$ we have $\|\zeta_{n_k}'\| \geq \epsilon$.

Since $(\mathcal{H}^A)^\perp$ is $\pi(G)$-invariant and the orthogonal projection is norm decreasing, we have that for all $g \in G$

$$\|\pi(g)\zeta_{n_k}' - \zeta_{n_k}'\| \leq \|\pi(g)\xi_{n_k} - \xi_{n_k}\|$$

from which, using Equation **2.2**, we obtain that

$$\left\| \pi(g) \left(\frac{\zeta_{n_k}'}{\|\zeta_{n_k}'\|} \right) - \frac{\zeta_{n_k}'}{\|\zeta_{n_k}'\|} \right\| \leq \frac{1}{\epsilon} \|\pi(g)\xi_{n_k} - \xi_{n_k}\|.$$

If now $K \subset G$ is any compact set, since the sequence $\{\xi_n\} \subset \mathcal{H}$ is asymptotically $\pi(G)$-invariant, we have

$$\lim_{k \to \infty} \sup_{k \in K} \left\| \pi(g) \left(\frac{\zeta_{n_k}'}{\|\zeta_{n_k}'\|} \right) - \frac{\zeta_{n_k}'}{\|\zeta_{n_k}'\|} \right\| \leq \lim_{k \to \infty} \sup_{k \in K} \frac{1}{\epsilon} \|\pi(g)\xi_{n_k} - \xi_{n_k}\| = 0,$$

from which we deduce that the sequence $\{\zeta_{n_k}'/\|\zeta_{n_k}'\|\}$ is also asymptotically $\pi(G)$-invariant. Hence $\pi(G)|_{(\mathcal{H}^A)^\perp}$ has non-zero $\pi(A)$-invariant vectors, which is a contradiction. $\square$

Proof of Proposition 2.4.

1) Let $(\pi, \mathcal{H})$ be a representation of L that almost has invariant vectors. Since weak containment is preserved under induction, the representation $\omega :=$ $\mathrm{Ind}_L^G(\pi)$ of G induced from π weakly contains the representation $\mathrm{Ind}_L^G(\mathbf{1}_L)$ induced to G from the identity representation of L; the latter in turn has invariant vectors since $L \backslash G$ has finite G-invariant measure and hence, by transitivity of weak containment, ω almost has invariant vectors. Thus, since (G, A) has relative property (T), the Hilbert space

$$\mathcal{L}_\omega := \{f : G \to \mathcal{H} : f \text{ is measurable}, f(\ell g) = \pi(\ell)(f(g)),$$

$$\text{for almost every } g \in G \text{ and for all } \ell \in L$$

$$\text{and } \int_{L\backslash G} \|f(g)\|^2 d\mu(g) < \infty\}$$

of the representation ω has nonzero A-invariant vectors with respect to the action $(\omega(g)f)(x) = f(xg)$ for all $g \in G$ and almost all $x \in G$. Let $f \in \mathcal{L}_\omega$ be such nonzero A-invariant vector. Then for all $a \in A \cap L$ and almost all $x \in G$ we have

$$\pi(a)(f(x)) = f(ax) = f(x(x^{-1}ax)) = f(x),$$

which shows that the space of $(L \cap A)$-invariant vectors in $\mathcal{H}$ is not trivial and hence shows (1).

2) Let $\pi : G \to \mathcal{U}(\mathcal{H})$ be a representation of G and $\{\xi_n\} \in \mathcal{H}$ be a sequence of asymptotically $\pi(G)$-invariant vectors.[2] Then $\{\xi_n\}$ is also asymptotically $\pi(L)$-invariant and since by hypothesis $(L, L \cap A)$ has relative property (T) and $L \cap A$ is normal in L, by Lemma 2.5 there exists a sequence $\{\zeta_n\} \in \mathcal{H}^{L \cap A}$ such that $\lim_{n \to \infty} \|\xi_n - \zeta_n\| = 0$. After rescaling if necessary, we may assume that $\|\zeta_n\| = 1$. For all $n \in \mathbb{N}$ the vectors

$$\eta_n := \int_{A/A \cap L} \pi(a)\zeta_n \, d\lambda(a) \in \mathcal{H},$$

where λ is the A-invariant probability measure on $A/A \cap L$, are obviously $\pi(A)$-invariant, and we only need to show that there exists an $n \in \mathbb{N}$ such that $\eta_n \neq 0$.

---

2. For the sake of simplicity we assume in the sequel that all locally compact groups are σ-compact. If not, the same arguments apply by replacing sequences with generalized sequences.

To see this, let $F \subset A \subset G$ be a compact fundamental domain for the action of $A \cap L$ on A. Then for all $a \in A$ there exists $f_a \in F$ and $\ell_a \in A \cap L$ such that $a = f_a \ell_a$, so that, by $\pi(L \cap A)$-invariance of ζ_n,

$$\|\eta_n - \zeta_n\| = \left\| \int_{A/A\cap L} \pi(a)\zeta_n \, d\lambda(a) - \zeta_n \right\| \leq \int_{A/A\cap L} \|\pi(a)\zeta_n - \zeta_n\| \, d\lambda(a)$$

$$\leq \sup_{f_a \in F} \|\pi(f_a)\zeta_n - \zeta_n\| .$$

Now we observe that $\{\zeta_n\}$ is asymptotically $\pi(G)$-invariant: indeed for all $g \in G$,

$$\|\pi(g)\zeta_n - \zeta_n\| \leq 2\|\zeta_n - \xi_n\| + \|\pi(g)\xi_n - \xi_n\| .$$

It follows that $\lim_{n\to\infty} \|\eta_n - \zeta_n\| = 0$ and thus in particular $\eta_n \neq 0$ for some $n \in \mathbb{N}$. $\square$

LEMMA 2.6. *Let $\Gamma \ltimes A$ be a semidirect product of discrete groups, with A abelian, and assume that $(\Gamma \ltimes A, A)$ has relative property (T). Then the $\mathbb{Z}[\Gamma]$-module A is finitely generated.*

Proof. Let $A_n \uparrow A$ be an increasing sequence of Γ-invariant subgroups finitely generated over $\mathbb{Z}[\Gamma]$ and $\cup A_n = A$. Consider the $\Gamma \ltimes A$-regular action on

$$\bigoplus_{n\geq 1} \ell^2(\Gamma \ltimes A / \Gamma \ltimes A_n) ;$$

the sequence $\delta_{e(\Gamma \ltimes A_n)}$ is asymptotically invariant, hence there is an A-invariant vector in some $\ell^2(\Gamma \ltimes A / \Gamma \ltimes A_n)$; this implies that there is a finite A-orbit in $(\Gamma \ltimes A / \Gamma \ltimes A_n) \cong A/A_n$ and hence $|A/A_n| < +\infty$. $\square$

3. Algebraic Actions and Measure Theory

We start by recalling the classical fact that if k is a local field and H_k consists of the k-points of an algebraic group defined over k, then any k-algebraic action of H_k on the k-points of a k-variety has orbits, which are locally closed in the Hausdorff topology [2]. Recall, moreover, that an action is almost effective if the intersection of the stabilizers is a finite group.

At the heart of what we are doing is the following theorem of Gromov: the proof we recall here is more elementary than the original one and appears in [1, théorème 6.5].

THEOREM 3.1. (GROMOV) *Let k be a local field, H a k-algebraic group, W a k-algebraic variety and $H \times W \to W$ a k-algebraic action. Let $\mu \in \mathcal{M}^1(W_k)$ be a probability measure on W_k such that the Zariski closure of its support is W, and assume that the action is almost effective. Then*

$$\mathrm{Stab}_{H_k}(\mu) := \{h \in H_k : h_*\mu = \mu\}$$

is compact.

Proof. Observe first of all that since $\mathrm{supp}(\mu) \subset W_k$, then $\overline{\mathrm{supp}(\mu)}^Z \subset \overline{W_k}^Z$, which, together with the hypothesis on the support of μ implies that W_k is Zariski dense in W. This and the fact that the action is almost effective imply that

3.1 $$\bigcap_{x \in W_k} \mathrm{Stab}_{H_k}(x) = \bigcap_{x \in W} \mathrm{Stab}_{H_k}(x) \text{ is finite}.$$

The Noetherian property for the k-algebraic group $\mathrm{Stab}_{H_k}(x)$ implies that there exist $x_1, \ldots, x_n \in W_k$ such that

3.2 $$\bigcap_{j=1}^{n} \mathrm{Stab}_{H_k}(x_j) = \bigcap_{x \in W_k} \mathrm{Stab}_{H_k}(x).$$

If we now let H act diagonally on W^n and define

$$\mathcal{O} := \{p \in W^n : \mathrm{Stab}_{H_k}(p) \text{ is finite}\},$$

we deduce immediately from Equations **3.1** and **3.2** that $\mathcal{O}$ is not empty. Likewise it is easy to see that $\mathcal{O}$ is Zariski open and hence, if $\nu := \mu^{\otimes n}$, $\nu(\mathcal{O}_k) > 0$. Since, moreover, $\mathcal{O}_k$ is H_k-invariant, then we conclude that $\nu|_{\mathcal{O}_k}$ is a finite measure left-invariant by $\mathrm{Stab}_{H_k}(\mu)$.

Since $\mathcal{O}_k$ is open in W_k^N and the H_k-orbits in W_k are locally closed, then also the $\mathrm{Stab}_{H_k}(\mu)$-orbits in $\mathcal{O}_k$ are locally closed in the Hausdorff topology. By decomposing $\nu|_{\mathcal{O}_k}$ if necessary into ergodic components and applying general considerations (see [26] or [12, 7]), we deduce that the finite Stab_{H_k}

(μ)-invariant measure on $\mathcal{O}_k$ is supported on an orbit $\text{Stab}_{H_k}(\mu) \cdot p_0$, with $p_0 \in \mathcal{O}_k$. But

$$\text{Stab}_{H_k}(\mu) \cdot p_0 \cong \text{Stab}_{H_k}(\mu)/\text{Stab}_{H_k}(p_0) \cap \text{Stab}_{H_k}(\mu)$$

and, by hypothesis, $\text{Stab}_{H_k}(p_0)$ is finite. It follows that $\text{Stab}_{H_k}(\mu)$ supports a finite measure that is invariant by translations and hence is compact. □

We can draw from this theorem some useful consequences on the structure of the stabilizer of a probability measure in projective space. The point is to be able to deal with measures whose support is not necessarily Zariski dense and with actions that are not necessarily effective. To apply the previous theorem and for further reference, let us set up some notation. If k is a local field, let E be a finite-dimensional vector space defined over k. If $\mu \in \mathcal{M}^1(\mathbb{P}(E_k))$, let us set

3.3
$$W := \overline{\text{supp}\,\mu}^Z \subseteq \mathbb{P}(E)$$

to be the projective subvariety defined as the Zariski closure of the support of μ and define the algebraic subgroups

3.4
$$N(W) := \{h \in \text{PGL}(E) : h(W) = W\}$$

$$I(W) := \{h \in \text{PGL}(E) : h|_W = Id_W\},$$

where $I(W)$ is normal in $N(W)$ and both are defined over k. The above notation will be typically applied when E is either a vector space as in the next corollary or the dual of a vector space (the only difference of course is lying in the action).

COROLLARY 3.2. *Let V_k be a finite-dimensional k-vector space seen as the set of k-points of the corresponding vector space V over an algebraic closure of k. The stabilizer $\text{Stab}_{\text{PGL}(V_k)}(\mu)$ of a probability measure $\mu \in \mathcal{M}^1(\mathbb{P}(V_k))$ on the projective space $\mathbb{P}(V_k)$ has a cocompact normal k-subgroup. More precisely, if $W \subset \mathbb{P}(V)$ and $I(W)$ are defined as in Equations (3.3) and (3.4), respectively, then $I(W)$ is a k-subgroup of $\text{PGL}(V)$ such that $I(W)_k$ is normal in $\text{Stab}_{\text{PGL}(V_k)}(\mu)$ and the quotient $\text{Stab}_{\text{PGL}(V_k)}(\mu)/I(W)_k$ is compact.*

Proof. By construction the group $H := N(W)/I(W)$ is defined over k and so is the algebraic action $H \times W \to W$, which is in addition effective by construction. Since $N(W)_k/I(W)_k$ is the $N(W)_k$-orbit of the identity in $(N(W)/I(W))_k = H_k$, it is locally closed in the Hausdorff topology and, being a topological group, is also closed; but $\text{Stab}_{\text{PGL}(V_k)}(\mu)/I(W)_k$ being closed in $N(W)_k/I(W)_k$ it is also closed in H_k. Since the quotient $\text{Stab}_{\text{PGL}(V_k)}(\mu)/I(W)_k$ is closed and contained in the compact $\text{Stab}_{H_k}(\mu)$, it is itself compact. $\qquad\square$

REMARK 3.3. *Furstenberg's lemma [9] asserts that either the stabilizer of a probability measure μ on $\mathbb{P}(V_k)$ is compact or the support of the measure is contained in the union of two proper subspaces. Using this lemma Zimmer has shown [26, theorem 3.2.4] (see also [18]) that under the same hypotheses as in Corollary 3.2 $\text{Stab}_{\text{PGL}(V_k)}(\mu)$ has a normal subgroup of finite index that contains a cocompact normal subgroup consisting of the k-points of a k-algebraic group.*

4. Applications to Group Pairs

Let V_k be as in §3 a k-vector space identified with the k-points of a vector space V over an algebraic closure of k, and let V_k^* be its dual. If $\mu \in \mathcal{M}^1(\mathbb{P}(V_k^*))$, define

4.1 $$G_\mu := \{h \in \text{GL}(V_k) : [h^*] \in \text{Stab}_{\text{PGL}(V_k^*)}(\mu)\}$$

where h^* denotes the transpose of h. Then we have

PROPOSITION 4.1. *Let $\mu \in \mathcal{M}^1(\mathbb{P}(V_k^*))$. Then the pair $(G_\mu \ltimes V_k, V_k)$ does not have relative property (T).*

Proof. Let us consider $W \subset \mathbb{P}(V^*)$ and $I(W)$ as defined in Equations (3.3) and (3.4) with $E = V^*$ and let us define

$$I_\mu := \{h \in G_\mu : [h^*] \in I(W)_k\}$$
$$= \{h \in \text{GL}(V_k) : [h^*] \in I(W)_k\}.$$

Let $[\lambda] \in \text{supp}\,\mu$, where $\lambda : V_k \to k$ is a nonzero linear form. Then, by definition of I_μ, there exists a continuous homomorphism $\chi_\lambda : I_\mu \to k^*$ such that for every $h \in I_\mu$

4.2 $$h^*\lambda = \chi_\lambda(h)\lambda.$$

But this is equivalent to saying that the map

$$q_\lambda : I_\mu \ltimes V_k \to \quad k^* \ltimes k$$

$$(h, v) \mapsto (\chi_\lambda(h), \lambda(v))$$

is a homomorphism. Since k is not compact and $k^* \ltimes k$ is amenable, then $(k^* \ltimes k, k)$ does not have relative property (T) (see, for instance, [8, lemma 8.3]), and since q_λ is continuous with $q_\lambda(V_k) = k$, then also $(I_\mu \ltimes V_k, V_k)$ does not have relative property (T) (Lemma 2.1). But I_μ is a normal subgroup of G_μ and G_μ/I_μ is compact by Corollary 3.2, so that Proposition 2.4(1) implies that also $(G_\mu \ltimes V_k, V_k)$ does not have relative property (T). □

Our source of examples is the following:

THEOREM 4.2. *Let k be a local field, V_k a finite-dimensional k-vector space endowed with its structure of (additive) locally compact group, H a locally compact group, and $\rho : H \to \mathrm{GL}(V_k)$ a continuous representation. Then the following are equivalent:*

i) *the pair $(H \ltimes V_k, V_k)$ has relative property (T); and*
ii) *there is no H-invariant probability measure on $\mathbb{P}(V_k^*)$.*

The proof of the implication (ii)⟹(i) was proven in [5, proposition 7 and examples] and is recalled below for completeness. The case in which $k = \mathbb{R}$ and H is a semisimple connected Lie group was established in [24, proposition 2.3]. The implication (i)⟹(ii) in the general case appears in [6, proposition 3.1.9], but it follows here immediately from the above proposition. In fact, assume that via the contragredient action ρ^* of ρ, the group H fixes a measure $\mu \in \mathcal{M}^1(\mathbb{P}(V_k^*))$. With G_μ as in Equation 4.1, consider the continuous homomorphism

$$H \ltimes V_k \overset{\rho \otimes \mathrm{Id}}{\longrightarrow} G_\mu \ltimes V_k \ .$$

Then Proposition 4.1 and the contrapositive of Lemma 2.1 imply that $(H \ltimes V_k, V_k)$ does not have relative property (T).

We recall now the proof of the implication (ii)⟹(i). Let $\rho : H \to \mathrm{GL}(V_k)$ be a continuous representation and let $G := H \ltimes V_k$. By fixing a nontrivial character $\chi \in \mathrm{Hom}_c (k, \mathbb{T})$ we get the usual isomorphism as locally compact groups of the Pontryagin dual $\widehat{V}_k$ of V_k with the dual V_k^*, via

$$V_k^* \longrightarrow \widehat{V}_k$$

$$\lambda \longmapsto \chi \circ \lambda .$$

Given a continuous unitary representation $\pi : G \to \mathcal{U}(\mathcal{H})$, we have the spectral measure

$$P : \mathcal{B}(V_k^*) \to \mathcal{P}(\mathcal{H})$$

of $\pi|_{V_k}$, that is, a map associating to every Borel set $B \subset V_k^*$ an orthogonal projection satisfying certain additional properties. Essential is the fact that for every $\xi \in \mathcal{H}$,

$$\mu_\xi(B) := \langle P(B)\xi, \xi \rangle$$

defines a bounded positive Radon measure $\mu_\xi \in \mathcal{M}^+(V_k^*)$ on V_k^* and

$$\langle \pi(v)\xi, \xi \rangle = \hat{\mu}_\xi(v) = \int_{V_k^*} \chi(\lambda(v)) \, d\mu_\xi(\lambda) .$$

In other words, μ_ξ is determined uniquely by the diagonal coefficients on V_k associated to ξ and it is easy to see that $P(B)$ is uniquely determined by the map $\xi \mapsto \mu_\xi(B)$; furthermore, we have for all $h \in H$ and $B \in \mathcal{B}(V_k^*)$

4.3 $$P(\rho(h)^* B) = \pi(h)^{-1} P(B) \pi(h) .$$

Now let $\xi \in \mathcal{H}$ and $\mu_\xi \in \mathcal{M}^+(V_k^*)$ and define

$$m_\xi := p_*(\mu_\xi|_{V_k^* \setminus \{0\}})$$

to be the push- forward of the measure $\mu_\xi|_{V_k^* \setminus \{0\}}$ under the projection $p : V_k^* \setminus \{0\} \to \mathbb{P}(V_k^*)$. If $B \in \mathcal{B}(\mathbb{P}(V_k^*))$ is a Borel subset in $\mathbb{P}(V_k^*)$ and we set $B' = p^{-1}(B)$, using Equation **4.3** we have

$$m_\xi(\rho^*(h)B) - m_\xi(B) = \langle P(\rho^*(h)B')\xi, \xi \rangle - \langle P(B')\xi, \xi \rangle$$

$$= \langle P(B')\pi(h)\xi, \pi(h)\xi \rangle - \langle P(B')\xi, \xi \rangle$$

$$= \mathrm{Re} \langle P(B')(\pi(h)\xi + \xi), \pi(h)\xi - \xi \rangle ,$$

so that

$$|m_\xi(\rho^*(h)B) - m_\xi(B))| \leq 2\|\xi\| \, \|\pi(h)\xi - \xi\| .$$

Now for any subset $K \subset H$, introduce the following quantity

$$\alpha(K, \rho) := \inf_{m \in \mathcal{M}^1(\mathbb{P}(V_k^*))} \sup_{B \in \mathcal{B}(\mathbb{P}(V_k^*))} \sup_{h \in K} |m(\rho^*(h)B) - m(B)|$$

which is somehow the extent to which "measures are moved by $\rho^*(h)$."

Since the total mass of the positive measure m_ξ is

$$\mu_\xi(\mathbb{P}(V_k^*)) = \mu_\xi(V_k^*) - \mu_\xi(\{0\}),$$

and since there are no $\rho(V_k)$-invariant vectors if and only if $\mu_\xi(\{0\}) = 0$ for every $\|\xi\| = 1$, we obtain the following:

PROPOSITION 4.3. *Let* $\rho : G \to \mathcal{U}(\mathcal{H})$ *be a continuous unitary representation of* $H \ltimes V_k$ *with no nonzero* V_k-*invariant vectors. Then for every compact subset* $K \subset H$ *and for every* $\xi \in \mathcal{H}$ *with* $\|\xi\| = 1$, *we have that*

$$\max_{h \in K} \|\rho(h)\xi - \xi\| \geq \frac{1}{2}\alpha(K, \rho).$$

The proof of Theorem 4.2(ii)$\Rightarrow$(i) will be complete if we show that H has no invariant measure on $\mathbb{P}(V_k^*)$ via ρ if and only if there exists a compact set $K \subset H$ with $\alpha(K, \rho) > 0$; but this is just an exercise.

4.1. Proof of the Results in §1.1

We adopt the same notation as in §1.1, namely, Γ is a discrete group, S a finite set of primes, $\mathbb{Z}[S]$ the ring obtained inverting the primes in S, $\rho : \Gamma \to \mathrm{GL}_N(\mathbb{Z}[S])$ a representation, and $\rho_p : \Gamma \to \mathrm{GL}_N(\mathbb{Q}_p)$ is the representation obtained by composing ρ with the embedding $\mathbb{Z}[S] \hookrightarrow \mathbb{Q}_p$, where $\mathbb{Q}_\infty := \mathbb{R}$. Since the diagonal embedding

$$\mathbb{Z}[S]^N \hookrightarrow \mathbb{R}^N \times \prod_{p \in S} \mathbb{Q}_p^N$$

realizes $\mathbb{Z}[S]^N$ as a cocompact lattice, Corollary 2.3 can be translated into the following:

COROLLARY 4.4. *The pair* $(\Gamma \ltimes \mathbb{Z}[S]^N, \mathbb{Z}[S]^N)$ *has relative property (T) if and only if the pair* $(\Gamma \ltimes (\mathbb{R}^N \times \prod_{p \in S} \mathbb{Q}_p^N), \mathbb{R}^N \times \prod_{p \in S} \mathbb{Q}_p^N)$ *has relative property (T).*

From this we deduce the equivalence of (i) and (iv) in Theorem 1.2, which we record here separately as it does not require the hypothesis of Zariski density of $\rho(\Gamma)$ in SL_N.

PROPOSITION 4.5. *With the above hypotheses, the following are equivalent:*

i) $(\Gamma \ltimes \mathbb{Z}[S]^N, \mathbb{Z}[S]^N)$ *has relative property (T); and*

ii) *for every $p \in S \cup \{\infty\}$, there is no $\rho_p^*(\Gamma)$-invariant probability measure on* $\mathbb{P}((\mathbb{Q}_p^N)^*)$.

Proof. Using Corollary 4.4 and Lemma 2.2 we deduce that the pair $(\Gamma \ltimes \mathbb{Z}[S]^N, \mathbb{Z}[S]^N)$ has relative property (T) if and only if $(\Gamma \ltimes \mathbb{Q}_p^N, \mathbb{Q}_p^N)$ has relative property (T) for every $p \in S \cup \{\infty\}$. Theorem 4.2 then implies the equivalence with (ii). $\qquad\square$

Proof of Theorem 1.2. The chain of implications that we shall prove is as follows: (iii)$\Rightarrow$(iv)$\Rightarrow$(i)$\Rightarrow$(ii)$\Rightarrow$(iii), where, however, the equivalence (iv)$\Leftrightarrow$(i) is the content of Proposition 4.5.

(iii)$\Rightarrow$(iv). We shall show the contrapositive statement; namely, we shall assume that for some $p \in S \cup \{\infty\}$ there exists a $\rho_p^*(\Gamma)$-invariant probability measure μ and we shall prove that then the image $\rho_p(\Gamma) \subset SL_N(\mathbb{Q}_p)$ is bounded.

Let $W \subset \mathbb{P}((\overline{\mathbb{Q}_p}^N)^*)$ and $I(W)$ be defined as in Equations **3.3** and **3.4** with $E = (\overline{\mathbb{Q}_p}^N)^*$ and $\overline{\mathbb{Q}_p}$ an algebraic closure of $\mathbb{Q}_p$, and let us consider $I'(W) = I(W) \cap PSL_N$. Since $I'(W)$ is normalized by $\rho(\Gamma)$ and $\rho(\Gamma)$ is Zariski dense in SL_N, then $I'(W)$ is a normal subgroup of PSL_N and hence either trivial or the full group. Obviously it cannot be the full group PSL_N since this cannot fix pointwise any nonempty subset in projective space. So since $I'(W)$ is trivial, then by Corollary 3.2 $\mathrm{Stab}_{PSL(\mathbb{Q}_p^N)}(\mu)$ is compact, so that $\rho_p(\Gamma)$ is bounded.

(i)$\Rightarrow$(ii). This follows from Lemma 2.6.

(ii)$\Rightarrow$(iii). If $p = \infty$, then it is easy to see that $\rho_\infty(\Gamma)$ cannot be bounded because otherwise it would be conjugate into a maximal compact subgroup that is a real algebraic group, thus contradicting the Zariski density of $\rho(\Gamma)$ in SL_N.

Let us assume that $\mathbb{Z}[S]^N$ is finitely generated as a module over $\mathbb{Z}[\Gamma]$, namely there exist $a_j \in \mathbb{Z}[S]^N$, with $1 \leq j \leq r$, such that $\mathbb{Z}[S]^N = \rho(\mathbb{Z}(\Gamma))a_1 + \cdots + \rho(\mathbb{Z}(\Gamma))a_r$. For $p \in S$, $n_i \in \mathbb{Z}$, and $\gamma_i \in \Gamma$, using the ultrametric inequality, we have

$$\left\| \sum_i^{<\infty} \sum_{j=1}^r n_i \rho_p(\gamma_i) a_j \right\| \leq \max_{j,i} \| n_i \rho_p(\gamma_i) a_j \|$$

$$= \max_{j,i} \| \rho_p(\gamma_i) a_j \| \leq \max_j \sup_{\gamma \in \Gamma} \| \rho_p(\gamma) a_j \|$$

so that if $\rho_p(\Gamma)$ were to be bounded for any of the p's in S, then $\mathbb{Z}[S]^N$ would also be bounded in $\mathbb{Q}_p^N$, which is not the case. $\qquad\square$

In the course of the proof of the implication (iii)⇒(i) in Theorem 1.2 we have proven the following fact, which we record as it might be of independent interest and which could be deduced also from Furstenberg's Lemma [9].

LEMMA 4.6. *Let* $\Lambda < \mathrm{PSL}_N(\mathbb{Q}_p)$ *be an unbounded closed subgroup that is Zariski dense in* PSL_N. *Then there exists no* Λ-*invariant probability measure on* $\mathbb{P}(\mathbb{Q}_p^N)$.

Proof of Corollary 1.5. Since $\mathrm{SL}_N(\mathbb{Z}[S])$ is an irreducible lattice in the product $\mathrm{SL}_N(\mathbb{R}) \times \prod_{\ell \in S} \mathrm{SL}_N(\mathbb{Q}_\ell)$, if $p \in S$ is a fixed prime, the projection

$$\alpha_p : \mathrm{SL}_N(\mathbb{Z}[S]) \to \mathrm{SL}_N(\mathbb{R}) \times \prod_{\ell \in S, \ell \neq p} \mathrm{SL}_N(\mathbb{Q}_\ell)$$

has dense image. We assume for the moment that we have proven that $\alpha_p(\mathrm{SL}_N(\mathbb{Z}[s]))$ cannot contain an open solvable subgroup, so that by the topological Tits alternative [4, theorem 1.6] there exists a free dense subgroup $\Lambda < \alpha_p(\mathrm{SL}_N(\mathbb{Z}[S]))$ of finite rank, which is also dense in $\mathrm{SL}_N(\mathbb{R}) \times \prod_{\ell \in S, \ell \neq p} \mathrm{SL}_N(\mathbb{Q}_\ell)$. This implies in particular that for all $\ell \in S$, $\ell \neq p$, the projection $\rho_\ell : \alpha_p^{-1}(\Lambda) \to \mathrm{SL}_N(\mathbb{Q}_\ell)$ is unbounded. If we show that also the projection $\rho_p : \Gamma \to \mathrm{SL}_N(\mathbb{Q}_p)$ is unbounded where $\Gamma := \alpha_p^{-1}(\Lambda) < \mathrm{SL}_N(\mathbb{Z}[S])$, then Theorem 1.2 will imply that the pair $(\Gamma \ltimes \mathbb{Z}[s]^N, \mathbb{Z}[s]^N)$ has relative property (T). In fact, if the projection $\rho_p : \Gamma \to \mathrm{SL}_N(\mathbb{Q}_p)$ were bounded with compact closure K, then Γ would be a discrete subgroup of $\mathrm{SL}_N(\mathbb{R}) \times \prod_{\ell \in S, \ell \neq p} \mathrm{SL}_N(\mathbb{Q}_\ell) \times K$, contradicting the fact that its projection in $\mathrm{SL}_N(\mathbb{R}) \times \prod_{\ell \in S, \ell \neq p} \mathrm{SL}_N(\mathbb{Q}_\ell)$ is dense.

To complete the proof we need to verify that $\alpha_p(\mathrm{SL}_N(\mathbb{Z}[S]))$ does not have open solvable subgroups. Let us start with the general observation that if $L < H < G$ are topological groups such that L is open in H and H is dense in G, then the closure $\overline{L}$ of L is an open subgroup of G. By applying this to $H = \alpha_p(\mathrm{SL}_N(\mathbb{Z}[S])) < \mathrm{SL}_N(\mathbb{R}) \times \prod_{\ell \in S, \ell \neq p} \mathrm{SL}_N(\mathbb{Q}_p)$, we have that if such open solvable subgroup $L < \alpha_p(\mathrm{SL}_N(\mathbb{Z}[S]))$ were to exist, then $\rho_\infty(\overline{L})$ would be an open solvable subgroup of $\mathrm{SL}_N(\mathbb{R})$, hence closed, and hence $\rho_\infty(\overline{L}) = \mathrm{SL}_N(\mathbb{R})$, since $\mathrm{SL}_N(\mathbb{R})$ is connected. This is not possible and the proof is completed. □

4.2. Proof of the results in §1.2
We now move on to the proof of the results in §1.2.

Proof of Theorem 1.6. In view of Theorem 4.2 it will be enough to prove the equivalence of the following statements:

1) There exists a $\rho^*(\Gamma)$-invariant probability measure on $\mathbb{P}(V_{\mathbb{R}}^*)$; and

2) The representation $\rho_{\mathbb{R}} : G_{\mathbb{R}} \to \mathrm{GL}(V_{\mathbb{R}})$ is not totally unbounded.

Remark first of all that one should avoid the temptation of trying to deduce Theorem 1.6 as an application of Theorem 1.2 (with $p = \infty$), as in order to do so one should require in addition to the hypotheses of Theorem 1.6 also the Zariski density of the image $\rho(\Gamma)$ in SL_N.

$(2) \Rightarrow (1)$. By hypothesis let $W_{\mathbb{R}} \subset V_{\mathbb{R}}$ be a $\rho_{\mathbb{R}}(G)$-invariant subspace such that $\rho_{\mathbb{R}}(G)|_{W_{\mathbb{R}}}$ is bounded. Then on $\mathbb{P}(W_{\mathbb{R}}) \subset \mathbb{P}(V_{\mathbb{R}})$ there exists a $\rho(\Gamma)$-invariant probability measure.

$(1) \Rightarrow (2)$. Let $\mu \in \mathcal{M}^1(\mathbb{P}(V_{\mathbb{R}}^*))$ be a $\rho^*(\Gamma)$-invariant probability measure and let $W \subset \mathbb{P}(V^*)$, $N(W)$, and $I(W)$ be as usual as defined in Equations **3.3** and **3.4** with $E = V^*$. If ρ^* is the contragredient representation, by hypothesis for all $g \in G$,

$$[\rho^*(g)] \in N(W).$$

If we define $N := \{g \in G : [\rho^*(g)] \in I(W)\}$, then we have an injective homomorphism

$$G/N \hookrightarrow N(W)/I(W),$$

and, by passing to the real points, we have an induced homomorphism

$$h : G_{\mathbb{R}}/N_{\mathbb{R}} \longrightarrow N(W)_{\mathbb{R}}/I(W)_{\mathbb{R}}$$

that is at most finite-to-one. If $p : G_{\mathbb{R}} \to G_{\mathbb{R}}/N_{\mathbb{R}}$ denotes the usual projection, then $h \circ p(\Gamma)$ is relatively compact since it is contained in $\mathrm{Stab}_{\mathrm{PGL}(V_{\mathbb{R}}^*)}$ $(\mu)/I(W)_{\mathbb{R}}$, which is compact by Corollary 3.2. Since h is at most finite-to-one, we infer that $p(\Gamma)$ is bounded. Since $(N_{\mathbb{R}})^\circ$ is of finite index in $N_{\mathbb{R}}$, if $p_1 : G_{\mathbb{R}} \to G_{\mathbb{R}}/(N_{\mathbb{R}})^\circ$, then $p_1(\Gamma)$ is also bounded in $G_{\mathbb{R}}/(N_{\mathbb{R}})^\circ$.

The rest of the proof will consist just in identifying the quotient $G_{\mathbb{R}}/(N_{\mathbb{R}})^\circ$ to deduce that $\rho_{\mathbb{R}}$ is not totally unbounded. To this purpose, observe that the connected component N° of N fixes pointwise every vector in the linear span E of $\pi^{-1}(W)$, where $\pi : V^* \to \mathbb{P}(V^*)$. In fact, for every $[\lambda] \in W$, $\lambda : V \to \mathbb{C}$, we have $\chi_\lambda : N \to \mathbb{C}^*$ given by

$$\rho^*(n)\lambda = \chi_\lambda(n)\lambda$$

for every $n \in N$. Thus, since N° is connected and a product of almost $\mathbb{R}$-simple factors of G, we get that $\chi_\lambda(N^\circ) = 1$. Since

$$N = \{g \in G : \rho^*(g)|_W = Id_W\}$$

$$\supset \{g \in G : \rho^*(g)|_E = Id_E\} \supset N^\circ$$

then $G_\mathbb{R}/(N_\mathbb{R})^\circ$ surjects onto $G_\mathbb{R}/\ker\{g \mapsto \rho^*(g)|_{E_\mathbb{R}}\}$. The fact that

$$G_\mathbb{R}/\ker\{g \mapsto \rho^*(g)|_{E_\mathbb{R}}\} \cong \text{im}\{g \mapsto \rho^*(g)|_{E_\mathbb{R}}\} \subset GL(E_\mathbb{R}),$$

together with the fact that $p_1(\Gamma)$ was bounded in $G_\mathbb{R}/(N_\mathbb{R})^\circ$, implies that $\rho^*(\Gamma)|_{E_\mathbb{R}}$ is bounded in $GL(E_\mathbb{R})$; again by Zariski density, $\rho^*_\mathbb{R}(G_\mathbb{R})|_{E_\mathbb{R}}$ is bounded, thus showing that $\rho_\mathbb{R}$ is not totally unbounded. $\square$

LEMMA 4.7. *Let H be a connected $\mathbb{Q}$-semisimple group that is either simply connected or adjoint and with no factors of $\mathbb{Q}$-rank $= 0$. Then there exists an irreducible representation $\rho : H \to GL(V)$ such that $\rho_\mathbb{R} : H_\mathbb{R} \to GL(V_\mathbb{R})$ is irreducible and unbounded.*

Proof. It will be enough to prove the assertion under the hypothesis that H is $\mathbb{Q}$-simple. In fact, since $H = H_1 \times \cdots \times H_n$, if a representation $\rho_j : H_j \to GL(V_j)$ with the desired properties exists for each factor H_i, then the Kronecker product $\rho := \bigotimes_{j=1}^n \rho_j : H \to GL\left(\bigotimes_{j=1}^n V_j\right)$ will have the desired properties for the group H.

So, recall (see for instance [17, pp. 47-48]) that if H is a connected almost $\mathbb{Q}$-simple group, there is a number field k and an absolutely simple k-group $\mathbb{L}$ such that $H = \text{Res}_{k/\mathbb{Q}} \mathbb{L}$. Thus, over $\mathbb{C}$, $H = \prod_{\sigma:k \to \mathbb{C}} \mathbb{L}^\sigma$, where the product is over all Archimedean places of k. If $\mathfrak{l}$ denotes the Lie algebra of $\mathbb{L}$, let $\text{Ad}_1 := \text{Ad} : \mathbb{L} \to \text{Aut}(\mathfrak{l})$ be the adjoint representation of $\mathbb{L}$ and $\text{Ad}_i : \mathbb{L}^{\sigma_i} \to \text{Aut}(\mathfrak{l}^i)$. Then by Weil's criterion $\rho := \bigotimes_{i=1}^n \text{Ad}_i : \prod \mathbb{L}^{\sigma_i} \to \text{Aut}(\bigotimes_{i=1}^n \mathfrak{l}^i)$ can be defined over $\mathbb{Q}$ and, setting $V := \bigotimes \mathfrak{l}^i$, we get an irreducible representation $\rho : H \to GL(V)$; then $\rho_\mathbb{R} : H_\mathbb{R} \to GL(V_\mathbb{R})$ is also irreducible and with unbounded image. $\square$

Proof of Corollary 1.7. By hypothesis, there exists a connected semisimple simply connected algebraic group H defined over $\mathbb{Q}$ and a surjective homomorphism $h : H_\mathbb{R} \twoheadrightarrow G$ such that $h(H_\mathbb{Z})$ is commensurable with Γ. By passing to a subgroup of finite index $\Gamma_0 \leq H_\mathbb{Z}$, we may assume that $p|_{\Gamma_0}$ is injective and has image $\Gamma' < \Gamma < G$ of finite index in Γ. If $\rho : H \to GL(V)$ is the representation in Lemma 4.7 and dim $V = N$, Theorem 1.6 implies that $(H_k \ltimes \mathbb{Z}^N, \mathbb{Z}^N)$ has relative property (T); hence $(\Gamma' \ltimes \mathbb{Z}^N, \mathbb{Z}^n)$ has relative property (T) where the action of Γ' on $\mathbb{Z}^N$ is via $\rho \circ h^{-1}$. $\qquad\square$

References

[1] N. A'Campo and M. Burger, *Réseaux arithmétiques et commensurateurs d'après G. A. Margulis*, Invent. Math. **116** (1994), 1–25.

[2] I. N. Bernstein and A. V. Zelevinskii, *Representations of the group gl(n, f), where f is a non-Archimedean local field*, Russian Math. Surveys **31** (1972), 1–68.

[3] A. Borel, *Density and maximality of arithmetic subgroups*, J. Reine Angew. Math. **224** (1966), 78–89.

[4] E. Breuillard and T. Gelander, *A topological Tits alternative*, Ann. of Math. **166** (2007), no. 2, 427–474.

[5] M. Burger, *Kazhdan constants for* SL(3, **Z**), J. Reine Angew. Math. **413** (1991), 36–67.

[6] Y. de Cornulier, *Relative Kazhdan property*, Ann. Sci. École Norm. Sup. (2) **39** (2006), 301–333.

[7] E. G. Effros, *Transformation groups and C*-algebras*, Ann. of Math. (2) **81** (1965), 38–55.

[8] T. Fernós, *Relative property (T) and linear groups*, Annales Institut Fourier **56** (2006), no. 6, 1767–1804.

[9] H. Furstenberg, *A Poisson formula for semisimple Lie groups*, Ann. of Math. **77** (1963), 335–383.

[10] D. Gaboriau, *Relative property (T) actions and trivial outer automorphisms groups*, preprint, 2008, arXiv:0804.0358v1.

[11] D. Gaboriau and S. Popa, *An uncountable family of nonorbit equivalent actions of* $\mathbb{F}_n$, J. Amer. Math. Soc. **18** (2005), no. 3, 547–559 (electronic).

[12] J. Glimm, *Locally compact transformation groups*, Trans. Amer. Math. Soc. **101** (1961), 124–138.

[13] Gh. Jaudon, *Notes on relative Kazhdan's property (T)*, personal notes, 2007, http://www.unige.ch/math/folks/jaudon/notes.pdf.

[14] D. A. Každan, *On the connection of the dual space of a group with the structure of its closed subgroups*, Funkcional. Anal. i Priložen. **1** (1967), 71–74.

[15] G. A. Margulis, *Explicit constructions of expanders*, Problemy Peredači Informacii **9** (1973), no. 4, 71–80, Problems in Information Transmission, Vol. 9 (1973) year 1975 pp. 325–332.

[16] _____, *Finitely-additive invariant measures on Euclidean spaces*, Erg. Th. Dyn. Sys. **2** (1982), 383–396.

[17] _____, *Discrete subgroups of semisimple Lie groups*, Springer-Verlag, New York, 1991.

[18] C. C. Moore, *Amenable subgroups of semisimple groups and proximal flows*, Israel J. Math. **34** (1979), no. 1–2, 121–138 (1980).

[19] A. Navas, *Quelques nouveaux phénomènes de rang 1 pour les groupes de difféomorphismes du cercle*, Comment. Math. Helv. **80** (2005), no. 2, 355–375.

[20] S. Popa, *On a class of type* II$_1$ *factors with Betti numbers invariants*, Ann. of Math. (2) **163** (2006), no. 3, 809–899.

[21] ———, *Some computations of 1-cohomology groups and construction of non-orbit-equivalent actions*, J. Inst. Math. Jussieu **5** (2006), no. 2, 309–332.

[22] ———, *Strong rigidity of II$_1$ factors arising from malleable actions of w-rigid groups. I*, Invent. Math. **165** (2006), no. 2, 369–408.

[23] ———, *Strong rigidity of II$_1$ factors arising from malleable actions of w-rigid groups. II*, Invent. Math. **165** (2006), no. 2, 409–451.

[24] A. Valette, *Old and new about Kazhdan's property (T)*, Representations of Lie groups and quantum groups (Trento, 1993), Pitman Res. Notes Math. Ser., Vol. 311, Longman Sci. Tech., Harlow, 1994, pp. 271–333.

[25] ———, *Group pairs with property (T), from arithmetic lattices*, Geom. Dedicata **112** (2005), 183–196.

[26] R. J. Zimmer, *Ergodic theory and semisimple groups*, Birkhäuser, Boston, 1984.

12

ANDERS KARLSSON[1] AND FRANÇOIS LEDRAPPIER[2]

NONCOMMUTATIVE ERGODIC THEOREMS

TO ROBERT J. ZIMMER ON THE OCCASION OF HIS 60TH BIRTHDAY

Abstract

We present recent results about the asymptotic behavior of ergodic products of isometries of a metric space X. If we assume that the displacement is integrable, then either there is a sublinear diffusion or there is, for almost every trajectory in X, a preferred direction at the boundary. We discuss the precise statement when X is a proper metric space [KL1] and compare it with classical ergodic theorems. Applications are given to ergodic theorems for nonintegrable functions, random walks on groups, and Brownian motion on covering manifolds.

In this note, we survey some recent results about the asymptotic behavior of ergodic products of 1-Lipschitz mappings of a metric space (X, d). If the mappings are translations on the real line $(\mathbb{R}, |\cdot|)$, then classical ergodic theorems apply, as we recall in Section 1. In more general settings, a suitable generalization of the convergence of averages is the ray approximation property: a typical orbit stays within a $O(\frac{1}{n})$ distance of some (random) geodesic ray ([Pa], [K3], and [KM], see Theorem 6 below). Most of this note is devoted to another generalization, valid in the case when the space $(X.d)$ is proper (see Theorem 7). It also says that there is a (random) direction followed by the typical trajectory, but now a direction is just a point in the metric compactification of (X, d). We discuss in Section 3 how Theorem 7 yields the ray approximation property when the space (X, d) is a CAT(0) metric space, and consequently Oseledets theorem (following [K3]). We give in Section 4 some applications when the space (X, d) is a Gromov hyperbolic space. In particular, by choosing different metrics on $\mathbb{R}$ we directly show some known ergodic theorems for nonintegrable functions. We prove Theorem 7 in Section 5 and give applications to Random Walks in Section 6. Section 6 comes from [KL2], with slightly simpler

1. Royal Swedish Academy of Sciences Research Fellow supported by a grant from the Knut and Alice Wallenberg Foundation. Supported also by the Swedish Research Council.
2. Supported in part by NSF grant DMS-0500630.

proofs. The gist of our results is that for a random walk with first moment on a locally compact group with a proper metric, the Liouville property implies that the linear drift of the random walk, if any, completely comes from a character on the group (see Section 6 for precise statements). This is to be compared with the results of Guivarc'h [G] in the case of connected Lie groups. Since our result applies to discrete groups, it can, through discretization, be applied to Brownian motion on Riemannian covers of finite-volume manifolds. We state in Section 7 the subsequent result from [KL3].

1. Classical Ergodic Theorems

We consider a Lebesgue probability space $(\Omega, \mathcal{A}, \mathbb{P})$, an invertible bimeasurable transformation T of the space $(\Omega, \mathcal{A})$ that preserves the probability $\mathbb{P}$, a function $f : \Omega \to \mathbb{R}$, and we define $S_n(\omega) := \sum_{i=0}^{n-1} f(T^i \omega)$. This setting occurs in particular in statistical mechanics and in mechanics, where Ω is the space of configurations, T the time-1 evolution and $\mathbb{P}$ is either the statistical distribution of states or the Liouville measure on the energy levels. The ergodic hypothesis led to assert that the ergodic averages

$$\frac{1}{n} S_n(\omega) = \frac{1}{n} \sum_{i=0}^{n-1} f(T^i \omega)$$

have some asymptotic regularity.

Around 1930, Koopman suggested that it might be useful to consider the operator U on functions f in $L^2(\Omega, \mathbb{P})$ defined by

$$(Uf)(\omega) = f(T\omega).$$

Since T is measure preserving, the operator U is unitary. The ergodic average then becomes

$$\frac{1}{n} \sum_{k=0}^{n-1} U^k f.$$

The system $(\Omega, \mathcal{A}, \mathbb{P}; T)$ is said to be ergodic if the only functions in L^2 that are invariant under the unitary operator U are the constant functions. In this text, for the sake of exposition, we assume that the system $(\Omega, \mathcal{A}, \mathbb{P}; T)$ is ergodic. Statements for nonergodic systems follow using the decomposition of the measure $\mathbb{P}$ into ergodic components. As an application of the spectral theorem, von Neumann indeed proved:

THEOREM 1. (VON NEUMANN ERGODIC THEOREM, 1931) *Assume that the transformation T is ergodic and that $\int f^2 d\mathbb{P} < \infty$, then*

$$\frac{1}{n} \sum_{k=0}^{n-1} U^k f \rightarrow \int_\Omega f d\mathbb{P}$$

in L^2.

This prompted Birkhoff to prove an almost everywhere convergence theorem:

THEOREM 2. (BIRKHOFF ERGODIC THEOREM, 1931) *Assume that the transformation T is ergodic and that $\int \max(f, 0) d\mathbb{P} < \infty$, then for $\mathbb{P}$-almost every ω, as $n \rightarrow \infty$:*

$$\frac{1}{n} S_n(\omega) \rightarrow \int f d\mathbb{P}.$$

A variant of the ergodic theorem applies to *subadditive* sequences. A sequence S_n of real functions on Ω is said to be subadditive if, for $\mathbb{P}$-almost every ω, all natural integers n, m:

$$S_{n+m}(\omega) \leq S_m(\omega) + S_n(T^m \omega).$$

THEOREM 3. (KINGMAN SUBADDITIVE ERGODIC THEOREM, 1968) *Assume that the transformation T is ergodic, and that $\int \max(S_1, 0) d\mathbb{P} < \infty$, then for $\mathbb{P}$-almost every ω, as $n \rightarrow \infty$:*

$$\frac{1}{n} S_n(\omega) \rightarrow \inf_n \frac{1}{n} \int S_n d\mathbb{P}.$$

Proofs of Theorems 2 and 3 often appeal to some combinatorics of the sequence $S_n(\omega)$ along individual orbits. The following technical lemma was proven by the first author and Margulis:

LEMMA 4. ([KM], PROPOSITION 4.2) *Let S_n be a subadditive sequence on an ergodic dynamical system $(\Omega, \mathcal{A}, \mathbb{P}; T)$. Assume that $\int \max(S_1, 0) d\mathbb{P} < \infty$ and that $\alpha := \inf_n \frac{1}{n} \int S_n d\mathbb{P} > -\infty$. Then, for $\mathbb{P}$ a.e. ω, all $\varepsilon > 0$, there exist $K = K(\omega)$ and an infinite number of instants n such that*

$$S_n(\omega) - S_{n-k}(T^k \omega) \geq (\alpha - \varepsilon)k \text{ for all } k, K \leq k \leq n.$$

In particular, it follows from subadditivity that $\liminf_k \frac{S_k(\omega)}{k} \geq \alpha$. Therefore Theorem 2 follows (in the case $\int f d\mathbb{P} > -\infty$) because in that case, both sequences S_k and $-S_k$ are subadditive and

$$\inf_n \frac{1}{n} \int S_n d\mathbb{P} = \sup_n \frac{1}{n} \int S_n d\mathbb{P} = \int f d\mathbb{P} = \alpha.$$

On the other hand, $\limsup_k \frac{S_k}{k}$ is a T-invariant function which, by sub-additivity, is not bigger than $\limsup_p \frac{1}{pk} \sum_{j=0}^{p-1} S_k(T^{jk}\omega)$. Thus the constant $\limsup_k \frac{S_k}{k}$ is not bigger than $\frac{1}{k} \int S_k d\mathbb{P}$. Theorem 3 follows in the case when $\alpha > -\infty$. To treat the case $\alpha = -\infty$ in both theorems, it suffices to replace S_n by $\max(S_n, -nM)$, and to let M go to infinity; see [Kr] for details.

2. Noncommutative Ergodic Theorems

Observe that Theorem 1 also holds true for any linear operator U of a Hilbert space assuming $\|U\| \leq 1$. One can take this a step further and define for any $g \in \mathcal{H}$, $\phi(g) := Ug + f$. Then ϕ is an isometry (or merely 1-Lipschitz in the case $\|U\| \leq 1$).

Note that

$$\phi^n(0) = \sum_{k=0}^{n-1} U^k f.$$

Pazy proved in [Pa] that more generally for any map $\phi : \mathcal{H} \to \mathcal{H}$ such that $\|\phi(x) - \phi(y)\| \leq \|x - y\|$, it holds that there is a vector $v \in \mathcal{H}$ such that

$$\frac{1}{n} \phi^n(0) \to v$$

in norm. This can be reformulated as follows: there is a unit speed geodesic $\gamma(t) = tv/\|v\|$ in $\mathcal{H}$ such that

1
$$\frac{1}{n} \left\| \phi^n(0) - \gamma(n\|v\|) \right\| = \frac{1}{n} \left\| \phi^n(0) - nv \right\| \to 0.$$

We call this property *ray approximation*. It turns out that this generalization of the ergodic theorem still holds for more general group actions than the actions of $\mathbb{Z}$. Let G be a second countable, locally compact, Hausdorff topological semigroup, and consider $g : \Omega \to G$ a measurable map. We form

$$Z_n(\omega) := g(\omega)g(T\omega) \dots g(T^{n-1}\omega)$$

and ask whether Z_n converges to infinity with some linear speed.

Assume G acts on a metric space (X, d) by 1-Lipschitz transformations. Then for a fixed $x_0 \in X$ we can define for $g \in G$, $|g| := d(x_0, gx_0)$. Clearly, up to a bounded error, $|Z_n(\omega)|$ does not depend on our choice of x_0. We have

PROPOSITION 5. *Assume the transformation T is ergodic, and $\int |g| d\mathbb{P} < \infty$. Then there is a nonnegative number α such that for $\mathbb{P}$-almost every ω, as $n \to \infty$:*

$$\frac{1}{n}|Z_n(\omega)| \to \alpha.$$

The number α is given by

2
$$\alpha = \inf_n \frac{1}{n} \int |Z_n(\omega)| d\mathbb{P}.$$

Proof. It suffices to observe that the sequence $|Z_n(\omega)|$ satisfies the hypotheses of Theorem 3. Our hypothesis says that $\int Z_1 < \infty$. The subaditivity follows from the 1-Lipschitz property:

$$
\begin{aligned}
|Z_{n+m}(\omega)| &= d(x_0, g(\omega) \ldots g(T^{n+m-1}\omega)x_0) \\
&\leq d(x_0, g(\omega) \ldots g(T^{m-1}\omega)x_0) + \\
&\quad + d(g(\omega) \ldots g(T^{m-1}\omega)x_0, g(\omega) \ldots g(T^{n+m-1}\omega)x_0) \\
&\leq |Z_m(\omega)| + d(x_0, g(T^m\omega) \ldots g(T^{n+m-1}\omega)x_0) \\
&= |Z_m(\omega)| + |Z_n(T^m\omega)|.
\end{aligned}
$$

Moreover we see that the limit α is given by $\inf_n \frac{1}{n} \int |Z_n(\omega)| d\mathbb{P}$. □

When $\alpha > 0$, Proposition 5 says that the points $Z_n(\omega)x$ go to infinity with a definite linear speed. The question arises of the convergence in direction of the points $Z_n(\omega)x$. Given Equation (1), we expect that an almost everywhere convergence theorem will say that $Z_n(\omega)x$ will stay at a sublinear distance of a geodesic. We present several results in that direction depending on different geometric hypotheses on the space X.

Assume X is a complete, Busemann nonpositively curved and uniformly convex (e.g., $CAT(0)$ or uniformly convex Banach space) metric space. Then,

THEOREM 6. [KM] *Under these assumptions, there is a constant $\alpha \geq 0$ and, for $\mathbb{P}$-almost every ω, a geodesic ray γ_ω such that*

$$\frac{1}{n}d(Z_n(\omega)x_0, \gamma_\omega(n\alpha)) \to 0.$$

We outline the proof (see [KM], section 5, for details). Let $a(n, \omega) = d(x_0, Z_n(\omega)x_0)$ for each n. Consider a triangle consisting of x_0, $Z_n(\omega)x_0$, and $Z_k(\omega)x_0$. Note that the side of this triangle has lengths $a(n, \omega)$, $a(k, \omega)$, and (at most) $a(n-k, T^k\omega)$. Given $\varepsilon > 0$ (and a.e. ω), for k large it holds that

$a(k, \omega) \leq (\alpha + \varepsilon)k$. Assume now in addition to k being large that n and k are as in Lemma 4. This implies that the triangle is thin in the sense that $Z_k(\omega)x_0$ lies close to the geodesic segment $[x_0, Z_n(\omega)x_0]$; more precisely, the distance is at most $\delta(\varepsilon)a(k, \omega)$, where δ depends only on the geometry. Thanks to the geometric assumptions this $\delta(\varepsilon)$ tends to 0 as ε tends to 0. Selecting ε tending to 0 fast enough we can, by selecting suitable n as in Lemma 4 and some simple geometric arguments, obtain a limiting geodesic. Finally, one has essentially from the construction that as $m \to \infty$, the points $Z_m(\omega)x_0$ lie at a sublinear distance from this geodesic ray.

This note is devoted to the generalization of the ergodic theorem to groups of isometries of a metric space (X, d). We assume that the space (X, d) is *proper* (closed bounded subsets are compact) and we consider the *metric compactification* of X. Define, for $x \in X$, the function $\Phi_x(z)$ on X by

$$\Phi_x(z) = d(x, z) - d(x, x_0).$$

The assignment $x \mapsto \Phi_x$ is continuous, injective, and takes values in a relatively compact set of functions for the topology of uniform convergence on compact subsets of X. The *metric compactification* $\overline{X}$ of X is the closure of X for that topology. The *metric boundary* $\partial X := \overline{X} \setminus X$ is made of Lipschitz continuous functions h on X such that $h(x_0) = 0$. Elements of ∂X are called *horofunctions*. Our main result is the following:

THEOREM 7. (ERGODIC THEOREM FOR ISOMETRIES [KL1]) *Let T be a measure-preserving transformation of the Lebesgue probability space $(\Omega, \mathcal{A}, \mathbb{P})$, G a locally compact group acting by isometries on a proper space X, and $g : \Omega \to G$ a measurable map satisfying $\int |g(\omega)| d\mathbb{P}(\omega) < \infty$. Then, for $\mathbb{P}$-almost every ω, there is some $h_\omega \in \partial X$ such that*

$$\lim_{n \to \infty} -\frac{1}{n} h_\omega(Z_n(\omega)x_0) = \lim_{n \to \infty} \frac{1}{n} d(x_0, Z_n(\omega)x_0).$$

For the convenience of the reader, the proof of Theorem 7 is given in Section 5. We explain in Section 3 why the convergence in Theorem 7 is equivalent to the ray approximation under the $CAT(0)$ assumption. Note that by Theorem 7 the former convergence holds for all norms on $\mathbb{R}^d$, but that Theorem 7 does not apply to infinite-dimensional Banach spaces. In this case, one can use Lemma 4 to prove a noncommutative ergodic theorem with linear functionals of norm 1, somewhat analogous to horofunctions. Namely,

THEOREM 8. [Ka] *Let $Z_n(\omega)$ be an ergodic integrable cocycle of 1-Lipschitz self-maps of a reflexive Banach space. Then for $\mathbb{P}$-almost every ω there is a linear functional f_ω of norm 1 such that*

$$\lim_{n \to \infty} \frac{1}{n} f_\omega(Z_n(\omega)0) = \alpha.$$

On the other hand, Kohlberg-Neyman [KN] found a counterexample to the norm convergence, or more precisely to (1), for general Banach spaces.

3. Case When X Is a $CAT(0)$ Proper Space

When the space (X, d) is a proper $CAT(0)$ metric space, both Theorems 6 and 7 apply. Because it is a direct generalization of the important case when G is a linear group, it is often called the Oseledets theorem. In this section we explain how to recover the ray approximation and other more familiar forms of Oseledets theorem from Theorem 7. Many of the geometric ideas in this section go back to Kaimanovich's extension of Oseledets theorem to more general semisimple groups [K3].

A metric geodesic space (X, d) is called a $CAT(0)$ space if its geodesic triangles are thinner than in the Euclidean space. Namely, consider four points A, B, C, and $D \in X$, D lying on a length-minimizing geodesic going from B to C. Draw four points A', B', C', and D' in the Euclidean plane with $AB = A'B'$, $BD = B'D'$, $DC = D'C'$, and $CA = C'A'$. The space is called $CAT(0)$ if for any such configuration, $AD \leq A'D'$. Simply connected Riemannian spaces with nonpositive curvature, locally finite trees, and Euclidean buildings are proper $CAT(0)$ spaces. If X is a $CAT(0)$ space, then the horofunctions $h \in \partial X$ are called Busemann functions, and for any $h \in \partial X$, there is a unique geodesic ray $\sigma_h(t), t \geq 0$ such that $\sigma_h(0) = x_0$ and $\lim_{t \to \infty} \Phi_{\sigma_h(t)} = h$. We have

COROLLARY 9. *Assume moreover that X is a $CAT(0)$ space and that $\alpha > 0$. Then, for $\mathbb{P}$-almost every ω, as n goes to ∞,*

$$\lim_n \frac{1}{n} d\left(Z_n(\omega)x_0, \sigma_{h_\omega}(\alpha n)\right) = 0,$$

where h_ω is given by Theorem 7.

Proof. Consider a geodesic triangle $A = Z_n(\omega)x_0$, $B = x_0$, $C_t = \sigma_{h_\omega}(t)$, for t very large, and choose $D = \sigma_{h_\omega}(n\alpha)$. We want to estimate the distance AD.

We have

$$AB = d(Z_n(\omega)x_0, x_0) = |Z_n(\omega)| =: n\alpha_n(\omega);$$

$$BC_t = t, \ BD = n\alpha; \text{ and}$$

$$C_t A = t + \Phi_{\sigma_{h_\omega}(t)}(Z_n(\omega)x_0) =: t - n\beta_n(\omega) + o_n(t).$$

For almost every ω, we have

- $\lim_n \alpha_n(\omega) = \alpha$ by Theorem 3;
- $\lim_n \beta_n(\omega) = \lim_n -\frac{1}{n}h_\omega(Z_n(\omega)x_0) = \alpha$ by Theorem 7; and
- for a fixed n, $\lim_{t\to\infty} o_n(t) = h_\omega(Z_n(\omega)x_0) - h_{\sigma_{h_\omega}(t)}(Z_n(\omega)x_0) = 0$.

Construct the comparison figure $A'B'C'_tD'$, and let t go to ∞. The point E'_t of $B'C'_t$ at the same distance from C'_t than A' converges to the orthogonal projection E'_∞ of A' on $B'C'_t$ and satisfies $B'E'_t = n\beta_n - o_n(t)$. Therefore, $B'E'_\infty = n\beta_n$. We have

$$(AE'_\infty)^2 = n^2(\alpha_n^2 - \beta_n^2), \quad (D'E'_\infty)^2 = n^2(\beta_n - \alpha)^2,$$

and therefore, as $n \to \infty$:

$$\lim_n \frac{1}{n^2}(A'D')^2 = \lim_n \left((\alpha_n^2 - \beta_n^2) + (\beta_n - \alpha)^2\right) = 0. \qquad \square$$

COROLLARY 10. *With the same assumptions, we have, for $\mathbb{P}$-almost every ω, $Z_n(\omega)x_0$ converges to h_ω in $\overline{X}$.*

In particular, when $\alpha > 0$ and X is proper $CAT(0)$, the direction h_ω given by Theorem 7 is unique.

Proof. In the above triangle, the geodesic σ_n joining x_0 to $Z_n(\omega)x_0$ converges to σ_{h_ω}. Therefore, all the accumulation points of $Z_n(\omega)x_0$ belong to the set seen from x_0 in the direction of h_ω. By the same proof, all the accumulation points of $Z_n(\omega)x_0$ belong to the set seen from $\sigma_{h_\omega}(K)$ in the direction of h_ω, for all K. As K goes to infinity, the intersection of those sets is reduced to the point h_ω. $\qquad \square$

In the case when G is a linear group, Corollary 9 is closely related to the well-known

THEOREM 11. (OSELEDETS MULTIPLICATIVE ERGODIC THEOREM [O], *1968) Let T be an ergodic transformation of the probability space $(\Omega, \mathcal{A}, \mathbb{P})$, and A :*

$\Omega \to GL(d, \mathbb{R})$ *a measurable map such that* $\int \max\{\ln \|A\|, \ln \|A^{-1}\|\} d\mathbb{P} < \infty$.
Then there exist

- *real numbers* $\lambda_1 \leq \lambda_2 \leq \cdots \leq \lambda_k$; *and*
- *integers* $m_i, i = 1, \ldots, k$ *with* $\sum_i m_i = d, \sum_i \lambda_i m_i = \int \ln |DetA| d\mathbb{P}$,
- *for* $\mathbb{P}$-*almost every* ω, *a flag of subspaces of* $\mathbb{R}^d$

$$\{0\} = V_{k+1}(\omega) \subset V_k(\omega) \subset \cdots \subset V_1(\omega) = \mathbb{R}^d$$

with, for all $i, 1 \leq i \leq k$, $DimV_i = \sum_{j \geq i} m_j$ *and a vector* v *belongs to* $V_i(\omega) \setminus V_{i+1}(\omega)$ *if and only if as n goes to* ∞,

$$\lim \frac{1}{n} \ln \|A(T^{n-1}\omega)A(T^{n-2}\omega)\ldots A(\omega)v\| = \lambda_i.$$

Observe that, automatically, the V_i depend measurably on ω and are invariant in the sense that $A(\omega)V_i(\omega) = V_i(T\omega)$. The usual complete form of Oseledets theorem follows by comparing the results of Theorem 11 for (T, A) and for $(T^{-1}, A^{-1} \circ T^{-1})$. Fix $\omega \in \Omega$, and let $e_i, i = 1, \ldots, d$ be an orthogonal base of $\mathbb{R}^d$ such that $e_\ell \in V_i(\omega)$ as soon as $\ell \leq \sum_{j \geq i} m_j$. Write $\mu_1 \geq \cdots \geq \mu_d$ for the exponents λ_j, each counted with multiplicity m_j, and consider $A^{(n)}(\omega) := A(T^{n-1}\omega)A(T^{n-2}\omega)\ldots A(\omega)$ in the base (e_i). To verify the statement of Theorem 11, it suffices to show that for all $\varepsilon > 0$ and for n large enough,

$$|A_{i,j}^{(n)}(\omega)| \leq e^{n(\mu_i + \varepsilon)} \text{ and } |\ln |DetA^{(n)}(\omega)| - \sum_j \mu_j| \leq \varepsilon.$$

With the notations of Section 2, consider the action by isometries of $GL(d, \mathbb{R})$ on the symmetric space $GL(d, \mathbb{R})/O(d, \mathbb{R})$ with origin $x_0 = O(d, \mathbb{R})$ and distance $|g| = \sqrt{\sum_{j=1}^d (\ln \tau_i)^2}$, where τ_i are the eigenvalues of gg^t. It is a $CAT(0)$ geodesic proper space. Set $g(\omega) = A^t(\omega)$. The moment hypothesis $\int |g| d\mathbb{P} < \infty$ is satisfied. We have $A^{(n)}(\omega) = (Z_n(\omega))^t$. If $\alpha = 0$, then the eigenvalues of $Z_n Z_n^t$ grow subexponentially and

$$\lim_n \frac{1}{n} \ln \|A^{(n)}(\omega)v\| = \frac{1}{2} \lim_n \frac{1}{n} \ln (\|Z_n^t v\|^2) = 0.$$

In this case $m_1 = d, \lambda_1 = 0$ and Theorem 11 holds. We may assume $\alpha > 0$, and apply Corollary 9.

Geodesics starting from the origin are of the form e^{tH}, where H is a nonzero symmetric matrix. Therefore, for $\mathbb{P}$-almost every ω, there is a nonzero symmetric matrix $H(\omega)$ such that $\frac{1}{n}d(exp(nH(\omega)), Z_n(\omega))$ goes to 0 as $n \to \infty$ (the constant α has been incorporated in H). In other words, $\frac{1}{n} \ln$ of the norm, and of the norm of the inverse, of the matrix $exp(-nH(\omega))(A^{(n)}(\omega))^t$, go to 0 as $n \to \infty$. We claim that this gives the conclusion of Theorem 11 with λ_i the eigenvalues of $H(\omega)$, m_i their respective multiplicities and V_i the sums of the eigenspaces corresponding to eigenvalues smaller than λ_i. Indeed, we write $exp(H(\omega)) = K(\omega)^t \Delta K(\omega)$ for K an orthogonal matrix and Δ a diagonal matrix with diagonal entries e^{μ_i}, and $A^{(n)}(\omega) = L_n(\omega)\Delta_n(\omega)K_n(\omega)$ a Cartan decomposition of $A^{(n)}$ with L_n, K_n orthogonal, Δ_n a diagonal matrix with nonincreasing diagonal entries $exp(n\delta_1^{(n)}(\omega)) \geq \cdots \geq exp(n\delta_d^{(n)}(\omega))$. The conclusion of Corollary 9 is therefore that, for $\mathbb{P}$-almost every ω, $\frac{1}{n}$ ln of the norm, and of the norm of the inverse, of the matrix $\Delta_n(\omega)K_n(\omega)K^t(\omega)exp(\quad n\Delta)$, go to 0 as $n \to \infty$.

It follows that, for such an ω, $|\ln|DetA^{(n)}(\omega)| - \sum_j \mu_j|$ goes to 0 as n goes to ∞. Furthermore, for n large enough, the entries $k_{i,j}^{(n)}(\omega)$ of the matrix $K_n(\omega)K^t(\omega)$ satisfy

$$|k_{i,j}^{(n)}(\omega)| \leq e^{n(\mu_j - \delta_i^{(n)} + \varepsilon)}.$$

We have

$$\|A^{(n)}(\omega)e_j\| = \|L_n(\omega)\Delta_n(\omega)K_n(\omega)e_j\| = \|\Delta_n(\omega)K_n(\omega)K^{-1}(\omega)f_j\|,$$

where f_i is the canonical base of $\mathbb{R}^d$. The components of this vector are $e^{n\delta_i^{(n)}(\omega)}k_{i,j}^{(n)}(\omega)$. Their absolute values are indeed smaller than $e^{n(\mu_j+\varepsilon)}$ for n large enough.

4. The Case When X Is a Gromov Hyperbolic Space (in Particular, $\mathbb{R}$).

Theorem 7 is due to Kaimanovich using an idea of Delzant when X is a Gromov hyperbolic geodesic space even without the condition that X is a proper space [K2]. As in the $CAT(0)$-case it is there formulated as Z_n lies on sublinear distance of a geodesic ray. From Theorem 7 one gets the following:

COROLLARY 12. *Assume moreover that X is a Gromov hyperbolic geodesic space and that $\alpha > 0$. Then, for $\mathbb{P}$-almost every ω, as n goes to ∞, there is a geodesic ray σ_ω such that*

$$\lim_n \frac{1}{n}d(Z_n(\omega)x_0, \sigma_{h_\omega}(\alpha n)) = 0.$$

Proof. Take h_ω given from Theorem 7. It is known, see [BH, p. 428], that for Gromov hyperbolic geodesic spaces it holds that there is a geodesic ray σ_ω such that $\sigma_\omega(x_0) = 0$ and

$$b_\omega(\cdot) = \lim_{t\to\infty} d(\cdot, \sigma_\omega(t)) - t$$

is a horofunction such that $\left|b_\omega(\cdot) - h_\omega(\cdot)\right| \le C$ for some constant C. This b_ω therefore clearly satisfies the conclusion of Theorem 7.

Now we use the notation and setup in the proof of Corollary 9. Consider the triangle ABC_t. By δ-hyperbolicity D must lie at most δ away from either AB or AC_t. Call the closest point X. By the triangle inequality we must have that

$$\alpha n - \delta \le XB \le \alpha n + \delta.$$

If X lies on AB, then it is clear that $XA = o(n)$ and hence $AD = o(n)$. If X lies on AC_t, then

$$t \quad \alpha n \quad \delta \le XC_t \le t \quad \alpha n \mid \delta.$$

In view of that $b_\omega(Z_n(\omega)) \approx -\alpha n$ we again reach the conclusion that X, and hence also D, lie on sublinear distance from A. $\qquad\square$

COROLLARY 13. *With the same assumptions, we have that for $\mathbb{P}$-almost every ω, $Z_n(\omega)x_0$ converges to the point $[\sigma_\omega]$ in the hyperbolic boundary $\partial_{hyp}X$.*

Proof. Clearly, the Gromov product $(Z_n(\omega), \sigma_\omega(\alpha n)) \to \infty$ as $n \to \infty$ in view of the previous corollary. $\qquad\square$

In the case when $G = \mathbb{R}$ and $X = (\mathbb{R}, |\cdot|)$, Corollaries 12 and 13 yield Theorem 2. Indeed, in this case the drift is

$$\alpha = \left| \int_\Omega f d\mathbb{P} \right|$$

and $\partial\mathbb{R} = \{h_+ = \Phi_{+\infty}(z) = -z, h_- = \Phi_{-\infty}(z) = z\}$. It follows from Corollary 13 that the index of h_ω is T invariant and is therefore almost everywhere constant. In other words, the existence of the h_ω with the required property amounts to the choice of the right sign:

$$h_\omega = \Phi_{\text{sign}\{\int_\Omega f(\omega)d\mathbb{P}(\omega)\}\infty}.$$

Then, Corollary 12 says exactly that if the function f is integrable, for $\mathbb{P}$-almost every ω

$$\frac{1}{n} S_n(\omega) \to \int_\Omega f(\omega)d\mathbb{P}(\omega).$$

The above observation is not a new proof of Theorem 2, because Theorem 2 is used in the proof of Theorem 7 (see Section 5). We want to only illustrate the meaning of the metric boundary on the simplest example. Nevertheless, it turns out that modifying the translation-invariant metric on $X = \mathbb{R}$ might have interesting consequences. The following discussion comes from [KMo], which in turn was inspired by [LL].

Let $D : \mathbb{R}_{\geq 0} \to \mathbb{R}_{\geq 0}$ be an increasing function, $D(t) \to \infty$ such that $D(0) = 0$ and $D(t)/t \to 0$ monotonically. From the inequality

$$\frac{1}{t+s} D(t+s) \leq \frac{1}{t} D(t)$$

we get the following subadditivity property:

$$D(t+s) \leq D(t) + \frac{s}{t}D(t) = D(t) + \frac{D(t)/t}{D(s)/s}D(s) \leq D(t) + D(s).$$

From all these properties of D, it follows that $(\mathbb{R}, D(|\cdot|))$ is a proper metric space, and clearly invariant under translations.

Now we determine $\partial\mathbb{R}$ with respect to this metric. Without loss of generality we may assume that $x_n \to \infty$. We claim that for any z

$$h(z) = \lim_{n\to\infty} D(x_n - z) - D(x_n) = 0.$$

Assume not. Then for some $s > 0$ and an infinite sequence of $t \to \infty$ that $D(t+s) - D(t) > c > 0$ (wlog). For such s and t with t large so that $D(t)/t < c/s$, we have

$$\frac{D(t+s)}{t+s} \geq \frac{D(t)+c}{t+s} \geq \frac{D(t) + \frac{D(t)}{t}s}{t+s} = \frac{D(t)}{t}$$

but this contradicts that $D(t)/t$ is strictly decreasing. Hence $\partial\mathbb{R} = \{h \equiv 0\}$.

Applying Theorem 7 in this setting yields a result already obtained by Aaronson with a different argument.

THEOREM 14. (AARONSON [A]) *Let $f : \Omega \to \mathbb{R}$ such that $\int_\Omega D(|f|)d\mu < \infty$. Then, for $\mathbb{P}$-almost every ω,*

$$\lim_{n\to\infty} \frac{1}{n} D\left(|S_n(\omega)|\right) = 0.$$

Proof. It was noted above that $\partial(\mathbb{R}, D(|\cdot|))$ only consisted of $h = 0$. The conclusion then follows from Theorem 7 since $h = 0$ forces $\alpha = 0$. $\square$

One can relax the conditions on D: for one thing, one can remove having $D(0) = 0$. More interestingly, the condition that $D(t)/t$ decreases to 0 can be weakened in the following way.

COROLLARY 15. (AARONSON-WEISS [A]) *Let $d(t)$ be an increasing positive function $d(t) \to \infty$, such that $d(t) = o(t)$, $d(t+s) \leq d(t) + d(s)$ and $\int_\Omega d(|f|)d\mu < \infty$ for some function $f : \Omega \to \mathbb{R}$. Then, for $\mathbb{P}$-almost every ω,*

$$\lim_{n \to \infty} \frac{1}{n} d\left(|S_n(\omega)|\right) = 0.$$

Proof. Define

$$D(t) = \sup\{d(ut)/u : u \geq 1\}.$$

Note that this satisfies all the assumptions made on D in Theorem 14. Moreover,

$$d(t) \leq D(t) \leq 2d(t),$$

since if $D(t) = d(tu)/u$, set $n = [u] + 1$ and then $D(u) \leq d(nt)/u \leq nd(t)/u \leq 2d(t)$. (See [A], p. 66, for more details.) This shows that Theorem 14 actually holds for d in place of D. $\qquad\square$

In particular, Corollary 15 applies to any metric $d(.,.)$ on $\mathbb{R}$ where balls grow superlinearly (where $d(t) := d(0,t)$). From this one obtains as a special case classical results like the one of Marcinkiewicz-Zygmund [MZ] and Sawyer [S]:

COROLLARY 16. *Let $0 < p < 1$. If $f \in L^p$, then for $\mathbb{P}$-almost every ω*

$$\lim_{n \to \infty} \frac{1}{n^{1/p}} S_n = 0.$$

Such moment conditions arise naturally in probability theory. These results are known to be best possible in certain ways (e.g., [S] and [A]). For the iid case the converse also holds [MZ]. Another example is as follows:

COROLLARY 17. *If f is log-integrable, then for $\mathbb{P}$-almost every ω*

$$\lim_{n \to \infty} |S_n|^{1/n} = 1.$$

One can modify the metric on any metric space X in the same way replacing $d(x, y)$ with $D(d(x, y))$, where $D(t)$ satisfies the assumptions for Theorem 14 or, more generally, the assumptions for Corollary 15. By estimating a subadditive by an additive cocycle in the obvious way

$$a(n, \omega) \leq \sum_{k=0}^{n-1} a(1, T^k \omega),$$

and Theorem 14 implies that

$$\frac{1}{n} D(d(Z_n x_0, x_0)) \rightarrow 0 \text{ a.e.,}$$

under the condition that $D(d(g(\omega)x_0, x_0))$ is integrable.

5. Proof of Theorem 7

We begin by a few observations: first, we can extend by continuity the action of G to $\overline{X}$, and write, for $h \in \overline{X}, g \in G$:

$$g.h(z) = h(g^{-1}z) - h(g^{-1}x_0).$$

Now define the skew product action on $\overline{\Omega} := \Omega \times \overline{X}$ by

$$\overline{T}(\omega, h) = (T\omega, g(\omega)^{-1} \cdot h).$$

Observe that $\overline{T}^n(\omega, h) = (T^n \omega, (Z_n(\omega))^{-1} \cdot h)$. Define the Furstenberg cocycle $\overline{F}(\omega, h)$ by $\overline{F}(\omega, h) := -h(g(\omega)x_0)$. We have

3
$$\overline{F}_n(\omega, h) := \sum_{i=0}^{n-1} \overline{F}(\overline{T}^i(\omega, h)) = -h(Z_n(\omega)x_0).$$

Relation (3) is proven by induction on n. We have $\overline{F}_1(\omega, h) := -h(g(\omega)x_0) = -h(Z_1(\omega)x_0)$ and

$$\overline{F}_n(\omega, h) = \overline{F}_{n-1}(\omega, h) + \overline{F}(\overline{T}^{n-1}(\omega, h))$$
$$= -h(Z_{n-1}(\omega)x_0) - (Z_{n-1}(\omega))^{-1} \cdot h(g(T^{n-1}\omega)x_0)$$
$$= -h(Z_{n-1}(\omega)x_0) - h(Z_{n-1}(\omega)g(T^{n-1}(\omega))x_0) + h(Z_{n-1}(\omega)x_0)$$
$$= -h(Z_n(\omega)x_0).$$

In particular, for any $\overline{T}$ invariant measure m on $\overline{\Omega}$ such that the projection on Ω is $\mathbb{P}$, we have $\int \overline{F}dm \leq \alpha$ because

$$\int \overline{F}(\omega, h)dm(\omega, h) = \frac{1}{n} \int -h(Z_n(\omega)x_0)dm(\omega, h) \leq \frac{1}{n} \int |Z_n(\omega)|d\mathbb{P}(\omega).$$

There is nothing to prove if $\alpha = 0$. To prove Theorem 7 in the case $\alpha > 0$, it suffices to construct a $\overline{T}$-invariant measure m on $\overline{\Omega}$ such that the projection on Ω is $\mathbb{P}$ and such that $\int \overline{F}(\omega, h)dm(\omega, h) = \alpha$. Indeed, since α is the largest possible value of $\int \overline{F}$, we still have the same equality for almost every ergodic component of m. By the ergodic Theorem 2, the set A of (ω, h) such that $-\frac{1}{n}h(Z_n(\omega)x_0) = \frac{1}{n}\overline{F}_n(\omega, h) \to \alpha$ as n goes to ∞ has full measure. Moreover, observe that if h is not a point in ∂X, $-\frac{1}{n}h_\gamma(Z_n(\omega)x_0)$ converges to $-\alpha$. Since $\alpha > 0$, this shows that $A \subset \partial X$. We get the conclusion of Theorem 7 by choosing for $\omega \mapsto h_\omega$ a measurable section of the set A.

We finally construct a measure m with those properties. We define a measure μ_n on $\overline{\Omega}$; for any measurable function Ξ on $\overline{\Omega}$ such that

$$\int \sup_{h \in \overline{X}} |\Xi(\omega, h)| \, d\mathbb{P}(\omega) < \infty,$$

we set:

$$\int_{\Omega \times \overline{X}} \Xi(\omega, h)d\mu_n(\omega, h) = \int_{\Omega} \Xi(\omega, \Phi_{Z_n(\omega)x_0})d\mathbb{P}(\omega).$$

The set of measures m on $\overline{\Omega}$ such that the projection on Ω is $\mathbb{P}$ is a convex compact subset of $L^\infty(\Omega, \mathcal{P}(\overline{X})) = (L^1(\Omega, C(\overline{X})))^*$ for the weak* topology. The mapping $m \mapsto (\overline{T})_*m$ is affine and continuous. We can take for m any weak* limit point of the sequence:

$$\eta_n = \frac{1}{n}\sum_{i=0}^{n-1}(\overline{T}^i)_*\mu_n.$$

The measure m is $\overline{T}$-invariant and, since

$$\|\overline{F}\|_{L^1(\Omega, C(\overline{X}))} = \int \sup_h |\overline{F}(\omega, h)|d\mathbb{P}(\omega) = \int \sup_h |h(g(\omega)x_0)|d\mathbb{P}(\omega) < +\infty,$$

we may write, using Relation (3) and Formula (2):

$$\int \overline{F}dm = \lim_{k \to \infty} \frac{1}{n_k} \int \sum_{i=0}^{n_k-1}(\overline{F} \circ \overline{T}^i)d\mu_{n_k}$$

$$= \lim_{k \to \infty} \frac{1}{n_k} \int \overline{F}_{n_k}(\omega, \Phi_{Z_{n_k}(\omega)x_0})d\mathbb{P}(\omega)$$

$$= \lim_{k \to \infty} \frac{1}{n_k} \int (-\Phi_{Z_{n_k}(\omega)x_0}(Z_{n_k}(\omega)x_0))d\mathbb{P}(\omega)$$

$$= \lim_{k \to \infty} \frac{1}{n_k} \int |Z_{n_k}(\omega)|d\mathbb{P}(\omega) = \alpha.$$

By the above discussion this achieves the proof of Theorem 7.

Observe that by putting together the discussions in Sections 5 and 3, we obtain a proof of Oseledets Theorem 11. As proofs of Theorem 11 go, this one is in some sense rather close to the original one [O], with the somewhat simplifying use of the geometric ideas from [K3] and invariant measures as in [W].

6. Random Walks

In this section we consider a probability v on a group G and apply the preceeding analysis to the random walk $Z_n = g_0 g_1 \ldots g_{n-1}$, where the g_i are independent with distribution v. We assume

- there is a proper left-invariant metric d on G that generates the topology of G (when G is second-countable locally compact, such a metric always exists, see [St]);
- $\int d(e, g) dv(g) < +\infty$ (we say that v has a *first moment*); and
- the closed subgroup generated by the support of v is the whole G (we say that v is *nondegenerate*).

Then there is a number $\ell(v) \geq 0$ such that, for almost every sequence $\{g_i\}$, $\lim_n \frac{1}{n} d(e, Z_n) = \ell(v)$. In the case when the group G is the group $SL_2(\mathbb{R})$ acting on the hyperbolic plane, $\ell(v)$ is twice the Lyapunov exponent of the independent product of matrices. In that case it is given by a formula involving the stationary measure on the circle, the Furstenberg-Khasminskii formula ([F1]; this appellation seems to be standard, cf. [Ar]). Seeing again the circle as the geometric boundary of the hyperbolic plane, we extend this formula to our general context:

THEOREM 18. (FURSTENBERG-KHASMINSKII FORMULA FOR THE LINEAR DRIFT, [KL2]) *Let (G, v) verify all the above assumptions, and let $\overline{G}$ be the metric compactification of (G, d). Then there exists a measure μ on $\overline{G}$ with the following properties:*

- *μ is stationary for the action of G, that is, μ satisfies $\mu = \int (g_* \mu) dv(g)$; and*
- *$\ell(v) = \int h(g^{-1}) d\mu(h) dv(g)$.*

Moreover, if $\ell(v) > 0$, then μ is supported on ∂G.

Proof. In the proof of Theorem 7, we constructed a measure m on $\Omega \times \overline{G}$. The measure μ can be seen as the projection on $\overline{G}$ of m, but it turns out that the measure μ can be directly constructed. Let $(\Omega^+, \mathcal{A}^+, \mathbb{P})$ be the space of sequences $\{g_0, g_1, \dots\}$ with product topology, σ-algebra, and measure $\mathbb{P} = \nu^{\otimes \mathbb{N}}$. For $n \geq 0$, let ν_n be the distribution of $Z_n(\omega)$ in $\overline{G}$. In other words, define, for any continuous function f on $\overline{G}$:

$$\int f d\nu_n = \int f(g_0 g_1 \cdots g_{n-1}) d\nu(g_0) d\nu(g_1) \cdots d\nu(g_{n-1}), \quad \nu_0 = \delta_e.$$

We claim that any weak* limit μ of the measures $\frac{1}{n} \sum_{i=0}^{n-1} \nu_i$ satisfies the conclusions of Theorem 18. Clearly, the measure μ is stationary: for any continuous function f on $\overline{G}$, we have

$$\int f(g.h) d\mu(h) d\nu(g) = \lim_{k \to \infty} \frac{1}{n_k} \sum_{i=0}^{n_k-1} \int f(gg_0 g_1 \cdots g_{i-1})$$

$$\times d\nu(g_0) d\nu(g_1) \cdots d\nu(g_{i-1}) d\nu(g)$$

$$= \lim_{k \to \infty} \frac{1}{n_k} \sum_{i=0}^{n_k-1} \int f d\nu_{i+1}$$

$$= \int f d\mu + \lim_{k \to \infty} \frac{1}{n_k} [\int f d\nu_{n_k} - f(e)]$$

$$= \int f d\mu.$$

In the same way, we get

$$\int h(g^{-1}) d\mu(h) d\nu(g) = \lim_{k \to \infty} \frac{1}{n_k} \sum_{i=0}^{n_k-1} \int [d(Z_i(\omega), g^{-1})$$

$$- d(Z_i(\omega), e)] d\mathbb{P}(\omega) d\nu(g)$$

$$= \lim_{k \to \infty} \frac{1}{n_k} \int d(Z_{n_k}, e) d\mathbb{P}(\omega)$$

$$= \ell(\nu).$$

This shows that the measure μ has the desired properties. Moreover, the measure $\mathbb{P} \times \mu$ on the space $\Omega^+ \times \overline{G}$ is $\overline{T}$-invariant. There is a unique $\overline{T}$-invariant measure m on $\Omega \times \overline{G}$ that extends $\mathbb{P} \times \mu$. The measure m satisfies all the properties we needed in the proof of Theorem 7. In particular, if $\ell(\nu)$ is positive,

$$\mu(\partial G) = (\mathbb{P} \times \mu)(\Omega^+ \times \partial G) = m(\Omega \times \partial G) = 1. \qquad \square$$

A bounded measurable $f : G \to \mathbb{R}$ is ν-*harmonic* if

$$f(g) = \int_G f(gh)d\nu(h)$$

for any $g \in G$. Constant functions are obviously ν-harmonic. If f is a bounded Harmonic function, then $f(Z_n)$ is a bounded Martingale and therefore converges almost surely. We say that (G, ν) satisfies the *Liouville property* (or (G, ν) is Liouville) if the constant functions are the only bounded ν-harmonic functions.

COROLLARY 19. [KL2] *Let G be a locally compact group with a left-invariant proper metric and ν be a nondegenerate probability measure on G with first moment. Then, if (G, ν) is Liouville, there is a 1-Lipschitz homomorphism $T : G \to \mathbb{R}$ such that for almost every trajectory Z_n of the corresponding random walk, we have*

$$\lim_{n \to \infty} \frac{1}{n} T(Z_n) = \int_G T(g)d\nu(g) = l(\nu).$$

Proof. The key observation is that if (G, ν) is Liouville and G acts continuously on a compact space Y, then every stationary measure μ is invariant. Indeed, for $f \in C(Y, \mathbb{R})$, the function $\varphi(g) := \int f d(g_* \mu)$ is harmonic and bounded, and therefore constant. In particular, the measure μ from Theorem 18 is invariant, and if we set

$$T(g) := \int h(g^{-1})\mu(dh),$$

the mapping T is Lipschitz continuous and is a group homomorphism because we have

$$T(g'g) = \int h(g^{-1}g'^{-1})\mu(dh)$$

$$= \int (g'.h)(g^{-1})\mu(dh) + \int h(g'^{-1})\mu(dh)$$

$$= \int h(g^{-1})(g'_*\mu)(dh) + T(g')$$

$$= T(g) + T(g'),$$

where we used the invariance of μ at the last line. Finally, by the Furstenberg-Khasminskii formula, we have

$$\ell(\nu) = \int T(g)d\mu(g). \qquad \square$$

A measure ν on G is called *symmetric* if it is invariant under the mapping $g \mapsto g^{-1}$. A measure is *centered* if every homomorphism of G into $\mathbb{R}$ is centered, meaning that the ν-weighted mean value of the image is 0. Every symmetric measure with first moment ν is centered, since for any homomorphism $T : G \to \mathbb{R}$, the mean value, which is

$$\int_G T(g)d\nu(g) = \int_G T(g^{-1})d\nu(g) = -\int_G T(g)d\nu(g),$$

must hence equal 0. By simple contraposition from Corollary 19, we get

COROLLARY 20. [KL2] *Let G be a locally compact group with a left-invariant proper metric and ν be a nondegenerate centered probability measure on G with first moment. Then, if $l(\nu) > 0$, there exist nonconstant bounded ν-harmonic functions.*

Corollary 20 was known in particular for ν with finite support ([Va], [M]) or in the continuous case, for ν with compact support and density [Al].

One case when all probability measures on G are centered is when there is no group homomorphism from G to $\mathbb{R}$. We can apply Corollary 20 to a countable finitely generated group. Let S be a finite symmetric generator for G, and endow G with the left-invariant metric $d(x, y) = |y^{-1}x|$ where $|z|$ is the shortest length of an S-word representing z. We say that G has subexponential growth if $\lim_n \frac{1}{n} \ln A_n = 0$, where A_n is the number of elements z of G with $|z| \le n$. Such a group has automatically the Liouville property [Av]. This yields

COROLLARY 21. [KL2] *Let G be a finitely generated group with subexponential growth and $H^1(G, \mathbb{R}) = 0$. Then for any nondegenerate ν on G with first moment, we have $\ell(\nu) = 0$.*

Observe that conversely, if there exists a nontrivial group homomorphism T from a finitely generated group into $\mathbb{R}$, then there exists a nondegenerate probability ν on G, with first moment and $\ell(\nu) > 0$. Indeed, there exists M such that $[-M, M]$ contains all the images of the elements of the generating set S. We can choose ν carried by all the elements of S with images in $[0, M]$. Since T is nontrivial, $\int T(g)d\nu(g) > 0$. The measure ν is nondegenerate, has finite support, and $\ell(\nu) > 0$ since for all $g \in G, |T(g)| \le M|g|$.

7. Riemannian Covers

In this section we consider a complete connected Riemannian manifold (M, g) with bounded sectional curvatures. In particular, if d_M is the Riemannian distance on M, (M, d) is a proper space. Associated to the metric is the Laplace-Beltrami operator Δ. A function f is *harmonic* if $\Delta f = 0$. We say that M is *Liouville* if all bounded and harmonic functions are constant.

Associated to Δ is a diffusion process B_t called *Brownian motion*. Since the curvature is bounded and M is complete, the Brownian motion is defined for all time. For all $x \in M$, there is a probability $\mathbb{P}_x$ on $C(\mathbb{R}_+, M)$ such that the process B_t given by the t coordinate is a Markov process with generator Δ and $B_0 = x$. We can define

$$\ell_g := \limsup_{t \to \infty} \frac{1}{t} d_M(x_0, B_t),$$

for any $x_0 \in M$.

THEOREM 22. [KL3] *Assume that (M, g) is a regular covering of a Riemannian manifold that has finite Riemannian volume and bounded sectional curvatures. Then M is Liouville if and only if*

$$\lim_{t \to \infty} \frac{1}{t} d(x_0, B_t) = 0 \ a.s.$$

The "if" part was proved by Kaimanovich, see [K1], and the converse is clear if the Brownian motion is recurrent on M. The proof of the new implication in Theorem 22 in the transient case uses the Furstenberg-Lyons-Sullivan discretization procedure. Let Γ be the covering group of isometries of M. This discretization consists in the construction of a probability measure ν on Γ, with the following properties:

- the restriction $f(\gamma) := F(\gamma x_0)$ is a one-to-one correspondence between bounded harmonic functions on M and bounded functions on Γ that satisfy

$$f(\gamma) = \sum_{g \in \Gamma} f(\gamma g) \nu(g);$$

- if $\gamma_1, \ldots, \gamma_n$ are chosen independent and with distribution ν, then $\lim_{n \to \infty} \frac{1}{n} d_M(x_0, \gamma_1 \ldots \gamma_n x_0)$ exists. It vanishes a.e. if and only if $\lim_{t \to \infty} \frac{1}{t} d_M(x_0, B_t) = 0$ a.s.; and
- in the case when the Brownian motion is transient, one can choose ν symmetric, that is, such that for all γ in Γ, $\nu(\gamma^{-1}) = \nu(\gamma)$.

The first property goes back to Furstenberg [F2] and has been systematically developed by Lyons and Sullivan [LS] and Kaimanovich [K5]. The second one was observed in certain situations by Guivarc'h [G] and Ballmann [Ba]. Babillot observed that the modified construction of [BL] has the symmetry property. Given the above, proving Theorem 22 mostly reduces to Corollary 20, if we can show that hypotheses of Corollary 20 are satisfied. We endow Γ with the metric defined by the metric of M on the orbit Γx_0. This defines a left-invariant and proper metric on Γ: bounded sets are finite because they correspond to pieces of the orbit situated in a ball of finite volume. The measure ν is nondegenerate because its support is the whole Γ. It is shown in [KL3] that the measure ν has a first moment. The proof uses the details of the construction, but the idea is that the distribution of ν is given by choosing some random time and looking at the point γx_0 close to the trajectory of the Brownian motion at that time. Since the curvature is bounded from below, if the expectation of the time is finite, the expectation of the distance of the Brownian point at that time is finite as well. It also follows that the rates of escape of the Brownian motion and of the random walk are proportional. Therefore, if the manifold (M, g) is Liouville, then the random walk (G, ν) is Liouville. By Corollary 20, $\lim_{n\to\infty} \frac{1}{n} d_M(x_0, \gamma_1 \ldots \gamma_n x_0) = 0$ and therefore, $\lim_{t\to\infty} \frac{1}{t} d_M(x_0, B_t) = 0$ a.s.

There are many results about the Liouville property for Riemannian covers of a compact manifold. Theorem 22 implies that the corresponding statements hold for the rate of escape of the Brownian motion. Guivarc'h [G] showed that if the group Γ is not amenable, then (M, g) is not Liouville, whereas when Γ is polycyclic, (M, g) is Liouville (Kaimanovich [K2]). Lyons and Sullivan ([LS], see also [Er] for a simply connected example) have examples of amenable covers without the Liouville property.

ACKNOWLEDGMENTS. This survey grew out from lectures given by both authors at the IIIème Cycle Romand de Mathématiques in Les Diablerets in March 2008 and from conversations there. We thank Tatiana Smirnova-Nagnibeda and Slava Grigorchuk for their invitation to this friendly and stimulating meeting.

References

[A] J. Aaronson, *An Introduction to Infinite Ergodic Theory*, AMS Math. Surv. Mon. 50, American Mathematical Society, Providence, RI, 1997.

[Al] G. Alexopoulos, On the mean distance of random walks on groups, *Bull. Sci. Math.* 111 (1987), 189–199.

[Ar] L. Arnold, *Random Dynamical Systems*, Springer Monographs in Mathematics, Springer-Verlag, Berlin, 1998.

[Av] A. Avez, Entropie des groupes de type fini, *C. R. Acad. Sci. Paris, Sér. A-B* 275 (1972), A1363–A1366.

[Ba]   W. Ballmann, On the Dirichlet problem at infinity for manifolds of nonpositive curvature, *Forum Math.* 1 (1989), no. 2, 201–213.

[BL]   W. Ballmann and F. Ledrappier, Discretization of positive harmonic functions on Riemannian manifolds and Martin boundary. Actes de la Table Ronde de Géométrie Différentielle (Luminy, 1992), 77–92, Sémin. Congr., 1, Soc. Math. France, Paris, 1996.

[BH]   M. R. Bridson and A. Haefliger, *Metric Spaces of Non-positive Curvature*, Grundl. Math. Wiss. 319, Springer-Verlag, Berlin, 1999.

[Er]   A. Erschler, Liouville property for groups and manifolds, *Invent. Math.* 155 (2004), no. 1, 55–80.

[F1]   H. Furstenberg, Non-commuting random products, *Trans. Amer. Math. Soc.* 108 (1963), 377–428.

[F2]   H. Furstenberg, Random walks and discrete subgroups of Lie groups, *Advances in Probability and Related Topics* 1 (1971), 1–63.

[G]   Y. Guivarc'h, Sur la loi des grands nombres et le rayon spectral d'une marche aléatoire, *Astérisque* 74 (1980), 47–98.

[K1]   V. A. Kaimanovich, Brownian motion and harmonic functions on covering manifolds: an entropic approach, *Soviet Math. Dokl.* 33 (1986), no. 3, 812–816.

[K2]   V. A. Kaimanovich, Boundaries of random walks on polycyclic groups and the law of large numbers for solvable Lie groups, (Russian) *Vestnik Leningrad. Univ. Mat. Mekh. Astronom.* (1987), vyp. 4, 93–95, 112.

[K3]   V. A. Kaimanovich, Lyapunov exponents, symmetric spaces and a multiplicative ergodic theorem for semisimple Lie groups, *J. Soviet Math.* 47 (1989), 2387–2398.

[K4]   V. A. Kaimanovich, Poisson boundaries of random walks on discrete solvable groups, in *Probability Measures on Groups*, X (Oberwolfach, 1990), Plenum, New York, 1991, 205–238.

[K5]   V. A. Kaimanovich, *Discretization of Bounded Harmonic Functions on Riemannian Manifolds and Entropy: Potential Theory* (Nagoya, 1990), 213–223, de Gruyter, Berlin, 1992.

[K6]   V.A. Kaimanovich, The Poisson formula for groups with hyperbolic properties, *Ann. Math.* 152 (2000), 659–692.

[KV]   V. Kaimanovich and A. Vershik, Random walks on discrete groups: boundary and entropy, *Ann. Prob.* 11 (1983), 457–490.

[Ka]   A. Karlsson, Linear rate of escape and convergence in directions, in *Proceedings of a Workshop at the Schrödinger Institute*, Vienna 2001, (ed. by V.A. Kaimanovich, in collab. with K. Schmidt and W. Woess), de Gruyter, Berlin, 2004.

[KL1]   A. Karlsson and F. Ledrappier, On laws of large numbers for random walks, *Ann. Prob.* 34 (2006), 1693–1706.

[KL2]   A. Karlsson and F. Ledrappier, Drift and entropy for random walks, *Pure Appl Math. Quarterly* 3 (2007), 1027–1036.

[KL3]   A. Karlsson and F. Ledrappier, Propriété de Liouville et vitesse de fuite du mouvement Brownien, *C. R. Acad. Sci. Paris, Sér. I*, 344 (2007), 685–690.

[KM]   A. Karlsson and G. Margulis, A multiplicative ergodic theorem and nonpositively curved spaces, *Comm. Math. Phys.* 208 (1999), 107–123.

[KMo]   A. Karlsson and N. Monod, Strong law of large numbers with concave moments, unpublished note, 2008.

[KN]   E. Kohlberg and A. Neyman, Asymptotic behavior of nonexpansive mappings in normed linear spaces, *Israel J. Math* 4 (1981), 269–275.

[Kr]   U. Krengel, *Ergodic Theorems*, de Gruyter Studies in Mathematics, 6, de Gruyter, Berlin, 1985.

[LL]   F. Ledrappier and S. Lim, A proof of a $L^{1/2}$ ergodic theorem, unpublished note, 2008.

[L]   T. Lyons, Instability of the Liouville property for quasi-isometric Riemannian manifolds and reversible Markov chains, *J. Diff. Geom* 26 (1987), 33–66.

[LS]   T. Lyons and D. Sullivan, Function theory, random paths and covering spaces, *J. Differential Geom.* 19 (1984), no. 2, 299–323.

[MZ]   J. Marcinkiewicz and A. Zygmund, Sur les fonctions indépendantes, *Fundam. Math.* 29 (1937), 60–90.

[M]   P. Mathieu, Carne-Varopoulos bounds for centered random walks, *Ann. Prob.* 34 (2006), 987–1011.

[O]   V. I. Oseledec, A multiplicative ergodic theorem: Lyapunov characteristic numbers for dynamical systems, *Trans. Moscow Math. Soc.* 19 (1968), 197–231.

[Pa]   A. Pazy, Asymptotic behavior of contractions in Hilbert space, *Israel J. Math.* 9 (1971), 235–240.

[S]   S. A. Sawyer, Maximal inequalities of weak type, *Ann. of Math.* (2) 84 (1966), 157–174.

[St]   R. A. Struble, Metrics in locally compact groups, *Comp. Math.* 28 (1974), 217–222.

[Va]   N. Th. Varopoulos, Long range estimates for Markov chains, *Bull. Sci. Math.* 109 (1985), 225–252.

[Ve]   A. Vershik, Dynamic theory of growth in groups: entropy, boundary, examples, *Russian Math. Surveys* 55 (2000), 667–733.

[W]   P. Walters, A dynamical proof of the multiplicative ergodic theorem, *Trans. Amer. Math. Soc.* 335 (1993), 245–257.

13

SORIN POPA[1] AND STEFAAN VAES[2]

COCYCLE AND ORBIT SUPERRIGIDITY FOR LATTICES IN SL$(n, \mathbb{R})$ ACTING ON HOMOGENEOUS SPACES

TO BOB ZIMMER AT THE OCCASION OF HIS 60TH BIRTHDAY

Abstract

We prove cocycle and orbit equivalence superrigidity for lattices in SL$(n, \mathbb{R})$ acting linearly on $\mathbb{R}^n$, as well as acting projectively on certain flag manifolds, including the real projective space. The proof combines operator algebraic techniques with the property (T) in the sense of Zimmer for the action SL$(n, \mathbb{Z}) \curvearrowright \mathbb{R}^n$, $n \geq 4$. We also show that the restriction of the orbit equivalence relation $\mathcal{R}(\mathrm{SL}(n, \mathbb{Z}) \curvearrowright \mathbb{R}^n)$ to a subset of finite Lebesgue measure provides a II_1 equivalence relation with property (T) and yet fundamental group equal to $\mathbb{R}_+$.

1. Introduction and Statement of Main Results

Over the last few years, operator algebraic methods were used to prove several orbit equivalence and cocycle superrigidity theorems: for Bernoulli actions of property (T) groups [19], of product groups [18], and for profinite actions of property (T) groups [13]. In this paper, we extend the scope of these methods to a more geometric class of actions, like the natural actions of lattices $\Gamma \subset \mathrm{SL}(n, \mathbb{R})$ on the vector space $\mathbb{R}^n$, on the projective space $P^{n-1}(\mathbb{R})$, and on certain flag manifolds, all of which can be viewed as SL$(n, \mathbb{R})$-homogeneous spaces.

None of these actions is probability measure preserving. Hence, property (T) of the acting group has to be replaced by Zimmer's notion of property (T) for a nonsingular action (see [25]), which plays a crucial role in this paper. It is shown that for any lattice $\Gamma \subset \mathrm{SL}(n, \mathbb{R})$ the linear action $\Gamma \curvearrowright \mathbb{R}^n$ has property (T) if and only if $n \geq 4$. We then deduce the following theorem; see Theorem 4.1.

1. Partially supported by NSF Grant DMS-0601082.

2. Partially supported by ERC Starting Grant VNALG-200749 and Research Programme G.0231.07 of the Research Foundation–Flanders (FWO).

THEOREM 1.1. *Let $\Gamma < SL(n, \mathbb{R})$ be a lattice and let $\mathcal{R}$ be the II_1 equivalence relation obtained by restricting the orbit equivalence relation $\mathcal{R}(\Gamma \curvearrowright \mathbb{R}^n)$ to a set of Lebesgue measure 1, for some $n \geq 4$. Then we have*

- *$\mathcal{R}$ has property (T), in the sense of Zimmer, yet the fundamental group of $\mathcal{R}$ equals $\mathbb{R}_+$; and*
- *$\mathcal{R}^t$ cannot be implemented by a free action of a group, $\forall t > 0$. Also, $\mathcal{R}^t$ cannot be implemented by a (not necessarily free) action of a discrete property (T) group, $\forall t > 0$.*

Every orbit equivalence relation $\mathcal{R}(\Gamma \curvearrowright X)$ of a probability measure preserving (p.m.p.) action of a property (T) group, has property (T) in the sense of Zimmer. Theorem 1.1 provides the first examples of property (T) equivalence relations that do not arise in this way.

We say that a Polish group is of finite type if it can be realized as the closed subgroup of the unitary group of some II_1 factor with separable predual. All countable and all second countable compact groups are Polish groups of finite type. In [19], the first author proved that every 1-cocycle for the Bernoulli action of a property (T) group with values in a Polish group of finite type is cohomologous to a group morphism. We say that actions with this property are $\mathcal{U}_{fin}$-cocycle superrigid. More precisely:

DEFINITION 1.2. The nonsingular action $G \curvearrowright (X, \mu)$ of the locally compact second countable group G on the standard measure space (X, μ) is called $\mathcal{U}_{fin}$-cocycle superrigid if every 1-cocycle for the action $G \curvearrowright (X, \mu)$ with values in a Polish group of finite type $\mathcal{G}$ is cohomologous to a continuous group morphism $G \to \mathcal{G}$.

Until now, the only known examples of $\mathcal{U}_{fin}$-cocycle superrigid actions were the Bernoulli and Gaussian actions of property (T) groups (cf. [19]) and, respectively, product groups (cf. [18]), all of which are p.m.p. actions. We exhibit here the first examples of nonsingular ergodic actions $G \curvearrowright (X, \mu)$ that are $\mathcal{U}_{fin}$-cocycle superrigid but for which there exist no invariant probability measure $\nu \sim \mu$ on X. As it happens, these examples are of genuine geometric nature.

THEOREM 1.3. *The following actions are $\mathcal{U}_{fin}$-cocycle superrigid.*

1) *For $n \geq 5$ and Γ any lattice in $SL(n, \mathbb{R})$, the linear action $\Gamma \curvearrowright \mathbb{R}^n$.*
2) *For $n \geq 5$ and Γ any finite index subgroup of $SL(n, \mathbb{Z})$, the affine action $\Gamma \ltimes \mathbb{Z}^n \curvearrowright \mathbb{R}^n$.*

3) For $n \geq 4k + 1$, Γ any lattice in SL$(n, \mathbb{R})$, and H any closed subgroup of GL$(k, \mathbb{R})$, the action

$$G \curvearrowright M_{n,k}(\mathbb{R}) \quad where \quad G := \begin{cases} \frac{\Gamma \times H}{\{\pm(1,1)\}} & if\,(-1, -1) \in \Gamma \times H, \\ \Gamma \times H & otherwise, \end{cases}$$

by left-right multiplication on the space $M_{n,k}(\mathbb{R})$ of $n \times k$ matrices equipped with the Lebesgue measure.

In [19], the first author introduced the notion of *malleability* for measure preserving actions $\Gamma \curvearrowright (X, \mu)$. Roughly speaking, this property requires the existence of a flow on $X \times X$, commuting with the diagonal Γ-action and connecting the identity map to the flip map on $X \times X$. Theorem 0.1 in [19] shows that every weakly mixing, malleable p.m.p. action of a property (T) group is $\mathcal{U}_{\mathrm{fin}}$-cocycle superrigid.

We prove here a cocycle superrigidity result for infinite measure preserving malleable actions. But, property (T) of the group Γ has to be replaced by property (T) of the diagonal action of $\Gamma \curvearrowright X \times X$. In the case of SL$(n, \mathbb{Z}) \curvearrowright \mathbb{R}^n$, this forces $n \geq 5$. Finally, weak mixing has to be replaced by the ergodicity of the 4-fold diagonal action $\Gamma \curvearrowright X \times X \times X \times X$, which in the case of SL$(n, \mathbb{Z}) \curvearrowright \mathbb{R}^n$ again holds exactly for $n \geq 5$.

Using the cocycle superrigidity of SL$(n, \mathbb{Z}) \curvearrowright \mathbb{R}^n$, we then derive a full classification of all 1-cocycles for the action SL$(n, \mathbb{Z}) \curvearrowright \mathbb{T}^n = \mathbb{R}^n/\mathbb{Z}^n$, with values in a Polish group of finite type. As such, our Example 5.12 below, complements Zimmer's celebrated cocycle superrigidity theorem [26]. Thus, Zimmer's result concerns arbitrary actions SL$(n, \mathbb{Z}) \curvearrowright (X, \mu)$, but makes certain restrictions on the target groups (which must be simple linear algebraic groups) and on the cocycles (Zariski density), while our result treats a very specific action, but allows rather general target groups and requires no conditions on the cocycles.

From the cocycle superrigidity Theorem 1.3, we derive orbit equivalence (OE) superrigidity results for the following concrete actions of lattices Γ in SL$(n, \mathbb{R})$ and PSL$(n, \mathbb{R})$:

1) The linear action $\Gamma \curvearrowright \mathbb{R}^n$.
2) The action $\Gamma \curvearrowright \mathbb{T}^n = \mathbb{R}^n/\mathbb{Z}^n$ for Γ a finite-index subgroup of SL$(n, \mathbb{Z})$.
3) The projective action $\Gamma \curvearrowright P^{n-1}(\mathbb{R})$.
4) The natural action $\Gamma \curvearrowright X$ of a lattice Γ in PSL$(n, \mathbb{R})$ on the real flag manifold X of signature $(d_1, \ldots, d_l, n)$. (Recall that points in X are flags $\{0\} \subset V_1 \subset \cdots \subset V_l \subset \mathbb{R}^n$ where V_i is a vector subspace of $\mathbb{R}^n$ with dimension d_i.)

The action in (1) has the Lebesgue measure as infinite invariant measure, while the actions in (3) and (4) do not have finite or infinite invariant measures. All the actions in (1)–(4) are essentially free and ergodic; see Lemma 5.7 for details.

The natural invariant measure class on the flag manifold X can be described as follows. Put $d_l = k$ and consider the set $M_{n,k}(\mathbb{R})$ of $n \times k$ matrices of rank k, equipped with the Lebesgue measure. Denote by $E = (E_1, \ldots, E_l)$ the standard flag of signature $(d_1, \ldots, d_l, n)$, that is $E_i = \text{span}\{e_1, \ldots, e_{d_i}\}$, where $e_1, \ldots, e_n$ are the standard basis vectors in $\mathbb{R}^n$. The group $GL(k, \mathbb{R})$ acts on $M_{n,k}(\mathbb{R})$ by right multiplication. This action is free and proper and

$$M_{n,k}(\mathbb{R})/H \to X : A \mapsto (AE_1, \ldots, AE_l)$$

is an isomorphism. Here, $H = \{g \in GL(k, \mathbb{R}) \mid gE_i \subset E_i \text{ for all } i = 1, \ldots, l\}$. Writing $k_1 = d_1$ and $k_i = d_i - d_{i-1}$ for $i \geq 2$, the group H can of course be written as

1.1
$$H = \begin{pmatrix} GL(k_1, \mathbb{R}) & * & \cdots & * \\ 0 & GL(k_2, \mathbb{R}) & \cdots & * \\ \vdots & \vdots & \ddots & \vdots \\ 0 & 0 & \cdots & GL(k_l, \mathbb{R}) \end{pmatrix}.$$

Before stating our OE superrigidity results, recall the following terminology.

DEFINITION 1.4. Let $\Gamma \overset{\alpha}{\curvearrowright} (X, \mu)$ and $\Lambda \overset{\beta}{\curvearrowright} (Y, \eta)$ be essentially free, ergodic, nonsingular actions of countable groups on standard measure spaces.

- A *stable orbit equivalence (SOE)* between α and β is a nonsingular isomorphism $\Delta : X_0 \to Y_0$ between nonnegligible subsets $X_0 \subset X$, $Y_0 \subset Y$, such that Δ is an isomorphism between the restricted orbit equivalence relations $\mathcal{R}(\Gamma \curvearrowright X)|_{X_0}$ and $\mathcal{R}(\Lambda \curvearrowright Y)|_{Y_0}$.
- We say that $\Gamma \curvearrowright X$ is *induced* from $\Gamma_1 \curvearrowright X_1$, if Γ_1 is a subgroup of Γ, X_1 is a nonnegligible subset of X and $g \cdot X_1 \cap X_1$ is negligible for all $g \in \Gamma - \Gamma_1$.

For the linear lattice actions and the quotient action $SL(n, \mathbb{Z}) \curvearrowright \mathbb{T}^n$, we get the following.

THEOREM 1.5. *Let $n \geq 5$ and $\Gamma \subset SL(n, \mathbb{R})$ a lattice. Let $\Lambda \curvearrowright (Y, \eta)$ be any essentially free, ergodic, nonsingular action of the countable group Λ.*

1) *The actions* $\Gamma \curvearrowright \mathbb{R}^n$ *and* $\Lambda \curvearrowright Y$ *are SOE if and only if* $\Lambda \curvearrowright Y$ *is conjugate to an induction of one of the following actions:*

- $\Gamma \curvearrowright \mathbb{R}^n$ *itself, and*
- *(only in case* $-1 \in \Gamma$*) the quotient action* $\Gamma/\{\pm 1\} \curvearrowright \mathbb{R}^n/\{\pm 1\}$.

2) *The actions* SL$(n, \mathbb{Z}) \curvearrowright \mathbb{T}^n$ *and* $\Lambda \curvearrowright Y$ *are SOE if and only if* $\Lambda \curvearrowright Y$ *is conjugate to an induction of one of the following actions*

- SL$(n, \mathbb{Z}) \curvearrowright \mathbb{T}^n$ *itself,*
- SL$(n, \mathbb{Z}) \ltimes \mathbb{Z}^n \curvearrowright \mathbb{R}^n$,
- SL$(n, \mathbb{Z}) \ltimes (\frac{\mathbb{Z}}{\lambda \mathbb{Z}})^n \curvearrowright \frac{\mathbb{R}^n}{\lambda \mathbb{Z}^n}$ *for some* $\lambda \in \mathbb{N} \setminus \{0, 1\}$, *and*
- *(only in case* n *is even) one of the actions*

$$\mathrm{PSL}\,(n, \mathbb{Z}) \curvearrowright \frac{\mathbb{T}^n}{\{\pm 1\}} \quad or \quad \mathrm{PSL}\,(n, \mathbb{Z}) \ltimes \left(\frac{\mathbb{Z}}{2\mathbb{Z}}\right)^n \curvearrowright \frac{\mathbb{R}^n/(2\mathbb{Z})^n}{\{\pm 1\}}\,.$$

Part (2) of the above theorem is a slightly more detailed version, with a very different proof, of a result of Furman [7, corollary B], who shows that for all $n \geq 3$, the actions SL$(n, \mathbb{Z}) \curvearrowright \mathbb{T}^n$ and $\Lambda \curvearrowright Y$ are SOE if and only if they are virtually conjugate. Furman's proof relies on Zimmer's cocycle superrigidity, while ours is based on the cocycle superrigidity Theorem 1.3.

To formulate easily the correct OE superrigidity statements for lattice actions on flag manifolds, make the following observations.

The real flag manifold of signature $(d_1, \ldots, d_l, n)$ has a natural 2^l-fold covering $\widetilde{X}$ consisting of oriented flags

$$\{0\} \subset (V_1, \omega_1) \subset \cdots \subset (V_l, \omega_l) \subset \mathbb{R}^n$$

where every V_i is a vector subspace of $\mathbb{R}^n$ with an orientation ω_i. Clearly, $\widetilde{X} = \mathrm{M}_{n,k}(\mathbb{R})/H_0$, where $H_0 = \{g \in H \mid \det(g|_{E_i}) > 0 \text{ for all } i = 1, \ldots, l\}$. In the Expression 1.1 above, H_0 consists of those matrices A that have on the diagonal $A_{ii} \in \mathrm{GL}\,(k_i, \mathbb{R})$ with $\det A_{ii} > 0$ for all i.

Denote $\Sigma_l = H/H_0$ and observe that $\Sigma_l \cong (\mathbb{Z}/2\mathbb{Z})^{\oplus l}$. Then, Σ_l acts on $\widetilde{X}$ by reversing orientations, but keeping the flags. We denote by $-1 \in \Sigma_l$ the multiplication by -1 and observe that $-1 = 1$ in Σ_l if all d_i are even. Clearly, $X = \widetilde{X}/\Sigma_l$.

THEOREM 1.6. *Let X be the real flag manifold of signature $(d_1, \ldots, d_l, n)$. Let $\Gamma \subset \mathrm{PSL}\,(n, \mathbb{R})$ be a lattice and assume that $n \geq 4d_l + 1$. Denote by $\widetilde{X}$ the 2^l-fold covering of X consisting of oriented flags, as explained before the theorem. Let $\Lambda \curvearrowright (Y, \eta)$ be any essentially free, ergodic, nonsingular action of the countable group Λ.*

The actions $\Gamma \curvearrowright X$ and $\Lambda \curvearrowright Y$ are SOE if and only if $\Lambda \curvearrowright Y$ is conjugate to an induction of one of the actions

- $\Gamma \times \dfrac{\Sigma_l}{\Sigma} \curvearrowright \tilde{X}/\Sigma$ *for some subgroup* $\Sigma < \Sigma_l$ *with* $-1 \in \Sigma$, *or*

- $\dfrac{\tilde{\Gamma} \times \Sigma_l/\Sigma}{\{\pm(1,1)\}} \curvearrowright \tilde{X}/\Sigma$ *for some* $\Sigma < \Sigma_l$ *with* $-1 \notin \Sigma$ *and with* $\tilde{\Gamma} = \{\pm 1\} \cdot \Gamma \subset GL(n, \mathbb{R})$.

EXAMPLE 1.7. If $n \geq 5$ and $\Gamma \subset PSL(n, \mathbb{R})$ is a lattice, the action $\Gamma \curvearrowright P^{n-1}(\mathbb{R})$ is a special case of the flag manifold action treated in Theorem 1.6. Hence, $\Gamma \curvearrowright P^{n-1}(\mathbb{R})$ and $\Lambda \curvearrowright Y$ are SOE if and only if $\Lambda \curvearrowright Y$ is conjugate to an induction of either $\Gamma \curvearrowright P^{n-1}(\mathbb{R})$ or its double cover $\tilde{\Gamma} \curvearrowright \mathbb{R}^n/\mathbb{R}_+$, where $\tilde{\Gamma} = \{\pm 1\} \cdot \Gamma \subset GL(n, \mathbb{R})$.

Finally, combining the work of [9] and the above OE superrigidity results, we classify up to stable orbit equivalence, the lattice actions on $\mathbb{R}^n$ and on flag manifolds; see Theorems 6.2 and 6.3. At the same time, we compute the outer automorphism group of the associated orbit equivalence relation.

2. Preliminaries

Let $\mathcal{R}$ be a countable measured equivalence relation on the standard measure space (X, μ), meaning that $\mathcal{R}$ is, on the one hand, a Borel subset of $X \times X$ and, on the other hand, an equivalence relation on X with countable equivalence classes, such that the σ-finite measures $\mu_l^{(1)}$ and $\mu_r^{(1)}$ on $\mathcal{R}$ defined by

$$\mu_l^{(1)}(Y) = \int_X \#\{y \in X \mid (x, y) \in Y\} \, d\mu(x) \, ,$$

$$\mu_r^{(1)}(Y) = \int_X \#\{x \in X \mid (x, y) \in Y\} \, d\mu(y) \, ,$$

are mutually absolutely continuous: $\mu_l^{(1)} \sim \mu_r^{(1)}$. One calls $\mathcal{R}$ ergodic if every $\mathcal{R}$-saturated measurable subset of X is negligible or conegligible. The measure μ is said to be invariant if $\mu_l^{(1)} = \mu_r^{(2)}$. A II_1 equivalence relation is a countable measured ergodic equivalence relation with invariant probability measure.

We now recall Zimmer's definition of property (T) for a countable measured equivalence relation $\mathcal{R}$. To this end, first define $\mathcal{R}^{(2)} = \{(x, y, z) \in X \times X \times X \mid x\mathcal{R}y \text{ and } y\mathcal{R}z\}$. Note that $\mathcal{R}^{(2)}$ comes equipped with a canonical measure class, induced by the σ-finite measure

$$\mu_l^{(2)}(Y) = \int_X \#\{(y, z) \in \mathcal{R} \mid (x, y, z) \in Y\} \, d\mu(x) \, .$$

Note that one defines absolutely continuous measures $\mu_m^{(2)}$, respectively, $\mu_r^{(2)}$, by integrating over y, respectively, z. But we only make use of the measure class given by $\mu_l^{(2)}$.

- A *1-cocycle of* $\mathcal{R}$ *with values in the unitary group* $\mathcal{U}(K)$ of a Hilbert space K is a Borel map $c : \mathcal{R} \to \mathcal{U}(K)$ satisfying $c(x, z) = c(x, y)c(y, z)$ for almost all $(x, y, z) \in \mathcal{R}^{(2)}$.
- Suppose that $c : \mathcal{R} \to \mathcal{U}(K)$ is a 1-cocycle of $\mathcal{R}$.
  - An *invariant unit vector* of c is a Borel map $\xi : X \to K$ satisfying $\xi(x) = c(x, y)\xi(y)$ for almost all $(x, y) \in \mathcal{R}$ and $\|\xi(x)\| = 1$ for almost all $x \in X$.
  - A *sequence of almost invariant unit vectors* of c is a sequence of Borel maps $\xi_n : X \to K$ satisfying

$$\|\xi_n(x) - c(x, y)\xi_n(y)\| \to 0 \quad \text{for almost all} \quad (x, y) \in \mathcal{R}$$

and $\|\xi_n(x)\| = 1$ for all $n \in \mathbb{N}$ and almost all $x \in X$.

DEFINITION 2.1. A countable measured equivalence relation $\mathcal{R}$ is said to have *property (T) in the sense of Zimmer* if the following holds: every 1-cocycle of $\mathcal{R}$ with values in the unitary group of a Hilbert space and admitting a sequence of almost invariant unit vectors admits a unit invariant vector.

We leave the proof of the following lemma as an exercise.

LEMMA 2.2. *Let* $\mathcal{R}$ *be a countable measured equivalence relation on* (X, μ) *and* $Y \subset X$ *a measurable subset such that the* $\mathcal{R}$-*saturation of* Y *is conegligible. Then, the restriction of* $\mathcal{R}$ *to* Y *has property (T) if and only if* $\mathcal{R}$ *has property (T).*

3. Property (T) for Actions of Locally Compact Groups

We recalled above Zimmer's definition of property (T) for a II_1 equivalence relation. In fact, one can define property (T) for measured groupoids in general; see [1]. We do not need this generality in this paper, but we do need the concept of property (T) for nonsingular actions of locally compact second countable (l.c.s.c.) groups on measure spaces. For a groupoid approach to this definition, we refer to [1]. For the convenience of the reader, we gather in this section the necessary concepts and results and present them in an operator algebra framework.

All von Neumann algebras are supposed to have separable predual and all locally compact groups are supposed to be second countable.

Given a von Neumann algebra $M \subset B(H)$ and a Hilbert space K, we make use of the W^*-module $M \overline{\otimes} K$, that can be defined as $\{T \in B(H, H \otimes K) \mid Tx = (x \otimes 1)T, \forall x \in M'\}$. By construction, $M \overline{\otimes} K$ is an $(M \overline{\otimes} B(K))$-$M$-bimodule and carries the strong*-topology inherited from $B(H, H \otimes K)$.

If M is a von Neumann algebra, we equip Aut (M) with the Polish topology making the functions Aut $(M) \to M_* : \alpha \mapsto \omega \circ \alpha$ continuous for all $\omega \in M_*$. An action α of an l.c.s.c. group G on a von Neumann algebra M, denoted $G \overset{\alpha}{\curvearrowright} M$, is a continuous group morphism $\alpha : G \to$ Aut (M).

- A 1-*cocycle* of an action $G \overset{\alpha}{\curvearrowright} M$ with values in the unitary group $\mathcal{U}(K)$ of a Hilbert space K, is a strongly continuous map $c : G \to \mathcal{U}(M \overline{\otimes} B(K))$ satisfying $c(gh) = c(g)(\alpha_g \otimes \mathrm{id})(c(h))$ for all $g, h \in G$. Note that by Theorem 3 in [16], it makes no difference to assume only that c is a measurable map, with the previous formula holding for almost all $(g, h) \in G \times G$.
- An *invariant unit vector* of the 1-cocycle c of $G \overset{\alpha}{\curvearrowright} M$, is an element ξ in the W^*-module $M \overline{\otimes} K$ satisfying $\xi^* \xi = 1$ and $c(g)(\alpha_g \otimes \mathrm{id})(\xi) = \xi$ for all $g \in G$.
- A *sequence of almost invariant unit vectors* of the 1-cocycle c of $G \overset{\alpha}{\curvearrowright} M$, is a sequence $\xi_n \in M \overline{\otimes} K$ satisfying $\xi_n^* \xi_n = 1$ for all n and $c(g)(\alpha_g \otimes \mathrm{id})(\xi_n) - \xi_n \to 0$ *-strongly, uniformly on compact subsets of G.
- The action $G \curvearrowright M$ is said to have *property (T)* if every 1-cocycle with values in the unitary group of a Hilbert space and admitting a sequence of almost invariant unit vectors, admits an invariant unit vector.

The relation between this notion of property (T) for an action and Zimmer's property (T) for a measured equivalence relation is given by the following straightforward lemma.

LEMMA 3.1. *Let $\Gamma \curvearrowright (X, \mu)$ be a nonsingular action of the countable group Γ and consider the orbit equivalence relation $\mathcal{R}(\Gamma \curvearrowright X)$. If the action $\Gamma \curvearrowright L^\infty (X)$ has property (T) in the sense above, then the equivalence relation $\mathcal{R}(\Gamma \curvearrowright X)$ has property (T) in the sense of Definition 2.1. The converse holds when $\Gamma \curvearrowright (X, \mu)$ is essentially free.*

Note that by combining Lemmas 2.2 and 3.1, we obtain the following: if $\Gamma \curvearrowright (X, \mu)$ is a nonsingular, essentially free ergodic action and $Y \subset X$ is non-negligible, then the restricted equivalence relation $\mathcal{R}(\Gamma \curvearrowright X)|_Y$ has property (T) iff the action $\Gamma \curvearrowright (X, \mu)$ has property (T).

The following result is proven for discrete groups in [25], for weakly mixing actions, and in [6], for arbitrary actions (see also [2, corollary 5.16]). These methods work as well in the locally compact case and for completeness, we give a proof in a von Neumann algebra setup.

PROPOSITION 3.2. *Let* $G \overset{\alpha}{\curvearrowright} M$ *be an action of the l.c.s.c. group G on the von Neumann algebra M. Suppose that τ is a faithful normal tracial state on M, invariant under α. Then, $G \curvearrowright M$ has property (T) if and only if the group G has property (T).*

Combining Proposition 3.2 and Lemma 3.1, it follows that for every p.m.p. action $\Gamma \curvearrowright (X, \mu)$ of a countable property (T) group Γ, the orbit equivalence relation $\mathcal{R}(\Gamma \curvearrowright X)$ has property (T) in the sense of Definition 2.1. Theorem 1.1 provides the first examples of property (T) equivalence relations that do not arise in this way.

Proof of Proposition 3.2. Suppose first that G has property (T). Let $c : G \to \mathcal{U}(M \overline{\otimes} B(K))$ be a 1-cocycle of $G \curvearrowright M$ having $\xi_n \in M \overline{\otimes} K$ as a sequence of almost invariant unit vectors. Define the unitaries u_g on $L^2(M, \tau)$ by extending α_g. Then,

$$\pi : G \to \mathcal{U}(L^2(M, \tau) \otimes K) : \pi(g) = c(g)(u_g \otimes 1)$$

is a unitary representation of G and we can view ξ_n as a sequence of almost invariant unit vectors. Since G has property (T), π admits an invariant unit vector. Even more, we find a sequence $\eta_n \in L^2(M, \tau) \otimes K$ of π-invariant vectors satisfying $\eta_n - \xi_n \to 0$.

Since $\|\xi_n\| = 1$, it follows that $\|\xi_n^* \eta_n - 1\|_2 \to 0$. Hence, the right support projection of $\xi_n^* \eta_n$ converges strongly to 1. A fortiori, the right support projection of η_n converges to 1. We view η_n as a closed operator from $L^2(M, \tau)$ to $L^2(M, \tau) \otimes K$. Taking the polar decomposition of η_n, we find $v_n \in M \overline{\otimes} K$ satisfying $c(g)(\alpha_g \otimes \mathrm{id})(v_n) = v_n$ for all $g \in G, n \in \mathbb{N}$ and such that $v_n^* v_n$ is a sequence of projections in M converging strongly to 1. Define the von Neumann algebra

$$N = \begin{pmatrix} M \overline{\otimes} B(K) & M \overline{\otimes} K \\ (M \overline{\otimes} K)^* & M \end{pmatrix}.$$

Define the action (γ_g) of G on N by

$$\gamma_g \begin{pmatrix} a & b \\ e & f \end{pmatrix} = \begin{pmatrix} c(g)(\alpha_g \otimes \mathrm{id})(a)c(g)^* & c(g)(\alpha_g \otimes \mathrm{id})(b) \\ (\alpha_g \otimes \mathrm{id})(e)c(g)^* & \alpha_g(f) \end{pmatrix}.$$

Define $p = \left(\begin{smallmatrix} 1 & 0 \\ 0 & 0 \end{smallmatrix}\right)$, $q = 1 - p$, and $w_n = \left(\begin{smallmatrix} 0 & v_n \\ 0 & 0 \end{smallmatrix}\right)$. Then, w_n is a sequence of partial isometries in the fixed-point algebra N^G, satisfying $w_n \in pN^G q$ and $w_n^* w_n \to q$ strongly. It follows that $q \prec p$ in the von Neumann algebra N^G. So, we find $v \in M \overline{\otimes} K$ satisfying $c(g)(\alpha_g \otimes \mathrm{id})(v) = v$ for all g and $v^* v = 1$. Hence, $G \curvearrowright M$ has property (T).

Suppose conversely that $G \curvearrowright M$ has property (T). Let $\pi : G \to \mathcal{U}(K)$ be a strongly continuous unitary representation of G admitting ξ_n as a sequence of almost invariant unit vectors. By [3, theorem 2.12.9], it is sufficient to prove that π has a nonzero finite-dimensional $\pi(G)$-invariant subspace. Define $c : G \to \mathcal{U}(M \overline{\otimes} B(K)) : c(g) = 1 \otimes \pi(g)$. Obviously, c is a 1-cocycle of $G \curvearrowright M$ having $1 \otimes \xi_n$ as a sequence of almost invariant unit vectors. By property (T) of $G \curvearrowright M$, we find $\xi \in M \overline{\otimes} K$ satisfying $\xi^* \xi = 1$ and $c(g)(\alpha_g \otimes \mathrm{id})(\xi) = \xi$ for all $g \in G$. Denoting again by $u : g \mapsto u_g$ the representation of G on $\mathrm{L}^2(M, \tau)$ obtained by extending α_g, we find that $u \otimes \pi$ admits an invariant unit vector ξ. Identify $\mathrm{L}^2(M, \tau) \otimes K$ with the Hilbert space of Hilbert-Schmidt operators from $\overline{\mathrm{L}^2(M, \tau)}$ to K. Then, $T := \xi \xi^*$ is a nonzero trace-class operator on K satisfying $\pi(g) T \pi(g)^* = T$ for all $g \in G$. So, for $\varepsilon > 0$ sufficiently small, the spectral projection $\chi_{[\varepsilon, +\infty)}(T)$ projects onto a nonzero finite-dimensional $\pi(G)$-invariant subspace of K. $\qquad\square$

The following is a slight generalization of [2, theorem 5.3]. When $H \curvearrowright M$ is an action, we denote by M^H the von Neumann algebra of H-fixed points.

LEMMA 3.3. *Let $G \overset{\alpha}{\curvearrowright} M$ be an action of the l.c.s.c. group G on the von Neumann algebra M. Let $H \lhd G$ be a closed normal subgroup and assume that there is a $*$-isomorphism $\theta : M \to \mathrm{L}^\infty(H) \overline{\otimes} M^H$ satisfying $\theta \circ \alpha_h = (\rho_h \otimes \mathrm{id}) \circ \theta$ for all $h \in H$, where ρ_h denotes the right translation by h on $\mathrm{L}^\infty(H)$.*

Then, $G \curvearrowright M$ has property (T) if and only if $G/H \curvearrowright M^H$ has property (T).

Proof. We say that two 1-cocycles c_1, c_2 of $G \overset{\alpha}{\curvearrowright} M$ with values in $\mathcal{U}(K)$ are unitarily equivalent if there exists a unitary $v \in \mathcal{U}(M \overline{\otimes} B(K))$ satisfying $c_1(g) = vc_2(g)(\alpha_g \otimes \mathrm{id})(v^*)$ for all $g \in G$. We denote by $\mathrm{H}^1(G \overset{\alpha}{\curvearrowright} M, \mathcal{U}(K))$ the set of equivalence classes of 1-cocycles.

In the first part of the proof, we show that the obvious map

$$\Theta : \mathrm{H}^1(G/H \curvearrowright M^H, \mathcal{U}(K)) \to \mathrm{H}^1(G \curvearrowright M, \mathcal{U}(K)) : \Theta(c) = c \circ \pi \quad \text{with}$$

$$\pi : G \to G/H \,,$$

is a bijection. In the second part of the proof, we show that this map and its inverse preserve the property of having invariant, respectively, almost invariant, vectors. Both parts together show that property (T) of $G \curvearrowright M$ is equivalent with property (T) of $G/H \curvearrowright M^H$.

It is straightforward to check that Θ is well defined and injective. Suppose that $c : G \to \mathcal{U}(M \,\overline{\otimes}\, \mathrm{B}(K))$ is a 1-cocycle of $G \curvearrowright M$. In order to prove that c is in the range of Θ, it suffices to prove that c is unitarily equivalent with c' satisfying $c'(h) = 1$ for all $h \in H$. Identify throughout $\mathcal{U}(M \,\overline{\otimes}\, \mathrm{B}(K))$ with $\mathcal{U}(\mathrm{L}^\infty(H) \,\overline{\otimes}\, M^H \,\overline{\otimes}\, \mathrm{B}(K))$ and view the latter as measurable functions $H \to \mathcal{U}(M^H \,\overline{\otimes}\, \mathrm{B}(K))$, modulo equality almost everywhere. By [16, theorem 1], take a measurable map $\varphi : H \times H \to \mathcal{U}(M^H \,\overline{\otimes}\, \mathrm{B}(K))$ such that $c(h) = \varphi(\cdot, h)$ for all $h \in H$. Since c is a cocycle, we find that

$$\varphi(k, hg) = \varphi(k, h)\varphi(kh, g) \quad \text{for almost all } (k, h, g) \in H \times H \times H.$$

By the Fubini theorem, take $k_0 \in H$ such that for almost all $(h, g) \in H \times H$, the previous equality holds for (k_0, h, g). Define the unitary $v \in \mathcal{U}(M \,\overline{\otimes}\, \mathrm{B}(K))$ as the function $H \to \mathcal{U}(M^H \,\overline{\otimes}\, \mathrm{B}(K))$ given by $h \mapsto \varphi(k_0, k_0^{-1}h)$ and set $c'(g) = vc(g)(\alpha_g \otimes \mathrm{id})(v^*)$. By construction, $c'(g) = 1$ for almost all $g \in H$ and hence for all $g \in H$ by continuity.

It is an exercise to check that the 1-cocycle $c \in \mathrm{H}^1(G/H \curvearrowright M^H, \mathcal{U}(K))$ has an invariant unit vector if and only if $\Theta(c)$ has. Also, a sequence of almost invariant unit vectors for c defines a sequence of almost invariant unit vectors for $\Theta(c)$. Finally, suppose that $\xi_n \in M \,\overline{\otimes}\, K$ is a sequence of almost invariant unit vectors for $\Theta(c)$. In order to conclude the proof of the lemma, it suffices to show that there exists a sequence $\eta_n \in M^H \,\overline{\otimes}\, K$ satisfying $\eta_n^* \eta_n = 1$ for all n and $\xi_n - \eta_n \to 0$ *-strongly.

Define the sets $X = \{v \in M \,\overline{\otimes}\, K \mid v^* v = 1\}$ and $Y = \{v \in M^H \,\overline{\otimes}\, K \mid v^* v = 1\}$. Through the isomorphism $M \cong \mathrm{L}^\infty(H) \,\overline{\otimes}\, M^H$, we identify X with the set of measurable functions from H to Y (modulo equality almost everywhere). Take a bounded metric d_0 on Y inducing the strong* topology. Let μ be a probability measure on H in the same measure class as the Haar measure. Following [16, p. 5], define the metric d on X by

$$d(v, w) = \int_H d_0(v(h), w(h)) d\mu(h).$$

Then, d induces the strong* topology on X. It is easy to check that when $v_n, w_n \in X$ such that $d(v_n, w_n)$ is summable, then $v_n(h) - w_n(h) \to 0$ *-strongly for almost every $h \in H$ (see [16, proposition 6]).

View $Y \subset X$ as constant functions. We have to prove that $d(\xi_n, Y) \to 0$. Suppose the contrary. Write H as an increasing union of compact subsets H_n. After passage to a subsequence, we find $\varepsilon > 0$ such that $d(\xi_n, Y) > \varepsilon$ for all n and such that

$$d(\xi_n, (\alpha_g \otimes \mathrm{id})(\xi_n)) < 2^{-n} \quad \text{for all } n \in \mathbb{N}, g \in H_n .$$

It follows that for all $g \in H$, we have

$$d_0(\xi_n(h), \xi_n(hg)) \to 0 \quad \text{for almost all } h \in H .$$

By the Fubini theorem, take $h_0 \in H$ such that $d_0(\xi_n(h_0), \xi_n(h_0 g)) \to 0$ for almost all $g \in H$. It follows that $d(\xi_n, \xi_n(h_0)) \to 0$, contradicting the assumption that $d(\xi_n, Y) > \varepsilon$ for all n. $\qquad\square$

If M is a commutative von Neumann algebra, an action $H \curvearrowright M$ satisfying the condition in Lemma 3.3 is called essentially free and proper. More precisely:

DEFINITION 3.4. The nonsingular action $G \curvearrowright (X, \mu)$ of the l.c.s.c group G on the standard measure space (X, μ) is called *essentially free and proper* if there exists a measurable map $\pi : X \to G$ such that $\pi(g \cdot x) = g\pi(x)$ for almost all $(g, x) \in G \times X$. Equivalently, there exists a *-isomorphism $L^\infty (X, \mu) \to L^\infty (G) \,\overline{\otimes}\, L^\infty (X, \mu)^G$ conjugating the G-action on $L^\infty (X, \mu)$ with the action $(\rho_g \otimes \mathrm{id})_{g \in G}$, where ρ_g denotes the right translation by g on $L^\infty (G)$. Indeed, suppose that we are given $\pi : X \to G$ as above. Denote by X/G the space of ergodic components of $G \curvearrowright X$ and define the map $\Theta : X \to G \times X/G : x \mapsto (\pi(x), \overline{x})$. Then, Θ is invertible with the inverse being well defined and given by $(g, \overline{x}) \mapsto g^{-1}\pi(x)^{-1} \cdot x$. Also, Θ conjugates the action $G \curvearrowright X$ with the action $\rho_g \times \mathrm{id}$.

PROPOSITION 3.5. *Let G be an l.c.s.c. group with closed subgroups H_1, H_2. Then, $H_1 \curvearrowright L^\infty (G/H_2)$ has property (T) if and only if $H_2 \curvearrowright L^\infty (G/H_1)$ has property (T).*

Proof. Set $M = L^\infty (G)$ and $G = H_1 \times H_2$ acting by left-right translations on M:

$$(\alpha_{(g,h)}(F))(x) = F(g^{-1}xh) \quad \text{for all} \quad F \in L^\infty (G), x \in G, g \in H_1, h \in H_2 .$$

We apply Lemma 3.3 to $G \curvearrowright M$ and the closed normal subgroups $H = H_i$, $i = 1, 2$ of G. By the Effros-Mackey theorem (see e.g., [14, theorem II.12.17]), the quotient map $G \to G/H$ admits a Borel lifting and hence, there exists an H-equivariant isomorphism $M \to L^\infty (H) \,\overline{\otimes}\, M^H$. So, by Lemma 3.3, property (T) of $G \curvearrowright M$ is equivalent with property (T) of $H_1 \curvearrowright L^\infty (G/H_2)$ as well as with property (T) of $H_2 \curvearrowright L^\infty (G/H_1)$. $\qquad\square$

4. The Lattice Actions $\Gamma \curvearrowright \mathbb{R}^n$ Have Property (T)

Recall from [6] that if $\Gamma \curvearrowright (X, \mu)$ is a free ergodic p.m.p. action, then property (T) of $\mathcal{R}(\Gamma \curvearrowright X)$ in the sense of Zimmer is equivalent with property (T) of the group Γ (see also [25, proposition 2.4] and Proposition 3.2 above).

If Γ is a property (T) group, the fundamental group of $\mathcal{R}(\Gamma \curvearrowright X)$ is countable for any free ergodic p.m.p. action (see [12, corollary 1.8] if, moreover, Γ has infinite conjugacy classes (ICC) and see [13, theorem 5.9] for the general case).

But more is true: we proved in [22, theorem 6.1] that the fundamental group of a II_1 equivalence relation $\mathcal{R}$ on (X, μ) is countable whenever the full group $[\mathcal{R}]$ contains a property (T) group that implements an ergodic action on (X, μ). As a result, the following theorem is rather surprising: we obtain a II_1 equivalence relation $\mathcal{R}$ with property (T) and fundamental group $\mathbb{R}_+$; hence, none of the $\mathcal{R}^t$ can be implemented by a free action of a group and none of the $\mathcal{R}^t$ can be implemented by a possibly nonfree action of a property (T) group. The following result covers Theorem 1.1 stated in the introduction.

THEOREM 4.1. *Let $\Gamma < SL(n, \mathbb{R})$ be a lattice and let $\mathcal{R}$ be the II_1 equivalence relation obtained by restricting the orbit equivalence relation $\mathcal{R}(\Gamma \curvearrowright \mathbb{R}^n)$ to a set of Lebesgue measure 1. If $n \geq 4$, the equivalence relation $\mathcal{R}$ has property (T) in the sense of Zimmer, but nevertheless $\mathcal{F}(\mathcal{R}) = \mathbb{R}_+$. In particular,*

- *none of the equivalence relations $\mathcal{R}^t$, $t > 0$ can be implemented by a free action of a group,*
- *none of the equivalence relations $\mathcal{R}^t$, $t > 0$ can be implemented by a possibly nonfree action of a property (T) group. In fact, $\mathcal{R}^t$ does not contain any ergodic subequivalence relation that can be implemented by a property (T) group.*

Proof. By the remark following Lemma 3.1, proving property (T) of $\mathcal{R}$ amounts to proving property (T) for the action $\Gamma \curvearrowright L^\infty (\mathbb{R}^n)$.

Define the l.c.s.c. group $\mathcal{G} = SL(n, \mathbb{R})$ and set $H_1 = \Gamma$. Consider the linear action $\mathcal{G} \curvearrowright \mathbb{R}^n$ and set $H_2 = \{A \in SL(n, \mathbb{R}) \mid Ae_1 = e_1\}$, where e_1 denotes the first basis vector of $\mathbb{R}^n$. By construction, the action $\Gamma \curvearrowright L^\infty (\mathbb{R}^n)$ can be viewed as $H_1 \curvearrowright L^\infty (\mathcal{G}/H_2)$. Hence, by Proposition 3.5, property (T) for this last action is equivalent with property (T) of $H_2 \curvearrowright L^\infty (\mathcal{G}/H_1)$. This action admits a finite invariant measure, because H_1 is a lattice in $\mathcal{G}$. Moreover, $H_2 \cong SL(n-1, \mathbb{R}) \ltimes \mathbb{R}^{n-1}$, which has property (T) for $n \geq 4$. So, it follows from Proposition 3.2 that $H_2 \curvearrowright L^\infty (\mathcal{G}/H_1)$ has property (T).

The action on $\mathbb{R}^n$ by multiples of the identity matrix scales the Lebesgue measure and commutes with the action of Γ. Hence, the fundamental group of $\mathcal{R}$ equals $\mathbb{R}_+$. The statements about implementing $\mathcal{R}^t$ by group actions follow from the discussion preceding the theorem. $\square$

REMARK 4.2. Note that a similar reasoning as in the proofs of Lemmas 3.1 and 3.3, and Proposition 3.2 show that those statements hold true if we replace everywhere "property (T)" by "amenable." By applying this to the action $SL(2, \mathbb{Z}) \curvearrowright \mathbb{R}^2$, one recovers a result from [2], showing that the action $SL(2, \mathbb{Z}) \curvearrowright \mathbb{R}^2$ is amenable and hence, $L^\infty(\mathbb{R}^2) \rtimes SL(2, \mathbb{Z})$ is isomorphic to the unique hyperfinite II_∞ factor, by Connes's theorem [4]. This also shows that the action $SL(3, \mathbb{Z}) \curvearrowright \mathbb{R}^3$ is not amenable, nor has property (T).

REMARK 4.3. Denote by $\mathcal{R}$ the II_1 equivalence relation constructed in Theorem 4.1. Since $\mathcal{R}$ has property (T), [20, theorem 4.4.1] and [17, corollaire 18] imply that $\mathcal{R}$ is finitely generated. Hence, $\mathcal{R}$ has finite cost and finite first ℓ^2-Betti number. Since the fundamental group of $\mathcal{R}$ equals $\mathbb{R}_+$, it follows from [10, proposition II.6], respectively [11, corollaire 5.7], that actually $\cot \mathcal{R} = 1$ and $\beta_1^{(2)}(\mathcal{R}) = 0$.

5. Cocycle and OE Superrigidity Theorems

5.1. Proof of Theorem 1.3

We prove in this section the cocycle superrigidity Theorem 1.3 as a consequence of the more general Theorem 5.3 below.

We do not know whether $SL(n, \mathbb{R}) \curvearrowright \mathbb{R}^n$ is $\mathcal{U}_{fin}$-cocycle superrigid for $n = 3, 4$. On the other hand, it is generally believed that whenever one has a $\mathcal{U}_{fin}$-cocycle superrigidity result, it actually holds true for more general target groups than the $\mathcal{U}_{fin}$-class. However, the following result, which we'll prove at the end of this subsection, shows that there even exist locally compact target groups for 1-cocycles for the action $SL(n, \mathbb{R}) \curvearrowright \mathbb{R}^n$, which cannot be untwisted.

PROPOSITION 5.1. Let $n \geq 3$. The action $SL(n, \mathbb{Z}) \curvearrowright \mathbb{R}^n$ admits a 1-cocycle with values in $SL(n-1, \mathbb{R}) \ltimes \mathbb{R}^{n-1}$ that is not cohomologous to a group morphism.

Recall from [19] the following definition of s-malleability of a measure preserving action.

DEFINITION 5.2. Let Γ be an l.c.s.c. group and $\Gamma \curvearrowright (X, \mu)$ an action preserving the finite or infinite measure μ. The action is called s-malleable if there exists

- a 1-parameter group $(\alpha_t)_{t\in\mathbb{R}}$ of measure preserving transformations of $X \times X$, and
- an involutive measure preserving transformation β of $X \times X$,

such that

- α_t and β commute with the diagonal action $\Gamma \curvearrowright X \times X$,
- $\alpha_1(x,y) \in \{y\} \times X$ for almost all $(x,y) \in X \times X$,
- $\beta(x,y) \in \{x\} \times X$ for almost all $(x,y) \in X \times X$, and
- $\alpha_t \circ \beta = \beta \circ \alpha_{-t}$ for all $t \in \mathbb{R}$.

Theorem 0.1 in [19] says the following. Let $\Gamma \curvearrowright (X,\mu)$ be an s-malleable, probability measure preserving action and $\Lambda < \Gamma$ a normal subgroup with the relative property (T) such that the restriction of $\Gamma \curvearrowright (X,\mu)$ to Λ is weakly mixing. Then every 1-cocycle of $\Gamma \curvearrowright (X,\mu)$ with values in a Polish group of finite type, is cohomologous to a group morphism.

Recall here that one of the equivalent formulations of weak mixing for a p.m.p. action $G \curvearrowright (X,\mu)$ of a countable group G is the ergodicity of the diagonal action $G \curvearrowright (X \times X, \mu \times \mu)$. If $G \curvearrowright (X,\mu)$ is a weakly mixing p.m.p. action and $G \curvearrowright (Y,\eta)$ is any ergodic p.m.p. action, then the diagonal action $G \curvearrowright X \times Y$ is ergodic. In particular, the diagonal action

$$G \curvearrowright \underbrace{X \times \cdots \times X}_{k \text{ times}}$$

is ergodic for every k, once it is ergodic for $k = 2$. For infinite measure preserving actions, things are more complicated and, for instance, the diagonal action

$$\mathrm{SL}(n,\mathbb{Z}) \curvearrowright \underbrace{\mathbb{R}^n \times \cdots \times \mathbb{R}^n}_{k \text{ times}}$$

is ergodic if and only if $k \leq n-1$. This partially explains the formulation of the following result.

THEOREM 5.3. *Let $\Gamma \curvearrowright (X,\mu)$ be an infinite measure preserving, s-malleable action. Assume that*

- *the diagonal action $\Gamma \curvearrowright X \times X$ has property (T),*
- *the 4-fold diagonal action $\Gamma \curvearrowright X \times X \times X \times X$ is ergodic.*

Then, $\Gamma \curvearrowright (X,\mu)$ is $\mathcal{U}_{\mathrm{fin}}$-cocycle superrigid.

The proof of Theorem 5.3 follows entirely the setup of the proof of [19, theorem 0.1]. But, one has to be careful at those places in [19] where weak mixing is applied. The proper way to deal with these issues lies in the following lemma distilled from the proof of [8, lemma 3.1].

LEMMA 5.4. *Let (Z, d) be a Polish space with separable complete metric d and $(\alpha_g)_{g \in G}$ a continuous action of a Polish group G by homeomorphisms of Z. Assume that d is $(\alpha_g)_{g \in G}$-invariant.*

Let Γ be an l.c.s.c. group and $\Gamma \curvearrowright (X, \mu)$, $\Gamma \curvearrowright (Y, \eta)$ nonsingular actions. Let $F : X \times Y \to Z$ be a measurable map satisfying

$$F(g \cdot x, g \cdot y) = \alpha_{\omega(g,x)}(F(x, y))$$

for almost all $(x, y) \in X \times Y$, $g \in \Gamma$, where $\omega : \Gamma \times X \to G$ is some measurable map.

Assume the diagonal action $\Gamma \curvearrowright X \times Y \times Y$ satisfies $L^\infty (X \times Y \times Y)^\Gamma = L^\infty (X)^\Gamma \otimes 1 \otimes 1$ (which holds in particular if the diagonal action $\Gamma \curvearrowright X \times Y \times Y$ is ergodic). Then, there exists a measurable $H : X \to Z$ with $F(x, y) = H(x)$ for almost all $(x, y) \in X \times Y$.

Proof. Define the map $G : X \times Y \times Y \to \mathbb{R} : G(x, y, z) = d(F(x, y), F(x, z))$. Since d is $(\alpha_g)_{g \in G}$-invariant, the map G is invariant under the diagonal Γ-action. By our assumption, $G(x, y, z) = G_0(x)$ for almost all (x, y, z) and some measurable map $G_0 : X \to \mathbb{R}_+$. We claim that $G_0(x) = 0$ for almost all $x \in X$. Let $\delta > 0$ and assume that $G_0(x) \geq \delta$ for all x in a nonnegligible subset $\mathcal{U}$ of X. Cover Z by a sequence $(B_n)_{n \in \mathbb{N}}$ of balls of diameter strictly smaller than δ. Write, for every $x \in X$, $F_x : y \mapsto F(x, y)$. By the Fubini theorem, for almost every $x \in \mathcal{U}$, we have $G(x, y, z) \geq \delta$ for almost all $(y, z) \in Y \times Y$. Hence, for almost every $x \in \mathcal{U}$, we have that $F_x^{-1}(B_n)$ is negligible for every n, which is absurd.

So, $G(x, y, z) = 0$ for almost all $(x, y, z) \in X \times Y \times Y$. Again by the Fubini theorem, take $z \in Y$ such that $d(F(x, y), F(x, z)) = 0$ for almost all $(x, y) \in X \times Y$. Putting $H(x) := F(x, z)$, we are done. $\square$

Proof of Theorem 5.3. Let N be a II$_1$ factor and $G \subset \mathcal{U}(N)$ a closed subgroup. Let $\omega : \Gamma \times X \to G$ be a 1-cocycle, meaning that for all $g, h \in \Gamma$, we have

$$\omega(gh, x) = \omega(g, h \cdot x) \omega(h, x) \quad \text{for almost all } x \in X .$$

Define the following 1-cocycles for the diagonal action $\Gamma \curvearrowright X \times X$.

$$\omega_0 : \Gamma \times X \times X \to \mathcal{G} : \omega_0(g, x, y) = \omega(g, x) \quad \text{and}$$

$$\omega_t : \Gamma \times X \times X \to \mathcal{G} : \omega_t(g, x, y) = \omega_0(g, \alpha_t(x, y)) \ .$$

Define the action $(\rho_g)_{g \in \Gamma}$ of Γ by automorphisms of $\mathrm{L}^\infty(X) \,\overline{\otimes}\, N$ by the formula

$$(\rho_{g^{-1}}(F))(x) = \omega(g, x)^* F(g \cdot x)\omega(g, x) \ .$$

Denote by B the von Neumann subalgebra of $(\rho_g)_{g \in \Gamma}$-fixed points.

CLAIM. Whenever p is a nonzero projection in B, there exists a measurable function $\varphi : X \times X \to N$ and a nonzero projection $q \in B$ such that $q \leq p$ and such that for all $g \in \Gamma$, we have

$$\omega(g, x)\varphi(x, y) = \varphi(g \cdot x, g \cdot y)\omega(g, y), \quad \varphi(x, y)\varphi(x, y)^* = q(x),$$

$$\varphi(x, y)^*\varphi(x, y) = q(y)$$

for almost all $(x, y) \in X \times X$.

PROOF OF THE CLAIM. Since p is $(\rho_g)_{g \in \Gamma}$-invariant, the function $x \mapsto \tau(p(x))$ is Γ-invariant and hence constantly equal to $0 < \lambda \leq 1$. Let $p_0 \in N$ be a projection with $\tau(p_0) = \lambda$. It follows that inside $\mathrm{L}^\infty(X) \,\overline{\otimes}\, N$, the projections p and $1 \otimes p_0$ are equivalent. Take a partial isometry $v \in \mathrm{L}^\infty(X) \,\overline{\otimes}\, N$ such that $v^*v = p$ and $vv^* = 1 \otimes p_0$. Define $\eta(g, x) = v(g \cdot x)\omega(g, x)v(x)^*$ and note that η is a 1-cocycle for $\Gamma \curvearrowright X$ with values in $\mathcal{U}(p_0 N p_0)$. Set $\eta_0(g, x, y) = \eta(g, x)$ and, for all $n \geq 1$, $\eta_n(g, x, y) = \eta_0(g, \alpha_{2^{-n}}(x, y))$.

Define the Hilbert space $K = \bigoplus_{k=1}^\infty p_0 \, \mathrm{L}^2(N)p_0$. We define the following 1-cocycle of $\Gamma \curvearrowright X \times X$ with values in the unitary group $\mathcal{U}(K)$ of K:

$$(c(g, x, y)\zeta)_k = \eta(g, x)\,\zeta_k\,\eta_k(g, x, y)^* .$$

Define the map $\xi_n : X \times X \to (K)_1$ by the formula

$$(\xi_n(x, y))_k = \begin{cases} \tau(p_0)^{-1/2}p_0 & \text{if } k = n \ , \\ 0 & \text{if } k \neq n \ . \end{cases}$$

One checks that ξ_n is a sequence of almost invariant unit vectors. Since $\Gamma \curvearrowright X \times X$ has property (T), we find an invariant unit vector, that is, a measurable map $\xi : X \times X \to K$ with $\|\xi(x, y)\| = 1$ for almost all (x, y) and for all $g \in \Gamma$,

$$\xi(g \cdot x, g \cdot y) = c(g, x, y)\,\xi(x, y)$$

almost everywhere. It follows that

$$\xi(g \cdot x, g \cdot y)_k = \eta(g, x)\, \xi(x, y)_k\, \eta_k(g, x, y)^*$$

almost everywhere. In particular, for every k, the function $(x, y) \mapsto \|\xi(x, y)_k\|$ is Γ-invariant and hence, constant. Since $\|\xi(x, y)\| = 1$ for almost all (x, y), we can pick k such that $\|\xi(x, y)_k\| \neq 0$ for almost all (x, y). Taking the polar decomposition of $\xi(x, y)_k$, we find a nonzero partial isometry $\psi \in L^\infty(X \times X) \overline{\otimes} p_0 N p_0$ satisfying

$$\psi(g \cdot x, g \cdot y)\, \eta_k(g, x, y) = \eta(g, x)\, \psi(x, y)$$

almost everywhere. We set $\varphi_0(x, y) := v(x)^* \psi(x, y) v_k(x, y)$, where $v_0(x, y) = v(x)$ and $v_k(x, y) = v_0(\alpha_{2^{-k}}(x, y))$. It follows that

$$\omega(g, x)\, \varphi_0(x, y) = \varphi_0(g \cdot x, g \cdot y)\, \omega_{t_0}(g, x, y)$$

where $t_0 = 2^{-k}$.

Set $r(x, y) = \varphi_0(x, y)\varphi_0(x, y)^*$. It follows that

$$r(g \cdot x, g \cdot y) = \omega(g, x)\, r(x, y)\, \omega(g, x)^*$$

almost everywhere. By Lemma 5.4, we find a projection $q \in L^\infty(X) \overline{\otimes} N$ such that $r(x, y) = q(x)$ almost everywhere. Then, q is a nonzero projection in B and $q \leq p$. Similarly, $\varphi_0(x, y)^* \varphi_0(x, y) = q_0'(\alpha_{t_0}(x, y))$ where q' is a projection in $L^\infty(X) \overline{\otimes} N$ and $q_0'(x, y) = q'(x)$.

Set $q_0(x, y) = q(x)$ and $q_t(x, y) = q_0(\alpha_t(x, y))$. We now construct $\varphi_1 : X \times X \to N$ such that $\varphi_1(x, y)\varphi_1(x, y)^* = q(x)$, $\varphi_1(x, y)^*\varphi_1(x, y) = q_{2t_0}(x, y)$ and

$$\omega(g, x)\, \varphi_1(x, y) = \varphi_1(g \cdot x, g \cdot y)\, \omega_{2t_0}(g, x, y)$$

almost everywhere. Continuing the same procedure k times (remember that $t_0 = 2^{-k}$), we will have found $\varphi = \varphi_k : X \times X \to N$ satisfying $\varphi(x, y)\varphi(x, y)^* = q(x)$, $\varphi(x, y)^*\varphi(x, y) = q_1(x, y) = q(y)$ and

$$\omega(g, x)\, \varphi(x, y) = \varphi(g \cdot x, g \cdot y)\, \omega_1(g, x, y) = \varphi(g \cdot x, g \cdot y)\, \omega(g, y),$$

hence proving the claim. In fact, it suffices to take

$$\varphi_1(x, y) = \varphi_0(x, y)\varphi_0(\beta(\alpha_{2t_0}(x, y)))^*$$

and to use that $(\alpha_t)_{t \in \mathbb{R}}$ is a 1-parameter group, $\beta \circ \alpha_t = \alpha_{-t} \circ \beta$ and $\beta(x, y) \in \{x\} \times Y$ for almost all (x, y). So, the claim above has been proven.

Using the claim and a maximality argument, we find a measurable function $\varphi : X \times X \to \mathcal{U}(N)$ such that

$$\omega(g, x)\, \varphi(x, y) = \varphi(g \cdot x, g \cdot y)\, \omega(g, y)$$

almost everywhere. Set $H(x, y, z) = \varphi(x, y)\varphi(y, z)$. It follows that

$$H(g \cdot x, g \cdot y, g \cdot z) = \omega(g, x) \, H(x, y, z) \, \omega(g, z)^*$$

for almost all (x, y, z). By Lemma 5.4 and because the 4-fold diagonal action $\Gamma \curvearrowright X \times X \times X \times X$ is ergodic, H is essentially independent of its second variable. So, we find a measurable $F : X \times X \to \mathcal{U}(N)$ such that $\varphi(x, y) = F(x, z)\varphi(y, z)^*$ for almost every (x, y, z). By the Fubini theorem, take $z \in X$ such that the previous formula holds for almost all (x, y). Set $\psi(x) = F(x, z)^*$ and $G(y) = \varphi(y, z)^*$. It follows that

$$\psi(g \cdot x) \, \omega(g, x) \, \psi(x)^* = G(g \cdot y) \, \omega(g, y) \, G(y)^*$$

almost everywhere. Hence, the left-hand side is independent of x and we have found a group morphism $\delta : G \to \mathcal{U}(N)$ and a measurable map $\psi : X \to \mathcal{U}(N)$ such that

$$\omega(g, x) = \psi(g \cdot x)^* \, \delta(g) \, \psi(x)$$

almost everywhere.

Consider the quotient Polish space $\mathcal{U}(N)/G$ with the induced metric, which is invariant under left multiplication by elements of $\mathcal{U}(N)$. Write $F : X \to \mathcal{U}(N)/G : F(x) = \psi(x)G$. It follows that $F(g \cdot x) = \delta(g)F(x)$ almost everywhere. By Lemma 5.4, F is essentially constant. So, we find a unitary $u \in \mathcal{U}(N)$ and a measurable map $w : X \to G$ such that $\psi(x) = uw(x)$ almost everywhere. Replacing δ by $u^*\delta(\cdot)u$, it follows that $\omega(g, x) = w(g \cdot x)^*\delta(g)w(x)$ almost everywhere. In particular, $\delta(g) \in G$ and we are done. $\qquad\square$

As a principle (cf. [19, proposition 3.6]), once the restriction of a 1-cocycle $\omega : G \times X \to G$ to a closed subgroup $H < G$ is cohomologous to a group morphism $H \to G$ and if H is sufficiently normal in G and acts sufficiently mixingly on X, the entire 1-cocycle ω is cohomologous to a group morphism $G \to G$. In our setting, we need the following:

LEMMA 5.5. *Let $G \curvearrowright (X, \mu)$ be a nonsingular action of the l.c.s.c. group G and $\omega : G \times X \to G$ a 1-cocycle with values in the Polish group of finite type G. Let $H < G$ be a closed subgroup and assume that $\omega|_H$ is cohomologous to a group morphism $H \to G$. If for every $g \in G$, the diagonal action of the group $H \cap gHg^{-1}$ on $X \times X$ is ergodic, then ω is cohomologous to a morphism $G \to G$.*

Proof. We may assume that for every $h \in H$, we have $\omega(h, x) = \delta(h)$ for almost every $x \in X$, where $\delta : H \to G$ is a continuous group morphism. Let $g \in G$ and put $F(x) = \omega(g, x)$. Using the cocycle equation, it follows that for all $h \in H \cap$

$g^{-1}Hg$, we have $F(h \cdot x) = \delta(ghg^{-1})F(x)\delta(h)^{-1}$ almost everywhere. By Lemma 5.4, F is essentially constant. So, we have shown that for every $g \in G$, the map $x \mapsto \omega(g, x)$ is essentially constant. It follows that $\omega(g, x) = \delta(g)$. $\qquad\square$

After showing the following lemma, we can prove Theorem 1.3 and Proposition 5.1.

LEMMA 5.6. *Let Γ be a lattice in $\mathrm{SL}\,(n, \mathbb{R})$ and consider the linear action of Γ on $\mathbb{R}^n$. Let*

$$\Gamma \curvearrowright X^{(k)} := \underbrace{\mathbb{R}^n \times \cdots \times \mathbb{R}^n}_{k \text{ times}}$$

be the k-fold diagonal action. Then, $\Gamma \curvearrowright X^{(k)}$

- *is ergodic if and only if $k \leq n - 1$, and*
- *has property (T) if and only if $k \leq n - 3$ or $k \geq n$ (the latter part being as interesting as the trivial group $\{e\}$ having property (T)).*

Proof. Writing the elements of $\mathbb{R}^n$ as column vectors, identify, up to measure 0, $X^{(n)}$ with $\mathrm{GL}\,(n, \mathbb{R})$, with the Γ-action given by left multiplication. The determinant function is invariant and not essentially constant, proving that $\Gamma \curvearrowright X^{(k)}$ is nonergodic for $k \geq n$. It also follows that for $k \geq n$, the action $\Gamma \curvearrowright X^{(k)}$ is essentially free and proper (see Definition 3.4). Hence, it has property (T) because the trivial group has property (T).

Now let $k \leq n - 1$. Denoting by $(e_i)_{i=1,\dots,n}$ the standard basis vectors in $\mathbb{R}^n$, the orbit of $(e_1, \dots, e_k) \in X^{(k)}$ under the diagonal $\mathrm{SL}\,(n, \mathbb{R})$-action has complement of measure 0, so that we can identify $\Gamma \curvearrowright X^{(k)}$ with $\Gamma \curvearrowright \mathrm{SL}\,(n, \mathbb{R})/H$, where $H = \{A \in \mathrm{SL}\,(n, \mathbb{R}) \mid Ae_i = e_i \text{ for all } i = 1, \dots, k\}$. Observe that $H \cong \mathrm{SL}\,(n - k, \mathbb{R}) \ltimes \mathrm{M}_{n-k}\,(\mathbb{R})$, where $\mathrm{M}_{n-k}\,(\mathbb{R})$ denotes the additive group of $(n - k) \times (n - k)$ matrices on which $\mathrm{SL}\,(n - k, \mathbb{R})$ acts by multiplication. By Moore's ergodicity theorem (see e.g., [26, theorem 2.2.6]), $H \curvearrowright \Gamma \backslash \mathrm{SL}\,(n, \mathbb{R})$ is ergodic and hence, $\Gamma \curvearrowright \mathrm{SL}\,(n, \mathbb{R})/H$ is ergodic.

By Proposition 3.5, property (T) of $\Gamma \curvearrowright \mathrm{SL}\,(n, \mathbb{R})/H$ is equivalent with property (T) of $H \curvearrowright \Gamma \backslash \mathrm{SL}\,(n, \mathbb{R})$. By Proposition 3.2, the latter is equivalent with the group H having property (T), which is in turn equivalent with $n - k \geq 3$. $\qquad\square$

Proof of Theorem 1.3. Observe that part (1) is a special case of part (3), by taking $k = 1$ and $H = \{1\}$. We start by proving part (3). Set $X = \mathrm{M}_{n,k}\,(\mathbb{R})$. The action

$\Gamma \curvearrowright X$ by left multiplication is s-malleable. It suffices to take

$$\alpha_t(A, B) = (\cos(\pi t/2)A + \sin(\pi t/2)B, -\sin(\pi t/2)A + \cos(\pi t/2)B) \quad \text{and}$$

$$\beta(A, B) = (A, -B)$$

whenever $t \in \mathbb{R}$ and $A, B \in M_{n,k}(\mathbb{R})$.

Moreover, $\Gamma \curvearrowright X$ can be viewed as the k-fold diagonal action $\Gamma \curvearrowright \mathbb{R}^n \times \cdots \times \mathbb{R}^n$. By Lemma 5.6 and because $n \geq 4k + 1$, the diagonal action $\Gamma \curvearrowright X \times X$ has property (T) and the 4-fold diagonal action $\Gamma \curvearrowright X \times X \times X \times X$ is ergodic. So, by Theorem 5.3, $\Gamma \curvearrowright X$ is $\mathcal{U}_{\text{fin}}$-cocycle superrigid. Since the diagonal action $\Gamma \curvearrowright X \times X$ is ergodic and Γ is a normal subgroup of G, Lemma 5.5 implies that $G \curvearrowright X$ is $\mathcal{U}_{\text{fin}}$-cocycle superrigid.

It remains to prove part (2) of the theorem. By part (1), we already know $\Gamma \curvearrowright \mathbb{R}^n$ is $\mathcal{U}_{\text{fin}}$-cocycle superrigid. Define, for every $x \in \mathbb{Z}^n$, $\Gamma_x = \{g \in \Gamma \mid gx = x\}$. We claim that the diagonal action $\Gamma_x \curvearrowright \mathbb{R}^n \times \mathbb{R}^n$ is ergodic for every $x \in \mathbb{Z}^n$. Once this claim is proven, Lemma 5.5 implies that $\Gamma \ltimes \mathbb{Z}^n \curvearrowright \mathbb{R}^n$ is $\mathcal{U}_{\text{fin}}$-cocycle superrigid.

For $x = 0$, the claim follows from Lemma 5.6. Now let $x \neq 0$. Define the closed subgroup $H = \{g \in SL(n, \mathbb{R}) \mid ge_1 = e_1\}$ of $SL(n, \mathbb{R})$. Exactly as in the proof of Lemma 5.6, when $n \geq 4$, the diagonal action $\Lambda \curvearrowright \mathbb{R}^n \times \mathbb{R}^n$ of any lattice $\Lambda \subset H$ is ergodic. Take $g_0 \in SL(n, \mathbb{Q})$ with $x = g_0e_1$. Since Γ is a finite index subgroup of $SL(n, \mathbb{Z})$, it follows that $g_0^{-1}\Gamma_x g_0$ contains a finite index subgroup of $\{g \in SL(n, \mathbb{Z}) \mid ge_1 = e_1\}$ and hence, is a lattice in H. So, its diagonal action on $\mathbb{R}^n \times \mathbb{R}^n$ is ergodic. Then, the same is true for the diagonal action of Γ_x on $\mathbb{R}^n \times \mathbb{R}^n$. $\qquad\square$

Proof of Proposition 5.1. Write $G = SL(n, \mathbb{R})$ and $\Gamma = SL(n, \mathbb{Z})$. As before, identify $\Gamma \curvearrowright \mathbb{R}^n$ with $\Gamma \curvearrowright G/H$, where $H \cong SL(n-1, \mathbb{R}) \ltimes \mathbb{R}^{n-1}$. Let $\theta : G/H \to G$ be a Borel lifting. Define the 1-cocycle $\omega : \Gamma \times G/H \to H$ by

$$g\theta(xH) = \theta(gxH)\omega(g, xH)$$

whenever $g \in \Gamma$ and $xH \in G/H$. Assume that ω is cohomologous to a group morphism $\delta : \Gamma \to H$. This means that we can choose the lifting θ such that $g\theta(xH) = \theta(gxH)\delta(g)$ for $g \in \Gamma$ and almost all $xH \in G/H$.

The image of any group morphism $SL(n, \mathbb{Z}) \to SL(n-1, \mathbb{R}) \ltimes \mathbb{R}^{n-1}$ is finite (see [23, theorem 6] for an elementary argument). So, we have found a finite index subgroup $\Gamma_0 \subset SL(n, \mathbb{Z})$ and a measurable map $\theta : \mathbb{R}^n \to SL(n, \mathbb{R})$ such that $\theta(gx) = g\theta(x)$ for all $g \in \Gamma_0$ and almost all $x \in \mathbb{R}^n$. It follows that the map $(x, y) \mapsto \theta(x)^{-1}\theta(y)$ is invariant under the diagonal Γ_0-action, which is ergodic by Lemma 5.6. Hence, θ is essentially constant, which is a contradiction with the formula $\theta(gx) = g\theta(x)$. $\square$

5.2. Proofs of Theorems 1.5 and 1.6

We will deduce Theorems 1.5 and 1.6 from the general Theorem 5.8 below, dealing with arbitrary actions of the form $\Gamma \curvearrowright M_{n,k}(\mathbb{R})/H$, where Γ is a lattice and $H < GL(k, \mathbb{R})$ a closed subgroup. First, observe that these actions are essentially free and ergodic.

LEMMA 5.7. *Let $n > k$ and $\Gamma < SL(n, \mathbb{R})$ any lattice. Let $H \subset GL(k, \mathbb{R})$ be a closed subgroup. If $(-1, -1) \in \Gamma \times H$, put $\Gamma_0 = \Gamma/\{\pm 1\}$, otherwise put $\Gamma = \Gamma_0$.*

The action $\Gamma_0 \curvearrowright M_{n,k}(\mathbb{R})/H$ is essentially free and ergodic. It never admits an invariant probability measure. It admits an infinite invariant measure if and only if H is unimodular and satisfies $\det g = \pm 1$ for all $g \in H$.

Proof. By Lemma 5.6, $\Gamma \curvearrowright M_{n,k}(\mathbb{R})$ is ergodic, because $k < n$. A fortiori, $\Gamma \curvearrowright M_{n,k}(\mathbb{R})/H$ is ergodic.

Denote $V = \text{span}\{e_1, \ldots, e_k\} \subset \mathbb{R}^n$ and define the closed subgroup $H_1 < SL(n, \mathbb{R})$ by

$$H_1 = \{g \in SL(n, \mathbb{R}) \mid gV = V \text{ and } g|_V \in H\}.$$

We can identify $\Gamma_0 \curvearrowright M_{n,k}(\mathbb{R})/H$ with $\Gamma \curvearrowright SL(n, \mathbb{R})/H_1$. From this description, essential freeness follows. Also the statement about invariant measures follows, because H_1 is unimodular if and only if H is unimodular and satisfies $\det g = \pm 1$ for all $g \in H$. $\square$

THEOREM 5.8. *Let $H \subset GL(k, \mathbb{R})$ be a closed subgroup and $\Lambda \curvearrowright (Y, \eta)$ any essentially free, ergodic, nonsingular action of the countable group Λ. Suppose $n \geq 4k + 1$.*

CASE $-1 \in H$. *Let $\Gamma \subset PSL(n, \mathbb{R})$ be a lattice and put $\widetilde{\Gamma} := \{\pm 1\} \cdot \Gamma \subset GL(n, \mathbb{R})$. The actions $\Gamma \curvearrowright M_{n,k}(\mathbb{R})/H$ and $\Lambda \curvearrowright Y$ are SOE if and only if $\Lambda \curvearrowright Y$ is conjugate to an induction of one of the following actions:*

1) $\Gamma \times H/N \curvearrowright M_{n,k}(\mathbb{R})/N$, where $N \triangleleft H$ is an open normal subgroup with $-1 \in N$, or

2) $\dfrac{\widetilde{\Gamma} \times H/N}{\{\pm(1, 1)\}} \curvearrowright M_{n,k}(\mathbb{R})/N$, where $N \triangleleft H$ is an open normal subgroup with $-1 \notin N$.

CASE $-1 \notin H$. Let $\Gamma \subset$ SL $(n, \mathbb{R})$ be a lattice. The actions $\Gamma \curvearrowright M_{n,k}(\mathbb{R})/H$ and $\Lambda \curvearrowright Y$ are SOE if and only if $\Lambda \curvearrowright Y$ is conjugate to an induction of one of the following actions:

1) $\Gamma \times H/N \curvearrowright M_{n,k}(\mathbb{R})/N$, where $N \lhd H$ is an open normal subgroup, or

2) (only when $-1 \in \Gamma$) $\dfrac{\Gamma}{\{\pm 1\}} \times \dfrac{H}{N} \curvearrowright M_{n,k}(\mathbb{R})/(\{\pm 1\} \cdot N)$, where $N \lhd H$ is an open normal subgroup.

REMARK 5.9. Given a stable orbit equivalence Δ . $M_{n,k}(\mathbb{R})/H \to Y$ between the actions $\Gamma \curvearrowright M_{n,k}(\mathbb{R})/H$ and $\Lambda \curvearrowright Y$, Theorem 5.8 provides a conjugacy Ψ between one of the listed actions and $\Lambda_1 \curvearrowright Y_1$ such that $\Lambda \curvearrowright Y$ is induced from $\Lambda_1 \curvearrowright Y_1$. In fact, moreover, one has $\Psi(\overline{x}) \in \Lambda \cdot \Delta(\overline{x})$ for almost every $x \in M_{n,k}(\mathbb{R})$.

If $\Delta : X_0 \to Y_0$ is a SOE between $\Gamma \curvearrowright X$ and $\Lambda \curvearrowright Y$, we can (and will tacitly) extend Δ to a countable to 1, measurable $\Delta : X \to Y$ satisfying $\Delta(g \cdot x) \in \Lambda \cdot \Delta(x)$ for all $g \in \Gamma$ and almost all $x \in X$.

If $\Delta : X \to Y$ is a SOE between the essentially free actions $\Gamma \curvearrowright X$ and $\Lambda \curvearrowright Y$, we get a *Zimmer 1-cocycle* ω for the action $\Gamma \curvearrowright X$ with values in Λ, determined by the formula

$$\Delta(g \cdot x) = \omega(g, x) \cdot \Delta(x)$$

almost everywhere. As a general principle, if the 1-cocycle ω is cohomologous to a group morphism $\Gamma \to \Lambda$, the stable orbit equivalence is "essentially" given by a conjugacy of the actions: see [26, proposition 4.2.11], [24, lemma 4.7], and [8, theorem 1.8]. In our framework of non-p.m.p. actions, we again need such a principle: see Lemma 5.10 below.

Recall from Definition 3.4 the notion of an essentially free and proper action.

LEMMA 5.10. *Let G be an l.c.s.c. group and $G \overset{\sigma}{\curvearrowright} (X, \mu)$ a nonsingular, essentially free, ergodic action. Assume that σ is $\mathcal{U}_{\mathrm{fin}}$-cocycle superrigid. Let $N \lhd G$ be an open normal subgroup such that the restricted action $\sigma|_N$ is proper. Let $\Lambda \curvearrowright (Y, \eta)$ be an essentially free, ergodic, nonsingular action.*

If $\Delta : X/N \to Y$ is an SOE between the actions $G/N \curvearrowright X/N$ and $\Lambda \curvearrowright Y$, then there exists

- *a subgroup $\Lambda_1 < \Lambda$ and a nonnegligible $Y_1 \subset Y$ such that $\Lambda \curvearrowright Y$ is induced from $\Lambda_1 \curvearrowright Y_1$, and*
- *an open normal subgroup $N_1 \lhd G$ such that $\sigma|_{N_1}$ is proper*

such that the actions $G/N_1 \curvearrowright X/N_1$ and $\Lambda_1 \curvearrowright Y_1$ are conjugate through the non-singular isomorphism $\Psi : X/N_1 \to Y_1$ and the group isomorphism $\delta : G/N_1 \to \Lambda_1$. Furthermore, $\Delta(\overline{x}) \in \Lambda \cdot \Psi(\overline{x})$ for almost all $x \in X$.

Proof. Let $\Delta : X/N \to Y$ be an SOE. By cocycle superrigidity of σ, take a measurable map $\varphi : X \to \Lambda$ such that, writing $\Theta(x) = \varphi(x)^{-1} \cdot \Delta(\overline{x})$, we have $\Theta(g \cdot x) = \delta(g) \cdot \Theta(x)$ almost everywhere, where $\delta : G \to \Lambda$ is a continuous group morphism. Put $N_1 = \operatorname{Ker} \delta$. So, N_1 is an open normal subgroup of G. Then, $N \cap N_1$ is still open in G and we consider $X/(N \cap N_1)$ with the quotient map $\pi : X/(N \cap N_1) \to X/N$. It follows that we can view Θ as a measurable map $\Theta : X/(N \cap N_1) \to Y$ such that $\Delta(\pi(x)) \in \Lambda \cdot \Theta(x)$ for almost all $x \in X/(N \cap N_1)$ and $\Theta(g \cdot x) = \delta(g) \cdot \Theta(x)$ almost everywhere.

Using the facts that the countable group $N/(N \cap N_1)$ acts freely and properly on $X/(N \cap N_1)$, that Λ is countable, and that $\Delta : X/N \to Y$ is locally a nonsingular isomorphism, it follows that $X/(N \cap N_1)$ can be partitioned in a sequence of nonnegligible subsets $(\mathcal{U}_n)_n$ such that for every n, $\Theta|_{\mathcal{U}_n}$ is a nonsingular isomorphism between $\mathcal{U}_n$ and a nonnegligible subset of Y. But then, for every nontrivial element $g \in N_1/(N \cap N_1)$ and every n, we conclude that $g \cdot \mathcal{U}_n \cap \mathcal{U}_n$ has measure 0. It follows that $N_1/(N \cap N_1)$ acts freely and properly on $X/(N \cap N_1)$. So, N_1 acts freely and properly on X. Hence, we can form the quotient space X/N_1 and view Θ as a measurable map $\Theta : X/N_1 \to Y$ such that $\Delta(\overline{x}) \in \Lambda \cdot \Theta(\overline{x})$ for almost all $x \in X$ and $\Theta(\overline{g} \cdot \overline{x}) = \delta(\overline{g}) \cdot \Theta(\overline{x})$ almost everywhere. Now, $\delta : G/N_1 \to \Lambda$ is an injective group morphism. Still, X/N_1 can be partitioned into a sequence of nonnegligible subsets $(\mathcal{U}_n)$ such that $\Theta|_{\mathcal{U}_n}$ is a nonsingular isomorphism between $\mathcal{U}_n$ and $\mathcal{V}_n \subset Y$.

We claim that $\mathcal{V}_n \cap \mathcal{V}_m$ has measure 0 for every $n \neq m$. If this is not the case, take $\mathcal{W} \subset \mathcal{U}_n$ and $\mathcal{W}' \subset \mathcal{U}_m$ nonnegligible and a nonsingular isomorphism $\rho : \mathcal{W} \to \mathcal{W}'$ such that $\Theta(\rho(x)) = \Theta(x)$ for $x \in \mathcal{W}$. Since Δ is an SOE, $\rho(x) \in (G/N_1) \cdot x$ for almost all $x \in \mathcal{W}$. Hence, making $\mathcal{W}$ smaller but still nonnegligible, we may assume that $\rho(x) = \overline{g} \cdot x$ for all $x \in \mathcal{W}$ and some $\overline{g} \in G/N_1$. Since $\mathcal{W} \cap \mathcal{W}'$ has measure 0 and the action of G/N_1 on X/N_1 is essentially free, we get $\overline{g} \neq e$. But also, $\delta(\overline{g}) \cdot \Theta(x) = \Theta(x)$ for almost all $x \in \mathcal{W}$. This is a contradiction with the injectivity of δ and the essential freeness of $\Lambda \curvearrowright Y$. This proves the claim and we have found that Θ is a nonsingular isomorphism between X/N_1 and a nonnegligible subset $Y_1 \subset Y$. Set $\Lambda_1 = \delta(G/N_1)$.

It remains to prove that $\Lambda \curvearrowright Y$ is induced from $\Lambda_1 \curvearrowright Y_1$. So, let $h \in \Lambda$ and assume that $h \cdot Y_1 \cap Y_1$ is nonnegligible. We have to prove that $h \in \Lambda_1$. By our assumption, take $\mathcal{W}, \mathcal{W}' \subset X/N_1$ nonnegligible and a nonsingular isomorphism $\rho : \mathcal{W} \to \mathcal{W}'$ such that $h \cdot \Theta(x) = \Theta(\rho(x))$ for all $x \in \mathcal{W}$. Since

Δ was an SOE, we can make $\mathcal{W}$ smaller but still nonnegligible and assume that $\rho(x) = \overline{g} \cdot x$ for all $x \in \mathcal{W}$ and some $\overline{g} \in G/N_1$. But then, $h \cdot \Theta(x) = \delta(\overline{g}) \cdot \Theta(x)$ for almost all $x \in \mathcal{W}$. Since $\Lambda \curvearrowright Y$ is essentially free, it follows that $h = \delta(\overline{g}) \in \Lambda_1$. $\qquad \square$

Proof of Theorem 5.8. We prove only the case $-1 \in H$; the case $-1 \notin H$ being analogous. By Theorem 1.3, the action of $G := (\widetilde{\Gamma} \times H)/\{\pm(1, 1)\}$ on $\mathrm{M}_{n,k}(\mathbb{R})$ is $\mathcal{U}_{\mathrm{fin}}$-cocycle superrigid. By Lemma 5.10, we have only to prove that the following subgroups of $\widetilde{\Gamma} \times H$ are the only open normal subgroups containing $(-1, -1)$ and acting properly on $\mathrm{M}_{n,k}(\mathbb{R})$:

- $\{\pm 1\} \times N$, where $N \lhd H$ is an open normal subgroup with $-1 \in N$, and
- $(\{1\} \times N) \cup (\{-1\} \times -N)$, where $N \lhd H$ is an open normal subgroup with $-1 \notin N$.

So, let $N \lhd (\widetilde{\Gamma} \times H)$ be a closed normal subgroup acting properly on $\mathrm{M}_{n,k}(\mathbb{R})$. It is sufficient to prove that $N \subset \{\pm 1\} \times H$. Suppose that $N \not\subset \{\pm 1\} \times H$ and take $(g, h) \in N$ with $g \neq \pm 1$. Take $k \in \widetilde{\Gamma}$ such that the commutator $t := kgk^{-1}g^{-1} \neq \pm 1$. It follows that $(t, 1) \in N$. By Margulis's normal subgroup theorem [15], we have $\Gamma_0 \times \{1\} \subset N$ for some finite index subgroup $\Gamma_0 < \widetilde{\Gamma}$. By Lemma 5.6, Γ_0 acts ergodically on $\mathrm{M}_{n,k}(\mathbb{R})$, contradicting the properness of $N \curvearrowright \mathrm{M}_{n,k}(\mathbb{R})$. $\square$

Proof of Theorem 1.5. Statement (1) follows immediately from Theorem 5.8.

We now prove statement (2). We claim that the following are the only normal subgroups N of SL$(n, \mathbb{Z}) \ltimes \mathbb{Z}^n$ that act properly on $\mathbb{R}^n$:

- $N = \{e\}$,
- $N = \lambda \mathbb{Z}^n$ for some $\lambda \in \mathbb{N} \setminus \{0\}$, and
- (only when n is even) $N = \{\pm 1\} \ltimes \lambda \mathbb{Z}^n$ for $\lambda \in \{1, 2\}$.

Statement (2) of Theorem 1.5 then follows from Lemma 5.10 and Theorem 1.3, where the $\mathcal{U}_{\mathrm{fin}}$-cocycle superrigidity of the affine action SL$(n, \mathbb{Z}) \ltimes \mathbb{Z}^n \curvearrowright \mathbb{R}^n$ was established.

So, let $N \lhd \mathrm{SL}(n, \mathbb{Z}) \ltimes \mathbb{Z}^n$ be a normal subgroup acting properly on $\mathbb{R}^n$. Suppose first that $N \not\subset \mathbb{Z}^n$. Taking the commutator of $(g, x) \in N$ with $g \neq 1$ and an arbitrary $(1, y)$, $y \in \mathbb{Z}^n$, it follows that $H := N \cap \mathbb{Z}^n \neq \{0\}$. Hence, H is a nonzero, globally SL$(n, \mathbb{Z})$-invariant subgroup of $\mathbb{Z}^n$. So, $H = \lambda \mathbb{Z}^n$ for some $\lambda \in \mathbb{N} \setminus \{0\}$. If $N \not\subset \{\pm 1\} \ltimes \mathbb{Z}^n$, it would follow that N has finite index in SL$(n, \mathbb{Z}) \ltimes \mathbb{Z}^n$, contradicting the properness of $N \curvearrowright \mathbb{R}^n$. So, we have shown that in all cases $N \subset \{\pm 1\} \ltimes \mathbb{Z}^n$. It is now straightforward to deduce the above list of possibilities for N. $\qquad \square$

Proof of Theorem 1.6. This theorem is a special case of Theorem 5.8. □

5.3. Describing All 1-Cocycles of Quotient Actions

Finally, cocycle superrigidity of $G \curvearrowright (X, \mu)$ allows us to describe all 1-cocycles for $G/N \curvearrowright X/N$ when $N \lhd G$ is a closed normal subgroup of G acting essentially freely and properly on X. We start with the following proposition, closely related to [21, lemma 5.3], and illustrate it with two examples.

PROPOSITION 5.11. *Let G be an l.c.s.c. group and $G \overset{\sigma}{\curvearrowright} (X, \mu)$ a nonsingular action. Let $N \lhd G$ be a closed, normal subgroup such that the restriction $\sigma|_N$ is essentially free and proper. Assume that σ is $\mathcal{U}_{\mathrm{fin}}$-cocycle superrigid.*

Choose a measurable map $\pi : X \to N$ satisfying $\pi(g \cdot x) = g\pi(x)$ for almost all $(g, x) \in N \times X$. Denote by $g \mapsto \bar{g}$ and $x \mapsto \bar{x}$ the quotient maps $G \to G/N$, respectively, $X \to X/N$. Then,

$$\omega : \frac{G}{N} \times \frac{X}{N} \to G : \omega(\bar{g}, \bar{x}) = \pi(g \cdot x)^{-1} g\pi(x)$$

is a well-defined 1-cocycle.

Every 1-cocycle for the action $G/N \curvearrowright X/N$ with values in a Polish group of finite type $\mathcal{G}$ is cohomologous to $\delta \circ \omega$ for a continuous group morphism $\delta : G \to \mathcal{G}$. If the diagonal action $G \curvearrowright X \times X$ is ergodic, δ is uniquely determined up to conjugacy by an element of $g \in \mathcal{G}$.

Proof. Let $\Omega : G/N \times X/N \to \mathcal{G}$ be a 1-cocycle with values in the Polish group of finite type $\mathcal{G}$. From cocycle superrigidity of σ, let $\varphi : X \to \mathcal{G}$ be a measurable map and $\delta : G \to \mathcal{G}$ a continuous group morphism such that

$$\Omega(\bar{g}, \bar{x}) = \varphi(g \cdot x)^{-1} \delta(g)\varphi(x)$$

almost everywhere. Replacing g by hg, $h \in N$, it follows that $\varphi(hg \cdot x)^{-1}\delta(h) = \varphi(g \cdot x)^{-1}$ and hence, $\varphi(h \cdot x) = \delta(h)\varphi(x)$ almost everywhere. So, we can define $\Psi(\bar{x}) = \delta(\pi(x))^{-1}\varphi(x)$. By construction, Ψ makes Ω cohomologous to $\delta \circ \omega$. The uniqueness of δ follows directly from Lemma 5.4. □

EXAMPLE 5.12. Let $\Gamma \subset \mathrm{SL}(n, \mathbb{Z})$ be a finite index subgroup and consider the action $\Gamma \curvearrowright \mathbb{T}^n$.

- Choosing a measurable map $p : \mathbb{R}^n \to \mathbb{Z}^n$ such that $p(x + y) = x + p(y)$ for all $x \in \mathbb{Z}^n$ and almost all $y \in \mathbb{R}^n$, the formula

$$\omega : \Gamma \times \mathbb{R}^n/\mathbb{Z}^n \to \Gamma \ltimes \mathbb{Z}^n : \omega(g, \bar{x}) = p(g \cdot x)^{-1} g p(x)$$

defines a 1-cocycle for $\Gamma \curvearrowright \mathbb{T}^n$ with values in $\Gamma \ltimes \mathbb{Z}^n$.

- Every 1-cocycle with values in a Polish group of finite type $\mathcal{G}$ is cohomologous with $\delta \circ \omega$ for a group morphism $\delta : \Gamma \ltimes \mathbb{Z}^n \to \mathcal{G}$, uniquely determined up to conjugacy by an element in $\mathcal{G}$.

Note that by Zimmer's cocycle superrigidity theorem [26, theorem 5.2.5], any Zariski-dense 1-cocycle for any ergodic p.m.p. action SL $(n, \mathbb{Z}) \curvearrowright (X, \mu)$, $n \geq 3$, taking values in a connected simple real algebraic noncompact center-free group, is cohomologous to a group morphism. For the specific action SL $(n, \mathbb{Z}) \curvearrowright \mathbb{R}^n / \mathbb{Z}^n$, $n \geq 5$, the previous example provides an explicit description of all 1-cocycles of SL $(n, \mathbb{Z}) \curvearrowright \mathbb{R}^n / \mathbb{Z}^n$ with values in an arbitrary Polish group of finite type.

EXAMPLE 5.13. Let X be the real flag manifold of signature $(d_1, \ldots, d_l, n)$, $\Gamma \subset \mathrm{PSL}(n, \mathbb{R})$ a lattice and $n \geq 4d_l + 1$. We obtain as follows all 1-cocycles for the action $\Gamma \curvearrowright X$ with values in a Polish group of finite type.

- Identify $X = \mathrm{M}_{n,k}(\mathbb{R}) / H$ and choose a measurable map $p : \mathrm{M}_{n,k}(\mathbb{R}) \to H$ satisfying $p(Ag) = p(A)g$ for almost all $A \in \mathrm{M}_{n,k}(\mathbb{R})$, $g \in H$. Let $\widetilde{\Gamma} = \{\pm 1\} \cdot \Gamma$ be the double cover of Γ in GL $(n, \mathbb{R})$ and define $G := (\widetilde{\Gamma} \times H) / \{\pm(1, 1)\}$. The formula

$$\omega : \Gamma \times X \to G : \omega(\overline{g}, \overline{x}) = (g, p(gx)p(x)^{-1}) \mod \{\pm(1, 1)\}$$

defines a 1-cocycle with values in G. Here $\widetilde{\Gamma} \to \Gamma : g \mapsto \overline{g}$ and $\mathrm{M}_{n,k}(\mathbb{R}) \to X : x \mapsto \overline{x}$ are the quotient maps.
- Every 1-cocycle with values in a Polish group of finite type $\mathcal{G}$ is cohomologous with $\delta \circ \omega$ for a continuous group morphism $\delta : G \to \mathcal{G}$, uniquely determined up to conjugacy by an element in $\mathcal{G}$.

6. Classification up to Orbit Equivalence

Combining the results of [9] with Theorem 5.8, we classify up to stable orbit equivalence, the linear lattice actions $\Gamma \curvearrowright \mathbb{R}^n$, as well as the natural lattice actions on flag manifolds. At the same time, we compute the outer automorphism group of the associated orbit equivalence relations.

We start with the following elementary lemma.

LEMMA 6.1. *Let $\Gamma \curvearrowright (X, \mu)$ be a nonsingular action of the countable group Γ and assume that the diagonal action $\Gamma \curvearrowright X \times X$ is ergodic. Then, $\Gamma \curvearrowright X$ is not induced, that is, if $\Gamma \curvearrowright X$ is induced from $\Gamma_1 \curvearrowright X_1$, then $\Gamma_1 = \Gamma$ and $\mu(X \setminus X_1) = 0$.*

Proof. Assume that $\Gamma \curvearrowright X$ is induced from $\Gamma_1 \curvearrowright X_1$. So, we find a quotient map $\pi : X \to \Gamma / \Gamma_1$ satisfying $\pi(g \cdot x) = g\pi(x)$ almost everywhere and $X_1 = \pi^{-1}(e\Gamma_1)$. Hence, the subset $\{(x, y) \in X \times X \mid \pi(x) = \pi(y)\}$ is nonnegligible and Γ-invariant. By the ergodicity of $\Gamma \curvearrowright X \times X$, it follows that $\pi(x) = \pi(y)$ for almost all $(x, y) \in X \times X$. This means that $\Gamma_1 = \Gamma$ and $\mu(X \setminus X_1) = 0$. $\square$

If $\mathcal{R}$ is a II_1 equivalence relation on (X, μ), we denote by $[\mathcal{R}]$ the *full group* of $\mathcal{R}$ consisting of nonsingular automorphisms $\Delta : X \to X$ satisfying $(x, \Delta(x)) \in \mathcal{R}$ for almost every $x \in X$. Then, $[\mathcal{R}]$ is a normal subgroup of the automorphism group $\mathrm{Aut}(\mathcal{R})$ of $\mathcal{R}$. The quotient group is denoted by $\mathrm{Out}(\mathcal{R})$ and called the *outer automorphism group* of $\mathcal{R}$. The *full pseudogroup* of $\mathcal{R}$ is denoted by $[[\mathcal{R}]]$ and consists of nonsingular partial isomorphisms $\phi : X_0 \subset X \to X_1 \subset X$ satisfying $(x, \phi(x)) \in \mathcal{R}$ for almost every $x \in X_0$. We denote $X_0 = D(\phi)$ and $X_1 = R(\phi)$.

THEOREM 6.2. *Let $n \geq 5$ and $\Gamma \subset \mathrm{SL}(n, \mathbb{R})$ a lattice. Let $n' \geq 2$ and $\Gamma' \subset \mathrm{SL}(n', \mathbb{R})$ a lattice. If the nonsingular isomorphism $\Delta : X_1 \subset \mathbb{R}^n \to X_1' \subset \mathbb{R}^{n'}$ is an SOE between $\Gamma \curvearrowright \mathbb{R}^n$ and $\Gamma' \curvearrowright \mathbb{R}^{n'}$, then*

- *$n = n'$ and there exists $A \in \mathrm{GL}(n, \mathbb{R})$ such that $\Gamma' = A\Gamma A^{-1}$, and*
- *there exists $\phi \in [[\mathcal{R}(\Gamma' \curvearrowright \mathbb{R}^{n'})]]$ with $R(\phi) = X_1'$ such that $\Delta(x) = \phi(A(x))$ for almost every $x \in X_1$.*

In particular, $\mathrm{Out}(\mathcal{R}(\Gamma \curvearrowright \mathbb{R}^n)) = \mathcal{N}_{\mathrm{GL}(n,\mathbb{R})}(\Gamma) / \Gamma$.

Proof. Let $\Delta : X_1 \subset \mathbb{R}^n \to X_1' \subset \mathbb{R}^{n'}$ be an SOE between $\Gamma \curvearrowright \mathbb{R}^n$ and $\Gamma' \curvearrowright \mathbb{R}^{n'}$. Since for $n' = 2$ the equivalence relation $\mathcal{R}(\Gamma' \curvearrowright \mathbb{R}^{n'})$ is hyperfinite, we have $n' \geq 3$ and hence the diagonal action $\Gamma' \curvearrowright \mathbb{R}^{n'} \times \mathbb{R}^{n'}$ is ergodic. By Lemma 6.1, $\Gamma' \curvearrowright \mathbb{R}^{n'}$ is not an induced action. By Theorem 1.5, the action $\Gamma' \curvearrowright \mathbb{R}^{n'}$ is conjugate with either $\Gamma \curvearrowright \mathbb{R}^n$ or, in case n is even, $\Gamma / \{\pm 1\} \curvearrowright \mathbb{R}^n / \{\pm 1\}$. By Mostow rigidity, the latter is impossible since $\Gamma' \not\cong \Gamma / \{\pm 1\}$. In the former case, we already conclude $n = n'$ and we have found a nonsingular isomorphism $\Theta : \mathbb{R}^n \to \mathbb{R}^n$ and a group isomorphism $\delta : \Gamma \to \Gamma'$ satisfying

- $\Theta(g \cdot x) = \delta(g) \cdot \Theta(x)$, and
- $\Delta(x) \in \Gamma' \cdot \Theta(x)$,

for all $g \in \Gamma$ and almost all $x \in \mathbb{R}^n$. Denoting by $B^\top$ the transpose of the matrix B, by Mostow rigidity, we find $A \in \mathrm{GL}(n, \mathbb{R})$ such that (a) $\delta(g) = AgA^{-1}$ for all $g \in \Gamma$, or (b) $\delta(g) = A(g^\top)^{-1}A^{-1}$ for all $g \in \Gamma$.

Define the subgroup $H \subset \mathrm{SL}(n, \mathbb{R})$ consisting of matrices g with $ge_1 = e_1$ and identify $\mathbb{R}^n = \mathrm{SL}(n, \mathbb{R})/H$. In case (b), we would get a conjugacy between the Γ-actions on $\mathrm{SL}(n, \mathbb{R})/H$ and $\mathrm{SL}(n, \mathbb{R})/H^\top$, which is ruled out by [9, theorem D]. In case (a), multiplying A by a nonzero scalar if necessary, [9, theorem D] implies that $\Theta(x) = Ax$ for almost all $x \in \mathbb{R}^n$.

Defining the partial isomorphism $\phi := \Delta \circ A^{-1}$ with $R(\phi) = X_1'$, it follows that $\phi \in [[\mathcal{R}(\Gamma' \curvearrowright \mathbb{R}^{n'})]]$ and that $\Delta(x) = \phi(A(x))$ for almost every $x \in X_1$. $\square$

Let X be the real flag manifold with signature $d := (d_1, \dots, d_l, n)$. Denote

$$d^\top := (n - d_l, n - d_{l-1}, \dots, n - d_1, n) .$$

If X' is the real flag manifold with signature $d^\top$, there is a natural diffeomorphism $X \to X' : x \mapsto \overline{x}$ satisfying $\overline{g \cdot x} = (g^\top)^{-1} \cdot \overline{x}$ for all $x \in X$, $g \in \mathrm{SL}(n, \mathbb{R})$.

THEOREM 6.3. *Let $\Gamma \subset \mathrm{PSL}(n, \mathbb{R})$ be a lattice and X the real flag manifold with signature $d := (d_1, \dots, d_l, n)$. Assume that $n \geq 4d_l + 1$. Let $\Gamma' \subset \mathrm{PSL}(n', \mathbb{R})$ be a lattice and X' the real flag manifold with signature $d' := (d_1', \dots, d_{l'}', n')$.*

If the nonsingular isomorphism $\Delta : X_1 \subset X \to X_1' \subset X'$ is an SOE between $\Gamma \curvearrowright X$ and $\Gamma' \curvearrowright X'$, then $n = n'$ and there exists $A \in \mathrm{PGL}(n, \mathbb{R})$, $\phi \in [[\mathcal{R}(\Gamma' \curvearrowright X')]]$ with $R(\phi) = X_1'$ such that either

- *$d' = d$, $\Gamma' = A\Gamma A^{-1}$, $\Delta(x) = \phi(A(x))$ for almost every $x \in X_1$, or*
- *$d' = d^\top$, $\Gamma' = A\Gamma^\top A^{-1}$, $\Delta(x) = \phi(A(\overline{x}))$ for almost every $x \in X_1$.*

In particular, $\mathrm{Out}(\mathcal{R}(\Gamma \curvearrowright X)) = \mathcal{N}_{\mathrm{PGL}(n, \mathbb{R})}(\Gamma)/\Gamma$.

Proof. Let $\Delta : X_1 \subset X \to X_1' \subset X'$ be an SOE between $\Gamma \curvearrowright X$ and $\Gamma' \curvearrowright X'$.

We first prove that the diagonal action $\Gamma' \curvearrowright X' \times X'$ is ergodic. If Y denotes the flag manifold of signature $(1, 2, \dots, n')$, the action $\Gamma' \curvearrowright X'$ is a quotient of the action $\Gamma \curvearrowright Y$. So, it suffices to prove ergodicity of $\Gamma' \curvearrowright Y \times Y$. Denoting by $D \subset \mathrm{SL}(n', \mathbb{R})$ the subgroup of diagonal matrices, $\Gamma' \curvearrowright Y \times Y$ can be identified with $\Gamma' \curvearrowright \mathrm{SL}(n', \mathbb{R})/D$, which follows ergodic because $D \curvearrowright \mathrm{SL}(n', \mathbb{R})/\Gamma'$ is ergodic by Moore's ergodicity theorem. So, by Lemma 6.1, $\Gamma' \curvearrowright X'$ is not an induced action.

Since Γ' has trivial center, Theorem 1.6 yields a group isomorphism $\delta : \Gamma \to \Gamma'$ and a nonsingular isomorphism $\Theta : X \to X'$ satisfying $\Theta(g \cdot x) = \delta(g) \cdot \Theta(x)$ and $\Delta(x) \in \Gamma' \cdot \Theta(x)$ almost everywhere. By Mostow rigidity, $n = n'$ and there exists $A \in \mathrm{PGL}(n, \mathbb{R})$ such that either $\delta(g) = AgA^{-1}$ or $\delta(g) = A(g^\top)^{-1}A^{-1}$ for all $g \in \Gamma$.

Set $k_1 = d_1$ and $k_i = d_i - d_{i-1}$ for $2 \leq i \leq l$. Put $k = d_l$ and define the closed subgroups H and H_1 of $\mathrm{GL}(k, \mathbb{R})$ by

$$
H := \left\{ \begin{pmatrix} A_{11} & * & \cdots & * \\ 0 & A_{22} & \cdots & * \\ \vdots & \vdots & \ddots & \vdots \\ 0 & 0 & \cdots & A_{ll} \end{pmatrix} \; \middle| \; A_{ii} \in \mathrm{GL}(k_i, \mathbb{R}) \; \text{ and } \; \det(A_{ii}) \neq 0 \right\},
$$

$$
H_1 := \left\{ \begin{pmatrix} A_{11} & * & \cdots & * \\ 0 & A_{22} & \cdots & * \\ \vdots & \vdots & \ddots & \vdots \\ 0 & 0 & \cdots & A_{ll} \end{pmatrix} \; \middle| \; A_{ii} \in \mathrm{GL}(k_i, \mathbb{R}) \; \text{ and } \; \det(A_{ii}) = \pm 1 \right\}.
$$

We identify $H/H_1 = \mathbb{R}_+^l$. From now on, we write X as $\mathrm{M}_{n,k}(\mathbb{R})/H$. Define analogously the subgroups H', H_1' of $\mathrm{GL}(k', \mathbb{R})$ and write X' as $\mathrm{M}_{n,k}(\mathbb{R})/H'$.

Whenever $N \lhd H$ is a closed normal subgroup containing -1, the quotient morphism $H \to H/N$ gives rise, as in Example 5.13, to a 1-cocycle ω_N for the action $\Gamma \curvearrowright X$ with values in H/N such that the action $\Gamma \curvearrowright \mathrm{M}_{n,k}(\mathbb{R})/N$ can be identified with the action $\Gamma \curvearrowright X \times H/N$ given by

$$
g \cdot (x, h) = (g \cdot x, \omega_N(g, x)h) .
$$

Since $\Gamma \curvearrowright \mathrm{M}_{n,k}(\mathbb{R})/N$ is ergodic, the 1-cocycle ω_N cannot be cohomologous to a 1-cocycle taking values in a proper closed subgroup of H/N. We similarly define the 1-cocycles $\omega_{N'}$ for $\Gamma' \curvearrowright X'$.

Since δ, Θ conjugate the actions $\Gamma \curvearrowright X$ and $\Gamma' \curvearrowright X'$, the map $\mu(g, x) = \omega_{H_1'}(\delta(g), \Theta(x))$ defines a 1-cocycle for $\Gamma \curvearrowright X$, with values in $\mathbb{R}_+^{l'}$ and with the property of not being cohomologous to a 1-cocycle taking values in a proper closed subgroup of $\mathbb{R}_+^{l'}$. We now apply Example 5.13, describing all $\mathcal{U}_{\mathrm{fin}}$-valued 1-cocycles for $\Gamma \curvearrowright X$, and note that $\mathbb{R}_+^{l'}$ belongs to $\mathcal{U}_{\mathrm{fin}}$. Every group morphism $\Gamma \to \mathbb{R}_+^{l'}$ is trivial and every continuous group morphism $H \to \mathbb{R}_+^{l'}$ is trivial on H_1. So, we find a continuous group morphism $\rho : \mathbb{R}_+^l \to \mathbb{R}_+^{l'}$ such that μ is cohomologous to $\rho \circ \omega_{H_1}$. Since μ cannot be cohomologous to a 1-cocycle taking values in a proper closed subgroup of $\mathbb{R}_+^{l'}$, it follows that ρ is onto.

Altogether, we find a closed normal subgroup $N \lhd H$, containing H_1, and a continuous isomorphism $\rho : H/N \to H'/H_1'$ such that μ is cohomologous to $\rho \circ \omega_N$. It follows that there exists a nonsingular isomorphism $\Theta_1 : \mathrm{M}_{n,k}(\mathbb{R})/N \to \mathrm{M}_{n',k'}(\mathbb{R})/H_1'$ satisfying $\Theta_1(g \cdot x) = \delta(g) \cdot \Theta_1(x)$ and $\Theta_1(x)H' = \Theta(xH)$ almost everywhere.

Since the action of Γ' on $M_{n',k'}(\mathbb{R})/H_1'$ is infinite measure preserving, Lemma 5.7 implies that $N = H_1$. We saw already that either $\delta(g) = AgA^{-1}$ or $\delta(g) = A(g^\top)^{-1}A^{-1}$ for all $g \in \Gamma$. In the former case, [9, theorem D] implies that $d' = d$ and that there exists $B \in H$ such that $\Theta_1(x) = AxB$ for almost every $x \in M_{n,k}(\mathbb{R})/H_1$. It follows that $\Theta(x) = A(x)$ for almost every $x \in X$. In the latter case, we prove analogously that $d' = d^\top$ and $\Theta(x) = A(\overline{x})$. $\qquad \square$

7. Implementation by Group Actions

In [5, p. 292], the question is raised whether every II_1 equivalence relation can be implemented by an essentially free action of a countable group. This question has been settled in the negative in [7, theorem D]. In Proposition 7.1 below, we give examples of II_1 equivalence relations $\mathcal{R}$ on (X, μ) with the following much stronger property: whenever $\Lambda \curvearrowright (Y, \eta)$ is a free, nonsingular action and $\Delta : X \to Y$ is a measurable map satisfying $\Delta(x) \in \Lambda \cdot \Delta(y)$ for almost all $(x, y) \in \mathcal{R}$, then there exists $y_0 \in Y$ such that $\Delta(x) \in \Lambda \cdot y_0$ for almost all $x \in X$.

Among other examples, [7, theorem D] proves that the restriction of the orbit equivalence relation $\mathrm{SL}(n, \mathbb{Z}) \curvearrowright \mathbb{R}^n/\mathbb{Z}^n$ to a subset of irrational measure, provides a II_1 equivalence relation that cannot be implemented by a free action of a group. By [19, theorem 0.3], if $\Gamma \curvearrowright [0, 1]^\Gamma$ is the Bernoulli action of a property (T) group Γ without finite normal subgroups, the restriction of its orbit equivalence relation to any subset of measure strictly between 0 and 1 is unimplementable by a free action.

But [7, theorem D] also provides examples of II_1 equivalence relations $\mathcal{R}$ such that none of the amplifications $\mathcal{R}^t$, $t > 0$, can be implemented by a free action. These equivalence relations are constructed using the following method. Suppose that G is an l.c.s.c. unimodular group and $G \curvearrowright (X, \mu)$ an essentially free, properly ergodic p.m.p. action. There exists a Borel set $Y \subset X$, a probability measure η on Y, and a neighborhood $\mathcal{U}$ of e in G that we equip with a multiple of the Haar measure, such that $\mathcal{U} \times Y \to X : (g, y) \mapsto g \cdot y$ provides a measure preserving isomorphism of $\mathcal{U} \times Y$ onto a nonnegligible subset of X. The restriction of the orbit equivalence relation of $G \curvearrowright (X, \mu)$ to Y is a II_1 equivalence relation on (Y, η). One calls $Y \subset X$ a *measurable cross section* for $G \curvearrowright (X, \mu)$. A different choice of measurable cross section yields a stably isomorphic II_1 equivalence relation.

By Theorem 4.1, the restriction of the orbit equivalence relation of $\mathrm{SL}(n, \mathbb{Z}) \curvearrowright \mathbb{R}^n$ to a subset of finite measure, provides other examples of II_1 equivalence relations $\mathcal{R}$ such that none of the finite amplifications can be implemented by a free action. In fact, one can show that $\mathcal{R}$ arises as the measurable cross section for the action of $SL(n-1, \mathbb{R}) \ltimes \mathbb{R}^{n-1}$ on $SL(n, \mathbb{R}) / SL(n, \mathbb{Z})$. Nevertheless, this example is not covered by [7, theorem D], since $SL(n-1, \mathbb{R}) \ltimes \mathbb{R}^{n-1}$ is not semisimple.

PROPOSITION 7.1. *Let G be an l.c.s.c. connected, unimodular group with normal closed subgroup G_0 having the relative property (T). Let $H_{\mathbb{R}}$ be a real Hilbert space and $\pi : G \to O(H_{\mathbb{R}})$ an orthogonal representation. Assume that π is injective and that the restriction of π to G_0 is weakly mixing (i.e., has no finite-dimensional invariant subspaces). Denote by $G \curvearrowright (X, \mu)$ the associated Gaussian action (see e.g., [8, section 2.7]). Choose a measurable cross-section $X_1 \subset X$ and denote by $\mathcal{R}$ the associated II_1 equivalence relation on X_1.*

If $\Lambda \curvearrowright (Y, \eta)$ is a free, nonsingular action and $\Delta : X_1 \to Y$ is a measurable map satisfying $\Delta(x) \in \Lambda \cdot \Delta(y)$ for almost all $(x, y) \in \mathcal{R}$, then there exists $y_0 \in Y$ such that $\Delta(x) \in \Lambda \cdot y_0$ for almost all $x \in X_1$.

Proof. Choose a measurable map $p : X \to X_1$ such that $p(x) \in G \cdot x$ for almost all $x \in X$. Define the 1-cocycle $\omega : G \times X \to \Lambda$ such that $\Delta(p(g \cdot x)) = \omega(g, x) \cdot \Delta(p(x))$ almost everywhere. As observed in [8], theorem 0.1 in [19] applies to $G \curvearrowright (X, \mu)$. Since G is connected, every group morphism from G to Λ is trivial and we find a measurable map $\varphi : X \to \Lambda$ such that $\omega(g, x) = \varphi(g \cdot x)^{-1} \varphi(x)$. So, the map $x \mapsto \varphi(x) \cdot \Delta(p(x))$ is G-invariant and hence, essentially constant. We therefore find $y_0 \in Y$ such that $\Delta(p(x)) \in \Lambda \cdot y_0$ for almost all $x \in X$. This concludes the proof of the proposition. $\square$

References

[1]   C. Anantharaman-Delaroche, Cohomology of property T groupoids and applications. *Ergod. Th. & Dynam. Sys.* **25** (2005), 977–1013.

[2]   P.L. Aubert, Deux actions de $SL(2, \mathbb{Z})$. In *Ergodic theory (Les Plans-sur-Bex, 1980)*, Monograph. Enseign. Math. **29**, Univ. Genève, Geneva, 1981, pp. 39–46.

[3]   B. Bekka, P. de la Harpe, and A. Valette, Kazhdan's property (T). *New Mathematical Monographs* **11**. Cambridge University Press, Cambridge, 2008.

[4]   A. Connes, Classification of injective factors. *Ann. Math.* **104** (1976), 73–115.

[5]   J. Feldman and C.C. Moore, Ergodic equivalence relations, cohomology, and von Neumann algebras I, II. *Trans. Amer. Math. Soc.* **234** (1977), 289–324, 325–359.

[6]   A. Furman, Gromov's measure equivalence and rigidity of higher rank lattices. *Ann. of Math.* **150** (1999), 1059–1081.

[7]   A. Furman, Orbit equivalence rigidity. *Ann. of Math.* **150** (1999), 1083–1108.

[8]   A. Furman, On Popa's cocycle superrigidity theorem. *Int. Math. Res. Not. IMRN* **19** (2007), Art. ID rnm073, 46 pp.

[9] A. Furman, Measurable rigidity of actions on infinite measure homogeneous spaces, II. *J. Amer. Math. Soc.* **21** (2008), 479–512.

[10] D. Gaboriau, Coût des relations d'équivalence et des groupes. *Invent. Math.* **139** (2000), 41–98.

[11] D. Gaboriau, Invariants l^2 de relations d'équivalence et de groupes. *Publ. Math. Inst. Hautes Études Sci.* **95** (2002), 93–150.

[12] S.L. Gefter and V.Ya. Golodets, Fundamental groups for ergodic actions and actions with unit fundamental groups. *Publ. Res. Inst. Math. Sci.* **24** (1988), 821–847.

[13] A. Ioana, Cocycle superrigidity for profinite actions of property (T) groups (2008), arXiv:0805.2998. To appear in *Duke Math. J.*

[14] A.S. Kechris, Classical descriptive set theory. *Graduate Texts in Mathematics* **156**. Springer-Verlag, New York, 1995.

[15] G.A. Margulis, Discrete subgroups of semisimple Lie groups. Ergebnisse der Mathematik und ihrer Grenzgebiete (3) **17**. Springer-Verlag, Berlin, 1991.

[16] C.C. Moore, Group extensions and cohomology for locally compact groups. III. *Trans. Amer. Math. Soc.* **221** (1976), 1–33.

[17] M. Pichot, Sur la théorie spectrale des relations d'équivalence mesurées. *Journal of the Inst. of Math. Jussieu* **6** (2007), 453–500.

[18] S. Popa, On the superrigidity of malleable actions with spectral gap. *J. Amer. Math. Soc.* **21** (2008), 981–1000.

[19] S. Popa, Cocycle and orbit equivalence superrigidity for malleable actions of w-rigid groups. *Invent. Math.* **170** (2007), 243–295.

[20] S. Popa, Correspondences, INCREST preprint No 56, 1986 (unpublished).

[21] S. Popa and S. Vaes, Strong rigidity of generalized Bernoulli actions and computations of their symmetry groups. *Adv. Math.* **217** (2008), 833–872.

[22] S. Popa and S. Vaes, On the fundamental group of II₁ factors and equivalence relations arising from group actions. To appear in *Quanta of Maths, Proceedings of the Conference in honor of Alain Connes*, Clay Math. Inst., 2010.

[23] R. Steinberg, Some consequences of the elementary relations in SL_n. In Finite groups— coming of age (Montreal, 1982). *Contemp. Math.* **45**, American Mathematical Society, Providence, RI, 1985, pp. 335–350.

[24] S. Vaes, Rigidity results for Bernoulli actions and their von Neumann algebras (after Sorin Popa). Séminaire Bourbaki, exp. no. 961. *Astérisque* **311** (2007), 237–294.

[25] R.J. Zimmer, On the cohomology of ergodic actions of semisimple Lie groups and discrete subgroups. *Amer. J. Math.* **103** (1981), 937–951.

[26] R.J. Zimmer, Ergodic theory and semisimple groups. Birkhäuser, Boston, 1984.

PART 3

Geometric Group Theory

HEIGHTS ON SL_2 AND FREE SUBGROUPS

DEDICTATED TO BOB ZIMMER ON THE OCCASION OF HIS 60TH BIRTHDAY

Abstract

We discuss and survey the strong uniform Tits Alternative [8] and give a complete proof of it in the special case of $GL_2(\mathbb{C})$. The main arithmetic ingredient, the height gap theorem [7], which can be seen as a nonabelian analogue to the Lehmer conjecture, is also given a complete treatment in that case. We then give several applications involving expansion properties of $SL_d(\mathbb{Z}/p\mathbb{Z})$, a uniform l^2 spectral gap, and diophantine properties of subgroups of $GL_d(\mathbb{C})$.

Contents

1. Introduction

1.1. Statement of the Main Results

Recall that the Tits alternative [31] asserts that any finitely generated subgroup of $GL_d(K)$, where K is some field, contains a nonabelian free subgroup on two generators unless it is amenable, or equivalently in this case, unless it contains a solvable subgroup of finite index (i.e., is virtually solvable). In [7] and [8], we showed the following uniform version of Tits's theorem :

The author acknowledges support from the IAS Princeton through NSF Grant DMS-0635607.

THEOREM 1.1. (STRONG UNIFORM TITS ALTERNATIVE [8])   *For every* $d \in \mathbb{N}$ *there is* $N = N(d) \in \mathbb{N}$ *such that if* K *is any field and* F *a finite symmetric subset of* $GL_d(K)$ *containing* 1, *either* F^N *contains two elements that freely generate a nonabelian free group, or the group generated by* F *is virtually solvable (i.e., contains a finite index solvable subgroup).*

We denote by $F^n = F \cdot \ldots \cdot F$ the product set of n copies of F. The purpose of this text is to discuss some consequences of this result and to give a self-contained proof of it in the special case of SL_2 (hence equivalently GL_2) and K any field of characteristic 0 (or, equivalently, for $K = \mathbb{C}$, since every finitely generated field of characteristic 0 embeds in $\mathbb{C}$). This case is already representative of the general case as it captures the main difficulty, namely, treat all number fields in a uniform way. The proof given in [7] and [8] is an elaboration of the proof for SL_2 that we are about to give.

Theorem 1.1 improves earlier refinements of Tits's theorem due to Eskin-Mozes-Oh (see [13]) and to T. Gelander and the author (see [9]). These two papers were concerned with the S-arithmetic version of Theorem 1.1, namely, they proved uniformity of N (the "freeness radius" of F) for sets F with coefficients inside a fixed finitely generated ring. While in [13] the main concern was to prove uniform exponential growth by constructing generators of a free semigroup in F^N, in [9] it was shown that the result of [13] could be pushed to get generators of a free group in F^N, where N was depending only on d and on the ring generated by the matrix coefficients of the elements of F. The main contribution of this text and of [7] and [8] is to remove the dependence on the ring of coefficients. As in Tits's proof or in [9], the proof of Theorem 1.1 can be divided into an arithmetic step, on the one hand, and a geometric step on the other. While in [9] (as well as in Tits's original theorem) the arithmetic step was the easier one and most of the work lied in showing that a certain geometric configuration (the so-called "ping-pong") did arise, roles are reversed in the proof of [8] in Theorem 1.1 and the arithmetic step is the harder step, while the geometric step routinely follows Tits's original proof after a careful check that all estimates are indeed uniform over all local fields. For SL_2, however, none of the usual difficulties of higher rank arise and as will become clear below, this geometric step is even more transparent (in that case or other rank-1 situations this geometric step can also be performed differently by acting directly on the hyperbolic space/tree as has been pointed out by T. Gelander).

The proof of Theorem 1.1 is effective in the sense that the constant N can, in principle, be made explicit. Examples due to Grigorchuk and de la Harpe [18] imply that $N(d)$ must tend to infinity with d. They exhibited a sequence $(\Gamma_n)_{n \geq 0}$ $(\Gamma_n \leq GL(n, \mathbb{Z}))$ of 4-generated subgroups whose rate of exponential growth exponent decays to 1. These groups arise by chopping the usual presentation of the Grigorchuk group after finitely many relators (see [18]).

The arithmetic step in Theorem 1.1 relies on the following result, proved by the author in [7], and for which we will give below a self-contained proof in the special case of SL_2. Before stating it, we need some preliminary definitions.

Let $\overline{\mathbb{Q}}$ be an algebraic closure of $\mathbb{Q}$. In [7] we introduced the *arithmetic spectral radius* (or normalized height) of F, defined as

$$\widehat{h}(F) = \lim_{n \to +\infty} \frac{1}{n} h(F^n),$$

where h is the (absolute) *height* defined for F a finite subset of $M_d(\overline{\mathbb{Q}})$ by

$$h(F) = \frac{1}{[K : \mathbb{Q}]} \sum_{v \in V_K} n_v \log^+ \|F\|_v$$

where $\log^+ = \max\{0, \log\}$, K is the number field generated by the matrix co-efficients of F, V_K is the set of all places of K, and $\|F\|_v = \max\{\|f\|_v, f \in F\}$ is the maximal operator norm of $f \in F$, where $\|f\|_v = \max_{x \neq 0} \|f(x)\|_v / \|x\|_v$ for the standard norm $\|x\|_v$ induced on K_v^d by the standard absolute value $|\cdot|_v$ on the completion K_v of K associated to $v \in V_K$. We have also set $n_v = [K_v, \mathbb{Q}_v]$, where $\mathbb{Q}_v$ is the field of p-adic numbers if $v|p$ is finite and is $\mathbb{R}$ if v is infinite. The normalization of the absolute value $|\cdot|_v$ is such that $|\lambda|_v^{n_v}$ is the modulus of the multiplication by λ on K_v.

The quantities $h(F)$ and $\widehat{h}(F)$ are well defined, that is, they are independent of the chosen number field. If $d = 1$ and F is just one element $\{x\}$, then $h(F) = h(x)$ is the classical (absolute, logarithmic) Weil height of x (see [3]). Moreover, $\widehat{h}(F)$ is invariant under conjugation by elements from $GL_d(\overline{\mathbb{Q}})$. The main statement is

THEOREM 1.2. (HEIGHT GAP THEOREM [7]) *There is a positive constant* $\varepsilon = \varepsilon(d) > 0$ *such that if F is a finite subset of $GL_d(\overline{\mathbb{Q}})$ generating a nonvirtually solvable subgroup Γ, then*

$$\widehat{h}(F) > \varepsilon.$$

Moreover, if the Zariski closure of Γ is semisimple, then

$$\widehat{h}(F) \leq \inf_{g \in GL_d(\overline{\mathbb{Q}})} h(gFg^{-1}) \leq C \cdot \widehat{h}(F)$$

for some absolute constant $C = C(d) > 0$.

This result can be seen as a *nonabelian* version of the well-known *Lehmer conjecture* from number theory. This conjecture states that the Weil height $h(x)$ of an algebraic number that is not a root of unity must be bounded below by $\frac{\varepsilon_0}{\deg(x)}$, where ε_0 is an absolute constant. If F consists of a single diagonal matrix $diag(x, x^{-1})$, then $\widehat{h}(F) = h(x)$.

When $d = 2$, that is, for SL_2, the first part of Theorem 1.2 has the following geometric interpretation: either there is a finite place v where F acts without global fixed point on the corresponding Bruhat-Tits tree, or after applying some Galois automorphism of $\mathbb{C}$, the set F, as it acts on the hyperbolic 3-space $\mathbb{H}^3$ via $SL_2(\mathbb{C})$, moves every point away from itself by a positive absolute constant ε. This is analogous to the Margulis lemma in hyperbolic geometry, according to which if F generates a discrete subgroup of $SL_2(\mathbb{C})$ that is not virtually nilpotent, then every point of $\mathbb{H}^3$ is moved away from itself by some element of F by some fixed constant (see [30]).

The main purpose of Theorem 1.2 is to yield in F, or a bounded power of F, a nice hyperbolic element, that is, a semisimple matrix whose eigenvalue is of modulus at least 2, say, in some local completion.

Finally, letime point out that the proof of Theorem 1.1 yields a free subgroup with the extra property that it is *uniformly undistorted* in the original subgroup, namely:

THEOREM 1.3. (UNIFORMLY QUASI-ISOMETRICALLY EMBEDDED FREE) SUBGROUP [8] *There is a constant $C = C(d) > 0$ such that if K is any field and Γ is any non virutally solvable subgroup of $GL_d(K)$ generated by a finite symmetric subset F (giving rise to a word metric $d_\Gamma(\cdot, \cdot)$ on Γ), there exists a free subgroup H of Γ generated by two elements (giving rise to a word metric $d_H(\cdot, \cdot)$) such that for all $h \in H$*

$$\frac{1}{C \cdot |F|^C} \cdot d_\Gamma(1, h) \leq d_H(1, h) \leq d_\Gamma(1, h).$$

1.2. Some Consequences for Uniform Growth, Spectral Gap, and Diophantine Properties

We will prove here these corollaries for $GL_d(\mathbb{C})$, $d \geq 2$.

COROLLARY 1.4. (STRONG UNIFORM EXPONENTIAL GROWTH) *There is $\varepsilon = \varepsilon(d) > 0$ such that if F is a finite subset of $GL_d(\mathbb{C})$ containing 1 and generating a nonamenable subgroup, then for all $n \geq 1$, $|F^n| \geq (1 + \varepsilon)^n$. In particular,*

$$\rho_F = \lim_{n \to +\infty} \frac{1}{n} \log |F^n| \geq \log(1 + \varepsilon) > 0.$$

Let us remark that it may be the case that ρ_F is bounded away from 0 assuming only that F generates a nonvirtually nilpotent subgroup of $GL_2(\mathbb{C})$. However, we observed in [6] that such an assertion, if true, would imply the Lehmer conjecture about the Mahler measure of algebraic numbers. We also observed there that although every linear solvable group of exponential growth contains a free semigroup, no analog of Theorem 1.1 (the UF property of [1]) holds for solvable groups: namely, one may find finite symmetric sets F_n in $GL_2(\mathbb{C})$ containing 1 and generating a solvable subgroup of exponential growth, such that no pair of elements in $(F_n)^n$ may generate a free semigroup.

Von Neumann showed that groups containing a free subgroup are nonamenable, that is, have a spectral gap in ℓ^2. The uniformity in Theorem 1.1 implies also a uniformity for the spectral gap (see [29] for this observation). More precisely:

COROLLARY 1.5. (STRONG UNIFORM SPECTRAL GAP IN ℓ^2) *There is $\varepsilon = \varepsilon(d) > 0$ with the following property. If F is a finite subset of $GL_d(\mathbb{C})$ containing the identity and generating a nonamenable subgroup and if Γ is a countable subgroup of $GL_d(\mathbb{C})$ containing F and $f \in \ell^2(\Gamma)$, then there is $\sigma \in F$ such that*

$$\sum_{x \in \Gamma} \left| f(\sigma^{-1}x) - f(x) \right|^2 \geq \varepsilon \cdot \sum_{x \in \Gamma} |f(x)|^2$$

In particular, if F in $GL_d(\mathbb{C})$ is a finite subset containing the identity and generating a nonamenable subgroup, then for every finite subset A in $GL_d(\mathbb{C})$, we have $|FA| \geq (1 + \varepsilon)|A|$.

In [24] Lubotzky, Phillips, and Sarnak showed that for the compact Lie group $G = SU(2)$, the spectral measure of the "Hecke operators" $T_\mu = \frac{1}{2k} \sum_{1 \leq i \leq k} g_i + g_i^{-1}$ acting on $\mathbb{L}_0^2(G)$ is supported on $[-m_\mu, m_\mu]$ where m_μ is the norm of T_μ viewed as an operator on $\ell^2(\Gamma)$, Γ being the abstract group generated by the g_i's. This spectral measure is by definition the limiting distribution of the eigenvalues of T_μ on the nth-dimensional representation of G. Corollary 1.5 implies that m_μ is bounded away from 1 independently of μ as soon as k is fixed and Γ is a nonamenable subgroup of G. In other words, there is $\varepsilon = \varepsilon(k) > 0$ such that the proportion of eigenvalues of T_μ lying in $[-1, -1 + \varepsilon] \cup [1 - \varepsilon, 1]$ tends to 0 as n tends to infinity. The analogous result for Cayley graphs of $SL_2(\mathbb{Z}/p\mathbb{Z})$ is also a direct consequence of Corollary 1.5, that is, the spectral measure of any limit of such Cayley graphs is supported on $[-1 + \varepsilon, 1 - \varepsilon]$. It is believed (*spectral gap conjecture*, see [24] and [28]) that 1 is never an accumulation point of eigenvalues of T_μ for any given μ (or at least almost any in $SU(2)$) whose support generates a nonamenable subgroup.

Since by Kesten's theorem m_μ is also the exponential rate of decay of the return probability (see [21]), we also have

COROLLARY 1.6. (STRONG UNIFORM DECAY OF RETURN PROBABILITY)
There is $\varepsilon = \varepsilon(k, d) > 0$ with the property that for any nonamenable k-generated subgroup Γ of $GL_d(\mathbb{C})$ we have

$$\mathbb{P}(S_n = 1) \leq (1 - \varepsilon)^n$$

for all $n \geq 1$, where S_n is the simple random walk on Γ.

The next corollary gives an upper bound on the cogrowth of subgroups of $GL_d(\mathbb{C})$.

COROLLARY 1.7. (CO-GROWTH GAP) *Given $m \in \mathbb{N}$, there is $n(m) > 0$ such that for every $n \geq n(m)$ the following holds: $F = \{a_1, ..., a_m\} \subset GL_d(\mathbb{C})$ generates a non virtually solvable subgroup, if and only if the number of elements w in the free group F_m of word length n such that $w(a_1, ..., a_m) = 1$ is at most $(2m - 1 - \frac{\varepsilon}{m^D})^n$. Here $\varepsilon, D > 0$ are constants depending on d only.*

This result can be paraphrased by saying that nonamenable subgroups of $GL_d(\mathbb{C})$ are very strongly nonamenable, that is, have few relations. This puts a purely group theoretical restriction on an abstract finitely generated group given in terms of generators and relations to admit an embedding in $GL_d(\mathbb{C})$.

The uniformity in Theorem 1.1 allows us to reduce mod p and we obtain a statement giving a lower bound on the girth of subgroups of GL_d in positive characteristic:

COROLLARY 1.8. (LARGE GIRTH) *Given $k, d \geq 2$, there is $N, M \in \mathbb{N}$ and $\varepsilon_0, C > 0$ such that for every prime p and every field K of characteristic p and any finite subset F with k elements generating a subgroup of $GL_d(K)$ that contains no solvable subgroup of index at most M, then F^N contains two elements a, b such that $w(a, b) \neq 1$ in $GL_d(K)$ for any nontrivial word w in F_2 of length at most $f(p) = C \cdot (\log p)^{\varepsilon_0}$.*

It was conjectured in [16] that the statement of Corollary 1.8 holds for generating subsets F of $SL_2(\mathbb{F}_p)$ with $\varepsilon_0 = 1$.

Theorem 1.3 on the uniform QI-embedding of the free subgroup yields a uniform bound for the distortion of the subgroup of large girth. This in turn gives uniform expansion for subsets of say $GL_2(\mathbb{F}_p)$ lying in a ball of radius $\leq (\log p)^{\varepsilon_0}$. Namely:

COROLLARY 1.9. (UNIFORM EXPANSION FOR SMALL SETS) *Given $k, d \geq 2$, there is $N, M \in \mathbb{N}$ and $\varepsilon_0, C, \alpha, \beta > 0$ such that for every prime p and every field K of characteristic p and any finite symmetric subset F with k elements, containing 1 and generating a subgroup of $GL_d(K)$ that contains no solvable subgroup of index at most M, then for any subset $A \subset F^{C \cdot (\log p)^{\varepsilon_0}}$ there is $s \in F$ such that $|sA \triangle A| \geq \alpha|A|$. In particular $\mu_F^{*n}(e) \leq (1 - \beta)^n$ for all $n \leq C \cdot (\log p)^{\varepsilon_0}$, where μ_F is the uniform probability measure on F.*

If one could get $\varepsilon_0 = 1$ in the above corollary, then applying the argument of Bourgain and Gamburd [4] would give a proof that the family of all Cayley graphs of $SL_2(\mathbb{Z}/p\mathbb{Z})$ for varying p but with a fixed number of generators forms an expander family. See Lubotzky's book [23] for background material on expanders. It was also proved in [16] that a random d-regular Cayley graph of $GL_2(\mathbb{F}_p)$ has girth at least $(1 - o(1)) \log_{d-1}(p)$. Here we obtain $\varepsilon_0 = 2^{-10}$ for GL_2, which is quite far.

In the same vein, one obtains the following two weak forms of "non-Liouvilleness" for subgroups of $GL_d(\mathbb{C})$. Let d be some Riemannian distance on $GL_2(\mathbb{C})$.

COROLLARY 1.10. (SHORT WORDS ARE NOT SIMULTANEOUSLY VERY LIOUVILLE) *Given d, there is $N_0 \in \mathbb{N}$ and $\varepsilon_1 > 0$ with the following property. For every finite set $F \subset GL_d(\mathbb{C})$ generating a nonvirtually solvable subgroup, there is $\delta_0(F) > 0$ such that for every $\delta \in (0, \delta_0)$ there are two short words $a, b \in F^N O$ such that $d(w(a, b), 1) \geq \delta$ for every reduced word w in the free group F_2 with length $\ell(w)$ at most $(\log \delta^{-1})^{\varepsilon_1}$.*

In [19] Kaloshin and Rodnianski proved that for $G = SU(2) \leq SL_2(\mathbb{C})$ almost every pair $(a, b) \in G \times G$ satisfies $d(w(a, b), 1) \geq \exp(- C(a, b) \cdot \ell(w)^2)$ for all $w \in F_2 \backslash \{e\}$ and some constant $C(a, b) > 0$. Besides, it is easy to see that if $a, b \in GL_2(\overline{\mathbb{Q}})$, then the pair (a, b) satisfies the stronger diophantine condition $d(w(a, b), 1) \geq \exp(- C(a, b) \cdot \ell(w))$. It is conjectured in [28] and [16] that this stronger condition also holds for almost every pair $(a, b) \in SU(2)$ with respect to Haar measure.

Our result also allows us to estimate the number of words of length $\leq n$ that fall in a shrinking neighborhood of 1 in $GL_d(\mathbb{C})$. More precisely,

COROLLARY 1.11. (WEAK DIOPHANTINE PROPERTY) *There are $\tau, \varepsilon_1, C >$ 0 with the following property. For every $\{a, b\} \leq GL_d(\mathbb{C})$ that generates a nonvirtually solvable subgroup, there is $\delta_0(a, b) > 0$ such that for every $\delta \in (0, \delta_0)$ and every $n \leq C(\log \delta^{-1})^{\varepsilon_1}$, the proportion of elements w in the free group F_2 of word length n such that $d(w(a, b), 1) \leq \delta$ is at most $\exp(-\tau n)$.*

In [15], Gamburd, Jacobson, and Sarnak showed for $G = SU(2)$ that if a pair $(a, b) \in G$ satisfies the conclusion of Corollary 1.11 with $\varepsilon_1 = 1$ and $C > C_0$ (for some explicit $C_0 > 0$), then (a, b) has a spectral gap on $\mathbb{L}^2(G)$. In [5], Bourgain and Gamburd showed that if a pair $(a, b) \in G$ satisfies the above condition with $\varepsilon_1 = 1$ and some $C = C(a, b) > 0$, perhaps small, then (a, b) has a spectral gap on $\mathbb{L}^2(G)$. This latter condition is automatically satisfied if (a, b) satisfies the stronger diophantine condition above, for instance, if $(a, b) \in GL_2(\overline{\mathbb{Q}})$. Hence these pairs have a spectral gap. It is unknown whether there are (topologically generating) pairs with no spectral gap.

REMARK 1.12. We emphasize here that all the constants in the above theorems and corollaries can be effectively computed. Only at one point in the proof do we use a compactness argument. This is in the proof of Lemma 2.1(b). However, this statement can be given an effective proof valid in $M_d(\mathbb{C})$ (available upon request) although the price to pay is a much lengthier argument.

Comment on the proof of Corollaries 1.8 to 1.11: Observe (see Proposition 9.2) that the condition on a finite subset $F = \{A, B\}$ of $GL_d(\mathbb{C})$ that it generates an amenable (or equivalently virtually solvable) subgroup of $GL_d(\mathbb{C})$ is an algebraic one, as is the condition that all short words in A and B satisfy a relation of length at most n. Thus the statement of Theorem 1.1 can be read as a union of countably many assertions of first-order logic. According to the "compactness theorem" from model theory, since each assertion holds for $\mathbb{C}$ it must also hold for an arbitrary field K of sufficiently large characteristic (depending on n). This readily gives this existence of some function $f(p)$ going to $+\infty$ with p in Corollary 1.8. To derive the bound $(\log p)^{\varepsilon_0}$, as well as the bounds $(\log \delta^{-1})^{\varepsilon_0}$ in Corollaries 1.10 and 1.11, we use instead a standard version of the effective Nullstellensatz due to Masser and Wustholz (see [25]).

1.3. Outline of the Paper and of the Proof of Theorems 1.2 and 1.1 for SL_2

In Section 3 we introduce our main objects, the height and normalized height of a finite set F of matrices and prove basic properties about them. One of the key properties, the comparison between $\widehat{h}$ and e, relies crucially on Section 2, which is devoted to the proof of a key lemma, the *spectral radius lemma*. This lemma says in substance that unless F can be conjugated in a bounded part of SL_2, one will find a short word with letters in F with a large eigenvalue. In Section 4 we prove the first part of Theorem 1.2 (the height gap). The proof makes crucial use of equidistribution properties of algebraic numbers of small height and in particular a result of Zhang (Theorem 4.9) and Bilu's equidistribution theorem for Galois orbits (Theorem 4.4 below). Using the Eskin-Mozes-Oh escape lemma (see Lemma 4.8 below) we first reduce to the 2-generated case, when $F = \{Id, A, B\}$. Conjugating if necessary, A can be assumed to be diagonal. By making local estimates at each place, and with the help of Bilu's theorem, we then show that if $\widehat{h}(F)$ is small, then the heights of b_{11}, b_{22}, and $b_{12}b_{21}$ are small. But as $b_{11}b_{22} - b_{12}b_{21} = 1$, Zhang's theorem quickly yields to a contradiction if $b_{12}b_{21} \neq 0$. So $b_{12}b_{21} = 0$ and F is made of upper or lower triangular matrices, that is, F generates a solvable subgroup and we are done.

Theorem 1.1 is of purely algebraic nature and we begin its proof by showing that its validity over $\overline{\mathbb{Q}}$ implies its validity over $\mathbb{C}$. In the case of SL_2, one needs to exhibit a place v where one can *play ping-pong* on the projective line $\mathbb{P}^1(K_v)$ for the local field K_v, as in Tits's proof of his alternative. These "ping-pong players" will be the generators of the desired free subgroup. To achieve this, one needs to be able to conjugate F in $SL_2(K_v)$ in such a way that three conditions are satisfied. First, the norm $||F||_v$ ought to be controlled (up to a fixed power) by the maximal eigenvalue say $|\lambda|_v$ of an element, say A, lying in F (or F^N for a bounded N). Second, $|\lambda|_v$ should be large enough, that is, bounded away from 1. And third, at least one element, say B, from F or F^N, must send the eigenvectors of A far apart from one another with a distance controlled by some negative power of $||F||_v$. This criterion for ping-pong is explained in Section 6.

In Section 7 we show that a place v with these properties does exist. This is done in two steps. First (Section 5), it is shown that the minimal height $\widehat{h}(F)$ can be almost achieved (up to multiplicative and additive constants) by the ordinary height $h(F)$ after possibly conjugating F inside $SL_2(\overline{\mathbb{Q}})$. This is the second half of Theorem 1.2: this step uses the estimates needed in the first part of Theorem 1.2 (i.e. the proof of the height gap). In a second step (Section 7) the product formula on $\mathbb{P}^1(\overline{\mathbb{Q}})$ is used to show that the distances between eigenvectors of A and their images under B are controlled in terms of $h(F)$, and hence $\widehat{h}(F)$, only. This implies the existence of a place v satisfying the first and third conditions. Theorem 1.2 ensures that v can be chosen to satisfy the second condition also.

Sections 8 and 9 are devoted to the applications.

2. Spectral Radius Lemma for Several Matrices

In this section we state and prove the crucial Lemma 2.1. It says that given a finite set of matrices with coefficients in a local field, one may always find a short word with letters in that finite set whose maximal eigenvalue is as large as the minimal norm of the finite set. Together with Proposition 2.5 it can be interpreted as an analogue for several matrices of the classical spectral radius lemma relating the maximal eigenvalue and the rate of growth of the powers of a matrix. This lemma expresses in a condensed form an idea from a key proposition of the work of Eskin-Mozes-Oh where the concept of an almost algebra was used to essentially achieve the same goal. We emphasize here that getting an equality in part (a) of Lemma 2.1 as opposed to a mere inequality like in part (b) of the same lemma is absolutely crucial in the whole proof and in particular in Theorem 1.2.

Let k be a local field of characteristic 0. Let $\|\cdot\|_k$ be the standard norm on k^2, that is, the canonical Euclidean (respectively, Hermitian) norm if $k = \mathbb{R}$ (respectively, $\mathbb{C}$) and the sup norm ($\|(x,y)\|_k = \max\{|x|_k, |y|_k\}$) if k is non Archimedean. We will also denote by $\|\cdot\|_k$ the operator norm induced on $M_2(k)$ by the standard norm $\|\cdot\|_k$ on k^2. Let Q be a bounded subset of matrices in $M_2(k)$. We set

$$\|Q\|_k = \sup_{g \in Q} \|g\|_k$$

and call it the *norm of* Q. Let $\bar{k}$ be an algebraic closure of k. It is well known (see Lang's Algebra [22]) that the absolute value on k extends to a unique absolute value on $\bar{k}$, hence the norm $\|\cdot\|_k$ also extends in a natural way to $\bar{k}^2$ and to $M_2(\bar{k})$. This allows us to define the *minimal norm* of a bounded subset Q of $M_2(k)$ as

$$E_k(Q) = \inf_{x \in GL_2(\bar{k})} \|xQx^{-1}\|_k.$$

We will also need to consider the *maximal eigenvalue of* Q, namely

$$\Lambda_k(Q) = \max\{|\lambda|_k,\ \lambda \in spec(q), q \in Q\}$$

where $spec(q)$ denotes the set of eigenvalues (the spectrum) of q in $\bar{k}$. Finally, let $R_k(Q)$ be the *spectral radius* of Q:

$$R_k(Q) = \lim_{n \to +\infty} \|Q^n\|_k^{\frac{1}{n}}.$$

These quantities are related to one another. The key property concerning them is given in the following assertion (which also holds in $M_d(k)$, $k \geq 2$; see [7]).

LEMMA 2.1. (SPECTRAL RADIUS LEMMA) *Let Q be a bounded subset of $M_2(k)$,*

  a) *if k is non-Archimedean, then $\Lambda_k(Q^2) = E_k(Q)^2$; and*
  b) *if k is Archimedean, there is a constant $c \in (0,1)$ independent of Q, such that*
    $\Lambda_k(Q^2) \geq c^2 \cdot E_k(Q)^2.$

Proof. We make use of the following easy lemmas. □

LEMMA 2.2. *Let L be a field and Q a subset of $M_2(L)$ such that Q and Q^2 consist of nilpotent matrices. Then there is a basis (u, v) of L^2 such that $Qu = 0$ and $Qv \subset Lu$.*

Proof. For any $A, B \in Q$, we have $A^2 = B^2 = (AB)^2 = 0$. It follows, unless A or B are 0, that $\ker A = \mathrm{Im}A$ and $\ker B = \mathrm{Im}B$. Also if $AB \neq 0$, we get $\ker B = \ker(AB) = \mathrm{Im}(AB) = \mathrm{Im}A$, while if $AB = 0$, then $\mathrm{Im}B = \ker A$. So at any case $\ker A = \mathrm{Im}A = \ker B = \mathrm{Im}B$. So we have proved that the kernels and images of nonzero elements of Q coincide and are equal to some line say Lu,. Pick $v \in L^2 \backslash \{Lu\}$, then (u, v) forms the desired basis. □

LEMMA 2.3. *Let k be a local field with ring of integers $\mathcal{O}_k$ and uniformizer π. Let $A = (a_{ij}) \in M_2(\mathcal{O}_k)$ such that $trace(A)$ and $\det(A)$ belong to (π^2) and $a_{11}, a_{22}, a_{21} \in (\pi)$, while $a_{12} \in \mathcal{O}_k^\times$. Then $a_{21} \in (\pi^2)$.*

Proof. We have $a_{12}a_{21} = a_{11}a_{22} - \det(A) \in (\pi^2)$ and $a_{12} \in \mathcal{O}_k^\times$, hence $a_{21} \in (\pi)^2$. □

When k is a non-Archimedean local field, if a set $Id + Q$ in $SL_2(k)$ and its square have only eigenvalues very close to 1, then it must fix pointwise the 1-neighborhood of some point in the Bruhat-Tits tree of $SL_2(k)$. This is essentially the content of the next lemma.

LEMMA 2.4. (SMALL EIGENVALUES IMPLIES LARGE FIXED-POINT SET)
Let k be a local field with ring of integers $\mathcal{O}_k$ and uniformizer π together with an absolute value $|\cdot|_k$, which is (uniquely) extended to an algebraic closure $\overline{k}$ of k. Let Q be a subset of $M_2(\mathcal{O}_k)$ such that $\Lambda_k(Q)$ and $\Lambda_k(Q^2)$ are both $\leq |\pi|_k^3$. Then there is $T \in GL_2(k)$ such that $TQT^{-1} \subset \pi M_2(\mathcal{O}_k)$.

Proof. We can write Q as the disjoint union $Q_1 \cup \pi Q_2$ where Q_1 does not intersect $\pi M_2(\mathcal{O}_k)$. Let $Q' = Q_1 \cup Q_2$. Then $\Lambda_k(Q')$ and $\Lambda_k(Q'^2)$ are both $\leq |\pi|_k$. Hence projecting to $M_2(L)$, where L is the residue field $L = \mathcal{O}_k/(\pi)$, the matrices from Q' and Q'^2 become nilpotent. According to Lemma 2.2, one may find a basis $(\overline{u}, \overline{v})$ of L^2 such that $Q'\overline{u} = 0$ and $Q'\overline{v} \subset L\overline{u}$. According to Nakayama's lemma, this basis is the projection of a basis (u, v) of $\mathcal{O}_k^2$. Up to conjugating by a matrix in $GL_2(\mathcal{O}_k)$, we may assume that (u, v) is the canonical basis of $\mathcal{O}_k^2$. Therefore Q' consists of matrices $A = (a_{ij}) \in M_2(\mathcal{O}_k)$ with $a_{11}, a_{22}, a_{21} \in (\pi)$. Moreover, matrices in Q_1 satisfy $a_{12} \in \mathcal{O}_k^\times$ and hence by Lemma 2.3, $a_{21} \in (\pi^2)$. But for the matrices coming from πQ_2 we also have $a_{21} \in (\pi^2)$. So we have $a_{21} \in (\pi^2)$ for all matrices in Q. Let $T = diag(\pi, 1) \in GL_2(k)$. Then clearly $TQT^{-1} \subset \pi M_2(\mathcal{O}_k)$.

We go back to the proof of Lemma 2.1. We first prove (b). By contradiction, if such a c did not exist, then we may find a sequence of Q_n such that $\frac{\Lambda_k(Q_n^2)}{E_k(Q_n)^2} \to 0$. We can change Q_n into $\frac{Q_n}{E_k(Q_n)}$ and thus obtain a sequence of compact sets in $M_2(k)$ such that $E_k(Q_n) = 1$ with $\Lambda_k(Q_n^2) \to 0$ and $\Lambda_k(Q_n) \to 0$. Passing to a limit, we obtain a compact subset Q of $M_2(k)$ such that $\Lambda_k(Q^2) = \Lambda_k(Q) = 0$ while $E_k(Q) = 1$. By Lemma 2.2, we can thus find a basis (u, v) where $Qu = 0$ and $Qv \subset Lu$. But then conjugating Q by a suitable diagonal matrix can shrink the norm of Q as much as we want, hence $E_k(Q) = 0$. This is a contradiction.

We now prove (a). Let π be a uniformizer for k. Let $x = \log E_k(Q)$ where the log is taken in base $|\pi|_k^{-1}$. Suppose that $\Lambda_k(Q^2) < E_k(Q)^2$ and let $\varepsilon = x - \frac{1}{2}\log \Lambda_k(Q^2) > 0$. Then as $\Lambda_k(Q) \leq \Lambda_k(Q^2)^{\frac{1}{2}}$, we have $x - \log \Lambda_k(Q) \geq \varepsilon > 0$. Note that with our choice of normalization, $\log \Lambda_k(Q^2) \in \frac{1}{2}\mathbb{Z}$. Let $n \in \mathbb{N}$ be such that $2n\varepsilon > 3$. Let k_0 be the extension $k(\sqrt[2n]{\pi})$ where $\sqrt[2n]{\pi}$ is some $2n$-root of π in $\overline{k}$. Since $x = \log E_k(Q) = \log\inf\{\|gQg^{-1}\|_k, g \in GL_2(\overline{k})\}$, we may assume after possibly conjugating Q inside $GL_2(k_1)$, for some finite extension k_1 of k_0, that $y := \log \|Q\|_k \leq x + \frac{1}{2n}$ and also that $y = \min\{\log \|gQg^{-1}\|_k, g \in GL_2(k_1)\}$. Let π_1 be a uniformizer in k_1. Then $\log |\pi_1|_k^{-1} \leq \frac{1}{2n}$ and $y = \log |\pi_1|^m$ for some $m \in \mathbb{Z}$. Let $Q_y = \pi_1^{-m}Q \subset M_2(\mathcal{O}_{k_1})$. We get $\log E_k(Q_y) = x - y \leq 0$ and $\log \Lambda_k(Q_y) \leq \frac{1}{2}\log \Lambda_k(Q_y^2) = x - y - \varepsilon \leq -\varepsilon \leq -\frac{3}{2n} \leq \log |\pi_1|_k^3$. We are thus in a position to apply Lemma 2.4, which implies that Q_y, and hence

Q itself, can be further conjugated inside $GL_2(k_1)$ so as to strictly reduce its norm. But this contradicts the minimality of γ. □

PROPOSITION 2.5. *Let Q be a bounded subset of $M_2(k)$. We have*

$$R_k(Q) = \lim_{n \to +\infty} E_k(Q^n)^{\frac{1}{n}} = \inf_{n \in \mathbb{N}} E_k(Q^n)^{\frac{1}{n}} = \lim_{n \to +\infty} \Lambda_k(Q^{2n})^{\frac{1}{2n}} = \sup_{n \in \mathbb{N}} \Lambda_k(Q^n)^{\frac{1}{n}}.$$

Moreover, if k is non-Archimedean, $R_k(Q) = E_k(Q)$, while if k is Archimedean, then $c \cdot E_k(Q) \leq R_k(Q) \leq E_k(Q)$, where c is the constant from Lemma 2.1(b).

Proof. We omit the proof: these identities follow either directly from the definitions or as a straightforward application of Lemma 2.1. □

Note that some periodicity phenomenon may arise if $Id \notin Q$, namely, it may be that $\Lambda_k(Q^{2n+1}) = 1$ for all n while $\Lambda_k(Q^{2n})$ tends to infinity (for instance, take for Q a set of symmetries around several points on a given geodesic in the hyperbolic plane). However, if $Id \in Q$, then we do have $\lim_{n \to +\infty} \Lambda_k(Q^n)^{\frac{1}{n}} = R_k(Q)$.

Note also that if Q belongs to $SL_2(k)$, then $E_k(Q) \geq R_k(Q) \geq \Lambda_k(Q) \geq 1$ and all three quantities remain unchanged if we add Id to Q. The following lemma explains what happens if these quantities are close or equal to 1.

LEMMA 2.6. (LINEAR GROWTH OF DISPLACEMENT SQUARED) *Suppose k is Archimedean (i.e., $k = \mathbb{R}$ or $\mathbb{C}$). Then we have for every $n \in \mathbb{N}$ and every bounded subset Q of $SL_2(k)$ containing Id,*

1
$$E_k(Q^n) \geq E_k(Q)^{\sqrt{\frac{n}{4}}}.$$

Moreover,

$$\log R_k(Q) \geq c_1 \cdot \log E_k(Q) \cdot \min\{1, \log E_k(Q)\}$$

for some constant $c_1 > 0$. In particular $E_k(Q) = 1$ iff $R_k(Q) = 1$.

Proof. We use nonpositive curvature of hyperbolic space $\mathbb{H}^3$. For $x \in \mathbb{H}^3$ set $L(Q, x) = \max_{g \in Q} d(gx, x)$ and $L(Q) = \inf_x L(Q, x)$. Fix $\varepsilon > 0$, set $\ell_n := L(Q^n) = 2 \log E_k(Q^n)$, and let r_n be the infimum over $x \in \mathbb{H}^3$ of the smallest radius of a closed ball containing $Q^n x$. Note first that $r_n \leq \ell_n \leq 2r_n$.

We now prove (1). Fix $\varepsilon > 0$ and let $x, y \in \mathbb{H}^3$ be such that $Q^{n+1}x$ is contained in a ball of radius $r_{n+1} + \varepsilon$ around y. Let $q \in Q$ be arbitrary. Since Q contains Id, we have $Q^n x \subset Q^{n+1} x$, and $qQ^n x$ lies in the two balls of radius $r_{n+1} + \varepsilon$ centered around qy and around y. By the CAT(0) inequality for the median, the intersection of the two balls is contained in the ball B of radius $t := \sqrt{(r_{n+1} + \varepsilon)^2 - d(qy, y)^2/4}$ centered around the midpoint m between y and qy. Translating by q^{-1}, we get that $Q^n x$ lies in the ball of radius t centered at $q^{-1}m$. In particular $r_n \leq t$. This means $d(qy, y)^2 \leq 4((r_{n+1} + \varepsilon)^2 - r_n^2)$. Since $q \in Q$ and $\varepsilon > 0$ were arbitrary, we obtain $\ell_1^2 \leq 4(r_{n+1}^2 - r_n^2)$. Summing over n, we get $n\ell_1^2 \leq 4r_n^2 \leq 4\ell_n^2$, hence (1). But by Lemma 2.1(b), $\Lambda_k(Q^{2n}) \geq c^2 E_k(Q^n)^2$, hence $R_k(Q) \geq \Lambda_k(Q^{2n})^{\frac{1}{2n}} \geq c^{\frac{1}{n}} E_k(Q)^{\sqrt{\frac{1}{4n}}}$. Optimizing in n, we obtain the desired bound. $\qquad\square$

3. Height, Arithmetic Spectral Radius and Minimal Height

For any rational prime p (or $p = \infty$) let us fix an algebraic closure $\overline{\mathbb{Q}_p}$ of the field of p-adic numbers $\mathbb{Q}_p$ (if $p = \infty$, set $\mathbb{Q}_\infty = \mathbb{R}$ and $\overline{\mathbb{Q}_\infty} = \mathbb{C}$). We take the standard normalization of the absolute value on $\mathbb{Q}_p$ (i.e., $|p|_p = \frac{1}{p}$). It admits a unique extension to $\overline{\mathbb{Q}_p}$, which we denote by $|\cdot|_p$. Let $\overline{\mathbb{Q}}$ be the field of all algebraic numbers and K a number field. Let V_K be the set of equivalence classes of valuations on K. For $v \in V_K$ let K_v be the corresponding completion. For each $v \in V_K$, K_v is a finite extension of $\mathbb{Q}_p$ for some prime p. We normalize the absolute value on K_v to be the unique one that extends the standard absolute value on $\mathbb{Q}_p$. Namely $|x|_v = |N_{K_v|\mathbb{Q}_p}(x)|_p^{\frac{1}{n_v}}$ where $n_v = [K_v : \mathbb{Q}_p]$. Equivalently K_v has n_v different embeddings in $\overline{\mathbb{Q}_p}$ and each of them gives rise to the same absolute value on K_v. We identify $\overline{K_v}$, the algebraic closure of K_v, with $\overline{\mathbb{Q}_p}$. Let V_f be the set of finite places and V_∞ the set of infinite places.

Let F be a finite subset in $M_2(K)$. For $v \in V_K$, in order not to surcharge notation, we will use the subscript v instead of K_v in the quantities $E_v(F) = E_{K_v}(F)$, $\Lambda_v(F) = \Lambda_{K_v}(F)$, and so on.

Recall that if $x \in K$ then its (absolute, logarithmic) Weil height is by definition (see, e.g., [3]) the following quantity:

$$h(x) = \frac{1}{[K : \mathbb{Q}]} \sum_{v \in V_K} n_v \log^+ |x|_v.$$

It is well defined (i.e., independent of the choice of $K \ni x$). Let us similarly define the height of a matrix $f \in M_2(K)$ by

$$h(f) = \frac{1}{[K : \mathbb{Q}]} \sum_{v \in V_K} n_v \log^+ \|f\|_v$$

and the height of a finite set F of matrices in $M_2(K)$ by

$$h(F) = \frac{1}{[K : \mathbb{Q}]} \sum_{v \in V_K} n_v \log^+ \|F\|_v,$$

where $n_v = [K_v : \mathbb{Q}_v]$. We also define the *minimal height* of F as

$$2 \qquad e(F) = \frac{1}{[K : \mathbb{Q}]} \sum_{v \in V_K} n_v \log^+ E_v(F)$$

and the *arithmetic spectral radius* (or normalized height) of F as

$$\widehat{h}(F) = \frac{1}{[K : \mathbb{Q}]} \sum_{v \in V_K} n_v \log^+ R_v(F).$$

For any height h, we also set $h = h_\infty + h_f$, where h_∞ is the infinite part of h (i e., the part of the sum over the infinite places of K) and h_f is the finite part of h (i.e., the part of the sum over the finite places of K).

Note that these heights are well defined independently of the number field K such that $F \subset M_2(K)$. The above terminology is justified by the following facts:

PROPOSITION 3.1. (BASIC PROPERTIES OF HEIGHTS I) *For any finite set F in $M_2(\overline{\mathbb{Q}})$, we have*

1) $\widehat{h}(F) = \lim_{n \to +\infty} \frac{1}{n} h(F^n) = \inf_{n \in \mathbb{N}} \frac{1}{n} h(F^n)$,
2) $e_f(F) = \widehat{h}_f(F)$ *and* $e(F) + \log c \le \widehat{h}(F) \le e(F)$ *where c is the constant in Lemma 2.1 (b), and*
3) $\widehat{h}(F^n) = n \cdot \widehat{h}(F)$ *and* $\widehat{h}(F \cup \{Id\}) = \widehat{h}(F)$.

Proof. This follows easily from Lemma 2.1 and Proposition 2.5. □

We also record the following simple observations:

PROPOSITION 3.2. (BASIC PROPERTIES OF HEIGHTS II) *We have, for a finite set F in $M_2(\overline{\mathbb{Q}})$,*

1) $e(xFx^{-1}) = e(F)$ *if $x \in GL_2(\overline{\mathbb{Q}})$,*
2) $e(F^n) \le n \cdot e(F)$, *and*
3) *if λ is an eigenvalue of an element of F, then $h(\lambda) \le \widehat{h}(F) \le e(F)$.*

Proof. This is clear. □

We can also compare $e(F)$ and $\widehat{h}(F)$ when $\widehat{h}(F)$ is small:

PROPOSITION 3.3. (BASIC PROPERTIES OF HEIGHTS III) *Let c_1 be the constant from Lemma 2.6, then*

$$\widehat{h}_\infty(F) \geq \frac{c_1}{4} \cdot e_\infty(F) \cdot \min\{1, e_\infty(F)\}$$

for any finite subset F in $SL_2(\overline{\mathbb{Q}})$. In particular $e(F)$ is small as soon as $\widehat{h}(F)$ is small.

Proof. From Lemma 2.6, $\widehat{h}_v(F) \geq c_1 \cdot e_v(F) \cdot \min\{1, e_v(F)\}$ for every $v \in V_K$. We may write $e_\infty(F) = \alpha e^+(F) + (1 - \alpha)e^-(F)$ where e^+ is the average of the e_v greater than 1 and e^- the average of the e_v smaller than 1. Applying Cauchy-Schwarz, we have $\widehat{h}_\infty(F) \geq c_1 \cdot (\alpha e^+ + (1 - \alpha)(e^-)^2)$. If $\alpha e^+(F) \geq \frac{1}{2}e_\infty(F)$, then $\widehat{h}_\infty(F) \geq \frac{c_1}{2}e_\infty(F)$, and otherwise $(1 - \alpha)e^- \geq \frac{e_\infty}{2}$, hence $\widehat{h}_\infty(F) \geq c_1(1 - \alpha)$ $(e^-)^2 \geq \frac{c_1}{4}e_\infty^2$. At any case $\widehat{h}_\infty(F) \geq \frac{c_1}{4} \cdot e_\infty(F) \cdot \min\{1, e_\infty(F)\}$. $\qquad\square$

4. Height Gap Theorem

In this section, we prove Theorem 1.2 from the introduction. First observe that according to Propositions 3.3 and 3.1(2), $\widehat{h}(F)$ is small if and only if $e(F)$ is small. So we may as well replace $\widehat{h}(F)$ by $e(F)$ in Theorem 1.2. We now assume that $F = \{Id, A, B\}$, with A semisimple. The general case follows from this as we will show in Lemma 4.7. Since $e(F)$ is invariant under conjugation by any element in $GL_2(\overline{\mathbb{Q}})$, we may assume that A is diagonal, that is,

$$\mathbf{3} \qquad\qquad A = \begin{pmatrix} \lambda & 0 \\ 0 & \lambda^{-1} \end{pmatrix}, \ B = \begin{pmatrix} a & b \\ c & d \end{pmatrix}.$$

Let $\deg(\lambda)$ be the degree of λ as an algebraic number over $\mathbb{Q}$. The main part of the argument consists in the following proposition:

PROPOSITION 4.1. (SMALL NORMALIZED HEIGHT IMPLIES SMALL HEIGHT OF MATRIX COORDINATES) *For every $\beta > 0$ there exists $d_0, \eta > 0$ such that, if $F = \{A, B\}$ are as in (3) and if $e(F) \leq \eta$ and $\deg(\lambda) \geq d_0$, then*

$$\max\{h(ad), h(bc)\} \leq \beta.$$

In order to prove this statement, we are first going to give local estimates at each place v, then use Bilu's equidistribution theorem to show that when these estimates are put together, the error terms give only a negligible contribution to the height.

Let K be the number field generated by the coefficients of A and B. Let $v \in V_K$ be a place of K. We set $s_v = \log E_v(F)$ and $\delta = \lambda - \lambda^{-1}$. We first show the following local estimate:

LEMMA 4.2. (LOCAL ESTIMATES) *For each $v \in V_K$ we have*

$$\max\{|a|_v, |d|_v, \sqrt{|bc|_v}\} \leq C_v e^{2s_v} \max\{1, |\delta^{-1}|_v\},$$

where C_v is a constant equal to 1 if v is a finite place and equal to a number $C_\infty > 1$ (independent of v) if v is infinite. Moreover, there are absolute constants $\varepsilon_0 > 0$ and $C_0 > 0$ such that if v is infinite and $s_v \leq \varepsilon_0$, then

$$\max\{|ad|_v, |bc|_v\} \leq 1 + C_0\left(\sqrt{s_v} + \frac{\sqrt{s_v}}{|\delta|_v} + \frac{s_v}{|\delta|_v^2}\right).$$

Proof. In order not to overburden notation in this proof we set s_v to be some number arbitrarily close but strictly bigger than $\log E_v(F)$ and we can let it tend to $\log E_v(F)$ at the end. If v is infinite, then $\overline{\mathbb{Q}}_v = \mathbb{C}$ and $SL_2(\mathbb{C}) = KAN$ where $\mathbf{K} = SU_2(\mathbb{C})$, A is the subgroup of diagonal matrices with real positive entries, and N is the subgroup of unipotent complex upper-triangular matrices. As K leaves the norm invariant, there must exist a matrix $P \in AN$ such that $\max\{\|PAP^{-1}\|, \|PBP^{-1}\|\} \leq e^{s_v}$. Since $P \in AN$, we may write $P = \begin{pmatrix} t & y \\ 0 & t^{-1} \end{pmatrix}$ with $t > 0$ and $y \in \mathbb{C}$. Then we have, setting $\delta = \lambda - \lambda^{-1}$,

$$PAP^{-1} = \begin{pmatrix} \lambda & ty\delta \\ 0 & \lambda^{-1} \end{pmatrix},$$

4

$$PBP^{-1} = \begin{pmatrix} a + cyt^{-1} & bt^2 + dyt - ayt - cy^2 \\ t^{-2}c & -yct^{-1} + d \end{pmatrix}.$$

If v is finite and K_v is the corresponding completion, with ring of integers $\mathcal{O}_v$ and uniformizer π, we have $SL_2(K_v) = K_v A_v N_v$ where $\mathbf{K}_v = SL_2(\mathcal{O}_v)$, $A_v = \{diag(\pi^n, \pi^{-n}), n \in \mathbb{Z}\}$ and N_v is the subgroup of unipotent upper-triangular matrices with coefficients in K_v. Hence, we also get a $P \in A_v N_v$ satisfying (4) with $y \in K_v$ and $t = \pi^n$ for some $n \in \mathbb{Z}$.

We first assume that v is finite. Recall that the operator norm in $SL_2(K_v)$ is given by the maximum modulus of each matrix coefficient. Hence we must have $|t^{-2}c|_v \leq e^{s_v}$ and $|ty\delta|_v \leq e^{s_v}$. It follows that $|cyt^{-1}|_v \leq e^{2s_v}|\delta^{-1}|_v$ and hence $|a|_v \leq \max\{e^{s_v}, e^{2s_v}|\delta^{-1}|_v\}$. Similarly, $|d|_v \leq \max\{e^{s_v}, e^{2s_v}|\delta^{-1}|_v\}$. Hence $|ad|_v \leq \max\{e^{2s_v}, e^{4s_v}|\delta^{-1}|_v^2\}$. Moreover $ad - bc = 1$, hence $|bc|_v \leq \max\{1, |ad|_v\} \leq \max\{e^{2s_v}, e^{4s_v}|\delta^{-1}|_v^2\}$.

Now we assume that v is infinite.

Claim: There is $u_0 > 0$ such that if $0 \leq u \leq u_0$ and $\|B\| \leq e^u$, then

5

$$\max\{|a - \overline{d}|, |b + \overline{c}|\} \leq 2\sqrt{u},$$

6 $$\max\{|a|^2+|b|^2,|d|^2+|c|^2\}\leq 1+6u+8\sqrt{u},\; and$$

7 $$\max\{|a|,|b|,|c|,|d|\}\leq 1+3u+4\sqrt{u}\leq 1+5\sqrt{u}.$$

To prove this, recall that the operator norm in $SL_2(\mathbb{C})$ satisfies $tr(B^*B)=|a|^2+|b|^2+|c|^2+|d|^2=\|B\|^2+\|B\|^{-2}$. Hence $|a|^2+...+|d|^2\leq 1+e^{2u}$, hence ≤ 4 if u is small enough (say, $u\leq .5$). On the other hand, for small u, $|a-\bar d|^2+|b+\bar c|^2=|a|^2+...+|d|^2-2\leq e^{2u}-1\leq 4u$. Hence (5). Now $|d|\leq |a|+2\sqrt{u}$ and since $|a|,|b|\leq 2$, we get $|d|^2\leq |a|^2+4u+8\sqrt{u}$ and vice versa and similarly for b and c. Hence (6) and (7), and the claim is proved.

Let now $\varepsilon>0$ and assume that $s_v\leq \varepsilon$. From (4) we get $|\lambda|^2+|\lambda^{-1}|^2+|ty\delta|^2\leq 1+e^{2\varepsilon}$, hence $|ty\delta|^2\leq e^{2\varepsilon}-1\leq 4\varepsilon$ if ε is small enough. So $|ty\delta|\leq 2\sqrt{\varepsilon}$. Now since $\|PBP^{-1}\|\leq e^\varepsilon$, we have $|t^{-2}c|\leq 2$ as soon as $\varepsilon\leq \frac{1}{2}$. Hence $|yct^{-1}|\leq \frac{4\sqrt{\varepsilon}}{|\delta|}$ and $\max\{|a|,|d|\}\leq 1+5\sqrt{\varepsilon}+\frac{4\sqrt{\varepsilon}}{|\delta|}$. Finally, for some absolute constant $C>0$ $|ad|\leq 1+C(\sqrt{\varepsilon}+\frac{\sqrt{\varepsilon}}{|\delta|}+\frac{\varepsilon}{|\delta|^2})$.

On the other hand, $|cy^2|=|t^{-2}c(ty)^2|\leq \frac{8\varepsilon}{|\delta|^2}$ and $|d-a||yt|\leq 2\max\{|a|,|d|\}$ $|yt|\leq \frac{24\sqrt{\varepsilon}}{|\delta|}+\frac{16\varepsilon}{|\delta|^2}$. Also by (5), $|bt^2+(d-a)yt-cy^2+t^{-2}\bar c|\leq 2\sqrt{\varepsilon}$, and $|bc+|t^{-2}c|^2|\leq 2|bt^2+t^{-2}\bar c|\leq 4\sqrt{\varepsilon}+\frac{48\sqrt{\varepsilon}}{|\delta|}+\frac{48\varepsilon}{|\delta|^2}$, and by (6), $|t^{-2}c|^2\leq 1+14\sqrt{\varepsilon}$, hence up to enlarging the absolute constant C, we also have $|bc|\leq 1+C(\sqrt{\varepsilon}+\frac{\sqrt{\varepsilon}}{|\delta|}+\frac{\varepsilon}{|\delta|^2})$.

Without the assumption that s_v is small, we can make a coarser estimate: $|t^{-2}c|^2\leq 1+e^{2s_v}$, $|ty\delta|^2\leq 1+e^{2s_v}$, hence $|cyt^{-1}|\leq \frac{1+e^{2s_v}}{|\delta|}$ and $\max\{|a|,|d|\}\leq \frac{1+e^{2s_v}}{|\delta|}+\sqrt{1+e^{2s_v}}\leq 4e^{2s_v}\max\{1,\frac{1}{|\delta|}\}$ and $|ad|\leq 16e^{4s_v}\max\{1,\frac{1}{|\delta|^2}\}$. Similarly, we compute $|bc|\leq 20e^{4s_v}\max\{1,\frac{1}{|\delta|^2}\}$. □

We now put together the local information obtained above to bound the heights. Let $n=[K:\mathbb{Q}]$ and V_f and V_∞ the set of finite and infinite places of K. Set ε_0, C_0, and C_∞ the constants obtained in the previous lemma. For $A>0$ and $x\in\overline{\mathbb{Q}}$, we set

8 $$h_\infty^A(x)=\frac{1}{[K:\mathbb{Q}]}\sum_{v\in V_\infty,|x|_v\geq A}n_v\cdot\log^+|x|_v$$

where the sum is limited to those $v\in V_\infty$, for which $|x|_v\geq A$. We have

LEMMA 4.3. *For some constant C_2 $(2\leq C_2\leq 2+(2\log C_\infty+4)/\log 2)$, we have for all $\varepsilon_1\in(0,\frac{1}{2})$ and all $\varepsilon\leq\min\{\varepsilon_0,\varepsilon_1^2\}$*

9 $$\max\{h(ad),h(bc)\}\leq C_{\varepsilon,\varepsilon_1}e(F)+6C_0\frac{\sqrt{\varepsilon}}{\varepsilon_1}+2h_f(\delta^{-1})+C_2\cdot h_\infty^{\varepsilon_1^{-1}}(\delta^{-1})$$

where $C_{\varepsilon,\varepsilon_1}=\left(4+\frac{2\log C_\infty}{\varepsilon}+\frac{2|\log\varepsilon_1|}{\varepsilon}\right)$ *and* $\delta=\lambda-\lambda^{-1}$.

Proof. Recall that $s_v = \log E_v(F)$. If $v \in V_\infty$ and $s_v \geq \varepsilon$, then according to Lemma 4.2 $\log^+ |ad|_v \leq 2 \log C_\infty + 4 s_v + 2 \log^+ |\delta^{-1}|_v$ hence

$$\frac{1}{n} \sum_{v \in V_\infty, s_v \geq \varepsilon} n_v \cdot \log^+ |ad|_v \leq \left(4 + \frac{2 \log C_\infty}{\varepsilon} \right) \frac{1}{n} \sum_{v \in V_\infty, s_v \geq \varepsilon} n_v s_v$$

$$+ \frac{2}{n} \sum_{v \in V_\infty, s_v \geq \varepsilon} n_v \cdot \log^+ |\delta^{-1}|_v.$$

Fix $\varepsilon_1 < \frac{1}{2}$. On the other hand, if $s_v \leq \varepsilon \leq \min\{\varepsilon_0, \varepsilon_1^2\}$ and $|\delta|_v \geq \varepsilon_1$, then $\log^+ |ad|_v \leq C_0 (\sqrt{s_v} + \frac{\sqrt{s_v}}{|\delta|_v} + \frac{s_v}{|\delta|_v^2}) \leq 3 C_0 \frac{\sqrt{\varepsilon}}{\varepsilon_1}$ and, as $n_v \leq 2$,

$$\frac{1}{n} \sum_{v \in V_\infty, s_v \leq \varepsilon, |\delta|_v \geq \varepsilon_1} n_v \cdot \log^+ |ad|_v \leq 6 C_0 \frac{\sqrt{\varepsilon}}{\varepsilon_1}.$$

While if $s_v < \varepsilon$ and $|\delta|_v \leq \varepsilon_1 \leq \frac{1}{2}$, then $\log^+ |ad|_v \leq C_2 \log^+ |\delta^{-1}|_v$ for some absolute constant C_2, $(2 \leq C_2 \leq 2 + (2 \log C_\infty + 4)/\log 2)$, hence

$$\frac{1}{n} \sum_{v \in V_\infty, s_v < \varepsilon, |\delta|_v \leq \varepsilon_1} n_v \cdot \log^+ |ad|_v \leq \frac{1}{n} \sum_{v \in V_\infty, s_v < \varepsilon, |\delta|_v \leq \varepsilon_1} C_2 n_v \cdot \log^+ |\delta^{-1}|_v.$$

When $v \in V_f$, from Lemma 4.2, we get

$$\sum_{v \in V_f} n_v \cdot \log^+ |ad|_v \leq \sum_{v \in V_f} 4 n_v s_v + \sum_{v \in V_f} 2 n_v \cdot \log^+ |\delta^{-1}|_v.$$

But

$$\frac{2}{n} \sum_{v \in V_\infty, s_v \geq \varepsilon, |\delta|_v \geq \varepsilon_1} n_v \cdot \log^+ |\delta^{-1}|_v \leq \frac{2 |\log \varepsilon_1|}{\varepsilon} \frac{1}{n} \sum_{v \in V_\infty, s_v \geq \varepsilon} n_v s_v.$$

Putting together the above estimates, we indeed obtain (9) for ad. The same computation works for bc. $\square$

It is now time to recall the following result (see also [12] and [26]):

THEOREM 4.4. (BILU'S EQUIDISTRIBUTION OF SMALL POINTS, [2])
Suppose $(\lambda_n)_{n \geq 1}$ is a sequence of algebraic numbers (i.e., in $\overline{\mathbb{Q}}$) such that $h(\lambda_n) \to 0$ and $\deg(\lambda_n) \to +\infty$ as $n \to +\infty$. Let $\mathcal{O}(\lambda_n)$ be the Galois orbit of λ_n in $\overline{\mathbb{Q}}$. Then we have the following weak convergence of probability measures on $\mathbb{C}$,

10
$$\frac{1}{\#\mathcal{O}(\lambda_n)} \sum_{x \in \mathcal{O}(\lambda_n)} \delta_x \underset{n \to +\infty}{\to} d\theta$$

where $d\theta$ is the normalized Lebesgue measure on the unit circle $\{z \in \mathbb{C}, |z| = 1\}$.

We now draw two consequences of this equidistribution statement:

LEMMA 4.5. (BOUNDING ERRORS TERMS VIA BILU'S THEOREM I) *For every $\alpha > 0$ there is $d_1, \eta_1 > 0$ and $\varepsilon_1 > 0$ with the following property. If $\lambda \in \overline{\mathbb{Q}}$ is such that $h(\lambda) \leq \eta_1$, $\deg(\lambda) \geq d_1$, then*

$$h_\infty^{\varepsilon_1^{-1}}\left(\frac{1}{1-\lambda}\right) \leq \alpha$$

where $h_\infty^{\varepsilon_1^{-1}}$ was defined in (8).

Proof. Let $P \in \mathbb{Z}[X]$ be the minimal polynomial of λ, that is, $P(X) = \sum_{0 \leq i \leq n} a_i X^i = a_n \prod_{x \in \mathcal{O}(\lambda)} (X - x)$. As $P(1) \in \mathbb{Z} \setminus \{0\}$, $\log|P(1)| = \log|a_n| + \sum_{x \in \mathcal{O}(\lambda)} \log|1 - x| \geq 0$. So

$$\sum_{|1-x| \leq \varepsilon_1} \log \frac{1}{|1-x|} \leq \sum_{|1-x| > \varepsilon_1} \log|1-x| + \log|a_n|.$$

Recall (see [22] III.1.) that $h(\lambda) = \frac{1}{n}\left(\sum_{x \in \mathcal{O}(\lambda)} \log^+|x| + \log|a_n|\right)$. Hence

11
$$\frac{1}{n}\sum_{|1-x| \leq \varepsilon_1} \log \frac{1}{|1-x|} \leq h(\lambda) + \frac{1}{n}\sum_{|1-x| > \varepsilon_1} \log|1-x|.$$

Consider the function $f_{\varepsilon_1}(z) = 1_{|z-1| > \varepsilon_1} \log|1 - z|$. It is locally bounded on $\mathbb{C}$. By Theorem 4.4, for every $\varepsilon_1 > 0$, there must exist $d_1, \eta_1 > 0$ such that, if $h(\lambda) \leq \eta_1$ and $\deg(\lambda) \geq d_1$, then $\left|\frac{1}{n}\sum_x f_{\varepsilon_1}(x) - \int_0^1 f_{\varepsilon_1}(e^{2\pi i\theta})d\theta\right| \leq \frac{\alpha}{3}$. On the other hand, we verify that $\theta \mapsto \log|1 - e^{2\pi i\theta}|$ is in $L^1(0,1)$ and $\int_0^1 \log|1 - e^{2\pi i\theta}|d\theta = 0$. Hence we can choose $\varepsilon_1 > 0$ small enough so that $\left|\int_0^1 f_{\varepsilon_1}(e^{2\pi i\theta})d\theta\right| \leq \frac{\alpha}{3}$. Combining these inequalities with (11) and choosing $\eta_1 \leq \frac{\alpha}{3}$, we get $h_\infty^{\varepsilon_1^{-1}}((1-\lambda)^{-1}) \leq \alpha$. $\qquad\square$

Combining this with Bilu's theorem, we get

LEMMA 4.6. (BOUNDING ERRORS TERMS VIA BILU'S THEOREM II) *For every $\alpha > 0$ there exists $\eta_0 > 0$ and $A_1 > 0$ such that for any $\lambda \in \overline{\mathbb{Q}}$, if $h(\lambda) \leq \eta_0$ and $d = \deg(\lambda) > A_1$, then*

$$h_f\left(\frac{1}{1-\lambda}\right) \leq 2\alpha.$$

Proof. We apply the *product formula* to $\mu = 1 - \lambda$, which takes the form $h(\mu) = h(\mu^{-1})$, hence $h_f(\mu^{-1}) = h_\infty(\mu) - h_\infty(\mu^{-1}) + h_f(\mu)$. But $h_f(\mu) = h_f(1 - \lambda) \leq$

$h_f(\lambda) \leq \eta_0$ and $h_\infty(\mu) - h_\infty(\mu^{-1}) = \frac{1}{[K:\mathbb{Q}]} \sum_{v \in V_\infty} n_v \cdot \log |\mu|_v$. Lemma 4.5 shows that the convergence in Equation (10) in Bilu's theorem not only holds for compactly supported functions on $\mathbb{C}$, but also for functions with logarithmic singularities at 1. In particular, it holds for the function $f(z) = \log |1 - z|$, which is exactly what we need, since $\int_0^1 f(e^{2\pi i \theta}) d\theta = 0$. Hence $\frac{1}{[K:\mathbb{Q}]} \sum_{v \in V_\infty} n_v \cdot \log |\mu|_v$ becomes small. We are done. $\qquad\square$

Proof of Proposition 4.1. Since $h_f(\frac{1}{\lambda - \lambda^{-1}}) \leq h_f(\lambda) + h_f(\frac{1}{1 - \lambda^2})$ and similarly $h_\infty^A(\frac{1}{\lambda - \lambda^{-1}}) \leq h_\infty^A(\lambda) + h_\infty^A(\frac{1}{1 - \lambda^2})$, it follows from the last two lemmas that we can find $\varepsilon_1 > 0$, $\eta > 0$ and $d_0 \in \mathbb{N}$ so that $2 h_f(\delta^{-1}) + C_2 \cdot h_\infty^{\varepsilon_1^{-1}}(\delta^{-1}) \leq \frac{\beta}{3}$ as soon as $h(\lambda) \leq e(F) \leq \eta$ and $\deg(\lambda) \geq d_0$. Then choose ε so the $2 C_1 \frac{\sqrt{\varepsilon}}{\varepsilon_1} \leq \frac{\beta}{3}$ and finally take η even smaller so that $C_{\varepsilon, \varepsilon_1} \eta \leq \frac{\beta}{3}$. Now apply Lemma 4.3 and we are done. $\qquad\square$

End of the proof of Theorem 1.2:. The following lemma allows us, when proving Theorem 1.2, to assume without loss of generality that $F = \{1, A, B\}$, where A and B are two semisimple elements in $SL_2(\overline{\mathbb{Q}})$ that do not satisfy some prescribed finite set of algebraic relations. More precisely: $\qquad\square$

LEMMA 4.7. *For every $d_1 \in \mathbb{N}$, there exists $N(d_1) \in \mathbb{N}$ with the following property. Let F be a finite subset of $SL_2(\overline{\mathbb{Q}})$ containing 1 and generating a nonvirtually solvable subgroup, then there exists $A, B \in F^{N(d_1)}$ such that A and B are semisimple, generate a nonvirtually solvable subgroup of SL_2, A is not of order at most d_1, and $bc \notin \{0, -1, e^{\frac{2i\pi}{3}}, e^{\frac{4i\pi}{3}}\}$ after we conjugate A and B in the form of Equation (3).*

Proof. This is a direct application of Lemma 4.8 below applied to $\Sigma = F \times F$ in $SL_2 \times SL_2 \leq GL_4$ with $X = X_1 \cup X_2 \cup X_3 \cup X_4$ where $X_1 = \{(A, B), A \text{ or } B \text{ has order at most } d_1\}$, $X_2 = \{(A, B), tr(A) \text{ or } tr(B) \text{ is } 2\}$, $X_3 = \{(A, B), A \text{ and } B \text{ generate a virtually solvable subgroup}\}$ and X_4 the Zariski closure of $\{((gAg^{-1}, gBg^{-1}), g \in SL_2, A \text{ diagonal}, bc \in \{0, -1, e^{\frac{2i\pi}{3}}, e^{\frac{4i\pi}{3}}\}\}$. For dimension reasons X_4 is a proper subvariety of $SL_2 \times SL_2$ and Propositions 9.2 and 9.1 show that so is X_3. $\qquad\square$

LEMMA 4.8. (ESKIN-MOZES-OH "ESCAPE FROM SUBVARIETIES", SEE [13] AND [9]) *Let K be a field, $d \in \mathbb{N}$. For every $m \in \mathbb{N}$, there is $N \in \mathbb{N}$ such that if X a K-algebraic subvariety of $GL_d(K)$ such that the sum of the degrees of the geometrically irreducible components of X is at most m, then for any subset*

$\Sigma \subset GL_d(K)$ *containing* Id *and generating a subgroup that is not contained in* $X(K)$, *we have* $\Sigma^N \not\subseteq X(K)$.

Observe (as follows from the irreducibility of cyclotomic polynomials, see [22] VI.3.) that for every $d_0 \in \mathbb{N}$ there is $\eta_0 > 0$ and $d_1 > 0$ such that if $h(\lambda) < \eta_0$ and λ is not a root of 1 of order at most d_1, then $\deg(\lambda) \geq d_0$. However, recall the following well-known result (which is also a straightforward corollary of Theorem 4.4),

THEOREM 4.9. (ZHANG'S THEOREM [32]) *There exists an absolute constant* $\alpha_0 > 0$ *such that for any* $x \in \overline{\mathbb{Q}}$, *we have*

$$h(x) + h(1+x) > \alpha_0$$

unless $x \in \{0, -1, e^{\frac{2i\pi}{3}}, e^{\frac{4i\pi}{3}}\}$.

Let $\beta = \frac{\alpha_0}{2}$ where α_0 is given by Theorem 4.9. Proposition 4.1 yields $d_0 > 0$ and $\eta = \eta(\frac{\alpha_0}{2}) > 0$ such that $\max\{h(ad), h(bc)\} \leq \beta$ as soon as $e(\{Id, A, B\}) \leq \eta$ and $\deg(\lambda) \geq d_0$. By Lemma 4.7, if we have some nice $A, B \in F^{N(d_1)}$. If $e(F) \leq \frac{\min\{\eta(\frac{\alpha_0}{2}), \eta_0\}}{N(d_1)}$, then $e(\{Id, A, B\}) \leq \min\{\eta, \eta_0\}$ and λ is not a root of 1 of order at most d_1. Hence $\deg(\lambda) \geq d_0$ and by Proposition 4.1, $h(ad) + h(bc) \leq 2\beta = \alpha_0$. Then according to Theorem 4.9, $bc \in \{0, -1, e^{\frac{2i\pi}{3}}, e^{\frac{4i\pi}{3}}\}$, which contradicts our choice of A, B. So $\frac{\min\{\eta(\frac{\alpha_0}{2}), \eta_0\}}{N(d_1)} > 0$ is the desired gap.

This ends the proof of Theorem 1.2.

Finally, observe that Theorem 1.2 combined with Propositions 3.1 to 3.3 implies

PROPOSITION 4.10. *There exists a constant* $c_0 > 0$ *such that if* F *is any finite subset of* $SL_2(\overline{\mathbb{Q}})$ *generating a nonvirtually solvable subgroup, then*

$$e(F) \geq \widehat{h}(F) \geq c_0 \cdot e(F).$$

5. Simultaneous Quasisymmetrization over $\overline{\mathbb{Q}}$

Here we are going to use our previous height estimates once again to show the following proposition. Observe that the minimal height $e(F)$ coincides with the infimum of $h(gFg^{-1})$ over all adelic points $g = (g_v)_v$. The lemma we are about to state essentially means that this infimum is attained (up to additive and multiplicative constants) with a conjugating matrix g lying already in $SL_2(\overline{\mathbb{Q}})$ as opposed to $SL_2(\mathbb{A})$ (the adelic group).

PROPOSITION 5.1. (SIMULTANEOUS QUASISYMMETRIZATION) *There is an absolute constant $C > 0$ such that if F is a finite subset of $SL_2(\overline{\mathbb{Q}})$ generating a nonvirtually solvable subgroup, then there is an element $g \in SL_2(\overline{\mathbb{Q}})$ such that*

$$h(gFg^{-1}) \leq C \cdot e(F) + C$$

When a matrix with real entries is symmetric, then its norm coincides with the modulus of its maximal eigenvalue. Thus the proposition amounts to conjugating F simultaneously (i.e., by a single $g \in SL_2(\overline{\mathbb{Q}})$) in a "quasisymmetric" position.

Proof. As we may replace F by a bounded power of it, Lemma 4.7 above allows us to assume that F contains a semisimple element. Let $F = \{Id, A, B_1, ..., B_k\}$ with A semisimple. Conjugating by some $g \in SL_2(\overline{\mathbb{Q}})$, we may assume that A is in diagonal form and we write each B_i in the form of Equation (3) with entries $a_i, b_i, c_i,$ and d_i. Changing F into F^2 if necessary, we may assume that both b_1 and c_1 are not 0 (otherwise F would be contained in the group of upper- or lower-triangular matrices). We may further conjugate F by the diagonal matrix $diag(t, t^{-1})$, where $t \in \overline{\mathbb{Q}}$ is a root of $t^4 = c_1/b_1$, so as to ensure $b_1 = c_1$. Then $h(B_1) \leq h(a_1) + h(d_1) + 2h(b_1) + \log 2$. On the one hand, since $a_1 d_1 - b_1 c_1 = 1$, we have $b_1^2 = a_1 d_1 - 1$ and $2h(b_1) = h(b_1^2) \leq h(a_1 d_1) + \log 2 \leq 2e(\{A, B\}) + \log 2 C_\infty$. On the other hand, by Lemma 4.2 applied to $\{A, B_i\}$, we have $\max\{|a_i|_v, |d_i|_v\} \leq C_v e^{2s_v} \max\{1, |\delta^{-1}|_v\}$, for every place v, where $\delta = \lambda - \lambda^{-1}$ and $s_v = s_v(\{A, B_i\}) = \log E_v(\{A, B_i\})$. Applying Lemma 4.2 to $\{A, B_1 B_i\}$ we get $\max\{|(B_1 B_i)_{11}|_v, |(B_1 B_i)_{22}|_v\} \leq C_v e^{2s_v} \max\{1, |\delta^{-1}|_v\}$ with $s_v = s_v(\{A, B_1 B_i\}) = \log E_v(\{A, B_1 B_i\})$. We compute the matrix entry $(B_1 B_i)_{11} = a_1 a_i + b_1 c_i$. We get

$$|c_i|_v = |[(B_1 B_i)_{11} - a_1 a_i]b_1^{-1}|_v \leq C_v e^{2s_v} \max\{1, |\delta^{-1}|_v\} \max\{1, |b_1^{-1}|_v\}.$$

Similarly for $|b_i|_v$. Hence,

$$\|F\|_v \leq C_v \max_{i=1,...,k} \{|\lambda|_v, |\lambda^{-1}|_v, |a_i|_v, |d_i|_v, |b_i|_v, |c_i|_v\}$$

$$\leq C_v \max_{i=1,...,k} E_v(\{A, B_1, B_1 B_i\})^2 \cdot \max\{1, |\delta^{-1}|_v\} \max\{1, |b_1^{-1}|_v\}.$$

In particular, this means that

$$h(F) \leq 2 \log C_\infty + 2e(F^2) + h(\delta) + h(b_1)$$

$$\leq 7e(F) + 4 \log 2 C_\infty.$$

So we are done. $\qquad \square$

COROLLARY 5.2. *There exists a constant $C_{qs} > 0$ such that if F is as in the proposition, then there is an element $g \in SL_2(\overline{\mathbb{Q}})$ such that*

$$h(gFg^{-1}) \leq C_{qs} \cdot e(F).$$

Proof. It is clear from the combination of the previous proposition and Theorem 1.2. □

6. Ping-pong

Here we state and prove a ping-pong criterion, which gives a sufficient condition on the finite set F for it, or a bounded power of it, to contain two free generators of a free subgroup. Let $k_1, k_2, k_3 \in \mathbb{N}$ be three positive integers and let k be a local field of characteristic 0 with its standard absolute value. We set $C_k = 2$ if k is Archimedean, and $C_k = 1$ if k is non-Archimedean. Let $F \subset SL_2(k)$ be a finite set containing 1 such that $\Lambda_k(F^{k_1}) > C_k||F||_k$ (see Section 2 for notation, it is important to require a strict inequality here when k is non-Archimedean). Let $A \in F^{k_1}$ be such that $\Lambda_k(A) = \Lambda_k(F^{k_1})$. Then of course A is semisimple and admits two distinct eigenvectors v^+ and v^- in k_q^2 where k_q is either k or some quadratic extension of k. Since we may always replace k by k_q, there is no loss of generality in assuming that v^+ and v^- lie in k^2. Let d_k be the canonical (Fubini-Study) projective distance on $\mathbb{P}^1(k)$, namely $d_k(u, v) = \frac{||u \wedge v||_k}{||u||_k ||v||_k}$.

LEMMA 6.1. (GEOMETRIC CONDITIONS FOR PING-PONG) *Assume that there is $B \in F^{k_2}$ such that $d_k(Bv^\varepsilon, v^{\varepsilon'}) \geq ||F||_k^{-k_3}$, and $d_k(v^\varepsilon, v^{\varepsilon'}) \geq ||F||_k^{-k_3}$ for each $\varepsilon, \varepsilon' \in \{\pm\}$. Then A^l and $BA^l B^{-1}$ play ping-pong on $\mathbb{P}^1(k)$ and generate a free subgroup of $SL_2(k)$ as soon as $l \geq (k_2 + 1)(k_3 + 1)$.*

Proof. Note that for all $u, v \in \mathbb{P}^1(k)$ we have $d_k(Bu, Bv) \leq ||B||^2 d_k(u, v)$ for $B \in SL_2(k)$. Note also that without loss of generality, we may assume that $||v^+||_k = ||v^-||_k = 1$. Let λ, λ^{-1} be the eigenvalues of A, where we have chosen $|\lambda|_k \geq 1$. By the assumption on A, $|\lambda| > C_k||F||_k \geq 1$. Since the roles of v^+ and v^- are interchangeable, we may assume that v^+ corresponds to λ and v^- to λ^{-1}. Let $P \in GL_2(k)$ be defined by $Pe_1 = v^+$ and $Pe_2 = v^-$. Note that $|\det P| = ||v^+ \wedge v^-|| = d_k(v^+, v^-)$. Also $||P|| = 1$ if k is non-Archimedean, and $||P||^2 \leq 2$ if k is Archimedean, so in general $||P||^2 \leq C_k$. Moreover, $||P^{-1}|| = ||P||/|\det P|_k \leq C_k||F||^{k_3}$. Set $A' = P^{-1}AP$, $B' = P^{-1}BP$, and $F' = P^{-1}FP$. Then $A' = diag(\lambda, \lambda^{-1})$.

For $u, v \in \mathbb{P}^1(k)$, $d_k(Pu, Pv) = \frac{\|Pu \wedge Pv\|}{\|Pu\| \cdot \|Pv\|} \leq |\det P| \|P^{-1}\|^2 d_k(u, v) \leq \frac{C_k \cdot d_k(u,v)}{|\det P|}$. Hence for $i, j \in \{1, 2\}$,

12
$$d_k(B'e_i, e_j) \geq \frac{1}{C_k} d_k(v^+, v^-) d_k(BPe_i, Pe_j) \geq \frac{1}{C_k} \frac{1}{\|F\|^{2k_3}}.$$

Observe also that $\|F'\| \leq \|F\| \cdot \|P\|^2 / |\det P| \leq C_k \|F\|^{k_3 + 1}$.

Let $m \leq 2l$ be positive integers to be determined shortly below. Let $\mathcal{U}_A^+ = \{x \in \mathbb{P}^1(k), d_k(x, e_1) \leq |\lambda|^{-2l}\}$, $\mathcal{U}_A^- = \{x \in \mathbb{P}^1(k), d_k(x, e_2) \leq |\lambda|^{-2l}\}$, $\mathcal{U}_C^+ = \{x \in \mathbb{P}^1(k), d_k(x, B'e_1) \leq |\lambda|^{-m}\}$, and $\mathcal{U}_C^- = \{x \in \mathbb{P}^1(k), d_k(x, B'e_2) \leq |\lambda|^{-m}\}$. We need to show that these four sets are disjoint, and that A'^l maps $(\mathcal{U}_A^-)^c$ into $\mathcal{U}_A^+$, A'^{-l} maps $(\mathcal{U}_A^+)^c$ into $\mathcal{U}_A^-$, $C' = B'A'^l B'^{-1}$ maps $(\mathcal{U}_C^-)^c$ into $\mathcal{U}_C^+$, and C'^{-1} maps $(\mathcal{U}_C^+)^c$ into $\mathcal{U}_C^-$.

If, for instance, $\mathcal{U}_A^+ \cap \mathcal{U}_C^- \neq \emptyset$, then $d(B'e_i, e_j) \leq \frac{C_k}{|\lambda|^m}$ for some i, j, which in turn would contradict (12) since $|\lambda|^m > C_k^2 \|F\|^{2k_3}$ as soon as $m \geq 2k_3$. The same holds in other situations as soon as $m \geq 2(k_3 + 1)$.

Now since A' is diagonal, A' maps $(\mathcal{U}_A^-)^c$ into $\mathcal{U}_A^+$, and A'^{-l} maps $(\mathcal{U}_A^+)^c$ into $\mathcal{U}_A^-$. Finally, let us check the last two conditions. If $x \in (\mathcal{U}_C^-)^c$, then $d_k(x, B'e_2) > |\lambda|^{-m}$ and $d_k(B'^{-1}x, e_2) \|B'\|^2 > |\lambda|^{-m}$. So $B'^{-1}x \in (\mathcal{U}_A^-)^c$ as long as $|\lambda|^{2l-m} \geq \|B'\|^2$. Then $A'^l B'^{-1}x \in \mathcal{U}_A^+$ and $d_k(C'x, B'e_1) \leq \|B'\|^2 / |\lambda|^{2l} \leq |\lambda|^{-m}$. And similarly if $x \in (\mathcal{U}_C^+)^c$.

So the above works as soon as $m \geq 2(k_3 + 1)$ (so that $|\lambda|^m > C_k^2 \|F\|^{2(k_3+1)}$) and $2l - m \geq 2k_2(k_3 + 1)$ (so that $|\lambda|^{2l-m} > C_k^{2k_2} \|F\|^{2k_2 k_3 + 2k_2} \geq \|F'\|^{2k_2} \geq \|B'\|^2$). $\qquad\square$

REMARK 6.2. A similar ping-pong lemma holds with the ping-pong players A^l and $BA^l B$ (instead of $BA^l B^{-1}$) if we assume similar lower bounds on $d_k(B^\delta v^\varepsilon, v^{\varepsilon'})$ for $\delta \in \{0, \pm 1, \pm 2\}$ and $\varepsilon, \varepsilon' \in \{\pm\}$. This allows us to find the ping-pong players in some F^n, that is, without having to take inverses of elements of F.

6.1. Quasi-Isometrically Embedded Free Subgroup

A free subgroup H generated by two free elements a and b in a group Γ with finite generating set F (assumed symmetric) is said to be C-quasi isometrically embedded if for all $h \in H$

$$\frac{1}{C} \cdot d_\Gamma(1, h) \leq d_H(1, h) \leq C \cdot d_\Gamma(1, h)$$

where d_Γ is the word metric in Γ associated to F and d_H the word metric in H corresponding to the generating set $\{a^{\pm 1}, b^{\pm 1}\}$. In the setting of Lemma 6.1 we have

LEMMA 6.3. (QI-EMBEDDING OF FREE SUBGROUP) *The two elements* A^l *and* $BA^l B^{-1}$ *generate a free subgroup* H, *which is* C-*quasi-isometrically embedded in the group* Γ *with generating set* F *with* $C = 2k_2 + k_1 l$. *More precisely,*

$$\frac{1}{C} \cdot d_\Gamma(1, h) \leq d_H(1, h) \leq 4 \cdot d_\Gamma(1, h).$$

Proof. The inequality on the left-hand side is clear as both $a := A^l$ and $b := BA^l B^{-1}$ belong to F^C. To prove the inequality on the right-hand side, observe that both a and b act on the complement of their repelling neighborhood by transformations that contract distances by a factor at least $\frac{1}{|\lambda|_k} \leq \frac{1}{||F||_k}$. This implies that any element h that can be written as $h = w(a, b)$ for some reduced word w of length $n = d_H(1, h)$ in the free group will act on some open subset of $\mathbb{P}^1(k)$ by contracting distances by a factor at least $\frac{1}{||F||_k^n}$, and in particular $Lip(h) \geq ||F||_k^n$, where $Lip(h)$ is the bi-Lipschitz constant of h acting on $\mathbb{P}^1(k)$, $Lip(h) = \sup\{\left(\frac{d(hx,hy)}{d(x,y)}\right)^{\pm 1} | x, y \in \mathbb{P}^1(k)\}$. On the other hand, one easily checks that for any $g \in SL_2(k)$, $Lip(g) \leq ||g||_k^4$, hence $Lip(F^n) \leq ||F||_k^{4n}$ for all n and hence $Lip(h) \leq ||F||_k^{4d_\Gamma(1,h)}$, which yields $d_H(1, h) \leq 4 \cdot d_\Gamma(1, h)$ as desired. $\square$

7. Proof of Theorems 1.1 and 1.3 for SL$_2$ (ℂ)

We first assume that F has coefficients in $\overline{\mathbb{Q}}$. We explain at the end of this section why this case implies the general case.

We will show that if F generates a nonvirtually solvable subgroup of $SL_2(K)$ for some number field K, then for at least one place $v \in V_K$ the conditions of the ping-pong Lemma 6.1 are satisfied, with k_1, k_2, and k_3 bounded and independent of K. This will be done by finding an appropriate prime and a place above it where F will satisfy the requirements of Lemma 6.1.

Let F be a finite subset of $SL_2(\overline{\mathbb{Q}})$, which generates a nonvirtually solvable subgroup and contains 1. According to Lemma 4.7, as one may change F into a bounded power of itself if necessary, we may assume that F contains two semisimple elements that generate a nonvirtually solvable subgroup. Now, from Corollary 5.2, after possibly conjugating F inside $SL_2(\overline{\mathbb{Q}})$, we may assume that $h(F) \leq C_{qs} \cdot e(F)$, where $C_{qs} > 0$ is the universal constant given by Corollary 5.2.

The last important ingredient in the proof of Theorem 1.1 is the product formula on the projective line $\mathbb{P}^1(\overline{\mathbb{Q}})$ (see [3, 2.8.21]), that is, $\forall (u, v) \in \mathbb{P}^1(\overline{\mathbb{Q}})^2$

13

$$\prod_{v \in V_K} d_v(u, v)^{\frac{n_v}{[K:\mathbb{Q}]}} = \frac{1}{H(u) \cdot H(v)}$$

where $\log H(u) = h(u) = \frac{1}{[K:\mathbb{Q}]} \sum_{v \in V_K} n_v \log \|(u_1, u_2)\|_v$ if $(u_1, u_2) \in K^2$ represents $u \in \mathbb{P}^1(K)$. This formula is straightforward from the usual product formula and the definition of the standard distance $d_v(u, v) = \frac{\|u \wedge v\|_{K_v}}{\|u\|_{K_v} \|v\|_{K_v}}$.

LEMMA 7.1. (HEIGHT OF f CONTROLS HEIGHTS OF EIGENOBJECTS)

Let $A \in SL_2(\overline{\mathbb{Q}})$ and $v \in \mathbb{P}^1(\overline{\mathbb{Q}})$ an eigendirection of A, then $h(v) \leq 3h(A) + \frac{3}{2} \log 2$.

Proof. Simply solve for v in $Av = \lambda v$ using Cramer's rule. $\qquad \square$

Let us introduce some notation. Suppose $A \in SL_2(\overline{\mathbb{Q}})$ is semisimple with eigendirections v_A^+ and v_A^- in $\mathbb{P}^1(\overline{\mathbb{Q}})$ and suppose $B \in SL_2(\overline{\mathbb{Q}})$. Then, assuming A and B have coefficients in a number field K, we set for each place $v \in V_K$:

$$\delta_v^{+,-}(B; A) = \log \frac{1}{d_v(Bv_A^+, v_A^-)}$$

where d_v is the standard distance on $\mathbb{P}^1(K_v)$ and K_v is the completion of K at v. Note that as $d_v \leq 1$, we have $\delta_v^{+,-}(B; A) \geq 0$. If $d_v(Bv_A^+, v_A^-) = 0$, we set $\delta_v^{+,-}(B; A) = 0$. We define similarly $\delta_v^{+,+}(B; A)$, $\delta_v^{-,+}(B; A)$, and $\delta_v^{-,-}(B; A)$ in the obvious manner and we set

$$\delta_v(B; A) = \delta_v^{+,-}(B; A) + \delta_v^{+,+}(B; A) + \delta_v^{-,+}(B; A) + \delta_v^{-,-}(B; A).$$

For a finite subset F of $SL_2(\overline{\mathbb{Q}})$, we also define

$$\delta_v(F) = \sum \delta_v(Id; A) + \delta_v(B; A)$$

where the sum runs over all pairs $\{A, B\}$ of elements of F with A semisimple and B in *"nice position"* with respect to A, namely, such that $Bv_A^+ \notin \{v_A^+, v_A^-\}$ and $Bv_A^- \notin \{v_A^+, v_A^-\}$. If this set of pairs is empty, we set δ to be 0. However, in our case, it will be nonempty if not for F itself, then for a bounded power of it (see Lemma 7.3 below). We also define the corresponding global quantity:

$$\delta(B; A) = \frac{1}{[K:\mathbb{Q}]} \sum_{v \in V_K} n_v \cdot \delta_v(B; A)$$

and

$$\delta(F) = \frac{1}{[K:\mathbb{Q}]} \sum_{v \in V_K} n_v \cdot \delta_v(F).$$

PROPOSITION 7.2. (HEIGHT OF f CONTROLS ADELIC DISTANCE BETWEEN EIGENOBJECTS)

With the above notation, for every $B \in SL_2(\overline{\mathbb{Q}})$ in nice position with respect to a semisimple $A \in SL_2(\overline{\mathbb{Q}})$ (or for $B = Id$), we have

$$\delta(B; A) \leq 24h(A) + 4h(B) + 12\log 2.$$

In particular, for any finite subset F containing 1 in $SL_2(\overline{\mathbb{Q}})$

$$\delta(F) \leq 12|F|^2(3h(F) + \log 2)$$

Proof. One the one hand, from the product formula (13) above we have $\delta^{+,-}(B; A) = h(Bv_A^+) + h(v_A^-)$. On the other hand, we easily compute $h(Bv_A^+) \leq h(B) + h(v_A^+) + \log 2$. From Lemma 7.1, we get $\delta^{+,-}(B; A) \leq h(B) + 6h(A) + 3\log 2$, hence the desired bounds. $\qquad\square$

Note that since we assume that F generates a nonvirtually solvable group, then according to Theorem 1.2, $h(F) \geq e(F) \geq \varepsilon$ for some fixed ε. Therefore, there exists a constant $D_{qs} > 0$ such that $\delta(F) \leq D_{qs}|F|^2 h(F)$.

LEMMA 7.3. *There is an integer $n_0 \geq 2$ such that if F is a finite subset of $SL_2(\mathbb{C})$ containing 1 and generating a nonvirtually solvable group, then for any semisimple $A \in F$ there exists $B \in F^{n_0}$, which is in nice position with respect to A.*

Proof. This is another occurrence of the escape trick described in Lemma 4.8. The subvarieties $X_A = \{B \in GL_2, Bv_A^+ \in \{v_A^\pm\}$ or $Bv_A^- \in \{v_A^\pm\}\}$ are conjugate to each other in GL_2. In particular, there is N as in Lemma 4.8 such that for each semisimple A in F, F^N is not contained in $X_A(\mathbb{C})$, as the group generated by F clearly cannot be contained in any $X_A(\mathbb{C})$ for it would otherwise be virtually solvable. $\qquad\square$

We have for all $n \in \mathbb{N}$

$$\delta(F^n) \leq D_{qs} \cdot |F^n|^2 \cdot h(F^n) \leq D_{qs} \cdot |F|^{2n} \cdot n \cdot h(F).$$

We may write with obvious notation

$$\delta = \sum_{p \in \{\infty\} \cup \mathcal{P}} \delta_p = \delta_\infty + \delta_f.$$

We fix $n = n_0$ as in Lemma 7.3 and let $D'_{qs} = D_{qs} \cdot n_0$ so that $\delta(F^{n_0}) \leq D'_{qs} \cdot |F|^{2n_0} \cdot h(F)$ and $h(F) \leq C_{qs} \cdot e(F)$. For each $p \in \{\infty\} \cup \mathcal{P}$ we set $e_p = e_p(F)$, $h_p = h_p(F)$ and $\delta_p = \delta_p(F^{n_0})$. We now claim:

Claim: There exists a constant $C'' > 0$ such that for any set F in $SL_2(\overline{\mathbb{Q}})$ containing 1 and generating a nonvirtually solvable subgroup, there exist $p \in \{\infty\} \cup \mathcal{P}$ and a place $v|p$ such that $\max\{\delta_v, h_v\} \leq C'' \cdot |F|^{n_0} \cdot e_v$ and $e_v > \frac{e_p}{2}$. Moreover, if $p = \infty$, we may assume that $e_\infty \geq \frac{1}{2}e$.

We now prove this claim. Suppose first that $e_\infty \geq \frac{1}{2}e$, then $\delta_\infty + h_\infty \leq 2C_{qs}(D'_{qs}|F|^{2n_0} + 1) \cdot e_\infty$. But

$$e_\infty \leq \frac{2}{[K:\mathbb{Q}]} \sum_{v \in V_\infty^+} n_v e_v$$

where $V_\infty^+ = \{v \in V_\infty, e_v \geq \frac{e_\infty}{2}\}$. Indeed

$$e_\infty = \frac{1}{[K:\mathbb{Q}]} \left(\sum_{v \in V_\infty^+} n_v e_v + \sum_{v \in V_\infty^-} n_v e_v \right) \leq \frac{1}{[K:\mathbb{Q}]} \sum_{v \in V_\infty^+} n_v e_v + \frac{e_\infty}{2}.$$

Hence $\sum_{v \in V_\infty^+} n_v(\delta_v + h_v) \leq 4C_{qs}(D'_{qs}|F|^{2n_0} + 1) \cdot \sum_{v \in V_\infty^+} n_v e_v$. So for at least one $v \in V_\infty^+$ we have $\max\{\delta_v, h_v\} \leq \delta_v + h_v \leq 4C_{qs}(D'_{qs}|F|^{2n_0} + 1) \cdot e_v$.

Now suppose $e_\infty < \frac{e}{2}$, then $e_f \geq \frac{e}{2} > 0$ and $\sum_{p \in \mathcal{P}} \delta_p + h_p \leq 2C_{qs}(D'_{qs}|F|^{2n_0} + 1) \cdot \sum_{p \in \mathcal{P}} e_p$, hence there must be one $p \in \mathcal{P}$ for which $e_p > 0$ and $\delta_p + h_p \leq 2C_{qs}(D'_{qs}|F|^{2n_0} + 1) \cdot e_p$. As this is an average over the places $v|p$, as before there must be some place $v|p$ for which $e_v \geq \frac{e_p}{2}$ and $\max\{\delta_v, h_v\} \leq \delta_v + h_v \leq 4C_{qs}(D'_{qs}|F|^{2n_0} + 1) \cdot e_v$. So we have justified the claim.

End of the proof of Theorems 1.3 and 1.1:. Let us recapitulate what we have so far. We started with a set F in $SL_2(\overline{\mathbb{Q}})$ containing 1 and generating a nonvirtually solvable subgroup. We found the constant $n_0 \geq 2$ as in Lemma 7.3. We also found a constant C'' such that for some prime p and a place $v|p$ one has $\max\{\delta_v(F^{n_0}), h_v(F)\} \leq C'' \cdot |F|^{2n_0} \cdot e_v(F)$, and $e_v(F) \geq \frac{1}{4}e_p(F) > 0$ (with $e_\infty \geq \frac{e}{2}$ in case $p = \infty$). Set $D''_F := C'' \cdot |F|^{2n_0}$.

Suppose first that $v \in V_f$. Recall that we had $\Lambda_v(F^2) \geq E_v(F)^2$ by Lemma 2.1.

Let $A_0 \in F^2$ be such that $\Lambda_v(A_0) = \Lambda_v(F^2)$. Then $\Lambda_v(A_0) \geq E_v(F)^2 \geq ||F||_v^{\frac{2}{D''_F}} > 1$ and hence if $k_1 \in \mathbb{N}$ is the first even integer strictly larger that D''_F, we have $\Lambda_v(A) > ||F||_v$ if $A = A_0^{k_1/2} \in F^{k_1}$. Moreover we have $\delta_v(F^{n_0}) \leq D''_F \cdot e_v(F)$, therefore for every $B \in F^{n_0}$ that is in nice position with respect to A_0 (and there are such B's according to Lemma 7.3) we have $\delta_v(Id; A_0) + \delta_v(B; A_0) \leq D''_F \cdot e_v(F)$. Fix one such B. We have $d_v(Bv_A^\varepsilon, v_A^{\varepsilon'}) \geq E_v(F)^{-D''_F} \geq ||F||_v^{-D''_F}$ and also $d_v(v_A^\varepsilon, v_A^{\varepsilon'}) \geq E_v(F)^{-D''_F} \geq ||F||_v^{-D''_F}$ for all $\varepsilon, \varepsilon' \in \{\pm\}$. Therefore we are in a position to apply the ping-pong Lemma 6.1 to the pair A and B with k_1 as above ($\leq D''_F + 2$), $k_2 = n_0$ and $k_3 = D''_F$. This ends the proof in the case when $v \in V_f$.

Suppose now that $v \in V_\infty$. We have $E_v(F) \geq \exp(\frac{e}{4}) \geq \exp(\frac{\varepsilon}{4})$ where ε is the constant from Theorem 1.2. Now Lemma 2.6 shows that there is a constant $n_1 = n_1(\varepsilon) \in \mathbb{N}$ such that $E_v(F^{n_1}) \geq \frac{2}{c^2}$ where c is the constant in Lemma 2.1.

Then by Lemma 2.1 $\Lambda_v(F^{2n_1}) \geq c^2 E_v(F^{n_1})^2 \geq 2E_v(F^{n_1}) \geq 2E_v(F) \geq 2||F||^{\frac{1}{D_F''}}$.
Observe that after possibly changing n_0 we may assume that it is larger than
$2n_1$. Pick $A_0 \in F^{2n_1}$ such that $\Lambda_v(A_0) = \Lambda_v(F^{2n_1})$. Finally, if k_1' is the smallest integer strictly larger than D_F'', we set $A = A_0^{k_1'} \in F^{k_1}$ where $k_1 = 2n_1 k_1'$.
We have $\Lambda_v(A) > 2||F||_v$. Moreover $\delta_v(F^{n_0}) \leq D_F'' \cdot e_v(F)$, therefore, for every
$B \in F^{n_0}$ that is in nice position with respect to A_0 (and there are such B's according to Lemma 7.3) we have $\delta_v(Id; A_0) + \delta_v(B; A_0) \leq D_F'' \cdot e_v(F)$. Fix one such B.
We have $d_v(Bv_A^\varepsilon, v_A^{\varepsilon'}) \geq E_v(F)^{-D_F''} \geq ||F||_v^{-D_F''}$ and also $d_v(v_A^\varepsilon, v_A^{\varepsilon'}) \geq E_v(F)^{-D_F''} \geq$
$||F||_v^{-D_F''}$ for all $\varepsilon, \varepsilon' \in \{\pm\}$. Therefore, we are in a position to apply the ping-
pong Lemma 6.1 to the pair A and B with k_1 as above ($\leq 2n_1(D_F''+1)$), $k_2 = n_0$
and $k_3 = D_F''$. $\qquad\square$

Theorem 1.3 on the quasi-isometric embedding of the free group (in the
case $F \subset SL_2(\overline{\mathbb{Q}})$) now follows readily by application of Lemma 6.3. To complete
the proof of Theorem 1.1, it remains to observe that we can reduce to the
situation where F has three elements $\{1, a, b\}$ by application of Lemma 4.7.
Note that we cannot do this reduction for Theorem 1.3 because there we need
to control the behavior of every element of F.

There are several ways to see that Theorems 1.3 and 1.1 for $SL_2(\overline{\mathbb{Q}})$ imply the
same theorems for $SL_2(\mathbb{C})$. One can use the remark made in the introduction
that both results are equivalent to a countable union of assertions expressible
in first-order logic. By elimination of quantifiers for algebraically closed fields,
we know that two algebraically closed fields of the same characteristic satisfy
the same statements of first-order logic (see, e.g., [14] ch. 9). Hence the validity
of Theorems 1.3 and 1.1 over $\overline{\mathbb{Q}}$ is equivalent to its validity over $\mathbb{C}$.

Another way to see it is to invoke Proposition 9.3 below and use the fact that
if V is an algebraic variety defined over $\mathbb{Q}$, then $V(\overline{\mathbb{Q}})$ is Zariski-dense in $V(\mathbb{C})$.
From Proposition 9.3 and the above proof over $\overline{\mathbb{Q}}$, we know that $W_n(\overline{\mathbb{Q}}) \subset V(\overline{\mathbb{Q}})$
for every $n \in \mathbb{N}$, which readily implies that $W_n(\mathbb{C}) \subset V(\mathbb{C})$ for every $n \in \mathbb{N}$ (W_n
is defined in Equation (15); see Section 9 below). And the theorem is proved
over $\mathbb{C}$ with the same constant N_0.

For Theorem 1.1, one could also use a specialization argument as is in [13]
for instance.

8. Uniform Spectral Gap in ℓ^2 and Cogrowth of Subgroups

We prove here Corollaries 1.4 to 1.7. Corollary 1.4 is a direct application of
Corollary 1.5, so we will not say more about it.

Proof of Corollary 1.5. We reproduce the argument given in [29] and [9]. Since the free group F_2 is nonamenable, there is a constant $\kappa > 0$ such that $\max\{\|a \cdot f - f\|_2, \|b \cdot f - f\|_2\} \geq \kappa \cdot \|f\|_2$ for every $f \in \ell^2(F_2)$ where a and b are the two free generators of F_2. Then according to Theorem 1.1, there are a and $b \in (F \cup F^{-1})^{N_0}$ such that a and b generate a free subgroup H. For $f \in \ell^2(\Gamma)$ and Hx a coset of H, let f_{Hx} denote the restriction of f to Hx. Let $\mathcal{A}$ (respectively, $\mathcal{B}$) be the subset of $H \backslash \Gamma$ of those cosets such that $\|a \cdot f_{Hx} - f_{Hx}\|_2 \geq \kappa \|f_{Hx}\|_2$ (respectively, $\|b \cdot f_{Hx} - f_{Hx}\|_2 \geq \kappa \|f_{Hx}\|_2$). And set $f_{\mathcal{A}} = \sum_{Hx \in \mathcal{A}} f_{Hx}$ and $f_{\mathcal{B}} = \sum_{Hx \in \mathcal{B}} f_{Hx}$. Since $\|f\|_2^2 \leq \|f_{\mathcal{A}}\|_2^2 + \|f_{\mathcal{B}}\|_2^2$ we may assume without loss of generality that $\|f_{\mathcal{A}}\|_2^2 \geq \|f\|_2^2/2$. Hence $\|a \cdot f - f\|_2^2 \geq \|a \cdot f_{\mathcal{A}} - f_{\mathcal{A}}\|_2^2 \geq \kappa^2 \|f_{\mathcal{A}}\|_2^2 \geq \frac{\kappa^2}{2} \|f\|_2^2$. Since $a \in (F \cup F^{-1})^{N_0}$ we have

$$\|a \cdot f - f\|_2 \leq \sum_{i=1}^{N_0} \|s_1 \ldots s_i \cdot f - s_1 \ldots s_{i-1} \cdot f\|_2 = \sum_{i=1}^{N_0} \|s_i \cdot f - f\|_2$$

where $a = s_1 \cdot \ldots \cdot s_{N_0}$ with $s_i \in F \cup F^{-1}$. Finally, for some i we have $\|s_i \cdot f - f\|_2 = \|s_i^{-1} \cdot f - f\|_2 \geq \frac{\kappa}{N_0 \sqrt{2}} \|f\|_2$. Hence we have proved the first assertion of Corollary 1.5 with $\varepsilon = \frac{\kappa}{N_0 \sqrt{2}}$.

To prove the second assertion, let F and A be as in the statement. Let Γ be the group generated by A and F and simply apply the above with f the indicator function of A in $\ell^2(\Gamma)$. $\square$

Proof of Corollary 1.6. Set $\Gamma = \langle F \rangle$ with $F = \{a_1^{\pm 1}, \ldots, a_m^{\pm 1}\}$ and as in the statement and $\mu = \frac{1}{2m} \sum_{1 \leq i \leq m} \delta_{a_i} + \delta_{a_i^{-1}}$. Then $\mathbb{P}(S_n = e) = \mu^n(e)$. But $\mu^{nN_0}(e) \leq \|\mu^{\frac{nN_0}{2}}\|^2 \leq \|\mu^{N_0}\|^n$ where $\| \cdot \|$ is the norm of the convolution operator. Theorem 1.1 shows that $\mu^{N_0} = \alpha \mu_{F_2} + (1 - \alpha)\nu$ for some probability measure ν, where $\mu_{F_2} = \frac{1}{4}(\delta_a + \delta_{a^{-1}} + \delta_b + \delta_{b^{-1}})$ and a, b are the free generators in F^{N_0}, and $\alpha = \frac{1}{(2m)^{N_0}}$. It follows that $\|\mu^{N_0}\| \leq 1 - \alpha\tau$ if $\|\mu_{F_2}\| = 1 - \tau < 1$. Hence the result. $\square$

Proof of Corollary 1.7. We keep the notation of the proof of Corollary 1.6. One can go from spectral gap to cogrowth in a one-to-one fashion, thanks to the following formula (see [17], [10], and [27])

14
$$(2m - 1)^\eta + (2m - 1)^{1-\eta} = (2m)^\theta$$

where $\eta = \lim_{n \text{ even}} \frac{1}{n} \log_{2m-1} |W_n'|$ and $\theta = \lim_{n \text{ even}} \frac{1}{n} \log_{2m} |W_n|$, with W_n the set of paths of length n in the free group F_{2m} whose projection to $\langle F \rangle$ goes from the identity to itself and W_n' is the set of elements in F_{2m} of length n that kill $(a_1, \ldots, a_m)$. Since $|W_{p+q+2}'| \geq |W_p'||W_q'|$, we must have $|W_n'| \leq$

$(2m-1)^{\eta(n+2)}$ for all $n \geq 1$. On the other hand, $\mu^n(e) = \frac{|W_n|}{(2m)^n}$ and hence $(2m)^\theta \leq 2m\|\mu^{N_0}\|^{\frac{1}{N_0}} \leq 2m(1-\alpha\tau)^{\frac{1}{N_0}}$ if $\|\mu_{F_2}\| = 1 - \tau$. Hence $(2m)^\theta \leq 2m(1 - \frac{\tau}{(2m)^{N_0}})^{\frac{1}{N_0}}$. Solving equation (14), we obtain $(2m-1)^\eta \leq 2m - 1 - \frac{\tau}{(2m)^{2N_0}}$. Hence $|W'_n| \leq (2m - 1 - \frac{\tau}{2(2m)^{2N_0}})^n$ for $n \geq n(m)$. We are done.

To see the converse, note that by Proposition 9.2 $F = \{a_1, ..., a_m\} \subset GL_d(\mathbb{C})$ generates a virtually solvable subgroup if and only if it contains a subgroup of index $< M$ that can be conjugated in the upper-triangular matrices, hence is of solvable length $\leq d$. In particular, $\Gamma = \langle F \rangle$ is a quotient of the free object on m generators, which we denote by S in this variety of groups. S is virtually solvable, hence amenable, and by the Kesten's criterion and Equation (14) must satisfy $|W'_n| \geq (2m - 1 - \varepsilon)^n$ for every $\varepsilon > 0$ and all $n \geq n(\varepsilon)$. Since every relation in S is also a relation in Γ we are done. $\qquad\square$

9. Large Girth

Here we prove Corollaries 1.8 to 1.11. Let $\overline{K}$ be an algebraically closed field, $\mathcal{F}_d$ the flag variety in $\overline{K}^d$ ($\mathcal{F}_d = \mathbb{P}^1(\overline{K})$ if $d = 2$) and let $\mathcal{V}_k$ be the set of k-tuples $(A_1, ..., A_k) \in GL_d(\overline{K})^k$ such that $\mathbf{A} = (A_1, ..., A_k)$ leaves invariant some subset $\{u_1, ..., u_M\}$ of M not necessarily distinct points of $\mathcal{F}_d$.

PROPOSITION 9.1. *Then $\mathcal{V}_k$ is a closed subscheme of $GL_d(\overline{K})^k$ defined over $\mathbb{Z}$.*

Proof. We write the proof for $k = 2$, the same argument works in general. Consider the map $\phi : GL_d \times GL_d \times \mathcal{F}_d^M \to \mathcal{F}_d^{3M}$, which maps $(A, B, u_1, ..., u_M)$ to $(Au_1, ..., Au_M, Bu_1, ..., Bu_M, u_1, ..., u_M)$. For every two permutations $\sigma, \eta \in S_M$ we set $\Delta_{\sigma,\eta} = \{(a_1, ..., a_M, b_1, ..., b_M, u_1, ..., u_M) \in \mathcal{F}_d^{3M}$ such that $a_i = u_{\sigma(i)}$ and $b_i = u_{\eta(i)}$ for each $i = 1, ..., M\}$ and let Δ the union of all $\Delta_{\sigma,\eta}$. Then Δ is a closed subvariety of $\mathcal{F}_d^{3M}$, therefore so is $\mathcal{V}_k = \pi \circ \phi^{-1}(\Delta)$, where π is the projection onto the $GL_d \times GL_d$ factor, which is a closed morphism since $\mathcal{F}_d^{3M}$ is complete. $\qquad\square$

PROPOSITION 9.2. (ZARISKI CLOSEDNESS OF VIRTUALLY SOLVABLE TUPLES) *There is $M = M(d) \in \mathbb{N}$ such that a k-tuple $\mathbf{A} = (A_1, ..., A_k)$ in $GL_d(\mathbb{C})$ generates a virtually solvable subgroup if and only if $F \in \mathcal{V}_k(\mathbb{C})$.*

Proof. The if part is clear. To show the converse observe that by induction on d we may assume that $\mathbb{G}$ acts irreducibly on $\mathbb{C}^d$. Since the connected component $\mathbb{G}_0$ is solvable, Borel's fixed-point theorem implies that it fixes a point on $\mathcal{F}_d$.

Let $\mathbb{U}$ be the unipotent radical of $\mathbb{G}_0$. If $\mathbb{U}$ is nontrivial, it must fix pointwise a nontrivial subspace of $\mathbb{C}^d$. As $\mathbb{G}$ normalizes $\mathbb{U}$, $\mathbb{G}$ also must fix that subspace, which contradicts the assumption of irreducibility. Hence $\mathbb{U}$ is trivial and $\mathbb{G}_0$ is a torus. Therefore $\mathbb{G}$ is contained in the normalizer $N(\mathbb{G}_0)$ and $N(\mathbb{G}_0)/Z(\mathbb{G}_0)$ embeds in the Weyl group of GL_d, hence has size at most $d!$. We may thus assume that $\mathbb{G}$ centralizes $\mathbb{G}_0$. As we may again assume that $\mathbb{G}$ acts irreducibly, this forces $\mathbb{G}_0$ to be trivial. Hence we are left with the case when $\mathbb{G}$ is finite and we invoke Jordan's theorem (see, e.g., [11]) to conclude: it gives $M \in \mathbb{N}$ such that $[\mathbb{G} : A] \leq M$ where A is an abelian subgroup of $GL_d(\mathbb{C})$ made of semisimple elements. Hence A is contained in a torus S, which fixes a flag. It follows that $\mathbb{G}$ stabilizes the $\mathbb{G}$-orbit of this flag, which is of cardinality at most M. $\square$

Let us now express the conclusion of Theorem 1.1 in terms of a countable family of algebraic conditions. Let N be the integer obtained in the statement of Theorem 1.1 and let $B_2(n)$ be the ball of radius n in the free group F_2 on two generators. For $n \geq 1$ let $\mathcal{W}_n$ be the set of k-tuples $\mathbf{A} = (A_1, ..., A_k) \in GL_d(\mathbb{C})^k$ such that for any words w_1 and w_2 in $B_k(N)$ there exists a word $w \in B_2(n)\backslash\{1\}$ such that $w(w_1(\mathbf{A}), w_2(\mathbf{A})) = 1$. Clearly $\mathcal{W}_n$ is a closed subvariety of $GL_d(\mathbb{C})^k$. Hence we obtain:

PROPOSITION 9.3. (REFORMULATION OF MAIN THEOREM IN TERMS OF EQUALITY OF ALGEBRAIC VARIETIES) *Theorem 1.1 is equivalent to the statement* $\forall n \geq 1\ \mathcal{W}_n \subset \mathcal{V}_k$.

REMARK 9.4. Clearly $\mathcal{W}_n \subset \mathcal{W}_{n+1}$. Also it is clear from Proposition 9.2 that $\mathcal{V}_k \subset \mathcal{W}_{n_0}$ for some $n_0 \geq 1$. Hence Theorem 1.1 is in fact equivalent to $\mathcal{W}_n = \mathcal{W}_{n_0} = \mathcal{V}_k$ for all $n \geq n_0$.

For $w \in F_k\backslash\{1\}$ let X_w be the word variety $X_w = \{\mathbf{A} \in GL_d(\mathbb{C})^k, w(\mathbf{A}) = 1\}$. Equivalently $X_w = \{\mathbf{A} \in GL_d(\mathbb{C})^k,\ Q_w^{ij} = 0$ for all $i,j = 1, ..., d\}$, where $Q_w^{ij} = P_w^{ij} - \delta_{ij}$ and $\{P_w^{ij}\}_{1 \leq i,j \leq d}$ is the matrix $w(\mathbf{A})$ with each $P_w^{ij} \in \mathbb{C}[((A_1)_{ij})_{ij}, ..., ((A_k)_{kl})_{kl}]$ a polynomial in the kd^2 variables of $\mathbf{A} = (A_1, ..., A_k)$. Let $\mathcal{A}$ be the set of couples (w_1, w_2) of words in $B_2(N)$. Let $\mathcal{B}_n$ be the set of words $w \in B_k(n)\backslash\{1\}$ and finally let $\mathcal{C}$ be the set of indices $\{ij\}_{1 \leq i,j \leq d}$. For $a = (w_1, w_2) \in \mathcal{A}$, $b = w \in \mathcal{B}_n$, and $c = \{ij\} \in \mathcal{C}$ set $Q_{a,b,c}$ to be the polynomial $Q_{w(w_1,w_2)}^{ij}$.

LEMMA 9.5. (DEGREE AND HEIGHT BOUNDS FOR WORD POLYNOMIALS) *For each* $a \in \mathcal{A}$, $b \in \mathcal{B}_n$, *and* $c \in \mathcal{C}$, *the polynomial* $Q_{a,b,c} \in \mathbb{Z}[((A_1)_{ij})_{ij}, ...,$

$((A_k)_{kl})_{kl}]$ has integer coefficients, has height at most $d^{nN}+1$, and degree at most nN.

Proof. Here the height is understood in the naive sense of maximal modulus of the coefficients. The proof is an easy induction on n and we omit the details. □

With this notation, we have $W_n = \cap_{a\in A} \cup_{b\in B_n} \cap_{c\in C}\{Q_{a,b,c}=0\}$, which we may rewrite as

15
$$W_n = \bigcup_{f\in B_n^A} W_{n,f}$$

where $W_{n,f} = \cap_{a\in A} \cap_{c\in C}\{Q_{a,f(a),c}=0\}$ where f ranges among all maps $f : A \to B_n$. Let I be the ideal of $\mathbb{Z}[(A_{ij})_{ij}, (B_{kl})_{kl}]$ associated to V. Let $I_{n,f}$ be the ideal of $\mathbb{Q}[(A_{ij})_{ij}, (B_{kl})_{kl}]$ generated by the $Q_{a,f(a),c}$ with $a \in A$ and $c \in C$. Let I_n^f be the ideal of all polynomials in $\mathbb{Q}[(A_{ij})_{ij}, (B_{kl})_{kl}]$ that vanish on $W_{n,f}$. Then Hilbert's Nullstellensatz asserts that $I_n^f = \sqrt{I_{n,f}}$, and Theorem 1.1 says that $I \subset I_n^f$ for every n and $f \in B_n^A$. Let $f_1, ..., f_m$ be generators of I with integer coefficients. The following effective version of the Nullstellensatz may be found in [25]:

THEOREM 9.6. (EFFECTIVE ARITHMETIC NULLSTELLENSATZ [25]) *Let* $r, d \in \mathbb{N}$, $h > 0$ *and* $f, q_1, ..., q_k$ *be polynomials in* $\mathbb{Z}[X_1, ..., X_r]$ *with logarithmic height at most* h *and degree at most* d. *Assume that* f *vanishes at all common zeros (if any) of* $q_1, ..., q_k$ *in* $\mathbb{C}[X_1, ..., X_r]$. *Then there exist* $a, e \in \mathbb{N}$ *and polynomials* $b_1, ..., b_k \in \mathbb{Z}[X_1, ..., X_r]$ *such that*

$$af^e = b_1 q_1 + ... + b_k q_k$$

with $e \le (8d)^{2^r}$, *the total degree of each* b_i *is at most* $(8d)^{2^r+1}$ *and the logarithmic height of each* b_i *as well as* a *is at most* $(8d)^{2^{r+1}+1}(h + 8d \log(8d))$.

Here, the logarithmic height is the log of the naive height used above. We can now finish the proof of Corollary 1.8. In our situation, Theorem 9.6 yields numbers $a_i \in \mathbb{N}$ and polynomials $b_{f,a,c}^i \in \mathbb{Z}[(A_{ij})_{ij}, (B_{kl})_{kl}]$ with logarithmic height $h_n = O_d(n^{2^{kd^2+2}})$ as well as numbers $e_i \in \mathbb{N}$ such that for each $i = 1, ..., m$

16
$$a_i f_i^{e_i} = \sum_{a\in A, c\in C} b_{f,a,c}^i Q_{a,f(a),c}$$

It follows that if $p > \exp(h_n)$ is a rational prime, then for any field K of characteristic p, and any $\mathbf{A} \in GL_d(K)^k$, if for any words w_1, w_2 in $B_k(N)$ there is a word $w \in B_2(n)\backslash\{1\}$ such that $w(w_1(\mathbf{A}), w_2(\mathbf{A})) = 1$, then $f_i(\mathbf{A}) = 0$ for all $i = 1, ..., m$. Since the f_i generate I, according to Proposition 9.1, this means that there must be a set $\{u_1, ..., u_M\}$ in $\mathcal{F}_d(\overline{K})$ of at most M points ($\overline{K}$ is an algebraic closure of K) that is fixed by $\mathbf{A}$. In particular the group Γ generated by $\mathbf{A}$ contains the solvable subgroup $\Gamma_0 = \{\gamma \in \Gamma, \gamma \cdot u_1 = u_1\}$ as a subgroup of index $\leq M$. Therefore Corollary 1.8 holds as soon as $p > \exp(h_n)$, that is, for all $n \leq O_d(\log p)^{2^{-kd^2 - 2}}$. This ends the proof of Corollary 1.8 with, for example, $\varepsilon_0 = 2^{-10}$ for $k = d = 2$.

Observe that in the above proof we may have replaced $\mathcal{W}_n$ by the larger subvariety $\mathcal{W}'_n$ equal to the subset of k-tuples $\mathbf{A} = (A_1, ..., A_k) \in GL_d(\mathbb{C})^k$ such that for any words w_1 and w_2 in $B_k(N)$ there exists a word $w \in B_2(n)\backslash\{1\}$ such that $w(w_1(\mathbf{A}), w_2(\mathbf{A})) = 1$ or there are words $w \in B_2(n)\backslash\{1\}$ and $w_0 \in B_k(\frac{1}{4d}\ell(w))$ such that $w(w_1(\mathbf{A}), w_2(\mathbf{A})) = w_0(\mathbf{A})$. The bounds on the height and degree of the polynomials defining $\mathcal{W}'_n$ are of the same magnitude as those of $\mathcal{W}_n$, hence the same conclusion holds, namely:

COROLLARY 9.7. (QI-EMBEDDED SUBGROUP OF LARGE GIRTH) *Given $d, k \geq 2$, there is $N, M \in \mathbb{N}$ and $\varepsilon_0, C > 0$ such that for every prime p and every field K of characteristic p and any finite subset F with k elements generating a subgroup G of $GL_d(K)$ that contains no solvable subgroup of index at most M, then F^N contains two elements a, b generating a subgroup H with Cayley graph $\mathcal{G}_H$ such that $\operatorname{girth}(\mathcal{G}_H) \geq f(p)$ and*

17
$$\frac{1}{C} \cdot d_G(1, h) \leq d_H(1, h) \leq C \cdot d_G(1, h)$$

for any h with $d_H(1, h) \leq f(p)$, where $f(p) = (\log p)^{\varepsilon_0}$. Note that C depends on d only.

We denoted by d_G the word distance on G induced by the generating set F, and by d_H the word distance on H induced by $\{a, b\}$. We are now ready to prove Corollary 1.9 from the introduction.

Proof of Corollary 1.9. This follows directly from Corollary 9.7 and the following proposition: □

PROPOSITION 9.8. (QI EMBEDDED SUBGROUP OF LARGE GIRTH IMPLIES UNIFORM EXPANSION ON SMALL SETS) *Suppose G is a k-generated group with Cayley graph $\mathcal{G}_G$ and word metric d_G and H is a finitely generated* subgroup with Cayley graph $\mathcal{G}_H$ and word metric d_H. Assume further that girth$(\mathcal{G}_H) \geq N$ and that (17) holds for all h such that $d_H(1, h) \leq N$. Let μ be the uniform symmetric probability measure on the generators of G. Then there exists an explicit constant $\beta = \beta(k, C) > 0$ such that

$$\|\mu * f\|_2 \leq (1 - \beta)\|f\|_2$$

for any function f supported on a ball of radius $\leq N/2C$ in $\mathcal{G}_G$.

We will see below that the uniform QI-embedding of H in G is used in a key way in the proof. This proposition also yields in a standard way the following corollary.

COROLLARY 9.9. In the setting of the Proposition, there is $\beta = \beta(C, k) > 0$ and $\alpha = \alpha(C, k) > 0$ such that $\mu^{*n}(e) \leq (1 - \beta)^n$ for all $n \leq N/2C$. Moreover, for any subset A of G lying in a ball of radius $\leq N/2C$, there is a generator s such that $|sA \triangle A| \geq \alpha|A|$.

Proof of Proposition 9.8. Let $B_G(N)$ be the ball of radius N in G centered at 1. Pick representatives $rep = \{\overline{x} \in B_G(N)\}$ for right cosets of H : $H \cdot B_G(N) = \cup_{\overline{x} \in rep} H\overline{x}$ and then split f as a sum of mutually orthogonal terms

$$f = \sum_{x \in rep} f_{\overline{x}}(\cdot \, \overline{x}^{-1})$$

where $f_{\overline{x}} : H \to \mathbb{R}$ send h to $f(h\overline{x})$. We have $\|f\|_{2,G}^2 = \sum_{rep} \|f_{\overline{x}}\|_{2,H}^2$. Let $S = B_H(1)$ the generating set for H. We know from the corresponding spectral gap estimate on the free group that for every $g : H \to \mathbb{R}$ such that $Supp(g) \subset B_H(N)$ there exists $s \in S$ such that $\|s \cdot g - g\|_{2,H} \geq \tau\|g\|_{2,H}$ where $\tau > 0$ is an absolute constant (independent of the rank of the free group). But if $f_{\overline{x}}(h) \neq 0$, then $h\overline{x} \in Supp(f) \subset B_G(N/2C)$, hence by (17), $h \in B_H(N)$, so $Supp(f_{\overline{x}}) \subset B_H(N)$. Hence there is $s_{\overline{x}} \in S$ such that $\|s_{\overline{x}} \cdot f_{\overline{x}} - f_{\overline{x}}\|_{2,H} \geq \tau\|f_{\overline{x}}\|_{2,H}$. We get

$$\sum_{s \in S} \|s \cdot f - f\|_{2,G}^2 = \sum_{\overline{x} \in rep} \sum_{s \in S} \|s \cdot f_{\overline{x}} - f_{\overline{x}}\|_{2,H} \geq \tau^2 \cdot \|f\|_{2,G}^2.$$

Now since $d_G(1, s) \leq C$ for each $s \in S$ by (17), we get

$$\|s \cdot f - f\|_{2,G}^2 \geq \frac{\tau^2}{|S|}\|f\|_{2,G}^2.$$

Note that $|S| \leq |B_G(C)| \leq (2k)^C$. From there it is straightforward to derive the bound:

$$\|\mu * f\|_2 \le \frac{1}{2}\left[1 + \sqrt{\left(1 - \frac{\tau^2}{32(2k)^{2C}}\right)}\right] \cdot \|f\|_2. \qquad \square$$

Proof of Corollary 1.10. Applying Lemma 4.7, we may assume that $Card(F) = 2$. If $F = \{A, B\}$ does not satisfy the conclusion of the corollary for N as in Theorem 1.1, then for arbitrarily small $\delta > 0$ there is a map $f : A \to B_{n(\delta)}$ where $n = n(\delta) \le (\log \delta^{-1})^{\varepsilon_1}$ (with ε_1 an absolute constant to be determined below) such that $d(f(a), 1) < \delta$ for all $a \in A$. This means that for some $C_1 > 0$ (depending on the choice of the Riemannian metric d), we have $\forall a \in A$ $\forall c \in C$, $|Q_{a, f(a), c}| < C_1 \cdot \delta$. Applying Theorem 9.6 we get Equation (16) as above. Moreover, the logarithmic height of the polynomials $b^i_{f, a, c}$ is at most $O_d(n^{2^{2d^2+2}})$, hence evaluated on $F = \{A, B\}$, $b^i_{f, a, c}$ is $O_{d, F}(n^{2^{2d^2+2}})$, and hence also $f^{e_i}_i = O_{F, d}(\delta \cdot \exp(n^{2^{2d^2+2}}))$. As $e_i = O(n^{2^{2d^2}})$, we see f_i is arbitrarily small when $\delta \to 0$ as soon as we take, for instance, $\varepsilon_1 = 2^{-(2d^2+3)}$. It follows that $(A, B) \in V_2$. This is a contradiction. $\qquad \square$

Proof of Corollary 1.11. With the notation of Corollary 1.7, define $\tau(m) > 0$ by $\frac{(2m - 1 - \frac{\varepsilon}{m^D})}{(2m - 1)} = e^{-\tau(m)}$ and let $\tau = \tau(2)$. By contradiction, suppose Corollary 1.11 does not hold for $(a, b) \in GL_d(\mathbb{C})$. Then for arbitrarily small δ, there is $n = n(\delta) \le (\log \delta^{-1})^{\varepsilon_1}$ for which one can find at least $3^n e^{-\tau n}$ reduced words w of length n such that $d(w, 1) < \delta$. Let $\mathcal{I}$ be the set of all subsets of cardinality $3^n e^{-\tau n}$ of reduced words of length n. For $I \in \mathcal{I}$, let V_I the subvariety of $GL_d(\mathbb{C})^2$ where all words in I vanish simultaneously. Then Corollary 1.7 can be reformulated as the statement $V_I \subset V_2$ for every $n \ge n(2)$ and every $I \in \mathcal{I}$, where V_2 is the closed subvariety above defined shortly before Proposition 9.1. As above, the effective Nullstellensatz applies and gives coefficients $a_i \in \mathbb{Z} \setminus \{0\}$, polynomials $b^k_{I, w, ij}$ with coefficients in $\mathbb{Z}$, of degree and logarithmic height $O\left(n^{2^{2d^2+2}}\right)$, and integers $e_i = O\left(n^{2^{2d^2}}\right)$ such that

$$a_k f^{e_k}_k = \sum_{w \in I, \{ij\} \in C} b^k_{I, w, ij} Q^{ij}_w.$$

It follows that $f^{e_k}_k = O(\delta \cdot \exp(n^{2^{2d^2+2}}))$, which again implies that f_k is arbitrarily small as $\delta \to 0$ if, say, $\varepsilon_1 = 2^{-2d^2-3}$. Hence $(a, b) \in V_2$, which is a contradiction. $\qquad \square$

ACKNOWLEDGMENTS. I am very grateful to P. Sarnak for his precious tips and his enthusiasm. I thank E. Bombieri, J-B. Bost, A. Chambert-Loir, and A.

Yafaev for sharing their insights about diophantine geometry. I am grateful to J. Bourgain and A. Gamburd for their encouragement and for pointing out the right-girth bound in the corollary about $SL_2(\mathbb{F}_p)$. I thank P. E. Caprace, G. Chenevier, E. Lindenstrauss, G. Prasad, and A. Salehi-Golsefidy for our stimulating conversations. Last but not least, I want to express my sincere and deep gratitude to T. Gelander for the time we spent doing mathematics together, which greatly contributed to my involvement in the questions addressed in this paper.

References

[1] Alperin, R., and Noskov, G., *Uniform exponential growth, actions on trees and GL_2*, AMS Contemporary Math. **298** (2002).

[2] Bilu, Y., *Limit distribution of small points on algebraic tori*, Duke Math. J. **89** (1997), no. 3, 465–476.

[3] Bombieri, E., and Gubler, W., *Heights in Diophantine geometry*, New Mathematical Monographs, **4**, Cambridge University Press, Cambridge, (2006).

[4] Bourgain, J., and Gamburd, A., *Uniform expansion bounds for Cayley graphs of $SL_2(\mathbb{F}_p)$*, Ann. of Math. (2) **167** (2008), no. 2, 625–642.

[5] Bourgain, J., and Gamburd, A., *On the spectral gap for finitely generated subgroups of $SU(2)$*, Invent. Math. **171** (2008), no. 1, 83–121.

[6] Breuillard, E., *On uniform exponential growth for solvable groups*, Pure and Applied Math. Quart. 3, no 4, Margulis Volume Part 1 (2007), 949–967.

[7] Breuillard, E., *A height gap theorem for the finite subsets of $SL_n(\overline{\mathbb{Q}})$ and non amenable linear groups*, preprint (April 2008), arXiv:0804.1391.

[8] Breuillard, E., *A Strong Tits alternative*, preprint (April 2008), arXiv:0804.1395.

[9] Breuillard, E., and Gelander, T., *Uniform independence in linear groups*, Invent. Math. **173** (2008), no. 2, 225–263.

[10] Cohen, J. M., *Cogrowth and amenability of Discrete Groups*, J. Funct. Anal. **48** (1982), 301–309.

[11] Curtis, C. W., and Reiner, I., *Representation theory of finite groups and associative algebras*, Interscience, New York, (1962).

[12] Erdös, P., and Turan, P., *On the distribution of roots of polynomials*, Ann. of Math. (2) **51** (1950), 105–119.

[13] Eskin, A., Mozes, S., and Oh, H., *On uniform exponential growth for linear groups*, Invent. Math. **160** (2005), no. 1, 1–30.

[14] Fried, M., and Jarden, M., *Field arithmetic*, 2nd edition, Ergebnisse der Math. Grenzg., **11**, Springer-Verlag, Berlin, (2005).

[15] Gamburd, A., Hoory, S., Shahshahani, M., Shalev, A., and Virag, B., *On the girth of random Cayley graphs*, Random Structures Algorithms **35** (2009), no. 1, 100–117.

[16] Gamburd, A., Jakobson D., and Sarnak, P., *Spectra of elements in the group ring of $SU(2)$*, J. Eur. Math. Soc. **1** (1999), no. 1, 51–85.

[17] Grigorchuk, R., *Symmetrical random walks on discrete groups*, in Multicomponent Random Systems, ed. Dobrushin, Y. Sinai, Adv. Prob. Related Topics **6**, Dekker, New York, (1980), 285-325.

[18] Grigorchuk, R., and de la Harpe, P., *Limit behaviour of exponential growth rates for finitely generated groups*, in Essays on geometry and related topics, Vol. 1, 2, 351–370, Monogr. Enseign. Math., **38**, (2001).

[19] Kaloshin, V., and Rodnianski, I., *Diophantine properties of elements of $SO(3)$*, Geom. Funct. Anal. **11** (2001), no. 5, 953–970.

[20] Kazhdan, D., and Margulis, G., *A proof of Selberg's hypothesis*, Mat. Sb. (N.S.) **75** (1968), 163–168.

[21] Kesten, H., *Symmetric walks on groups*, Trans. Amer. Math. Soc. **92** (1959), 336–354.

[22] Lang, S., *Fundamentals of Diophantine geometry*, Springer-Verlag, New York, (1983).

[23] Lubotzky, A., *Discrete groups, expanding graphs and invariant measures*. Progress in Math., Birkhauser, Basel (1994).

[24] Lubotzky, A., Phillips, R., and Sarnak, P., *Hecke operators and distributing points on the sphere I*, CPAM **39** (1987), 149–186.

[25] Masser, D., and Wustholz, G., *Fields of large transcendence degree generated by values of elliptic functions*, Invent. Math. **72** (1983), no. 3, 407–464.

[26] Mignotte, M., *Sur un théorème de M. Langevin*, Acta Arith. **54** (1989), 81–86.

[27] Ollivier, Y., *Cogrowth and spectral gap of generic groups*, Annales de l'institut Fourier **55** (2005), no. 1, 289–317

[28] Sarnak, P., *Applications of modular forms*, Cambridge University Press, Cambridge, (1990).

[29] Shalom, Y., *Explicit Kazhdan constants for representations of semisimple and arithmetic groups*, Ann. Inst. Fourier **50** (2000), no. 3, 833–863.

[30] Thurston, W., *Three-dimensional geometry and topology*, Vol. 1. Ed. Silvio Levy, Princeton Mathematical Series, **35**, Princeton University Press, Princeton, NJ (1997).

[31] Tits, J., *Free subgroups of linear groups*, Journal of Algebra **20** (1972), 250–270.

[32] Zhang, S., *Small points and adelic metrics*, J. Algebraic Geom. **4** (1995), no. 2, 281–300.

15

THOMAS DELZANT, OLIVIER GUICHARD,

FRANÇOIS LABOURIE, AND SHAHAR MOZES

DISPLACING REPRESENTATIONS
AND ORBIT MAPS

FOR BOB ZIMMER, WITH ADMIRATION

Contents

O. Guichard et F. Labourie ont benéficié du soutien de l'ANR Repsurf : ANR-06-BLAN-0311, F. Labourie and S. Mozes were partially support by the Israeli-French grant 0387610

1. Introduction

Let γ be an isometry of a metric space X. We recall that the *displacement* of γ is

$$d_X(\gamma) = \inf_{x \in X} d(x, \gamma(x)).$$

In the case of the Cayley graph of a group Γ with set of generators S and word length $\| \ \|_S$, the displacement function is called the *translation length*—or the *stable translation length*—and is denoted by $\ell_S : \ell_S(\gamma) = \inf_\eta \|\eta\gamma\eta^{-1}\|_S$. We finally say the action by isometries on X of a group Γ is *displacing*, if given a set S of generators of Γ, there exist positive constants A and B such that

$$d_X(\gamma) \geqslant A\ell_S(\gamma) - B.$$

This definition does not depend on the choice of S. As first examples, it is easy to check that cocompact groups are displacing, as well as convex cocompact whenever X is *Hadamard* (i.e., complete, nonpositively curved, and simply connected). We recall that a cocompact action is by definition a properly discontinuous action whose quotient is compact, and a convex cocompact is an action such that there exists a convex invariant on which the action is cocompact.

The notion naturally arose in [7] where it is shown that for displacing representations of surface groups, the energy functional is proper on Teichmüller space and that, moreover, a large class of representations of surface groups are displacing.

This definition is a cousin to a more well-known one. Assume Γ acts by isometries on a space X. We say the *orbit maps are quasi-isometric embedding*—or in short the action is QI—if for every x in X there exist constants A and B so that we have

$$\forall \gamma \in \Gamma, \quad d(x, \gamma(x)) \geq A\|\gamma\|_S - B.$$

The purpose of this modest note is to collect some elementary observations about the relation between the two notions in order to complete the circle of ideas discussed in [7]. For hyperbolic groups the two notions turn out to be equivalent. In general, however, this relation is slightly more involved than expected.

Does displacing imply QI? In general, the answer is no: this follows immediately from the existence proved by Osin in [9] of infinite groups with finitely many conjugacy classes. However, we isolate a class of groups for which displacing implies QI. We say a finitely generated group is *undistorted in its conjugacy classes* (see Section 2)—or satisfies the *U*-property if there exists finitely many elements $g_1, \ldots, g_p$ of Γ, positive constants A and B such that

$$\forall \gamma \in \Gamma, \quad \|\gamma\|_S \leqslant A \sup_{1 \leqslant i \leqslant n} \ell_S(g_i \gamma) + B.$$

We prove in Theorem 2.1.1 that some class of linear groups—in particular lattices—enjoy the U-property. We also prove that hyperbolic groups have the U-property. For all groups enjoying the U-property displacing implies QI.

Does QI imply displacing? In general, again, the answer is no: in Corollary 3.2.3, we show that for a residually finite group (see Section 3.1) any linear that representation contains a unipotent is not displacing. It follows that $SL(n, \mathbb{Z})$ acting on $SL(n, \mathbb{R})/SO(n, \mathbb{R})$ is QI but not displacing. However, again a simple argument using the stable length shows that for hyperbolic groups QI implies displacing (see Section 4).

We expand in the article the discussion of this introduction and start by discussing the U-property.

2. Groups Whose Displacing Actions Have Orbit Maps That Are Quasi-Isometric Embeddings

We say a finitely generated group is *undistorted in its conjugacy classes*—in short, has the *U-property*,—if there exists finitely many elements $g_1, \ldots, g_p$ of Γ, positive constants A and B such that

$$\forall \gamma \in \Gamma, \quad \|\gamma\| \leqslant A \sup_{1 \leqslant i \leqslant n} \ell(g_i \gamma) + B.$$

REMARKS:

- This property is clearly independent of S. (Hence we omitted the subscript S.)
- This property is satisfied by free groups and commutative groups. On the other hand, the groups constructed by D. Osin [9] described in the above paragraph do not have this U-property.
- Note also that this property is very similar to the statement of Abels, Margulis, and Soifer's result [1, theorem 4.1] and it is no surprise that their result plays a role in the proof of Theorem 2.1.1.
- Finally, by the conjugacy invariance of the translation length, $\ell(g_i \gamma) = \ell(\gamma g_i)$, so we will indifferently write this property with left or right multiplication by the finite family (g_i).

LEMMA 2.0.1. *If Γ has the U-property, then every displacing action has orbit maps that are quasi-isometric embeddings.*

Proof. Indeed, assume that Γ acts on X by isometries and that the action is displacing. In particular, we have positive constants α and β so that

$$\forall x \in X, \ \gamma \in \Gamma, \ d(x, \gamma(x)) \geqslant \alpha \ell_S(\gamma) - \beta.$$

Moreover, there exists finitely many elements $g_1, \ldots, g_p$ of Γ, positive constants A and B such that

$$\forall \gamma \in \Gamma, A\|\gamma\| - B \leqslant \sup_{1 \leqslant i \leqslant n} \ell(g_i \gamma).$$

Hence, let $x \in X$, then

$$d(x, \gamma(x)) \geqslant \sup_{1 \leqslant i \leqslant n} d(x, g_i \cdot \gamma(x)) - \sup_{1 \leqslant i \leqslant n} d(x, g_i(x))$$

$$\geqslant \alpha \sup_{1 \leqslant i \leqslant n} \ell(g_i \gamma) - \beta - \sup_{1 \leqslant i \leqslant n} d(x, g_i(x))$$

$$\geqslant \alpha . A\|\gamma\| - B\alpha - \beta - \sup_{1 \leqslant i \leqslant n} d(x, g_i(x)).$$

Hence, the orbit map is a quasi-isometric embedding. $\qquad\square$

We will prove in the next section,

THEOREM 2.0.2. *Every uniform lattice—and nonuniform lattice in higher rank—in characteristic 0 has the U-property. In particular, every surface group has the U-property.*

Moreover, we show that hyperbolic groups have the U-property.

2.1. Linear Groups Having the U-Property

We prove the following result that implies Theorem 2.0.2.

THEOREM 2.1.1. *Let Γ be a finitely generated group and $\mathbf{G}$ a reductive group defined over a field F, suppose that*

- *there exists a homomorphism $\Gamma \to \mathbf{G}(F)$ with Zariski-dense image, and*
- *there are a finitely many field homomorphisms $(i_v)_{v \in S}$ of F in local fields F_v such that the diagonal embedding $\Gamma \to \Pi_{v \in S} \mathbf{G}(F_v)$ is a quasi-isometric embedding.*

Then the group Γ has the U-property.

Since lattices are Zariski dense (Borel theorem) and that higher-rank irreducible lattices in characteristic 0 are quasi-isometrically embedded [8], this theorem implies Theorem 2.0.2 for higher-rank lattices. The same holds for all uniform lattices. A specific corollary is the following:

COROLLARY 2.1.2. *Let Γ be a finitely generated group that is quasi-isometrically embedded and Zariski dense in a reductive group $G(F)$ where F is a local field. Then Γ has the U-property.*

2.1.1. GENERALITIES FOR THE U-PROPERTY We prove two lemmas for the U-property.

LEMMA 2.1.3. *Let Γ be a finitely generated group. Let $\Gamma_0 \lhd \Gamma$ be a normal subgroup of finite index. If Γ_0 has the U-property, so has Γ.*

We do not know whether the converse statement holds, in other words, whether the U-property is a property of commensurability classes.

Proof. We first observe that every finite-index subgroup of a finitely generated group is finitely generated. Let S_0 be a generating set for Γ_0 and write Γ as the union of left cosets for Γ_0

$$\Gamma = \bigcup_{t \in T} \Gamma_0 \cdot t.$$

We assume that T is symmetric. Clearly $S = S_0 \cup T$ is a generating set for Γ.

We denote $\| \cdot \|_{\Gamma_0}$ and ℓ_{Γ_0} the word and translation lengths, respectively, for Γ_0.

We observe that Γ_0 is quasi-isometrically embedded in Γ. Hence there exist positive constants α and β such that

1
$$\forall \gamma \in \Gamma_0, \quad \|\gamma\|_{\Gamma_0} \geqslant \|\gamma\|_\Gamma \geqslant \alpha \|\gamma\|_{\Gamma_0} - \beta.$$

For any γ in Γ, we write $\gamma = \gamma_0 t_0$ with $t_0 \in T$ and $\gamma_0 \in \Gamma_0$. Hence

$$\|\gamma\|_\Gamma \leqslant \|\gamma_0\|_{\Gamma_0} + 1 \leqslant A \sup \ell_{\Gamma_0}(\gamma_0 g_i) + B + 1$$

since Γ_0 has the U-property.

Next, we need to compare ℓ_Γ and ℓ_{Γ_0}. Let δ in Γ_0, then

$$\ell_\Gamma(\delta) = \inf_{t \in T, \eta \in \Gamma_0} \|t\eta t \delta \eta^{-1} t^{-1}\|_\Gamma$$

$$\geqslant \inf_{\eta \in \Gamma_0} \|\eta \delta \eta^{-1}\|_\Gamma - 2$$

$$\geqslant \alpha \inf_{\eta \in \Gamma_0} \|\eta \delta \eta^{-1}\|_{\Gamma_0} - \beta - 2$$

2
$$\geqslant \alpha \ell_{\Gamma_0}(\delta) - \beta - 2.$$

Finally, combining Inequality (2) and the one preceding, we have

$$\|\gamma\|_\Gamma \leqslant \frac{A}{\alpha} \sup_{t \in T, i \in I} \ell_\Gamma(\gamma t^{-1} \beta_i) + B + 1 + \frac{\beta+2}{\alpha}.$$

This is exactly the U-property for Γ. □

Also,

LEMMA 2.1.4. *Let Γ be a finitely generated group. Suppose that $\Gamma \to \Gamma_0$ is onto with finite kernel. Then the group Γ has the U-property if and only if Γ_0 has the U-property.*

Proof. We choose a generating set S for Γ that contains the kernel of $\Gamma \to \Gamma_0$. We choose the generating set S_0 for Γ_0 to be the image of S. Then we have, using surjectivity, for all γ projecting to γ_0

$$\|\gamma\|_\Gamma \geqslant \|\gamma_0\|_{\Gamma_0} \geqslant \|\gamma\|_\Gamma - 1,$$

$$\ell_\Gamma(\gamma) \geqslant \ell_{\Gamma_0}(\gamma_0) \geqslant \ell_\Gamma(\gamma) - 1.$$

These two inequalities enable us to transfer the U-property from Γ to Γ_0 and vice versa. □

2.1.2. PROXIMALITY We recall the notion of proximality and a result of Abels, Margulis, and Soifer.

Let k be a local field. Let V be a finite-dimensional k-vector space equipped with a norm. Let d be the induced metric on $\mathbb{P}(V)$. Let r and ϵ be positive numbers such that
$$r > 2\epsilon.$$

An element g of $\mathrm{SL}(V)$ is said to be (r, ε)-*proximal*, if there exist a point x_+ in $\mathbb{P}(V)$ and a hyperplane H in V such that

- $d(x_+, \mathbb{P}(H)) \geqslant r$, and
- $\forall x \in \mathbb{P}(V), \quad d(x, \mathbb{P}(H)) \geqslant \varepsilon \implies d(g \cdot x, x_+) \leqslant \varepsilon.$

In particular, a proximal element has a unique eigenvalue of highest norm. Conversely, if an element g admits a unique eigenvalue of highest norm, then some power of g is proximal (for some (r, ϵ)).

We cite the needed result from [1] and [2].

THEOREM 2.1.5. ([1], THEOREM 5.17) *Let* $\mathbf{G}$ *a semisimple group over a field* F. *Let* $(i_v)_{v \in V}$ *be finitely many field homomorphisms of* F *in local fields* F_v. *Let* $\rho_v : \mathbf{G}(F_v) \to \mathrm{GL}(n_v, F_v)$ *be an irreducible representation of* $\mathbf{G}(F_v)$ *for each* v.

Suppose that Γ *is a Zariski-dense subgroup of* $\mathbf{G}(F)$. *Suppose that for every* v, $\rho_v(\Gamma)$ *contains proximal elements. Then there exist*

- $r > 2\varepsilon > 0$, *and*
- *a finite subset* $\Delta \subset \Gamma$,

such that for every γ *in* Γ *there is some* δ *in* Δ *such that* $\rho_v(\gamma s)$ *is* (r, ε)-*proximal for every* v *in* V.

This result is usually stated with *one* local field but the proof of the above extension and the following is straightforward.

We shall also need the following Lemma.

LEMMA 2.1.6. ([2], COROLLAIRE P.13) *Let* Γ *be a Zariski-dense subgroup in* $\mathbf{G}(k)$, $\mathbf{G}$ *a reductive group over a local field* k. *Then* Γ *is unbounded if and only if there exists an irreducible representation of* $\mathbf{G}(k)$, *such that* $\rho(\Gamma)$ *contains a proximal element.*

The special case of $k = \mathbb{R}$ was proved in [3].

2.1.3. PROXIMAL ELEMENTS AND TRANSLATION LENGTHS

We recall some facts on length and translation length in $\mathbf{G}(k)$ where $\mathbf{G}$ is a semisimple group over a local field k.

Let K be the maximal compact subgroup of $\mathbf{G}(k)$. This defines a norm $\|g\|_G = d_{G/K}(K, gK)$ in $\mathbf{G}(k)$, which satisfies

$$\|gh\|_G \leqslant \|g\|_G + \|h\|_G.$$

We also consider the translation length ℓ_G in $\mathbf{G}(k)$:

$$\ell_G(g) = \inf_{h \in G} \|hgh^{-1}\|_G.$$

Observe that the translation norm is actually independent of the choice of the maximal compact subgroup since they are all conjugated. The translation length and norm of (r, ϵ)-proximal elements can be compared:

LEMMA 2.1.7. (COMPARE [2] §4.5) *Let* $\mathbf{G}$ *be a semisimple group over a local field* k. *Let* $\rho : \mathbf{G}(k) \to \mathrm{GL}(n, k)$ *be an irreducible representation. Let* $\varepsilon > 0$.

Then there exist positive constants α and β such that if $\rho(g)$ is (r, ε)-proximal, then

3
$$\ell_G(g) \geqslant \alpha \|g\|_G - \beta.$$

Proof. We use classical notation and refer to [2, pp. 6–8] for precise definitions.

Any g in $\mathbf{G}(k)$ is contained in a unique double coset $K\mu(g)K$. The element $\mu(g) \in A^+ \subset A$ is called the Cartan projection of g. Here we see A^+ as a subset of a cone $A^\times$ in some $\mathbf{R}$-*vector space*.

For some integer n ($n = 1$ in the Archimedean case) the element g^n admits a Jordan decomposition $g^n = g_e g_h g_u$ with g_e, g_h, g_u commuting, g_e elliptic, g_u unipotent, and g_h hyperbolic, that is, conjugated to a unique element $a \in A^+$. We set $\lambda(g) = \frac{1}{n} a \in A^\times$.

If we fix some norm on the vector space containing the cone $A^\times$, then (up to quasi-isometry constants) the norm of $\mu(g)$ is $\|g\|_G$ in $\mathbf{G}(k)$ and the norm of $\lambda(g)$ is $\ell_G(g)$.

Then by a result of Y. Benoist [2], there exists a compact subset N_ε of the vector space containing $A^\times$ such that for every g such that $\rho(g)$ is (r, ε)-proximal we have
$$\lambda(g) - \mu(g) \in N_\varepsilon. \qquad \square$$

The lemma follows.

2.1.4. PROOF OF THEOREM 2.1.1

By taking a finite-index normal subgroup and projecting (Lemmas 2.1.3 and 2.1.4) we can make the hypothesis that $\mathbf{G}$ is the product of a semisimple group $\mathbf{S}$ and a torus $\mathbf{T}$ and Γ is a subgroup of $\mathbf{S}(F) \times \mathbf{T}(F)$.

Moreover, since length and translation length for elements in $\mathbf{T}(F_v)$ are equal, we only need to work with the semisimple part $\mathbf{S}$.

Finally, it suffices to prove the existence of a finite family $F \subset \Gamma$ and constants A, B such that, for any $\gamma \in \Gamma$

4
$$\|\gamma\|_S \leqslant A \sup_{f \in F} \ell_S(\gamma f) + B,$$

where $S = \Pi_{v \in V} \mathbf{S}(F_v)$. Indeed, since Γ is quasi-isometrically embedded in G, $\|\gamma\|_\Gamma$ is less than $\|\gamma\|_G = \|\gamma\|_T + \|\gamma\|_S$ and $\ell_G(\gamma) = \ell_T(\gamma) + \ell_S(\gamma) = \|\gamma\|_T + \ell_S(\gamma)$ is less than $\ell_\Gamma(\gamma)$—up to quasi-isometry constants—and Inequality (4) implies that Γ has the U-property.

Note that we can forget any completion F_v where the subgroup $\Gamma \subset \mathbf{S}(F_v)$ is bounded without changing the fact that Γ is quasi-isometrically embedded. So by Lemma 2.1.6, for each v there is an irreducible representation $\rho_v : \mathbf{S}(F_v) \to$ GL(n_v, F_v) such that $\rho_v(\Gamma)$ contains proximal elements.

Applying Theorem 2.1.5 we find a finite family $F \subset \Gamma$ and r, ε such that for every γ in Γ there is some $f \in F$ such that $\rho_v(\gamma f)$ is (r, ε)-proximal for every v. Hence, as a consequence of Lemma 2.1.7, we have for such γ and f:

$$\|\gamma f\|_S \leqslant A\ell_S(\gamma f) + B.$$

This implies Inequality 4 and concludes the proof.

2.2. Hyperbolicity and the U-Property

Let Γ be a finitely generated group and S a set of generators. Let d be its word distance and $\|g\| = d(e, g)$. We denote by

$$\langle g, h \rangle_u = \frac{1}{2}(d(g, u) + d(h, u) - d(g, h)),$$

the *Gromov product*—based at U—on Γ. We abbreviate $\langle g, h \rangle_e$ by $\langle g, h \rangle$. Observe that

5 $$\langle gu, hu \rangle_u = \langle g, h \rangle_e.$$

Recall that Γ is called δ-*hyperbolic* if for all g, h, k in Γ we have

6 $$\langle g, k \rangle \geqslant \inf \left(\langle g, h \rangle; \langle h, k \rangle \right) - \delta$$

and Γ is called *hyperbolic* if it is δ-hyperbolic for some δ. A hyperbolic group is called *nonelementary* if it is not finite and does not contain $\mathbb{Z}$ as a subgroup of finite index.

Then,

PROPOSITION 2.2.1. *Hyperbolic groups have the U-property.*

We recall the *stable translation length of an element* g:

$$[g]_\infty = \lim_{n \to \infty} \frac{\|g^n\|}{n}.$$

We remark that obviously

$$[g]_\infty \leqslant \ell(g).$$

We shall actually prove

PROPOSITION 2.2.2. *Let Γ be hyperbolic. There exist a pair $u, v \in \Gamma$ and a constant α such that for every g one has*

$$\|g\| \leqslant 3 \sup \left([g]_\infty, [gu]_\infty, [gv]_\infty \right) + \alpha.$$

In particular Γ has the U-property.

REMARK:

- Let Γ be a free group generated by some elements $u, v, w_1, \ldots, w_n$. Then G is 0 hyperbolic. If $[g] \neq \|g\|$ the first letter of g must be equal to the inverse of the last one. Multiplying either by u or by v we find a new element that is cyclically reduced; for this element the stable translation length and the length are equal. The proof of Proposition 2.2.2 is a generalization of this remark.

2.2.1. ALMOST CYCLICALLY REDUCED ELEMENTS An element g in Γ is said to be *almost cyclically reduced* if $\langle g, g^{-1} \rangle \leqslant \frac{\|g\|}{3} - \delta$. We prove in this paragraph

LEMMA 2.2.3. *If g is almost cyclically reduced, then*

$$[g]_\infty \geqslant \frac{\|g\|}{3}.$$

The following result [5, lemma 1.1] will be useful.

LEMMA 2.2.4. *Let (x_n) be a finite or infinite sequence in G. Suppose that*

$$d(x_{n+2}, x_n) \geqslant \sup \left(d(x_{n+2}, x_{n+1}), d(x_{n+1}, x_n) \right) + a + 2\delta,$$

or equivalently that

$$\langle x_{n+2}, x_n \rangle_{x_{n+1}} \leqslant \frac{1}{2} \inf \left(d(x_{n+2}, x_{n+1}), d(x_{n+1}, x_n) \right) - \frac{a}{2} - \delta.$$

Then

$$d(x_n, x_p) \geqslant |n - p| a.$$

This implies Lemma 2.2.3.

Proof. Let $x_n = g^n$. By left invariance and since g is almost cyclically reduced,

$$\langle x_{n+2}, x_n \rangle_{x_{n+1}} = \langle g, g^{-1} \rangle \leqslant \frac{\|g\|}{2} - \frac{a}{2} - \delta,$$

for $a = \frac{\|g\|}{3}$. By Lemma 2.2.4,

$$\|g^n\| \geqslant n \frac{\|g\|}{3}. \qquad \square$$

The result follows:

2.2.2. PING-PONG PAIRS A *ping-pong pair* in Γ is a pair of elements u, v such that

1) $\inf(\|u\|, \|v\|) \geqslant 100\delta$,
2) $\langle u^{\pm 1}, v^{\pm 1} \rangle \leqslant \frac{1}{2} \inf(\|u\|, \|v\|) - 20\delta$, and
3) $\langle u, u^{-1} \rangle \leqslant \frac{\|u\|}{2} - 20\delta$ and $\langle v, v^{-1} \rangle \leqslant \frac{\|v\|}{2} - 20\delta$.

REMARKS:

- A ping-pong pair generates a free subgroup. This is an observation from [5]. To prove this, consider a reduced word w on the letter u, v, u^{-1}, v^{-1}. If x_n is the prefix of length n of w, the sequence x_n satisfies the hypothesis of Lemma 2.2.4.
- In the present proof, the third property will not be used.

We shall prove

LEMMA 2.2.5. *If Γ is hyperbolic nonelementary, there exists a ping-pong pair.*

Proof. In [6] explicit ping-pong pairs are constructed. Here is a construction whose idea goes back to F. Klein. Let f be some hyperbolic element (an element of infinite order). Replacing f by a conjugate of some power, we may assume that

$$\|f\| = [f] > 1000\delta.$$

As Γ is not elementary, there exists a generator a of Γ that does not fix the pair of fixed points f^+, f^- of f on the boundary $\partial\Gamma$: otherwise, since the action of Γ is topologically transitive, $\partial\Gamma$ would be reduced to these two points and Γ would be elementary. Now, let us prove that for some integer N, $(f, af^N a^{-1})$ is a ping-pong pair. We have

$$\lim_{N \to +\infty} f^N = f^+ \neq af^+ = \lim_{N \to +\infty} af^N a^{-1}.$$

It follows that the Gromov product $\langle f^N, af^N a^{-1} \rangle$ remains bounded, by the very definition of the boundary. Hence, for N large enough, we have

$$(f, af^N a^{-1}) \leqslant \frac{1}{2} \inf(\|f^N\|, \|af^N a^{-1}\|) - 20\delta.$$

A similar argument also yields that $\langle f^N, af^{-N} a^{-1} \rangle$ remains bounded. Therefore, $(f^N, af^N a^{-1})$ is a ping-pong pair for $N \gg 1$. $\square$

2.2.3. PROOF OF PROPOSITION 2.2.2 We first reduce this proof to the following lemma.

LEMMA 2.2.6. *Let (u, v) be a ping-pong pair. Let $g \in \Gamma$ such that*

$$\|g\| \geqslant 3 \sup (\|u\|, \|v\|) + 100\delta.$$

Then one of the three elements g, gu, gv is almost cyclically reduced.

We observe at once that Proposition 2.2.2 follows from Lemmas 2.2.3, 2.2.5, and 2.2.6; choose a ping-pong pair u, v and take

$$\alpha = 3 \sup (\|u\|, \|v\|) + 100\delta.$$

Proof. Assume g is not almost cyclically reduced. Then

7 $$\langle g, g^{-1} \rangle \geqslant \frac{\|g\|}{3} - \delta \geqslant \sup (\|u\|, \|v\|) + 30\delta.$$

Moreover, one of the following pair of inequalities holds:

8 $$\langle g^{-1}, u^{\pm 1} \rangle \leqslant \frac{\|u\|}{2} - 10\delta,$$

or

9 $$\langle g^{-1}, v^{\pm 1} \rangle \leqslant \frac{\|v\|}{2} - 10\delta.$$

Otherwise, by the definition of hyperbolicity we would have for some $\varepsilon, \varepsilon' \in \{\pm 1\}$,

$$\langle u^{\varepsilon}, v^{\varepsilon'} \rangle \geqslant \frac{1}{2} \inf (\|u\|, \|v\|) - 10\delta - \delta,$$

contradicting the second property of the definition of ping-pong pairs.

So we may assume that Inequality (8) holds. We will show that gu is almost cyclically reduced. Let $k = gu$. By the triangle inequality, $\langle u^{-1}, g \rangle \leqslant \|u\|$. Thus from Inequality (7) we deduce that

$$\inf (\langle g, u^{-1} \rangle, \langle g, g^{-1} \rangle) = \langle g, u^{-1} \rangle.$$

Then, by the definition of hyperbolicity, we get

10 $$\langle g, u^{-1} \rangle \leqslant \langle u^{-1}, g^{-1} \rangle + \delta.$$

Using Inequality (8) now, we have

11 $$\langle g, u^{-1} \rangle \leqslant \frac{\|u\|}{2} - 9\delta.$$

Note that

$$\langle g, k \rangle = \frac{1}{2} (\|g\| + \|gu\| - \|u\|) \geqslant \frac{1}{2} (2\|g\| - 2\|u\|) \geqslant 2\|u\| + 100\delta.$$

By the triangle inequality again,

$$\langle k, u^{-1} \rangle \leqslant \|u\|.$$

Therefore,

$$\inf \left(\langle g, k \rangle, \langle k, u^{-1} \rangle \right) = \langle k, u^{-1} \rangle.$$

From hyperbolicity and Inequality (11), we have

12 $$\langle k, u^{-1} \rangle \leqslant \langle g, u^{-1} \rangle + \delta \leqslant \frac{\|u\|}{2} - 8\delta.$$

Applying successively Inequalities (8) and (12), we get that

13 $$\langle k^{-1}, u^{-1} \rangle = \|u\| - \langle u, g^{-1} \rangle \geqslant \frac{\|u\|}{2} + 10\delta \geqslant \langle k, u^{-1} \rangle + 18\delta.$$

By hyperbolicity,

$$\inf \left(\langle k, k^{-1} \rangle, \langle k^{-1}, u^{-1} \rangle \right) \leqslant \langle k, u^{-1} \rangle + \delta.$$

Therefore, Inequalities (12) and (13) imply that

14 $$\langle k, k^{-1} \rangle \leqslant \langle k, u^{-1} \rangle + \delta \leqslant \frac{\|u\|}{2} - 7\delta.$$

Since $\|k\| \geqslant \|g\| - \|u\| \geqslant 3\|u\| - \|u\| + 100\delta$, we finally obtain that k is almost cyclically reduced. $\qquad \square$

3. Nondisplacing Actions Whose Orbit Maps are Quasi-Isometric Embeddings

We prove in particular

PROPOSITION 3.0.7. *The action of* $SL(n, \mathbb{Z})$ *on* $X_n = SL(n, \mathbb{R})/SO(n, \mathbb{R})$ *is not displacing, although, for* $n \geqslant 3$, *the orbit maps are quasi-isometric embeddings.*

The second part of this statement is a theorem of Lubotzky, Mozes, and Ragunathan [8]. Note that for the action of $SL(2, \mathbb{Z})$ on the hyperbolic plane $\mathbb{H}_2 = X_2$ the orbit maps are not quasi-isometries so it is obviously not displacing since $SL(2, \mathbb{Z})$ is a hyperbolic group (see Corollary 4.0.6).

3.1. Infinite Contortion

We say a group Γ has *infinite contortion*, if the set of conjugacy classes of powers of every nontorsion element is infinite. In other words, for every nontorsion element γ, for every finite family $g_1, \ldots, g_q$ of conjugacy classes of elements of Γ, there exists $k > 0$ such that

$$\forall i \in \{1, \ldots, q\}, \gamma^k \notin g_i.$$

We prove

LEMMA 3.1.1. *Every residually finite group has infinite contortion.*

Proof. Let γ be a nontorsion element. Let $g_1, \ldots, g_n$ be finitely many conjugacy classes. We want to prove that there exists $k > 0$ such that γ^k belongs to no g_i. Since γ is not a torsion element, we can assume that all the g_i are nontrivial. Let $h_i \in g_i$. Since all h_i are nontrivial, by residual finiteness there exist a homomorphism ϕ in a finite group H, such that

$$\forall i, \varphi(h_i) \neq 1.$$

Let $k = \|H\|$, hence $\varphi(\gamma^k) = 1$. This implies that $\gamma^k \notin g_i$. $\qquad \square$

3.2. Displacement Function and Infinite Contortion
We will prove

LEMMA 3.2.1. *Assume Γ has infinite contortion. Assume Γ acts cocompactly and properly discontinuoulsy by isometry on a space X. Assume furthermore that every closed bounded set in X is compact. Then, for every nontorsion element γ in Γ, we have*

$$\limsup_{p \to \infty} d_X(\gamma^p) = \infty.$$

REMARKS:

- We should notice that the conclusion immediately fails if Γ does not have infinite contortion. Indeed there exists an element γ such that its powers describe only finitely many conjugacy classes of elements $g_1, \ldots, g_q$, and hence

$$\limsup_{p \to \infty} d_X(\gamma^p) \leq \sup_{i \in \{1, \ldots, q\}} d_X(g_i) < \infty.$$

- It is also interesting to notice that there are groups with infinite contortion that possess elements γ such that

$$\liminf_{p \to \infty} d_X(\gamma^p) < \infty.$$

Indeed, there are finitely generated linear groups that contain elements γ that are conjugated to infinitely many of its powers. Hence, for such γ

we have

$$\liminf_{p \to \infty} d_X(\gamma^p) \leqslant d_X(\gamma).$$

Here is a simple example. We take $\Gamma = \mathrm{SL}(2, \mathbb{Z}[\frac{1}{p}])$ and

$$\gamma = \begin{pmatrix} 1 & 1 \\ 0 & 1 \end{pmatrix}.$$

Then for all n, γ^{p^n} is conjugated to γ.

- However, in Section 3.4, we shall give a condition—*bounded depth roots* (satisfied, for example, by any group commensurable to a subgroup of $\mathrm{SL}(n, \mathbb{Z})$) so that together with the hypothesis of the previous lemma,

$$\lim_{p \to \infty} d_X(\gamma^p) = \infty.$$

Proof. We want to prove that

$$\limsup_{p \to \infty} \inf_{x \in X} d(x, \gamma^p x) = \infty.$$

Assume the contrary, then there exists

- a constant R, and
- a sequence of points x_i of points in X,

such that for every p,

$$d(x_i, \gamma^p x_i) \leqslant R.$$

Now let K be a compact in X such that $\Gamma.K = X$. Let $f_i \in \Gamma$ such that $y_i = f_i^{-1}(x_i) \in K$. Then

$$d(y_i, f_i^{-1} \gamma^p f_i(y_i)) \leqslant R.$$

Let $K_R = \{z \in X, d(z, K) \leqslant R\}$. It follows that

$$\forall p, \quad f_i^{-1} \gamma^p f_i(K_R) \cap K_R \neq 0.$$

Observe that K_R is compact. By the properness of the action of Γ, we conclude that the family $\{g_i^{-1} \gamma_i^p g_i\}$ is finite. Hence the family of conjugacy classes of the sequence γ^p is finite. But this contradicts infinite contortion for Γ. $\quad\square$

COROLLARY 3.2.2. *Assume Γ has infinite contortion. Let C be its Cayley graph, then for γ a nontorsion element*

$$\limsup_{p \to \infty} \ell(\gamma^p) = \infty.$$

COROLLARY 3.2.3. *Assume Γ has infinite contortion. Let ρ be a representation of dimension n. Assume $\rho(\Gamma)$ contains a nontrivial unipotent, then ρ is not displacing on $X_n = \mathrm{SL}(n, \mathbb{R})/SO(n, \mathbb{R})$.*

Proof. Assume ρ is displacing. Let γ such that $\rho(\gamma)$ is a nontrivial unipotent. Then for all p, $d_{X_n}(\gamma^p) = 0$. However, γ is not a torsion element. We obtain the contradiction using the previous lemma. $\qquad\square$

3.3. Nonuniform Lattices

LEMMA 3.3.1. *For $n \geqslant 3$, the action of $\mathrm{SL}(n, \mathbb{Z})$ on X_n is such that the orbit maps are quasi-isometric embeddings. But it is not displacing.*

Proof. The group $\mathrm{SL}(n, \mathbb{Z})$ is residually finite. Hence it has infinite contortion by Lemma 3.1.1. The standard representation ρ contains a nontrivial unipotent; hence it is not displacing by Corollary 3.2.3.

By a theorem of Lubotzky, Mozes, and Ragunathan [8] irreducible higher-rank lattices Λ are quasi-isometrically embedded in the symmetric space. $\qquad\square$

3.4. Bounded Depth Roots

This section is complementary. We say that group Γ has *bounded depth roots property*, if for every γ in Γ is a nontorsion element, there exists some integer p, such that we have

$$q \geqslant p, \eta \in \Gamma \implies \eta^q \neq \gamma.$$

Observe that $\mathrm{SL}(2, \mathbb{Z}[\frac{1}{p}])$ does not have bounded depth roots. Note that this property is well behaved by taking subgroups and is a property of commensurability (see Lemma 3.4.4).

We prove

PROPOSITION 3.4.1. *The following groups have bounded depth roots property:*

- *the group $\mathrm{SL}(n, \mathbb{Z})$,*
- *$\mathrm{SL}(n, \mathcal{O})$ where $\mathcal{O}$ is the ring of integers of a number field F,*
- *any subgroup of a group having bounded depth roots property or any group commensurable to a group having this property, and*
- *in particular, any arithmetic lattice in an Archimedian Lie group.*

LEMMA 3.4.2. (BOUNDED DEPTH ROOT) *Let Γ be a group with bounded depth roots property. Assume Γ acts cocompactly and properly discontinuously by* *isometry on a space X. Assume furthermore that every closed bounded set in X is compact. Then, for every nontorsion element γ in Γ, we have*

$$\lim_{p \to \infty} d_X(\gamma^p) = \infty.$$

For the proof see Section 3.4.3. The lemma and proposition above again imply that the action of $\mathrm{SL}(n, \mathbb{Z})$ on X_n is not displacing.

3.4.1. BOUNDED DEPTH ROOTS PROPERTY FOR SL$(n, \mathbb{Z})$ We prove the above proposition.

LEMMA 3.4.3. $\mathrm{SL}(n, \mathbb{Z})$ *has bounded depth roots.*

Proof. Let $A \in \mathrm{SL}(n, \mathbb{Z})$. Let $B \in \mathrm{SL}(n, \mathbb{Z})$. We assume there exists k such that $B^k = A$. Let $\{\lambda_j^A\}$ and $\{\lambda_j^B\}$ be the eigenvalues of A and B, respectively. Let

$$K = \sup_j |\lambda_j^A|.$$

Then,

$$\sup_j |\lambda_j^B| \leqslant K^{\frac{1}{k}} \leqslant K.$$

Hence, all the coefficients of the characteristic polynomial of B have a bound K_1 that depends only on A. Therefore, since these coefficients only take values in $\mathbb{Z}$, it follows the characteristic polynomials of B that belongs to the finite family

$$\mathcal{P} = \{P(x) = x^n + \sum_{k=0}^{k=n-1} a_k x^k \ : \ a_k \in \mathbb{Z}, |a_k| \leqslant K_1\}.$$

Since $\mathcal{P}$ is finite, there exists a constant $b > 1$ such that for every root λ of a polynomial $P \in \mathcal{P}$,

$$|\lambda| > 1 \implies |\lambda| \geqslant b.$$

Let $q \in N$ be such that b^q is greater than K. It follows that if $B^q = A$, then all eigenvalues of B have complex modulus 1. Therefore, the same holds for A.

It follows from this discussion that we can reduce to the case where

$$\forall i, j, |\lambda_j^A| = 1.$$

We say such an element has *trivial hyperbolic part*. Note that necessarily also B has trivial hyperbolic part.

We first prove that there exists an integer M depending only on n, such that if $C \in SL(n, \mathbb{Z})$ has a trivial hyperbolic part, then C^M is unipotent. The same argument as above shows that the characteristic polynomials of elements with a trivial hyperbolic part belong to a finite family of the form

$$\mathcal{P} = \{P(x) = x^n + \sum_{k=0}^{k=n-1} a_k x^k \ : \ a_k \in \mathbb{Z}, |a_k| \leqslant K_2\},$$

where K_2 depends only on n. Note that roots of polynomials belonging to $\mathcal{P}$ that are of complex modulus 1 are roots of unity. Thus we may take M to be a common multiple of the orders of those roots of unity and deduce that C^M is unipotent if $C \in SL(n, \mathbb{Z})$ has a trivial hyperbolic part.

Returning to our setting we can replace A by A^M and B by B^M and consider $A = B^k$ where both A and B are unipotents. There is some rational matrix $g_0 \in SL(n, \mathbb{Q})$ depending on A such that $A_0 = g_0 A g_0^{-1}$ is in a Jordan form. We claim that $B_0 = g_0 B g_0^{-1}$ is made of blocks that correspond to the Jordan blocks of A_0 and each such block of B_0 is an upper-triangular unipotent matrix (this follows from observing that for unipotent matrices a matrix and its powers have the same invariant subspaces). Moreover, note that the denominators of the entries of B_0 are bounded by some $L \in \mathbb{N}$ depending only on g_0 (and thus on A). By considering (some of) the entries just above the main diagonal it is easily seen that $k \leq L$. $\square$

3.4.2. COMMENSURABILITY We observe

LEMMA 3.4.4. *If Γ is commensurable to a subgroup of a group that has bounded depth roots, then Γ has bounded depth root.*

Proof. By definition a subgroup of a group having bounded depth roots has bounded depth roots. Let G be a group and H a subgroup having finite index k. Observe that for every element g of G, we have $g^k \in H$. It follows that if H has bounded depth roots, then G has bounded depth roots. $\square$

3.4.3. PROOF OF LEMMA 3.4.2.

Proof. Let K be a compact in X. We first prove that

$$\lim_{p \to \infty} \inf_{x \in K, \eta \in \Gamma} d(x, \eta^{-1} \gamma^p \eta x) = \infty.$$

Assume the contrary, then there exists

- a constant R,
- a sequence of integers p_i going to infinity,
- a sequence of points x_i in K, and
- a sequence of elements η_i of Γ,

such that $d(x_i, \eta_i^{-1}\gamma^{p_i}\eta_i x_i) \leqslant R$. It follows that

$$\forall i, \, (\eta_i^{-1}\gamma\eta_i)^{p_i} K_R \cap K_R \neq 0.$$

By the properness of the action of Γ, we conclude that the family $\{(\eta_i^{-1}\gamma\eta_i)^{p_i}\}$ is finite. But this contradicts the bounded depth root property.

We now choose the compact K such that $\Gamma.K = X$. It follows that

$$\lim_{p\to\infty}\inf_{x\in X} d(x, \gamma^p x) = \lim_{p\to\infty}\inf_{x\in K, \eta\in\Gamma} d(\eta x, \gamma^p \eta x) = \infty.$$

This is what we wanted to prove. $\qquad\qquad\qquad\qquad\qquad\qquad\square$

4. Stable Norm, Quasi-Isometric Embedding of Orbits, and Displacing Action

If a group Γ acts by isometries on a metric space X, we define the *stable norm with respect to X* by

$$[g]_\infty^X = \lim_{n\to\infty}\frac{1}{n}d(x_0, g^n(x_0)).$$

We observe that this quantity does not depend on the choice of the base point x_0. The *stable norm* $[g]_\infty$ is the stable norm with respect to the Cayley graph of Γ.

We now prove the following easy result

PROPOSITION 4.0.5. *Let Γ be a group. Assume that there exists $\alpha > 0$ such that*

$$\forall g \in \Gamma, \, [g]_\infty \geqslant \alpha.\ell(g).$$

Then every action of Γ on (X, d) for which the orbit map is a quasi-isometric embedding is displacing.

Proof. By definition, if the orbit map is a quasi-isometric embedding, for every $x \in X$ there exists some constant A and B such that

$$A\|\gamma\| + B \geqslant d(x, \gamma(x)) \geqslant A^{-1}\|\gamma\| - B.$$

It follows that

$$A[\gamma]_\infty \geqslant [\gamma]_\infty^X \geqslant A^{-1}[\gamma]_\infty.$$

Now we remark that

$$[\gamma]_\infty^X \leqslant d_X(\gamma). \qquad\qquad \square$$

The result follows:

REMARKS:

- We will show later that this inequality fails for $SL(n, \mathbb{Z})$ for $n \geqslant 3$.
- On the other hand, we observe following [4, p. 119], that for Γ a hyperbolic group, the stable norm of an element coincides up to a constant with its translation length; there exists a constant K such that $|\ell(g) - [g]_\infty| \leqslant K$.

Therefore, we have

COROLLARY 4.0.6. *Let Γ be a hyperbolic group. If an isometric action is such that the orbit maps are quasi-isometries, then this action is displacing.*

5. Infinite Groups Whose Actions Are Always Displacing

We have

PROPOSITION 5.0.7. *There exists infinite finitely generated groups whose actions are always displacing. Hence there exists action for which the orbit maps are not quasi-isometric embeddings, but that are displacing.*

Proof. Denis Osin [9] has constructed infinite finitely generated groups with exactly n conjugacy classes. Any action of such a group is displacing. For the second part, we just take the trivial action on a point. $\qquad \square$

References

[1] Herbert Abels, Gregory A. Margulis, and Gregory A. Soĭfer, *Semigroups containing proximal linear maps*, Israel J. Math. **91** (1995), no. 1–3, 1–30.
[2] Yves Benoist, *Propriétés asymptotiques des groupes linéaires*, Geom. Funct. Anal. **7** (1997), no. 1, 1–47.
[3] Yves Benoist and François Labourie, *Sur les difféomorphismes d'Anosov affines à feuilletages stable et instable différentiables*, Invent. Math. **111** (1993), no. 2, 285–308.
[4] Michel Coornaert, Thomas Delzant, and Athanase Papadopoulos, *Géométrie et théorie des groupes*, Lecture Notes in Mathematics, Vol. 1441, Springer-Verlag, Berlin, 1990.
[5] Thomas Delzant, *Sous-groupes à deux générateurs des groupes hyperboliques*, Group theory from a geometrical viewpoint (Trieste, 1990), World Science Publications, River Edge, NJ, 1991, 177–189.

[6] Malik Koubi, *Croissance uniforme dans les groupes hyperboliques*, Ann. Inst. Fourier (Grenoble) **48** (1998), no. 5, 1441–1453.

[7] François Labourie, *Cross ratios, Anosov representations and the energy functional on Teichmüller space*, arXiv math.DG/0512070, 2005.

[8] Alexander Lubotzky, Shahar Mozes, and M. S. Raghunathan, *The word and Riemannian metrics on lattices of semisimple groups*, Inst. Hautes Études Sci. Publ. Math. **91** (2000), 5–53.

[9] Denis Osin, *Small cancellations over relatively hyperbolic groups and embedding theorems*, arXiv math.GR/0411039, 2004.

16

BENSON FARB, CHRIS HRUSKA, AND ANNE THOMAS

PROBLEMS ON AUTOMORPHISM GROUPS OF NONPOSITIVELY CURVED POLYHEDRAL COMPLEXES AND THEIR LATTICES

TO BOB ZIMMER ON HIS 60TH BIRTHDAY

Contents

The first author is supported in part by the NSF. The second author is supported by the NSF under Grant Nos. DMS-0505659 and DMS-0731759. This work of the third author was supported by NSF Grant No. DMS-0805206. The third author is currently supported by EPSRC Grant No. EP/D073626/2.

1. Introduction

The goal of this paper is to present a number of problems about automorphism groups of nonpositively curved polyhedral complexes and their lattices. This topic lies at the juncture of two slightly different cultures. In geometric group theory, universal covers of 2-complexes are studied as geometric and topological models of their fundamental groups, and an important way of understanding groups is to construct "nice" actions on cell complexes, such as cubical complexes. From a different perspective, automorphism groups of connected, simply connected, locally finite simplicial complexes may be viewed as locally compact topological groups, to which we can hope to extend the theory of algebraic groups and their discrete subgroups. In the classification of locally compact topological groups, these automorphism groups are the natural next examples to study after algebraic groups. In this paper we pose some problems meant to highlight possible directions for future research.

Let G be a locally compact topological group with left-invariant Haar measure μ. A *lattice* (respectively, uniform lattice) in G is a discrete subgroup $\Gamma < G$ with $\mu(\Gamma \backslash G) < \infty$ (respectively, $\Gamma \backslash G$ compact). The classical study of Lie groups and their lattices was extended to algebraic groups G over non-Archimedean local fields K by Ihara, Bruhat-Tits, Serre, and many others. This was done by realizing G as a group of automorphisms of the Bruhat-Tits (Euclidean) building X_G, which is a $\mathrm{rank}_K(G)$-dimensional, nonpositively curved (in an appropriate sense) simplicial complex. More recently, Kac-Moody groups G have been studied by considering the action of G on the associated (twin) Tits buildings (see, for example, Carbone-Garland [CG] and Rémy-Ronan [RR]).

The simplest example in the algebraic case is $G = \mathrm{SL}\,(n, K)$, where one can take $K = \mathbf{Q}_p$ (where $\mathrm{char}\,(K) = 0$) or $K = \mathbf{F}_p((t))$ (where $\mathrm{char}\,(K) = p > 0$).

When $n = 2$, that is, $\text{rank}_K(G) = 1$, the building X_G is the regular simplicial tree of degree $p + 1$. One can then extend this point of view to study the full group of simplicial automorphisms of a locally finite tree as a locally compact topological group, and investigate the properties of the lattices it contains. This leads to the remarkably rich theory of "tree lattices" to which we refer the reader to the book of Bass-Lubotzky [BL] as the standard reference.

One would like to build an analogous theory in dimensions 2 and higher, with groups like $\text{SL}(n, \mathbf{Q}_p)$ and $\text{SL}(n, \mathbf{F}_p((t)))$, for $n \geq 3$, being the "classical examples." The increase in dimension makes life much harder, and greatly increases the variety of phenomena that occur.

Now, let X be a locally finite, connected, simply connected simplicial complex. The group $G = \text{Aut}(X)$ of simplicial automorphisms of X naturally has the structure of a locally compact topological group, where a decreasing neighborhood basis of the identity consists of automorphisms of X that are the identity on bigger and bigger balls. With the right normalization of the Haar measure μ, due to Serre [Se], there is a useful combinatorial formula for the covolume of a discrete subgroup $\Gamma < G$:

$$\mu(\Gamma \backslash G) = \sum_{v \in A} \frac{1}{|\Gamma_v|}$$

where the sum is taken over vertices v in a fundamental domain $A \subseteq X$ for the Γ-action, and $|\Gamma_v|$ is the order of the Γ-stabilizer of v. A discrete subgroup Γ is a *lattice* if and only if this sum converges, and Γ is a *uniform lattice* if and only if the fundamental domain A is compact.

In this paper we concentrate on the case when $\dim(X) = 2$. Most questions also make sense in higher dimensions, where even less is understood. When X is a product of trees much is known (see, for example, Burger-Mozes [BM]). However, the availability of projections to trees makes this a special (but deep) theory; we henceforth assume also that X is not a product. There are several themes we wish to explore, many informed by the classical (algebraic) case and the theory of tree lattices in [BL]. We also hope that classical cases may be re-understood from a new, more geometric point of view. Part of our inspiration for this paper came from Lubotzky's beautiful paper [Lu], where he discusses the theory of tree lattices in relation to the classical (real and p–adic) cases.

This paper is not meant to be encyclopedic. It presents a list of problems from a specific and biased point of view. An important criterion in our choice of problem is that it presents some new phenomenon, or requires some new technique or viewpoint in order to solve it. After some background in Section 2, we describe the main known examples of polyhedral complexes and their lattices in Section 3. We have grouped problems on the structure of the complex X itself together with basic group-theoretic and topological properties of Aut (X) in Section 4. Section 5 focuses on whether important properties of linear groups and their lattices hold in this setting, while Section 6 discusses group-theoretic properties of lattices in Aut (X) themselves.

We would like to thank Noel Brady and John Crisp for permission to use Figure 2, and Laurent Saloff-Coste for helpful discussions. We would also like to thank Frédéric Haglund for making many useful comments, which greatly improved the exposition of this paper.

2. Some Background

This preliminary material is mostly drawn from Bridson-Haefliger [BH]. We give the key definitions for polyhedral complexes in Section 2.1. (Examples of polyhedral complexes are described in Section 3.) Conditions for a polyhedral complex X to have nonpositive curvature, and some the consequences for X, are recalled in Section 2.2. The theory of complexes of groups, which is used to construct both polyhedral complexes and their lattices, is sketched in Section 2.3.

2.1. Polyhedral Complexes

Polyhedral complexes may be viewed as generalizations of (geometric realizations of) simplicial complexes. The quotient of a simplicial complex by a group acting by simplicial automorphisms is not necessarily simplicial, and so we work in this larger category. Roughly speaking, a polyhedral complex is obtained by gluing together polyhedra from some constant curvature space by isometries along faces.

More formally, let $\mathbf{X}^n$ be S^n, $\mathbf{R}^n$, or $\mathbf{H}^n$, endowed with Riemannian metrics of constant curvature 1, 0, and -1, respectively. A *polyhedral complex* X is a finite-dimensional CW-complex such that

1) each open cell of dimension n is isometric to the interior of a compact convex polyhedron in $\mathbf{X}^n$; and
2) for each cell σ of X, the restriction of the attaching map to each open codimension one face of σ is an isometry onto an open cell of X.

A polyhedral complex is said to be (piecewise) *spherical, Euclidean,* or *hyperbolic* if $\mathbf{X}^n$ is S^n, $\mathbf{R}^n$, or $\mathbf{H}^n$, respectively. Polyhedral complexes are usually not thought of as embedded in any space. A 2-dimensional polyhedral complex is called a *polygonal complex.*

Given a polyhedral complex X, we write $G = \text{Aut}\,(X)$ for the group of auto-morphisms, or cellular isometries, of X. A subgroup $H \leq G$ is said to act *without inversions* on X if for every cell σ of X, the setwise stabilizer of σ in H is equal to its pointwise stabilizer. Note that any subgroup $H \leq G$ acts without inversions on the barycentric subdivision of X. The quotient of a polyhedral complex by a group acting without inversions is also a polyhedral complex so that the quotient map is a local isometry.

Let x be a vertex of an n-dimensional polyhedral complex X. The *link* of x, denoted $\text{Lk}\,(x, X)$, is the spherical polyhedral complex obtained by intersecting X with an n-sphere of sufficiently small radius centered at x. For example, if X has dimension 2, then $\text{Lk}\,(x, X)$ may be identified with the graph having vertices (the 1-cells of X containing x) and edges (the 2-cells of X containing x); two vertices in the link are joined by an edge in the link if the corresponding 1-cells in X are contained in a common 2-cell. The link may also be thought of as the space of directions, or of germs of geodesics, at the vertex x. By rescaling so that for each x the n-sphere around x has radius, say, 1, we induce a metric on each link, and we may then speak of isometry classes of links of X.

2.2. Nonpositive Curvature

In this section, we recall conditions under which the metrics on the cells of X, a Euclidean or hyperbolic polyhedral complex, may be pieced together to obtain a global metric that is, respectively, CAT(0) or CAT(-1). Some of the consequences for X are then described.

Any polyhedral complex X has an intrinsic pseudometric d, where for $x, y \in X$, the value of $d(x, y)$ is the infimum of lengths of paths Σ from x to y in X, such that the restriction of Σ to each cell of X is geodesic. Bridson [BH] showed that if X has only finitely many isometry types of cells, for example, if $G = \text{Aut}\,(X)$ acts cocompactly, then (X, d) is a complete geodesic metric space.

Now assume X is a Euclidean (respectively, hyperbolic) polyhedral complex such that (X, d) is a complete geodesic space. By the Cartan-Hadamard theorem, if X is locally CAT(0) (respectively, locally CAT(-1)), then the universal cover $\widetilde{X}$ is CAT(0) (respectively, CAT(-1)). Thus to see whether a simply connected X has a CAT(0) metric, we need only check a neighborhood of each point $x \in X$.

If $\dim(X) = n$ and x is in the interior of an n-cell of X, then a neighborhood of x is isometric to a neighborhood in Euclidean (respectively, hyperbolic) n-space. If x is not a vertex but is in the intersection of two n-cells, then it is not hard to see that a neighborhood of x is also CAT(0) (respectively, CAT(-1)).

Hence, the condition that X be CAT(0) comes down to a condition on the nieghborhoods of the vertices of X, that is, on their links.

There are two special cases in which it is easy to check whether neighborhoods of vertices are CAT(0) or CAT(-1). These are when dim$(X) = 2$ and when X is a cubical complex (defined below, and discussed in Section 3.8).

THEOREM 1. (GROMOV LINK CONDITION) *A 2-dimensional Euclidean (respectively, hyperbolic) polyhedral complex X is locally CAT(0) (respectively, CAT(-1)) if and only if for every vertex x of X, every injective loop in the graph* Lk(x, X) *has length at least 2π.*

Let $I^n = [0, 1]^n$ be the cube in $\mathbf{R}^n$ with edge lengths 1. A *cubical complex* is a Euclidean polyhedral complex with all n-cells isometric to I^n. Let L be a simplicial complex. We say L is a *flag* complex if whenever L contains the 1-skeleton of a simplex, it contains the simplex ("no empty triangles").

THEOREM 2. (GROMOV) *A finite-dimensional cubical complex X is locally CAT(0) if and only if the link L of each vertex of X is a flag simplicial complex.*

In general, let X be a polyhedral complex of piecewise constant curvature κ (so $\kappa = 0$ for X Euclidean, and $\kappa = -1$ for X hyperbolic).

THEOREM 3. (GROMOV) *If X is a polyhedral complex of piecewise constant curvature κ, and X has finitely many isometry types of cells, then X is locally CAT(κ) if and only if for all vertices x of X, the link* Lk(x, X) *is a CAT(1) space.*

The condition that a metric space be nonpositively curved has a number of implications, described, for example, in [BH]. We highlight the following results:

- Any CAT(0) space X is contractible.
- Let X be a complete CAT(0) space. If a group Γ acts by isometries on X with a bounded orbit, then Γ has a fixed point in X.

In particular, suppose X is a locally finite CAT(0) polyhedral complex and $\Gamma < $ Aut(X) is a finite group acting on X. Then Γ is contained in the stabilizer of some cell of X.

2.3. Complexes of Groups

The theory of complexes of groups, due to Gersten-Stallings [St] and Haefliger [Hae, BH], generalizes Bass-Serre theory to higher dimensions. It may be used to construct both polyhedral complexes and lattices in their automorphism groups. We give here only the main ideas and some examples, and refer the reader to [BH] for further details.

Throughout this section, if Y is a polyhedral complex, then Y' will denote the first barycentric subdivision of Y. This is a simplicial complex with vertices $V(Y')$ and edges $E(Y')$. Each $a \in E(Y')$ corresponds to cells $\tau \subset \sigma$ of Y, and so may be oriented from $i(a) = \sigma$ to $t(a) = \tau$. Two edges a and b of Y' are *composable* if $i(a) = t(b)$, in which case there exists an edge $c = ab$ of Y' such that $i(c) = i(b)$, $t(c) = t(a)$ and a, b, and c form the boundary of a 2-simplex in Y'.

A *complex of groups* $G(Y) = (G_\sigma, \psi_a, g_{a,b})$ over a polyhedral complex Y is given by

1) a group G_σ for each $\sigma \in V(Y')$, called the *local group* at σ;
2) a monomorphism $\psi_a: G_{i(a)} \to G_{t(a)}$ for each $a \in E(Y')$; and
3) for each pair of composable edges a, b in Y', an element $g_{a,b} \in G_{t(a)}$, such that

$$\text{Ad}\,(g_{a,b}) \circ \psi_{ab} = \psi_a \circ \psi_b$$

where $\text{Ad}\,(g_{a,b})$ is conjugation by $g_{a,b}$ in $G_{t(a)}$, and for each triple of composable edges a, b, c the following cocycle condition holds

$$\psi_a(g_{b,c})\,g_{a,bc} = g_{a,b}\,g_{ab,c}.$$

If all $g_{a,b}$ are trivial, the complex of groups is *simple*. To date, most applications have used only simple complexes of groups. In the case Y is 2-dimensional, the local groups of a complex of groups over Y are often referred to as face, edge, and vertex groups.

EXAMPLE

Let P be a regular right-angled hyperbolic p-gon, $p \geq 5$, and let q be a positive integer ≥ 2. Let $G(P)$ be the following *polygon of groups* over P. The face group is trivial, and each edge group is the cyclic group $\mathbf{Z}/q\mathbf{Z}$. The vertex groups are the direct products of adjacent edge groups. All monomorphisms are natural inclusions, and all $g_{a,b}$ are trivial.

Let G be a group acting without inversions on a polyhedral complex X. The action of G induces a complex of groups, as follows. Let $Y = G \backslash X$ with $p: X \to Y$ the natural projection. For each $\sigma \in V(Y')$, choose $\tilde{\sigma} \in V(X')$ such that $p(\tilde{\sigma}) = \sigma$. The local group G_σ is the stabilizer of $\tilde{\sigma}$ in G, and the ψ_a and $g_{a,b}$ are defined using further choices. The resulting complex of groups $G(Y)$ is unique (up to isomorphism).

Let $G(Y)$ be a complex of groups. Then one defines the *fundamental group of* $G(Y)$, denoted by $\pi_1(G(Y))$, as well as the universal cover of $G(Y)$, denoted by $\widetilde{G(Y)}$, and an action of $\pi_1(G(Y))$ without inversion on $\widetilde{G(Y)}$. The quotient of $\widetilde{G(Y)}$ by this action is naturally isomorphic to Y, and for each cell σ of Y the stabilizer of any lift $\tilde{\sigma} \subset \widetilde{G(Y)}$ is a homomorphic image of G_σ. The complex of groups is called *developable* whenever each homomorphism $G_\sigma \longrightarrow \mathrm{Stab}_{\pi_1(G(Y))}(\sigma)$ is injective. Equivalently, a complex of groups is developable if it is isomorphic to the complex of groups associated as above to an action without inversion on a simply connected polyhedral complex.

Unlike graphs of groups, complexes of groups are not in general developable:

EXAMPLE (K. BROWN)

Let $G(Y)$ be the triangle of groups with trivial face group and edge groups infinite cyclic, generated by, say, a, b, and c. Each vertex group is isomorphic to the Baumslag-Solitar group $BS(1,2) = \langle x, y \mid xyx^{-1} = y^2 \rangle$, where the generators x and y are identified with the generators of the adjacent edge groups. The fundamental group of $G(Y)$ then has presentation $\langle a, b, c \mid aba^{-1} = b^2, bcb^{-1} = c^2, cac^{-1} = a^2 \rangle$. It is an exercise that this is the trivial group. Thus $G(Y)$ is not developable.

We now describe a local condition for developability. Let Y be a connected polyhedral complex and let $\sigma \in V(Y')$. The *star* of σ, written $\mathrm{St}(\sigma)$, is the union of the interiors of the simplices in Y' that meet σ. If $G(Y)$ is a complex of groups over Y, then even if $G(Y)$ is not developable, each $\sigma \in V(Y')$ has a *local development*. That is, we may associate to σ an action of G_σ on the star $\mathrm{St}(\tilde{\sigma})$ of a vertex $\tilde{\sigma}$ in some simplicial complex, such that $\mathrm{St}(\sigma)$ is the quotient of $\mathrm{St}(\tilde{\sigma})$ by the action of G_σ. To determine the local development, its link may be computed in combinatorial fashion.

EXAMPLE

Suppose $G(Y)$ is a simple polygon of groups, with $G_\sigma = V$ a vertex group, with adjacent edge groups E_1 and E_2, and with face group F. We identify the groups E_1, E_2, and F with their images in V. The link L of the local development at σ is then a bipartite graph. The two sets of vertices of L correspond to the cosets of E_1 and E_2, respectively, in V, and the edges of L correspond to cosets of F in V. The number of edges between vertices $g_1 E_1$ and $g_2 E_2$ is equal to the number of cosets of F in the intersection $g_1 E_1 \cap g_2 E_2$. In the polygon of groups $G(P)$ given above, the link of the local development at each vertex of P will be the complete bipartite graph $K_{q,q}$.

If $G(Y)$ is developable, then for each $\sigma \in V(Y')$, the local development St $(\tilde{\sigma})$ is isomorphic to the star of each lift $\tilde{\sigma}$ of σ in the universal cover $\widetilde{G(Y)}$. The local development has a metric structure induced by that of the polyhedral complex Y. We say that a complex of groups $G(Y)$ is *nonpositively curved* if for all $\sigma \in V(Y')$, the star St $(\tilde{\sigma})$ is CAT(0) in this induced metric. The importance of this condition is given by

THEOREM 4. (STALLINGS (ST), HAEFLIGER (HAE, BH)) *A nonpositively curved complex of groups is developable.*

EXAMPLE

The polygon of groups $G(P)$ above is nonpositively curved and thus developable. The links are the complete bipartite graph $K_{q,q}$ with edge lengths $\frac{\pi}{2}$, and so Gromov's link condition (Theorem 1) is satisfied.

Let $G(Y)$ be a developable complex of groups, with universal cover a locally finite polyhedral complex X, and fundamental group Γ. We say that $G(Y)$ is *faithful* if the action of Γ on X is faithful. If so, Γ may be regarded as a subgroup of Aut (X). Moreover, Γ is discrete if and only if all local groups of $G(Y)$ are finite, and Γ is a uniform lattice if and only if Y is compact.

EXAMPLE

Let $G(P)$ be the (developable) polygon of groups above, with fundamental group, say, Γ and universal cover, say, X. Then $G(P)$ is faithful since its face group is trivial. As all the local groups are finite and P is compact, Γ may be identified with a uniform lattice in Aut (X).

3. Examples of Polyhedral Complexes and Their Lattices

In this section we present the most studied examples of locally finite polyhedral complexes X and their lattices. For each case, we give the key definitions, and sketch known constructions of X and of lattices in Aut (X). There is some overlap between examples, which we describe. We will also try to indicate the distinctive flavor of each class. While results on existence of X are recalled here, we defer questions of uniqueness of X, given certain local data, to Section 4.1. Existence of lattices is also discussed further in Section 6.1.

Many of the examples we discuss are buildings, which form an important class of nonpositively curved polyhedral complexes. Roughly, buildings may be thought of as highly symmetric complexes, which contain many flats, and often have algebraic structure. Classical buildings are those associated to groups such as $SL(n, \mathbf{Q}_p)$, and play a similar role for these groups to that of symmetric spaces for real Lie groups. The basic references for buildings are Ronan [Ron2] and Brown [Br]. A much more comprehensive treatment by Abramenko-Brown [AB] is to appear shortly. These works adopt a combinatorial approach. For our purposes we present a more topological definition, from [HP2].

Recall that a *Coxeter group* is a group W with a finite-generating set S and presentation of the form

$$W - \langle s \in S \mid (s_i s_j)^{m_{ij}} = 1 \rangle$$

where $s_i, s_j \in S$, $m_{ii} = 1$, and if $i \neq j$, then m_{ij} is an integer ≥ 2 or $m_{ij} = \infty$, meaning that there is no relation between s_i and s_j. The pair (W, S), or (W, I) where I is the finite-indexing set of S, is called a *Coxeter system*. A spherical, Euclidean, or hyperbolic *Coxeter polytope* of dimension n is an n-dimensional compact convex polyhedron P in the appropriate space, with every dihedral angle of the form π/m for some integer $m \geq 2$ (not necessarily the same m for each angle). The group W generated by reflections in the codimension-1 faces of a Coxeter polytope P is a Coxeter group, and its action generates a tesselation of the space by copies of P.

DEFINITION: Let P be an n-dimensional spherical, Euclidean, or hyperbolic Coxeter polytope. Let $W = (W, S)$ be the Coxeter group generated by the set of reflections S in the codimension-1 faces of P. A *spherical, Euclidean,* or *hyperbolic building* of type (W, S) is a polyhedral complex X equipped with a maximal family of subcomplexes, called *apartments*, each polyhedrally isometric to the tesselation of, respectively, S^n, $\mathbf{R}^n$, or $\mathbf{H}^n$ by the images of P under W (called *chambers*), such that

1) any two chambers of X are contained in a common apartment; and
2) for any two apartments A and A', there exists a polyhedral isometry from A onto A' that fixes $A \cap A'$.

The links of vertices of n-dimensional buildings are spherical buildings of dimension $n - 1$, with the induced apartment and chamber structure. Using this and Theorem 3, it follows that Euclidean (respectively, hyperbolic) buildings are CAT(0) (respectively, CAT(−1)). Since buildings are such important examples in the theory, we will spend some time describing them in detail.

3.1. Euclidean Buildings

Euclidean buildings are also sometimes known as *affine buildings*, or *buildings of affine type*. A simplicial tree X is a 1-dimensional Euclidean building of type (W, S), where W is the infinite dihedral group, acting on the real line with fundamental domain P an interval. The chambers of X are the edges of the tree, and the apartments X are the geodesic lines in the tree. Since the product of two buildings is also a building, it follows that products of trees are higher-dimensional (reducible) Euclidean buildings (see Section 3.2). In this section we consider Euclidean buildings X of dimension $n \geq 2$ that are not products.

3.1.1. CLASSICAL EUCLIDEAN BUILDINGS
Classical Euclidean buildings are those Euclidean buildings that are associated to algebraic groups, as we now outline. We first construct the building for $G = \mathrm{SL}\,(n, K)$ where K is a non-Archimedean local field, in terms of lattices in K^n and then in terms of BN-pairs (defined below). We then indicate how the latter construction generalizes to other algebraic groups. Our treatment is based upon [Br].

Let K be a field. We recall that a *discrete valuation* on K is a surjective homomorphism $v\colon K^* \longrightarrow \mathbf{Z}$, where K^* is the multiplicative group of nonzero elements of K, such that

$$v(x + y) \geq \min\{v(x), v(y)\}$$

for all $x, y \in K^*$ with $x + y \neq 0$. We set $v(0) = +\infty$, so that v is defined and the above inequality holds for all of K. A discrete valuation induces an absolute value $|x| = e^{-v(x)}$ on K, which satisfies the non-Archimedean inequality

$$|x + y| \leq \max\{|x|, |y|\}.$$

A metric on K is obtained by setting $d(x, y) = |x - y|$. The set $\mathcal{O} = \{\, x \in K \mid |x| \leq 1 \,\}$ is a subring of K called the *ring of integers*. The ring $\mathcal{O}$ is compact and open in the metric topology induced by v. Pick an element $\pi \in K$ with $v(\pi) = 1$, called a *uniformizer*. Every $x \in K^*$ is then uniquely expressible in the form $x = \pi^n u$ where $n \in \mathbf{Z}$ and u is a unit of $\mathcal{O}^*$ (so $v(u) = 0$). The ideal $\pi\mathcal{O}$ generated by π is a maximal ideal, since every element of $\mathcal{O}$ not in $\pi\mathcal{O}$ is a unit. Hence $k = \mathcal{O}/\pi\mathcal{O}$ is a field, called the *residue field*.

EXAMPLES

1) For any prime p the p-adic valuation v on the field of rationals $\mathbf{Q}$ is defined by $v(x) = n$, where $x = p^n a/b$ and a and b are integers not divisible by p. The field of p-adics $K = \mathbf{Q}_p$ is the completion of $\mathbf{Q}$ with respect to the metric induced by v, and the valuation v extends to $\mathbf{Q}_p$ by continuity. The ring of integers is the ring of p-adic integers $\mathbf{Z}_p$, and we may take $\pi = p$ as uniformizer. The residue field of $\mathbf{Q}_p$ is then the finite field $k = \mathbf{F}_p$.

2) Let q be a power of a prime p. The field $K = \mathbf{F}_q((t))$ of formal Laurent series with coefficients in the finite field $\mathbf{F}_q$ has valuation v given by

$$v\left(\sum_{j=-m}^{\infty} a_j t^j \right) = -m$$

where $a_{-m} \neq 0$, a uniformizer is t, and the ring of integers is the ring of formal power series $\mathbf{F}_q[[t]]$. The residue field is $k = \mathbf{F}_q$.

A *local non-Archimedean field* is a field K that is complete with respect to the metric induced by a discrete valuation, and whose residue field is finite. Examples are $K = \mathbf{Q}_p$, which has char $(K) = 0$, and $K = \mathbf{F}_q((t))$, which has char $(K) = p > 0$. In fact, all local non-Archimedean fields arise as finite extensions of these examples.

We now fix K to be a local non-Archimedean field, $\mathcal{O}$ its ring of integers, π a uniformizer, and k its residue field. The *Euclidean building* associated to the group $G = \mathrm{SL}\,(n, K)$ is the geometric realization $|\Delta|$ of the abstract simplicial complex Δ, which we now describe.

Let V be the vector space K^n. A *lattice* in V is an $\mathcal{O}$-submodule $L \subset V$ of the form $L = \mathcal{O}v_1 \oplus \cdots \oplus \mathcal{O}v_n$ for some basis $\{v_1, \ldots, v_n\}$ of V. If L' is another lattice, then we may choose a basis $\{v_1, \ldots, v_n\}$ for L such that L' admits the basis $\{\lambda_1 v_1, \ldots, \lambda_n v_n\}$ for some $\lambda_i \in K^*$. The λ_i may be taken to be powers of π. Two lattices L and L' are *equivalent* if $L = \lambda L'$ for some $\lambda \in K^*$. We write $[L]$ for the equivalence class of L, and $[v_1, \ldots, v_n]$ for the equivalence class of the lattice with basis $\{v_1, \ldots, v_n\}$.

The abstract simplicial complex Δ is defined to have vertices the set of equivalence classes of lattices in V. To describe the higher-dimensional simplices of Δ, we introduce the following incidence relation. (An *incidence relation* is a relation that is reflexive and symmetric.) Two equivalence classes of lattices Λ and Λ' are *incident* if they have representatives L and L' such that

$$\pi L \subset L' \subset L.$$

This relation is symmetric, since $\pi L' \in \Lambda'$ and $\pi L \in \Lambda$ satisfy

$$\pi L' \subset \pi L \subset L'.$$

The simplices of Δ are then defined to be the finite sets of pairwise incident equivalence classes of lattices in V.

By the definition of incidence, every top-dimensional simplex of Δ has vertex set

$$[v_1, \ldots, v_i, \pi v_{i+1}, \ldots, \pi v_n] \text{ for } i = 1, \ldots, n,$$

for some basis $[v_1, \ldots, v_n]$ of V. Hence Δ is a simplicial complex of dimension $n - 1$. The geometric realization $X = |\Delta|$ is thus a Euclidean polyhedral complex of dimension $n - 1$. We note that $n - 1$ is equal to the K-rank of $G = \text{SL}(n, K)$.

We now construct a simplicial complex isomorphic to Δ, using certain subgroups B and N of $G = \text{SL}(n, K)$. For now, we state without proof that, with the correct Euclidean metrization, $X = |\Delta|$ is indeed a Euclidean building, with chambers its $(n-1)$-cells, and that the vertex set of an apartment of X is the set of equivalence classes

$$[\pi^{m_1} v_1, \ldots, \pi^{m_n} v_n]$$

where the m_i are integers ≥ 0, and $\{v_1, \ldots, v_n\}$ is a fixed basis for V.

Observe that the group $G = \text{SL}(n, K)$ acts on the set of lattices in V. This action preserves equivalence of lattices and the incidence relation, so G acts without inversions on X. Let $\{e_1, \ldots, e_n\}$ be the standard basis of V. We define the *fundamental chamber* of X to be the simplex with vertices

$$[e_1, \ldots, e_i, \pi e_{i+1}, \ldots, \pi e_n], \text{ for } i = 1, \ldots, n,$$

and the *fundamental apartment* of X to be the subcomplex with vertex set

$$[\pi^{m_1} e_1, \ldots, \pi^{m_n} e_n], \text{ where } m_i \geq 0.$$

Define B to be the stabilizer in G of the fundamental chamber, and N to be the stabilizer in G of the fundamental apartment. There is a surjection $\text{SL}(n, \mathcal{O}) \longrightarrow \text{SL}(n, k)$ induced by the surjection $\mathcal{O} \longrightarrow k$. It is not hard to verify that B is the inverse image in $\text{SL}(n, \mathcal{O})$ of the upper-triangular subgroup of $\text{SL}(n, k)$, and that N is the monomial subgroup of $\text{SL}(n, K)$ (that is, the set of matrices with exactly one nonzero entry in each row and each column). We say that a subgroup of G is *special* if it contains a coset of B.

Now, from the set of cosets in G of special subgroups, we form a partially ordered set, ordered by opposite inclusion. There is an abstract simplicial complex $\Delta(G, B)$ associated to this poset. The vertices of $\Delta(G, B)$ are cosets of special subgroups, and the simplices of $\Delta(G, B)$ correspond to chains of opposite inclusions. Using the action of G on Δ and the construction of $\Delta(G, B)$, it is not hard to see that $\Delta(G, B)$ is isomorphic to (the barycentric subdivision of) Δ.

We now generalize the construction of $\Delta(G, B)$ to algebraic groups besides $G = SL(n, K)$. Let G be an absolutely almost-simple, simply connected linear algebraic group defined over K. Examples other than $SL(n, K)$ include $Sp(2n, K)$, $SO(n, K)$, and $SU(n, K)$. All such groups G have a Euclidean BN-pair, which we now define. A BN-pair is a pair of subgroups B and N of G, such that

- B and N generate G;
- the subgroup $T = B \cap N$ is normal in N, and
- the quotient $W = N/T$ admits a set of generators S satisfying certain (technical) axioms, which ensure that (W, S) is a Coxeter system.

A BN-pair is *Euclidean* if the group W is a Euclidean Coxeter group. The letter B stands for the Borel subgroup, T for the torus, N for the normalizer of the torus, and W for the Weyl group.

For $G = SL(n, K)$, the B and N defined above, as G-stabilizers of the fundamental chamber and fundamental apartment of Δ, are a BN-pair. Their intersection T is the diagonal subgroup of $SL(n, \mathcal{O})$. The group W acts on the fundamental apartment of Δ with quotient the fundamental chamber, and is in fact isomorphic to the Coxeter group generated by reflections in the codimension one faces of a Euclidean $(n-1)$-simplex (with certain dihedral angles).

For any group G with a Euclidean BN-pair, one may construct the simplicial complex $\Delta(G, B)$ from the poset of cosets of special subgroups, as described above. The geometric realization of $\Delta(G, B)$ is a Euclidean polyhedral complex, of dimension equal to the K-rank of G. To prove that the geometric realization of $\Delta(G, B)$ is a building, one uses the axioms for a BN-pair, results about Coxeter groups, and the Bruhat-Tits decomposition of G.

For classical Euclidean buildings X, there is a close relationship between the algebraic group G to which this building is associated and the group $\mathrm{Aut}(X)$, so long as $\dim(X) \geq 2$.

THEOREM 5. (TITS [Ti3]) *Let G be an absolutely almost-simple, simply connected linear algebraic group defined over a non-Archimedean local field K. Let X be the Euclidean building for G. If $\mathrm{rank}_K(G) \geq 2$, then G has finite index* in Aut (X) when char $(K) = 0$, and is cocompact in Aut (X) when char $(K) = p > 0$.

Thus the lattice theory of Aut (X) is very similar to that of G. Existence and construction of lattices in groups G as in Theorem 5 are well understood. If char $(K) = 0$, then G does not have a nonuniform lattice (Tamagawa [Ta]), but does admit a uniform lattice, constructed by arithmetic means (Borel-Harder [BHar]). If char $(K) = p > 0$, then G has an arithmetic nonuniform lattice, and an arithmetic uniform lattice if and only if $G = SL(n, K)$ (Borel-Harder [BHar]). In real rank at least 2 (for example if $G = SL(n, K)$, for $n \geq 3$) every lattice of G is arithmetic (Margulis [Ma]).

3.1.2. NONCLASSICAL EUCLIDEAN BUILDINGS Nonclassical Euclidean buildings are those Euclidean buildings (see Definition (3) that are not the building for any algebraic group G over a non-Archimedean local field. Tits constructed uncountably many isometry classes of nonclassical Euclidean buildings [Ti4]. Nonclassical buildings may also be constructed as universal covers of finite complexes, a method developed by Ballmann-Brin [BB1], and examples of this kind were obtained by Barré [Ba] as well. Ronan [Ron1] used a construction similar to the inductive construction of Ballmann-Brin, described in Section 3.7, to construct 2-dimensional nonclassical Euclidean buildings. Essert [Es] has constructed Euclidean buildings (of type $\tilde{A}_2$ and $\tilde{C}_2$), both classical and nonclassical, as universal covers of finite complexes of finite groups.

Very few lattices are known for nonclassical buildings. In [CMSZ], exotic lattices that act simply transitively on the vertices of various classical and non-classical Euclidean buildings (of type $\tilde{A}_2$) are constructed by combinatorial methods. The fundamental groups of the complexes of groups constructed by Essert [Es] are uniform lattices that act simply transitively on the panels of the universal cover.

3.2. Products of Trees

Let T_1 and T_2 be locally finite simplicial trees. The product space $T_1 \times T_2$ is a polygonal complex, where each 2-cell is a square (edge × edge), and the link at each vertex is a complete bipartite graph. Products of more than two trees may also be studied.

The group $G = \text{Aut}(T_1 \times T_2)$ is isomorphic to $\text{Aut}(T_1) \times \text{Aut}(T_2)$ (with a semidirect product with $\mathbf{Z}/2\mathbf{Z}$ if $T_1 = T_2$). Thus any subgroup of G may be projected to the factors. Because of this availability of projections, the theory of lattices for products of trees is a special (but deep) theory. See, for example, the work of Burger-Mozes [BM]. Many of the problems listed below may be posed in this context, but we omit questions specific to this case.

3.3. Hyperbolic Buildings

The simplest example of a hyperbolic building is *Bourdon's building* $I_{p,q}$, defined and studied in [B1]. Here p and q are integers, $p \geq 5$ and $q \geq 2$. The building $I_{p,q}$ is the (unique) hyperbolic polygonal complex such that each 2-cell (chamber) is isometric to a regular right-angled hyperbolic p-gon P, and the link at each vertex is the complete bipartite graph $K_{q,q}$. The apartments of $I_{p,q}$ are hyperbolic planes tesselated by copies of P. Bourdon's building is CAT(-1), and may be regarded as a hyperbolic version of the product of two q-regular trees, since it has the same links. However, $I_{p,q}$ is not globally a product space. The example of a polygon of groups $G(P)$ given in Section 2.3 has universal cover $I_{p,q}$, and the fundamental group Γ of this polygon of groups is a uniform lattice in Aut $(I_{p,q})$.

Bourdon's building is a *Fuchsian building*, that is, a hyperbolic building of dimension 2. More general Fuchsian buildings have all chambers hyperbolic k-gons, $k \geq 3$, with each vertex angle of the form π/m, for some integer $m \geq 2$ (depending on the vertex). The link at each vertex with angle π/m is a 1-dimensional spherical building L that is a generalized m-gon, that is, a graph with diameter m and girth $2m$. For example, a complete bipartite graph is a generalized 2-gon.

Unlike Euclidean buildings, hyperbolic buildings do not exist in arbitrary dimension. This is because there is a bound ($n \leq 29$), due to Vinberg [Vi], on the dimension n of a compact convex hyperbolic Coxeter polytope. Gaboriau-Paulin [GP] broadened the definition of building given above (Definition 3) to allow hyperbolic buildings with noncompact chambers, in which case there are examples in any dimension, with chambers for ideal hyperbolic simplexes.

Various constructions of hyperbolic buildings are known. In low dimensions, right-angled buildings (see Section 3.4) may be equipped with the structure of a hyperbolic building. In particular, Bourdon's building is a right-angled building. Certain hyperbolic buildings arise as Kac-Moody buildings (see Section 3.5), and some Davis-Moussong complexes may also be metrized as hyperbolic buildings (see Section 3.6). Vdovina constructed some Fuchsian buildings as universal covers of finite complexes [Vd]. Fuchsian buildings were constructed as universal covers of polygons of groups by Bourdon [B1, B2] and by Gaboriau-Paulin [GP]. Haglund-Paulin [HP2] have constructed

3–dimensional hyperbolic buildings using "treelike" decompositions of the corresponding Coxeter systems.

Many of these constructions of hyperbolic buildings X also yield lattices in Aut (X). When a hyperbolic building X is a Kac-Moody building then a few lattices in Aut (X) are known from Kac-Moody theory (see, for example, [Rem1]), and when X is a Davis-Moussong complex for a Coxeter group W then W may be regarded as a uniform lattice in Aut (X). If X is the universal cover of a finite complex, the fundamental group of that complex is a uniform lattice in Aut (X). As described in Section 2.3, if X is the universal cover of a finite complex of finite groups, such as a polygon of finite groups, then the fundamental group of the complex of groups is a uniform lattice in Aut (X). More elaborate complexes of groups were used by Thomas to construct both uniform and nonuniform lattices for certain Fuchsian buildings in [Th3]. In [B2] Bourdon was able to "lift" lattices for affine buildings to uniform and nonuniform lattices for certain Fuchsian buildings.

3.4. Right-Angled Buildings

Recall that (W, I) is a *right-angled* Coxeter system if all the m_{ij} with $i \neq j$ equal 2 or ∞. A building X of type (W, I) is then a *right-angled building*. Products of trees are examples of right-angled buildings, with associated Coxeter group the direct product of infinite dihedral groups.

Bourdon's building $I_{p,q}$, discussed in Section 3.3, is another basic example of a right-angled building. The Coxeter group W here is generated by reflections in the sides of a regular right-angled hyperbolic p-gon. Right-angled Coxeter polytopes exist only in dimensions $n \leq 4$, and this bound is sharp (Potyagailo-Vinberg [PV]). Thus right-angled buildings may be metrized as hyperbolic buildings (with compact chambers) only in dimensions ≤ 4.

We may broaden the definition of building given above (Definition 3) to allow apartments that are Davis-Moussong complexes for W (see Section 3.6), rather than just the manifold S^n, $\mathbf{R}^n$, or $\mathbf{H}^n$ tesselated by the action of W. With this definition, Gromov-hyperbolic right-angled buildings, equipped with a piecewise Euclidean metric, exist in arbitrary dimensions (Januszkiewicz-Świątkowski [JŚ1]).

The following construction of a right-angled building X and a uniform lattice in Aut (X) appears in [HP1]; this construction was previously known to Davis and Meier. It is a generalization of the polygon of groups $G(P)$ in Section 2.3. Let (W, I) be a right-angled Coxeter system, and $\{q_i\}_{i \in I}$ a set of cardinalities with $q_i \geq 2$. Let N be the finite nerve of W, with first barycentric subdivision N', and let K be the cone on N'. For example, if W is generated by reflections in the sides of a right-angled hyperbolic p-gon P, then N is a circuit of p edges, and K is isomorphic to the barycentric subdivision of P. For each $i \in I$, let G_i be a group of order q_i. Each vertex of K has a type J, where $J \subset I$ is such that the group W_J generated by $\{s_i\}_{i \in J}$ is finite. For each i, let K_i be the subcomplex of K, which is the closed star of the vertex of type $\{i\}$ in N'. Let $G(K)$ be the complex of groups where the vertex of K with type J has local group the direct product

$$\coprod_{i \in J} G_i$$

and all monomorphisms are natural inclusions. This complex of groups is developable, with universal cover a right-angled building X of type (W, I). The copies of K in X are called *chambers*, and each copy of K_i in X is contained in q_i distinct chambers. Moreover, the fundamental group of this complex of groups may be viewed as a uniform lattice in $\mathrm{Aut}\,(X)$ (if all q_i are finite).

Many other lattices for right-angled buildings (in any dimension) were obtained by promoting tree lattices, using complexes of groups, in Thomas [Th2].

3.5. Kac-Moody Buildings

Kac-Moody groups over finite fields $\mathbf{F}_q$ may be viewed as infinite-dimensional analogues of Lie groups. See, for example, Carbone-Garland [CG] and Rémy-Ronan [RR]. For any Kac-Moody group Λ there are associated (twin) buildings X_+ and X_-, constructed using twin BN-pairs (B^+, N) and (B^-, N) (see Section 3.1.1). The group Λ acts diagonally on the product $X_+ \times X_-$, and for q large enough Λ is a nonuniform lattice in $\mathrm{Aut}\,(X_+ \times X_-)$ (see [Rem1]). A *Kac-Moody building* is a building that appears as one of the twin buildings for a Kac-Moody group. Kac-Moody buildings are buildings, but unlike classical Euclidean buildings (see Theorem 5), nonisomorphic Kac-Moody groups may have the same building (Rémy [Rem2]). One may also study a *complete Kac-Moody group* G, which is the completion of Λ with respect to some topology. Very few lattices in complete Kac-Moody groups are known.

3.6. Davis-Moussong Complexes

Given any Coxeter system (W, S), the associated Davis-Moussong complex is a locally finite, CAT(0), piecewise Euclidean polyhedral complex on which W acts properly discontinuously and cocompactly. We describe a special case of this construction in dimension 2.

Let L be a connected, finite simplicial graph with all circuits of length at least 4, and let $k \geq 2$ be an integer. The Coxeter system corresponding to this data has a generator s_i of order 2 for each vertex v_i of L, and a relation $(s_i s_j)^k = 1$ if and only if the vertices v_i and v_j are connected by an edge in L. The Coxeter group defined by this Coxeter system is denoted $W = W(k, L)$. If $k = 2$, then W is a right-angled Coxeter group.

For any such $W = W(k, L)$, Davis-Moussong constructed a CAT(0) piecewise Euclidean complex $X = X(2k, L)$ (see [D, Mo]). The cells of X correspond to cosets in W of spherical subgroups of W, and in particular the 0-cells of X correspond to the elements of W, viewed as cosets of the trivial subgroup. Recall that a *spherical subgroup* of W is a subgroup W_T generated by some subset $T \leq S$, such that W_T is finite.

The Davis-Moussong complex may be identified with (the first barycentric subdivision of) a polygonal complex X with all links L and all 2-cells regular Euclidean $2k$-gons. The group W has a natural left action on X that is properly discontinuous, cellular, and simply transitive on the vertices of X. Thus W may be viewed as a uniform lattice in Aut (X). This construction can also be carried out in higher dimensions, provided L is a CAT(1) spherical simplicial complex. In dimension 2, where L is a graph, this is equivalent to all circuits having length at least 4, by the Gromov link condition (Theorem 1). If W is right angled, then each apartment of a right-angled building of type W is isomorphic to the Davis-Moussong complex for W.

Davis-Moussong also found easy-to-verify conditions on L such that $X(2k, L)$ may be equipped with a CAT(-1) piecewise hyperbolic structure. In this way, some hyperbolic buildings (or rather, their first barycentric subdivisions) may be constructed as Davis-Moussong complexes, with the graph L a 1-dimensional spherical building.

3.7. (k, L)-Complexes

Let L be a finite graph and k an integer ≥ 3. A (k, L)-*complex* is a polygonal complex X such that the link of each vertex of X is L, and each 2-cell of X is a regular k-gon (usually but not necessarily Euclidean).

Many polygonal complexes already described are (k, L)-complexes. For example, 2-dimensional Euclidean or hyperbolic buildings, with all links the same, are (k, L)-complexes where L is a 1-dimensional spherical building. The 2-dimensional Davis-Moussong complexes described in Section 3.6 are barycentric subdivisions of (k, L)-complexes with $k \geq 4$ even. An example of a (k, L)-complex that is not a building or a Davis-Moussong complex is where k is odd and L is the Petersen graph (Figure 1).

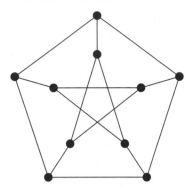

Fig. 1. Petersen graph.

There are simple conditions on the pair (k, L) ensuring that a (k, L)-complex satisfies Gromov's link condition (Theorem 1) and thus has nonpositive curvature. Ballmann-Brin [BB1] showed that any (k, L)-complex where k and L satisfy these conditions may be constructed in an inductive manner, by adding k-gons to the previous stage without obstructions. This construction is discussed in more detail in Section 4.1. Some (k, L)-complexes may also be constructed as universal covers of triangles of groups, as done in [JLVV]. In this case the fundamental group of the triangle of groups is a uniform lattice. Constructions of uniform and nonuniform lattices as fundamental groups of complexes of groups are carried out for certain highly symmetric (k, L)-complexes, including those with Petersen graph links, as in Thomas [Th4].

3.8. CAT(0) Cubical Complexes

Recall that a *cubical complex* is a Euclidean polyhedral complex with all n-cells isometric to the Euclidean n-cube, and that a cubical complex X is locally CAT(0) if and only if each vertex of X is a flag simplicial complex (Theorem 2). Trees and products of trees are examples of CAT(0) cubical complexes.

Groups of automorphisms of CAT(0) cubical complexes are different in many ways from groups acting on the Euclidean buildings discussed in Section 3.1. Examples of discrete groups that act properly on CAT(0) cube complexes include finitely generated Coxeter groups [NRe2], many small cancellation groups [W4], 1-relator groups with torsion [LW], many diagram groups, including Thompson's group F [Far], and groups acting properly on products of trees.

In this setting, the main geometric objects of study are hyperplanes, defined as follows. Consider two edges of a CAT(0) cube complex X to be *equivalent* if they are opposite edges of some 2-cube. This relation generates an equivalence relation whose equivalence classes are the *combinatorial hyperplanes* of X. One can also define *geometric hyperplanes* of X as unions of *midplanes* of cubes, where a midplane of the cube $C = [0, 1]^n$ is a subset of the form

$$[0, 1] \times \cdots \times [0, 1] \times \frac{1}{2} \times [0, 1] \times \cdots \times [0, 1].$$

Thus C has n midplanes, which intersect transversely at the barycenter of C. Given a combinatorial hyperplane H, the corresponding geometric hyperplane is the union of all midplanes meeting the barycenters of the edges of H. Each geometric hyperplane is itself a CAT(0) cubical complex, whose cubes are midplanes of cubes of X. Each geometric hyperplane separates X into two complementary components, called *half-spaces*. The properties of hyperplanes generalize the separation properties of edges in a tree. The main new feature in higher dimensions, not present in trees, is that hyperplanes can have transverse intersections. In fact, CAT(0) cubical complexes have a rich combinatorial structure arising from the incidence and nesting properties of hyperplanes.

Geometrically, the most significant subgroups in a group acting on a CAT(0) cubical complex are the codimension-1 subgroups, which typically arise as stabilizers of hyperplanes. If a group Γ has a finite-generating set S, a subgroup $H \leq \Gamma$ is *codimension*-1 provided that some neighborhood of H separates Cayley(G, S) into at least two "deep" complementary components, where a component is *deep* if it contains elements arbitrarily far away from H. For instance, if M is a 3-manifold with an immersed, incompressible surface S, then $\pi_1(S)$ is a codimension-1 subgroup of $\pi_1(M)$.

Sageev has shown (together with a result proved independently by Gerasimov and Niblo-Roller) that a finitely generated group Γ has a codimension-1 subgroup if and only if Γ acts on a CAT(0) cube complex with no global fixed point [Sa, Ger, NRo]. The cube complex produced by Sageev's theorem is sometimes infinite-dimensional and sometimes locally infinite.

Several representation-theoretic aspects of actions on trees extend naturally to actions on CAT(0) cubical complexes. If a topological group with property (T) acts on a CAT(0) cubical complex, then the action must have a global fixed point. On the other hand, if a topological group G acts metrically properly on a CAT(0) cubical complex X, then G is a-T-menable [NRo]. In particular, if X is locally finite then any discrete subgroup $\Gamma \leq \text{Aut}(X)$ is a-T-menable. Niblo-Reeves have also shown that if X is any CAT(0) cube complex, then every uniform lattice $\Gamma \leq \text{Aut}(X)$ is biautomatic [NRe1].

3.9. Systolic Complexes

Systolic complexes are a family of simplicial complexes whose geometry exhibits many aspects of nonpositive curvature, yet which are not known to be CAT(0). A *systolic complex* is a flag simplicial complex that is connected and simply connected, such that the link of each vertex does not contain an isometric edge cycle of length 4 or 5. In [Ch], Chepoi proved that a graph is the 1-skeleton of a systolic complex if and only if it is a *bridged graph*, which is a connected graph having no isometric edge cycles of length at least 4.

Bridged graphs were introduced by Soltan-Chepoi [SC] and independently by Farber-Jamison [FJ] where they were shown to share certain convexity properties with CAT(0) spaces. Their geometric and algorithmic properties were studied for many years from the point of view of graph theory. Systolic complexes were rediscovered independently by Januszkiewicz-Świątkowski [JŚ2], Haglund [H4], and Wise, and have subsequently been the subject of much study in geometric group theory.

A simplicial complex can be metrized in many ways, but the most natural metric, called the *standard piecewise Euclidean metric*, is given by declaring each simplex to be isometric to a regular Euclidean simplex with all side lengths equal to 1. In dimension 2, a simplicial complex is systolic exactly when the standard piecewise Euclidean metric is CAT(0). In higher dimensions being systolic is neither stronger nor weaker than the standard piecewise Euclidean metric being CAT(0). A much more subtle question is whether a systolic complex admits any piecewise Euclidean metric that is CAT(0). No answer is known, but the answer is generally expected to be negative.

Systolic complexes do share many properties with CAT(0) spaces. For example, any finite-dimensional systolic simplicial complex is contractible. As with CAT(0) cubical complexes, any group acting properly discontinuously and cocompactly on a systolic complex is biautomatic. An interesting question is whether all systolic groups are in fact CAT(0) groups.

Systolic complexes are constructed in [JŚ2] as universal covers of simplices of groups, using the result that a locally 6-large complex of groups is developable. The fundamental groups of these simplices of groups are uniform lattices in the automorphism group of the universal cover.

4. Properties of X and Aut(X)

The goal of this section is to understand the general structure of a polyhedral complex X and its full automorphism group Aut (X). For instance, how much local data is required in order to uniquely determine X? What are the basic topological and group-theoretic properties of Aut (X)?

4.1. When Do Local Data Determine X?

As seen in many examples in Section 3, polyhedral complexes X are often constructed as universal covers of complexes of groups, and lattices in Aut (X) are often fundamental groups of complexes of groups. In each case, the local structure of the universal cover is determined by the local structure of the quotient space, together with the attached local groups of the complex of groups. Thus it is critical to know how much local data is needed in order to uniquely specify a desired polyhedral complex X. To simplify matters, we focus on the special case when X is a (k, L)-complex (see Section 3.7).

Question 6.

For a fixed (k, L), is there a unique (k, L)-complex X? If not, then what additional local data is needed to determine X uniquely?

If L is the complete bipartite graph $K_{m,n}$, then in many cases there is a unique (k, L)-complex X. If $k = 4$, this complex is the product of an m-valent and an n-valent tree [W1]. If $k > 4$ and either k is even or $n = m$, the unique (k, L)-complex is isomorphic to Bourdon's building $I_{p,q}$, a right-angled Fuchsian building, with $k = p$ and $L = K_{q,q}$ ([B1, Św1]; $I_{p,q}$ is discussed in Section 3.3). If $k > 4$ is odd and $n \neq m$, then there does not exist a (k, L)-complex.

On the other hand, when L is the complete graph K_n for $n \geq 4$, Ballmann-Brin [BB1] and Haglund [H1] independently constructed uncountably many nonisometric (k, L)-complexes. We now discuss these constructions. As mentioned in Section 3.7, simply connected nonpositively curved complexes can be constructed "freely" by building successive balls outward from a given cell. Provided that certain obvious local obstructions do not occur, we can glue in cells arbitrarily at each stage. Ballmann-Brin showed that for many choices of k and L, every nonpositively curved (k, L)-complex can be constructed in this manner [BB1].

In this inductive construction of a (k, L)-complex, choices may or may not arise. Let us consider the case when $k = 6$ and $L = K_4$. Then each 2-cell of a (k, L)-complex X is a regular hexagon. Each 1-cell of X is contained in three distinct hexagons. Fix a 2-cell A of X, and consider the 12 surrounding 2-cells that contain one of the six 1-cells bounding A. These 2-cells are arranged locally in two sheets, whose union is a band surrounding A. However, if one

Fig. 2. If the union of the 12 hexagons surrounding A is a Möbius band, then the holonomy around the boundary of A is nontrivial.

follows the sheets around the boundary of A, there are two cases, depending on whether the union of the 12 hexagons is an annulus, or is the Möbius band shown in Figure 2.

To describe and analyze this phenomenon, Haglund [H3] introduced the notion of *holonomy*, which measures the twisting of the 2-cells neighboring a given 2-cell C as one traverses the boundary cycle of C. In many cases, the choices of holonomies around each 2-cell uniquely determine the isomorphism type of a nonpositively curved (k, L)-complex. The existence of holonomies depends on combinatorial properties of the graph L.

For instance, when $n \geq 4$, the complete graph $L = K_n$ admits nontrivial holonomies. Roughly speaking, Ballmann-Brin and Haglund constructed uncountably many (k, K_n)-complexes by showing that, at each stage, a countable number of holonomies can be specified arbitrarily. In particular, K_4 has a unique nontrivial holonomy, which is illustrated in Figure 2. The unique $(6, K_4)$-complex with trivial holonomies around every 2-cell is the Cayley complex for the presentation $\langle a, b \mid ba^2 = ab^2 \rangle$, which defines the Geisking 3-manifold group. The unique $(6, K_4)$-complex with nontrivial holonomies around every 2-cell is the Cayley complex for the presentation $\langle a, b \mid aba^2 = b^2 \rangle$, which is δ-hyperbolic (see [BC] for more details). On the other hand, the complete bipartite graph $L = K_{m,n}$ admits only the trivial holonomy, which explains why there is a unique (k, L)-complex in this case.

Świątkowski [Św1] considered (k, L)-complexes X where L is a trivalent graph and X has *Platonic symmetry*, that is, $\mathrm{Aut}(X)$ acts transitively on the set of flags (vertex, edge, face) in X. He found elementary graph-theoretic conditions on L that imply that such an X is unique. Januszkiewicz-Leary-Valle-Vogeler [JLVV] classify Platonic (k, L)-complexes X in which L is a complete graph. Their main results are for finite complexes X.

In general, holonomies are not enough to uniquely determine a (k, L)-complex. For instance, Haglund has observed that the Euclidean buildings for SL$(3, \mathbf{Q}_p)$ and SL$(3, \mathbf{F}_p((t)))$ are $(3, L)$-complexes with the same link L and the same holonomies. Yet the buildings are not isomorphic, by Theorem 5.

Nonclassical buildings with given local structures have been studied by Gaboriau-Paulin and Haglund-Paulin, who proved results analogous to those for (k, K_n)-complexes and $(k, K_{m,n})$-complexes discussed above. If $q > 4$ is a prime power, Gaboriau-Paulin [GP] proved that for every hyperbolic Coxeter polygon P with all vertex angles $\pi/6$, and for every prime power $q > 4$, there exist uncountably many hyperbolic buildings with chambers P such that the links of vertices are all isomorphic to the building for the projective plane over the finite field $\mathbf{F}_q$. On the other hand, Haglund-Paulin [HP2] showed that if (W, I) is a right-angled Coxeter system and $(q_i)_{i \in I}$ is a collection of cardinalities, then there exists a unique building X of type (W, I) such that for each $i \in I$, each codimension 1-cell containing a vertex of type $\{i\}$ in X is a face of q_i distinct chambers. This generalizes the result that Bourdon's building $I_{p,q}$ is the unique $(p, K_{q,q})$-complex.

In many cases it is still unknown how much local data is required to uniquely specify a (k, L)-complex.

4.2. Nondiscreteness of Aut(X)

Let X be a locally finite, nonpositively curved polyhedral complex. The most basic question about the locally compact group $G = \text{Aut}(X)$ is whether it is discrete. Recall that in the compact-open topology, the group $G = \text{Aut}(X)$ is nondiscrete exactly when, for each positive integer n, there is an element $g_n \in G$, with g_n fixing pointwise the ball of radius n in X, and $g_n \neq \text{Id}$. The theory of lattices in a discrete group is trivial, hence this issue is of crucial importance. We again focus on the case of (k, L)-complexes (see Section 3.7).

Question 7.

Given a (k, L)-complex X, is $G = \text{Aut}(X)$ discrete?

The answer is known in certain cases, and is closely related to the notion of a flexible complex.

DEFINITION: *A complex X is flexible if there exists $\phi \in \text{Aut}(X)$ such that ϕ fixes the star of some vertex in X but $\phi \neq \text{Id}$.*

Flexibility was introduced by Ballmann-Brin in [BB1]. If X is locally finite and not flexible, then the stabilizer of any vertex $v \in X$ is finite, since an automorphism of X that fixes v is uniquely determined by its action on the link of v. In particular, Aut(X) is discrete if X is not flexible. The following result is nearly immediate from the definition of flexibility.

THEOREM 8. (DISCRETENESS CRITERION) *If the graph L is not flexible, then no (k, L)-complex X is flexible, and* Aut(X) *is discrete.*

Theorem 8 has the following converse when $X = X(2k, L)$ is the Davis-Moussong complex for the Coxeter group $W = W(k, L)$, discussed in Section 3.6. The result was proved independently by Haglund and Świątkowski in the case that X is 2-dimensonal [H2, Św1], and was extended to arbitrary Coxeter systems by Haglund-Paulin [HP1].

THEOREM 9. (NONDISCRETENESS CRITERION) *Suppose L is a finite simplicial graph and $k \geq 2$. Let $X = X(2k, L)$ be the Davis-Moussong complex for the Coxeter group $W = W(k, L)$. If L is flexible, then* Aut(X) *is nondiscrete.*

The proof of Theorem 9 relies on the fact that Davis-Moussong complexes have numerous symmetries. For other (k, L)-complexes, particularly those with k odd, much less is known. It is not clear whether this reflects the limitations of our techniques, or actual differences in behavior for k odd and k even.

4.3. Simplicity and Nonlinearity

Let X be a locally finite, nonpositively curved polyhedral complex, with locally compact automorphism group $G = $ Aut(X). In this section we discuss whether two basic group-theoretic properties, simplicity and (non)linearity, hold for G. We assume for this section that $G = $ Aut(X) is nondiscrete.

Question 10.

When is $G = $ Aut(X) a simple group?

For X a locally finite regular or biregular tree, Tits [Ti1] proved simplicity of the group Aut$_0$ (X) of type-preserving automorphisms of X (which is finite index in the full automorphism group $G = $ Aut(X)). Haglund-Paulin [HP1] showed that various type-preserving automorphism groups in several higher-dimensional cases are simple. We note that the method of proof of these results lies in geometric group theory.

We say that a group G is *linear* if it has a faithful representation $G \longrightarrow$ GL (n, K) for some field K. On the question of linearity, suppose X is a classical

Euclidean building, associated to the algebraic group G over a local non-Archimedean field K (see Section 3.1.1). Theorem 5 says that if char$(K) = 0$, then G is finite index in Aut(X) (and if char$(K) = p > 0$, then G is cocompact in Aut(X)). By inducing, we see in particular that when char$(K) = 0$, the group Aut(X) has a faithful linear representation over K. On the other hand, for several higher-dimensional complexes X that are not classical buildings, Haglund-Paulin [HP1] proved that the full automorphism group Aut(X) has no such faithful linear representation. For dim$(X) = 2$, we pose the following problem:

PROBLEM 11. Find conditions on the link L so that a (k, L)-complex X has linear automorphism group.

Haglund [H5] has recently shown that Aut(X) is nonlinear for certain Fuchsian buildings X (see Section 3.3). Is it possible that linearity of Aut(X) characterizes those X that are classical Euclidean buildings among all nonpositively curved X?

5. Comparisons with Linear Groups

While one expects some of the phenomena and results from the theory of linear groups $G \subset$ GL$(n, \mathbf{C})$ to hold for the group $G =$ Aut(X) and its lattices, most of the methods from that theory are unavailable in this new context. There are no eigenvalues or traces. There are no vectors to act on. It therefore seems important to attack such questions, as they will (hopefully) force us to come up with new methods and tools.

One new approach to the study of automorphism groups of nonpositively curved polyhedral complexes is the structure theory of totally disconnected locally compact groups (see the survey [W]). An example of this approach is the computation of the flat rank of automorphism groups of buildings with sufficiently transitive actions [BRW].

5.1. Some Linear-Type Properties

One of the basic properties of linear groups G is the *Tits alternative*: any finitely generated linear group either contains a nonabelian free group or has a solvable subgroup of finite index (see [Ti2]). The following problem is well known.

PROBLEM 12. Let X be a nonpositively curved polyhedral complex. Prove that finitely generated subgroups of $G =$ Aut(X) satisfy the Tits alternative.

When X is a CAT(-1) space, uniform lattices in $G = \mathrm{Aut}(X)$ are word hyperbolic, and thus satisfy the Tits alternative (Gromov [Gr]). The usual ping-pong argument for the Tits alternative requires strong expanding/contracting behavior for the action of isometries of X on the visual boundary ∂X. The difficulty with Problem 12 lies in the fact that if X is just nonpositively curved, rather than negatively curved, this behavior on ∂X is not strong enough to immediately allow for the usual ping-pong argument to work.

The Iwasawa decomposition KAN of a semisimple Lie group G plays a fundamental role in the representation theory of G. Here, K is a compact subgroup, A is abelian, and N is nilpotent. In the topology on $G = \mathrm{Aut}(X)$, where X is a locally finite polyhedral complex, the stabilizers of vertices are maximal compact subgroups.

Question 13.
For which X does $G = \mathrm{Aut}(X)$ have a KAN structure?

Answering this question might be a first step toward investigating various analytic properties of X, the group $G = \mathrm{Aut}(X)$, and its lattices. For instance, random walks on classical buildings have been studied using the representation theory of the associated algebraic group (see, for example, Cartwright-Woess [CW] and Parkinson [Pa]), but for more general complexes X this machinery is not available.

Kazhdan proved that simple Lie groups G have *property (T)*: the trivial representation is isolated in the unitary dual of G (see, for example, [Ma]). Ballmann-Świątkowski [BŚ], Żuk [Żu], and Dymara-Januszkiewicz [DJ] have proven that many $G = \mathrm{Aut}(X)$ satisfy this important property.

Question 14.
For which X does $\mathrm{Aut}(X)$ have property (T)?

We remark that a locally compact topological group G has property (T) if and only if any of its lattices has property (T).

One of the deepest theorems about irreducible lattices Γ in higher-rank semisimple Lie groups is Margulis's normal subgroup theorem (see [Ma]), which states that any normal subgroup of Γ is finite or has finite index in Γ.

Question 15.
For which X does a normal subgroup theorem hold for $\mathrm{Aut}(X)$?

Such a theorem has been shown for products of trees by Burger-Mozes [BM].

Recall that the *Frattini subgroup* $\Phi(\Gamma)$ of a group Γ is the intersection of all maximal subgroups of Γ. Platonov [Pl] proved that $\Phi(\Gamma)$ is nilpotent for every finitely generated linear group. Ivanov [I] proved a similar result for mapping class groups. Kapovich [K] proved that $\Phi(\Gamma)$ is finite for finitely generated subgroups of finitely generated word-hyperbolic groups.

PROBLEM 16. Compute the Frattini subgroup $\Phi(\Gamma)$ for finitely generated subgroups $\Gamma < \mathrm{Aut}\,(X)$.

Part of the fascination of lattices in $\mathrm{Aut}\,(X)$ is that they exhibit a mixture of rank 1 and higher-rank behavior. Ballmann-Eberlein (see [Eb]) defined an invariant rank (Γ), called the *rank* of Γ, which is defined for any finitely generated group Γ as follows. Let Γ_i denote the set of elements $g \in \Gamma$ so that the centralizer of g contains $\mathbf{Z}^d$ for some $d \leq i$ as a finite-index subgroup. Let $r(\Gamma)$ be defined to be the smallest i so that Γ is a finite union of translates

$$\Gamma = g_1 \Gamma_i \cup \cdots \cup g_n \Gamma_i$$

for some $g_j \in \Gamma$. Then define rank (Γ) to be the maximum of $r(\Gamma')$, where Γ' runs over all finite-index subgroups of Γ.

Work of Prasad-Raghunathan shows that this notion of rank agrees with the classical one for arithmetic lattices. Ballmann-Eberlein [BE] proved that the rank of the fundamental group of a complete, finite-volume, nonpositively curved manifold M equals the geometric rank of the universal cover of M. Since centralizers of infinite-order elements in word-hyperbolic groups Γ are virtually cyclic, it is clear that rank $(\Gamma) = 1$ in these cases. Thus for nonpositively curved, connected, simply connected 2-complexes X, lattices in $\mathrm{Aut}\,(X)$ can have rank 1 and also rank 2 (the latter, for example, when X is a classical Euclidean building, discussed in Section 3.1.1).

PROBLEM 17. Compute rank (Γ) for lattices $\Gamma < \mathrm{Aut}\,(X)$.

A basic property of any finitely generated linear group is that it is residually finite. In contrast, there are lattices Γ in $G = \mathrm{Aut}\,(X)$ that are not residually finite. Indeed, Burger-Mozes [BM] have constructed, in the case when X is a product of simplicial trees, lattices that are simple groups. Wise had earlier constructed lattices for such X that are not residually finite [W2]. Kac-Moody lattices are also simple, and their buildings have arbitrarily large dimension (see [CapRem]).

PROBLEM 18. Construct a lattice Γ in $G = \text{Aut}(X)$ that is a simple group, and where X is not a product of trees.

For residual finiteness, a key case is Bourdon's building $I_{p,q}$ (see Section 3.3) whose 2-cells are right-angled hyperbolic p-gons. Wise [W3] has shown that fundamental groups of polygons of finite groups, where the polygon has at least 6 sides, are residually finite. Thus there are residually finite uniform lattices for $I_{p,q}$, $p \geq 6$, but the question of whether every uniform lattice in $I_{p,q}$ is residually finite is completely open for $p = 5$, that is, for pentagons. The question of residual finiteness of uniform lattices is open even for triangular hyperbolic buildings (see [KV]).

Question 19.

Which lattices $\Gamma < G = \text{Aut}(X)$ are residually finite?

A related but broader problem is as follows. Most of the known CAT(0) groups are residually finite, hence virtually torsion free. As remarked in Section 2.3, to date most applications of the theory of complexes of groups have used only simple complexes of groups. Now, if the fundamental group Γ of a complex of groups $G(Y)$ is virtually torsion free, then $G(Y)$ has a finite cover $G(Y') \longrightarrow G(Y)$ where all local groups of $G(Y')$ are trivial, hence $G(Y')$ is a simple complex of groups.

PROBLEM 20. (HAGLUND) Find a nonpositively curved complex of groups $G(Y)$ that is not finitely covered by a simple complex of groups. Do this in the negatively curved setting as well. Is there a CAT(0) group Γ that is not virtually the fundamental group of *any* (nonpositively curved) simple complex of groups?

5.2. Rigidity

Automorphism groups G of nonpositively curved polyhedral complexes X, and lattices $\Gamma < G$, are natural places in which to study various rigidity phenomena, extending what we know in the classical, algebraic cases. A first basic problem is to prove strong (Mostow) rigidity. In other words, one wants to understand the extent to which a lattice Γ in G determines G.

PROBLEM 21. (STRONG RIGIDITY) Let X_1 and X_2 be nonpositively curved polyhedral complexes, and let Γ_i be a lattice in $G_i = \text{Aut}(X_i)$, $i = 1, 2$. Find conditions on the X_i that guarantee that any abstract group isomorphism

$\phi \colon \Gamma_1 \longrightarrow \Gamma_2$ extends to an isomorphism $G_1 \longrightarrow G_2$. Further, determine when any two copies of Γ_i in G_i are conjugate in G_i.

Some assumptions on the X_i, for example, that every 1-cell is contained in a 2-cell, are needed to rule out obvious counterexamples.

A harder, more general problem is to prove quasi-isometric rigidity.

PROBLEM 22. (QUASI-ISOMETRIC RIGIDITY) Compute the quasi-isometry groups of nonpositively curved polyhedral complexes X. Prove *quasi-isometric rigidity* theorems for these complexes; that is, find conditions on X for which

1) Any quasi-isometry of X is a bounded distance from an isometry (automorphism), and
2) Any finitely generated group quasi-isometric to X is (a finite extension of) a cocompact lattice in Aut (X).

A standard trick due to Cannon-Cooper shows that (1) implies (2). It is also immediate from Mostow's original argument that (1) implies strong rigidity. Quasi-isometric rigidity was proven in the case of Euclidean buildings by Kleiner-Leeb [KL]. Bourdon-Pajot [BP] proved quasi-isometric rigidity for Bourdon's building $I_{p,q}$, and Xie [X] generalized this to Fuchsian buildings (see Section 3.3). One would expect that higher-dimensional buildings would be more rigid, and indeed they seem to be harder to construct, so they might be a good place to look for rigidity phenomena.

Another kind of rigidity problem follows:

PROBLEM 23. Suppose X_1 and X_2 are locally finite, connected, simply connected 2-complexes, such that for $i = 1, 2$, the group Aut (X_i) acts cocompactly on X_i. If Aut (X_1) is isomorphic to Aut (X_2), is X_1 isometric to X_2?

A variety of other rigidity phenomena from Riemannian geometry have natural analogues in this context. Examples include rank rigidity, hyperbolic rank rigidity, minimal entropy rigidity, and marked-length-spectrum rigidity. A rank rigidity theorem for nonpositively 2-complexes was proven by Ballmann-Brin in [BB2].

5.3. Geometry of the Word Metric

One of the few results about the geometry of the word metric for nonuniform lattices is the theorem of Lubotzky-Mozes-Raghunathan [LMR], which we now discuss.

Let G be a semisimple Lie group over $\mathbf{R}$ (respectively, over a non-archimedean local field K), and let X be the associated symmetric space (respectively, Euclidean building). Thus X is a nonpositively curved Riemannian manifold (respectively, simplicial complex) on which G acts by isometries. Let Γ be a lattice in G. If K is non-archimedean and G has rank 1 over K, then nonuniform lattices in G are not finitely generated. On the other hand, when G is either a real Lie group or a non-archimedean group with K-rank at least 2, then all lattices Γ in G are finitely generated. In this section, we consider only the finitely generated case, and we endow Γ with the word metric for a finite-generating set.

If Γ is uniform, then the natural map $\psi \colon \Gamma \longrightarrow X$ sending Γ to any of its orbits is a quasi-isometry. When Γ is nonuniform, the orbit map is never a quasi-isometry, since the quotient $\Gamma \backslash X$ is noncompact. When G has real rank 1, the map ψ is not even a quasi-isometric embedding, as can be seen by considering any nonuniform lattice acting on real hyperbolic space. In this case, the maximal parabolic subgroups of Γ are exponentially distorted in X.

The theorem of Lubotzky-Mozes-Raghunathan [LMR] states that, when G has real rank (respectively, K-rank) at least 2, then ψ is indeed a quasi-isometric embedding. Each of the known proofs of this result is heavily algebraic, depending on the structure of matrix groups. Thus the following problem presents an interesting challenge, even in terms of giving a geometric proof in the (non-archimedean) algebraic case.

QUESTION 24. Let Γ be a finitely generated nonuniform lattice in the automorphism group of a nonpositively curved polyhedral complex X. When is the natural map $\psi \colon \Gamma \longrightarrow X$, sending Γ to any of its orbits, a quasi-isometric embedding?

When X is a product of trees, ψ need not be a quasi-isometric embedding. When X is not a product of trees, is ψ always a quasi-isometric embedding?

5.4. Dynamics

Let G be (any) locally compact topological group, equipped with Haar measure, and let Γ be a lattice in G. Then G acts on the left on G/Γ, preserving the finite measure on G/Γ induced by the Haar measure on G. We thus obtain an action of every closed subgroup $H < G$ on G/Γ. It is a basic question in understanding these dynamical systems, in particular to determine when the action of H on G/Γ is *ergodic*; that is, when every H-invariant set has 0 or full measure. When G is a semisimple Lie group with no compact factors, and Γ

is an irreducible lattice in G, *Moore's ergodicity theorem* (see [Zi]) states that the H-action on G/Γ is ergodic if and only if H is noncompact.

Now let X be a simply connected, locally finite polyhedral complex of nonpositive curvature. Equip $G = \mathrm{Aut}\,(X)$ with left-invariant Haar measure, and let Γ be a lattice in G.

PROBLEM 25. Determine which closed subgroups of $G = \mathrm{Aut}\,(X)$ act ergodically on G/Γ.

One reason we consider Problem 25 to be worthwhile is that the usual method of proving Moore's ergodicity theorem uses the unitary representation theory of G. We thus believe that, apart from being interesting in its own right, attempts to solve Problem 25 will require us either to find new approaches to Moore's theorem, or to develop the unitary representation theory of $G = \mathrm{Aut}\,(X)$.

6. Lattices in Aut(X)

In this section we consider properties of the lattices in $\mathrm{Aut}\,(X)$ themselves. Some lattice properties have already been mentioned in Section 5, on comparisons with linear groups. Here, we discuss topics where new phenomena, contrasting with classical cases, have already been observed, and where the known techniques of proof are combinatorial or geometric in flavor.

6.1. Existence and Classification Theorems

Given a locally compact group G, the most basic question in the lattice theory of G is whether G admits a uniform or nonuniform lattice.

For algebraic groups, the existence of both uniform and nonuniform lattices was settled by Borel and others, using arithmetic constructions (see the final paragraph of Section 3.1.1). For automorphism groups of trees, precise conditions are known for the existence of both uniform lattices (Bass-Kulkarni [BK]) and nonuniform lattices (Bass-Carbone-Rosenberg, in [BL]). In Section 3, for each example X of a polyhedral complex, we described known constructions of lattices in $G = \mathrm{Aut}\,(X)$. These constructions are nonarithmetic, for X not a classical building. The following question is still largely open.

Question 26.

When does $G = \mathrm{Aut}\,(X)$ admit a uniform lattice? A nonuniform lattice?

A special case of this question is

Question 27.
For which positive integers $k \geq 3$ and finite simplicial graphs L does the automorphism group of a (k, L)-complex X admit lattices?

Once one establishes the existence of lattices in a given $G = \text{Aut}(X)$, the next problem is to classify all such lattices. We discuss commensurability of lattices in Section 6.2. An even more fundamental problem is

PROBLEM 28. Classify lattices in $G = \text{Aut}(X)$ up to conjugacy.

We note that in the case of real Lie groups, classification theorems are difficult. For SO $(3, 1)$, for example, the classification is precisely the classification of all finite-volume, complete hyperbolic orbifolds. On the other hand, for higher-rank real (and p-adic) semisimple Lie groups, Margulis's arithmeticity theorem (see [Ma]) states that all lattices are arithmetic, and arithmetic lattices can in some sense be classified (although this is also not easy). So, even solving Problem 28 in any special case, for example, for specific hyperbolic buildings, would be of great interest.

6.2. Commensurability
One of the basic problems about a locally compact topological group G is to classify its lattices up to commensurability. Recall that two lattices $\Gamma_1, \Gamma_2 \leq G$ are *commensurable in* G if there exists $g \in G$ so that $g\Gamma_1 g^{-1} \cap \Gamma_2$ has finite index in both $g\Gamma_1 g^{-1}$ and Γ_2. Since covolume is multiplicative in index, two commensurable lattices have covolumes that are commensurable real numbers, that is, they have a rational ratio.

PROBLEM 29. Classify lattices in $G = \text{Aut}(X)$ up to commensurability. As a subproblem, find commensurability invariants of lattices.

If G is an algebraic group of rank at least 2 over a non-Archimedean local field K, then there exist noncommensurable arithmetic lattices in G. If G is a rank-1 simple real Lie group, then lattices are again not all commensurable, as there exist both arithmetic and nonarithmetic lattices.

For $G = \text{Aut}(X)$, commensurability of uniform lattices is strikingly different. When X is a locally finite tree, Leighton proved in [Lei] that all torsion-free uniform lattices in $G = \text{Aut}(X)$ are commensurable. The torsion-free hypothesis was removed by Bass–Kulkarni in [BK], establishing that there is at most one commensurability class of uniform lattices in the tree case. Haglund [H5] has shown the same result for many Fuchsian buildings (see Section 3.3). He has also found a sufficient condition for a uniform lattice in the automorphism group of a Davis-Moussong complex X (see Section 3.6) to be commensurable to the corresponding Coxeter group W. As specific instances of Problem 29, we have

PROBLEM 30. Suppose X is a (k, L)-complex. Find conditions on L such that all uniform lattices in Aut (X) are commensurable, and find examples of such L.

And on the other hand:

PROBLEM 31. (HAGLUND) Find a Gromov-hyperbolic CAT(0) complex X such that Aut (X) admits two noncommensurable uniform lattices.

For nonuniform lattices in $G = $ Aut (X), the situation seems much wilder. Even in the tree case, there seems to be a great deal of flexibility in the construction of nonuniform lattices. For instance, Farb-Hruska [FH] have shown that when X is the biregular tree there are uncountably many commensurability classes of nonuniform lattices in $G = $ Aut (X) with any given covolume $v > 0$. To prove this result, they construct several new commensurability invariants, and then evaluate them on lattices constructed using graphs of groups.

A similar result holds when X is a right-angled building (see Section 3.4), by work of Thomas [Th2]. Lattices in right-angled hyperbolic buildings, such as Bourdon's building $I_{p,q}$, are known to exhibit higher-rank phenomena, such as quasi-isometric rigidity (see [BP] and Section 5.2). In contrast, Thomas's theorem indicates a similarity of these lattices with tree lattices. In fact, Thomas proves this theorem by constructing a functor that takes tree lattices to lattices in right-angled buildings. This functor preserves many features of the lattice.

The most important commensurability invariant of a group Γ inside a group G is the *commensurator* $\text{Comm}_G (\Gamma)$ of Γ in G, defined by

$$\text{Comm}_G (\Gamma) := \{ g \in G \mid \Gamma \cap g\Gamma g^{-1} \text{ has finite index in both } \Gamma \text{ and } g\Gamma g^{-1} \}.$$

Margulis proved that a lattice Γ in a semisimple Lie group G is arithmetic if and only if $\text{Comm}_G (\Gamma)$ is dense in G (see [Zi]). Lubotzky proposed this density property as a definition of "arithmeticity" when $G = $ Aut (X).

PROBLEM 32. For lattices Γ in $G = \text{Aut}(X)$, compute $\text{Comm}_G(\Gamma)$. Determine whether $\text{Comm}_G(\Gamma)$ is dense in G.

When X is a tree, density of commensurators of uniform lattices was proved by Bass-Kulkarni [BK] and Liu [Liu]. Haglund established density of commensurators of uniform lattices for many Davis-Moussong complexes in[H2], and Haglund [H6] and independently Barnhill-Thomas [KBT] have recently shown the same result for right-angled buildings. For commensurators of nonuniform lattices, however, even for trees very little is known (see [BL]).

6.3. Finiteness Properties of Lattices

Uniform lattices in $G = \text{Aut}(X)$ are always finitely generated, for obvious reasons. However, nonuniform lattices need not be finitely generated.

QUESTION 33. For which $G = \text{Aut}(X)$ are all nonuniform lattices non-finitely generated? Do there exist G that admit both finitely generated and nonfinitely generated nonuniform lattices?

Higher-rank algebraic groups, such as $G = \text{SL}(3, (\mathbf{F}_q((t)))$, have Kazhdan's property (T) (see Section 5.1). Furthermore, property (T) is inherited by lattices, and all countable groups with property (T) are finitely generated. Therefore, lattices in higher-rank groups are all finitely generated.

On the other hand, if X is a tree, every nonuniform lattice in $\text{Aut}(X)$ is non-finitely generated [BL]. Thomas's functor mentioned in Section 6.2 implies that many nonuniform lattices in right-angled hyperbolic buildings are nonfinitely generated as well.

CONJECTURE 34. *Let Γ be a nonuniform lattice in $G = \text{Aut}(X)$, where X is any right-angled hyperbolic building. Then Γ is not finitely generated.*

We are starting to believe that finite generation of nonuniform lattices in 2-complexes is actually a miracle, and could even characterize the remarkable nonuniform lattices in algebraic groups in characteristic $p > 0$. Even these lattices are not finitely presentable, and so we make the following:

CONJECTURE 35. *If Γ is a nonuniform lattice in $G = \text{Aut}(X)$, where X is a locally finite polyhedral complex, then Γ is not finitely presentable.*

6.4. Covolumes

One of the more striking ways in which the study of lattices in Aut (X) diverges from the case of lattices in semisimple Lie groups is the study of covolumes of lattices in a fixed Aut (X). New phenomena are seen to occur right away, and much remains to be understood.

PROBLEM 36. *Given* $G = \text{Aut}(X)$ *with Haar measure* μ, *describe the set of covolumes*

$$\mathcal{V}(G) := \{\, \mu(\Gamma \backslash G) \mid \Gamma \text{ is a lattice in } G \,\}.$$

Note that $\mathcal{V}(G)$ is a set of positive real numbers.

If G is a noncompact simple real Lie group, such as PSL $(n, \mathbf{R})$, then the set $\mathcal{V}(G)$ has positive lower bound (Kazhdan-Margulis, [KM]) and in most cases is discrete (see [Lu] and the references therein). If G is a higher-rank algebraic group over a non-Archimedean local field, such as PSL $(n, \mathbf{Q}_p)$ with $n \geq 3$, the strong finiteness result of Borel-Prasad [BPr] implies that for any $c > 0$, there are only finitely many lattices in G with covolume less than c. Hence $\mathcal{V}(G)$ is discrete, has positive lower bound, and for any $v \in \mathcal{V}(G)$ there are only finitely many lattices of covolume v.

The set of covolumes for tree lattices is very different. Suppose G is the group of automorphisms of a regular locally finite tree. Then, for example, Bass-Kulkarni [BK] showed that $\mathcal{V}(G)$ contains arbitrarily small elements, by constructing a tower of uniform lattices (see Section 6.5). Bass-Lubotzky [BL] showed that the set of nonuniform covolumes is $(0, \infty)$.

A few higher-dimensional nonclassical cases have been studied. In [Th2] and [Th3], Thomas considered covolumes for, respectively, right-angled buildings and certain Fuchsian buildings (see Sections 3.4 and 3.3, respectively). In both these settings, $\mathcal{V}(G)$ shares properties, such as nondiscreteness, with covolumes of tree lattices, even though such buildings also have some rigidity properties typical of classical cases (see Section 5.2). Little is known about covolumes for X not a building. In [Th4], the class of (k, L)-complexes X of Platonic symmetry (introduced by Świątkowski [Św1]; see Section 3.7) is considered. A sample result is that if $k \geq 4$ is even and L is the Petersen graph, then $\mathcal{V}(G)$ is nondiscrete. Many cases are completely open.

From a different point of view, Prasad [Pr] gave a computable formula for the covolumes of lattices Γ in algebraic groups G over non-archimedean local fields. This formula is in terms of discriminants of field extensions and numbers of roots. If Γ is viewed instead as a lattice in Aut (X), where X is the building associated to the algebraic group G, we also have Serre's more geometrically flavored formula for the covolume of Γ, stated in the introduction.

Question 37.

Can Serre's geometric formula for covolumes tell us anything new about lattices in classical cases?

More generally, using Serre's geometric formula, in [Th1] Thomas established a computable number-theoretic restriction on the set of covolumes of uniform lattices, for all locally finite X with $G = \text{Aut}(X)$ acting cocompactly, in all dimensions.

PROBLEM 38. Suppose $\nu > 0$ satisfies the restriction of [Th1]. Construct a uniform lattice in G of covolume ν, or show that such a lattice does not exist. Also, find the cardinality of the set of uniform lattices of covolume ν. For nonuniform lattices, the same questions for any $\nu > 0$.

This problem was solved for right-angled buildings (see Section 3.4) in [Th2].

The properties of the set of volumes of hyperbolic 3-manifolds are well understood (see [Thu]), and one could investigate whether similar properties hold for volumes of lattices in $\text{Aut}(X)$. For instance, for every nonuniform lattice Γ in $\text{SO}(3, 1)$, there is a sequence of uniform lattices with covolumes converging to that of Γ, obtained by Dehn surgery. This gives a surjective homomorphism from Γ to each of these uniform lattices. It is not known whether any nonuniform lattices in $\text{Aut}(X)$ surject onto uniform lattices.

6.5. Towers

The study of towers of lattices is closely related to covolumes (Section 6.4). A *tower* of lattices in a locally compact group G is an infinite strictly ascending sequence

$$\Gamma_1 < \Gamma_2 < \cdots < \Gamma_n < \cdots < G$$

where each Γ_n is a lattice in G.

Question 39.

Does $G = \text{Aut}(X)$ admit a tower of (uniform or nonuniform) lattices?

If G admits a tower, then the covolumes of lattices in this tower tend to 0, hence the set $\mathcal{V}(G)$ of covolumes does not have positive lower bound. It follows that in classical (algebraic) cases, G does not admit any towers, by the Kazhdan-Margulis theorem in Section 6.4.

The first examples of towers of tree lattices are due to Bass-Kulkarni [BK]. Generalizing these constructions, Rosenberg [Ros] proved that if X is a tree such that $\operatorname{Aut}(X)$ is nondiscrete and admits a uniform lattice, then $\operatorname{Aut}(X)$ admits a tower of uniform lattices. Carbone-Rosenberg [CR] considered nonuniform lattice towers in $\operatorname{Aut}(X)$ for X a tree, showing that, with one exception, if $\operatorname{Aut}(X)$ admits a nonuniform lattice, then it admits a tower of nonuniform lattices.

In higher dimensions, for X a right-angled building (see Section 3.4) Thomas [Th2] constructed a tower of uniform and of nonuniform lattices. Other higher-dimensional cases are open. In particular, it is not known whether the automorphism groups of any Fuchsian buildings that are not right angled (see Section 3.3) admit towers.

A finer version of Question 6.5 is the following:

Question 40.
Does G admit a tower of *homogeneous* lattices, that is, lattices acting transitively on cells of maximum dimension in X?

For $X = T_{p,q}$ the (p, q)-biregular tree, if p or q is composite, there is a homogeneous tower in $G = \operatorname{Aut}(X)$ (Bass-Kulkarni [BK]). When X is the 3-regular tree, a deep theorem of Goldschmidt [Go] implies that G does not admit such a tower, since G contains only finitely many conjugacy classes of edge-transitive lattices. The Goldschmidt-Sims conjecture (see [Gl]), which remains open, is that if p and q are both prime, then there are only finitely many conjugacy classes of homogeneous lattices in $\operatorname{Aut}(T_{p,q})$. If X is the product of two trees of prime valence, Glasner [Gl] has shown that there are only finitely many conjugacy classes of (irreducible) homogeneous lattices in $G = \operatorname{Aut}(X)$. For all other higher-dimensional X, the question is open.

Question 41.
Does G admit maximal lattices?

In the algebraic setting, lattices of minimal covolume are known in many cases (see [Lu] and its references), and so these lattices are maximal. Examples of maximal lattices in $G = \operatorname{Aut}(X)$ are some of the edge-transitive lattices for X the 3-regular tree, classified by Goldschmidt [Go].

A coarse version of the question of towers is

Question 42. (Lubotzky)

Let Γ be a uniform lattice in $G = \text{Aut}(X)$. Define

$$u_\Gamma(n) = \#\{\,\Gamma' \mid \Gamma' \text{ is a lattice containing } \Gamma, \text{ and } [\Gamma' : \Gamma] = n\}.$$

By similar arguments to [BK], $u_\Gamma(n)$ is finite. What are the asymptotics of $u_\Gamma(n)$?

The case X a tree was treated by Lim [Lim]. If X is the building associated to a higher-rank algebraic group, then for any Γ, we have $u_\Gamma(n) = 0$ for $n >> 0$, since $\mathcal{V}(G)$ has positive lower bound. In contrast, if $\text{Aut}(X)$ admits a tower of lattices (for example, if X is a right-angled building), there is a Γ with $u_\Gamma(n) > 0$ for arbitrarily large n. Lim-Thomas [LT], by counting coverings of complexes of groups, found an upper bound on $u_\Gamma(n)$ for very general X, and a lower bound for certain right-angled buildings X. It would be interesting to sharpen these bounds for particular cases.

6.6. Biautomaticity of Lattices

The theory of automatic and biautomatic groups is closely related to nonpositive curvature. All word-hyperbolic groups are biautomatic [ECHLPT]. Yet it is not known whether an arbitrary group acting properly, cocompactly, and isometrically on a CAT(0) space is biautomatic, or even automatic. Indeed, the following special case is open:

Question 43

Suppose a group Γ acts properly, cocompactly, and isometrically on a CAT(0) piecewise Euclidean 2-complex. Is Γ biautomatic? Is Γ automatic?

Biautomaticity is known in several cases for groups acting on complexes built out of restricted shapes of cells. Gersten-Short established biautomaticity for uniform lattices in CAT(0) 2-complexes of type $\tilde{A}_1 \times \tilde{A}_1$, $\tilde{A}_2$, $\tilde{B}_2$, and $\tilde{G}_2$ in [GS1] and [GS2]. In particular, Gersten-Short's work includes CAT(0) square complexes, 2-dimensional systolic complexes, and 2-dimensional Euclidean buildings.

Several special cases of Gersten-Short's theorem have been extended. For instance, Świątkowski proved that any uniform lattice in a Euclidean building is biautomatic [Św2]. Niblo-Reeves [NRe1] proved biautomaticity of all uniform lattices acting on CAT(0) cubical complexes. In particular, this result includes all finitely generated right-angled Coxeter groups and right-angled Artin groups. Systolic groups, that is, uniform lattices acting on arbitrary systolic simplicial complexes, are also biautomatic by work of Januszkiewicz-Świątkowski [JŚ1].

Gersten-Short's work applies only to 2-complexes with a single shape of 2-cell. Levitt has generalized Gersten-Short's theorem to prove biautomaticity of any uniform lattice acting on a CAT(0) triangle-square complex, that is, a 2-complex each of whose 2-cells is either a square or an equilateral triangle [Lev].

Epstein proved that all nonuniform lattices in $SO(n, 1)$ are biautomatic [ECHLPT, 11.4.1]. Rebbechi [Reb] showed more generally that a relatively hyperbolic group is biautomatic if its peripheral subgroups are biautomatic. Finitely generated virtually abelian groups are biautomatic by [ECHLPT, §4.2]. It follows from work of Hruska-Kleiner [HK] that any uniform lattice acting on a CAT(0) space with isolated flats is biautomatic.

By a theorem of Brink-Howlett, all finitely generated Coxeter groups are automatic [BHo]. Biautomaticity has been considerably harder to establish, and remains unknown for arbitrary Coxeter groups. Biautomatic structures exist when the Coxeter group is affine, that is, virtually abelian, and also when the Coxeter group has no affine parabolic subgroup of rank at least 3 by a result of Caprace-Mühlherr [CM]. Coxeter groups whose Davis-Moussong complex has isolated flats are also biautomatic by [HK]. The Coxeter groups with isolated flats have been classified by Caprace [Cap].

Let W be a Coxeter group, and let X be a building of type W. Świątkowski [Św2] has shown that any uniform lattice Γ in $G = \text{Aut}(X)$ is automatic. If W has a geodesic biautomatic structure, he shows that Γ is biautomatic as well. Together with Caprace's work mentioned above it follows that if W is a Coxeter group with isolated flats, then Γ is biautomatic [Cap]. This consequence can be seen in two ways: using the fact that W is biautomatic, or alternately using the fact, established by Caprace, that Γ is relatively hyperbolic with respect to uniform lattices in Euclidean buildings.

References

[AB]   P. Abramenko and K. Brown, *Buildings: Theory and Applications*, Springer Science+Business Media, New York, 2008.

[BB1]   W. Ballmann and M. Brin, *Polygonal complexes and combinatorial group theory*, Geom. Dedicata **50** (1994), 165–191.

[BB2]   W. Ballmann and M. Brin, *Orbihedra of nonpositive curvature*, Inst. Hautes Études Sci. Publ. Math. **82** (1995), 169–209.

[BE]   W. Ballmann and P. Eberlein, *Fundamental groups of manifolds of nonpositive curvature*, J. Differential Geom. **25** (1987), 1–22.

[BŚ]   W. Ballmann and J. Świątkowski, *On L^2-cohomology and property (T) for auto-morphism groups of polyhedral cell complexes*, Geom. Funct. Anal. **7** (1997), 615–645.

[Ba]   S. Barré, *Immeubles de Tits triangulaires exotiques*, Ann. Fac. Sci. Toulouse Math. (6) **9** (2000), 575–603.

[BK]   H. Bass and R. Kulkarni, *Uniform tree lattices*, J. Amer. Math. Soc **3** (1990), 843–902.

[BL]   H. Bass and A. Lubotzky, *Tree Lattices*. Prog. in Math., **176**, Birkhäuser, Boston.

[BRW]   U. Baumgartner, B. Rémy, and G. A. Willis, *Flat rank of automorphism groups of buildings*, Transform. Groups **12** (2007), 413–436.

[BHar]   A. Borel and G. Harder, *Existence of discrete cocompact subgroups of reductive groups over local fields*, J. Reine Angew. Math. **298** (1978), 53–64.

[BPr]   A. Borel and G. Prasad, *Finiteness theorems for discrete subgroups of bounded covolume in semi-simple groups*, Inst. Hautes Études Sci. Publ. Math., 69, 1989, 119–171.

[B1]   M. Bourdon, *Immeubles hyperboliques, dimension conforme et rigidité de Mostow*, Geom. Funct. Anal. **7** (1997), 245–268.

[B2]   M. Bourdon, *Sur les immeubles fuchsiens et leur type de quasi-isométrie*, Ergod. Th. & Dynam. Sys. **20** (2000), 343–364.

[BP]   M. Bourdon and H. Pajot, *Rigidity of quasi-isometries for some hyperbolic buildings*, Comment. Math. Helv. **75** (2000), 701–736.

[BC]   N. Brady and J. Crisp, *CAT(0) and CAT(-1) dimensions of torsion free hyperbolic groups*, Comment. Math. Helv. **82** (2007), 61–85.

[BH]   M.R. Bridson and A. Haefliger, *Metric Spaces of Non-Positive Curvature*, Springer-Verlag, Berlin, 1999.

[BHo]   B. Brink and R.B. Howlett, *A finiteness property and an automatic structure for Coxeter groups*, Math. Ann. **296** (1993), 179–190.

[Br]   K.S. Brown, *Buildings*, Springer-Verlag, New York, 1989.

[BM]   M. Burger and S. Mozes, *Lattices in products of trees*, Inst. Hautes Études Sci. Publ. Math. **92** (2000), 151–194.

[Cap]   P.-E. Caprace, *Buildings with isolated subspaces*, arXiv:math.GR/0703799.

[CM]   P.-E. Caprace and B. Mühlherr, *Reflection triangles in Coxeter groups and biautomaticity*, J. Group Theory **8** (2005), 467–489.

[CapRem] P.-E. Caprace and B. Rémy, *Simplicity and superrigidity of twin building lattices*, Invent. Math. **176** (2009), 169–221.

[CG]   L. Carbone and H. Garland, *Existence of lattices in Kac-Moody groups over finite fields*, Commun. Contemp. Math. **5** (2003), 813–867.

[CR]   L. Carbone and G. Rosenberg, *Infinite towers of tree lattices*, Math. Res. Lett. **8** (2001), 469–477.

[CMSZ]   D.I. Cartwright, A.M. Mantero, T. Steger, and A. Zappa, *Groups acting simply transitively on the vertices of a building of type $\tilde{A}_2$*, Geom. Dedicata **47** (1993), 143–166.

[CW]   D.I. Cartwright and W. Woess, *Isotropic random walks in a building of type $\tilde{A}_d$*, Math. Z. **247** (2004), 101–135.

[Ch]   V. Chepoi, *Graphs of some CAT(0) complexes*, Adv. in Appl. Math. **24** (2000), 125–179.

[D]   M.W. Davis, *Groups generated by reflections and aspherical manifolds*, Ann. of Math. **117** (1983), 293–324.

[DJ]   J. Dymara and T. Januszkiewicz, *Cohomology of buildings and their automorphism groups*, Invent. Math. **150** (2002), 579–627.

[Eb]   P. Eberlein, *Geometry of Nonpositively Curved Manifolds*, University of Chicago Press, Chicago, 1996.

[ECHLPT] D. Epstein, J. Cannon, D. Holt, S. Levy, M. Paterson, and W. Thurston, *Word Processing in Groups*, Jones and Bartlett Publishers, Boston, 1992.

[Es]   J. Essert, *A geometric construction of panel-regular lattices in buildings of types $\tilde{A}_2$ and $\tilde{C}_2$*, arXiv:0908.2713v2.

[FE]   B. Farb and A. Eskin, *Quasi-flats and rigidity in higher rank symmetric spaces*, J. Amer. Math. Soc. **10** (1997), 653–692.

[FH]   B. Farb and G.C. Hruska, *Commensurability invariants for nonuniform tree lattices*, Israel J. of Math. **152** (2006), 125–142.

[FJ]   M. Farber and R. E. Jamison, *On local convexity in graphs*, Discrete Math. **66** (1987), 231–247.

[Far]   D.S. Farley, *Finiteness and CAT(0) properties of diagram groups*, Topology **42** (2003), 1065–1082.

[GP]   D. Gaboriau and F. Paulin, *Sur les immeubles hyperboliques*, Geom. Dedicata **88** (2001), 153–197.

[Ger]   V.N. Gerasimov, *Semi-splittings of groups and actions on cubings*, in Izdat. Ross. Akad. Nauk Sib. Otd. Inst. Mat., Novosibirsk, 1997. Eng. Trans. *Fixed-point-free actions on cubings*, Siberian Adv. Math. **8** (1998), 36–58.

[GS1]   S. Gersten and H. Short, *Small cancellation theory and automatic groups*, Invent. Math. **102** (1990), 305–334.

[GS2]   S. Gersten and H. Short, *Small cancellation theory and automatic groups II*, Invent. Math. **105** (1991), 641–662.

[Gl]   Y. Glasner, *A two-dimensional version of the Goldschmidt–Sims conjecture*, J. Algebra **269** (2003), 381–401.

[Go]   D.M. Goldschmidt, *Automorphisms of trivalent graphs*, Ann. of Math. (2) **111** (1980), 377–406.

[Gr]   M. Gromov, *Hyperbolic groups*, in Essays in Group Theory, Math. Sci. Res. Inst. Publ. **8**, Springer, New York, 1987, 75–263.

[Hae]   A. Haefliger, *Complexes of groups and orbihedra*, in Group Theory from a Geometrical Viewpoint (E. Ghys, A. Haefliger, A. Verjovsky, eds)., Proc. ICTP Trieste 1990, World Scientific, Singapore, 1991, 504–540.

[H1]   F. Haglund, *Les polyèdres de Gromov*, C. R. Acad. Sci. Paris, Série I **313** (1991), 603–606.

[H2]   F. Haglund, *Réseaux de Coxeter-Davis et commensurateurs*, Ann. Inst. Fourier (Grenoble) **48** (1998), 649–666.

[H3]   F. Haglund, *Existence, unicité et homogénéité de certains immeubles hyperboliques*, Math. Z. **242** (2002), 97–148.

[H4]   F. Haglund, *Complexes simpliciaux hyperboliques de grande dimension*, Prépublication d'Orsay, November 2003.

[H5]   F. Haglund, *Commensurability and separability of quasiconvex subgroups*, Algebr. Geom. Topol. **6** (2006), 949–1024.

[H6]   F. Haglund, *Finite index subgroups of graph products*, Geom. Dedicata **135** (2008), 167–209.

[HP1]   F. Haglund and F. Paulin, *Simplicité de groupes d'automorphismes d'espaces à courbure négative*, The Epstein birthday schrift, 181–248 (electronic), Geom. Topol. Monogr., 1, Geom. Topol. Publ., Coventry, 1998.

[HP2]   F. Haglund and F. Paulin, *Constructions arborescentes d'immeubles*, Math. Ann. **325** (2003), 137–164.

[HK]   G.C. Hruska and B. Kleiner, *Hadamard spaces with isolated flats*, with an appendix by M. Hindawi, G.C. Hruska, and B. Kleiner, Geom. Topol. **9** (2005), 1501–1538.

[I]   N.V. Ivanov, *Subgroups of Teichmüller modular groups and their Frattini subgroups*, Funktsional. Anal. i Prilozhen. **21** (1987), 76–77.

[JLVV]   T. Januszkiewicz, I. Leary, R. Valle, and R. Vogeler, *Simple Platonic polygonal complexes*, in preparation.

[JŚ1]   T. Januszkiewicz and J. Świątkowski, *Hyperbolic Coxeter groups of large dimension*, Comment. Math. Helv. **78** (2003), 555–583.

[JŚ2]   T. Januszkiewicz and J. Świątkowski, *Simplicial nonpositive curvature*, Publ. Math. Inst. Hautes Études Sci. **104** (2006), 1–85.

[KV]   R. Kangaslampi and A. Vdovina, *Triangular hyperbolic buildings*, C. R. Acad. Sci. Paris, Ser. I **342** (2006), 125–128.

[K]   I. Kapovich, *The Frattini subgroups of subgroups of hyperbolic groups*, J. Group Theory **6** (2003), 115–126.

[KM]   D. Kazhdan and G. Margulis, *A proof of Selberg's hypothesis*, Mat. Sbornik **75** (1968), 162–168.

[KL]   B. Kleiner and B. Leeb, *Rigidity of quasi-isometries for symmetric spaces and Euclidean buildings*, Inst. Hautes Études Sci. Publ. Math. **86** (1997), 115–197.

[KBT]   A. Kubena Barnhill and A. Thomas, *Density of commensurators for uniform lattices of right-angled buildings*, arXiv:0812.2280.

[LW]   J. Lauer and D.T. Wise, *Cubulating one-relator groups with torsion*, in preparation.

[Lei]   F.T. Leighton, *Finite common coverings of graphs*, J. Combin. Theory Ser. B **33** (1982), 231–238.

[Lev]   R. Levitt, *Biautomaticity and triangle-square complexes*, in preparation.

[Lim]   S. Lim, *Counting overlattices in automorphism groups of trees*, Geom. Dedicata **118** (2006), 1–21.

[LT]   S. Lim and A. Thomas, *Counting overlattices for polyhedral complexes*, submitted.

[Liu]   Y.-S. Liu, *Density of the commensurability groups of uniform tree lattices*, J. Algebra **165** (1994), 346–359.

[Lu]   A. Lubotzky, *Tree-lattices and lattices in Lie groups*, Combinatorial and geometric group theory (Edinburgh, 1993), 217–232, London Math. Soc. Lecture Note Ser., 204, Cambridge University Press, Cambridge, 1995.

[LMR]   A. Lubotzky, S. Mozes, and M.S. Raghunathan, *The word and Riemannian metrics on lattices of semisimple groups*, Inst. Hautes Études Sci. Publ. Math. **91** (2000), 5–53.

[Ma]   G.A. Margulis, *Discrete Subgroups of Semi-Simple Lie Groups*, Springer-Verlag, New York, 1991.

[Mo]   G. Moussong, *Hyperbolic Coxeter groups*, Ph.D. Thesis, Ohio State University, 1988.

[NRe1]   G.A. Niblo and L.D. Reeves, *The geometry of cube complexes and the complexity of their fundamental groups*, Topology 37 (1998), 621–633.

[NRe2]   G.A. Niblo and L.D. Reeves, *Coxeter groups act on CAT(0) cube complexes*, J. Group Theory 6 (2003), 399–413.

[NRo]   G.A. Niblo and M.A. Roller, *Groups acting on cubes and Kazhdan's property (T)*, Proc. Amer. Math. Soc. 126 (1998), 693–699.

[Pa]   J. Parkinson, *Isotropic random walks on affine buildings*, Ann. Inst. Fourier (Grenoble) 57 (2007), 379–419.

[Pl]   V.P. Platonov, *Frattini subgroup of linear groups and finitary approximability*, Dokl. Akad. Nauk SSSR 171 (1966), 798–801.

[PV]   L. Potyagailo and E. Vinberg, *On right-angled reflection groups in hyperbolic spaces*, Comment. Math. Helv. 80 (2005), 63–73.

[Pr]   G. Prasad, *Volumes of S-arithmetic quotients of semi-simple groups*, Inst. Hautes Études Sci. Publ. Math. 69 (1989), 91–114.

[Reb]   D.Y. Rebbechi, *Algorithmic properties of relatively hyperbolic groups*, Ph.D. thesis, Rutgers University, 2001, arXiv:math.GR/0302245.

[Rem1]   B. Rémy, *Constructions de réseaux en theorie de Kac-Moody*, C. R. Acad. Sci. Paris 329 (1999), 475–478.

[Rem2]   B. Rémy, *Immeubles de Kac-Moody hyperboliques. Isomorphismes abstraits entre groupes de même immeuble*, Geom. Dedicata 90 (2002), 29–44.

[RR]   B. Rémy and M. Ronan, *Topological groups of Kac-Moody type, right-angled twinnings and their lattices*, Comment. Math. Helv. 81 (2006), 191–219.

[Ron1]   M.A. Ronan, *A construction of buildings with no rank 3 residues of spherical type*, Buildings and the geometry of diagrams (Como, 1984), 242–248, Lecture Notes in Math. 1181, Springer, Berlin, 1986.

[Ron2]   M.A. Ronan, *Lectures on Buildings*, Perspectives in Mathematics, 7, Academic Press, Boston, 1989.

[Ros]   G.E. Rosenberg, *Towers and Covolumes of Tree Lattices*, Ph.D. Thesis, Columbia University, 2001.

[Sa]   M. Sageev, *Ends of group pairs and non-positively curved cube complexes*, Proc. London Math. Soc. 71 (1995), 585–617.

[Se]   J.-P. Serre, *Cohomologie des groupes discrets*, Ann. of Math. Studies, no. 70, Princeton University Press, Princeton, NJ, 1971, 77–169.

[SC]   V.P. Soltan and V.D. Chepoi, *Conditions for invariance of set diameters under d-convexification in a graph*, Cybernetics 19 (1983), 750–756.

[St]   J. Stallings, *Non-positively curved triangles of groups*, in Group Theory from a Geometrical Viewpoint (Trieste, 1990), 491–503, World Science Publications 1991.

[Św1]   J. Świątkowski, *Trivalent polygonal complexes of nonpositive curvature and Platonic symmetry*, Geom. Dedicata 70 (1998), 87–110.

[Św2]   J. Świątkowski, *Regular path systems and (bi)automatic groups*, Geom. Dedicata 118 (2006), 23–48.

[Ta]   T. Tamagawa, *On discrete subgroups of p-adic algebraic groups*, 1965 Arithmetical Algebraic Geometry (Proc. Conf. Purdue University, 1963), pp. 11–17, Harper & Row, New York.

[Th1]   A. Thomas, *Covolumes of uniform lattices acting on polyhedral complexes*, Bull. London Math. Soc. 39 (2007), 103–111.

[Th2]   A. Thomas, *Lattices acting on right-angled buildings*, Algebr. Geom. Topol. **6** (2006), 1215–1238.

[Th3]   A. Thomas, *On the set of covolumes of lattices for Fuchsian buildings*, C. R. Acad. Sci. Paris Sér. I **344** (2007), 215–218.

[Th4]   A. Thomas, *Lattices acting on Platonic polygonal complexes*, in preparation.

[Thu]   W. Thurston, *The Geometry and Topology of Three-Manifolds*, Princeton Notes.

[Ti1]   J. Tits, *Sur le groupe d'automorphismes d'un arbre*, in Essays on Topology (Mémoires dédiés à Georges de Rham), Springer-Verlag (1970), 188–211.

[Ti2]   J. Tits, *Free subgroups in linear groups*, J. Algebra **20** (1972), 250–270.

[Ti3]   J. Tits, *Buildings of Spherical Type and Finite BN-pairs*, Lecture Notes in Math. 386, Springer-Verlag, New York, 1974.

[Ti4]   J. Tits, *Endliche Spielungsgruppen, die als Weylgruppen auftreten*, Invent. Math (1977), 283–295.

[Vd]   A. Vdovina, *Combinatorial structure of some hyperbolic buildings*, Math Z. **241** (2002), 471–478.

[Vi]   E.B. Vinberg, *The non existence of crystallographic reflection groups in Lobachevskii spaces of large dimension*, Trudy Moskov. Mat. Obshch. **47** (1984), 68–102.

[W]   G.A. Willis, *A canonical form for automorphisms of totally disconnected locally compact groups*, Random walks and geometry, 295–316, Walter de Gruyter GmbH & Co. KG, Berlin, 2004.

[W1]   D.T. Wise, *Non-positively curved square complexes, aperiodic tilings, and non-residually finite groups*, Ph.D. Thesis, Princeton University, 1996.

[W2]   D.T. Wise, *A non-Hopfian automatic group*, J. Algebra **180** (1996), 845–847.

[W3]   D.T. Wise, *The residual finiteness of negatively curved polygons of finite groups*, Invent. Math. **149** (2002), 579–617.

[W4]   D.T. Wise, *Cubulating small cancellation groups*, Geom. Funct. Anal. **14** (2004), 150–214.

[X]   X. Xie, *Quasi-isometric rigidity of Fuchsian buildings*, Topology **45** (2006), 101–169.

[Zi]   R. J. Zimmer, *Ergodic Theory and Semisimple Groups*, Birkhäuser, Boston, 1984.

[Żu]   A. Żuk, *La propriété (T) de Kazhdan pour les groupes agissant sur les polyèdres*, C. R. Acad. Sci. Paris Sér. I Math. **323** (1996), 453–458.

17

JENNIFER TABACK AND PETER WONG

THE GEOMETRY OF TWISTED CONJUGACY
CLASSES IN WREATH PRODUCTS

TO ROBERT J. ZIMMER ON THE OCCASION OF HIS 60TH BIRTHDAY

Abstract

We present a geometric proof based on recent work of Eskin, Fisher, and Whyte that the lamplighter group L_n has infinitely many twisted conjugacy classes for any automorphism φ only when n is divisible by 2 or 3, originally proved by Gonçalves and Wong. We determine when the wreath product $G \wr \mathbb{Z}$ has this same property for several classes of finite groups G, including symmetric groups and some nilpotent groups.

1. Introduction

We say that a group G has *property* R_∞ if any automorphism φ of G has an infinite number of φ-twisted conjugacy classes. Two elements $g_1, g_2 \in G$ are φ-twisted conjugate if there is an $h \in G$ so that $h g_1 \varphi(h)^{-1} = g_2$. The study of this property is motivated by topological fixed point theory, and is discussed below. Groups with property R_∞ include

1) Baumslag-Solitar groups $BS(m, n) = \langle a, b | ba^m b^{-1} = a^n \rangle$ except for $BS(1, 1)$; [FG2].
2) Generalized Baumslag-Solitar (GBS) groups, that is, finitely generated groups that act on a tree with all edge and vertex stabilizers infinite cyclic as well as any group quasi-isometric to a GBS group; [L, TWo2].
3) Lamplighter groups $\mathbb{Z}_n \wr \mathbb{Z}$ iff $2|n$ or $3|n$; [GW1].
4) Some groups of the form $\mathbb{Z}^2 \rtimes \mathbb{Z}$; [GW2].
5) Nonelementary Gromov hyperbolic groups; [LL, F].

2000 *Mathematics Subject Classification.* Primary: 20E45; Secondary: 20E08, 20F65, 55M20
Key words and Phrases. Reidemeister number, twisted conjugacy class, lamplighter group, Diestel-Leader graph, wreath product.

The first author acknowledges support from NSF grant DMS-0604645, and would like to thank David Fisher and Kevin Whyte for many useful conversations about this paper, and Kevin Wortman for comments.

6) The solvable generalization Γ of $BS(1, n)$, which is defined by the short exact sequence $1 \to \mathbb{Z}[\frac{1}{n}] \to \Gamma \to \mathbb{Z}^k \to 1$ as well as any group quasi-isometric to Γ; [TWo1].

7) The mapping class group; [FG1].

8) Relatively hyperbolic groups; [FG1].

9) Any group with a *characteristic* and nonelementary action on a Gromov hyperbolic space; [TWh, FG1].

These results fall into two categories: those that show the property holds using a group presentation, and those that show that property R_∞ is geometric in some way, whether invariant under quasi-isometry, or dependent on an action of a group on a particular space. The final statement on the list is the most general, relying only on the existence of a certain group action. This was proven independently in [TWh] and in the appendix of [FG1] using similar methods, and provides a wealth of examples of groups with this property.

The study of the finiteness of the number of twisted conjugacy classes arises in Nielsen fixed point theory. (see, for example, [B]) Given a self-map $f : X \to X$ of a compact connected manifold X, the nonvanishing of the classical Lefschetz number $L(f)$ guarantees the existence of fixed points of f. Unfortunately, $L(f)$ yields no information about the size of the set of fixed points of f. However, the Nielsen number $N(f)$, a more subtle homotopy invariant, provides a lower bound on the size of this set. For $\dim X \geq 3$, a classical theorem of Wecken [We] asserts that $N(f)$ is a sharp lower bound on the size of this set, that is, $N(f)$ is the minimal number of fixed points among all maps homotopic to f. Thus the computation of $N(f)$ is a central issue in fixed point theory.

Given an endomorphism $\varphi : \pi \to \pi$ of a group π, the φ-twisted conjugacy classes are the orbits of the (left) action of π on π via $\sigma \cdot \alpha \mapsto \sigma \alpha \varphi(\sigma)^{-1}$. Given a self-map $f : X \to X$ of a compact connected polyhedron, the fixed point set $Fixf = \{x \in X | f(x) = x\}$ is partitioned into fixed point classes, which are identical to the φ-twisted conjugacy classes where $\varphi = f_\sharp$ is the homomorphism induced by f on the fundamental group $\pi_1(X)$. The Reidemeister number $R(f)$ is the number of φ-twisted conjugacy classes, and is an upper bound for $N(f)$. When $R(f)$ is finite, this provides additional information about the cardinality of the set of fixed points of f.

For a certain class of spaces, called *Jiang spaces*, the vanishing of the Lefshetz number implies that $N(f) = 0$ as well, and a nonzero Lefshetz number, combined with a finite Reidemeister number, implies that $N(f) = R(f)$. As the Reidemeister number is much easier to calculate than the Lefshetz number, this provides a valuable tool for computing the cardinality of the set of fixed points of the map f. Jiang's results [J] have been extended to self-maps of simply connected spaces, generalized lens spaces, topological groups, orientable coset spaces of compact connected Lie groups, nilmanifolds, certain C-nilpotent spaces where C denotes the class of finite groups, certain solvmanifolds, and infrahomogeneous spaces (see, e.g., [Wo2, Wo3]). Groups G, which satisfy property R_∞, that is, every automorphism φ has $R(\varphi) = \infty$, will never be the fundamental group of a manifold, which satisfies the conditions above.

In this paper, we study the (in)finiteness of the number of φ-twisted conjugacy classes for automorphisms $\varphi : G \wr \mathbb{Z} \to G \wr \mathbb{Z}$ where $G \wr \mathbb{Z}$ is the restricted wreath product of a finite group G with $\mathbb{Z}$. When $G = \mathbb{Z}_n$, so that $G \wr \mathbb{Z}$ is a "lamplighter" group, we prove that $G \wr \mathbb{Z}$ has property R_∞ if and only if 2 or 3 divides n using the geometry of the Cayley graph of these groups. Our proofs rely on recent work of A. Eskin, D. Fisher, and K. Whyte identifying all quasi-isometries of L_n, and classifying these groups up to quasi-isometry; [EFW1]. Our theorem was originally proven in [GW1] using algebraic methods. The geometric techniques used below extend to certain other groups of this form, but combining geometric and algebraic methods we determine several larger classes of finite groups G for which $G \wr \mathbb{Z}$ has property R_∞. In particular, $S_n \wr \mathbb{Z}$ has this property for $n \geq 5$, yielding the corollary that every group of the form $G \wr \mathbb{Z}$ can be embedded into a group that has property R_∞.

We note that property R_∞ is not geometric in the sense that it is preserved under quasi-isometry. Namely, $\mathbb{Z}_4 \wr \mathbb{Z}$ is quasi-isometric to $(\mathbb{Z}_2)^2 \wr \mathbb{Z}$ but according to [GW1] the former has property R_∞ while the latter does not. It also fails to be geometric when one considers cocompact lattices in Sol. Let $A, B \in GL(2, \mathbb{Z})$ be matrices whose traces have absolute value greater than 2. We know that $\mathbb{Z}^2 \rtimes_A \mathbb{Z}$ and $\mathbb{Z}^2 \rtimes_B \mathbb{Z}$ are always quasi-isometric, as they are both cocompact lattices in Sol, but they may not both have property R_∞. However, there are classes of groups for which this property is invariant under quasi-isometry; these include the solvable Baumslag-Solitar groups, their solvable generalization Γ given above, and the generalized Baumslag-Solitar groups.

2. Background on Twisted Conjugacy

2.1. Twisted Conjugacy

We say that a group G has *property R_∞* if for any $\varphi \in Aut(G)$, we have $R(\varphi) = \infty$. If G is abelian, then $R(\varphi) = \#Coker(Id - \varphi)$. For general groups, the main technique we use for computing $R(\varphi)$ is as follows. We consider groups that can be expressed as group extensions, for example, $1 \to A \to B \to C \to 1$. Suppose that an automorphism $\varphi \in Aut(B)$ induces the following commutative diagram, where the vertical arrows are group homomorphisms, that is, $\varphi|_A = \varphi'$ and $\overline{\varphi}$ is the quotient map induced by φ on C:

1

$$
\begin{array}{ccccccccc}
1 & \longrightarrow & A & \overset{i}{\longrightarrow} & B & \overset{p}{\longrightarrow} & C & \longrightarrow & 1 \\
 & & \varphi' \downarrow & & \varphi \downarrow & & \overline{\varphi} \downarrow & & \\
1 & \longrightarrow & A & \overset{i}{\longrightarrow} & B & \overset{p}{\longrightarrow} & C & \longrightarrow & 1
\end{array}
$$

Then both φ' and $\overline{\varphi}$ are automorphisms and we obtain a short exact sequence of sets and corresponding functions $\hat{i}$ and $\hat{p}$:

2
$$
\mathcal{R}(\varphi') \overset{\hat{i}}{\to} \mathcal{R}(\varphi) \overset{\hat{p}}{\to} \mathcal{R}(\overline{\varphi})
$$

where if $\overline{1}$ is the identity element in C, we have $\hat{i}(\mathcal{R}(\varphi')) = \hat{p}^{-1}([\overline{1}])$, and $\hat{p}$ is onto. The following result is straightforward and follows from more general results discussed in [Wo1].

LEMMA 2.1. *Given the commutative diagram labeled* 1 *above,*

1) *if* $R(\overline{\varphi}) = \infty$, *then* $R(\varphi) = \infty$,
2) *if* C *is finite or* $Fix(\overline{\varphi}) = 1$, *and* $R(\varphi') = \infty$ *then* $R(\varphi) = \infty$.

When $C \cong \mathbb{Z}$, *we have one of two situations:*

3) *the map* $\overline{\varphi}$ *is the identity, in which case* $R(\overline{\varphi}) = \infty$ *and hence* $R(\varphi) = \infty$, *or*
4) $\overline{\varphi}(t) = t^{-1}$ *and* $R(\overline{\varphi}) = 2$. *In this case,* $R(\varphi) < \infty$ *iff* $R(\varphi') < \infty$ *and* $R(t \cdot \varphi') < \infty$, *in which case* $R(\varphi) = R(\varphi') + R(t \cdot \varphi')$.

To show that a group G does not have property R_∞, it suffices to produce a single example of $\varphi \in Aut(G)$ with $R(\varphi) < \infty$. When G additionally has a semidirect product structure, such an automorphism is often constructed as follows. Write $G = A \rtimes B$ as $1 \to A \to G \to B \to 1$, with $\Theta : B \to Aut(A)$. We can use automorphisms $\varphi' : A \to A$ and $\overline{\varphi} : B \to B$ to construct the following commutative diagram defining $\varphi \in Aut(G)$, provided that

3
$$
\varphi'(\Theta(b)(a)) = \Theta(\overline{\varphi}(b))(\varphi'(a))
$$

for $b \in B$ and $a \in A$:

4

$$
\begin{array}{ccccccccc}
1 & \longrightarrow & A & \overset{i}{\longrightarrow} & G & \overset{p}{\longrightarrow} & B & \longrightarrow & 1 \\
 & & \varphi' \downarrow & & \varphi \downarrow & & \overline{\varphi} \downarrow & & \\
1 & \longrightarrow & A & \overset{i}{\longrightarrow} & G & \overset{p}{\longrightarrow} & B & \longrightarrow & 1
\end{array}
$$

Through careful selection of the maps φ' and $\overline{\varphi}$, we can sometimes create an example of an automorphism of G with finite Reidemeister number using the above diagram and Lemma 2.1. We should point out, however, that the map φ so constructed using φ' and $\overline{\varphi}$ is not unique.

3. Lamplighter Groups and Diestel-Leader Graphs

We begin with a discussion of the geometry of the lamplighter groups $L_n = \mathbb{Z}_n \wr \mathbb{Z}$ and later address more general groups $G \wr \mathbb{Z}$ where G is any finite group.

The standard presentation of L_n is $\langle a, t | a^n = 1, [a^{t^i}, a^{t^j}] = 1 \rangle$ where x^y denotes the conjugate yxy^{-1}. Equivalently, we recall that a wreath product is simply a certain semidirect product, and write L_n as $\left(\bigoplus_{i=-\infty}^{\infty} A_i \right) \rtimes \mathbb{Z}$, with each $A_i \cong \mathbb{Z}_n$, fitting into the split short exact sequence

$$0 \to \bigoplus_{i=-\infty}^{\infty} A_i \to L_n \to \mathbb{Z} \to 0$$

where the generator of $\mathbb{Z}$ is taken to be t from the presentation given above.

The "lamplighter" picture of elements of this group is the following. Consider a bi-infinite string of lightbulbs placed at integer points along the real line, each of which has n states corresponding to the powers of the generator a of $\mathbb{Z}_n$, and a cursor, or lamplighter, indicating the particular bulb under consideration. The action of the group on this picture is such that t moves the cursor one position to the right, and a changes the state of the current bulb under consideration. Thus each element of L_n can be interpreted as a series of instructions to illuminate a finite collection of lightbulbs to some allowable states, leaving the lamplighter at a fixed integer.

Understanding group elements via this picture, the generator $a_j = t^j a t^{-j}$ of A_j in $\bigoplus_{i=-\infty}^{\infty} A_i$ has a single bulb in position j illuminated to state a, and the cursor at the origin of $\mathbb{Z}$. This leads to two possible normal forms for $g \in L_n$, as described in [CT], separating the word into segments corresponding to bulbs indexed by negative and non-negative integers:

$$rf(g) = a_{i_1}^{e_1} a_{i_2}^{e_2} \ldots a_{i_k}^{e_k} a_{-j_1}^{f_1} a_{-j_2}^{f_2} \ldots a_{-j_l}^{f_l} t^m$$

or

$$lf(g) = a_{-j_1}^{f_1} a_{-j_2}^{f_2} \ldots a_{-j_l}^{f_l} a_{i_1}^{e_i} a_{i_2}^{e_i} \ldots a_{i_k}^{e_k} t^m$$

with $i_k > \ldots i_2 > i_1 \geq 0$ and $j_l > \ldots j_2 > j_1 > 0$ and e_i, f_j in the range $\{-h, \ldots, h\}$, where h is the integer part of $\frac{n}{2}$. When n is even, we omit a^{-h} to ensure uniqueness, since $a^h = a^{-h}$ in $\mathbb{Z}_{2h}$. If we consider the normal form as a series of instructions for acting on the lamplighter picture to create these group elements, then the normal form $rf(g)$ first illuminates bulbs at or to the right of the origin, and $lf(g)$ first illuminates the bulbs to the left of the origin. It is proven in [CT] that the word length of $g \in L_n$ with respect to the finite generating set $\{a, t\}$, using the notation of either normal form given above, is

$$\sum e_i + \sum f_j + min\{2j_l + i_k + |m - i_k|, 2i_k + j_l + |m + j_l|\}.$$

We note that the lamplighter picture of L_n does not yield the Cayley graph of the group; to obtain a useful and understandable Cayley graph for this group we introduce Diestel-Leader graphs in §3.1. However, to use a Diestel-Leader graph as the Cayley graph of L_n, we must consider the generating set $\{t, ta, ta^2, \ldots, ta^{n-1}\}$ for L_n.

There is another generating set for L_2 worth noting. R. Grigorchuk and A. Zuk in [GZ] show that this group is an example of an automata group. The natural generating set arising from this interpretation is $\{a, ta\}$. Grigorchuk and Zuk compute the spectral radius with respect to this generating set and find that it is a discrete measure.

3.1. Diestel-Leader Graphs

We now describe explicitly the Cayley graph for the lamplighter group L_m with respect to one particular generating set. This Cayley graph is an example of a Diestel-Leader graph, which we define in full generality as follows.

For positive integers $m \leq n$, the Diestel-Leader graph $DL(m, n)$ is a subset of the product of the regular trees of valence $m + 1$ and $n + 1$, which we denote, respectively, as T_1 and T_2. We orient these trees so that each vertex has m (respectively, n) outgoing edges. By fixing a base point, we can use this orientation to define a height function $h_i : T_i \to \mathbb{Z}$. The Diestel-Leader graph $DL(m, n)$ is defined to be the subset of $T_1 \times T_2$ for which $h_1 + h_2 = 0$. This definition lends itself to the following pictorial representation. Each tree has a distinguished point at infinity, which determines the height function. If we place these points for the two trees at opposite ends of a page, the Diestel-Leader graph can be seen as those pairs of points, one from each tree, lying on the same horizontal line. See Figure 1 for a portion of the graph $DL(3, 3)$.

When $n = m$, we will see that $DL(m, m)$ is the Cayley graph of the lamplighter group L_m with respect to a particular generating set. This fact was first noted by Moeller and Neumann [MN]. Both Woess [Woe] and Wortman [Wor] describe slightly different methods of understanding this model of the

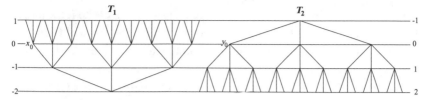

Fig. 1. Part of the Cayley graph $DL(3,3)$ of L_3 with the point $Id = (x_0, y_0)$ labeled. The integers on the edges of the diagram represent the height in each tree.

lamplighter groups; our explanation is concrete in a different way than either of theirs.

The group L_m is often presented by $L_m \cong \langle a, t | a^n = 1, [a^{t^i}, a^{t^j}] = 1 \rangle$. However, the Cayley graph resulting from this presentation is rather intractable. When we take the generating set $\{t, ta, ta^2, \ldots, ta^{m-1}\}$ we obtain the Cayley graph $DL(m, m)$ for L_m. Thus a group element $g \in L_m$ corresponds to a pair of points (c, B) where $c \in T_1$ and $B \in T_2$ so that $h_1(c) + h_2(B) = 0$. Our notational convention will be to use lowercase letters for vertices in T_1 and capital letters for vertices in T_2, with the exception of the identity, which we always denote (x_0, y_0).

The action of the group L_m on $DL(m, m)$ is as follows:

1) t translates up in height in T_1 and down in height in T_2, so that the condition $h_1 + h_2 = 0$ is preserved, and

2) if (v, W) is a point in $DL(m, m)$, let $v_0, v_1, \ldots, v_{m-1}$ be the vertices in T_1 adjacent to v with $h_1(v_i) = h_1(v) + 1$. Then a performs a cyclic rotation among these vertices, so that $a \cdot (v_i, W) = (v_{(i+1)(\text{mod } m)}, W)$. Conjugates of a by powers of t perform analogous rotations on the vertices with a common parent in T_2 while fixing the coordinate in T_1.

We now describe how to identify group elements with vertices of $DL(m, m)$. Express $g \in L_m$ using the generating set $\{t, ta, ta^2, \ldots, ta^{m-1}\}$ and inverses of these elements. For example, let $m = 3$ and take $a = t^{-1}(ta)$. We describe a path in $DL(m, m)$ from the identity to a, using the expression $a = t^{-1}(ta)$ and the action of the group on the graph. We introduce coordinates on the trees as needed, but always designate the identity as (x_0, y_0). From a given vertex $(x, Y) \in DL(m, m)$, there are edges emanating from this vertex labeled by all these generators and their inverses. The edge labeled by a generator of the form ta^i for $i = 0, 1, \ldots, m - 1$ leads to a point (x', Y') with $h_1(x') = h_1(x) + 1$ and $h_2(Y') = h_2(Y) - 1$, and the edge labeled by a generator of the form $(ta^i)^{-1}$ leads to a point (x', Y') with $h_1(x') = h_1(x) - 1$ and $h_2(Y') = h_2(Y) + 1$.

We now describe how a sequence of generators of L_m of the form $(ta^i)^{\pm 1}$ for $i \in \{0, 1, 2, \ldots, m - 1\}$ determines a path to a particular vertex of $DL(m, m)$. The first generator in the sequence corresponds to an edge in the Cayley graph $DL(m, m)$ beginning at the identity with that label. Each successive generator in the sequence extends the path by an edge in $DL(m, m)$, that is, an edge in each of T_1 and T_2, one of which increases height and the other decreases height. When an edge in T_i terminates in a vertex at a lesser height, there is no choice in that tree of which edge to traverse, and we label that edge in T_i by the generator. This immediately labels the other $m - 1$ edges sharing a common parent with this edge using the other $m - 1$ generators in a cyclic order; in fact, to ensure consistency under the action of both a and t on this Cayley graph, all sets of m edges with a common parent at this same height inherit this same labeling by the m generators. In the tree in which the edge traversed increases height, one of two situations can occur: if the edges at this height have already been labeled corresponding to a previous generator in the sequence, traverse the edge corresponding to the generator just read in the sequence; when the edges at this height have not been previously labeled, use the default labeling in which the leftmost edge has label t, the next has label ta, and so on. It is possible for the edges, say, initiating at height k and terminating at height $k + 1$ in T_i, to be relabeled as one proceeds along this sequence of generators, if there are two distinct generators in the sequence with traverse an edge in T_i on which the height function decreases from height $k + 1$ to height k.

These rules allow one to read a path of generators of L_m and obtain the corresponding point in the Cayley graph $DL(m, m)$. Notice that the labeling of edges in each tree within $DL(m, m)$ will vary depending on the sequence of generators. So they are not "permanent" labels on the edges of the two tress. This is because what we are doing in this process is labeling the edges of $DL(m, m)$ and not simply of T_1 and T_2; the reason the labels are not "permanent" is that the generator labeling a single edge in T_1 (for example) will differ depending on which edge of T_2 it is paired with to form an edge in the graph $DL(m, m)$. We illustrate this procedure in several examples below.

We begin by finding the group element $a = t^{-1}(ta)$ as a vertex in $DL(3, 3)$. We read the generators from left to right as a series of instructions for forming a path in $DL(3, 3)$ from the identity to a. Using the group action on $DL(3, 3)$ described above, the generator t always increases the height function in T_1 while decreasing it in T_2, and the action of a (respectively, a^{-1}) is to cyclically rotate the edge we traverse in T_1 (respectively, T_2).

For example, in $DL(3, 3)$ we trace a path to $a = t^{-1}(ta)$ in each tree as follows. The edge emanating from $Id = (x_0, y_0)$ labeled t^{-1} must decrease the height in

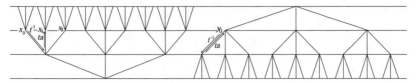

Fig. 2. The path in the Cayley graph $DL(3,3)$ of L_3 from $Id = (x_0, y_0)$ to the group element $a = t^{-1}(ta)$.

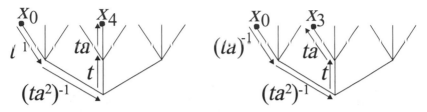

Fig. 3. The T_1 coordinates of the paths in $DL(3,3)$ leading from the identity (x_0, y_0) to the group elements $t^{-1}(ta^2)^{-1}t(ta)$ and $(ta)^{-1}(ta^2)^{-1}t(ta)$, respectively. The T_2 coordinates of the final point in each path is y_0.

T_1 and increase it in T_2; this determines a vertex x' in T_1 and Y' in T_2. The edge emanating from (x', Y') labeled by ta must increase the height function in T_1 and decrease it in T_2. There is a unique vertex in T_2 satisfying the conditions necessary for a coordinate in the terminus of this edge, namely y_0. In T_1, since the initial edge was labeled t^{-1}, we do not traverse up that edge, but rather the cyclically adjacent edge, ending at a point whose coordinates we label (x_1, y_0). Using the coordinates in Figure 2, we see that $a^2 = t^{-1}(ta^2) = (x_2, y_0)$.

For a more involved example, consider the group elements $t^{-1}(ta^2)^{-1}t(ta)$ and $(ta)^{-1}(ta^2)^{-1}t(ta)$. In Figure 3, we trace out the paths in T_1 leading from x_0 to the first coordinate of these two elements in $DL(3,3)$. This example illustrates the cyclic labeling of edges determined by the downward path of edges and the action of a on the graph. It is not hard to see that the T_2 coordinate of both of these group elements is y_0.

Writing L_m as a group extension:

5
$$0 \to \bigoplus_{i=-\infty}^{\infty} A_i \to L_n \to \mathbb{Z} \to 0$$

where $A_i \cong \mathbb{Z}_m$, we see that the map onto $\mathbb{Z}$ is determined by the exponent sum on all instances of the generator t in any word representing $g \in L_m$, using the generating set $\{t, a\}$. Thus the kernel of this map is simply those group elements in which there are equal numbers of t and t^{-1}. It is easy to see that when we change to the generating set $\{t, ta, \dots, ta^{n-1}\}$ these elements can be

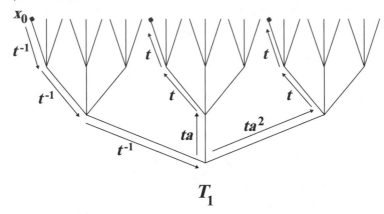

Fig. 4. Paths from the identity to the T_1 coordinates of the points of A_3 in $\bigoplus_{i=-\infty}^{\infty} A_i$ where $A_i \cong \mathbb{Z}_3$ for all i.

characterized in the same manner. Thus in $DL(m, m)$ the points corresponding to the group elements in the kernel of this map again have coordinates with height 0 in both trees.

To understand this picture more thoroughly, we identify in $DL(m, m)$ the points corresponding to the different factors of $A_i \cong \mathbb{Z}_m$. First suppose that $i \leq 0$. The points in A_i all have the form $t^{-i}a^k t^i$ with respect to the generating set $\{t, a\}$ for L_m, for $0 \leq k \leq m - 1$, and $t^{-(i+1)}(ta^k)t^i$ with respect to the generating set $\{t, ta, ta^2, \ldots, ta^{m-1}\}$. Thus they are easy to find in $DL(m, m)$: the coordinate in T_2 will be y_0, and in T_1, follow the unique path that takes $i + 1$ edges decreasing in height, cyclically rotate over k edges, proceed up in height one edge, then regardless of k, proceed up the identical path in each subtree labeled by t at each step until you reach height 0. The factors of A_i for $i > 0$ are reversed in that points in such a factor all have the first coordinate in T_1 equal to x_0 and the second coordinate is found in T_2 in an analogous manner (see Figure 4).

Using the lamplighter picture to understand group elements, a natural subset to identify is the collection of elements with bulbs illuminated (to any state) within a given set of positions, say, between $-i$ and 0, for some $i \geq 0$. The situation we describe below is completely analogous for $i < 0$, with the two trees interchanged. Let l be the line in $DL(m, m)$, which is the orbit of the identity under the group generator t. Viewing T_1 and T_2 simply as trees, and not as part of $DL(m, m)$, the line l in $DL(m, m)$ determines a line $l_k \subset T_k$ for each k, for $k = 1, 2$.

When $-i \leq 0$, to determine this set of group elements, we recall that in the generating set corresponding to $DL(m, m)$, any instance of a generator

$(ta^k)^{\pm 1}$ for k strictly greater than 0 in a word will illuminate a bulb in the lamplighter picture for that element. The position of this bulb is determined by the exponent sum on all instances of the generator t, whether as part of ta^k or not, within the word up until the string a^k is read. Thus all group elements of this form can be expressed as follows:

6
$$t^{-(i+1)} \prod_{j=1}^{i+1} \alpha_j$$

where α_j lies in $\{t, ta, ta^2, \ldots, ta^{m-1}\}$. It is not hard to see that all of these group elements have y_0 for their T_2 coordinate. The set of elements with bulbs illuminated only in positions between 0 and i, for $i > 0$, will share x_0 as their T_1 coordinate. We return to the set of elements whose only bulbs are illuminated between positions $-i$ and 0, and identify the T_1 coordinates of these elements as follows. Let x_{i+1} be the point on l_1 at height $-(i+1)$. Then the m^{i+1} points that can be reached in T_1 by a path of length $i + 1$, which increases in height at each step, are precisely all the points of the form given in Expression (b). The points at height 0 in the tree T_1 that are shown in Figure 1, when paired with $y_0 \in T_2$, give the complete collection of elements of L_3 with bulbs illuminated in all combinations of positions between -2 and 0, inclusive.

3.2. Results of A. Eskin, D. Fisher, and K. Whyte

In our proofs below, we rely on results of A. Eskin, D. Fisher and K. Whyte concerning the classification of the Diestel-Leader graphs up to quasi-isometry. [EFW1] These results are relevant since any automorphism of $\mathbb{Z}_n \wr \mathbb{Z}$ can be viewed as a quasi-isometry of the Cayley graph $DL(n, n)$.

Introduce the following coordinates on $DL(m, n)$: a point in $DL(m, n)$ can be uniquely identified by the triple (x, y, z) where $x \in \mathbb{Q}_m$, $y \in \mathbb{Q}_n$ and $z \in \mathbb{Z}$. Here, $\mathbb{Q}_m$ denotes the m-adic rationals, which we view as the ends of the tree T_1 once the height function has been fixed. The n-adic rationals correspond in an analogous way to the ends of T_2. Restricting to T_1, any $x \in \mathbb{Q}_m$ determines a unique line in T_1 on which the height function is strictly increasing. This line has a unique point at height z. Similarly, the line in T_2 determined by $y \in \mathbb{Q}_n$ contains a unique point at height $-z$. This pair of points together forms a single point in $DL(m, n)$ that is uniquely identified in this way.

Following [EFW1], we define a *product map* $\hat{\varphi} : DL(m, n) \to DL(m', n')$ as follows. Namely, $\hat{\varphi}$ is a product map if it is within a bounded distance of a map of the form $(x, y, z) \mapsto (f(x), g(y), q(z))$ or $(x, y, z) \mapsto (g(y), f(x), q(z))$ where $f : \mathbb{Q}_m \to \mathbb{Q}_{m'}$ (or $\mathbb{Q}_{n'}$), $g : \mathbb{Q}_n \to \mathbb{Q}_{n'}$ (or $\mathbb{Q}_{m'}$) and $q : \mathbb{R} \to \mathbb{R}$.

A product map is called *standard* if it is the composition an isometry and a product map in which q is the identity, and f and g are bilipshitz.

Eskin, Fisher, and Whyte determine the form of any quasi-isometry between Diestel-Leader graphs in the following theorem, which is stated in [EFW1] and proven there for $m \neq n$, and proven in the remaining case in [EFW2].

THEOREM 3.1. ([EFW1], THEOREM 2.3). *For any $m \leq n$, any (K, C)-quasi-isometry φ from $DL(m, n)$ to $DL(m', n')$ is within bounded distance of a height respecting quasi-isometry φ'. Furthermore, the bound is uniform in K and C.*

Below, we use the fact that any height-respecting quasi-isometry is at a bounded distance from a standard map. Eskin, Fisher and Whyte describe the group of standard maps of $DL(m, n)$ to itself; this group is isomorphic to $(Bilip(\mathbb{Q}_m) \times Bilip(\mathbb{Q}_n)) \rtimes \mathbb{Z}/2\mathbb{Z}$ when $m = n$ and $(Bilip(\mathbb{Q}_m) \times Bilip(\mathbb{Q}_n))$ otherwise. They deduce that this is the quasi-isometry group of $DL(m, n)$. The extra factor of $\mathbb{Z}_2$ in the case $m = n$ reflects the fact that the trees in this case may be interchanged.

This result has the following implications for our work below. The product map structure implies that a quasi-isometry takes a line in T_1 and maps it to within a uniformly bounded distance of a line in either T_1 or T_2, depending on whether the quasi-isometry interchanges the two tree factors or not. The same is true when we begin with a line in T_2. This is a geometric fact not dependent on the group structure, and we use it repeatedly below.

4. Counting Twisted Conjugacy Classes in $L_n = \mathbb{Z}_n \wr \mathbb{Z}$

In this section, we present a geometric proof of the following theorem, first proven in [GW1]. We follow the outline of the lemmas in [GW1], providing proofs based on the geometry of the group and the results of A. Eskin, D. Fisher, and K. Whyte in [EFW1].

THEOREM 4.1. ([GW1], THEOREM 2.3) *Let $m \geq 2$ be a positive integer with its prime decomposition*

$$m = 2^{e_1} 3^{e_2} \Pi_i p_i^{e_i}$$

where each p_i is a prime greater than or equal to 5. Then there exists an automorphism φ of $\mathbb{Z}_m \wr \mathbb{Z}$ with $R(\varphi) < \infty$ if and only if $e_1 = e_2 = 0$.

Let $\varphi \in Aut(L_n)$. We first must show that the infinite direct sum $\bigoplus A_i$ where each $A_i \cong \mathbb{Z}_n$ is characteristic in L_n regardless of whether $(n, 6) = 1$, allowing us to obtain "vertical" maps on the short exact sequence defining L_n given in (5). Let $DL(n, n)$ be the Cayley graph of L_n with respect to the generating set $\{t, ta, ta^2, \ldots, ta^{n-1}\}$. We label the identity using the coordinates (x_0, y_0) and label other points as needed. In general, we use lowercase letters for coordinates in the first tree T_1, and capital letters to denote coordinates in the second tree T_2.

LEMMA 4.2. *Let* $\varphi \in Aut(L_n)$. *Then* $\varphi(\bigoplus A_i) \subset \bigoplus A_i$.

Proof. Let $a = t^{-1}(ta) - (b, y_0)$. Since $\psi(Id) - \psi(x_0, y_0) = (x_0, y_0)$, we have one of two situations: a quasi-isometry of $DL(n, n)$, which interchanges the tree factors will take any line in T_1 through x_0 to within a uniformly bounded distance of a line in T_2 passing through y_0, and a quasi-isometry that does not interchange the tree factors will coarsely preserve the set of lines in T_1 through x_0.

Respectively, we have that $\varphi(a) = \varphi(b, y_0) = (x_0, C)$ or $\varphi(b, y_0) = (c, y_0)$. In either case it is clear that $\varphi(a)$ must lie at height 0 in $DL(n, n)$, that is, $\varphi(a) \in \bigoplus A_i$. If a_i is the generator of the i-th copy of A in $\bigoplus A_i$, then $a_i = t^i a_0 t^{-i}$, and it is easy to see that a_i either has coordinates (x_0, D) or (d, x_0) in $DL(n, n)$. Thus the same argument shows that $\varphi(a_i) \in \bigoplus A_i$ and the lemma follows. $\square$

Algebraically, the validity of this lemma, as well as Lemma 5.1, lies in the fact that $\bigoplus A_i$ represents the set of torsion elements that must be mapped to torsion elements under any automorphism. Lemma 4.2 guarantees the following commutative diagram:

$$
\begin{array}{ccccccccc}
1 & \longrightarrow & \bigoplus A_i & \longrightarrow & L_n & \longrightarrow & \mathbb{Z} & \longrightarrow & 1 \\
 & & \varphi' \downarrow & & \varphi \downarrow & & \overline{\varphi} \downarrow & & \\
1 & \longrightarrow & \bigoplus A_i & \longrightarrow & L_n & \longrightarrow & \mathbb{Z} & \longrightarrow & 1
\end{array}
$$

In particular, if $\overline{\varphi}(t) = t$, it follows immediately from Lemma 2.1 that $R(\varphi) = \infty$. In all that follows, we assume that $\overline{\varphi}(t) = -t$, which necessarily means that φ interchanges the tree factors in $DL(n, n)$. Since the fixed point set of $\overline{\varphi}$ is trivial, if we can show that $R(\varphi') = \infty$, it will follow from Lemma 2.1 that $R(\varphi) = \infty$ as well.

When n is prime, we are able to determine the image of the generator a_i of the i-th copy of A_i in $\bigoplus A_i$ under any automorphism $\varphi \in Aut(L_n)$ using only the geometry of the group. The simplest proof of this lemma uses the "lamplighter" picture describing group elements. The lemma is proven for all n in [GW1].

LEMMA 4.3. *Let $G = \mathbb{Z}_p \wr \mathbb{Z} = \left(\bigoplus \mathbb{Z}_p\right) \rtimes \mathbb{Z}$ where p is prime, and let φ be any automorphism of G. Denote by a_i the generator of the i-th copy of $\mathbb{Z}_p$ in $\bigoplus \left(\mathbb{Z}_p\right)_i$. For each $j \in \mathbb{Z}$, $\exists i \in \mathbb{Z}$ and $k \in \{1, 2, \dots, n-1\}$ so that $\varphi'(a_i^k) = a_j$.*

Proof. Let $\varphi'(a_0) = a_{i_1}^{c_1} a_{i_2}^{c_2} \dots a_{i_k}^{c_k} a_{-j_1}^{d_1} a_{-j_2}^{d_2} \dots a_{-j_l}^{d_l}$. This expression yields a particular collection of bulbs that are illuminated in the lamplighter picture corresponding to this element, with a_{min} and a_{max}, respectively, the left- and right-most of these illuminated bulbs. Then $\varphi'(a_i) = \varphi'(t^i \cdot a_0) = \overline{\varphi}(t^i) \cdot \varphi'(a_0)$. Since $\overline{\varphi}(t^i)$ subtracts i from the indices of all terms in $\varphi'(a_0)$, we see that $\varphi'(a_i)$ is just the same series of illuminated bulbs as in $\varphi'(a_0)$ shifted i units to the left.

Suppose that $\varphi'(x) = a_j$, where $x = a_{n_1}^{e_1} a_{n_2}^{e_2} \dots a_{n_k}^{e_k} a_{-m_1}^{f_1} a_{-m_2}^{f_2} \dots a_{-m_l}^{f_l}$. Since p is prime, no exponents in $\varphi'(a_{n_i}^{e_i})$ or $\varphi'(a_{-m_i}^{f_i})$ become congruent to 0 mod p, and thus each element of the form $\varphi'(a_{n_i}^{e_i})$ or $\varphi'(a_{-m_i}^{f_i})$ has the same number of illuminated bulbs in its lamplighter picture as $\varphi(a_0)$.

If we assume that $a_{min} \neq a_{max}$, it is clear that $\varphi'(x)$ has illuminated bulbs in positions $min - m_l$ and $max + n_k$ and thus cannot be equal to a_j. We conclude that both $a_{min} = a_{max}$ and $x = a_i^k$ for some $i \in \mathbb{Z}$. $\square$

Since p is prime, by taking powers of a_i^k it follows from Lemma 4.3 that $\varphi'(a_i) = a_j^r$. Thus, if $\varphi'(a_0) = a_j^k$, then $\varphi'(a_i) = a_{j-i}^k$ and $\varphi'(a_{j-i}) = a_i^k$ for the same value of k, and we obtain by restriction a map $\varphi' : A_i \bigoplus A_{j-i} \to A_i \bigoplus A_{j-i}$.

We now recall the definition of φ'-twisted conjugacy classes in $\bigoplus A_i$. If g_1, $g_2 \in \bigoplus A_i$ are φ'-twisted conjugate, then there exists $h \in \bigoplus A_i$ so that $hg_1\varphi'(h)^{-1} = g_2$. Since $\bigoplus A_i$ is abelian, we rewrite this equation as $h\varphi'(h)^{-1} = g_1^{-1}g_2$. Switching to additive notation, we see that if $Id - \varphi' : A_i \bigoplus A_{j-i} \to A_i \bigoplus A_{j-i}$ is invertible, then $R(\varphi') < \infty$. In particular, if we can find a fixed point of φ', we guarantee that $Id - \varphi'$ is not surjective and hence not invertible. In the proof of Theorem 4.1, we prove that φ' must have a fixed point if n is 2 or 3, and if $(n, 2) = (n, 3) = 1$, then it is possible to construct automorphisms φ' so that $Id - \varphi'$ is surjective. This geometric method will allow us to prove that L_2 and L_3 have property R_∞, as well as prove that when $(n, 6) = 1$, L_n does not have this property. When $(n, 6) \neq 1$, it is a consequence of the following lemma that L_n does have property R_∞.

LEMMA 4.4. *If $(n, 6) \neq 1$, then there is a characteristic subgroup $H_i \lhd L_n$ with quotient $\mathbb{Z}_{\frac{n}{i}}$ where $i | n$, for $i = 2$ or 3, and $H_i \cong L_i$.*

Proof. Suppose that $2|n$. Let $i = 2$, and if τ is the generator for $\mathbb{Z}_n$, then the element $\tau^{\frac{n}{2}}$ is the unique element of order 2 in $\mathbb{Z}_n$. Consider the subgroup $H_2 = \langle \tau^{\frac{n}{2}} \rangle \wr \mathbb{Z}$ of L_n, where $\mathbb{Z} = \langle t \rangle$. It is clear that H_2 is isomorphic to L_2 and that H_2 contains $\bigoplus_j (\langle \tau^{\frac{n}{2}} \rangle)_j$, the subgroup of L_n consisting of elements of order 2. It is easy to see that every automorphism $\varphi \in Aut(L_n)$ sends $\bigoplus_j (\langle \tau^{\frac{n}{2}} \rangle)_j$ to itself, and the subgroup generated by $(1, t)$ to itself, where $1 \in \bigoplus_j (\langle \tau^{\frac{n}{2}} \rangle)_j$ and $t \in \mathbb{Z}$. Since H_2 is generated by $(\tau^{\frac{n}{2}}, 1)$ and $(1, t)$, it follows that H_2 is characteristic in L_n. Now $L_n/H_2 \cong \mathbb{Z}_{\frac{n}{2}}$. The case when $3|n$ and we let $i = 3$ is similar. $\qquad\square$

One can also show that if $(n, 6) \neq 1$, then there is a characteristic subgroup K_i of L_n such that $L_n/K_i \cong L_i$ for $i = 2$ or 3.

Proof of Theorem 4.1. Consider $\varphi \in Aut(L_n)$, and suppose that $\varphi'(a_0) = a_j^k$. Since φ is an automorphism, it is also a quasi-isometry. By [EFW1] it is a uniformly bounded distance from a product map. Let K and C be the quasi-isometry constants of this product map. We note that since we are assuming that $\overline{\varphi}(t) = -t$, this product map interchanges the two tree factors, that is, it maps lines in the first tree factor in $DL(n, n)$ to within a uniformly bounded distance D of lines in the second tree factor, where the constant D depends on K and C.

We know that $\varphi' : \bigoplus A_i \to \bigoplus A_i$, and our assumption that $\varphi'(a_0) = a_j^k$ yields the restriction $\varphi' : A_i \bigoplus A_{j-i} \to A_i \bigoplus A_{j-i}$. We call this subset of $\bigoplus A_i$ a *block*. We now explain the effect on these blocks of the geometric fact that φ is a uniformly bounded distance from a product map. Recall that the value of j is determined by the image of a_0, that is, $\varphi'(a_0) = a_j^k$. We always consider $i > j$, so that one factor in the block has a positive index, and the other a negative index. This ensures that coordinates can be introduced on the n^2 points in this block so that if $a_i^{m_1} = (x_0, B)$ and $a_{j-i}^{m_2} = (c, y_0)$, then $a_i^{m_1} a_{j-i}^{m_2}$ will have coordinates (c, B).

Since φ is within a uniformly bounded distance D of a product map, we additionally choose $i > KD$, where K is the maximum of the bilipshitz constants of the maps on the tree factors. (Note that this lower bound is perhaps much larger than required.) This ensures that when we consider the coordinates in T_2 of the points a_i^k and in T_1 of the points a_{j-i}^k for $0 \leq k \leq n-1$, these coordinates are at least distance $2KD$ apart, using the standard metric on the tree, which assigns each edge length 1. This is easy to see by writing $a_i^k = t^{(i-1)}ta^k t^{-i}$; when $i < 0$, the second coordinates of these points, as k varies, is always y_0. The first coordinates, pairwise, are distance $2i > 2KD$

apart in T_1. More importantly, if l is any line in T_2 through the second coordinate of a_i^k on which the height function is strictly decreasing, this product map takes l to within D of a line in T_1; the D-neighborhood of this line, by construction, can contain a unique first coordinate of one of the points in A_{j-i}.

Thus the fact that lines in T_1 are mapped to within a uniformly bounded distance of lines in T_2 tells us that there is a surjective function f_1 from the set of T_1 coordinates of the points in A_{j-i} to the set of T_2 coordinates of the points in A_i. In addition, there is a surjective function f_2 from the set of T_2 coordinates of the points in A_i to the set of T_1 coordinates of the points in A_{j-i}. Moreover, these maps combine in the following way, as follows from the definition of a product map given in [EFW1]. If $(x, Y) \in A_i \oplus A_{j-i}$, then $\varphi'(x, Y) = (f_2(Y), f_1(x))$.

We begin with the case $n = 2$. Consider the restriction $\varphi' : A_i \oplus A_{j-i} \to A_i \oplus A_{j-i}$, where j is defined by the image of a_0 under φ'. Then the set $A_i \oplus A_{j-i}$ consists of four points. Assume that $i > j$ and that i was chosen to be sufficiently large, as described above. The first choice ensures that the index of one factor is negative, and the other is positive; the second choice guarantees that the points within the block $A_i \oplus A_{j-i}$ have coordinates in the appropriate trees that are at least D units apart from each other. We describe the points in $A_i \oplus A_{j-i}$ using both the lamplighter picture of L_2 as well as giving coordinates in the graph $DL(2, 2)$, as follows:

1) the identity (x_0, y_0),
2) a single bulb illuminated in position i, where we introduce the coordinates $a_i = (x_0, B)$,
3) a single bulb illuminated in position $j - i$, where we introduce the coordinates $a_{j-i} = (c, y_0)$, and
4) bulbs illuminated in positions i and $j - i$, which necessarily has coordinates (c, B).

Note that it is exactly the fact that we ensured that i and $j - i$ have opposite signs that determines the coordinates of $a_i a_{j-i}$ to be (c, B).

We use the geometric result of [EFW1] that φ is a uniformly bounded distance from a product map, which is bilipshitz on each tree and then interchanges the tree factors, to show that (c, B) is a fixed point for φ'. Let f_1 and f_2 be the functions on the coordinates of the points of $A_i \oplus A_{j-i}$ defined above. We must have $f_1(x_0) = y_0$ and $f_2(y_0) = x_0$ since the identity (x_0, y_0) is preserved under any automorphism. Lemma 4.3 implies that $f_1(B) = c$ and $f_2(c) = B$. This forces $\varphi'(c, B) = (c, B)$; the existence of a fixed point for φ' says that

$Id - \varphi'|_{A_i \oplus A_{j-i}}$ is not invertible when $n = 2$. Thus within each block of the form $A_i \oplus A_{j-i}$, for $i > j$, there are at least two distinct φ'-twisted conjugacy classes. By choosing group elements in $\oplus A_i$ with entries in the different φ'-twisted conjugacy classes within each block, we can create infinitely many different φ'-twisted conjugacy classes in $\oplus A_i$. We conclude that $R(\varphi')$ and hence $R(\varphi)$ are infinite.

In the case $n = 3$, the blocks $A_i \oplus A_{j-i}$ have nine elements, and we have the additional algebraic fact that $a_i^2 = a_i^{-1}$. We have two choices for the maps f_1 and f_2, as follows:

1) If $\varphi'(a_i) = a_{j-i}$ and $\varphi'(a_{j-i}) = a_i$, then it is clear that $f_1(c_1) = B_1$ and $f_2(B_1) = c_1$, resulting in (c_1, B_1) being fixed under φ'.

2) If $\varphi'(a_i) = a_{j-i}^2$ and $\varphi'(a_{j-i}) = a_i^2$, then we know that $f_1(c_1) = B_2$ and $f_2(B_1) = c_2$. This forces $f_1(c_2) = B_1$ and $f_2(B_2) = c_1$. We see immediately that (c_1, B_2) is fixed under φ'.

In either case, the existence of a nontrivial fixed point ensures that L_3 has property R_∞.

When $n > 3$ is odd and $(n, 3) = 1$, we use the blocks $A_{j-i} \oplus A_i$ to construct $\varphi' : A_i \oplus A_{j-i} \to A_i \oplus A_{j-i}$ with no nontrivial fixed points, under the assumption that $\varphi'(a_0) = a_j^k$. Thus the map $Id - \varphi'$ will be surjective, hence invertible, on each block, and we see that there must be a single φ'-twisted conjugacy class in $\oplus A_i$. This then implies that L_n does not have property R_∞. We begin by assuming that $\varphi' : \oplus A_i \to \oplus A_i$ is an automorphism with $\varphi'(a_0) = a_j^k$, and that $\overline{\varphi}(t) : \mathbb{Z} \to \mathbb{Z}$ is given by $\overline{\varphi}(t) = t^{-1}$. The compatibility condition given in Equation (3) ensures that these automorphisms together induce an automorphism of $L_n = \mathbb{Z}_n \wr \mathbb{Z}$.

Since $\varphi'(a_0) = a_j^k$, we again obtain a map on blocks, where the block $A_i \oplus A_{j-i}$ consists of n^2 points; we introduce coordinates on these points as follows:

	a_i^0	a_i	a_i^2	$\cdots$	a_i^{n-1}
a_{j-i}^0	(x_0, y_0)	(x_0, B_1)	(x_0, B_2)	$\cdots$	(x_0, B_{n-1})
a_{j-i}	(c_1, y_0)	(c_1, B_1)	(c_1, B_2)	$\cdots$	(c_1, B_{n-1})
a_{j-i}^2	(c_2, y_0)	(c_2, B_1)	(c_2, B_2)	$\cdots$	(c_2, B_{n-1})
$\vdots$	$\vdots$	$\vdots$	$\vdots$	$\vdots$	$\vdots$
a_{j-i}^{n-1}	(c_{n-1}, y_0)	(c_{n-1}, B_1)	(c_{n-1}, B_2)	$\cdots$	(c_{n-1}, B_{n-1})

These points determine n points in each tree, all at height 0, namely $\{x_0, c_1, \ldots, c_{n-1}\}$ in T_1 and $\{y_0, B_1, \ldots, B_{n-1}\}$ in T_2. Choose $i > \max\{j, KD\}$ as above.

We now show that when $(n, 2) = (n, 3) = 1$, we can construct maps f_1 and f_2 so that the induced automorphism $\varphi' : A_i \bigoplus A_{j-i} \to A_i \bigoplus A_{j-i}$ has no non-trivial fixed points, and thus $Id - \varphi' : A_i \bigoplus A_{j-i} \to A_i \bigoplus A_{j-i}$ is invertible, which implies that $R(\varphi') < \infty$. Suppose that $\varphi'(a_i) = a_{j-i}^2$ and $\varphi'(a_{j-i}) = a_i^2$. To obtain a fixed point, we must solve the equations $\varphi'(a_i^k) = a_{j-i}^{2q(\text{mod } n)}$ and $\varphi'(a_{j-i}^{2q(\text{mod } n)}) = a_i^q$ for q, that is, $3q \equiv 0(\text{mod } n)$, which has a nontrivial solution exactly when $(3, n) \neq 1$. Thus when $(3, n) = 1$, we find no nontrivial fixed points, and it follows that the map $1 - \varphi' : A_i \to A_i$ is invertible on each block, hence invertible on $\bigoplus A_i$.

Since we are trying to show that $R(\varphi) < \infty$, we refer to part (4) of Lemma 2.1. We know that $R(\overline{\varphi}) = \#Coker(1 - \overline{\varphi}) = 2$, so $\overline{\varphi}$ has two twisted conjugacy classes: $[1]$ and $[t]$. Recalling the short exact sequence, we must count the twisted conjugacy classes lying over the classes $[1]$ and $[t]$. Over $[1]$ we have $R(\varphi')$ twisted conjugacy classes, and over $[t]$ we have $R(t \cdot \varphi')$. Since $Fix\overline{\varphi} = \{1\}$, it follows that $R(\varphi) = R(\varphi') + R(t \cdot \varphi')$. The action of t shifts the block $A_i \bigoplus A_{j-i}$ by increasing all the indices by 1, and thus the geometry of the lamplighter picture for L_n easily implies that $R(\varphi') = R(t \cdot \varphi')$ and hence $R(\varphi) = 2R(\varphi') < \infty$.

When $2|n$, we construct the following short exact sequence of groups:

$$1 \to \mathbb{Z}_2 \wr \mathbb{Z} \to \mathbb{Z}_n \wr \mathbb{Z} \to \mathbb{Z}_{\frac{n}{2}} \to 0$$

It is proven in Lemma 4.2 that $\mathbb{Z}_2 \wr \mathbb{Z}$ is characteristic in $\mathbb{Z}_n \wr \mathbb{Z}$, allowing us to obtain a commutative diagram based on the above short exact sequence and any $\varphi \in Aut(\mathbb{Z}_n \wr \mathbb{Z})$. We proved above that $\mathbb{Z}_2 \wr \mathbb{Z}$ has property R_∞ and it follows from Lemma 2.1(2) that $\mathbb{Z}_n \wr \mathbb{Z}$ has property R_∞ as well. The argument is analogous when $3|n$. $\quad\square$

When G is a finite abelian group, $G \wr \mathbb{Z}$ has as its Cayley graph $DL(|G|, |G|)$, with respect to the generating set $\{t, tg_1, tg_2, \ldots, tg_{n-1}\}$ where $\{g_i\}$ represents all elements of G. Gonçalves and Wong [GW1] prove the following general result concerning which groups of this form have property R_∞.

THEOREM 4.5. ([GW1], THEOREM 3.7) *Let*

$$G = \bigoplus_j (\mathbb{Z}_{p_j^{k_j}})^{r_j}$$

be a finite abelian group, where the p_j are distinct primes. Then for all automorphisms $\varphi \in Aut(G \wr \mathbb{Z})$, $R(\varphi) = \infty$ if and only if $p_j = 2$ or 3 for some j and $r_j = 1$.

5. Counting Twisted Conjugacy Classes in $G \wr \mathbb{Z}$

We end this paper by studying property R_∞ for general lamplighter groups of the form $G \wr \mathbb{Z}$, where G is an arbitrary finite group. The Cayley graph of $G \wr \mathbb{Z}$ is again the Diestel-Leader graph $DL(n, n)$, where $n = |G|$, with respect to the generating set $\{tg | g \in G\}$, where $<t> \cong \mathbb{Z}$.

We prove several results about when these general lamplighter groups have property R_∞, but do not give a complete classification of which groups of this form have or do not have the property. The first two results are algebraic, using various expressions of G as a group extension, and we end with geometric results relying on $DL(n, n)$. We first note that as for $\mathbb{Z}_n \wr \mathbb{Z}$, we have the following short exact sequence in which $\bigoplus G_i$ is characteristic:

$$1 \to \bigoplus G_i \to G \wr \mathbb{Z} \to \mathbb{Z} \to 1.$$

LEMMA 5.1. Let $\varphi \in Aut(G \wr \mathbb{Z})$. Then $\varphi(\bigoplus G_i) \subset \bigoplus G_i$.

Proof. Let $|G| = n$ and put coordinates on $DL(n, n)$ as we did in the case of the lamplighter group $L_n = \mathbb{Z}_n \wr \mathbb{Z}$: denote the identity by (x_0, y_0) and the generators of G_0, the copy of G in $\bigoplus G$ indexed by 0, by (x_0, B_i).

Let $\varphi \in Aut(G \wr \mathbb{Z})$. The argument given in Lemma 4.2 quoting the result of [EFW1] again applies, and we conclude that $\varphi(x_0, B_i)$ lies at height 0 in $DL(n, n)$, that is, $\varphi(x_0, B_i) \in \bigoplus G_i$. Since $\varphi(G_0) \subset \bigoplus G_i$ and the action of t is to translate between the factors of G in the sum $\bigoplus G_i$, we see that $\bigoplus G_i$ is characteristic in $G \wr \mathbb{Z}$. $\square$

Applying Lemma 5.1, we see that the following diagram is commutative, for $\phi \in Aut(G \wr \mathbb{Z})$.

$$
\begin{array}{ccccccccc}
1 & \longrightarrow & \bigoplus G_i & \longrightarrow & G \wr \mathbb{Z} & \longrightarrow & \mathbb{Z} & \longrightarrow & 1 \\
 & & \varphi' \downarrow & & \varphi \downarrow & & \bar{\varphi} \downarrow & & \\
1 & \longrightarrow & \bigoplus G_i & \longrightarrow & G \wr \mathbb{Z} & \longrightarrow & \mathbb{Z} & \longrightarrow & 1
\end{array}
$$

Let $G = \{Id, g_1, g_2, \ldots g_n\}$. To denote the elements of the i-th copy of G in $\bigoplus G_i$, we use the notation $g_{i,j}$ where i is the index of G_i in $\bigoplus G_i$ and $j \in \{1, 2, \ldots, n\}$. The symbol Id_i will denote the identity in G_i.

If G itself can be written as a group extension $1 \to A \to G \to C \to 1$ then one always obtains the following short exact sequences, although in both cases the kernel is not necessarily a characteristic subgroup of $G \wr \mathbb{Z}$:

9 $1 \to \bigoplus A_i \to G \wr \mathbb{Z} \to C \wr \mathbb{Z} \to 1$ and $1 \to A \wr \mathbb{Z} \to G \wr \mathbb{Z} \to C \to 1$

where $A_i \cong A$ for all i.

Given an arbitrary finite group G, we have two natural short exact sequences:

$$1 \to [G, G] \to G \to G^{Ab} \to 1 \quad \text{and} \quad 1 \to Z(G) \to G \to G/Z(G) \to 1$$

where $[G, G]$ and $Z(G)$ denote the commutator subgroup and the center of G, respectively. Both of these subgroups are characteristic in G. To obtain the related commutative diagrams, we must first prove the following lemma.

LEMMA 5.2. *The subgroups $\bigoplus ([G, G])_i$ and $Z(G) \wr \mathbb{Z}$ of $G \wr \mathbb{Z}$ are both characteristic.*

Proof. We show that $\bigoplus ([G, G])_i$ is characteristic in $\bigoplus Gi$ and thus in $G \wr \mathbb{Z}$. The commutator subgroup $[\bigoplus G_i, \bigoplus G_i]$ has the following form. Let $a, b \in \bigoplus G_i$. Then

$$a = \sum_s g_{\sigma(s), j_s} \text{ and } b = \sum_t g_{\tau(t), j_t}$$

where σ and τ are injective functions from $\{1, 2, 3, \ldots, k\}$ and $\{1, 2, \ldots, l\}$, respectively, to $\mathbb{Z}$ so that $\sigma(1) < \sigma(2) < \cdots < \sigma(k)$ and $\tau(1) < \tau(2) < \cdots < \tau(l)$. When we form $[a, b]$, one of two things must occur:

1) If $\sigma(s) = \tau(t)$ for some t, then the coordinate of $[a, b]$ in $G_{\sigma(s)}$ is $[g_{\sigma(s), j_s}, g_{\tau(t), j_t}]$.

2) If $\sigma(s) \neq \tau(t)$ for any t, then the coordinate of $[a, b]$ in $G_{\sigma(s)}$ is $g_{\sigma(s), j_s} g_{\sigma(s), j_s}^{-1} = Id_{\sigma(s)}$. The same is true for indices $\tau(t)$, which are not equal to $\sigma(s)$ for any s.

Since each non-identity coordinate of $[a, b]$ is a commutator of G, we see that $[\bigoplus G_i, \bigoplus G_i] \subset \bigoplus ([G, G])_i$, and the other inclusion is clear. Thus $\bigoplus ([G, G])_i$ must be a characteristic subgroup.

To show that $Z(G) \wr \mathbb{Z}$ is characteristic in $G \wr \mathbb{Z}$, we first show that $\bigoplus Z(G)_i$ is characteristic in $\bigoplus G_i$. Since the group operation between elements of $\bigoplus G_i$ involves componentwise multiplication, it follows immediately that $Z(\bigoplus G_i) = \bigoplus Z(G)_i$, and thus this subgroup is characteristic.

Let $\varphi \in Aut(G \wr \mathbb{Z})$. We obtain automorphisms $\varphi' : \bigoplus G_i \to \bigoplus G_i$ and $\overline{\varphi} : \mathbb{Z} \to \mathbb{Z}$; we assume the latter is given by $\overline{\varphi}(t) = t^{-1}$. Since $\bigoplus Z(G)_i$ is characteristic in $\bigoplus G_i$, we can restrict to obtain $\varphi' : \bigoplus Z(G)_i \to \bigoplus Z(G)_i$. Since φ' and $\overline{\varphi}$ satisfy the compatibility condition given in Equation (3), we induce a map on $Z(G) \wr \mathbb{Z}$, which makes the following diagram commute and must be the restriction of φ to $Z(G) \wr \mathbb{Z}$:

10

$$
\begin{array}{ccccccccc}
1 & \longrightarrow & \bigoplus Z(G)_i & \longrightarrow & Z(G) \wr \mathbb{Z} & \longrightarrow & \mathbb{Z} & \longrightarrow & 1 \\
& & \varphi' \downarrow & & \varphi \downarrow & & \overline{\varphi} \downarrow & & \\
1 & \longrightarrow & \bigoplus Z(G)_i & \longrightarrow & Z(G) \wr \mathbb{Z} & \longrightarrow & \mathbb{Z} & \longrightarrow & 1
\end{array}
$$

From this we conclude that $Z(G) \wr \mathbb{Z}$ is a characteristic subgroup of $G \wr \mathbb{Z}$. $\square$

It follows from Lemma 5.2 that given any $\varphi \in Aut(G \wr \mathbb{Z})$, we obtain the following commutative diagrams:

11

$$
\begin{array}{ccccccccc}
1 & \longrightarrow & \bigoplus ([G, G])_i & \longrightarrow & G \wr \mathbb{Z} & \longrightarrow & G^{Ab} \wr \mathbb{Z} & \longrightarrow & 1 \\
& & \varphi' \downarrow & & \varphi \downarrow & & \overline{\varphi} \downarrow & & \\
1 & \longrightarrow & \bigoplus ([G, G])_i & \longrightarrow & G \wr \mathbb{Z} & \longrightarrow & G^{Ab} \wr \mathbb{Z} & \longrightarrow & 1
\end{array}
$$

and

12

$$
\begin{array}{ccccccccc}
1 & \longrightarrow & Z(G) \wr \mathbb{Z} & \longrightarrow & G \wr \mathbb{Z} & \longrightarrow & G/Z(G) & \longrightarrow & 1 \\
& & \varphi' \downarrow & & \varphi \downarrow & & \overline{\varphi} \downarrow & & \\
1 & \longrightarrow & Z(G) \wr \mathbb{Z} & \longrightarrow & G \wr \mathbb{Z} & \longrightarrow & G/Z(G) & \longrightarrow & 1
\end{array}
$$

Here, the projection $G \wr \mathbb{Z} \to G/Z(G)$ is given by

$$
\left(\sum_k^m a_{i_k}, t^j \right) \mapsto \prod_k^m [a_{i_k}]
$$

where $[a]$ is the image of $a \in G$ in $G/Z(G)$ and $i_1 < \cdots < i_m$.

We will always use φ' and $\overline{\varphi}$ for the induced homomorphisms on the kernel and on the quotient, respectively. We note that the two maps in the diagrams above labeled φ' (respectively, $\overline{\varphi}$) denote different maps in the two diagrams.

Denote by $\mathfrak{A}$ the family of finite abelian groups A such that $A \wr \mathbb{Z}$ has property R_∞. This family was completely determined in [GW1], and is listed above as Theorem 4.5. Similarly, let $\mathfrak{L}$ be the family of finite groups G such that $G \wr \mathbb{Z}$ has property R_∞. We use both the algebra and the geometry of these lamplighter groups to determine some conditions under which $G \in \mathfrak{L}$.

THEOREM 5.3. *Let G be a finite group. If $G^{Ab} \in \mathfrak{A}$ or $Z(G) \in \mathfrak{A}$, then $G \in \mathfrak{L}$.*

Proof. First suppose that $G^{Ab} \in \mathfrak{A}$. It follows from commutative diagram **11** and Lemma 2.1 that $G \wr \mathbb{Z}$ has property R_∞.

Next, suppose that $Z(G) \in \mathfrak{A}$, which yields $R(\varphi') = \infty$. The inclusion $i : Z(G) \wr \mathbb{Z} \to G \wr \mathbb{Z}$ induces a function $\hat{i} : \mathcal{R}(\varphi') \to \mathcal{R}(\varphi)$ where $\mathcal{R}(\psi)$ denotes the set of ψ-twisted conjugacy classes. Since $G/Z(G)$ is finite, so is the number of fixed points of $\overline{\varphi}$. The reasoning in part (4) of Lemma 2.1 shows that $\mathcal{R}(\varphi)$ is in 1-1 correspondence with $\mathcal{R}(\varphi')$ modulo the action of $Fix\overline{\varphi}$. Thus $R(\varphi) = \infty$ as well. □

Using Theorem 4.5 one can easily determine when $Z(G) \in \mathfrak{A}$. We now list some examples of groups with property R_∞, which follow immediately from Theorem 5.3.

EXAMPLE 5.4. (DIHEDRAL GROUPS D_{2n} FOR ODD n) Let $G = D_6 \cong \mathbb{Z}_3 \rtimes \mathbb{Z}_2$ be the dihedral group of order 6. In this case, $[G, G] \cong \mathbb{Z}_3$ while $Z(G) = \{1\}$. Since $G^{ab} \cong \mathbb{Z}_2 \in \mathfrak{A}$, it follows from Theorem 5.3 that $D_6 \in \mathfrak{L}$, that is, $D_6 \wr \mathbb{Z}$ has property R_∞. Moreover, if n is odd and D_{2n} is the dihedral group of order $2n$, then $\mathbb{Z}_n$ is a characteristic subgroup of $D_{2n} \cong \mathbb{Z}_n \rtimes \mathbb{Z}_2$. Using the commutative diagram **8** and the fact that $\mathbb{Z}_2 \in \mathfrak{A}$, it follows from Theorem 5.3 and part (1) of Lemma 2.1 that $D_{2n} \in \mathfrak{L}$ when n is odd.

EXAMPLE 5.5. (DIHEDRAL GROUPS D_{2n} FOR EVEN n) When n is even, $D_{2n} \cong \mathbb{Z}_n \rtimes \mathbb{Z}_2$, then $D_{2n} \wr \mathbb{Z}$ also has property R_∞. If we take $\mathbb{Z}_n \cong \langle t \rangle$, then the center of D_{2n} is isomorphic to $\mathbb{Z}_2 \cong \langle t^{\frac{n}{2}} \rangle$. Thus $Z(D_{2n}) \in \mathfrak{A}$ and it follows from Theorem 5.3 that $D_{2n} \wr \mathbb{Z}$ has property R_∞ for all $n > 0$.

EXAMPLE 5.6. (QUATERNION GROUP OF ORDER 8) Let $G = Q_8$ be the quaternion group of order 8. In this case, $[G, G] = Z(G) \cong \mathbb{Z}_2$. Since $\mathbb{Z}_2 \in \mathfrak{A}$, it follows from Theorem 5.3 that $Q_8 \in \mathfrak{L}$.

Combining Theorem 4.1 with Examples 5.4 and 5.5, we have proven the following:

PROPOSITION 5.7. *If G is any finite group with order $2p$ where p is an odd prime, then $G \wr \mathbb{Z}$ has property R_∞.*

Proof. It is an elementary theorem in algebra that any group of order $2p$, where p is an odd prime, must be either cyclic or dihedral. $\qquad\square$

If $Z(G)$ or G^{ab} is unknown, the following special condition may be applicable, which allows us to extend our results to some nilpotent groups, dependent on order.

THEOREM 5.8. *Let G be a finite group with a unique Sylow 2-group S_2. If $Z(S_2) \in \mathfrak{A}$ then $G \in \mathfrak{L}$. Similarly, if G has a unique Sylow 3-group S_3 with $Z(S_3) \in \mathfrak{A}$, then $G \subset \mathfrak{L}$.*

Before proving Theorem 5.8 we state two corollaries.

COROLLARY 5.9. *Let G be a finite nilpotent group whose order is divisible by either 2 or 3, and let S_2 and/or S_3 denote, respectively, the unique Sylow 2- and/or 3-subgroups. If $Z(S_i) \in \mathfrak{A}$ for $i = 2$ or $i = 3$, then $G \wr \mathbb{Z}$ has property R_∞.*

Some elementary group theory leads to an additional corollary.

COROLLARY 5.10. *Let p and q be prime, and let G be a finite nonabelian group whose order is one of:*

1) $2p^n$, *where* $2 < p$,
2) $3q^m$, *where* $3 < q$,
3) $2p^n q^m$, *where* $2 < p < q$, *or*
4) $3p^n q^m$, *where* $3 < p < q$.

Then $G \wr \mathbb{Z}$ has property R_∞.

Proof. It is an exercise in group theory to check that groups with the above orders have either a unique Sylow 2-subgroup of order 2 or a unique Sylow 3-subgroup of order 3. The conclusion then follows directly from Theorem 5.8. $\square$

We now return to the proof of Theorem 5.8.

Proof of Theorem 5.8. We prove the theorem in the case that G has a unique Sylow 2-subgroup S_2; the case for the unique Sylow 3-subgroup is analogous. It follows immediately from Theorem 5.3 that $S_2 \wr \mathbb{Z}$ has property R_∞. We will show that $S_2 \wr \mathbb{Z}$ is a characteristic subgroup of $G \wr \mathbb{Z}$ and the theorem then follows from Lemma 2.1, the short exact sequence $1 \to S_2 \wr \mathbb{Z} \to G \wr \mathbb{Z} \to G/S_2 \to 1$, and the fact that a unique Sylow subgroup is normal.

Let $\varphi \in Aut(G \wr \mathbb{Z})$, φ' its restriction to $\bigoplus G_i$ and $\overline{\varphi}$ its projection to $\mathbb{Z}$. Since S_2 contains all elements of G whose order is a power of 2, the same is true for $\bigoplus (S_2)_i \subset \bigoplus G_i$. Since order is preserved under automorphism, $\varphi'(\bigoplus (S_2)_i) \subset \bigoplus (S_2)_i$ and this subgroup is characteristic. We now have vertical maps from the short exact sequence $1 \to \bigoplus (S_2)_i \to S_2 \wr \mathbb{Z} \to \mathbb{Z} \to 1$ where the map on the kernel is the restriction of φ' and the map on the quotient is $\overline{\varphi}$. These maps induce a map on $S_2 \wr \mathbb{Z}$, which must be the restriction of φ to $S_2 \wr \mathbb{Z}$. Thus $S_2 \wr \mathbb{Z}$ is a characteristic subgroup of $G \wr \mathbb{Z}$. $\qquad \square$

We extend these results further for simple groups, and as a consequence of Theorem 5.12, determine that $A_n \wr \mathbb{Z}$ and $S_n \wr \mathbb{Z}$ have property R_∞ for all $n \geq 5$.

LEMMA 5.11 *Let G be a finite simple group, and $\varphi \in Aut(G \wr \mathbb{Z})$. For each $j \in \mathbb{Z}$, there exists $i \in \mathbb{Z}$ so that $\varphi'(G_i) = G_j$.*

Proof. Since G is finite, there is an $r \in \mathbb{Z}^+$ so that $\varphi'(G_0) \subset \cup_{k=1}^r G_{i_k}$, where the i_k are distinct indices. Consider the homomorphism defined by composing $\varphi'|_{G_0}$ with projection onto the first factor of $\cup_{k=1}^r G_{i_k}$. Since G is simple, there can be no kernel, so this is an automorphism of G. This argument holds for any factor in $\cup_{k=1}^r G_{i_k}$. We obtain a set $\{\xi_k\}$ of automorphisms of G so that $\xi_k : G \to G_{i_k}$. We observe that since each ξ_k is an automorphism, for all nontrivial $g \in G_0$ and for all values of k, the element $\xi_k(g)$ is nontrivial.

As the action of t on $\bigoplus G_i$ is by translation, the same set of automorphisms can be used to determine the image of any element of G_i in $\bigoplus G_i$.

Since we began with a group automorphism, and $\bigoplus G_i$ is a characteristic subgroup of $G \wr \mathbb{Z}$, there is some $x \in \bigoplus G_i$ so that $\varphi'(x) = g_j$. Now the proof of Lemma 4.3 allows us to conclude that $x \in G_i$ for some i. Since the action of t on $\bigoplus G_i$ is by translation, there is a single automorphism $\xi : G \to G$, which is used to determine the image of any element in $\bigoplus G_i$ under $\varphi \in Aut(G \wr \mathbb{Z})$. $\qquad \square$

Using Lemma 5.11, we can now prove the following theorem.

THEOREM 5.12. *Let G be a finite simple group whose outer automorphism group is trivial. Then $G \in \mathcal{L}$, that is, $G \wr \mathbb{Z}$ has property R_∞.*

Proof. It follows from Lemma 5.11 that $\varphi' : \bigoplus G_i \to \bigoplus G_i$ preserves blocks of the form $G_i \bigoplus G_{j-i}$. We mimic the proof of Theorem 4.1, and consider the restriction of φ' to each block. If G has no outer automorphisms, then $\varphi'|_{G_i}$ must be conjugation by the same group element on each block. Since conjugation by $g \in G$ always has g as a nontrivial fixed point, the element $(g, g) \in G_i \oplus G_{j-i}$ will be a nontrivial fixed point for $\varphi' : G_i \oplus G_{j-i} \to G_i \oplus G_{j-i}$. Note that the number of φ-twisted conjugacy classes need not be the cardinality of the cokernel $Coker(Id - \varphi)$ when the group is nonabelian. Instead, the number of such classes is given by the number of ordinary conjugacy classes $[x]$ for which $[x] = [\varphi(x)]$ (see, e.g., [FH, theorem 5]). In particular, a nontrivial fixed point of φ yields a class other than that of the identity element. Thus there are at least two φ-twisted conjugacy classes on each block, and the theorem follows. □

Since both $\mathbb{Z}_2$ and $\mathbb{Z}_3$ are finite simple groups with trivial outer automorphism groups, Theorem 5.12 yields the immediate corollary that L_2 and L_3 have property R_∞ [GW1].

It follows immediately that many of the finite simple groups lie in $\mathfrak{L}$, including:

1) the Matthieu groups M_{11}, M_{23}, and M_{24},
2) the Conway groups C_1, C_2, and C_3,
3) the Janko groups J_1 and J_4,
4) the baby monster B and the Fischer-Griess monster M, and
5) other sporadic groups: the Fischer group Fi_{23}, the Held group He, the Harada-Norton group HN, the Lyons group Ly, and the Thompson group Th.

In the case of A_n for $n = 5$ and $n \geq 7$, which has outer automorphism group isomorphic to $\mathbb{Z}_2$, we obtain the following corollary.

COROLLARY 5.13. *The alternating group A_n for $n \geq 5$, $n \neq 6$, is in $\mathfrak{L}$, that is, $A_n \wr \mathbb{Z}$ has property R_∞.*

Proof. For $n \neq 6$, we have $Out(A_n) = \mathbb{Z}_2$. This single outer automorphism is conjugation by an odd permutation, and thus preserves a conjugacy class within A_n. Since this outer automorphism is defined up to inner automorphisms, we can choose it so that there is a fixed point in A_n. Then the argument in the proof of Theorem 5.12 works verbatim, that is, the automorphism φ' when restricted to the blocks $G_j \oplus G_{i-j}$ again must have a nontrivial fixed point. □

Since A_n is a subgroup of index 2 in S_n, we can apply reasoning similar to the proof of Lemma 5.11 to obtain the following proposition.

PROPOSITION 5.14. *For $n \geq 5$, the symmetric group $S_n \in \mathfrak{L}$, that is, $S_n \wr \mathbb{Z}$ has property R_∞.*

Proof. Since A_n is a subgroup of index 2 in S_n, we obtain the following short exact sequence:

$$1 \to \bigoplus (A_n)_i \to S_n \wr \mathbb{Z} \to \mathbb{Z}_2 \wr \mathbb{Z} \to 1.$$

Therefore, if we can show that $\bigoplus (A_n)_i$ is a characteristic subgroup of $\bigoplus (S_n)_i$, then the proposition follows from Lemma 2.1 and Theorem 4.1.

Following the proof of Lemma 5.11, let $G_0 = (S_n)_0$ and suppose that $\varphi'(G_0) \subset \cup_{k=1}^{r} G_{i_k}$. When we compose with the projection $pr_{i_j} : \cup_{k=1}^{r} G_{i_k} \to G_{i_j}$ onto any of these G_{i_k} factors, we obtain one of two possible images for $G \cong S_n$: either S_n or $\mathbb{Z}_2$, where the latter arises when the kernel of the homomorphism is A_n. When we restrict $pr_{i_j} \circ \varphi'$ for each j to $A_n \subset G_0$, we see that the image must either be the identity or all of A_n. As φ' is an automorphism and A_n is simple, there must be at least one j for which $pr_{i_j} \circ \varphi'(A_n) = A_n$. We now restrict the map φ' to $\bigoplus (A_n)_i$ initially and it follows that $\bigoplus (A_n)_i$ is a characteristic subgroup of $\bigoplus (S_n)_i$. $\qquad \square$

The following corollary follows immediately from Proposition 5.14, since any finite group can be embedded into S_n for some value of n.

COROLLARY 5.15. *Every group of the form $G \wr \mathbb{Z}$ where G is a nontrivial finite group can be embedded into a group that has property R_∞.*

References

[B]   R. Brown et al. (eds.), Handbook of topological fixed point theory, Springer, Dordrecht, 2005.

[CT]   S. Cleary and J. Taback, Dead end words in lamplighter groups and other wreath products, *Quarterly J. Math.* 56, no. 2 (2005), 165–178.

[EFW1]   A. Eskin, D. Fisher, and K. Whyte, Coarse differentiation of quasi-isometries I: Spaces not quasi-isometric to Cayley graphs, preprint, 2007.

[EFW2]   A. Eskin, D. Fisher, and K. Whyte, Coarse differentiation of quasi-isometries II: Rigidity for Sol and Lamplighter groups, preprint, 2007.

[F]   A.L. Fel'shtyn, The Reidemeister number of any automorphism of a Gromov hyperbolic group is infinite, *Zap. Nauchn. Sem. POMI* 279 (2001), 229–241.

[FG1]   A.L. Fel'shtyn and D. Gonçalves, Twisted conjugacy classes in Symplectic groups, mapping class groups and Braid groups (including an appendix: Geometric group theory and R_∞ property for mapping class groups, written with Francois Dahmani), preprint, 2007.

[FG2]   A.L. Fel'shtyn and D. Gonçalves, Twisted conjugacy classes of automorphisms of Baumslag-Solitar groups, *Algebra Discrete Math.* no. 3 (2006), 36–48.

[FH]   A.L. Fel'shtyn and R. Hill, The Reidemeister zeta function with applications to Nielsen theory and a connection with Reidemeister torsion, *K-Theory* 8, no. 4 (1994), 367–393.

[GW1]   D. Gonçalves and P. Wong, Twisted conjugacy classes in wreath products, *Internat. J. Alg. Comput.* 16 (2006), 875–886.

[GW2]   D. Gonçalves and P. Wong, Twisted conjugacy in exponential growth groups, *Bull. London Math. Soc.* 35 (2003), 261–268.

[GZ]   R. Grigorchuk and A. Zuk, The lamplighter group as a group generated by a 2-state automaton, and its spectrum, *Geom. Dedicata*, 87, no. 1–3 (2001) 209–244.

[J]   B. Jiang, Lectures on Nielsen fixed point theory, *Contemp. Math.* 14, American Mathematical Society, Providence, RI, 1983.

[L]   G. Levitt, On the automorphism group of generalized Baumslag-Solitar groups, *Geom. Topol.* 11 (2007), 473–515.

[LL]   G. Levitt and M. Lustig, Most automorphisms of a hyperbolic group have simple dynamics, *Ann. Sci. Ecole Norm. Sup.* 33 (2000), 507–517.

[MN]   Letter from R. Moeller to W. Woess, 2001.

[TWh]   J. Taback and K. Whyte, Twisted conjugacy and group actions, preprint, 2005.

[TWo1] J. Taback and P. Wong, Twisted conjugacy and quasi-isometry invariance for generalized solvable Baumslag-Solitar groups, *Journal London Math. Soc. (2)* 75 (2007), 705–717.

[TWo2] J. Taback and P. Wong, A note on twisted conjugacy and generalized Baumslag-Solitar groups, arXiv:math.GR/0606284, preprint, 2006.

[We]   F. Wecken, Fixpunktklassen. III. Mindestzahlen von Fixpunkten, (German) *Math. Ann.* 118 (1942), 544–577.

[Woe]   W. Woess, Lamplighters, Diestel-Leader graphs, random walks, and harmonic functions, *Combinatorics, Probability & Computing* 14 (2005), 415–433.

[Wo1]   P. Wong, Reidemeister number, Hirsch rank, coincidences on polycyclic groups and solvmanifolds, *J. Reine Angew. Math.* 524 (2000), 185–204.

[Wo2]   P. Wong, Fixed point theory for homogeneous spaces—a brief survey, Handbook of topological fixed point theory, 265–283, Springer, Dordrecht, 2005.

[Wo3]   P. Wong, Fixed point theory for homogeneous spaces, II, *Fund. Math.* 186, no. 2 (2005), 161–175.

[Wor]   K. Wortman, A finitely presented solvable group with small quasi-isometry group, *Michigan Math. J.* 55, no. 1 (2007), 3–24.

PART 4

*Group Actions on
Representations Varieties*

18

WILLIAM M. GOLDMAN AND EUGENE Z. XIA

ERGODICITY OF MAPPING CLASS GROUP ACTIONS ON SU(2)-CHARACTER VARIETIES

TO BOB ZIMMER ON HIS 60TH BIRTHDAY

Abstract

Let Σ be a compact orientable surface with genus g and n boundary components $\partial_1, \ldots, \partial_n$. Let $b = (b_1, \ldots, b_n) \in [-2, 2]^n$. Then the mapping class group $\mathrm{Mod}(\Sigma)$ acts on the relative SU(2)-character variety $\mathfrak{X}_b := \mathrm{Hom}_b(\pi, \mathrm{SU}(2))/\mathrm{SU}$, comprising conjugacy classes of representations ρ with $\mathrm{tr}(\rho(\partial_i)) = b_i$. This action preserves a symplectic structure on the open dense smooth submanifold of $\mathrm{Hom}_b(\pi, \mathrm{SU}(2))/\mathrm{SU}$ corresponding to irreducible representations. This subset has full measure and is connected. In this note we use the symplectic geometry of this space to give a new proof that this action is ergodic.

1. Introduction

Let $\Sigma = \Sigma_{g,n}$ be a compact oriented surface of genus g with n boundary components $\partial_1(\Sigma), \ldots, \partial_n(\Sigma)$. Choose base points $p_0 \in \Sigma$ and $p_i \in \partial_i(\Sigma)$. Let $\pi = \pi_1(\Sigma, p_0)$ denote the fundamental group of Σ. Choosing arcs from p_0 to each p_i identifies each fundamental group $\pi_1(\partial_i(\Sigma), p_i)$ with a subgroup $\pi_1(\partial_i) \hookrightarrow \pi$. The orientation on Σ induces orientations on each $\partial_i(\Sigma)$. For each i, denote the positively oriented generator of $\pi_1(\partial_i \Sigma)$ also by ∂_i.

The *mapping class group* $\mathrm{Mod}(\Sigma)$ consists of isotopy classes of orientation-preserving homeomorphisms of Σ, which pointwise fix each ∂_i. The Dehn-Nielsen theorem (see, for example, Farb-Margalit [1] or Morita [19]), identifies $\mathrm{Mod}(\Sigma)$ with a subgroup of $\mathrm{Out}(\pi) := \mathrm{Aut}(\pi)/\mathrm{Inn}(\pi)$.

Consider a connected compact semisimple Lie group G. Its complexification $G^{\mathbb{C}}$ is the group of complex points of a semisimple linear algebraic group defined over $\mathbb{R}$. Fix a conjugacy class $B_i \subset G$ for each boundary component ∂_i. Then the *relative representation variety* is

Goldman gratefully acknowledges partial support from National Science Foundation grant DMS070781 and the Oswald Veblen Fund at the Institute for Advanced Study. Xia gratefully acknowledges partial support by the National Science Council, Taiwan, with grants 96-2115-M-006-002 and 97-2115-M-006-001-MY3.

$$\mathrm{Hom}_B(\pi, G) := \{\rho \in \mathrm{Hom}(\pi, G) \mid \rho(\partial_j) \in B_j, \text{ for } 1 \le j \le n\}.$$

The action of the automorphism group $\mathrm{Aut}(\pi)$ on π induces an action on $\mathrm{Hom}_B(\pi, G^{\mathbb{C}})$ by composition. Furthermore, this action descends to an action of $\mathrm{Mod}(\Sigma) \subset \mathrm{Out}(\pi)$ on the categorical quotient or the *relative character variety*

$$\mathfrak{X}_B^{\mathbb{C}}(G) := \mathrm{Hom}_B(\pi, G^{\mathbb{C}}) /\!/ G^{\mathbb{C}}.$$

The moduli space $\mathfrak{X}_B^{\mathbb{C}}(G)$ has an invariant dense open subset that is a smooth complex submanifold. This subset has an invariant complex symplectic structure $\omega^{\mathbb{C}}$, which is algebraic with respect to the structure of $\mathfrak{X}_B^{\mathbb{C}}(G)$ as an affine algebraic set. The pull-back ω of the real part of this complex symplectic structure under

$$\mathfrak{X}_B(G) := \mathrm{Hom}_B(\pi, G)/G \longrightarrow \mathfrak{X}_B^{\mathbb{C}}(G)$$

defines a symplectic structure on a dense open subset, which is a smooth submanifold. The smooth measure defined by the symplectic structure is finite [3, 13, 11] and $\mathrm{Mod}(\Sigma)$-invariant. The main result of Goldman [6] (when G has $\mathrm{SU}(2)$- and $\mathrm{U}(1)$-factors) and Pickrell-Xia [24] (when $g > 1$) is

THEOREM. *The action of* $\mathrm{Mod}(\Sigma)$ *on each component of* $\mathfrak{X}_B(G)$ *is ergodic with respect to the measure induced by* ω.

The goal of this note is to give a short proof in the case that $G = \mathrm{SU}(2)$.

Recently, F. Palesi [23] proved ergodicity of $\mathrm{Mod}(\Sigma)$ on $\mathfrak{X}_B(\mathrm{SU}(2))$ when Σ is a compact connected *nonorientable* surface with $\chi(\Sigma) \le -2$. When Σ is nonorientable, the character variety fails to possess a symplectic structure (in fact its dimension may be odd) and it would be interesting to adapt the proof given here to the nonorientable case.

The proof given here arose from our investigation [10] of ergodic properties of subgroups of $\mathrm{Mod}(\Sigma)$ on character varieties. The closed curves on Σ play a central role. Namely, every closed curve defines a conjugacy class of elements in π, and hence a regular function

$$\mathrm{Hom}(\pi, G^{\mathbb{C}}) \xrightarrow{f_\alpha} \mathbb{C}$$

$$\rho \longmapsto \mathrm{tr}(\rho(\alpha))$$

for some representation $G^{\mathbb{C}} \longrightarrow \mathrm{GL}(N, \mathbb{C})$. These trace functions f_α are $G^{\mathbb{C}}$-conjugate invariant and results of Procesi [25] imply that such functions generate the coordinate ring $\mathbb{C}[\mathfrak{X}_B(\mathrm{SL}(2, \mathbb{C}))]$ of $\mathfrak{X}_B(\mathrm{SL}(2, \mathbb{C}))$.

Simple closed curves α determine elements of $\mathrm{Mod}(\Sigma)$, namely, the *Dehn twists* τ_α. Let S be a set of simple closed curves on Σ. Our methods apply to the subgroup $\Gamma_S \subset \mathrm{Mod}(\Sigma)$ generated by τ_α, where $\alpha \in S$. Our proof may be summarized: *if the trace functions f_α generate $\mathbb{C}[\mathfrak{X}_B(\mathrm{SL}(2,\mathbb{C}))]$, then the action of Γ_S on each component of $\mathfrak{X}_B(\mathrm{SU}(2))$ is ergodic.*

The original proof [6] decomposes Σ along a set $\mathfrak{P}$ of $3g - 3 + 2n$ disjoint curves into

$$2g - 2 + n = -\chi(\Sigma)$$

3-holed spheres (a *pants decomposition*.) The subgroup $\Gamma_\mathfrak{P}$ of $\mathrm{Mod}(\Sigma)$ stabilizing $\mathfrak{P}$ is generated by Dehn twists along curves in $\mathfrak{P}$. The corresponding trace functions define a map

$$\mathfrak{X}_b \xrightarrow{f_\mathfrak{P}} [-2,2]^\mathfrak{P},$$

which is an ergodic decomposition for the action of $\Gamma_\mathfrak{P}$. Thus any measurable function invariant under $\Gamma_\mathfrak{P}$ must factor through $f_\mathfrak{P}$. Changing $\mathfrak{P}$ by elementary moves on 4-holed spheres, and a detailed analysis in the case of $\Sigma_{0,4}$ and $\Sigma_{1,1}$, implies ergodicity under all of $\mathrm{Mod}(\Sigma)$. The present proof uses the commutative algebra of the character ring (in particular the work of Horowitz [12], Magnus [17], and Procesi [25]), and the identification of the twist flows with the Hamiltonians of trace functions [4]. Although it is not used in [6], the map $f_\mathfrak{P}$ is the *moment map* for the $\mathbb{R}^\mathfrak{P}$-action by twist flows, as well as the ergodic decomposition for $\Gamma_\mathfrak{P}$. Finding sets S of simple curves whose trace functions generate the character ring promises to be useful to prove ergodicity of the subgroup of $\mathrm{Mod}(\Sigma)$ generated by Dehn twists along elements of S (Goldman-Xia [10].)

In a similar direction, Sean Lawton has pointed out that this method of proof (combined with [15, 16]) implies in at least some cases ergodicity of $\mathrm{Mod}(\Sigma)$ on the relative $\mathrm{SU}(3)$-character varieties (except when $\Sigma \approx \Sigma_{0,3}$, where it is not true).

We are grateful to Sean Lawton, David Fisher, and the anonymous referee for helpful suggestions on this manuscript.

With great pleasure we dedicate this paper to Bob Zimmer. Goldman first presented this result in Zimmer's graduate course at Harvard University in fall 1985, and would like to express his warm gratitude to Zimmer for the friendship, support, and mathematical inspiration he has given over many years.

2. Simple Generators for the Character Ring

In this note we restrict to the case $G = SU(2)$ and $G^{\mathbb{C}} = SL(2, \mathbb{C})$. Conjugacy classes in $G = SU(2)$ are level sets of the *trace function* $SU(2) \overset{\text{tr}}{\to} [-2, 2]$. Thus a collection $B = (B_1, \ldots, B_n)$ of conjugacy classes in $SU(2)$ corresponds to an n-tuple

$$b = (b_1, \ldots, b_n) \in [-2, 2]^n.$$

We denote the *relative representation variety* by

$$\text{Hom}_b(\pi, SU(2)) := \{\rho \in \text{Hom}(\pi, SU(2)) \mid \text{tr}(\rho(\partial_i)) = b_i\}$$

and its quotient, the *relative character variety*, by

$$\mathfrak{X}_b :- \text{Hom}_b(\pi, SU(2))/SU(2).$$

THEOREM 2.1. *There exists a finite subset $S \subset \pi$ corresponding to simple closed curves on Σ such that the set of their trace functions $\{f_\gamma : \gamma \in S\}$ generates the coordinate ring $\mathbb{C}[\mathfrak{X}_b]$.*

We prove this theorem in §2.1 and §2.2.

2.1. Magnus-Horowitz-Procesi Generators

The following well-known proposition is a direct consequence of the work of Horowitz [12] and Procesi [25]. Compare also Magnus [17], Newstead [22], and Goldman [8].

PROPOSITION 2.2. *Let F_N be the free group freely generated by $A_1, \ldots, A_N$, and let*

$$\mathfrak{X}(N) := \text{Hom}(F_N, SL(2, \mathbb{C}))/\!\!/SL(2, \mathbb{C})$$

be its $SL(2, \mathbb{C})$-character variety. Denote by $\mathfrak{I}_N$ the collection of all

$$I = (i_1, \ldots, i_k) \in \mathbb{Z}^k$$

where

$$1 \le i_1 < \cdots < i_k \le N$$

and $k \le 3$. For $I \in \mathfrak{I}_N$, define

$$A_I := A_{i_1}, \ldots, A_{i_k}$$

and let

$$\mathfrak{X}(N) \xrightarrow{f_I} \mathbb{C}$$

$$[\rho] \longmapsto \operatorname{tr}(\rho(A_I))$$

the corresponding trace functions. Then the collection

$$\{f_I \mid I \in \mathfrak{I}_N\}$$

generates the coordinate ring $\mathbb{C}[\mathfrak{X}(N)]$.

We shall refer to the coordinate ring $\mathbb{C}[\mathfrak{X}(N)]$ as the *character ring*. Recall that by definition it is the subring of the ring of regular functions

$$\mathsf{SL}(2, \mathbb{C})^N \longrightarrow \mathbb{C}$$

consisting of $\mathsf{Inn}(\mathsf{SL}(2, \mathbb{C}))$-invariant functions.

2.2. Constructing Simple Loops

Suppose that Σ has genus $g \geq 0$ and $n > 0$ boundary components. (We postpone the case when Σ is closed, that is $n = 0$, to the end of this section.) We suppose that $\chi(\Sigma) = 2 - 2g - n < 0$. Then $\pi_1(\Sigma)$ is free of rank $N = 2g + n - 1$. We describe a presentation of $\pi_1(\Sigma)$ such that the above elements A_I, for $I \in \mathfrak{I}_N$, can be represented by simple closed curves on Σ (compare Figures 1–4). We also identify I with the *subset*

$$\{i_1, \ldots, i_k\} \subset \{1, \ldots, N\}.$$

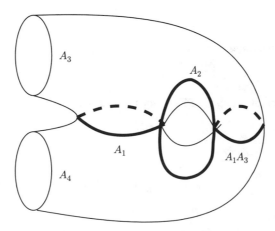

Fig. 1. Simple loops on $\Sigma_{1,2}$ corresponding to words A_1, A_2, A_3, A_1A_3 and $A_4^{-1} = A_1A_2A_1^{-1}A_2^{-1}A_3$ in free generators $\{A_1, A_2, A_3\}$.

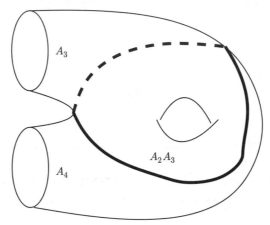

Fig. 2. Simple loop corresponding to $A_2 A_3$.

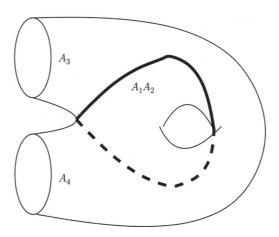

Fig. 3. Simple loop corresponding to $A_1 A_2$.

The fundamental group $\pi_1(\Sigma)$ admits a presentation with generators

$$A_1, \ldots, A_{2g}, A_{2g+1}, \ldots, A_{2g+n}$$

subject to the relation

$$A_1 A_2 A_1^{-1} A_2^{-1} \ldots A_{2g-1} A_{2g} A_{2g-1}^{-1} A_{2g}^{-1} \ldots A_{2g+1} \ldots A_{2g+n} = 1.$$

Then

$$\pi = \pi_1(\Sigma) \cong \mathsf{F}_{2g+n-1},$$

freely generated by the set $\{A_1, \ldots, A_{2g+n-1}\}$.

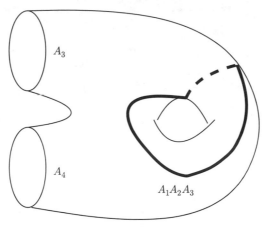

Fig. 4. Simple loop corresponding to $A_1 A_2 A_3$.

To represent the elements $A_I \in \pi_1(\Sigma)$ explicitly as *simple loops,* we realize Σ as the union of a planar surface P and g *handles* $H_1, \ldots, H_g$. In the notation of [8], $P \approx \Sigma_{0,g+n}$ has $g + n$ boundary components

$$\alpha_1, \ldots, \alpha_g, \alpha_{g+1}, \ldots, \alpha_{g+n}$$

and each handle $H_j \approx \Sigma_{1,1}$ is a 1-holed torus. The original surface Σ is obtained by attaching H_j to P along α_j for $j = 1, \ldots, g$.

We construct the curves A_i, for $i = 1, \ldots, 2g + n$ as follows. Choose a pair of base points p_j^+, p_j^- on each α_j for $j = 1, \ldots, g + n$. Let α_j^- be the oriented subarc of α_j from p_j^- to p_j^+, and α_j^+ the corresponding subarc from p_j^+ to p_j^-. Thus $\alpha_j \simeq \alpha_j^- * \alpha_j^+$ is a boundary component of P.

Choose a system of disjoint arcs β_j from p_j^+ to p_{j+1}^-, where β_{g+n} runs from p_{g+n}^+ to p_1^- in the *cyclic indexing* of $\{1, 2, \ldots, g + n\}$. Compare Figure 5.

For $I \in \mathfrak{I}_N$, the curve A_I will be the concatenation $E_1^I * \ldots E_{g+n}^I$ of simple arcs E_j^I running from p_j^- to p_{j+1}^-. Define

$$E_j^\emptyset := \alpha_j^- * \beta_j,$$

so that

$$A^\emptyset := E_1^\emptyset * \cdots * E_N^\emptyset$$

is a contractible loop.

Suppose first that $i > 2g$. Then the curve A_i will be freely homotopic to the oriented loop α_i^{-1}, corresponding to a component of $\partial \Sigma$. The arc

$$E_i^+ := (\alpha_i^+)^{-1} * \beta_i$$

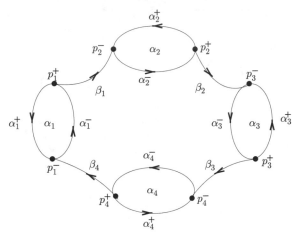

Fig. 5. A planar surface $P \approx \Sigma_{0,4}$.

goes from p_i^- to p_{i+1}^- (cyclically). Then A_i corresponds to the arc

$$A_i := E_1^\emptyset * \cdots * E_{2g}^\emptyset$$

$$* E_{2g+1}^\emptyset * \cdots * E_i^+ * \ldots E_{2g+n-1}^\emptyset.$$

For $i \leq 2g$, the curves A_i will lie on the handles H_j. The curves A_{2j-1} and A_{2j} define a basis for the relative homology of H_j and the relative homology class of the curve

$$A_{2j-1,2j} := A_{2j-1}A_{2j}$$

is their sum. Compare Figures 6 and 7.

As above, we define three simple arcs $\gamma_j, \delta_j, \eta_j$ running from p_j^- to p_j^+ to build these three curves, respectively.

The boundary ∂H_j identifies with α_j for $j = 1, \ldots, g$. The two points on ∂H_j, which identify to

$$p_j^\pm \in \alpha_j \subset \partial P,$$

divide ∂H_j into two arcs. Without danger of confusion, denote these arcs by $\alpha_j^\pm$ as well. On the handle H_j, choose disjoint simple arcs γ_j, δ_j, and η_j running from p_i^+ to p_i^- such that the

$$H_j \setminus (\gamma_j \cup \delta_j)$$

is a hexagon. Two of its edges correspond to the arcs $\alpha_j^\pm$. Its other four edges are the two pairs obtained by splitting γ_j and δ_j. (Compare Figure 6.) Let η_j to

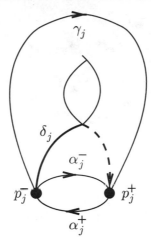

Fig. 6. A handle $H_j \approx \Sigma_{1,1}$.

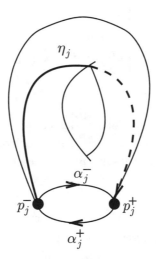

Fig. 7. A $(1, 1)$-curve η_j on the handle H_j.

be a simple arc homotopic to $\gamma_j * (\alpha_j^+)^{-1} * \delta_j$, where $*$ denotes concatenation. For each $j \leq g$, the arcs

$$E_j^\gamma = \gamma_j * \beta_j$$

$$E_j^\delta = \delta_j * \beta_j$$

$$E_j^\eta = \eta_j * \beta_j$$

run from p_j^- to p_j^+ and define

$$A_{2j-1} = E_1^\emptyset * \cdots * E_j^\gamma * \ldots E_g^\emptyset * \cdots * E_{g+n}^\emptyset$$

$$A_{2j} = E_1^\emptyset * \cdots * E_j^\delta * \ldots E_g^\emptyset * \cdots * E_{g+n}^\emptyset$$

$$A_{2j-1,2j} = E_1^\emptyset * \cdots * E_j^\eta * \ldots E_g^\emptyset * \cdots * E_{g+n}^\emptyset.$$

In general, suppose that $I \in \mathfrak{I}_N$. Define

$$A_I := E_1^I * \cdots * E_{g+n}$$

where

$$E_j^I = \begin{cases} E_j^\emptyset & \text{if } j \notin I \\ E_j^+ & \text{if } j \in I \end{cases}$$

if $j > g$ and

$$E_j^I = \begin{cases} E_j^\emptyset & \text{if } 2j-1, 2j \notin I \\ E_j^\gamma & \text{if } 2j-1 \in I, 2j \notin I \\ E_j^\delta & \text{if } 2j-1 \notin I, 2j \in I \\ E_j^\eta & \text{if } 2j-1, 2j \in I \end{cases}$$

if $j \le g$.

Now each A_I is *simple:* Each of the oriented arcs $\alpha_i^\pm$, β_i, γ_i, δ_i, η_i are embedded and intersect only along $p_i^\pm$. In particular, each of the above oriented arcs begins at some $p_i^\pm$ and ends at some $p_i^\mp$. Thus each

$$E_j^\emptyset, E_j^+, E_j^\gamma, E_j^\delta, E_j^\eta$$

is a simple arc running from p_j^- to p_{j+1}^-, cyclically. The loop A_I concatenates these arcs, which only intersect along the p_i^-. Each of these endpoints occurs exactly twice, once as the initial endpoint and once as the terminal endpoint. Therefore, the loop A_I is simple.

This collection A_I, for $I \in \mathfrak{I}_N$, of simple curves determines a collection of regular functions f_I on $\mathfrak{X}^{\mathbb{C}}$, which generate the character ring. Since the inclusion

$$\mathfrak{X}_b^{\mathbb{C}} \hookrightarrow \mathfrak{X}^{\mathbb{C}}$$

is a morphism of algebraic sets, the restrictions of f_I to $\mathfrak{X}_b^{\mathbb{C}}$ generate the coordinate ring of $\mathfrak{X}_b^{\mathbb{C}}$.

The case $n = 0$ remains. To this end, the character variety of $\Sigma_{g,0}$ appears as the relative character variety of $\Sigma_{g,1}$ with boundary condition $b_1 = 2$. As above, the restrictions of the f_I from $\Sigma_{g,1}$ to the character variety of $\Sigma_{g,0}$ generate its coordinate ring. The proof of Theorem 2.1 is complete.

3. Infinitesimal Transitivity

The application of Theorem 2.1 involves several lemmas to deduce that the flows of the Hamiltonian vector fields $\mathsf{Ham}(f_\gamma)$, where $\gamma \in \mathcal{S}$, generate a transitive action on $\mathfrak{X}_b$.

LEMMA 3.1. *Let X be an affine variety over a field k. Suppose that $\mathcal{F} \subset k[X]$ generates the coordinate ring $k[X]$. Let $x \in X$. Then the differentials $df(x)$, for $f \in \mathcal{F}$, span the cotangent space $T_x^*(X)$.*

Proof. Let $\mathfrak{M}_x \subset k[X]$ be the maximal ideal corresponding to x. Then the functions $f - f(x)1$, where $f \in \mathcal{F}$, span $\mathfrak{M}_x$. The correspondence

$$\mathfrak{M}_x \longrightarrow T_x^*(X)$$

$$f \longmapsto df(x)$$

induces an isomorphism $\mathfrak{M}_x/\mathfrak{M}_x^2 \xrightarrow{\cong} T_x^*(X)$. In particular, it is onto. Therefore, the covectors $df(x)$ span $T_x^*(X)$ as claimed. $\square$

LEMMA 3.2. *Let X be a connected symplectic manifold and $\mathcal{F}$ be a set of functions on X such that at every point $x \in X$, the differentials $df(x)$, for $f \in \mathcal{F}$, span the cotangent space $T_x^*(X)$. Then the group $\mathfrak{G}$ generated by the Hamiltonian flows of the vector fields $\mathsf{Ham}(f)$, for $f \in \mathcal{F}$, acts transitively on X.*

Proof. The nondegeneracy of the symplectic structure implies that the vector fields $\mathsf{Ham}(f)(x)$ span the tangent space $T_x X$ for every $x \in X$. By the inverse function theorem, the $\mathfrak{G}$-orbit $\mathfrak{G} \cdot x$ of x is open. Since the orbits partition X and X is connected, $\mathfrak{G} \cdot x = X$ as claimed. $\square$

PROPOSITION 3.3. *Let $b = (b_1, \ldots, b_m) \in [-2, 2]^n$. Then $\mathfrak{X}_b$ is either empty or connected.*

The proof follows from Newstead [21] and Goldman [5]. Alternatively, apply Mehta-Seshadri [18] to identify $\mathfrak{X}_b$ with a moduli space of semistable parabolic bundles, and apply their result that the corresponding moduli space is irreducible.

COROLLARY 3.4. *Let $\mathfrak{G}$ be the group generated by the flows of the Hamiltonian vector fields* $\mathsf{Ham}(f_\gamma)$, *where* $\gamma \in S$. *Then $\mathfrak{G}$ acts transitively on* $\mathfrak{X}_b$.

Proof. By Theorem 2.1,

$$\{f_\gamma \mid \gamma \in S\}$$

generates $\mathbb{C}[\mathfrak{X}_b]$. Lemma 3.1 implies that at every point $x \in \mathfrak{X}_b$ the differentials $df_\gamma(x)$ span $T_x^*(\mathfrak{X}_b)$. Proposition 3.3 implies that $\mathfrak{X}_b$ is connected. Now apply Lemma 3.2. $\qquad\qquad\square$

4. Hamiltonian Twist Flows

We briefly review the results of Goldman [4], describing the flows generated by the Hamiltonian vector fields $\mathsf{Ham}(f_\alpha)$, when α represents a *simple* closed curve. In that case the local flow of this vector field on $\mathfrak{X}(G)$ lifts to a flow ξ_t on the representation variety $\mathsf{Hom}_B(\pi, G)$. Furthermore this flow admits a simple description [4] as follows:

4.1. Invariant Functions and Centralizing 1-Parameter Subgroups
Let Ad be the adjoint representation of G on its Lie algebra $\mathfrak{g}$. We suppose that Ad preserves a nondegenerate symmetric bilinear form $\langle , \rangle$ on $\mathfrak{g}$. In the case $G = \mathsf{SU}(2)$, this will be

$$\langle X, Y \rangle := \mathsf{tr}(XY).$$

Let $G \xrightarrow{f} \mathbb{R}$ be a function invariant under the inner automorphisms $\mathsf{Inn}(G)$. Following [4], we describe how f determines a way to associate to every element $x \in G$ a 1-parameter subgroup

$$\zeta^t(x) = \exp\left(t\mathsf{F}(x)\right)$$

centralizing x. Given f, define its *variation function* $G \xrightarrow{\mathsf{F}} \mathfrak{g}$ by

$$\langle \mathsf{F}(x), \upsilon \rangle = \frac{d}{dt}\bigg|_{t=0} f(x \exp\left(t\upsilon\right))$$

for all $\upsilon \in \mathfrak{g}$. Invariance of f under $\mathsf{Ad}(G)$ implies that F is G-equivariant:

$$\mathsf{F}(gxg^{-1}) = \mathsf{Ad}(g)\mathsf{F}(x).$$

Taking $g = x$ implies that the 1-parameter subgroup

4.1 $$\zeta^t(x) := \exp\left(t\mathsf{F}(x)\right)$$

lies in the centralizer of $x \in G$.

Intrinsically, $F(x) \in \mathfrak{g}$ is dual (by $\langle , \rangle$) to the element of $\mathfrak{g}^*$ corresponding to the left-invariant 1-form on G extending the covector $df(x) \in T_x^*(G)$.

There are two cases, depending on whether α is *nonseparating* or *separating*. Let $\Sigma|\alpha$ denote the surface with boundary obtained by *splitting* Σ along α. The boundary of $\Sigma|\alpha$ has two components, denoted by $\alpha_\pm$, corresponding to α. The original surface Σ may be reconstructed as a quotient space under the identification of α_- with α_+.

4.2. Nonseparating Loops

If α is nonseparating, then $\pi = \pi_1(\Sigma)$ can be reconstructed from the fundamental group $\pi_1(\Sigma|\alpha)$ as an HNN-extension:

4.2
$$\pi \cong \Big(\pi_1(\Sigma|\alpha) \amalg \langle \beta \rangle \Big) \Big/ \Big(\beta \alpha_- \beta^{-1} = \alpha_+ \Big).$$

A representation ρ of π is determined by

- the restriction ρ' of ρ to the subgroup $\pi_1(\Sigma|\alpha) \subset \pi$, and
- the value $\beta' = \rho(\beta)$,

which satisfies

4.3
$$\beta' \rho'(\alpha_-) \beta'^{-1} = \rho'(\alpha_+).$$

Furthermore, any pair (ρ', β') where ρ' is a representation of $\pi_1(\Sigma|\alpha)$ and $\beta' \in G$ satisfies Equation **4.3** determines a representation ρ of π.

The *twist flow* ξ_α^t, for $t \in \mathbb{R}$ on $\mathrm{Hom}(\pi, \mathsf{SU}(2))$, is then defined as follows:

4.4
$$\xi_\alpha^t(\rho) : \gamma \longmapsto \begin{cases} \rho(\gamma) & \text{if } \gamma \in \pi_1(\Sigma|\alpha) \\ \rho(\beta) \zeta^t(\rho(\alpha_-)) & \text{if } \gamma = \beta. \end{cases}$$

where ζ^t is defined in Equation **4.1**. This flow covers the flow generated by $\mathrm{Ham}(f_\alpha)$ on $\mathfrak{X}_b$ (see [4]).

4.3. Separating Loops

If α separates, then $\pi = \pi_1(\Sigma)$ can be reconstructed from the fundamental groups $\pi_1(\Sigma_i)$ of the two components Σ_1, Σ_2 of $\Sigma|\alpha$, as an amalgam

4.5
$$\pi \cong \pi_1(\Sigma_1) \amalg_{\langle \alpha \rangle} \pi_1(\Sigma_2).$$

A representation ρ of π is determined by its restrictions ρ_i to $\pi_1(\Sigma_i)$. Furthermore, any two representations ρ_i of π satisfying $\rho_1(\alpha) = \rho_2(\alpha)$ determines a representation of π.

The *twist flow* is defined by

4.6 $$\xi_\alpha^t(\rho) : \gamma \longmapsto \begin{cases} \rho(\gamma) & \text{if } \gamma \in \pi_1(\Sigma_1) \\ \zeta^t(\rho(\alpha))\, \rho(\gamma)\, \zeta^{-t}(\rho(\alpha)) & \text{if } \gamma \in \pi_1(\Sigma_2) \end{cases}$$

where ζ^t is defined in Equation **4.1**.

4.4. Dehn Twists

Let $\alpha \subset \Sigma$ be a simple closed curve. The *Dehn twist* along α is the mapping class $\tau_\alpha \in \mathsf{Mod}(\Sigma)$ represented by a homeomorphism $\Sigma \longrightarrow \Sigma$ supported in a tubular neighborhood $N(\alpha)$ of α defined as follows. In terms of a homeomorphism $S^1 \times [0,1] \overset{h}{\to} N(\alpha)$, which takes α to $S^1 \times \{0\}$, the Dehn twist is

$$\tau_\alpha \circ h(\zeta, t) = h(e^{2t\pi i}\zeta, t).$$

If α is essential, then τ_α induces a nontrivial element of $\mathsf{Out}(\pi)$ on $\pi = \pi_1(\Sigma)$.

If α is nonseparating, then $\pi = \pi_1(\Sigma)$ can be reconstructed from the fundamental group $\pi_1(\Sigma|\alpha)$ as an HNN-extension as in Equation **4.2**. The Dehn twist τ_α induces the automorphism $(\tau_\alpha)_* \in \mathsf{Aut}(\pi)$ defined by

$$(\tau_\alpha)_* : \gamma \longmapsto \begin{cases} \gamma & \text{if } \gamma \in \pi_1(\Sigma|\alpha) \\ \gamma\alpha & \text{if } \gamma = \beta. \end{cases}$$

The induced map $(\tau_\alpha)^*$ on $\mathsf{Hom}(\pi, G)$ maps ρ to

4.7 $$(\tau_\alpha)^*(\rho) : \gamma \longmapsto \begin{cases} \rho(\gamma) & \text{if } \gamma \in \pi_1(\Sigma|\alpha) \\ \rho(\gamma)\rho(\alpha)^{-1} & \text{if } \gamma = \beta. \end{cases}$$

If α separates, then $\pi = \pi_1(\Sigma)$ can be reconstructed from the fundamental groups $\pi_1(\Sigma_i)$ as an amalgam as in Equation **4.5**. The Dehn twist τ_α induces the automorphism $(\tau_\alpha)_* \in \mathsf{Aut}(\pi)$ defined by

$$(\tau_\alpha)_* : \gamma \longmapsto \begin{cases} \gamma & \text{if } \gamma \in \pi_1(\Sigma_1) \\ \alpha\gamma\alpha^{-1} & \text{if } \gamma \in \pi_1(\Sigma_2). \end{cases}$$

The induced map $(\tau_\alpha)^*$ on $\mathsf{Hom}(\pi, G)$ maps ρ to

4.8 $$(\tau_\alpha)^*(\rho) : \gamma \longmapsto \begin{cases} \rho(\gamma) & \text{if } \gamma \in \pi_1(\Sigma_1) \\ \rho(\alpha)^{-1}\rho(\gamma)\rho(\alpha) & \text{if } \gamma \in \pi_1(\Sigma_2). \end{cases}$$

5. The Case $G = \text{SU}(2)$

Now we specialize the preceding theory to the case $G = \text{SU}(2)$. Its Lie algebra $\mathfrak{su}(2)$ consists of 2×2 traceless skew-Hermitian matrices over $\mathbb{C}$.

5.1. One-Parameter Subgroups

The trace function

$$\text{SU}(2) \xrightarrow{f} [-2, 2]$$

$$x \longmapsto \text{tr}(x)$$

induces the variation function

$$\text{SU}(2) \xrightarrow{F} \mathfrak{su}(2)$$

$$x \longmapsto x - \frac{\text{tr}(x)}{2}\mathbb{I},$$

the projection of $x \in \text{SU}(2) \subset M_2(\mathbb{C})$ to $\mathfrak{su}(2)$. Explicitly, if $x \in \text{SU}(2)$, there exists $g \in \text{SU}(2)$ such that

$$x = g \begin{bmatrix} e^{i\theta} & 0 \\ 0 & e^{-i\theta} \end{bmatrix} g^{-1}.$$

Then $f(x) = 2\cos(\theta)$,

$$F(x) = g \begin{bmatrix} 2i\sin(\theta) & 0 \\ 0 & -2i\sin(\theta) \end{bmatrix} g^{-1} \in \mathfrak{su}(2),$$

and the corresponding 1-parameter subgroup is

$$\zeta^t(x) = g \begin{bmatrix} e^{2i\sin(\theta)t} & 0 \\ 0 & e^{-2i\sin(\theta)t} \end{bmatrix} g^{-1} \in \text{SU}(2).$$

Except in two exceptional cases this 1-parameter subgroup is isomorphic to S^1. Namely, if $f(x) = \pm 2$, then $x = \pm\mathbb{I}$. These two elements comprise the *center* of $\text{SU}(2)$. In all other cases, $-2 < f(x) < 2$ and $\zeta^t(x)$ is a circle subgroup. Notice that this circle subgroup contains x:

5.1
$$x = \zeta^{s(x)}(x)$$

where

5.2
$$s(x) := \frac{2}{\sqrt{4 - f(x)^2}} \cos^{-1}\left(\frac{f(x)}{2}\right).$$

Furthermore,

5.3
$$\zeta^t(x) = \mathbb{I}$$

if and only if

$$t \in \frac{4\pi}{\sqrt{4 - f(x)^2}} \, \mathbb{Z}.$$

(Compare Goldman [6].)

PROPOSITION 5.1. *Let $\alpha \in \pi$ be represented by a simple closed curve, and ξ_α^t be the corresponding twist flow on $\mathrm{Hom}(\pi, G)$ as defined in flows 4.4 and 4.6. Let $\rho \in \mathrm{Hom}(\pi, G)$. Then*

$$(\tau_\alpha)^*(\rho) = \xi_\alpha^{s(\rho(\alpha))}$$

where s is defined in Equation 5.2.

Proof. Combine Equation 5.1 with flow 4.4 when α is nonseparating and flow 4.6 when α separates. $\qquad\qquad\square$

The basic dynamical ingredient of our proof, (like the original proof in [6]) is the ergodicity of an irrational rotation of S^1. There is a unique translation-invariant probability measure on S^1 (Haar measure). Furthermore, this measure is ergodic under the action of any infinite cyclic subgroup. Recall that an action of group Γ of measure-preserving transformations of a measure space $(X, \mathcal{B}, \mu)$ is *ergodic* if and only if every invariant measurable set either has measure 0 or has full measure (its complement has measure 0).

LEMMA 5.2. *If $\cos^{-1}(f(x)/2)/\pi$ is irrational, then the cyclic group $\langle x \rangle$ is a dense subgroup of the 1-parameter subgroup*

$$\{\zeta^t(x) \mid t \in \mathbb{R}\} \cong S^1$$

and acts ergodically on S^1 with respect to Lebesgue measure.

For these basic facts see Furstenberg [2], Haselblatt-Katok [14], Morris [20], or Zimmer [26].

COROLLARY 5.3. *Let α, ξ_α^t and τ_α be as in Proposition 5.1. Then for almost every $b \in [-2, 2]$, $(\tau_\alpha)^*$ acts ergodically on the orbit*

$$\{\xi_\alpha^t([\rho])\}_{t \in \mathbb{R}},$$

when $f_\alpha(\rho) = b$.

Proof. Combine Proposition 5.1 with Lemma 5.2. $\qquad\qquad\square$

PROPOSITION 5.4. *Let* $\alpha \in S$ *be a simple closed curve, with twist vector field* ξ_α *and Dehn twist* τ_α. *Let* $\mathfrak{X}_b \xrightarrow{\psi} \mathbb{R}$ *be a measurable function invariant under the cyclic group* $\langle(\tau_\alpha)^*\rangle$. *Then there exists a nullset* $\mathcal{N}$ *of* $\mathfrak{X}_b$ *such that the restriction of* ψ *to the complement of* $\mathcal{N}$ *is constant on each orbit of the twist flow* ξ_α.

Proof. Disintegrate the symplectic measure on $\mathfrak{X}_b$ over the quotient map

$$\mathfrak{X}_b \longrightarrow \mathfrak{X}_b/\xi_\alpha$$

as in Furstenberg [2] or Morris [20], 3.3.3, 3.3.4]. By Equation **5.3** almost all fibers of this map are circles.

The subset

$$\mathcal{N} := f_\alpha^{-1}(2\cos(\mathbb{Q}\pi)) \subset \mathfrak{X}_b$$

has measure 0. By Corollary 5.3, the action of $(\tau_\alpha)^*$ is ergodic on each circle in the complement of $\mathcal{N}$. In particular, ψ factors through the quotient map, as desired. □

Conclusion of proof of main theorem. Suppose that $\mathfrak{X}_b \xrightarrow{\psi} \mathbb{R}$ is a measurable function invariant under $\mathrm{Mod}(\Sigma)$; we show that ψ is almost everywhere constant.

To this end let S be the collection of simple closed curves in Theorem 2.1. Then, for each $\alpha \in S$, the function ψ is invariant under the mapping $(\tau_\alpha)^*$ induced by the Dehn twist along $\alpha \in S$. By Proposition 5.4, ψ is constant along almost every orbit of the Hamiltonian flow ξ_α of $\mathrm{Ham}(f_\alpha)$. Thus, up to a nullset, ψ is constant along the orbits of the group $\mathfrak{G}$ generated by these flows. By Corollary 3.4, $\mathfrak{G}$ acts transitively on $\mathfrak{X}_b$. Therefore ψ is almost everywhere constant, as claimed. The proof is complete. □

References

[1] Farb, B. and Margalit, D., *A primer on the mapping class group*, Princeton University Press, Princeton, NJ (to appear).

[2] Furstenberg, H., *Recurrence in Ergodic Theory and Combinatorial Number Theory*, Princeton University Press, Princeton, NJ (1981).

[3] Goldman, W., *The symplectic nature of fundamental groups of surfaces*, Adv. Math. **54** (1984), 200–225.

[4] ———, *Invariant functions on Lie groups and Hamiltonian flows of surface group representations*, Inv. Math. **85** (1986), 1–40.

[5] ———, *Topological components of spaces of representations*, Inv. Math. **93** (1988), no. 3, 557–607.

[6] ———, *Ergodic theory on moduli spaces*, Ann. Math. **146** (1997), 475–507.

[7] _____, The complex symplectic geometry of SL(2, C)-characters over surfaces, in Algebraic Groups and Arithmetic, Proceedings of the celebration of the sixtieth birthday of M. S. Raghunathan, December 17–22, 2001, Tata Institute of Fundamental Research, Mumbai (2004), 375–407, math.DG/0304307.

[8] _____, Trace coordinates on Fricke spaces of some simple hyperbolic surfaces, chapter 15, pp. 611–684, in Handbook of Teichmüller theory, vol. II (A. Papadopoulos, ed.), IRMA Lectures in Mathematics and Theoretical Physics, volume 13, European Mathematical Society (2008), math.GM.0402103.

[9] Goldman, W. and Millson, J., Eichler-Shimura homology and the finite generation of cusp forms by hyperbolic Poincaré series, Duke Math. J. 53 (1986), 1081–1091.

[10] Goldman, W. and Xia, E., Action of the Johnson-Torelli group on SU(2)-character varieties (in preparation).

[11] Guruprasad, K., Huebschmann, J., Jeffrey, L., and Weinstein, A., Group systems, groupoids, and moduli spaces of parabolic bundles, Duke Math. J. 89 (1997), 377–412.

[12] Horowitz, R., Characters of free groups represented in the two-dimensional linear group, Comm. Pure Appl. Math. 25 (1972), 635–649.

[13] Huebschmann, J., Symplectic and Poisson structures of certain moduli spaces I, Duke Math. J. 80 (1995), 737–756.

[14] Katok, A. and Hasselblatt, B., Introduction to the modern theory of dynamical systems, in Encyclopedia of Mathematics and Its Applications 54, Cambridge University Press, Cambridge (1995).

[15] Lawton, S., Minimal affine coordinates for SL(3, C)-character varieties of free groups, J. Algebra 320 (2008), no. 10, 3773–3810.

[16] _____, Algebraic independence in SL(3, C)-character varieties of free groups, arXiv:0807.0798v1.

[17] Magnus, W., Rings of Fricke characters and automorphism groups of free groups, Math. Zeit. 170 (1980), 91–103.

[18] Mehta, V. B. and Seshadri, C. S., Moduli of vector bundles on curves with parabolic structures, Math. Ann. 248 (1980), no. 3, 205–239.

[19] Morita, S., Introduction to mapping class groups of surfaces and related groups, chapter 7 in Handbook of Teichmüller theory, vol. I (A. Papadopoulos, ed.), IRMA Lectures in Mathematics and Theoretical Physics 11, European Mathematical Society, Zürich, (2007), 353–386.

[20] Morris, D. W., Ratner's theorems on unipotent flows, Chicago Lectures in Mathematics. University of Chicago Press, Chicago, (2005).

[21] Newstead, P. E., Topological properties of some spaces of stable bundles, Topology 6 (1967), 241–262.

[22] _____, Introduction to moduli problems and orbit spaces, Tata Institute of Fundamental Research, Bombay (1978).

[23] Palesi, F., Ergodic actions of mapping class groups on moduli spaces of representations of non-orientable surfaces, math.GT.0807.1615.

[24] Pickrell, D. and Xia, E., Ergodicity of mapping class group actions on representation varieties I: Closed surfaces, Comment. Math. Helv. 77 (2002), 339–362.

[25] Procesi, C., The invariants of n × n matrices, Adv. Math. 19 (1976), 306–381

[26] Zimmer, R., Ergodic theory and semisimple groups, Monographs in Mathematics 81, Birkhäuser, Basel (1984).

19

ALEXANDER LUBOTZKY

DYNAMICS OF Aut(F_N) ACTIONS ON GROUP PRESENTATIONS AND REPRESENTATIONS

TO BOB ZIMMER

Abstract

Several different areas of group theory, topology, and geometry have led to the study of the action of Aut(F_n)—the automorphism group of the free group on n generators—on Hom(F_n, G) when G is either finite, compact, or simple Lie group. In this survey, we describe these topics and results, with special emphasis on some similarities and with an effort to give a somewhat uniform treatment. This perspective sometimes suggests new questions, conjectures, and methods borrowed from one area to another.

Contents

1. Introduction

In this paper we will survey a collection of results originated from various different contexts but all have the following common form:

Let $F = F_n$ be the free group on $n \geq 2$ generators and G a group. Denote

1.1 $\qquad H_n(G) = \mathrm{Hom}\,(F_n, G)$

1.2 $\qquad E_n(G) = \mathrm{Epi}\,(F_n, G) = \{\varphi \in H_n(G) | \varphi \text{ is onto}\}.$

The group $\text{Aut}(G) \times \text{Aut}(F_n)$ acts on $H_n(G)$ by

1.3 $$(\beta, \alpha)(\varphi) = \beta \circ \varphi \circ \alpha^{-1}$$

where $\alpha \in \text{Aut}(F_n)$, $\beta \in \text{Aut}(G)$, and $\varphi \in H_n(G)$.

The action clearly preserves $E_n(G)$. We denote by $\bar{H}_n(G)$ (respectively, $\bar{E}_n(G)$) the quotient set $\text{Aut}(G) \backslash H_n(G)$ (respectively, $\text{Aut}(G) \backslash E_n(G)$). So $\text{Aut}(F_n)$ acts on $\bar{H}_n(G)$ and on $\bar{E}_n(G)$. What really acts on $\bar{E}_n(G)$ is $\text{Out}(F_n)$, as $\text{Inn}(F_n)$ acts trivially. The orbits of $\bar{E}_n(G)$ under $\text{Aut}(F_n)$ are called "T-systems of G" in some of the old literature about the subject, but we will not use this term. What we will present here is a collection of results and methods to study these sets and these actions. The motivations come from various quite different areas of research: presentation of groups, actions of groups on handlebodies, computational group theory and the product replacement algorithm, the theory of linear groups, Fuchsian and Kleinian groups, compact groups, and more.

Here is a brief outline of the content of this survey, section by section:

The set $\bar{E}_n(G)/\text{Aut}(F_n)$ is in a one-to-one correspondence with equivalence classes of presentations of G using n generators. While for an infinite groups G this can be a very large and complicated set, it is known in some cases, and conjectured in others, that it is a "tame set" when G is finite. In Section 2, we describe some of the known results and open problems. The orbits of $\text{Aut}(F_n)$ on $E_n(G)$ can be illustrated as the connected components of a graph—the PRA graph.

The product replacement algorithm (PRA) is a probabilistic algorithm providing a pseudorandom element in a finite group given by a set of generators. It is really a random walk on the PRA graph whose vertices are the n-generating sets, that is, the set $E_n(G)$.

Furthermore, the vertices of the graph and its connected components are also in correspondence with actions of G on handlebodies and their equivalence classes. So, the topological question and the computational group theory problem discussed in Section 3, are different forms of the questions raised in Section 2 on the presentation theory of G.

In Section 4, we take the opportunity to present in passing another graph associated with G—the Andrews-Curtis graph. Its origin is the classical Andrews-Curtis conjecture in combinatorial group theory and topology, but it also got a new interest from computational group theory. In Sections 5 and 6, respectively, we discuss the connectivity of the PRA graphs for solvable and simple groups. The case of finite simple groups is of special interest. A long-standing conjecture (Wiegold conjecture) suggests various extensions to compact and Lie groups as well as potential applications to the representation theory of Aut (F_n).

In Section 7, we will compare the situation to the case where F is replaced by its profinite completion $\hat{F}$ and explain why profinite presentations "behave nicer" than discrete presentations: all is due to a beautiful lemma of Gashütz, which seems to be not as well known as it should.

In Section 8, we will treat the case where G is a semisimple compact Lie group. Here Hom (F_n, G) and Epi (F_n, G) are the same form a measure theoretical point of view (where by Epi (F_n, G) we mean now, the homomorphisms φ for which $\varphi(F_n)$ is dense in G). We study the ergodicity of the action of Aut (F_n) on $H_n(G) = \mathrm{Hom}\,(F_n, G) = G^n$.

Let now G be a general semisimple Lie group. Here almost nothing is known, except for the cases $G = \mathrm{PSL}_2\,(\mathbb{R})$ and PSL $(\mathbb{C})$. We present in Section 9 some basic questions and few results.

In the last section we will replace F_n by $T_g = \pi_1(S_g)$ the fundamental group of a closed surface of genus g. The case of surface groups deserves a separate survey but we will touch it only briefly, suggesting along the way a possible program for proving that the mapping class groups are not linear. The group Aut (T_g) acts on Hom (T_g, G) and Out (T_g) on the equivalence classes mod Inn (G). Note that Out (T_g) is the mapping class group M_g and by restricting ourselves to a subset of Hom $(T_g, \mathrm{PSL}_2\,(\mathbb{R}))$ we recover the action of M_g on the Fricke-Teichmüller space. The case of G-compact has also been studied in the literature and even the case of G-finite came up recently in the work of Dunfield and Thurston on finite covers of random 3-manifolds.

These notes are based on a series of lectures given at Yale University in February 2008. I would like to thank the participants for their remarks and questions and, in particular, to Yair Minsky for information about Kleinian groups. I am also grateful to Nir Avni, Shelly Garion, Tsachik Gelander, and Yair Glasner for several discussions.

This paper is dedicated to Bob Zimmer from whom I learned to search for the "big picture." Bob has always been a friend as well as a source of inspiration in his leadership as a scientist and as an administrator.

2. Presentations

Let G be a finitely generated group. By $d(G)$ we denote the minimal number of generators of G. A presentation of G is an exact sequence

2.1
$$1 \rightarrow R \rightarrow F_n \xrightarrow{\varphi} G \rightarrow 1$$

where $F = F_n$ is the free group on n generators $x_1, \ldots, x_n$. Clearly, $n \geq d(G)$. We denote by $d_F(R)$ the minimal number of generators of R as a *normal* subgroup of F.

A basic question in the theory of presentations of groups is to what extent a presentation as in sequence **2.1** is unique. Clearly, we can "twist" Sequence **2.1** by the action of $\mathrm{Aut}\,(G) \times \mathrm{Aut}\,(F_n)$ in the following way

2.2 $\qquad (\beta, \alpha)(\varphi) = \beta \circ \varphi \circ \alpha^{-1} \quad \text{where } (\beta, \alpha) \in \mathrm{Aut}\,(G) \times \mathrm{Aut}\,(F_n).$

Another way to change a presentation is by enlarging n. Let $\pi : F_{n+1} \to F_n$ be the natural epimorphism sending x_i (as an element of F_{n+1}) to x_i (as an element of F_n) for $i = 1, \ldots, n$ and x_{n+1} to the identity. Then

2.3 $\qquad\qquad\qquad 1 \to N \to F_{n+1} \xrightarrow{\varphi \circ \pi} G \to 1$

is also a presentation of G, which we call a lifting of Sequence **2.1**.

2.4. Waldhausen's Problem
Waldhausen (see [LS77, p. 92]) raised the question whether every presentation of a group G with $n > d(G)$ can be obtained from a presentation with $n = d(G)$ by a sequence of liftings. An equivalent formulation is

Question
Let F_m be the free group on m generators and $N \lhd F_m$. Assume $d(G) < m$ where $G = F_m/N$. Does N contain a primitive element of F_m?

Recall that $\gamma \in F_m$ is primitive if it is part of some basis of F_m or equivalently, it is in the orbit of x_1 under the action of $\mathrm{Aut}\,(F_m)$.

2.5. Gruenberg's Questions
It was probably B. H. Neumann who already in the 1930s considered in a systematic way the connection between various presentations of the same group G. Let us follow Gruenberg's treatment in [Gru76], who presented three fundamental questions:

Question (2.5a)
Given two presentations

$$1 \to R_i \to F_n \xrightarrow{\varphi_i} G \to 1$$

$i = 1, 2$. Is $d_{F_n}(R_1) = d_{F_n}(R_2)$?

Question (2.5b)

Let $r(G)$ be the minimum number of relations needed to define G, that is, the minimum of $d_{F_n}(R)$ over all possible presentations of G as in Sequence **2.1**. Is $r(G)$ realized in a minimal presentation (i.e., one in which $n = d(F_n) = d(G)$)?

Question (2.5c)

Is $d_{F_n}(R) - d(F_n)$ independent of the presentation in sequence **2.1** and therefore an invariant of G?

Of course, a positive answer to Question 2.5 would imply a similar one for 2.5 and 2.5.

Unfortunately, the answer to these questions in general is negative. For example, Dunwoody and Pietrowski [DP73] showed that the group $G = \langle a, b : a^2 = b^3 \rangle$ is a 1 relator group that has also a presentation with 2 generators, which needs more than 1 relater (see also [Gru76] and references therein).

Another example, attributed to G. Higiman, is given in [LS77, p. 93] where it is shown that the Baumslag-Solitar group $G = \langle x, y : x^{-1}y^2x = y^3 \rangle$ is also generated by x and $z = y^4$, but has no presentation with these two generators with a single defining relation.

This shows that the answers to Questions 2.5 and 2.5 are negative. Noskov [Nos81] (see also Evans [Eva93a]) showed that the answer to Waldenhaus's question 2.4 is negative. We do not know a counterexample to Question 2.5, but one very likely exists.

One can easily see that if for every n, the action of $\text{Aut}(G) \times \text{Aut}(F_n)$ on the set of possible presentations of G (with n generators) is transitive, then the answer to the first three questions raised in this section would be positive. But this is far from being the case if G is infinite. The case of finite G is more delicate and more interesting, and this will be the subject of the next sections. Before elaborating on it, we will introduce the convenient language of the product replacement graphs.

3. The Product Replacement Algorithm and Its Graph

3.1.

The *product replacement algorithm* (PRA) is a practical algorithm for generating random elements of a finite group. The algorithm was introduced and analyzed in [CLGM$^+$95]. Although it has no rigorous justification, practical experiments have shown excellent performance. It quickly became a popular algorithm for generating random group elements, and was included in two frequently used group computation packages: GAP and MAGMA.

The PRA can be described as a random walk on a graph, called the *product replacement graph* (or the *PRA Graph*). It will be more convenient for us to look at the following extended graph. For any $n \geq d(G)$, let

$$V_n(G) = \{(g_1, \ldots, g_n) \in G^n : \langle g_1, \ldots, g_n \rangle = G\}$$

be the set of all *generating n-tuples* of G.

The *extended PRA graph*, denoted $\tilde{X}_n(G)$, has $V_n(G)$ as its set of vertices. The edges correspond to the following so-called *Nielsen moves* $R_{i,j}^\pm, L_{i,j}^\pm, P_{i,j}, I_i$ for $1 \leq i \neq j \leq n$, where

$$R_{i,j}^\pm : (g_1, \ldots, g_i, \ldots, g_n) \rightarrow (g_1, \ldots, g_i \cdot g_j^{\pm 1}, \ldots, g_n)$$

$$L_{i,j}^\pm : (g_1, \ldots, g_i, \ldots, g_n) \rightarrow (g_1, \ldots, g_j^{\pm 1} \cdot g_i, \ldots, g_n)$$

$$P_{i,j} : (g_1, \ldots, g_i, \ldots, g_j, \ldots, g_n) \rightarrow (g_1, \ldots, g_j, \ldots, g_i, \ldots, g_n)$$

$$I_i : (g_1, \ldots, g_i, \ldots, g_n) \rightarrow (g_1, \ldots, g_i^{-1}, \ldots, g_n).$$

Strictly speaking, the PRA is a random walk on a subgraph $X_n(G)$ of $\tilde{X}_n(G)$, which is obtained by removing the edges corresponding to the $P_{i,j}$ and I_i. The output of the algorithm is a random entry chosen from the tuple at the end of the random walk. As observed in [Pak01, proposition 2.2.1], when $n \geq d(G) + 1$, the graph $X_n(G)$ is connected if and only if $\tilde{X}_n(G)$ is connected. The connected components of $\tilde{X}_n(G)$ are also called Nielsen equivalence classes of n-generating sets of G.

3.2.

Recall now the well-known result of Nielsen (cf. [LS77]) that Aut (F_n) is generated by the Nielsen moves $\{R_{i,j}^\pm, L_{i,j}^\pm, P_{i,j}, I_i\}_{1 \leq i,j \leq n}$ viewed as automorphisms of F_n acting on the n-tuple of generators $(x_1, \ldots, x_n)$. Moreover, $V_n(G)$ is naturally identified with $E_n(G)$ the set of epimorphisms from F_n, the free group on $x_1, \ldots, x_n$, onto G, since every such epimorphism φ is uniquely defined by the generating vector $(\varphi(x_1), \ldots, \varphi(x_n))$. It follows, therefore, that the connected components of the graph $\tilde{X}_n(G)$ are exactly the orbits of Aut (F_n) acting on $E_n(G)$. This observation plays a crucial role in [LP01] when the mixing rate of the random walk on $\tilde{X}_n(G)$ is related to the possibility of Aut (F_n) (or some variants of it) have property (T) or (τ). We will not elaborate on this issue here—referring the reader to [LP01] for more information and for some interesting open problems.

We denote by $\bar{X}_n(G)$ the quotient graph $\mathrm{Aut}\,(G)\backslash\tilde{X}_n(G)$. Its vertices are in one-to-one correspondence with the points of $\bar{E}_n(G)$. These in turn are in one-to-one correspondence with the set of normal subgroups N of F_n with F_n/N isomorphic to G. The action of $\mathrm{Aut}\,(F_n)$ on $\bar{E}_n(G) = \tilde{X}_n(G)$ factor through $\mathrm{Out}\,(F_n) = \mathrm{Aut}\,(F_n)/\mathrm{Inn}\,(F_n)$.

The connectivity of $\tilde{X}_n(G)$ (respectively, $\bar{X}_n(G)$) is equivalent to $|E_n(G)/\mathrm{Aut}\,(F_n)| = 1$ (respectively, $|\bar{E}_n(G)/\mathrm{Aut}\,(F_n)| = 1$) and if $\tilde{X}_n(G)$ is connected, so is $\bar{X}_n(G)$. But we do not know if the converse is true in general. In [Pak01, proposition 2.4.1] it is observed that this is the case if $n \geq 2d(G)$. Anyway, the recent interest in the product replacement algorithm put forward the question of the connectivity of $\tilde{X}_n(G)$, which is essentially equivalent to the question we started with in Section 2, that is, whether G has an essentially unique presentation on n generators. From now on we will use both languages.

There are very few general results in this context that hold for every finite group. Here is one:

PROPOSITION 3.1. *Let G be a finite group and $\mu(G)$ the maximal size of a minimal set of generators of G (i.e., if $m > \mu(G)$, every set S of generators of G with $|S| \geq m$ contains a proper subset of generators). Then $\tilde{X}_n(G)$ is connected for every $n \geq \mu(G) + d(G)$.*

Proof. Let $g_1, \ldots, g_d$ be a set of generators of G, $d = d(G)$ and $(\underline{g}) = (1, \ldots, 1, g_1, \ldots, g_d) \in \tilde{X}_n(G)$. Let $(\underline{x}) = (x_1, \ldots, x_n) \in \tilde{X}_n(G)$ be an arbitrary vector. By assumption it contains a subset S of size $\mu = \mu(G)$ of generators. After changing order we can assume that these are $x_1, \ldots, x_\mu$. We can now move $(\underline{x})$ a series of Nielsen moves to $(x_1, \ldots, x_\mu, 1, \ldots, 1, g_1, \ldots, g_d)$ and the latter can be moved to $(\underline{g})$. Thus $\tilde{X}_n(G)$ is connected. $\qquad\square$

Very few results seem to be known on $\mu(G)$: Whiston [Whi00] and Saxl–Winston [WS02] estimate it for S_n and $PSL_2(q)$, respectively. Nikolov (unpublished) showed that for every finite simple group of Lie-type G of rank at most r over a finite field of order at most p^e, $\mu(G) \leq f(r, e)$ where f depends on r and e but not on p.

Anyway, for every finite group G, one can easily see that $d(G)$ and $\mu(G)$ are both bounded by $\log_2(|G|)$. So we can deduce that $\tilde{X}_n(G)$ is connected for any $n \geq 2\log_2(|G|)$. But as observed in [MW03] one can do even better: Let $l = l(G)$ be the maximum length of a chain of strictly decreasing nontrivial subgroups of G. It is easy to see that $d(G) \leq \mu(G) \leq l(G)$.

PROPOSITION 3.2. *If $n > l(G)$, then $\tilde{X}_n(G)$ is connected.*

Proof. Fix a generating vector $(s_1, \ldots, s_d)$ with $d = d(G)$. Let $(\underline{t}) = (t_1, \ldots, t_n) \in \tilde{X}_n(G)$. We will show that $(\underline{t})$ is connected to $(s_1, \ldots, s_d, 1, \ldots, 1)$.

Put $G_i = \langle t_1, \ldots, t_i \rangle$, so $G = G_n \geq G_{n-1} \geq \cdots \geq G_1 \geq \{1\}$. Since $n > l(G)$, $G_j = G_{j-1}$ for some $j > 0$, so t_j is a word in $t_1, \ldots, t_{j-1}$. So $(\underline{t})$ is connected to $(t_1, \ldots, t_{j-1}, 1, t_{j+1}, \ldots, t_n)$. We can (by changing names) assume that t_1 of $(\underline{t})$ satisfies $t_1 = 1$. Thus $G = \langle t_2, \ldots, t_n \rangle$ and hence $(\underline{t})$ is connected to $(s_1, t_2, \ldots, t_n)$. Define now $G_1 = \langle s_1 \rangle$ and $G_i = \langle s_1, t_2, \ldots, t_i \rangle$ for $i \geq 2$. Again, as $n > l(G)$, we must have $G_j = G_{j-1}$ for some j and since $s_1 \neq 1$, $j > 1$. So $t_j \in \langle s_1, t_2, \ldots, t_{j-1} \rangle$ and $(\underline{t})$ is equivalent to $(s_1, t_2, \ldots, t_{j-1}, 1, t_{j+1}, \ldots, t_n)$ and hence to $(s_1, s_2, t_3, \ldots, t_m)$. We continue by induction to deduce that $(\underline{t})$ is connected to $(s_1, \ldots, s_d, t_{d+1}, \ldots, t_n)$. As $\langle s_1, \ldots, s_d \rangle = G$, it implies that $(\underline{t})$ is connected to $(s_1, \ldots s_d, 1, \ldots, 1)$ as promised. □

Since $l(m) \leq \log_2(|G|)$ we deduce

COROLLARY 3.3. *$\tilde{X}_n(G)$ is connected for $n > \log_2(|G|)$.*

3.3. Free Action of Finite Groups on Handlebodies

The Neilsen equivalence classes of generating sets of G parametrize free actions of the group G on handlebodies as we will now explain, following [MW03] and the references therein.

Let $\mathcal{H}$ be an orientable 3-dimensional handlebody of genus $g \geq 1$. Two (effective) actions $\rho_1, \rho_2 : G \to \mathrm{Homeo}(\mathcal{H})$ are said to be *equivalent* if there is a homeomorphism $h : \mathcal{H} \to \mathcal{H}$ such that $h\rho_1(g)h^{-1} = \rho_2(g)$ for each $g \in G$. They are *weakly equivalent* if their images are conjugate, that is, there is $h \in \mathrm{Homeo}(\mathcal{H})$ such that $h\rho_1(G)h^{-1} = \rho_2(G)$. Equivalently, there is $\alpha \in \mathrm{Aut}(G)$ such that $h\rho_1(g)h^{-1} = \rho_2(\alpha(g))$, that is, ρ_1 and $\rho_2 \circ \alpha$ are equivalent.

From now on when we talk about actions of G on $\mathcal{H}$ we mean orientation-preserving *free* actions. We will assume $g \geq 1$, as the only free action on the handlebody of genus 0, the 3-ball, is by the trivial group.

If G acts on $\mathcal{H}$ freely, then the quotient map $\mathcal{H} \to \mathcal{H}/G$ is a covering map that induces an extension

3.4 $$1 \to \pi_1(\mathcal{H}) \to \pi_1(\mathcal{H}/G) \to G \to 1.$$

Note that $\pi_1(\mathcal{H})$ is a free group on g generators and by [Hem76, theorem 5.2] $\mathcal{H}/G$ is also a handlebody, so $\pi_1(\mathcal{H}/G)$ is also a free group on, say, n generators. Nielsen-Schreier theorem implies that $g = 1 + |G|(n-1)$, that is, $n = 1 + \frac{1}{|G|}(g-1)$. Conversely, if we start with a handlebody $\mathcal{H}'$ of genus n, so $\pi_1(\mathcal{H}') = F_n$, every epimorphism onto G gives rise to a covering $\mathcal{H}$ that is a handlebody of genus g, on which G acts. This sets up a surjective map from the family of (free) actions of G on the handlebody $\mathcal{H}$ of genus g onto the set of epimorphisms from F_n onto G. The latter is exactly the set of vertices of $\tilde{X}_n(G)$. Now, every automorphism of $F_n = \pi_1(\mathcal{H}/G)$ is induced by an homeomorphism of $\mathcal{H}/G$. The following theorem is deduced in [MW03, theorem 2.3] using elementary arguments from covering theory.

THEOREM 3.4. *Let G be a finite group, $n \geq 1$ and $g = 1 + |G|(n-1)$. The equivalence classes of actions of G on a genus g handlebody correspond bijectively to the Nielsen equivalence classes of n-generating sets of G, that is, to the connected components of $\tilde{X}_n(G)$. The weak equivalence classes of these actions correspond to the connected components of $\bar{X}_n(G)$.*

The theorem shows that all the results discussed in this survey on the connected components of $\tilde{X}_n(G)$ or $\bar{X}_n(G)$ have direct topological applications and in fact are equivalent to such topological statements. The correspondence goes even further. An important notion in the study of actions of G on $\mathcal{H}$ is "stabilization": two actions ρ_1 and ρ_1 can be equivalent after "adding" one (or more) handle(s) to $\mathcal{H}$. This is equivalent to the question whether the two corresponding n-generating sets $(\underline{t}) = (t_1, \ldots, t_n)$ and $(\underline{s}) = (s_1, \ldots, s_n)$ are on the same connected component when considered as $(s_1, \ldots, s_n, 1)$ and $(t_1, \ldots, t_n, 1)$ in $\tilde{X}_{n+1}(G)$. So, all these seemingly pure algebraic questions, to be discussed later, carry a significant amount of topological information. We will usually stick to the algebraic language, leaving the reader the translation to this topological setting.

Later in the paper we will show various connectivity results. It is a highly nontrivial problem to show that two n-generating sets are *not* Nielsen equivalent (except for $n = 2$; see Section 6.2). Some powerful methods using Fox calculus were developed in [Lus91] and [LM93] but as far as we know these methods have not been applied as of now to yield nonequivalence for n-generating sets of finite groups.

4. The Andrews-Curtis Conjecture and Its Graph

Before going to a more detailed study of $\tilde{X}_n(G)$, let us mention in passing another graph associated with a group—the Andrews-Curtis graph $AC_n(G)$. This graph also has its roots in a deep problem in topology and in presentation theory, but the interest in it has revived recently from the point of view of computational group theory.

Let G be a group generated by a finite symmetric set $h_1, \ldots, h_d$, and N a normal subgroup of G. As usual $d_G(N)$ denotes the minimal number of elements of N generating N as a normal subgroup of G. For $n \geq d_G(N)$ we define the graph $AC_n(G, N)$ as follows:

Its vertices are the n-tuples $(g_1, \ldots, g_n) \in N^n$ with $\langle\langle g_1, \ldots, g_n \rangle\rangle_G = N$, that is, those which generate N as a normal subgroup of G.

A vector $(g_1, \ldots, g_n)$ will be connected to its image under the moves $L_{i,j}^{\pm}$, $R_{i,j}^{\pm}$, $P_{i,j}$, I_i as in 3.1 as well as $(g_1, \ldots, g_n) \rightarrow (g_1, \ldots, h_j g_i h_j^{-1}, \ldots, g_n)$ for every $1 \leq i \leq n$ and $1 \leq j \leq d$.

Note that all moves indeed take an n-tuple of normal generating set of N to another one. The case when $N = G$ is of special interest. In this case we write $AC_n(G)$ instead of $AC_n(G, G)$.

The famous Andrew-Curtis conjecture is equivalent to:

CONJECTURE 4.1. (ANDREWS-CURTIS [AC65]) *The graph $AC_n(F_n)$ is connected.*

The conjecture is usually expressed in a different language. Note that a vector of $AC_n(F_n)$ amounts to a vector of n elements of the free group F_n, that is, n words in $x_1, \ldots, x_n$, which normally generate F_n. In other words, this is a presentation of the trivial group by n generators and n relations. The Andrews-Curtis conjecture predicts that any such presentation is obtained from the standard presentation $\langle x_1, \ldots, x_n; x_1, \ldots, x_n \rangle$ using a finite series of Nielsen moves or conjugations. This is one of the most outstanding conjectures in combinatorial group theory, with various potential applications to topology (see [AC65]).

Let us mention that in general $AC_n(G)$ is not connected. For example, if $G = \mathbb{F}_p^n$ (or any abelian group with $d(G) = n$), then $AC_n(G)$ is the same as $\tilde{X}_n(G)$ and, as will be shown in the next section (in a different language!), the number of the connected component of $\tilde{X}_n(G)$ is $(p-1)/2$ and the graph is not connected.

The topic has received a new interest in recent years from computational group theory [BKM03]: A natural generalization of the PRA is the following algorithm to produce a pseudorandom element in the normal closure N of a given set of elements $g_1, \ldots, g_n$ inside a group G generated by given generators $h_1, \ldots, h_d$. The algorithm starts with the vector $(g_1, \ldots, g_n)$ of $AC_n(G, N)$ and takes a random walk on the graph. The output is a random component from the end vector of the random walk.

Many researchers believe that the Andrews-Curtis conjecture is false and some (see [BKM03] and the references therein) have tried to disprove it by using computer calculations. The idea is that if $\pi : H \to G$ is an epimorphism of groups, then π induces a graph theoretic map $\tilde{\pi} : AC_n(H) \to AC_n(G)$ (not necessarily onto!). If one could find one finite group G and an epimorphism $\pi : F_n \to G$ such that $\tilde{\pi}(AC_n(F_n))$ is not connected, then the conjecture is false. Various calculations have been performed till it was shown in [BLM05].

THEOREM 4.2. *Let G be a finite group and $n \geq \max\{d_G(G), 2\}$. Then two vectors of $AC_n(G)$ are connected by a path iff their images in $AC_n(G/[G, G])$ are connected.*

The connected components of the Andrews-Curtis graph of an abelian group are easy to understand and one can deduce

COROLLARY 4.3. *For every epimorphism $\pi : F_n \to G$ where G is a finite group, the image $\tilde{\pi}(AC_n(F_n))$ is connected.*

This "finitary Andrews-Curtis Conjecture" does not give much insight on the original conjecture, but shows that the computational efforts carried out in order to disprove it have all been in vain.

We end this section by giving a sketch of the proof for a special case of the Theorem — the case when G is perfect. In this case $G/[G, G] = \{1\}$ and the theorem claims that $AC_n(G)$ is connected. Let's prove it:

First, denote by $M(G)$ the intersection of all maximal normal subgroups of G. An easy observation is that a subset $\{y_1, \ldots, y_k\}$ of G normally generates G if and only if it generates $\bmod M(G)$. For infinite groups, this is usually a useless observation (e.g., $M(F) = \{e\}$ if F is a free group). But for a finite group G, $G/M(G)$ is always a direct product of finite simple groups, and if G is also perfect, all the simple groups are nonabelian. We can replace G by $G/M(G) = \prod_{i=1}^{r} S_i$, S_i nonabelian finite simple groups and every $z = (z_1, \ldots, z_r) \in \prod_{i=1}^{r} S_i$ with $z_i \neq e$ for all $1 \leq i \leq r$, generates $G/M(G)$ normally. To show now that $AC_n(G)$ is connected, we show that every vector in $AC_n(G) \subseteq G^n$ is connected to $(z, 1, \ldots, 1)$, where z is an element as above.

So let $(y_1, \ldots, y_n) \in AC_n(G)$. Look at y_{n-1} and y_n as elements of $G/M(G) = \prod_{i=1}^{r} S_i$. We can conjugate y_{n-1} by some $g \in G$ such that $\tilde{y}_{n-1} = y_{n-1}^{g} y_n$ is not the identity in every component, unless *both* y_{n-1} and y_n are identities at that component: to do so, simply conjugate such that the conjugation of y_{n-1} in component i is different from the component of y_n there. Then use the normal closure of the new $\tilde{y}_{n-1}$, which is the product of all components in which $\tilde{y}_{n-1}$ is a nonidentity, to "clean" y_n, that is, to make it the identity in these components and hence altogether to replace y_n by the identity.

One can continue like that to get that $(y_1, \ldots, y_n)$ is connected to $(z', 1, \ldots, 1) \in G^n$ where all components of z' as an element of $\prod_{i=1}^r S_i$ are nonidentity. As $n \geq 2$, we can change $(z', 1, \ldots, 1)$ to $(z', z, 1, \ldots, 1)$, then switch to $(z, z', 1, \ldots, 1)$ and then "clean" z' by z to get $(z, 1, \ldots, 1)$ as promised.

5. Finite Solvable Groups

While the theory of presentations of infinite groups seems wild and one expects very few general results to hold, the theory of presentations of finite groups may have some pleasant properties. In particular, the question we are interested in, that is, the number of orbits of $\mathrm{Aut}\,(G) \times \mathrm{Aut}\,(F_n)$ acting on the n-generators presentations of G, is quite well understood for finite solvable groups due to the following two results of Dunwoody.

THEOREM 5.1A. (DUN 70) *Let G be a finite solvable group with $d(G) < n$. Then $\mathrm{Aut}\,(F_n)$ acts transitively on $E_n(G)$, that is, $\tilde{X}_n(G)$ (and hence also $\bar{X}_n(G)$) is connected.*

The assumption $d(G) < n$ is crucial. Note that for $G = \mathbb{F}_p^n$ the n-dimensional vector space over the field $\mathbb{F}_p$, $E_n(G)$ is the set of all bases of G, so can be identified with $GL_n(\mathbb{F}_p)$. The action of $\mathrm{Aut}\,(F_n)$ on it is via $SL_n^{\pm}(\mathbb{F}_p) = \{A \in GL_n(\mathbb{F}_p) | \det(A) = \pm 1\}$ and so $|E_n(G)/\mathrm{Aut}\,(F_n)| = (p-1)/2$. On the other hand, $\mathrm{Aut}\,(G) = GL_n(\mathbb{F}_p)$ and so $|\bar{E}_n(G)/\mathrm{Aut}\,(F_n)| = 1$. But there are groups where even the second set is large:

THEOREM 5.1B. (DUN 63) *For every $2 \leq n \in \mathbb{N}$, every $k \in \mathbb{N}$, and every prime p, there exists a finite p-group of nilpotency class 2 with $|\bar{E}_n(G)/\mathrm{Aut}\,(F_n)| \geq k$. In particular, the PRA graph $\tilde{X}_n(G)$ has at least k components.*

Theorem 5.1a shows that for finite solvable groups, the question in section 2.4 and 2.5 have an affirmative answer. The same is true for Question 2.5, provided $n > d(G)$.

The proof of Theorem 5.1a is nontrivial but quite elementary. As this is the main positive result known, let us sketch its proof:

Let $\{e\} = M_0 \lhd M_1 \lhd \cdots \lhd M_r = G$ be a chief series of G, that is, $M_i \lhd G$ and M_{i+1}/M_i is a minimal normal subgroup of G/M_i. It is, therefore, an irreducible $\mathbb{F}_p[G/M_{i+1}]$ module, for some prime p. Let $h_1, \ldots, h_{n-1}$ be a finite set of generators for G and $(g_1, \ldots, g_n) \in V_n(G)$. We will argue by induction on r, so assume the theorem for $r - 1$. This means that we can move by Nielsen transformations from $(g_1, \ldots, g_n)$ to $(m, m_1 h_1, \ldots, m_{n-1} h_{n-1})$ where $m, m_1, \ldots, m_{n-1} \in M_1$. We should show that the last vector is connected to $(e, h_1, \ldots, h_{n-1})$.

We can assume $m \neq e$. Indeed, if $m = e$, then $m_1 h_1, \ldots, m_{n-1} h_{n-1}$ generate G (since $(e, m_1 h_1, \ldots, m_{n-1} h_{n-1}) \in V_n(G)$) and so we can create any word of them in the first component.

So assume that $m \neq e$. Think of M_1 as an additive group, a G/M_1 module. Note also that as M_1 is abelian, the action of $m_i h_i$ on M_1 by conjugation is the same as that of h_i. As M_1 is irreducible, one deduces that every $m' \in M_1$ can be written as a sum of conjugates of m by words in $m_1 h_1, \ldots, m_{n-1} h_{n-1}$ and this is so for every $m \neq e$. Now for each $i = 1, \ldots, n-1$, write m_i as a sum $m_i = m_{i1} + \cdots + m_{it}$ where m_{ij} is a conjugate of m. Then we can change m to m_{ij} to clean m_{ij} out of m_i and gradually clean m_i. Do it for all $i = 1, \ldots, n-1$ to get $(\tilde{m}, h_1, \ldots, h_{n-1})$ and then eliminate $\tilde{m}$, which is possible since $h_1, \ldots, h_{n-1}$ generate G.

6. Finite Simple Groups

6.1.

We are coming to the most interesting case, with some very interesting open problems and some potential applications—the case when G is a finite non-abelian simple group. For such a group G the leading long-standing conjecture is the following one, which is attributed to Jim Wiegold:

CONJECTURE 6.1. *Let G be a finite simple group. Then $\bar{X}_n(G)$ is connected for every $n \geq 3$.*

Note that by the classification of the finite simple groups, it is known that for such G, $d(G) \leq 2$, so the conjecture combined with Dunwoody's Theorem 5.1a made Pak [Pak01] ask

QUESTION 6.2. *Is it true that for every finite group G, $\bar{X}_n(G)$ is connected if $n > d(G)$?*

It is interesting to mention that Dunwoody in a review on [Gil77] (see Math. Review MR0435226 in 1977) wrote: "It seems unlikely that this result is true for an arbitrary finite group G". But in the years since then no counter example has been found, so maybe the answer to Question 6.2 is indeed positive. As an intermediate step one can suggest the following conjecture that looks more feasible, as it is known to be true for simple groups (see Theorem 6.6) and for solvable groups (Theorem 5.1a).

CONJECTURE 6.3. *Let G be a finite group, $(\underline{t}) = (t_1, \ldots, t_n)$ and $(\underline{s}) = (s_1, \ldots, s_n) \in \tilde{X}_n(G)$. Then $(t_1, \ldots, t_n, 1)$ and $(s_1, \ldots, s_n, 1)$ are connected in $\bar{X}_{n+1}(G)$.*

Note that it is easy to see that $(\underline{s})$ and $(\underline{t})$ become connected to each other in $\tilde{X}_{n+d}(G)$ when $d = d(G)$ (see [MW03, proposition 6.1]). Note also that Conjecture 6.3 has a topological equivalent formulation, as hinted in Section 3.3, it asserts that any two (free) actions of G on a handlebody become equivalent after adding one handle (see [MW03] for more on stabilizations of actions).

We will come again to the case where $n > d(G)$, but let us first clear up the situation when $n = d(G)$.

6.2. G Simple and n = 2

When G is a finite simple group and $n = d(G) = 2$, the situation is very much different than what is predicted by Conjecture 6.1.

THEOREM 6.4. (GARION-SHALEV [GS]) *Let G be a nonabelian finite simple group. Then $|\bar{E}_2(G)/\operatorname{Aut}(F_2)| \to \infty$ when $|G| \to \infty$, or equivalently the number of connected components of $\bar{X}_2(G)$ is going to infinity with G.*

The special case of $G = \mathrm{PSL}_2(q)$ was proved by Guralnick and Pak [GP03], who also conjectured the general case.

Let us sketch the proof: A classical result of Nielsen asserts that if $\alpha \in \operatorname{Aut} F_2$, with $F_2 = F(x, y)$ the free group in x and y, then the commutator $[\alpha(x), \alpha(y)]$ is conjugate in F_2 to either $[x, y]$ or $[x, y]^{-1}$. This implies that if $\varphi, \psi \in \operatorname{Epi}(F_2, G)$ are on the same $\operatorname{Aut}(F_2) \times \operatorname{Aut}(G)$ orbits, then $[\varphi(x), \varphi(y)]$ is conjugate to $[\psi(x), \psi(y)]^{\pm 1}$ in $\operatorname{Aut}(G)$. So to prove the theorem, Garion and Shalev showed that "almost all the elements of G are commutators of pairs of generators of G," that is, the proportion of these elements in G goes to one when the order of G goes to infinity. Once this is proved, it follows that the number of components of $\bar{X}_2(G)$ grows at least as fast as the number of the conjugacy classes of G. The claim above is proved by combining two methods. First, note that the function $F(g) = \#\{(a, b) \in G \times G | [a, b] = g\}$ is a class function on G, that is, constant on conjugacy classes. So, by harmonic analysis on finite groups, it can be expressed using characters of G. Moreover, a classical result of Frobenius from 1896 gives an explicit formula:

$$F(g) = |G| \cdot \sum_{\chi} \frac{\chi(g)}{\chi(1)}.$$

Thus one can estimate $F(g)$ by estimating the normalized characters' values $\frac{\chi(g)}{\chi(1)}$. A lot has been done on this issue in recent years and Garion and Shalev used it to prove that almost all elements of G are commutators (a well-known conjecture of Ore asserting that in finite nonabelian simple group every element is a commutator has been proved recently [LOST]). But one needs more: g should be equal to $[a, b]$ when the pair $\{a, b\}$ generates G. A well-known result of Dixon [Dix69], Kantor-Lubotzky [KL90], and Liebeck-Shalev [LS95] says

THEOREM 6.5. *Almost all pairs* $(a, b) \in G \times G$ *generate* G.

Garion and Shalev show that the distribution of the commutators $[a, b]$ over generating pairs (a, b) is approximately the same as over all pairs and Theorem 6.4 then follows.

6.3. G simple and $n \geq 3$:

We saw in Theorem 6.5 that almost all pairs of elements of G generate G. This implies that for $n \geq 3$, almost all $n - 1$ n-tuples of generators of G are *redundant*, that is, a proper subset of the n-set already generates G.

The following important result was proved by Gilman [Gil77] for $n \geq 4$ and extended by Evans [Eva93b] to $n \geq 3$:

THEOREM 6.6. *Let* G *be a finite simple group and* $3 \leq n \in \mathbb{N}$. *Then*

a) *all the redundant vectors in* $\tilde{X}_n(G)$ *lie in the same connected component* Y *of* $\tilde{X}_n(G)$, *and*

b) *the group* Aut (F_n) *acts on* $\bar{Y}$, *the projection of* Y *to* $\bar{E}_n(G)$, *as the alternating or the symmetric group of degree* $|\bar{Y}|$.

COROLLARY 6.7. *Let* G *be a finite simple group. For n large enough,* Aut (F_n) *acts on* $\bar{E}_n(G)$ *as the alternating or the symmetric group of degree* $|\bar{E}_n(G)|$.

As observed by Pak [Pak01], part (a) of the theorem together with Theorem 6.5 imply that the graph $\tilde{X}_n(G)$ has a huge connected component Y whose size is at least $(1 - \varepsilon)|\tilde{X}_n(G)|$ for every $\varepsilon > 0$ when $|G| \to \infty$. So, the Wiegold conjecture is essentially true. For some questions this is enough, but for others (see Section 6.4) it is crucial to know that there are no very small connected components in $\tilde{X}_n(G)$.

Part (b) of the theorem is also very interesting. We will see in Section 8 its analogues when G is a compact group.

Part (a) of the theorem is proved using the notion of "spread." (A notion that was first introduced in [BW75]).

DEFINITION 6.8. A 2-generated group G is said to have spread r if for every nonidentity elements $y_1, \ldots, y_r \in G$, there exists $z \in G$ such that $G = \langle y_i, z \rangle$ for every $i = 1, \ldots, r$.

THEOREM 6.9. (BREUER-GURALNICK-KANTOR [BGK08]) *All finite simple groups have spread 2.*

Many (but not all) of them have even spread 3, but we need only 2 in order to prove Theorem 6.6(a) as follows: Let $z = (z_1, \ldots, z_n)$ and $y = (y_1, \ldots, y_n)$ be two redundant generating vectors. We can assume $z_i = y_j = 1$ for some i and j and after permuting the elements $z_n = y_n = 1$.

We can further assume that $z_1 \neq 1 \neq y_2$. As G has spread 2, there exists $w \in G$ with $\langle z_1, w \rangle = \langle w, y_2 \rangle = G$. Now, as $\langle z_1, \ldots, z_{n-1} \rangle = G$ we can move $z = (z_1, \ldots, z_{n-1}, 1)$ to $(z_1, \ldots, z_{n-1}, w)$ and then, using the fact that $\langle z_1, w \rangle = G$, to $(z_1, y_2, \ldots, y_{n-1}, w)$. But we also have $\langle y_2, w \rangle = G$, so the latter can be transformed to $(y_1, y_2, \ldots, y_{n-1}, w)$ and finally to $(y_1, \ldots, y_{n-1}, 1)$ as $\langle y_1, \ldots, y_{n-1} \rangle = G$.

The proof of part (b) of Theorem 6.6 is more involved: One first shows that the action of $\text{Aut}(F_n)$ on $\bar{Y}$ is double transitive. Then it is shown that there exists $\beta \in \text{Aut}(F_n)$ acting nontrivially on $\bar{Y}$ but moves at most $|G|$ elements. Now, an old result of Bochert (from 1897) asserts that a double transitive permutation subgroup of $\text{Sym}(N)$ with a nonidentity element that moves less then $\frac{1}{3}(N - 2\sqrt{N})$ elements, must contain $\text{Alt}(N)$. From Theorem 6.5 we know that $|Y|$ grows like $|G|^{n-1}$ so one can deduce that the action $\text{Aut}(F_n)$ on $\bar{Y}$ contains $\text{Alt}(\bar{Y})$ at least when G is large and $n \geq 4$ (and with little more precision one sees that this is true for every G and $n \geq 3$).

Another corollary of Theorem 6.6a is that the following is an equivalent reformulation of Wiegold's Conjecture 6.1 (a formulation that is easier to generalize to infinite groups—see Sections 8 and Sections 9).

CONJECTURE 6.10. *Let $n \geq 3$ and $\Psi : F_n \twoheadrightarrow G$ an epimorphism onto a finite simple group. Then F_n has a proper free factor H such that $\Psi(H) = G$.*

Conjecture 6.10 asserts that for some set of free generators $(g_1, \ldots, g_n)$ of F_n, $\varphi(g_1), \ldots, \varphi(g_{n-1})$ generate G and this is the same as saying that the n-generating vector $(\varphi(x_1), \ldots, \varphi(x_n))$ is connected to the redundant vector $(\varphi(g_1), \ldots, \varphi(g_n))$.

Let us denote by $E_n'(G)$ the set of all epimorphisms $\Psi : F_n \twoheadrightarrow G$ for which F_n has a proper factor H with $\Psi(H) = G$. So Wiegold's Conjecture 6.1 predicts that for finite simple group G, $E_n'(G) = E_n(G)$ and $\bar{Y} = \bar{X}_n(G)$ for $n \geq 3$. For the purpose of the PRA, the fact that $\bar{Y}$ is almost all of $\bar{X}_n(G)$ is just as good. One starts the algorithm with a redundant vector and so the random walk can take it to almost every other generating vector. (For the rate of mixing see [Pak01], [LP01], and [DSC98]). But there are several good reasons to want to know that $\bar{Y} = \bar{X}_n(G)$. This would imply that every presentation of G is equivalent to a minimal one (i.e., one with minimal numbers of generators— see Section 2) and that the minimal possible of relations can be obtained with a minimal number of generators. Another application to the representation theory of Aut (F_n) will be described below, but let us first summarize the very few partial results known toward the Wiegold conjecture.

Essentially all the known results are elaborations of the seminal paper of Gilman [Gil77]. Here is the current state of affairs:

THEOREM 6.11. *Let G be a finite simple group and $n \in \mathbb{N}$, then $\tilde{X}_n(G)$ is connected in the following cases:*

i) $G = PSL_2(p)$, *p prime and $n \geq 3$ [Gil77].*

ii) $G = S_z(2^n)$, *the Suzuki groups, or $G = PSL_2(2^n)$ and $n \geq 3$ [Eva93b].*

iii) $G = PSL_2(3^p)$, *p prime and $n \geq 3$ [MW03].*

iv) $G = PSL_2(p^r)$, *p prime, $r \in \mathbb{N}$, and $n \geq 4$ [Gar].*

v) *G is a finite simple group of Lie rank at most r and $n \geq f(r)$ for a suitable function f depending only on r [AG08].*

vi) $G = A_k$, *$k \leq 10$, and $n = 3$ (see [Pak01, theorem 2.5.6.]).*

In light of Theorem 6.6a, proving a connectivity result for $\tilde{X}_n(G)$ amounts to showing that every nonredundant vector $(g) = (g_1, \ldots, g_n) \in X_n(G)$ is connected to a redundant one. We can therefore assume that $g_1, \ldots, g_{n-1}$ generate a proper subgroup of G.

The proof of (i) by Gilman is heavily based on the explicit known list of subgroups of G and the same remark is true for the works of Evans,

McCullough-Wanderley, and Garion proving (ii), (iii), and (iv), respectively. For (v), Avni and Garion are using the work of Larsen and Pink [LP], which gives some quantitative description of the possible subgroups of G.

Part (v) should be compared with Proposition 3.1 and the result of Nikolov thereafter. The point here is that the function depends only on r and not on the defining field. But there is a price for it, the function $f(r)$ in the proof in [AG08] grows quite fast (exponentially) with r.

Part (vi) was proved using ad hoc arguments and by help of computer calculations (see [Pak01] and the references therein).

So altogether, Wiegold conjecture is known only in very limited cases. In the next subsection, we will give further motivation to prove the conjecture or even a weak form of it.

6.4. Representations of Aut (F_n)

For many years, it has not been known if Aut (F_n), the automorphism group of the free group $F_n (n \geq 2)$, or B_n, the braid group on n strands $(n \geq 4)$, or M_g—the mapping class group of a closed surface of genus $g \geq 2$, are linear groups, that is, whether they have faithful linear representations over the field $\mathbb{C}$ of complex numbers. It has been felt anyway that all the problems are similar and a solution to one of them would lead to a solution of all others. Moreover, it was shown in [DFG82] that B_4 is linear iff Aut (F_2) is linear. In [FP92], Formanek and Processi showed that Aut (F_n) is *not* linear for $n \geq 3$, leaving the case of $n = 2$ open. The proof was very special for these groups and did not shed any light on B_n (which is a subgroup of Aut (F_n)). Moreover, [BHT01] shows that their method of "poison subgroup" cannot be applied at all to M_g. On the other hand, Bigelow [Big01] and Krammer [Kra02] showed that B_n are all linear and so Aut (F_2) is linear in spite of the fact that Aut (F_n), $n \geq 3$ are not.

A new method to produce representations of Aut (F_n) onto arithmetic groups has been developed recently in [GL], but none of these representations is faithful.

We want now to explain a way of looking at this problem that suggests that the difference between $n = 2$ and $n \geq 3$ in the linearity question is related to the difference between Wiegold Conjecture 6.1 for $n \geq 3$ on the connectivity of $\bar{X}_n(G)$ and $n = 2$ where Garion-Shalev showed a strong nonconnectivity (Theorem 6.4).

In fact, even more can be said. The proof of the nonlinearity of Aut (F_n), $n \geq 3$ in [FP92] strongly suggests the following stronger statement:

CONJECTURE 6.12. *Let $n \geq 3$ and $\rho : \mathrm{Aut}\,(F_n) \to GL_k(\mathbb{C})$ be a linear represen-*
tation. Then $\rho(\mathrm{Inn}\,(F_n))$ is virtually solvable, where $\mathrm{Inn}\,(F_n)$ is the group of inner
automorphisms of F_n.

Let us now show:

CLAIM 6.13. *If Weigold's Conjecture 6.1 is true, then Conjecture 6.12 is true when*
$H = \overline{\rho(\mathrm{Aut}\,(F_n))}$, *the Zariski closure of $\rho(\mathrm{Aut}\,(F_n))$, is connected.*

As this is only a conditional result, we shall only sketch the proof:

Assume there is such a ρ with $\rho(\mathrm{Inn}\,F_n)$ not virtually solvable, then by
dividing H by its solvable radical we can assume H is semisimple, and even
simple, by choosing a suitable factor. Furthermore, by [LL04, theorem 4.1]
there is a specialization of ρ so that $\rho(\mathrm{Aut}\,(F_n))$ is in $GL_r(k)$ for some number
field k and the Zariski closure L is defined over k but it is still isomorphic to
the simple group H over $\mathbb{C}$.

As $\mathrm{Aut}\,(F_n)$ is finitely generated, $\rho(\mathrm{Aut}\,(F_n))$ is a subgroup of $L(O_S)$, the S-
integers of k, where S is a finite set of primes of k and O is the ring of integers
of k. We can further arrange that L is simply connected and then to apply
the strong approximation theorem for linear groups of [LS03, window 9] to
deduce that $\rho(\mathrm{Aut}\,(F_n))$ is almost dense in the congruence completion $L(\hat{O}_S)$.
The same applies also to $\rho(\mathrm{Inn}\,(F_n))$, since it is also Zariski dense in L. This
implies that for almost every prime ideal $\mathcal{P}$ in O_S, the projection of $\rho(\mathrm{Inn}\,(F_n))$
to the finite semisimple group $M = L(O_S/\mathcal{P})/Z$ is onto (where Z is the center).
Let $N_{\mathcal{P}}$ be the kernel of the map from $\mathrm{Inn}\,(F_n)$ to M. This subgroup $N_{\mathcal{P}}$ is also
normal in $\mathrm{Aut}\,(F_n)$ as it is equal to $\mathrm{Inn}\,(F_n)$ intersected with the kernel of the
map from $\mathrm{Aut}\,(F_n)$ to M. This means that $N_{\mathcal{P}}$ is a characteristic subgroup of
$\mathrm{Inn}\,(F_n) \simeq F_n$.

Now, M is a product of a bounded number of finite simple groups (a bound
independent of $\mathcal{P}$). Thus, for infinitely many finite simple groups G, when
$\mathrm{Aut}\,(F_n)$ acts on the set of kernels of epimorphisms from F_n onto G, it has
orbits of a bounded length. This means that the graph $\bar{X}_n(G)$ has a component
of bounded size in contradiction to its connectivity predicted by the Wiegold
conjecture.

The proof shows that much less than the Wiegold conjecture is needed.
For example, the following would suffice: given a Chevalley group scheme G,
prove that $X_n(G(\mathbb{F}_q))$ cannot have components of bounded size when $q \to \infty$.
(In fact, using the Chebotarev density theorem, it would suffice to assume q
is a prime.)

Here is a "baby version": prove that F_n has no characteristic subgroup N such that F_n/N is a finite simple group.

Another version that would make it: prove that for every epimorphism Ψ : $F_n \to G(\mathbb{F}_q)$ (as above), $\Psi(\Phi)$ contains an unbounded number of conjugacy classes in $G(\mathbb{F}_q)$ when $q \to \infty$. Here Φ is the set of primitive elements of F_n (i.e., those belonging to a basis of F_n). Clearly, $\Psi(\Phi)$ is a union of conjugacy classes. It is not difficult to see that if Ψ corresponds to a redundant generating vector in $\tilde{X}_n(G(\mathbb{F}_q))$, then $\Psi(\Phi) = G(\mathbb{F}_q)$. So, again one should only look at nonredundant vectors.

We end this section by remarking that the connection we observed above goes also in the opposite direction: if $\tilde{X}_n(G(\mathbb{F}_{q_i}))$ has a bounded size, say l, component for infinitely many primes q_i, then we get maps

$$\rho_{q_i} : \text{Aut}\,(F_n) \to \text{Aut}\,(G(\mathbb{F}_{q_i})^l) = \text{Aut}\,(G(\mathbb{F}_{q_i})^l) \rtimes \text{Sym}\,(l).$$

From these one can cook up a characteristic 0 representation ρ of Aut (F_n), with $\rho(\text{Inn}\,(F_n))$ being Zariski dense in $G(\mathbb{C})$, that is, contradicting Conjecture 6.12.

7. Profinite Groups

In this section we discuss the analogous problem in the category of profinite groups. We will show a strong positive result to all the questions mentioned above, in the context of profinite groups.

The main technical tool that is responsible for it is the following result of Gashütz. Since it is so important, we give an elegant proof due to Roquette (see [FJ05, lemma 17.7.2]).

LEMMA 7.1. (GASHÜTZ LEMMA) *Let* $\pi : G \twoheadrightarrow H$ *be an epimorphism between two finite groups. Assume* $d(G) \leq d$ *and let* $z_1, \dots, z_d \in H$ *be a set of d elements with* $\langle z_1, \dots, z_d \rangle = H$. *Then there exist* $y_1, \dots, y_d \in G$ *with* $\pi(y_i) = z_i$ *for* $i = 1, \dots, d$ *and* $G = \langle y_1, \dots, y_d \rangle$. *In other words, any generating d-vector of H can be lifted to a generating d-vector of G.*

Proof. For every subgroup $B \leq G$ and every $(\underline{t}) = (t_1, \dots, t_d) \in H$ with $\langle t_1, \dots, t_d \rangle = H$, we denote

$$\varphi_B((\underline{t})) = \#\{(b_1, \dots, b_d) \in B^d \,|\, \pi(b_i) = t_i \text{ for } i = 1, \dots, d, \text{ and } \langle b_1, \dots, b_d \rangle = B\}.$$

CLAIM. *The function* $\varphi_B((\underline{t}))$ *depends only on B and not on* $(\underline{t})$, *that is, it is constant on* $(\underline{t}) \in X_d(H)$.

This is proved by induction on the size of B. Assume it is true for every proper subgroup A of B and we will prove it for B: now, if $\pi(B) \lneqq H$, then $\varphi_B((\underline{t})) = 0$ for every $(\underline{t})$ and we are done. Otherwise,

$$\varphi_B((\underline{t})) = |K_B|^d - \sum_{A \lneqq B} \varphi_A((\underline{t}))$$

where $K_B = \mathrm{Ker}\,(\pi : B \twoheadrightarrow H)$. By induction, $\varphi_A((\underline{t}))$ is independent of $(\underline{t})$ and so $\varphi_B((\underline{t}))$ is also independent of $(\underline{t})$. This proves the claim.

We deduce now that $\varphi_G((\underline{t}))$ is independent of $(\underline{t})$. Let $x_1, \ldots, x_d \in G$ be elements such that $\langle x_1, \ldots, x_d \rangle = G$ (such elements exist since $d(G) \leq d$). Thus $\varphi_G((\pi(x_i))) > 0$, hence also $\varphi_G((z_i)) > 0$, which is exactly what the lemma says. $\qquad\square$

The Gashütz lemma does not hold if G is infinite. When $G = F_d$, its failure is "measured" by the number of connected components of $\tilde{X}_d(H)$. The lemma has the following corollary for the PRA graphs:

COROLLARY 7.2. *Let $\pi : G \twoheadrightarrow H$ be an epimorphism between two finite groups. Then the induced map $\tilde{\pi} : \tilde{X}_n(G) \to \tilde{X}_n(H)$ is onto for every $n \geq d(G)$.*

Standard inverse limit arguments imply that Lemma 7.1 holds also when G and H are profinite groups and "generating" means generating in the topological sense, that is, generating a dense subgroup. We can now deduce

PROPOSITION 7.3. *Let $F = \hat{F}_d$ be the free profinite group on $d \in \mathbb{N}$ generators. If G is a profinite group and π_1, π_2 two epimorphisms from F onto G, then there exists a (continuous) automorphism α of F, $\alpha \in \mathrm{Aut}\,(F)$ such that $\pi_1 \circ \alpha = \pi_2$.*

Proof. Say $\hat{F}_d = \hat{F}(x_1, \ldots, x_n)$ and denote $z_i = \pi_2(x_i)$ for $i = 1, \ldots, d$. Let $y_1, \ldots, y_d \in \hat{F}_d$ with $\pi_1(y_i) = z_i$ for $i = 1, \ldots, d$, and $\overline{\langle y_1, \ldots, y_d \rangle} = \hat{F}_d$. Such y_i's exist by Lemma 7.1.

Let α be the homomorphism from $\hat{F}_d$ to $\hat{F}_d$ sending x_i to y_i. Then, α is an epimorphism and hence an automorphism since every epimorphism from a finitely generated profinite group onto itself is an automorphism. Moreover, $\pi_1 \circ \alpha(x_i) = \pi_1(y_i) = z_i = \pi_2(x_i)$ and so $\pi_1 \circ \alpha = \pi_2$ as claimed. $\qquad\square$

It follows that profinite presentations satisfy all the good properties discussed in Section 2. For example, the Waldhausen conjecture: that is, if $N \lhd \hat{F}_d$ and $d(\hat{F}_d/N) < d$, then N contains a primitive element of $\hat{F}_d$ (i.e., an element that belongs to a basis of $\hat{F}_d$).

Recall that $\langle X; R \rangle$ is a profinite presentation for a profinite group G, if R is a subset of the free profinite group $\hat{F}_d$ on $X = \{x_1, \ldots, x_d\}$ and $G \simeq \hat{F}_d / \overline{\langle\langle R \rangle\rangle}$ where $\overline{\langle\langle R \rangle\rangle}$ is the topological closure of the normal closure of R in $\hat{F}_d$. We denote by $\hat{r}(G)$ the minimal possible size of R over all possible profinite presentations of G. Proposition 7.3 now implies that $\hat{r}(G)$ is obtained with a presentation on $d = d(G)$ generators and in all such presentations (as they are all equivalent by the proposition). Moreover, in [Lub01b] it is shown that it is obtained only in these representations. We mention in passing the long-standing open problem:

PROBLEM 7.4. *Let G be a finite group. Is $r(G) = \hat{r}(G)$ where $r(G)$ is the minimal number of relations needed to define G in the discrete category (see Section 2) and $\hat{r}(G)$ the number needed in the profinite category?*

Clearly $\hat{r}(G) \leq r(G)$ but there is no single example of a finite group where a strict inequality is known. The potential difference between $\hat{r}(G)$ and $r(G)$ was used in [Lub01a] to prove the Mann-Pyber conjecture on the normal subgroup growth of free group. For $\hat{r}(G)$ one has an exact formula in terms of the cohomology of G (see [GK99] and [Lub01b]) while estimating $r(G)$ is highly nontrivial. In fact, we do not know any lower bound on $r(G)$ for any group that is not, at the same time, also a lower bound for $\hat{r}(G)$. In [GKKL08] and [GKKL07] presentations of finite simple group are studied and the connections and differences between discrete and profinite presentations are discussed in length.

8. Compact Lie Groups

In this section $F = F_n$ will denote again the discrete free group on n generators, while G will be a connected compact Lie group. In this case, for every $n \geq 2$, the set of n-tuple $(\gamma_1, \ldots, \gamma_n) \in G^n$, for which $\overline{\langle \gamma_1, \ldots, \gamma_n \rangle} = G$, is open dense and of full measure in G^n. Thus measurewise the set G^n and the set Epi $(F, G) = \{\varphi : F \to G | \varphi(\bar{F}) = G\}$ are indistinguishable. We shall therefore look at the action of Aut (F_n) on G^n.

The main result here is due to Gelander [Gel08].

THEOREM 8.1. *Let G be a compact connected Lie group and let $n \geq 3$. The action of Aut (F_n) on G^n is ergodic.*

Theorem 8.1 was conjectured by Goldman [Gol07] who proved it for $G = SU(2)$. He also showed that $n \geq 3$ is necessary (compare Theorem 6.4; the reason is similar) and that a proof for the semisimple case would imply the general case.

The space G^n can be thought, also, as the space of n-generated marked subgroups of G (i.e., the subgroup is marked by an ordered n-generating set of it, so each subgroup appears many times in this space). A result of the kind of Theorem 8.1 implies that for every measurable property of subgroups of G we have a 0-1 law, that is, either the property is true for almost all subgroups or it is false for almost all of them, since the subset of marked subgroups with this property is a measurable subset of G^n that is invariant under Aut (F_n). (Note that the action of Aut (F_n) changes the generating set but not the generated subgroup of G.) An interesting example of such a property is the *spectral gap property*: Let $\Gamma \leq G$ be a dense subgroup. The left-translation action of Γ on G induces a unitary representation of Γ on $\mathcal{L}^2(G)$. The complement to the constant functions $\mathcal{L}_0^2(G) = \{f \in \mathcal{L}^2(G)| \int f d\mu = 0\}$ is Γ invariant. We say that the action of Γ on G has a *spectral gap* if the action of Γ on $\mathcal{L}_0^2(G)$ does not weakly contain the trivial representation. It is well known that this happens, for example, if Γ has Kazhdan property (T) (see [Lub94] and the references therein). It also happens for $G = SU(2)$ with very special choices of Γ (see [Lub94]) based on Deligne's solution to the Ramanujan conjecture. Altogether, for every semisimple connected compact Lie group G, there is such Γ. Such Γ is responsible for the affirmative answer to the Ruziewicz problem (see [Lub94]). But it is not known what is the behavior of the generic group with respect to the spectral gap property. (But see [LPS86, theorem 1.4].) In [Fis06], Fisher pointed out that Theorem 8.1 implies that either almost all subgroups of G have the spectral gap or almost all do not. In any event it implies that the set of n-tuples $(n \geq 3)$ in G^n, which generate a group with the spectral gap property, is dense in G^n. An analogous problem in the finite groups world is: essentially all finite simple groups G have a subset of k generators Σ, with respect to which the Cayley graph $Cay(G; \Sigma)$ is an ε-expander (k and ε are independent of G), and this is also a spectral gap property (see [KLN06]). But it is not known what is the behavior of the random set of generators of finite simple groups, except of the case of the family $\{PSL_2(p)|p \text{ prime}\}$ where Bourgain and Gamburd [BG06] showed that almost all k-tuple of elements ($k \geq 2$) give rise to expanders.

Let us now sketch the proof of Theorem 8.1 for the case G is a semisimple group. Assume the contrary; let $A \subset G^n$ be an Aut (F_n)-almost-invariant measurable subset that is neither null nor conull. Since Aut (F_n) is countable we can assume, by replacing A by $\bigcap_{\alpha \in \text{Aut}(F_n)} \alpha(A)$, that A is Aut (F_n)-invariant. Now, the action of G^n on itself is clearly ergodic, so at least one of the components, say the first one $G = G_1$, does not preserve A. For $(\underline{g}) = (g_2, \ldots, g_n) \in G^{n-1}$ denote $A_{(\underline{g})} = \{g \in G | (g, g_2, \ldots, g_n) \in A\}$.

CLAIM. *For a set of positive measure of* $(\underline{g}) = (g_2, \ldots, g_n) \in G^{n-1}$, *the set* $A_{(\underline{g})}$ *is neither null nor conull.*

Proof. By Fubini, $\mu(A) = \int_{(\underline{g})} \mu(A_{(\underline{g})}) d\underline{g}$. Now, $\mu(A) > 0$. We can throw out those $(\underline{g})$ with $\mu(A_{(\underline{g})}) = 0$ (they contribute measure 0). So, if for almost all the rest $\mu(A_{(\underline{g})}) = 1$, then for every $h \in G_1$, $h \cdot A$ is almost A (since $hA_{(\underline{g})} \sim A_{(\underline{g})}$). But we assumed that A is not G_1-almost-invariant. Thus for a positive measure of $(\underline{g})$, $0 < \mu(A_{(\underline{g})}) < 1$ as claimed.

Fix now a point $(\underline{g}) = (g_2, \ldots, g_n)$ in the subset of the claim, such that $\{g_2, g_3\}$ generates a dense subgroup of G (recall that we noticed that the set of such pairs is open dense and of full measure in G^2—so such $(\underline{g})$ does exist!). The orbits of the action of $\langle g_2, g_3 \rangle$ by left translation on $G_1 = G$ coincides with the (projection to the first factor of the) action of the Nielsen moves $\langle L(1, 2), L(1, 3) \rangle$ on $\{(g, g_2, g_3, \ldots, g_n) | g \in G\}$ (where $L(1, i)$ sends $(z_1, z_2, \ldots, z_n)$ to $(z_i z_1, z_2, \ldots, z_n)$). Let $A_1 = A_{(\underline{g})}$ for the $(\underline{g})$ chosen above. By our assumption A_1 is neither null nor conull. But on the other hand, it is invariant under $\langle g_2, g_3 \rangle$, a dense subgroup of G, a contradiction, since every dense subgroup acts ergodically on G. The theorem is now proven. $\square$

We conclude by mentioning another result of Gelander that is proved by similar methods.

THEOREM 8.2. *Let* $n \geq 3$ *and* G *a connected compact Lie group. Assume* $\Gamma \leq G$ *is an* $(n - 1)$-*generated dense subgroup. Then every* n *elements* $s_1, \ldots, s_n \in G$ *admit an arbitrary small deformation* $t_1, \ldots, t_n$ *with* $\Gamma = \langle t_1, \ldots, t_n \rangle$. *In other words, the set* $\{f \in \mathrm{Hom}\,(F_n, G) | f(F_n) = \Gamma\}$ *is dense in* $\mathrm{Hom}\,(F_n, G)$.

This theorem can be used to prove that given a simple compact Lie group G containing a dense Kazhdan subgroup, then for some n, any n elements can be ε-deformed (for every $\varepsilon > 0$) to generate a Kazhdan subgroup of G.

9. Non-compact Simple Lie Groups

Now let G be a noncompact simple real Lie group. In this case one cannot expect to have an ergodic action of $\mathrm{Aut}\,(F_n)$ on $G^n = \mathrm{Hom}\,(F_n, G)$. The representations with discrete image on one hand, and those with dense image on the other hand, form two disjoint Aut (F_n)-invariant subsets with nontrivial interior and so the action is not ergodic. Very little seems to be known about the decomposition of G^n under the Aut (F_n)-action in the general case. But, recently Minsky has revealed the picture for $G = \mathrm{PSL}_2$ ($\mathbb{R}$) and PSL_2 ($\mathbb{C}$). His (somewhat surprising) description shows that this decomposition can be quite delicate, but very interesting.

In the rest of this chapter let G be either PSL_2 ($\mathbb{R}$) or PSL_2 ($\mathbb{C}$) and $n \geq 3$. (For the case $n = 2$, the situation is similar to what we show in Section 6.2 for finite groups and Section 8 for compact groups: the trace of the commutator is an invariant that is preserved by Aut (F_n) and hence the action is far from being ergodic.) It will be more convenient to talk about the character variety $X_n(G) = \mathrm{Hom}\,(F_n, G)/G$. (We are ignoring the difference between this quotient and the geometric-invariant category quotient—see [LM85]—as anyway the representations that are not Zariski dense in G form a measure 0 set.) We will describe now the Out (F_n) decomposition of $X_n(G)$ following [Min]. The reader is referred to that paper and the references thereof, for unexplained notions and proofs.

Let $\mathcal{D}$ be the subset of the (equivalent classes) of *faithful discrete representations*. It contains $\mathcal{S}$, the Schottky representations. In fact, it is known (and by no means trivial) that $\mathcal{S}$ is precisely the interior of $\mathcal{D}$ and $\bar{\mathcal{S}} = \mathcal{D}$. The action of Out (F_n) on $\mathcal{S}$ is properly discontinuous.

At the other side we have $\mathcal{E} = \bar{E}_n(G)$—the set of representations with dense image, which is an open subset of $X_n(G)$. The complement of $\mathcal{D} \cup \mathcal{E}$ in $X_n(G)$ is the set of all representations that are either discrete but not faithful or are nondiscrete but not dense. This is a measure 0 set, so can be ignored for our purpose. The naive expectation has been that while Out (F_n) acts properly discontinuously on $\mathcal{S}$, it would act ergodically on $\mathcal{E}$—a phenomenon that might be seen as an extension of Wiegold's Conjecture 6.1 for finite simple groups and Gelander's Theorem (Goldman's conjecture) 8.1 for compact groups. But this is *not* the case! In fact, Minsky's main result in [Min] is

THEOREM 9.1. *There is an open subset of $X_n(G)$, strictly larger than $\mathcal{S}$, the set of Schottky representations, which is Out (F_n) invariant and on which Out (F_n) acts properly discontinuous.*

The set promised in the theorem is $\mathcal{PS}$ the primitive-stable representations, to be defined below. While it has a nonempty open intersection with $\mathcal{E}$, the set of dense representations, it has an empty intersection with R- the set of redundant representations, that is, those representations $\rho : F_n \to G$, for which there exists a proper free factor A of F_n with $\rho(A)$ dense in G. (Compare to Theorem 6.6 and Conjecture 6.10. Note that for G compact R is conull in G^n—see Section 8.) The set R is open. One is tempted to suggest:

CONJECTURE 9.2.

  a) *The action of* Out (F_n) *on R is ergodic.*
  b) $R \cup \mathcal{PS}$ *is conull in* $X(G)$.

If true, this conjecture gives a nice satisfactory picture: $X_n(G)$ is, up to a set of measure 0, a union of two Out (F_n)-invariant open subsets $\mathcal{PS}$ and R. On the first Out (F_n) acts properly discontinuously and on the second it acts ergodically. But, at this point this is just wishful thinking. (See a remark added in proof at the end of this section).

Let us now define $\mathcal{PS}$ and describe Minsky's main ingredients.

Let C denote the Cayley graph of F_n with respect to the free generators $x_1, \ldots, x_n$, and $\partial C = \partial F_n$ the boundary of F_n, that is, the rays from an initial vertex to infinity on the graph. Let $\partial^2 F_n = (\partial F_n \times \partial F_n) \setminus \Delta$ where Δ is the diagonal. Thus $\partial^2 F_n$ is the set of bi-infinite (oriented) lines on C.

To each $w \neq 1$ in F_n we associate a bi-infinite line, that is, a point $\bar{w} = (\infty_1, \infty_2)$ in $\partial^2 F_n$, that is, a bi-infinite word obtained by concatenating infinitely many copies of a representative of w. If $g \in F_n$, then the point of $\partial^2 F_n$ associated with gwg^{-1} is $(g\infty_1, g\infty_2)$. We denote by $\bar{\bar{w}}$ the F_n-orbit of $\bar{w}$. Let $\mathcal{P}$ be the subset of $\partial^2 F_n$ of all points associated with primitive elements of F_n. It is clearly invariant under the action of F_n, and Out (F_n) acts on the set $\mathcal{B}$ of F_n-orbits.

A representation $\rho : F_n \to G$ and a base point x_0 in the symmetric space $\mathbb{H}$ (which is either $\mathbb{H}^2$ if $G = \mathrm{PSL}_2(\mathbb{R})$ or $\mathbb{H}^3$ if $G = \mathrm{PSL}_2(\mathbb{C})$) gives rise to a unique map $\tau_{\rho, x_0} : C \to \mathbb{H}$ mapping the origin of C to x_0, which is ρ-equivariant and maps each edge to a geodesic. Every element of $\mathcal{B}$ is represented by an F_n-invariant set of infinite lines, which is mapped to a family of broken geodesic paths in $\mathbb{H}$.

DEFINITION 9.3. A representation $\rho : F_n \to G$ is *primitive stable* if there are constants K, δ in $\mathbb{R}_+$ and a base point $x_0 \in \mathbb{H}$ such that $\tau_{\rho, x}$ takes the lines representing the primitive elements $\mathcal{P}$ to (K, δ)-quasigeodesic. This means that for some $K, \delta \in \mathbb{R}_+$, for any two vertices v_1, v_2 on a line in $\mathcal{P}$, $\frac{1}{K} \mathrm{dist}_{\mathbb{H}}(v_1, v_2) - \delta \leq \mathrm{dist}_C(v_1, v_2) \leq K \, \mathrm{dist}_{\mathbb{H}}(v_1, v_2) + \delta$.

If there is one such base point, then any base point will do, at the expense of increasing δ. The set of primitive-stable representations is Aut (F_n)-invariant. Its image in $X_n(G)$ will be denoted $\mathcal{PS}$.

Schottky representations give rise to quasi-isometric embeddings of C in $\mathbb{H}$, so $\mathcal{S} \subset \mathcal{PS}$. The converse is not true, but Minsky showed that if $\rho \in \mathcal{PS}$, then for every proper free factor A of F_n, $\rho(A)$ is Schottky group. The set $\mathcal{PS}$ like $\mathcal{S}$ is also open; and the action of Out (F_n) is properly discontinuous.

The crucial point in proving this last claim is that the image of the set $\{\alpha \in \text{Aut}(F_n) | \|\alpha(w)\| \leq c\|w\| \ \forall \text{ primitive } w\}$ in Out (F_n) is finite, where c is any finite constant and for $g \in F_n$ we denote by $\|g\|$ the length of its cyclically reduced word (i.e., the minimal length in its conjugacy class).

The above facts are relatively simple to deduce from the basic definitions. The nontrivial fact is that $\mathcal{PS}$ is indeed larger than $\mathcal{S}$. To this end Minsky shows that one representation ρ_0 at the boundary of $\mathcal{S}$ (i.e., ρ_0 in $\mathcal{D} \setminus \mathcal{S}$) is primitive stable (in fact, he gives a method to produce many such examples, but one suffices!). Since $\mathcal{PS}$ is open it implies that some open neighborhood of ρ_0 is also in $\mathcal{PS}$—such a neighborhood has a nontrivial open intersection with $\mathcal{E}$. So indirectly one deduces the existence of many primitive-stable representations with dense images—even though only a discrete one is explicitly constructed!

To construct the discrete non-Schottky primitive-stable representation, Minsky appeals to a result of Whitehead that, using what nowadays is called the Whitehead graph, gives a necessary criterion for a word in F_n to be primitive. Given $g \in F_n$ we define the graph $Wh(g)$ to be the graph with $2n$ vertices denoted by the generators $\{x_i\}_{i=1}^n$ and their inverses $\{x_i^{-1}\}_{i=1}^n$. A pair (a, b) of vertices is an edge if ab^{-1} appears in g or in a cyclic permutation of g (which is the same as saying it appears in g or g starts with b^{-1} and ends with a); call this last edge the additional edge if it does not appear anyway in g.

THEOREM 9.4. (WHITEHEAD) *Let g be a cyclically reduced primitive element in F_n. Then by eliminating (at most) one vertex, $Wh(g)$ becomes a nonconnected graph.*

Whitehead's result gives a simple sufficient criterion for a word w to be "blocking"—that is, one which cannot appear as a subword of any cyclically reduced primitive element. This is the case if $Wh(w)$, minus the additional edge, contains a cycle that passes through all the vertices of the graph. It is easy to see that this is the case for $\beta^2 = ([x_1, x_2][x_3, x_4] \cdots [x_{2m-1}, x_{2m}])^2$ as an element of F_n, $n = 2m$. It also follows now that β^2 is not inside any proper free factor of F_n.

Now let Σ be a surface of genus m with one boundary that is a curve represented by β in $\pi_1(\Sigma) = F_{2m}$. Let $\rho : \pi_1(\Sigma) \to \mathrm{PSL}_2(\mathbb{R})$ be a discrete representation with $\rho(\beta)$ being parabolic. It is wellknown that such ρ exists in this case. It is a very special case of a general result asserting that every simple curve γ on the boundary of a 3-dimensional handlebody gives rise to a geometrically finite representation into $\mathrm{PSL}_2(\mathbb{C})$ for which $\rho(\gamma)$ is parabolic (see [Min] and the references therein). As $\rho(F_{2m})$ contains a parabolic element, it is not Schottky.

Minsky then proves that this ρ is primitive stable. This is done as follows: let Y be the convex hull of the limit set of $\rho(\pi_1(\Sigma))$. In our case, as $\rho(\pi_1(\Sigma))$ is a nonuniform lattice in $\mathrm{PSL}_2(\mathbb{R})$, Y is actually equal to $\mathbb{H}^2$, but this is not crucial for the general case. Let $Z = Y/\rho(\pi_1(\Sigma))$—the convex core of ρ. This is a surface with a unique cusp. Minsky shows that all the primitive elements of $F_n = \pi_1(\Sigma)$ are represented by geodesics in a fixed compact set $K \subset Z$. The idea is that in order to leave a compact set, a primitive element must wind around the cusp and this is prohibited by the blocking property deduced from Whitehead's lemma. The existence of this compact K implies the quasi-isometric condition for primitive elements, in a way similar to the standard argument that a group acting cocompactly is quasi-isometric to the space upon which it acts.

This finishes the sketch of the proof of Theorem 9.1 for n even and some modifications give the general case.

The theorem leaves various interesting problems. Define $\mathcal{PS}'(G)$ to be the set of all (equivalent classes of) representations of F_n where restrictions to proper free factors are Schottky. So $\mathcal{PS} \subset \mathcal{PS}'$ and Minsky shows that this is a proper inclusion. He asks whether $\mathcal{PS}$ is the interior of $\mathcal{PS}'$. He also shows that no point outside $\mathcal{PS}'$ can be in the domain of discontinuity of $\mathrm{Out}(F_n)$ acting on $X_n(G)$. Thus a positive answer to this question will show that $\mathcal{PS}$ is exactly the domain of discontinuity for the action of $\mathrm{Out}(F_n)$ on $X_n(G)$. Together with Conjecture 9.2 this will give a nice picture of the action of $\mathrm{Out}(F_n)$ on $X_n(G)$ for these two cases of G. One can speculate to suggest that a similar picture holds also for $G = \mathrm{PSO}(r, 1)$ for $r \geq 2$. For these G's, at least the definitions make sense. We do not even know what to expect the situation to be for higher-rank simple Lie groups G. The work of Minsky shows that the naive extension of Weigold-Goldman Conjectures 8.1 and 6.1 is false. But it still seems somewhat likely that the action of $\mathrm{Aut}(F_n)$ on $R(G)$—the set of redundant representations—is always ergodic. This will be a beautiful analogue of the theorem of Gilman and Evans (Theorem 6.6) and Gelander's Theorem 8.1.

We end this section by describing a recent work of Glasner [Gla] that shows that this is indeed the case for two families of simple locally compact groups. So let us now switch notations and assume that G is either $PSL_2(K)$ where K is a non-Archimedean local field of characteristic 0 or $G = \text{Aut}^+(T_k)$—the group of orientation-preserving automorphisms of the k-regular tree T_k, $k \geq 3$. (It is a simple group of index 2 in the full automorphism group of T_k.) Note also that $PSL_2(K)$ is acting on a tree; the Bruhat-Tits tree associated with it.

The Schottky subgroups of these G's were studied in detail in [Lub91]. It is shown there that the subset S of the Schottky representations is an open and closed subset of $H_n(G) = G^n$ and of $X_n(G) = \text{Hom}(F_n, G)/G$. The action of $\text{Out}(F_n)$ on $\text{Hom}(F_n, G)/G$ is not studied there, but from the discussion it is not difficult to see that $\text{Out}(F_n)$ acts properly discontinuously on S. Now let $\bar{E}_n(G)$ denote the subset of $X_n(G)$ of all the dense representations.

THEOREM 9.5. (GLASNER [GLA]) *Let G be either* $PSL_2(K)$ *or* $\text{Aut}^+(T_k)$. *Then for every $n \geq 3$,* $\text{Out}(F_n)$ *acts ergodically on* $\bar{E}_n(G)$.

In fact, he shows that $\text{Aut}(F_n)$ acts ergodically on the set of all dense representations in $\text{Hom}(F_n, G) = G^n$. Before sketching the proof, let us first mention that for these G's, $S \cup \bar{E}_n(G)$ is far from covering the whole space. We also have an open subset of all the representations of F_n whose images lie in the compact open subgroup (the stabilizer of a vertex).

Glasner's proof is based on two main ingredients. The first is a result of Weidman [Wei02] asserting that if $\rho(F_n)$ is dense in G, then $\rho(w)$ is elliptic (i.e., fixes a vertex) for some *primitive* element w of F_n. This implies that every n-tuple in $E_n(G)$ is conjugate mod $\text{Aut}(F_n)$ to an n-tuple of type $(w, g_2, \ldots, g_n)$ with $\rho(w)$ elliptic. Glasner shows further that $\rho(g_2)$ can be made to be hyperbolic. Then he uses another result (proved in [AGa] for $\text{Aut}^+(T_k)$ and in [Gla] for $PSL_2(K)$): for almost every elliptic element a and almost every hyperbolic element b the group generated by a and b is dense in G. From this, he applies some arguments of a similar nature to the Gelander proof of Theorem 8.1 to deduce the theorem. Along the way he shows that $R(G)$ is conull in $E_n(G)$.

All these results of Minsky and Glasner seem to indicate that only the tip of the iceberg has been revealed. It looks like a rich and interesting theory should be explored here for general noncompact Lie groups (or other locally compact groups).

Added in Proof. Conjecture 9.2(a) has been proved recently by Gelander and Minsky [GeMi]. In fact, they proved it for every simple k-group defined over a characteristic 0 local field k. Their work explains the difference between $G = PSL_2(\mathbb{R})$ or $PSL_2(\mathbb{C})$ for which the action of Out (F_n) on $\bar{E}_n(G)$ is not ergodic (see Theorem 9.1) and the group $G = PSL_2(\mathbb{Q}_p)$ for which it is ergodic (Theorem 9.5). The crucial difference is that for the latter, almost every dense representation of $F_n(n \geq 3)$ to $G = PSL_2(\mathbb{Q}_p)$ is redundant (a fact whose proof by Glasner uses Weidman [Wei02] in a crucial way).

10. The Mapping Class Group Action on Surface Group Representations

In the previous sections we studied the action of Aut (F_n) on Hom (F_n, G) (and of Out (F_n) on Hom $(F_n, G)/G$ for various groups G. In principle, one can do this not only for F_n but also for any finitely generated group Γ. A case of special interest is $\Gamma = \Pi_g$—the fundamental group of a closed surface Σ_g of genus $g \geq 2$. Indeed, this case has been studied in the literature in great detail as it is related to classical geometric and topological topics such as Fricke-Teichmüller spaces. A comprehensive survey is given by Goldman [Gol06], who is responsible, to a large extent, for the modern systematic development of the theory. In this section we mention only a few points out of this theory. Our main goal is to call attention to a particular direction that is not covered in [Gol06]; the study of the action of Out (Π_g), the mapping class group, on Epi $(\Pi_g, G)/G$ when G is a finite group. This issue came out in a recent paper of Dunfield and Thurston [DT06] where finite sheeted covers of random 3-manifolds are studied. It suggests developing a theory of the kind described in Sections 5 and 6, for Π_g instead of F_n. One may, for example, suggest an analogous conjecture to Weigold's, a proof of which (or even of a weak form of it) would imply that the mapping class groups are not linear.

But let us start with G being infinite. It is of interest to note that for $\Gamma = F_n$, the study of the Aut (Γ) action on Hom (Γ, G), has started with G-finite in presentation theory, as described in Sections 2–6, and only later a systematic study for G-compact or semisimple has emerged. On the other hand, for $\Gamma = \Pi_g$ the most classical case is the study of Hom $(\Pi_g, PSL_2(\mathbb{R}))/PSL_2(\mathbb{R})$. The faithful discrete representations form a connected component that is exactly the space classifying the equivalent classes of conformal structures on Σ_g, or also equivalence classes of hyperbolic structures on Σ_g. But there are more components that are indexed by the Euler class $e : \mathrm{Hom}(\Pi_g, PSL_2(\mathbb{R}))/PSL_2(\mathbb{R}) \to H^2(\Sigma_g; \mathbb{Z}) \simeq \mathbb{Z}$ whose image is $\{2 - 2g, \ldots, 2g - 2\}$, that is, $4g - 3$ connected components. The components $e^{-1}(\pm(2 - 2g))$ are two copies of the Teichmüller space that differ by the choice of orientation. On these two, $M_g = \mathrm{Out}(\Pi_g)$, which is classically known as the mapping class group of Σ_g, is acting properly discontinuously and a lot of study has been devoted to this action by many authors (see [Gol06] and the references therein). Much less is known about the action on the other $4g - 5$ components. Goldman conjectures that M_g acts ergodically on each of these. If $PSL_2(\mathbb{R})$ is replaced by a connected compact Lie group, then it was indeed proved by Pickrell and Xia [PX02] that M_g acts ergodically on every component of $\mathrm{Hom}(\Pi, G)/G$. The special case $G = SU(2)$ was proved by Goldman [Gol97], who conjectured the general case for $\Gamma = \Pi_g$ as well as for $\Gamma = F_n$, as discussed in Section 8. If G is semisimple compact group, then the number of the connected components of $\mathrm{Hom}(\Pi_g, G)/G$ is equal to the order of the fundamental group of G. The same applies for complex semisimple groups G, but is not true in general; for example, $\mathrm{Hom}(\Pi_g, \widetilde{SL_3(\mathbb{R})})/\widetilde{SL_3(\mathbb{R})}$ is not connected.

A wealth of additional information is given in [Gol06], but we will move now to the case when G is a finite group, which is not discussed there.

In [DT06], Dunfield and Thurston suggest an interesting model to produce random 3-manifolds. It briefly goes that it is well known that every closed 3-manifold M has an Heegard splitting, that is, it can be presented as a union of two handlebodies of genus g, H_1, and H_2 that are glued along their boundaries, each of which is a genus g closed surface. Their idea is to use this as a way to produce closed 3-manifolds of Heegard genus (at most) g in the following way: Fix g and fix a set of generators S for the mapping class group M_g of Σ_g. Take a random walk along the Cayley graph of M_g with respect to S. This will produce a random element $\varphi \in M_g$. Use this random φ to glue the boundary of H_1—a handlebody of genus g—to a copy of it, H_2, along the boundary. This will give the resulting "random" 3-manifold.

They were interested in finite covers of such random manifolds and in questions of the following type: given a finite group G, what is the probability that a random 3-manifold M of genus g as above, has a finite sheeted cover M' with a cover group isomorphic to G? This is really the question: What is the probability that there is an epimorphism from $\pi_1(M)$ onto G? Now, $\pi_1(M)$ can be described in the following way: Start with $\pi_1(\Sigma_g) = \langle a_1, b_1, \ldots, a_g, b_g \mid \prod_{i=1}^{g}[a_i, b_i] = 1 \rangle$ the fundamental group of the surface Σ_g. Gluing H_1 to it "kills" $a_1, \ldots, a_g$ and we get the free group on $b_1, \ldots, b_g$. Then gluing H_2, the second copy, amounts to dividing $\pi_1(\Sigma_g)$ further by $\varphi(a_1), \ldots, \varphi(a_g)$ to get $\pi_1(M)$. (Note that $\varphi \in M_g$ gives an element of $\mathrm{Aut}\,\pi_1(\Sigma_g)$ that is well defined only up to inner automorphism but the normal closure of $\varphi(a_1), \ldots, \varphi(a_g)$ is well defined.)

Now, let $\rho : \Pi_g \twoheadrightarrow G$ be an epimorphism, then it "survives" in the above process if and only if a_i and $\varphi(a_i)$ are in $\mathrm{Ker}\,\rho$ for every $i = 1, \ldots, g$. There are many such epimorphisms ρ (a very good estimate is given in [LS04], at least for the most interesting case, when G is a finite simple group). We can ask the above question in a different way now: start with an epimorphism $\rho : \Pi_g \to G$ with $\rho(a_i) = 1$ for every $i = 1, \ldots, g$. What is the probability that for a random $\varphi \in M_g$, $\varphi^{-1}(\mathrm{Ker}\,\rho)$ still contain $a_1, \ldots, a_g$? For the discussion of this question and the interesting answer(s) we refer the reader to [DT06]. For our context, what is relevant is the steps taken in [DT06] to study the action of M_g on the set of all kernels of epimorphisms from Π_g onto G.

This last action cannot be expected to be transitive in general. In fact, the epimorphism $\rho : \Pi_g \twoheadrightarrow G$ induces a map $H_2(\Pi_g, \mathbb{Z}) \to H_2(G; \mathbb{Z})$ and thus to every kernel $\Sigma_g \twoheadrightarrow G$ one associates an invariant $[c] \in H_2(G, \mathbb{Z})/\mathrm{Out}\,(G)$. By using a "stabilization" result of Livingston [Liv85], it is shown in [DT06]:

THEOREM. *Let G be a nonabelian finite simple group. Then for all sufficiently large g, the orbits of* $\mathrm{Epi}\,(\Pi_g, G)$ *under* $M_g = \mathrm{Out}\,(\Pi_g)$ *correspond bijectively to* $H_2(G, \mathbb{Z})/\mathrm{Out}\,(G)$. *Moreover, the action of M_g on each orbit is by the full alternating group of that orbit.*

This theorem is the analogue of Corollary 6.7. It will be of interest to give a quantitative estimate of the g needed for a given G, as in Corollary 3.3. It will be even more remarkable if one can prove a "Wiegold's conjecture" in this context, that is, that for $g \geq 3$ (and actually maybe even $g \geq 2$) M_g acts transitively on all the kernels of $\Pi_g \twoheadrightarrow G$ with the same invariant in $H_2(G, \mathbb{Z})/\mathrm{Out}\,(G)$. One can then imitate the discussion in Section 6.4 (and just like there, a weaker statement suffices: there are no bounded size orbits) to deduce that $\mathrm{Aut}\,(\Pi_g)$ is not linear. From this last statement one can conclude that M_{g+1} is not linear. As of now, this is a long-standing open problem.

References

[AGa]   M. Abert and Y. Glasner, *Generic groups acting on regular trees*, Trans. Amer. Math. Soc. **361** (2009), no. 7, 3597–3610.

[AC65]   J. J. Andrews and M. L. Curtis, *Free groups and handlebodies*, Proc. Amer. Math. Soc. **16** (1965), 192–195. MR0173241 (30 #3454)

[AG08]   N. Avni and S. Garion, *Connectivity of the product replacement graph of simple groups of bounded Lie rank*, J. Algebra **320** (2008), 945–960.

[Big01]   S. J. Bigelow, *Braid groups are linear*, J. Amer. Math. Soc. **14** (2001), no. 2, 471–486.

[BKM03]   A. V. Borovik, E. I. Khukhro, and A. G. Myasnikov, *The Andrews-Curtis conjecture and black box groups*, Internat. J. Algebra Comput. **13** (2003), no. 4, 415–436. MR2022117 (2004k:20050)

[BLM05]   A. V. Borovik, A. Lubotzky, and A. G. Myasnikov, *The finitary Andrews-Curtis conjecture*, Infinite groups: geometric, combinatorial and dynamical aspects, Progr. Math., vol. 248, Birkhäuser, Basel, 2005, pp. 15–30. MR2195451 (2007f:20045)

[BG06]   J. Bourgain and A. Gamburd, *New results on expanders*, C. R. Math. Acad. Sci. Paris **342** (2006), no. 10, 717–721. MR2227746 (2007f:20084)

[BHT01]   T. E. Brendle and H. Hamidi-Tehrani, *On the linearity problem for mapping class groups*, Algebr. Geom. Topol. **1** (2001), 445–468 (electronic). MR1852767 (2002h:57003)

[BW75]   J. L. Brenner and J. Wiegold, *Two-generator groups. I*, Michigan Math. J. **22** (1975), 53–64. MR0372033 (51 #8250)

[BGK08]   T. Breuer, R. Guralnick, and W. M. Kantor, *Probablistic generation of finite simple groups II*, J. Algebra **320** (2008), 443–494.

[CLGM+95]   F. Celler, C. R. Leedham-Green, S. H. Murray, A. C. Niemeyer, and E. A. O'Brien, *Generating random elements of a finite group*, Comm. Algebra **23** (1995), 4931–4948.

[DSC98]   P. Diaconis and L. Saloff-Coste, *Walks on generating sets of groups*, Invent. Math. **134** (1998), 251–299.

[Dix69]   J. D. Dixon, *The probability of generating the symmetric group*, Math. Z. **110** (1969), 199–205.

[DT06]   N. M. Dunfield and W. P. Thurston, *Finite covers of random 3-manifolds*, Invent. Math. **166** (2006), 457–521.

[Dun63]   M. J. Dunwoody, *On T-systems of groups*, J. Austral. Math. Soc. **3** (1963), 172–179. MR0153745 (27 #3706)

[Dun70]   ———, *Nielsen transformations*, Computational Problems in Abstract Algebra (Proc. Conf., Oxford, 1967), Pergamon, Oxford, 1970, pp. 45–46. MR0260852 (41 #5472)

[DP73]   M. J. Dunwoody and A. Pietrowski, *Presentations of the trefoil group.*, Canad. Math. Bull. **16** (1973), 517–520.

[DFG82]   J. L. Dyer, E. Formanek, and E. K. Grossman, *On the linearity of automorphism groups of free groups*, Arch. Math. (Basel) **38** (1982), no. 5, 404–409. MR666911 (84a:20047)

[Eva93a]   M. J. Evans, *Presentations of groups involving more generators than are necessary*, Proc. London Math. Soc. (3) **67** (1993), no. 1, 106–126. MR1218122 (94f:20062)

[Eva93b]   ———, *T-systems of certain finite simple groups*, Math. Proc. Cambridge Philos. Soc. **113** (1993), no. 1, 9–22. MR1188815 (93m:20022)

[Fis06]   D. Fisher, *Out(F_n) and the spectral gap conjecture*, Int. Math. Res. Not. (2006), Art. ID 26028, 9.

[FP92]   E. Formanek and C. Procesi, *The automorphism group of a free group is not linear*, J. Algebra **149** (1992), 494–499.

[FJ05]   M. D. Fried and M. Jarden, *Field arithmetic*, 2nd ed., Ergebnisse der Mathematik und ihrer Grenzgebiete. 3. Folge. A Series of Modern Surveys in Mathematics, vol. 11, Springer-Verlag, Berlin, 2005. MR2102046 (2005k:12003)

[Gar]   S. Garion, *Connectivity of the product replacement algorithm graph of* PSL(2, q), J. Group Theory **11** (2008), no. 6, 765–777.

[GS]   S. Garion and A. Shalev, *Commutator maps, measure preservation, and T-systems*, Trans. Amer. Math. Soc. **361** (2009), no. 9, 4631–4651.

[Gel08]   T. Gelander, *On deformations of F_n in compact Lie groups*, Israel J. of Math. **167** (2008), 15–26.

[GeMi]   T. Gelander and Y. Minsky, *Ergodicity of the Aut(F_n) action on spaces of redundant representations*, in preparation.

[Gil77]   R. Gilman, *Finite quotients of the automorphism group of a free group*, Canad. J. Math. **29** (1977), 541–551.

[Gla]   Y. Glasner, *A zero-one law for random subgroups of some totally disconnected groups*, Transform. Groups **14** (2009), no. 4, 787–800.

[Gol97]   W. M. Goldman, *Ergodic theory on moduli spaces*, Ann. of Math. (2) **146** (1997), no. 3, 475–507. MR1491446 (99a:58024)

[Gol06]   ———, *Mapping class group dynamics on surface group representations*, Problems on mapping class groups and related topics, Proc. Sympos. Pure Math., vol. 74, American Mathematical Society, Providence, RI, 2006, pp. 189–214. MR2264541 (2007h:57020)

[Gol07]   ———, *An ergodic action of the outer automorphism group of a free group*, Geom. Funct. Anal. **17** (2007), no. 3, 793–805. MR2346275 (2008g:57001)

[Gru76]   K. W. Gruenberg, *Relation modules of finite groups*, American Mathematical Society, Providence, RI, 1976, Conference Board of the Mathematical Sciences Regional Conference Series in Mathematics, no. 25. MR0457538 (56 #15743)

[GK99]   K. W. Gruenberg and L. G. Kovács, *Proficient presentations and direct products of finite groups*, Bull. Austral. Math. Soc. **60** (1999), 177–189.

[GL]   F. Grunewald and A. Lubotzky, *Linear representations of the automorphism group of a free group*, Geom. Funct. Anal. **18** (2009), no. 5, 1564–1608.

[GKKL07]   R. M. Guralnick, W. M. Kantor, M. Kassabov, and A. Lubotzky, *Presentations of finite simple groups: profinite and cohomological approaches*, Groups Geom. Dyn. **1** (2007), no. 4, 469–523. MR2357481

[GKKL08]   ———, *Presentations of finite simple groups: a quantitative approach*, J. Amer. Math. Soc. **21** (2008), no. 3, 711–774. MR2393425

[GP03]   R. M. Guralnick and I. Pak, *On a question of B. H. Neumann*, Proc. Amer. Math. Soc. **131** (2003), 2021–2025.

[Hem76]   J. Hempel, *3-manifolds*, Ann. of Math. Studies, Vol. 86, Princeton University Press, Princeton, NJ; University of Tokyo Press, Tokyo, 1976.

[KL90]   W. M. Kantor and A. Lubotzky, *The probability of generating a finite classical group*, Geom. Dedicata **36** (1990), 67–87.

[KLN06]   M. Kassabov, A. Lubotzky, and N. Nikolov, *Finite simple groups as expanders*, Proc. Natl. Acad. Sci. USA **103** (2006), no. 16, 6116–6119. MR2221038 (2007d:20025)

[Kra02]   D. Krammer, *Braid groups are linear*, Ann. of Math. (2) **155** (2002), 131–156.

[LL04]   M. J. Larsen and A. Lubotzky, *Normal subgroup growth of linear groups: the (G_2, F_4, E_8)-theorem*, Algebraic groups and arithmetic, Tata Institute of Fundamental Research, Mumbai, 2004, pp. 441–468. MR2094120 (2005k:20061)

[LP]   M. J. Larsen and R. Pink, *Finite subgroups of algebraic groups*, J. Amer. Math. Soc., to appear, available at http://www.math.ethz.ch/~pink/preprints .html.

[LOST]   M. W. Liebeck, E. A. O'Brien, A. Shalev, and P. Tiep, J. European Math. Soc. **12** (2010), 939–1008.

[LS95]   M. W. Liebeck and A. Shalev, *The probability of generating a finite simple group*, Geom. Dedicata **56** (1995), 103–113.

[LS04]   _____, *Fuchsian groups, coverings of Riemann surfaces, subgroup growth, random quotients and random walks.*, J. Algebra **276** (2004), 552–601.

[Liv85]   C. Livingston, *Stabilizing surface symmetries*, Michigan Math. J. **32** (1985), no. 2, 249–255. MR783579 (86h:57002)

[Lub91]   A. Lubotzky, *Lattices in rank one Lie groups over local fields*, Geom. Funct. Anal. **1** (1991), no. 4, 406–431. MR1132296 (92k:22019)

[Lub94]   _____, *Discrete groups, expanding graphs and invariant measures*, Progress in Mathematics, Vol. 125, Birkhäuser Verlag, Basel, 1994.

[Lub01a]   _____, *Enumerating boundedly generated finite groups*, J. Algebra **238** (2001), 194–199.

[Lub01b]   _____, *Pro-finite presentations*, J. Algebra **242** (2001), 672–690.

[LM85]   A. Lubotzky and A. R. Magid, *Varieties of representations of finitely generated groups*, Mem. Amer. Math. Soc. **58** (1985), no. 336, xi+117. MR818915 (87c:20021)

[LP01]   A. Lubotzky and I. Pak, *The product replacement algorithm and Kazhdan's property (T)*, J. Amer. Math. Soc. **14** (2001), 347–363.

[LPS86]   A. Lubotzky, R. Phillips, and P. Sarnak, *Hecke operators and distributing points on the sphere. I*, Comm. Pure Appl. Math. **39** (1986), no. S, suppl., S149–S186, Frontiers of the mathematical sciences: 1985 (New York, 1985). MR861487 (88m:11025a)

[LS03]   A. Lubotzky and D. Segal, *Subgroup growth*, Progress in Mathematics, Vol. 212, Birkhäuser Verlag, Basel, 2003. MR1978431 (2004k:20055)

[Lus91]   M. Lustig, *Nielsen equivalence and simple-homotopy type*, Proc. London Math. Soc. (3) **62** (1991), 537–562.

[LM93]   M. Lustig and Y. Moriah, *Generating systems of groups and Reidemeister-Whitehead torsion*, J. Algebra **157** (1993), no. 1, 170–198.

[LS77]   R. C. Lyndon and P. E. Schupp, *Combinatorial group theory*, Springer-Verlag, Berlin, 1977, Ergebnisse der Mathematik und ihrer Grenzgebiete, Band 89. MR0577064 (58 #28182)

[MW03]   D. McCullough and M. Wanderley, *Free actions on handlebodies*, J. Pure Appl. Algebra **181** (2003), no. 1, 85–104. MR1971807 (2004c:57031)

[Min]   Y. Minsky, *On dynamics of Out(F_n) on PSL(2,C) characters*, arXiv:0906.3491, Israel J. of Math., to appear.

[Nos81]   G. A. Noskov, *Primitive elements in a free group*, (Russian) Mat. Zametki **30** (1981), no. 4, 497–500, 636. MR638422 (83e:20039)

[Pak01]   I. Pak, *What do we know about the product replacement algorithm?* Groups and computation, III (Columbus, OH, 1999), Ohio State Univ. Math. Res. Inst. Publ., vol. 8, de Gruyter, Berlin, 2001, pp. 301–347. MR1829489 (2002d:20107)

[PX02]   D. Pickrell and E. Z. Xia, *Ergodicity of mapping class group actions on representation varieties. I. Closed surfaces*, Comment. Math. Helv. **77** (2002), 339–362.

[Wei02]   R. Weidmann, *The Nielsen method for groups acting on trees*, Proc. London Math. Soc. (3) **85** (2002), no. 1, 93–118. MR1901370 (2003c:20029)

[Whi00]   J. Whiston, *Maximal independent generating sets of the symmetric group*, J. Algebra **232** (2000), no. 1, 255–268. MR1783924 (2001e:20004)

[WS02]   J. Whiston and J. Saxl, *On the maximal size of independent generating sets of* $PSL_2(q)$, J. Algebra **258** (2002), no. 2, 651–657. MR1943940 (2003k:20016)

Contributors

Emmanuel Breuillard
Laboratoire de Mathématiques
Université Paris-Sud 11
91405 Orsay cedex
France

Marc Burger
Departement Mathematik
ETH Zentrum
CH-8092 Zürich
Switzerland

Michael G. Cowling
School of Mathematics and Statistics
University of New South Wales
UNSW Sydney 2052
Australia

Thomas Delzant
IRMA, Université de Strasbourg et CNRS
67084 Strasbourg
France

Sorin Dumitrescu
Laboratoire J.-A. Dieudonné
Université de Nice-Sophia Antipolis
Parc Valrose
06108 Nice Cedex 02
France

Benson Farb
Department of Mathematics
University of Chicago
Chicago, IL 60637-1514
USA

Renato Feres
Department of Mathematics
Washington University
St. Louis, MO 63130
USA

David Fisher
Department of Mathematics
Indiana University Bloomington
Bloomington, IN 47405
USA

Alex Furman
Department of Mathematics, Statistics and
Computer Science
University of Illinois at Chicago
Chicago, IL 60607-7045
USA

William M. Goldman
Department of Mathematics
University of Maryland
College Park, MD 20742-4015
USA

Olivier Guichard
CNRS, Laboratoire de Mathématiques
Université Paris-Sud
F-91405 Orsay Cedex
France

Chris Hruska
Department of Mathematical Sciences
University of Wisconsin–Milwaukee
Milwaukee, WI 53201-0413
USA

Alessandra Iozzi
Departement Mathematik
ETH Zentrum
CH-8092 Zürich
Switzerland

Anders Karlsson
Section de mathématiques
Université de Genève
1211 Genève 4
Switzerland

François Labourie
Université Paris-Sud 11
F-91405 Orsay Cedex
France

François Ledrappier
LPMA
Université Paris 6,
75252 Paris Cedex 05
France

Alexander Lubotzky
Einstein Institute of Mathematics
The Hebrew University of Jerusalem
Jerusalem, 91904
Israel

Dave Witte Morris
Department of Mathematics and Computer
Science
University of Lethbridge
Lethbridge, Alberta
Canada T1K 3M4

Shahar Mozes
Einstein Institute of Mathematics
The Hebrew University of Jerusalem
Jerusalem, 91904
Israel

Sorin Popa
Mathematics Department
University of California at Los Angeles
Los Angeles, CA 90095-1555
USA

Pierre Py
Department of Mathematics
University of Chicago
Chicago, IL 60637-1514
USA

Raul Quiroga-Barranco
Cimat A.C.
Colonia Mineral de Valenciana
Guanajuato, Guanajuato
C.P. 36240
Mexico

Emily Ronshausen
Department of Mathematics
Washington University
St. Louis, MO 63130
USA

Jennifer Taback
Department of Mathematics
Bowdoin College
Brunswick, ME 04011
USA

Anne Thomas
Mathematical Institute
Oxford OX1 3LB
United Kingdom

Stefaan Vaes
Department of Mathematics
K.U.Leuven
B-3001 Leuven
Belgium

Shmuel Weinberger
Department of Mathematics
University of Chicago
Chicago, IL 60637-1514
USA

Peter Wong
Department of Mathematics
Bates College
Lewiston, ME 04240
USA

Eugene Z. Xia
Department of Mathematics
National Cheng Kung University
Tainan 701
Taiwan